"十二五"普通高等教育本科国家级规划教材
国家级来华留学英语授课品牌课程建设项目
"十三五"江苏省高等学校重点教材(编号：2019-1-097)

Engineering Mechanics (Third Edition)

工程力学（第三版）

Kaifu WANG
王开福　著

U0248641

Science Press
科学出版社
北京

Synopsis

内 容 简 介

This book consists of statics, kinematics, kinetics, and mechanics of materials. The main contents of the book include statics of particles, reduction of force system, statics of rigid bodies, friction, kinematics of particles, kinematics of rigid bodies in plane motion, resultant motion of particles, kinetics of particles, kinetics of rigid bodies in plane motion, analytical mechanics, stresses and strains, tension and compression, torsion, bending internal forces, bending stresses, bending deformation, stress analysis and strength theories, combined loadings, stability of columns, and energy methods.

The book can be used as an English, Chinese, or bilingual textbook of engineering mechanics for the engineering students majoring in aeronautical engineering, engineering mechanics, mechanical engineering, materials science, civil engineering, etc.

本书由静力学、运动学、动力学和材料力学组成。本书主要内容包括质点静力学、力系简化、刚体静力学、摩擦、质点运动学、刚体平面运动学、质点合成运动、质点动力学、刚体平面动力学、分析力学、应力与应变、拉伸与压缩、扭转、弯曲内力、弯曲应力、弯曲变形、应力分析与强度理论、组合载荷、压杆稳定和能量方法。

本书可作为高等院校航空工程、工程力学、机械工程、材料科学和土木工程等工科专业的英文、中文或双语工程力学教材。

图书在版编目(CIP)数据

工程力学=Engineering Mechanics：汉、英/王开福著.—3版.—北京：科学出版社，2021.9
“十二五”普通高等教育本科国家级规划教材 “十三五”江苏省高等学校重点教材
ISBN 978-7-03-067398-5

Ⅰ.①工… Ⅱ.①王… Ⅲ.①工程力学-高等学校-教材-汉、英 Ⅳ.①TB12

中国版本图书馆 CIP 数据核字(2020)第 266941 号

责任编辑：许 蕾/责任校对：杨聪敏
责任印制：张 伟/封面设计：许 瑞

科学出版社 出版
北京东黄城根北街 16 号
邮政编码：100717
http://www.sciencep.com

北京九州迅驰传媒文化有限公司印刷
科学出版社发行 各地新华书店经销
*
2012 年 5 月第 一 版 开本：787×1092 1/16
2016 年 9 月第 二 版 印张：31 1/2
2021 年 9 月第 三 版 字数：900 000
2024 年 7 月第十次印刷

定价：129.00 元
(如有印装质量问题，我社负责调换)

Preface

Engineering mechanics is a required course for engineering students majoring in aeronautical engineering, engineering mechanics, mechanical engineering, materials science, and civil engineering, which is usually taught during the sophomore year. This book is intended to provide the students of these specialties with the basic theories, fundamental principles, calculation methods and engineering applications of engineering mechanics.

Engineering mechanics usually includes statics, kinematics, kinetics, and mechanics of materials. Statics deals with the equilibrium of bodies subjected to the action of forces, i.e., it discusses the relation between the forces acting on the bodies that are balanced. Statics usually consists of statics of particles, reduction of force system, statics of rigid bodies, friction, principle of virtual work, etc. Kinematics investigates the motion of bodies without regard to the forces acting on these bodies, i.e., it only deals with the geometry of motion. Kinematics usually includes kinematics of particles, kinematics of rigid bodies, resultant motion of particles in different systems of coordinates, etc. Kinetics relates the forces acting on bodies and the motion of the bodies subjected to these forces. It mainly involves kinetics of particles, kinetics of rigid bodies, Lagrange's Equations, etc. Mechanics of materials studies the strength and rigidity of structural members subjected to axial forces, shearing forces, torsional moments, bending moments, or combined loadings, and the stability of structural members subjected to axial compressive loads. It usually includes tension and compression, torsion, bending internal forces, bending stresses, bending deformation, stress analysis, strength theories, combined loadings, stability of a column, energy methods, etc.

The book can be used as an English, Chinese, or bilingual textbook of engineering mechanics for the student majoring in aeronautical engineering, engineering mechanics, mechanical engineering, materials science, civil engineering, etc.

Special thanks to Senior Engineer of NUAA College of Materials Science and Technology, Minghui GAO, for her valuable work in typing and proofreading.

Kaifu WANG

Nanjing, September 2021

前　言

工程力学是航空工程、工程力学、机械工程、材料科学和土木工程等工科学生的必修课程，通常在大二学年讲授。本书旨在向学生传授工程力学基础理论、基本原理、计算方法和工程应用。

工程力学通常包括静力学、运动学、动力学和材料力学。静力学研究受力物体的平衡，即研究作用在平衡物体上的力之间的关系。静力学通常包含质点静力学、力系简化、刚体静力学、摩擦和虚功原理等。运动学研究物体运动而不考虑作用在物体上的力，即仅研究运动几何。运动学通常包含质点运动学、刚体运动学和质点在不同坐标系的合成运动等。动力学研究作用在物体上的力与受力作用而产生的物体运动之间的关系。动力学通常包含质点动力学、刚体动力学和拉格朗日方程等。材料力学研究构件受轴力、剪力、扭矩、弯矩或组合载荷作用时的强度和刚度，以及构件受轴向压缩载荷作用时的稳定性。材料力学通常包含拉压、扭转、弯曲内力、弯曲应力、弯曲变形、应力分析、强度理论、组合载荷、压杆稳定和能量方法等。

本书可作为航空工程、工程力学、机械工程、材料科学和土木工程等工科学生的英文、中文或双语工程力学教材。

特别感谢南京航空航天大学材料科学与技术学院高级工程师高明慧在打字和校对方面所做的宝贵工作。

王开福

2021 年 9 月于南京

Contents

目　录

Chapter 1 Statics of Particles

If the size of a body does not affect the solution to the problem under consideration, then this body can be idealized as a particle. A particle has a finite mass, but a negligible size. All the forces acting on a particle can be assumed to be applied at a common point in space, and will thus form a concurrent force system.

1.1 Resultant of Concurrent Coplanar Force System

A concurrent coplanar force system is one in which the lines of action of all the forces lie in a common plane and intersect at a common point.

1. Graphical Solution

Two forces acting on a particle can be replaced by a resultant force obtained by drawing the diagonal, passing through the intersection point of the two forces, of the parallelogram which has sides equal to the given forces. For example, two forces $\boldsymbol{F}_1$ and $\boldsymbol{F}_2$ acting on particle O, Fig. 1.1a, can be replaced by a single force $\boldsymbol{R}$, Fig. 1.1b, which is called the resultant force of the forces $\boldsymbol{F}_1$ and $\boldsymbol{F}_2$. The resultant force $\boldsymbol{R}$ can be obtained by drawing a parallelogram using $\boldsymbol{F}_1$ and $\boldsymbol{F}_2$ as two adjacent sides of the parallelogram. The diagonal that passes through particle O represents the resultant force $\boldsymbol{R}$, i.e., $\boldsymbol{R} = \boldsymbol{F}_1 + \boldsymbol{F}_2$. This is known as the parallelogram law.

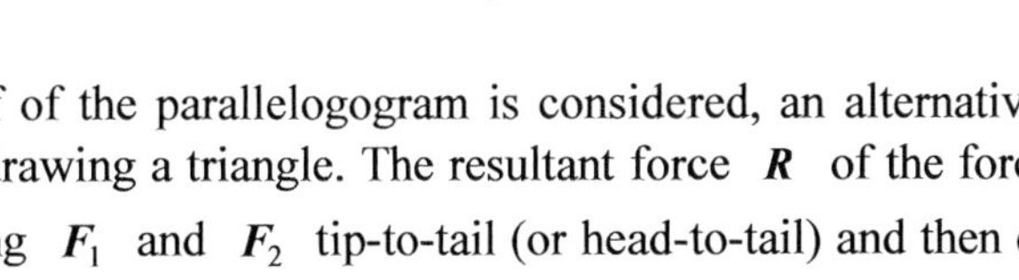

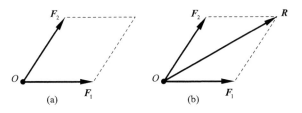

Fig. 1.1

When only half of the parallelogogram is considered, an alternative method, Fig. 1.2b, can be obtained by drawing a triangle. The resultant force $\boldsymbol{R}$ of the forces $\boldsymbol{F}_1$ and $\boldsymbol{F}_2$ can be found by arranging $\boldsymbol{F}_1$ and $\boldsymbol{F}_2$ tip-to-tail (or head-to-tail) and then connecting the tail of $\boldsymbol{F}_1$ with the tip of $\boldsymbol{F}_2$, i.e., $\boldsymbol{R} = \boldsymbol{F}_1 + \boldsymbol{F}_2$. This is known as the triangle rule (or trigonometry).

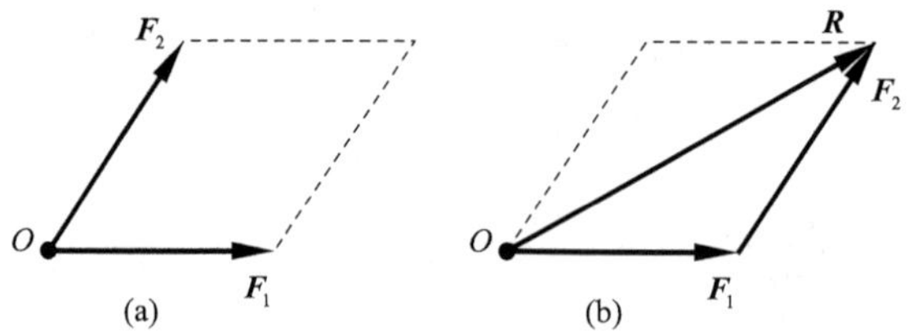

Fig. 1.2

If a particle is acted upon by three or more concurrent coplanar forces, the resultant force can be obtained by repeated applications of the triangle rule. Considering that particle O is acted upon by concurrent coplanar forces $\boldsymbol{F}_1$, $\boldsymbol{F}_2$, and $\boldsymbol{F}_3$, Fig. 1.3a, the resultant force $\boldsymbol{R}$ can be obtained graphically by arranging all the given forces tip-to-tail and connecting the tail of the first force with the tip of the last one, Fig. 1.3b, i.e., $\boldsymbol{R} = \boldsymbol{F}_1 + \boldsymbol{F}_2 + \boldsymbol{F}_3$. This is known as the polygon rule.

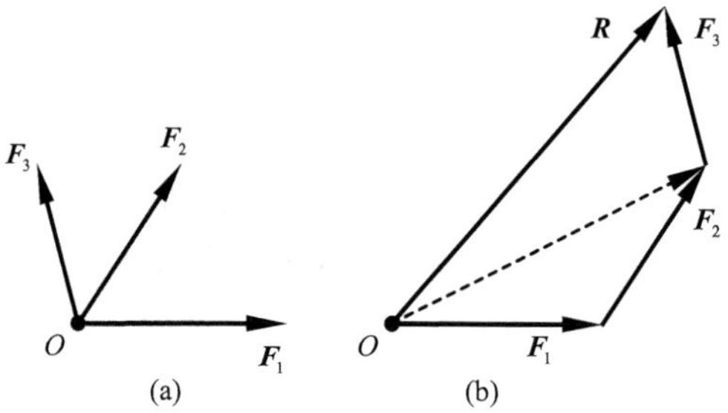

Fig. 1.3

We thus conclude that a concurrent coplanar system of forces acting on a particle can be replaced by a resultant force through the intersection point, and that the resultant force is equal to the vector sum of the given concurrent coplanar forces, i.e.,

$$\boldsymbol{R} = \sum \boldsymbol{F} \tag{1.1}$$

Example 1.1 Two rods AC and AD are attached at A to column AB, Fig. E1.1a. Knowing that $F_1 = 150\ \text{N}$, $\theta_1 = 30°$, and $\theta_2 = 15°$, determine (a) F_2 if the resultant force is directed vertically upward, (b) the magnitude of the resultant force.

Solution $\boldsymbol{F}_1$ and $\boldsymbol{F}_2$ can be replaced by $\boldsymbol{R}$ from the parallelogram law, Fig. E1.1b. Considering the shaded triangle shown in Fig. E1.1b and using the law of sines, we have

$$\frac{F_1}{\sin(90° - \theta_2)} = \frac{F_2}{\sin(90° - \theta_1)} = \frac{R}{\sin(\theta_1 + \theta_2)}$$

Substituting $F_1 = 150\ \text{N}$, $\theta_1 = 30°$, and $\theta_2 = 15°$ into the above equations, we get

$$F_2 = \frac{\sin(90° - \theta_1)}{\sin(90° - \theta_2)} F_1 = 134.49\ \text{N},\ R = \frac{\sin(\theta_1 + \theta_2)}{\sin(90° - \theta_2)} F_1 = 109.81\ \text{N}$$

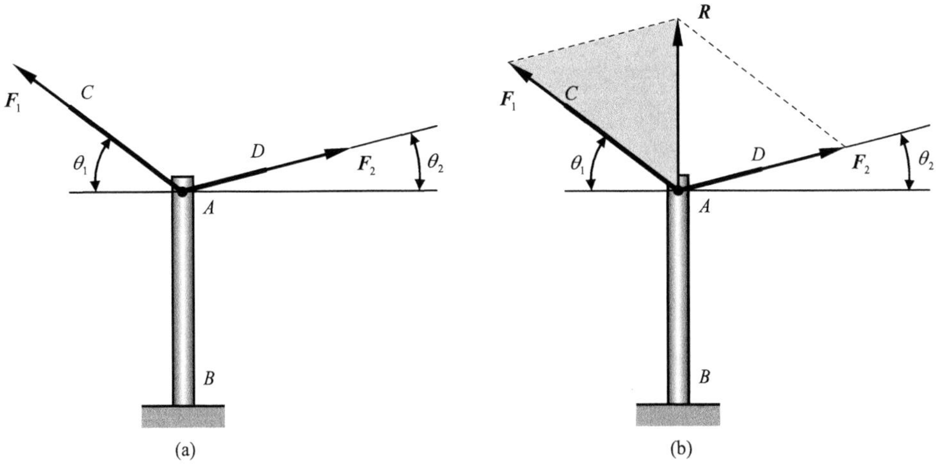

Fig. E1.1

Example 1.2 Two rods *AC* and *AD* are attached at *A* to column *AB*, Fig. E1.2a. Knowing that $F_1 = 120\ \text{N}$, $F_2 = 100\ \text{N}$, $\theta_1 = 35°$, and $\theta_2 = 20°$, determine the resultant force.

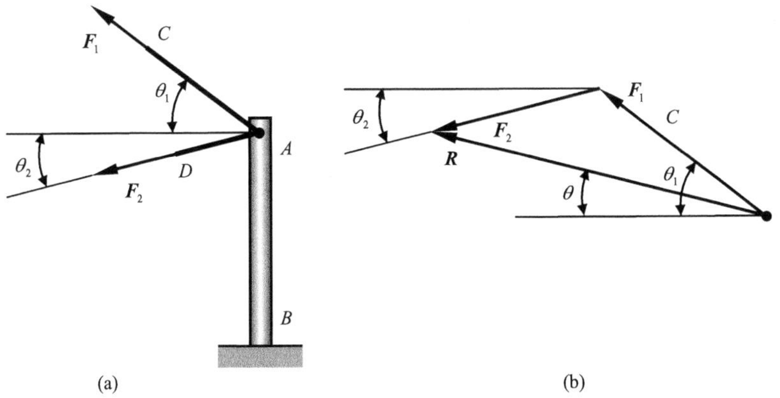

Fig. E1.2

Solution The force triangle is shown in Fig. E1.2b. Using the laws of cosines and of sines, we have

$$R^2 = F_1^2 + F_2^2 - 2F_1F_2\cos[180° - (\theta_1 + \theta_2)],\quad \frac{F_2}{\sin(\theta_1 - \theta)} = \frac{R}{\sin[180° - (\theta_1 + \theta_2)]}$$

Substituting $F_1 = 120\ \text{N}$, $F_2 = 100\ \text{N}$, $\theta_1 = 35°$, and $\theta_2 = 20°$ into the above equations yields

$$R = \sqrt{F_1^2 + F_2^2 + 2F_1F_2\cos(\theta_1 + \theta_2)} = 195.36\ \text{N},\ \theta = \theta_1 - \arcsin[\frac{F_2}{R}\sin(\theta_1 + \theta_2)] = 10.21°$$

2. Analytical Solution

Two or more forces acting on a particle can be replaced by a resultant force. Conversely, one force acting on a particle can also be replaced by two or more component forces which, together, have the same effect on the particle. For example, $\boldsymbol{F}$ can be replaced by $\boldsymbol{F}_1$ and $\boldsymbol{F}_2$, Fig. 1.4a, where $\boldsymbol{F}_1$ and $\boldsymbol{F}_2$ are the vector components of $\boldsymbol{F}$. Substituting $\boldsymbol{F}_1$ and $\boldsymbol{F}_2$ for $\boldsymbol{F}$ is called the resolution of a force into components. Clearly, for $\boldsymbol{F}$ there exist infinite sets of vector components, Fig. 1.4b.

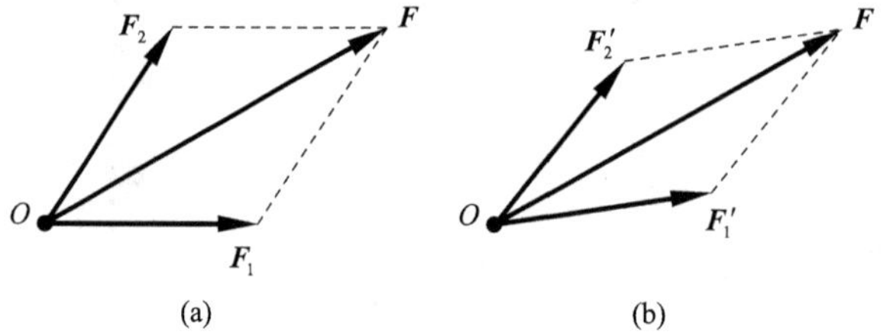

Fig. 1.4

It is often convenient to resolve a force into components perpendicular to each other. For example, $\boldsymbol{F}$ can be resolved into two vector components $\boldsymbol{F}_x$ and $\boldsymbol{F}_y$, Fig. 1.5a.

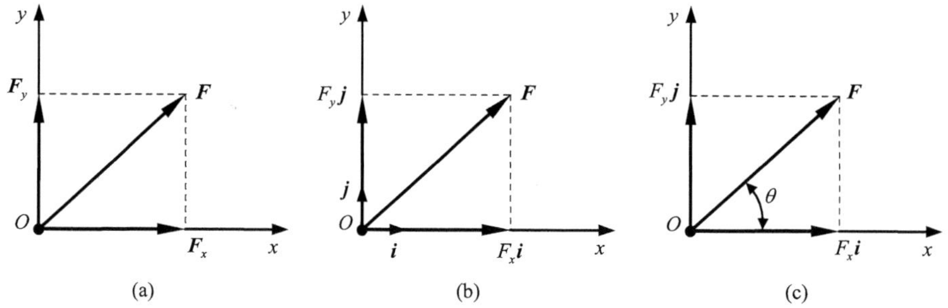

Fig. 1.5

By introducing two unit vectors $\boldsymbol{i}$ and $\boldsymbol{j}$, Fig. 1.5b, $\boldsymbol{F}$ can also be expressed as $\boldsymbol{F} = F_x\boldsymbol{i} + F_y\boldsymbol{j}$, where F_x and F_y are the scalar components of $\boldsymbol{F}$.

Denoting by F the magnitude of $\boldsymbol{F}$ and by θ the angle of $\boldsymbol{F}$ from the positive x axis, Fig. 1.5c, we can express the scalar components F_x and F_y as follows: $F_x = F\cos\theta,\ F_y = F\sin\theta$.

Using the graphical solution to determine the resultant force often requires extensive geometric or trigonometric calculation, especially for finding the resultant force of three or more forces. Instead, problems of this type are easily solved by using the analytical solution.

Considering $\boldsymbol{F}_1$, $\boldsymbol{F}_2$, and $\boldsymbol{F}_3$ acting on particle O, Fig. 1.6, then the resultant force $\boldsymbol{R}$ of these forces can be expressed, using the graphical solution, as $\boldsymbol{R} = \boldsymbol{F}_1 + \boldsymbol{F}_2 + \boldsymbol{F}_3$. Resolving each force, including the resultant force, into its rectangular components, we write

$$R_x\boldsymbol{i} + R_y\boldsymbol{j} = (F_{1x} + F_{2x} + F_{3x})\boldsymbol{i} + (F_{1y} + F_{2y} + F_{3y})\boldsymbol{j} \tag{1.2}$$

from which it follows that

$$R_x = F_{1x} + F_{2x} + F_{3x},\ R_y = F_{1y} + F_{2y} + F_{3y} \tag{1.3}$$

We thus conclude that the scalar component along an arbitrary axis of the resultant of forces acting on a particle is equal to the algebraic sum of the scalar components on the same axis of the given forces, i.e.,

$$R_x = \sum F_x,\ R_y = \sum F_y \tag{1.4}$$

Therefore, the magnitude R of the resultant force and the angle θ that the resultant force forms with the positive x axis can be written as

$$R = \sqrt{R_x^2 + R_y^2},\ \theta = \arctan\frac{R_y}{R_x} \tag{1.5}$$

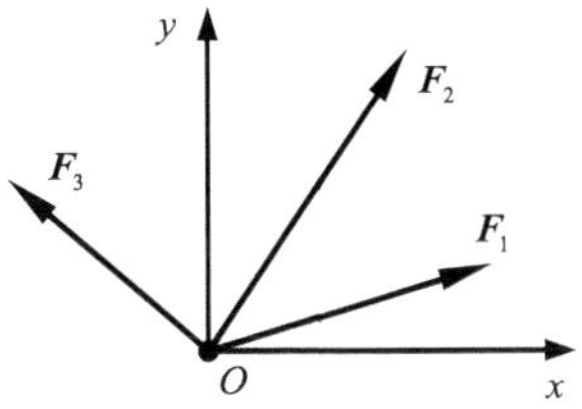

Fig. 1.6

Example 1.3 Two rods *AC* and *AD* are attached at *A* to column *AB*, Fig. E1.3a. Knowing that $F_1 = 150$ N, $\theta_1 = 30°$, and $\theta_2 = 15°$, determine (a) F_2 if the resultant force is directed vertically upward, (b) the magnitude of the resultant force.

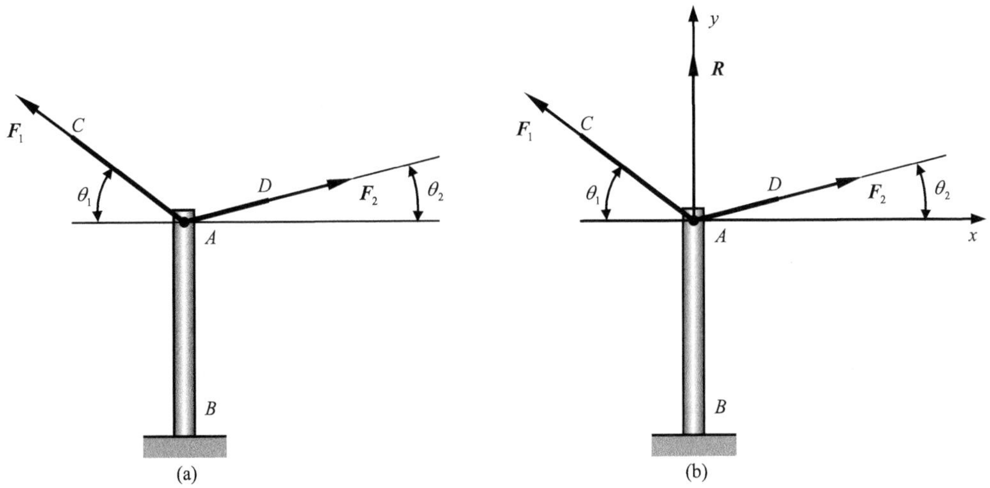

Fig. E1.3

Solution Establishing the coordinate system as shown in Fig. E1.3b, then the scalar components of the resultant force can be expressed as

$$R_x = -F_1 \cos\theta_1 + F_2 \cos\theta_2,\ R_y = F_1 \sin\theta_1 + F_2 \sin\theta_2$$

Since the resultant force is vertical, i.e., $R_x = 0$, we get

$$F_2 = \frac{\cos\theta_1}{\cos\theta_2} F_1 = 134.49 \text{ N},\ R = \sqrt{R_x^2 + R_y^2} = F_2 \sin\theta_2 + F_1 \sin\theta_1 = 109.81 \text{ N}$$

Example 1.4 A block subjected to three forces lies on the inclined plane for which the angle of inclination $\alpha = 25°$, Fig. E1.4a. Knowing that $\theta = 40°$, $F_1 = 150$ N, $F_2 = 250$ N, and $F_3 = 200$ N, determine the resultant of the forces acting on the block.

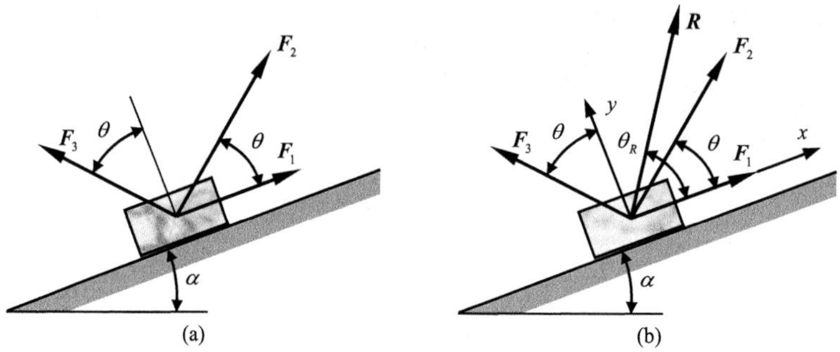

Fig. E1.4

Solution Establishing the coordinate system as shown in Fig. E1.4b, then the scalar components of the resultant force can be expressed as

$$R_x = F_1 + F_2 \cos\theta - F_3 \sin\theta = 212.95 \text{ N},\ R_y = F_2 \sin\theta + F_3 \cos\theta = 313.91 \text{ N}$$

from which it follows that the magnitude and direction of the resultant force can be written as

$$R = \sqrt{R_x^2 + R_y^2} = 379.32 \text{ N},\ \theta_R = \arctan\frac{R_y}{R_x} = 55.85°$$

Thus the resultant force is $R = 379.32$ N in magnitude and $\alpha + \theta_R = 80.85°$ in the angle of inclination.

1.2 Equilibrium of Concurrent Coplanar Force System

In solving problems concerning the equilibrium of a particle, it is necessary to consider all the forces acting on the particle. This can be done by choosing the particle as a free body and drawing a separate free-body diagram showing all the forces acting on the particle.

1. Graphical Solution

A particle is said to be in equilibrium if the resultant of the forces acting on the particle is zero. Thus the condition of equilibrium for a particle subjected to concurrent coplanar forces can be expressed as

$$\sum \boldsymbol{F} = \mathbf{0} \tag{1.6}$$

It can be seen from Eq. (1.6) that a particle is in equilibrium if all the forces acting on the particle can form a closed polygon. For example, considering that particle O is acted upon by forces $\boldsymbol{F}_1$, $\boldsymbol{F}_2$, and $\boldsymbol{F}_3$, Fig. 1.7a, the resultant force can be obtained by the polygon rule. Starting from point O with $\boldsymbol{F}_1$ and arranging all the forces tip-to-tail, the tip of $\boldsymbol{F}_3$ will coincide with the starting point O, Fig. 1.7b. Thus the resultant force is zero, and the particle is in equilibrium.

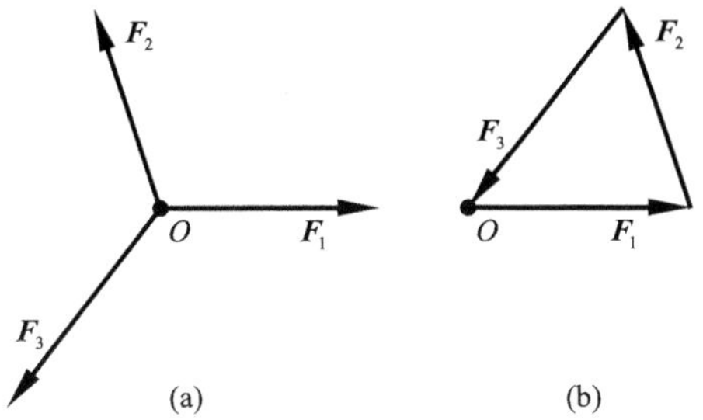

Fig. 1.7

Example 1.5 Three cables are tied together at A and are loaded as shown in Fig. E1.5a. Determine the tension (a) in AB, (b) in AC.

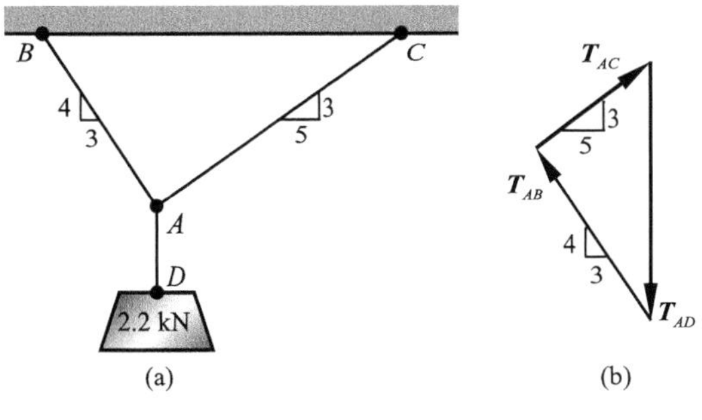

Fig. E1.5

Solution Since point A is in equilibrium, all the forces acting on A will form a closed triangle, Fig. E1.5b. Using the law of sines, we get

$$\frac{T_{AB}}{\sin[\arctan(5/3)]} = \frac{T_{AC}}{\sin[\arctan(3/4)]} = \frac{T_{AD}}{\sin[\arctan(3/5) + \arctan(4/3)]}$$

Substituting $T_{AD} = 2.2$ kN into the above equations gives

$$T_{AB} = 1.90 \text{ kN}, \ T_{AC} = 1.33 \text{ kN}$$

2. Analytical Solution

The condition of equilibrium for a particle subjected to concurrent coplanar forces can be expressed as $\sum \boldsymbol{F}=\mathbf{0}$. Resolving each force into its rectangular components, we write

$$(\sum F_x)\boldsymbol{i}+(\sum F_y)\boldsymbol{j}=\mathbf{0} \tag{1.7}$$

from which it follows that the condition of equilibrium for a particle subjected to concurrent coplanar forces can be expressed as

$$\sum F_x=0,\ \sum F_y=0 \tag{1.8}$$

which are the equations of equilibrium.

Example 1.6 Three cables are tied together at *A* and are loaded as shown in Fig. E1.6a. Determine the tension (a) in *AB*, (b) in *AC*.

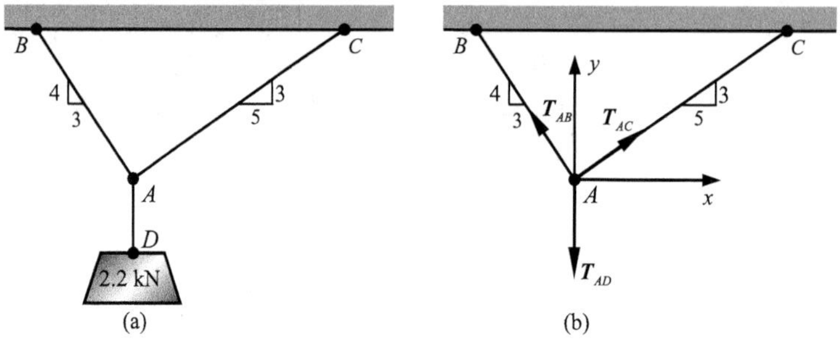

Fig. E1.6

Solution Establishing the coordinate system and considering the equilibrium of point *A* as shown in Fig. E1.6b, then we have

$$\sum F_x=0,\ -T_{AB}\frac{3}{\sqrt{3^2+4^2}}+T_{AC}\frac{5}{\sqrt{5^2+3^2}}=0$$

$$\sum F_y=0,\ T_{AB}\frac{4}{\sqrt{3^2+4^2}}+T_{AC}\frac{3}{\sqrt{5^2+3^2}}-T_{AD}=0$$

Solving the above equations for T_{AB} and T_{AC}, we obtain

$$T_{AB}=1.90\ \text{kN},\ T_{AC}=1.33\ \text{kN}$$

1.3 Resultant of Concurrent Noncoplanar Force System

Force $\boldsymbol{F}$ in space, Fig. 1.8a, can be resolved into three vector components $\boldsymbol{F}_x$, $\boldsymbol{F}_y$, and $\boldsymbol{F}_z$, i.e., $\boldsymbol{F}=\boldsymbol{F}_x+\boldsymbol{F}_y+\boldsymbol{F}_z$.

By introducing three unit vectors $\boldsymbol{i}$, $\boldsymbol{j}$, and $\boldsymbol{k}$, Fig. 1.8b, $\boldsymbol{F}$ can be expressed as $\boldsymbol{F}=F_x\boldsymbol{i}+F_y\boldsymbol{j}+F_z\boldsymbol{k}$, where F_x, F_y, and F_z are the scalar components of $\boldsymbol{F}$.

Denoting by F the magnitude of $\boldsymbol{F}$ and by θ_x, θ_y, and θ_z the angles between $\boldsymbol{F}$ and the positive *x*, *y*, and *z* axes, Fig. 1.8c, we can express the scalar components as follows:

$$F_x=F\cos\theta_x,\ F_y=F\cos\theta_y,\ F_z=F\cos\theta_z \tag{1.9}$$

where $\cos\theta_x$, $\cos\theta_y$, and $\cos\theta_z$ are the direction cosines of $\boldsymbol{F}$. These direction cosines satisfy the following relation:

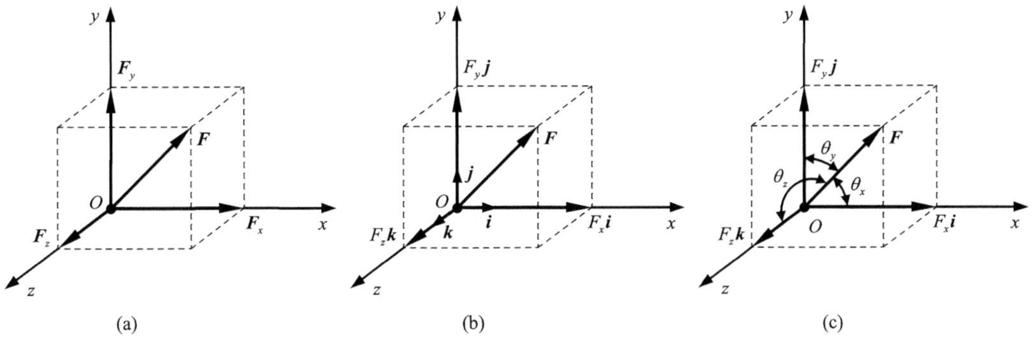

Fig. 1.8

$$\cos^2\theta_x + \cos^2\theta_y + \cos^2\theta_z = 1 \tag{1.10}$$

If we have known the angle γ between $\boldsymbol{F}$ and the positive y axis, and the angle φ between the positive z axis and the plane containing $\boldsymbol{F}$ and the y axis, Fig. 1.9, the corresponding scalar components can be expressed as

$$F_x = F_{xz}\sin\varphi = F\sin\gamma\sin\varphi,\ F_y = F\cos\gamma,\ F_z = F_{xz}\cos\varphi = F\sin\gamma\cos\varphi \tag{1.11}$$

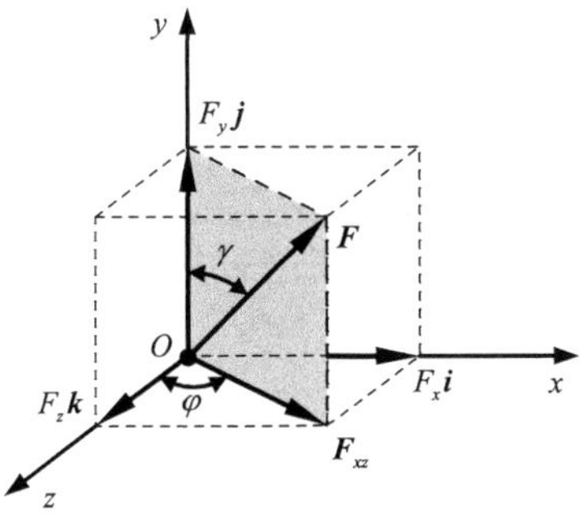

Fig. 1.9

Considering three forces $\boldsymbol{F}_1$, $\boldsymbol{F}_2$, and $\boldsymbol{F}_3$ acting on particle O, Fig. 1.10, then, using the graphical solution, the resultant force $\boldsymbol{R}$ of these forces can be expressed as $\boldsymbol{R} = \boldsymbol{F}_1 + \boldsymbol{F}_2 + \boldsymbol{F}_3$. Resolving each force, including the resultant force, into its rectangular components, we write

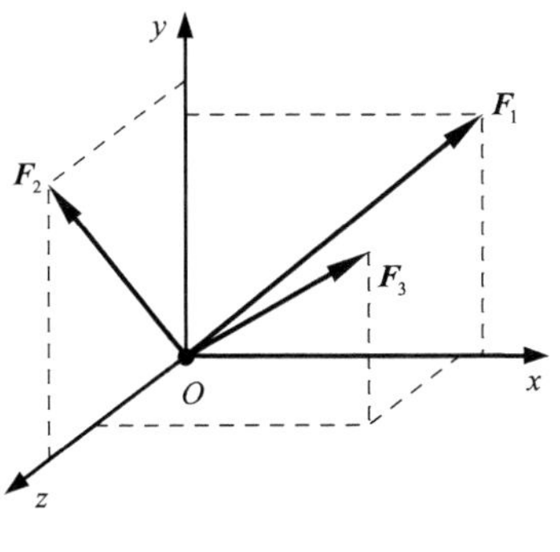

Fig. 1.10

$$R_x\boldsymbol{i}+R_y\boldsymbol{j}+R_z\boldsymbol{k}=(F_{1x}+F_{2x}+F_{3x})\boldsymbol{i}+(F_{1y}+F_{2y}+F_{3y})\boldsymbol{j}+(F_{1z}+F_{2z}+F_{3z})\boldsymbol{k} \quad (1.12)$$

from which it follows that

$$R_x=F_{1x}+F_{2x}+F_{3x},\ R_y=F_{1y}+F_{2y}+F_{3y},\ R_z=F_{1z}+F_{2z}+F_{3z} \quad (1.13)$$

We therefore conclude that the scalar component along an arbitrary axis of the resultant of the forces acting on a particle is equal to the algebraic sum of the scalar components on the same axis of the given forces, i.e.,

$$R_x=\sum F_x,\ R_y=\sum F_y,\ R_z=\sum F_z \quad (1.14)$$

Therefore, the magnitude R of the resultant force and the angles θ_x, θ_y, and θ_z that the resultant force forms with the positive *x*, *y*, and *z* axes can be written as

$$R=\sqrt{R_x^2+R_y^2+R_z^2},\ \theta_x=\arccos\frac{R_x}{R},\ \theta_y=\arccos\frac{R_y}{R},\ \theta_z=\arccos\frac{R_z}{R} \quad (1.15)$$

Example 1.7 Determine the magnitude and direction of the resultant of the three forces as shown in Fig. E1.7, knowing that $F_1=300\ \text{N}$, $F_2=200\ \text{N}$, and $F_3=100\ \text{N}$.

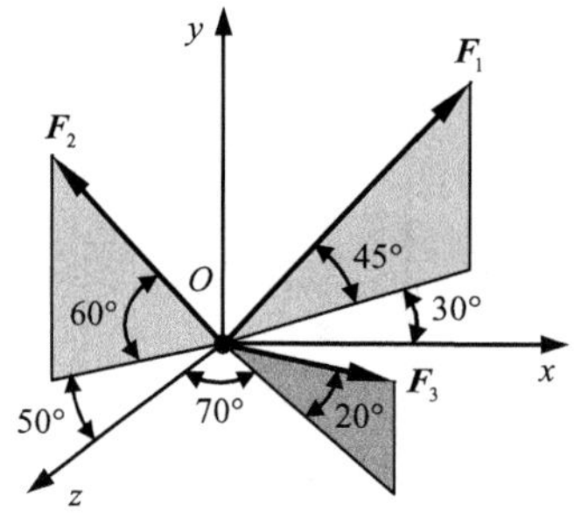

Fig. E1.7

Solution The three scalar components of the resultant force can be given by

$$R_x=F_1\cos45°\cos30°-F_2\cos60°\sin50°+F_3\cos20°\sin70°=195.41\ \text{N}$$

$$R_y=F_1\sin45°+F_2\sin60°+F_3\sin20°=419.54\ \text{N}$$

$$R_z=-F_1\cos45°\sin30°+F_2\cos60°\cos50°+F_3\cos20°\cos70°=-9.65\ \text{N}$$

Therefore, the magnitude and direction of the resultant force are

$$R=\sqrt{R_x^2+R_y^2+R_z^2}=462.92\ \text{N}$$

$$\theta_x=\arccos\frac{R_x}{R}=65.0°,\ \theta_y=\arccos\frac{R_y}{R}=25.0°,\ \theta_z=\arccos\frac{R_z}{R}=91.2°$$

1.4 Equilibrium of Concurrent Noncoplanar Force System

The condition of equilibrium for a particle subjected to concurrent noncoplanar forces can be expressed as $\sum \boldsymbol{F}=\boldsymbol{0}$. Resolving each force into its rectangular components, we get

$$(\sum F_x)\boldsymbol{i} + (\sum F_y)\boldsymbol{j} + (\sum F_z)\boldsymbol{k} = \boldsymbol{0} \tag{1.16}$$

from which it follows that the condition of equilibrium for a particle subjected to concurrent noncoplanar forces can be expressed as

$$\sum F_x = 0, \sum F_y = 0, \sum F_z = 0 \tag{1.17}$$

which are the equations of equilibrium.

Example 1.8 Knowing that a horizontal homogeneous circular plate having a weight of $W = 300$ N is suspended from three wires which form $30°$ angles with the vertical, Fig. E1.8a, determine the tension in each wire.

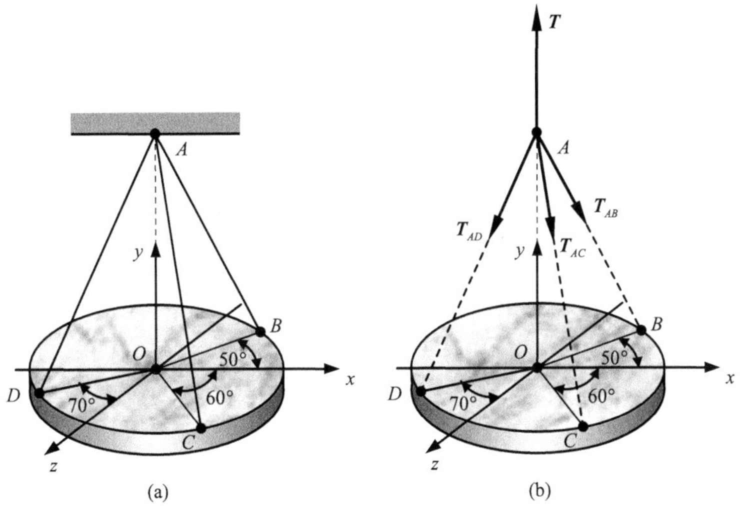

Fig. E1.8

Solution Since the whole system is in equilibrium, Fig. E1.8b, we have

$$T = W = 300 \text{ N}$$

Taking point A as a free body, and drawing its free-body diagram, then we have

$$\sum F_x = 0,\ T_{AB} \sin 30° \cos 50° + T_{AC} \sin 30° \cos 60° - T_{AD} \sin 30° \sin 70° = 0$$

$$\sum F_y = 0,\ T - T_{AB} \cos 30° - T_{AC} \cos 30° - T_{AD} \cos 30° = 0$$

$$\sum F_z = 0,\ -T_{AB} \sin 30° \sin 50° + T_{AC} \sin 30° \sin 60° + T_{AD} \sin 30° \cos 70° = 0$$

Solving the above equations for T_{AB}, T_{AC}, and T_{AD} yields

$$T_{AB} = 140.71 \text{ N},\ T_{AC} = 71.44 \text{ N},\ T_{AD} = 134.26 \text{ N}$$

Problems

1.1 Knowing that $F_2 = 100$ N, $\theta_1 = 20°$ and $\theta_2 = 10°$, determine (a) F_1 if the resultant force is directed vertically upward, (b) the magnitude of the resultant force.

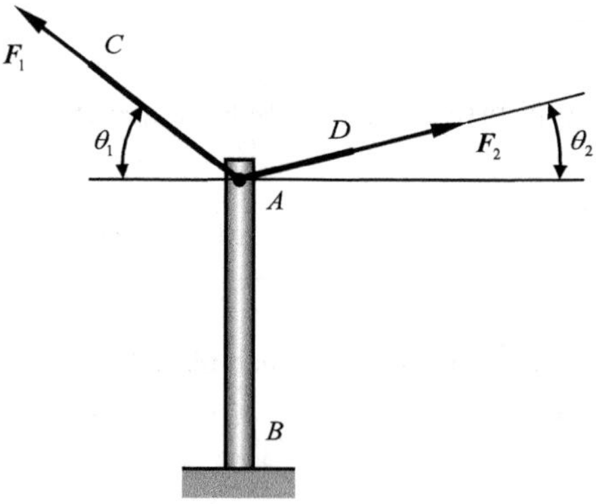

Fig. P1.1

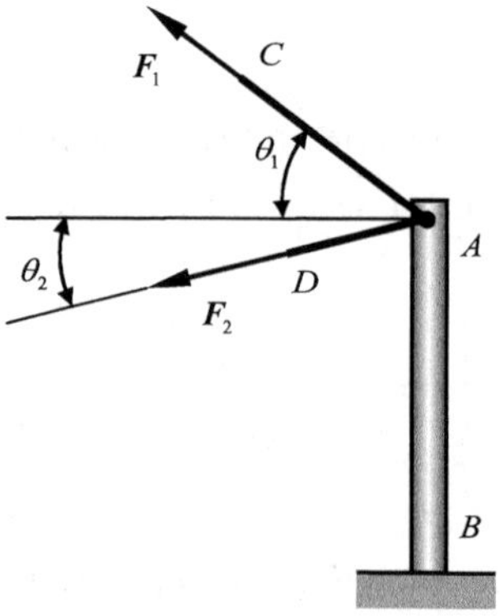

Fig. P1.2

1.2 Knowing that $F_1 = 150\ \text{N}$, $\theta_1 = 30°$ and $\theta_2 = 15°$, determine (a) F_2 if the resultant force is horizontal to the left, (b) the magnitude of the resultant force.

1.3 Knowing that $F_1 = 110\ \text{N}$, $F_2 = 90\ \text{N}$, $\theta_1 = 40°$ and $\theta_2 = 25°$, determine the resultant force.

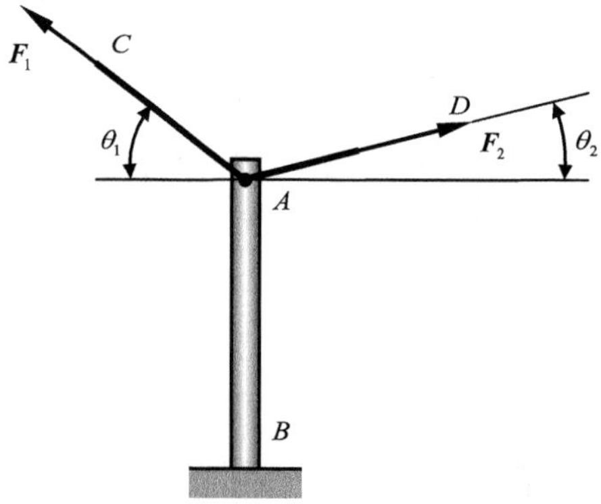

Fig. P1.3

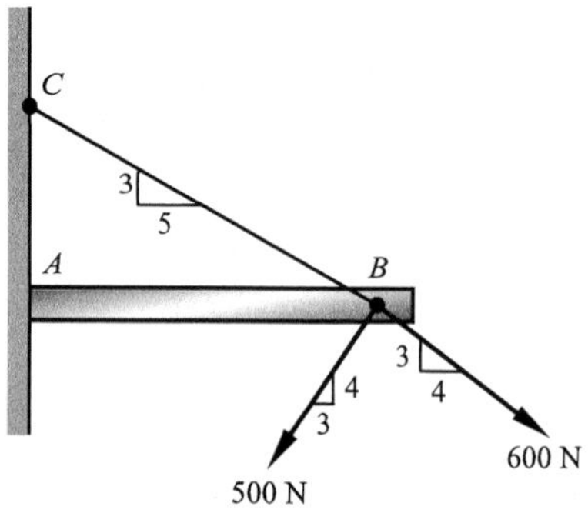

Fig. P1.4

1.4 Knowing that the tension in cable *BC* is 650 N, determine the resultant of the three forces exerted at point *B* of beam *AB*.

1.5 Knowing that three cables are tied together at *A* and are loaded as shown, determine the tension (a) in *AB*, (b) in *AC*.

1.6 Assuming the lever is in equilibrium as shown, and knowing that $F_1 = 200\ \text{N}$, $F_2 = 175\ \text{N}$ and $\theta_1 = 30°$, determine (a) θ_2, (b) the resultant of the forces exerted by the rods on the lever.

1.7 Collar *A* can slide on a frictionless vertical rod and is attached to the spring *C* through a frictionless fixed pulley *B*. Assuming that the spring is unstretched when $h = 0.3\ \text{m}$ and that the collar is in equilibrium when $h = 0.4\ \text{m}$, and knowing that $a = 0.4\ \text{m}$ and the spring stiffness $k = 500\ \text{N/m}$, determine the weight of the collar and the force exerted by the rod on the collar.

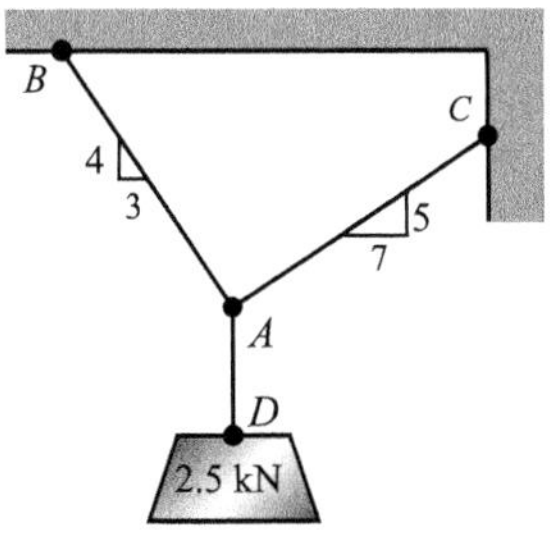

Fig. P1.5

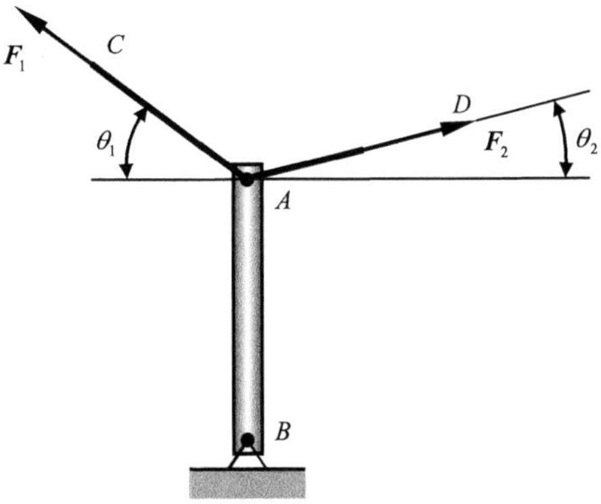

Fig. P1.6

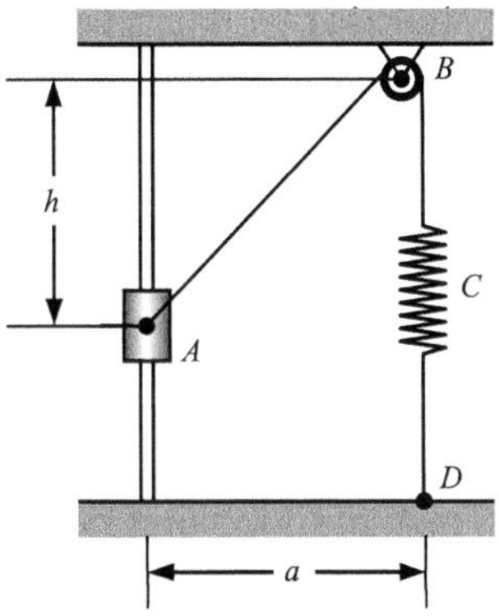

Fig. P1.7

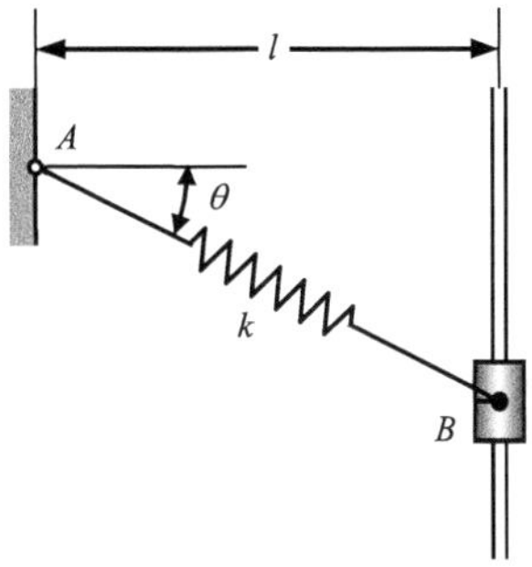

Fig. P1.8

1.8 Collar B of weight W can move freely along the vertical rod. Assuming that the spring stiffness is k and that the spring is unstretched when $\theta=0$, and knowing that W = 13.5 N, l = 150 mm, and k = 120 N/m, determine the value of θ corresponding to equilibrium.

1.9 A horizontal circular plate is suspended from three wires which form 30° angles with the vertical. Knowing that the x component of the force exerted by wire AB on the plate is 50 N, determine (a) the tension in wire AB, (b) the angles θ_x, θ_y, and θ_z that the force exerted at B forms with the coordinate axes.

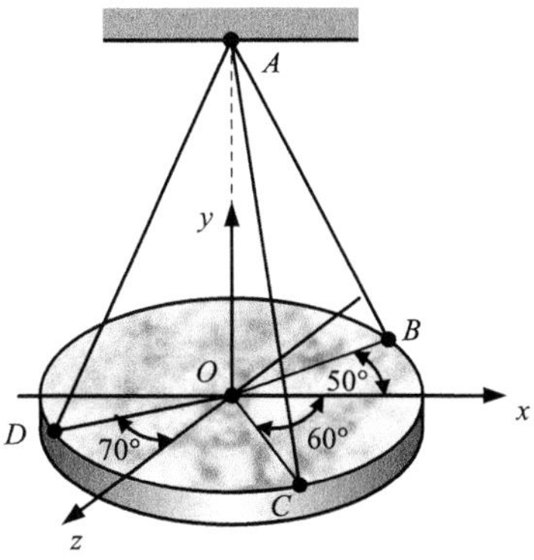

Fig. P1.9

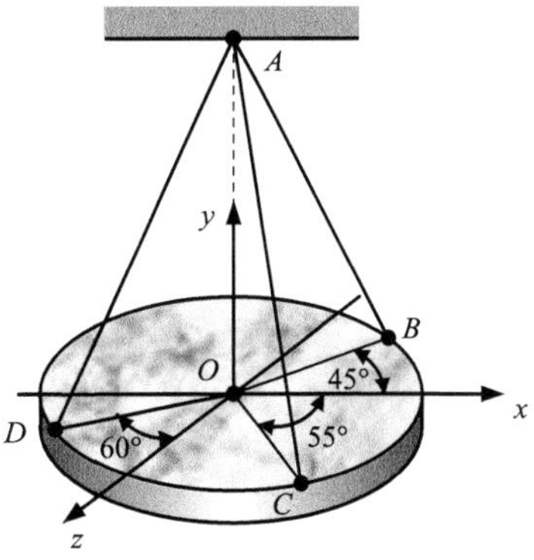

Fig. P1.10

1.10　Assuming that a horizontal homogeneous circular plate having a weight of $W = 200$ N is suspended from three wires which form $35°$ angles with the vertical, determine the tension in each wire.

Chapter 2 Reduction of Force System

2.1 Moment of Force about Point

Assuming that O is a reference point and that A is an arbitrary point on the line of action of force $\boldsymbol{F}$ as shown in Fig. 2.1, then the moment $\boldsymbol{M}_O(\boldsymbol{F})$ of $\boldsymbol{F}$ about O is defined as the vector product of $\boldsymbol{r}$ and $\boldsymbol{F}$, i.e.,

$$\boldsymbol{M}_O(\boldsymbol{F}) = \boldsymbol{r} \times \boldsymbol{F} \tag{2.1}$$

where $\boldsymbol{r}$ is a position vector directed from O to A, and $\boldsymbol{M}_O(\boldsymbol{F})$ satisfies:

(1) The magnitude of the moment is equal to $rF\sin\theta = Fd$, where θ is the angle between $\boldsymbol{r}$ and $\boldsymbol{F}$, and d is the moment arm (i.e., the perpendicular distance from point O to the line of action of $\boldsymbol{F}$).

(2) The direction of the moment is obtained from the right-hand rule, which can be stated as follows: If your right-hand fingers are curled from $\boldsymbol{r}$ to $\boldsymbol{F}$, then your right-hand thumb is in the direction of the moment.

Establishing a rectangular coordinate system $Oxyz$ and denoting by $\boldsymbol{i}$, $\boldsymbol{j}$, $\boldsymbol{k}$ the unit vectors along three axes x, y, z respectively, as shown in Fig. 2.2, then the moment $\boldsymbol{M}_O(\boldsymbol{F})$ of $\boldsymbol{F}$ about O can also be expanded as

$$\boldsymbol{M}_O(\boldsymbol{F}) = \begin{vmatrix} \boldsymbol{i} & \boldsymbol{j} & \boldsymbol{k} \\ x & y & z \\ F_x & F_y & F_z \end{vmatrix} = (yF_z - zF_y)\boldsymbol{i} + (zF_x - xF_z)\boldsymbol{j} + (xF_y - yF_x)\boldsymbol{k} \tag{2.2}$$

where x, y, z are the coordinates of $\boldsymbol{r}$ and F_x, F_y, F_z are the scalar components of $\boldsymbol{F}$.

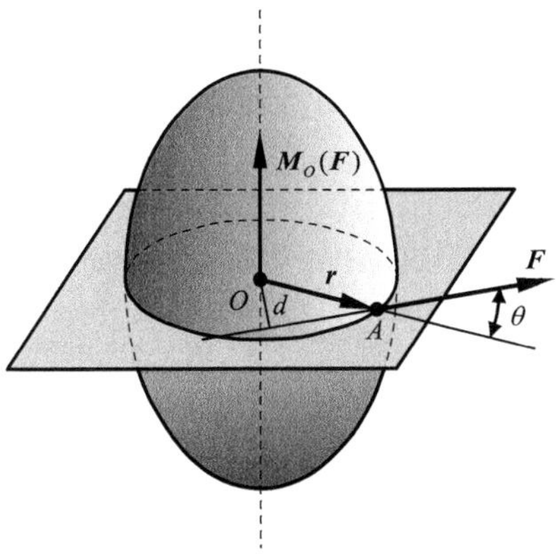

Fig. 2.1

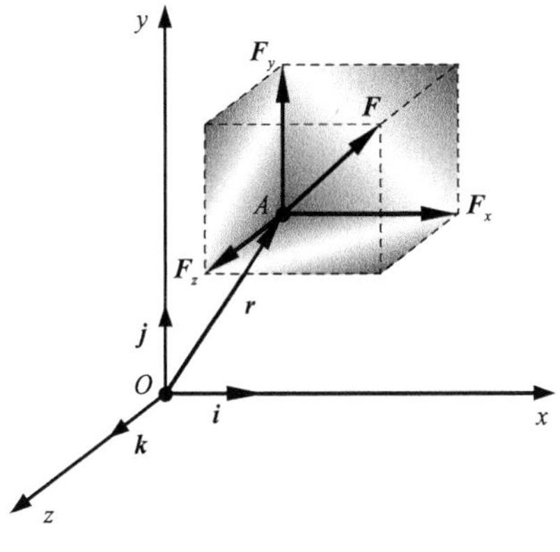

Fig. 2.2

Example 2.1 Two cables *AB* and *AC* are attached to a concrete column, Fig. E2.1a. Knowing that the tensions in cables *AB* and *AC* are 800 N and 500 N respectively, determine the moment about *O* of the resultant force exerted by the cables at *A* on the column.

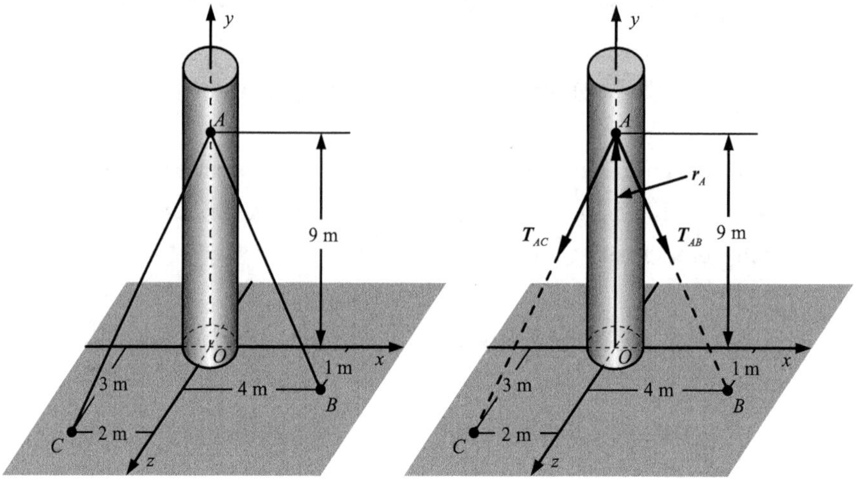

Fig. E2.1

Solution Denoting by $\boldsymbol{i}$, $\boldsymbol{j}$, and $\boldsymbol{k}$ the unit vectors along *x*, *y*, and *z* axes respectively, Fig. E2.1b, then we can write

$$\boldsymbol{r}_A = (9\boldsymbol{j})\ \text{m},\ \boldsymbol{r}_{AB} = (4\boldsymbol{i} - 9\boldsymbol{j} + \boldsymbol{k})\ \text{m},\ \boldsymbol{r}_{AC} = (-2\boldsymbol{i} - 9\boldsymbol{j} + 3\boldsymbol{k})\ \text{m}$$

$$\boldsymbol{T}_{AB} = T_{AB}\frac{\boldsymbol{r}_{AB}}{r_{AB}} = (800\ \text{N})\frac{4\boldsymbol{i} - 9\boldsymbol{j} + \boldsymbol{k}}{\sqrt{4^2 + (-9)^2 + 1^2}} = 80.81(4\boldsymbol{i} - 9\boldsymbol{j} + \boldsymbol{k})\ \text{N}$$

$$\boldsymbol{T}_{AC} = T_{AC}\frac{\boldsymbol{r}_{AC}}{r_{AC}} = (500\ \text{N})\frac{-2\boldsymbol{i} - 9\boldsymbol{j} + 3\boldsymbol{k}}{\sqrt{(-2)^2 + (-9)^2 + 3^2}} = 51.57(-2\boldsymbol{i} - 9\boldsymbol{j} + 3\boldsymbol{k})\ \text{N}$$

Using $\boldsymbol{R} = \boldsymbol{T}_{AB} + \boldsymbol{T}_{AC}$, we have

$$\boldsymbol{R} = (220.1\boldsymbol{i} - 1191.4\boldsymbol{j} + 235.5\boldsymbol{k})\ \text{N}$$

Therefore, the moment about *O* of the resultant force exerted by the cables at *A* on the column is equal to

$$\boldsymbol{M}_O(\boldsymbol{R}) = \boldsymbol{r}_A \times \boldsymbol{R} = \begin{vmatrix} \boldsymbol{i} & \boldsymbol{j} & \boldsymbol{k} \\ 0 & 9 & 0 \\ 220.1 & -1191.4 & 235.5 \end{vmatrix}\ \text{N}\cdot\text{m} = (2119.5\boldsymbol{i} - 1980.9\boldsymbol{k})\ \text{N}\cdot\text{m}$$

2.2 Moment of Force about Axis

Assuming that OL is an axis through O, Fig. 2.3, then the moment M_{OL} of $\boldsymbol{F}$ about OL is defined as the projection of $\boldsymbol{M}_O(\boldsymbol{F})$ onto OL, i.e.,

$$M_{OL} = \boldsymbol{\lambda} \cdot \boldsymbol{M}_O(\boldsymbol{F}) = \boldsymbol{\lambda} \cdot (\boldsymbol{r} \times \boldsymbol{F}) \tag{2.3}$$

where $\boldsymbol{\lambda}$ is the unit vector along OL.

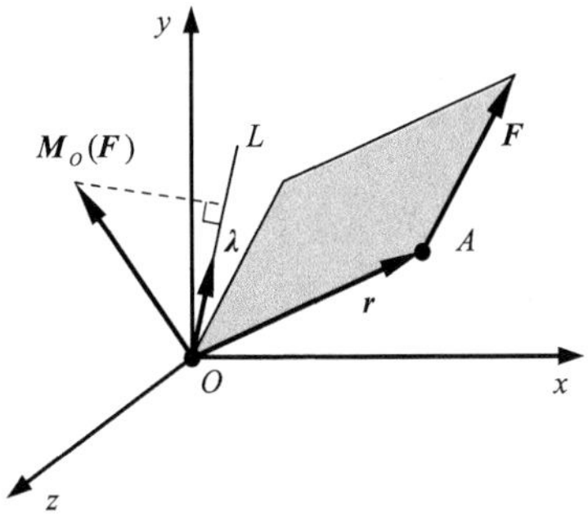

Fig. 2.3

2.3 Principle of Moments

Assuming that forces $\boldsymbol{F}_1$, $\boldsymbol{F}_2$, and $\boldsymbol{F}_3$ are applied at a common point A and that the resultant of these forces is denoted by $\boldsymbol{R}$, Fig. 2.4, then we have

$$\boldsymbol{M}_O(\boldsymbol{R}) = \sum \boldsymbol{M}_O(\boldsymbol{F}) \tag{2.4}$$

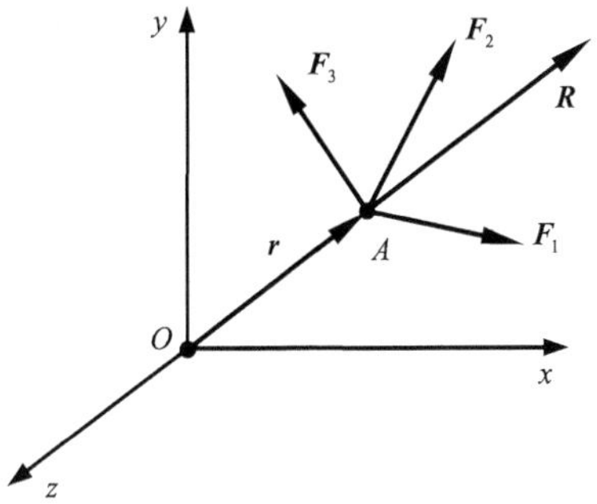

Fig. 2.4

We thus conclude that the moment of the resultant of concurrent forces about a point is equal to the vector sum of the moments of the various concurrent forces about the same point. This relation is known as the principle of moments or Varignon's theorem.

Example 2.2 A plate is suspended from two chains *AG* and *BH*, Fig. E2.2a. Knowing that the tension in *BH* is 200 N, determine (a) the moment about *A* of the force exerted by the chain *BH*, (b) the smallest force applied at *E* which creates the same moment about *A*.

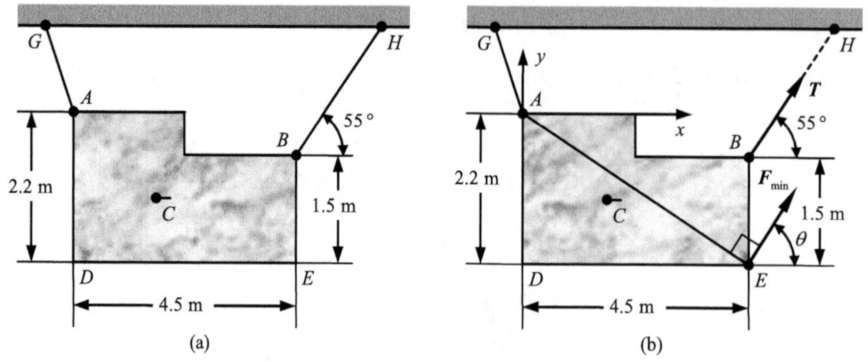

Fig. E2.2

Solution (a) Establishing the coordinate system Axy shown in Fig. E2.2b, then we have

$$x_B = 4.5\text{ m}, y_B = -0.7\text{ m}; T_x = T\cos 55° = 114.72\text{ N}, T_y = T\sin 55° = 163.83\text{ N}$$

Using the principle of moments yields

$$M_A(\boldsymbol{T}) = M_A(\boldsymbol{T}_x) + M_A(\boldsymbol{T}_y) = -y_B T_x + x_B T_y = 817.54\text{ N}\cdot\text{m}$$

(b) Using $M_A(\boldsymbol{F}_{\min}) = r_E F_{\min}\sin 90° = M_A(\boldsymbol{T})$ gives

$$F_{\min} = \frac{M_A(\boldsymbol{T})}{r_E \sin 90°} = 163.21\text{ N}, \ \theta = \arctan\frac{DE}{AD} = 63.95°$$

2.4 Moment of Couple

Two forces $\boldsymbol{F}$ and $\boldsymbol{F}'$ having equal magnitude, parallel lines of action, and opposite directions will form a couple, Fig. 2.5. Clearly, the sum of the components of the two forces in any direction is zero. The sum of the moments of the two forces about an arbitrary point, however, is not zero. Therefore, the two forces do not translate the body, but have tendency to rotate it.

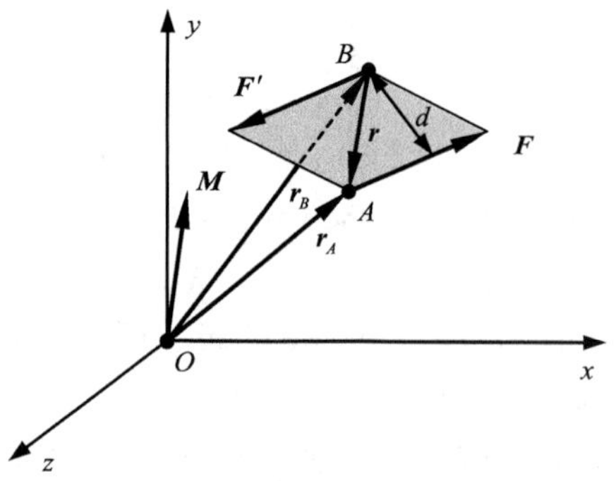

Fig. 2.5

Assuming that $\boldsymbol{r}_A$ and $\boldsymbol{r}_B$ are the position vectors of the points of application of $\boldsymbol{F}$ and $\boldsymbol{F}'$, then the sum of the moments of $\boldsymbol{F}$ and $\boldsymbol{F}'$ about O can be expressed as $\boldsymbol{M}_O(\boldsymbol{F},\boldsymbol{F}') = \boldsymbol{r}_A \times \boldsymbol{F} + \boldsymbol{r}_B \times \boldsymbol{F}' = \boldsymbol{r}_A \times \boldsymbol{F} + \boldsymbol{r}_B \times (-\boldsymbol{F}) = (\boldsymbol{r}_A - \boldsymbol{r}_B) \times \boldsymbol{F} = \boldsymbol{r} \times \boldsymbol{F}$. $\boldsymbol{M}_O(\boldsymbol{F},\boldsymbol{F}')$ is not dependent on the choice of O, thus it can be rewritten as

$$\boldsymbol{M} = \boldsymbol{r} \times \boldsymbol{F} \tag{2.5}$$

where $\boldsymbol{M}$ is called a couple moment.

Since a couple moment is independent of the choice of a reference point, it is a free vector which can be applied at any point of a rigid body as long as its direction remains unchanged.

Example 2.3 A block is acted upon by three couples, Fig. E2.3. Knowing that $M_1 = 10\ \text{N}\cdot\text{m}$, $M_2 = 15\ \text{N}\cdot\text{m}$, and $M_3 = 8\ \text{N}\cdot\text{m}$, replace these couples with a single equivalent couple, and specify its magnitude and direction.

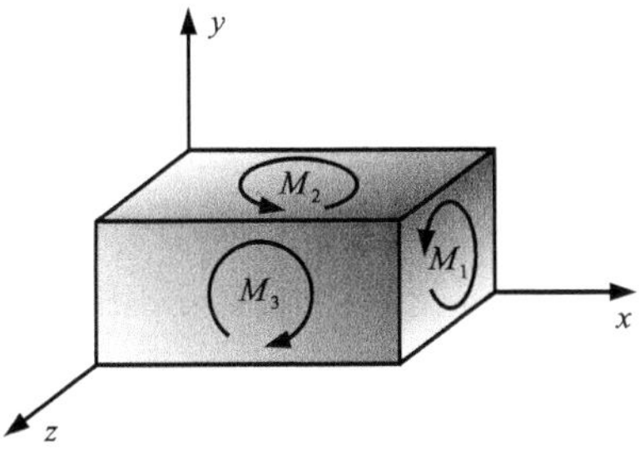

Fig. E2.3

Solution Assuming that the unit vectors along *x*, *y*, and *z* axes are denoted by $\boldsymbol{i}$, $\boldsymbol{j}$, and $\boldsymbol{k}$ respectively, then the moments of the three couples can be written as

$$\boldsymbol{M}_1 = (10\boldsymbol{i})\ \text{N}\cdot\text{m},\ \boldsymbol{M}_2 = (15\boldsymbol{j})\ \text{N}\cdot\text{m},\ \boldsymbol{M}_3 = (-8\boldsymbol{k})\ \text{N}\cdot\text{m}$$

Thus the moment of the equivalent couple can be expressed as

$$\boldsymbol{M} = \boldsymbol{M}_1 + \boldsymbol{M}_2 + \boldsymbol{M}_3 = (10\boldsymbol{i} + 15\boldsymbol{j} - 8\boldsymbol{k})\ \text{N}\cdot\text{m}$$

And the magnitude and direction of the equivalent couple moment are, respectively, given by

$$M = \sqrt{10^2 + 15^2 + (-8)^2} = 19.72\ \text{N}\cdot\text{m}$$

$$\theta_x = \arccos(\frac{10}{19.72}) = 59.53°,\ \theta_y = \arccos(\frac{15}{19.72}) = 40.48°,\ \theta_z = \arccos(\frac{-8}{19.72}) = 113.93°$$

2.5 Equivalence of Force

A force can be applied at any point on its line of action without changing its effect on the rigid body. Force $\boldsymbol{F}$ acting at point *O*, Fig. 2.6a, can be replaced by force $\boldsymbol{F}'$, Fig. 2.6b, of the same magnitude and the same direction, but acting at a different point O' on the same line of action. $\boldsymbol{F}$ and $\boldsymbol{F}'$ are equivalent since they have the same effect on the rigid body.

This is the principle of transmissibility which shows that the effect of a force on a rigid body remains unchanged provided the force acting on the rigid body is moved along its line of action. Thus a force acting on a rigid body is a sliding vector.

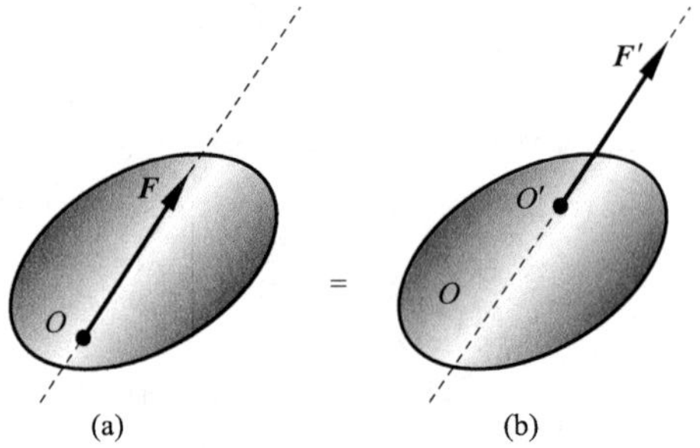

Fig. 2.6

Force $\boldsymbol{F}$ acts on a rigid body at point A defined by position vector $\boldsymbol{r}$, Fig. 2.7a. From the principle of transmissibility, $\boldsymbol{F}$ cannot be moved to O which does not lie on the line of action without modifying the action of $\boldsymbol{F}$ on the rigid body. We can, however, attach two forces $\boldsymbol{F}'$ and $\boldsymbol{F}''$ at O, $\boldsymbol{F}'=\boldsymbol{F}$ and $\boldsymbol{F}''=-\boldsymbol{F}$, without modifying the action of $\boldsymbol{F}$ on the rigid body, Fig. 2.7b. As a result of this transformation, $\boldsymbol{F}'$ is now applied at O, and $\boldsymbol{F}$ and $\boldsymbol{F}''$ will form a couple of moment $\boldsymbol{M}_O=\boldsymbol{r}\times\boldsymbol{F}$, Fig. 2.7c.

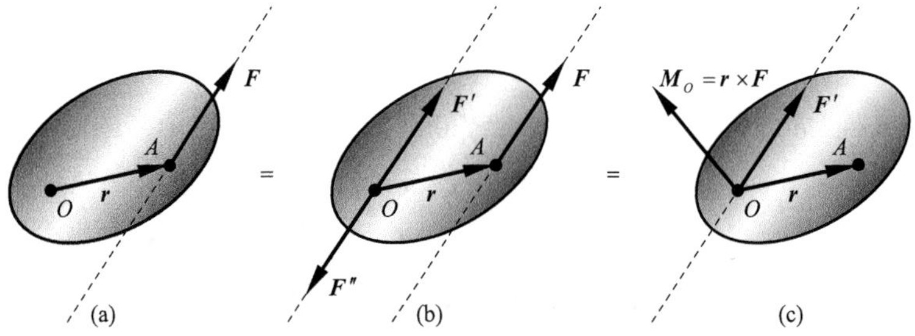

Fig. 2.7

We thus conclude that any force $\boldsymbol{F}$ acting on a rigid body can be moved to an arbitrary point O provided that a couple is added whose moment is equal to the moment $\boldsymbol{M}_O$ of $\boldsymbol{F}$ about O. Since $\boldsymbol{M}_O$ is a free vector, it may be applied anywhere; for convenience, however, the couple vector is usually attached at O.

Example 2.4 A vertical force $\boldsymbol{F}$ acts at C of a planar truss, Fig. E2.4a. Knowing that $F=80\text{ N}$, replace $\boldsymbol{F}$ with an equivalent force-couple system at G.

Solution The equivalent force-couple system, Fig. E2.4b, can be expressed as

$$\boldsymbol{F}'=\boldsymbol{F}=(-80\boldsymbol{j})\text{ N},\quad M_G(\boldsymbol{F})=480\text{ N}\cdot\text{m}$$

where $\boldsymbol{j}$ is the unit vector along the positive y axis.

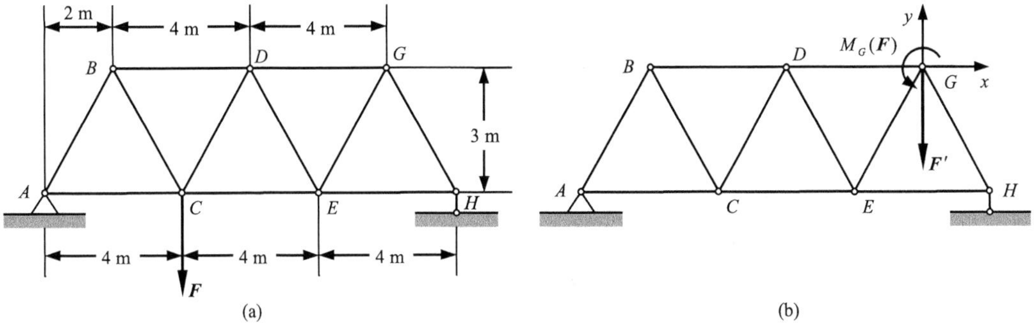

Fig. E2.4

2.6 Reduction of Force System to Force-Couple System

Forces $\boldsymbol{F}_1$, $\boldsymbol{F}_2$, $\boldsymbol{F}_3$ act on a rigid body at points A_1, A_2, A_3 defined by position vectors $\boldsymbol{r}_1$, $\boldsymbol{r}_2$, $\boldsymbol{r}_3$, respectively, Fig. 2.8a. $\boldsymbol{F}_1$ can be moved from A_1 to O if a couple moment $\boldsymbol{M}_O(\boldsymbol{F}_1)=\boldsymbol{r}_1\times\boldsymbol{F}_1$ is added to the system. Repeating this with $\boldsymbol{F}_2$ and $\boldsymbol{F}_3$, we will obtain a new force system consisting of $\boldsymbol{F}_1'$, $\boldsymbol{F}_2'$, $\boldsymbol{F}_3'$, $\boldsymbol{M}_O(\boldsymbol{F}_1)=\boldsymbol{r}_1\times\boldsymbol{F}_1$, $\boldsymbol{M}_O(\boldsymbol{F}_2)=\boldsymbol{r}_2\times\boldsymbol{F}_2$, $\boldsymbol{M}_O(\boldsymbol{F}_3)=\boldsymbol{r}_3\times\boldsymbol{F}_3$, Fig. 2.8b. In this new force system, $\boldsymbol{F}_1'$, $\boldsymbol{F}_2'$, $\boldsymbol{F}_3'$ can be replaced by force $\boldsymbol{R}'$, and $\boldsymbol{M}_O(\boldsymbol{F}_1)$, $\boldsymbol{M}_O(\boldsymbol{F}_2)$, $\boldsymbol{M}_O(\boldsymbol{F}_3)$ can also be replaced by a couple moment $\boldsymbol{M}_O$, Fig. 2.8c. $\boldsymbol{R}'$ and $\boldsymbol{M}_O$ can be written as

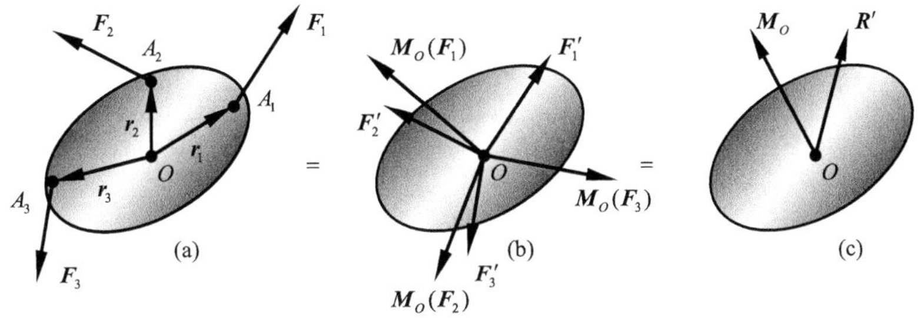

Fig. 2.8

$$\boldsymbol{R}'=\sum\boldsymbol{F}'=\sum\boldsymbol{F},\ \boldsymbol{M}_O=\sum\boldsymbol{M}_O(\boldsymbol{F}) \tag{2.6}$$

which shows that $\boldsymbol{R}'$ is obtained by adding all the forces, and that $\boldsymbol{M}_O$ is obtained by adding the moments about O of all the forces.

2.7 Reduction of Force-Couple System

Any system of forces acting on a rigid body can be reduced to an equivalent force-couple system at O consisting of force $\boldsymbol{R}'$ and couple moment $\boldsymbol{M}_O$. The simplified results of a force system are analyzed as follows:

(1) If $\boldsymbol{R}'=0$ and $\boldsymbol{M}_O=0$, the system of forces is in equilibrium.

(2) If $\boldsymbol{R}'=0$ but $\boldsymbol{M}_O\neq 0$, the system of forces can be reduced to a single couple of moment $\boldsymbol{M}=\boldsymbol{M}_O$, called the resultant couple of the system.

(3) If $\boldsymbol{R}'\neq 0$ but $\boldsymbol{M}_O=0$, the system of forces can be reduced to a single force $\boldsymbol{R}=\boldsymbol{R}'$, called the resultant force of the system. If $\boldsymbol{R}'\neq 0$ and $\boldsymbol{M}_O\neq 0$, where $\boldsymbol{R}'$ and $\boldsymbol{M}_O$ are perpendicular, the system of forces can be further reduced to a resultant force $\boldsymbol{R}=\boldsymbol{R}'$. $\boldsymbol{R}'$ and $\boldsymbol{M}_O$ are always perpendicular for the system of concurrent forces, coplanar forces, or parallel forces.

(4) If $\boldsymbol{R}'\neq 0$ and $\boldsymbol{M}_O\neq 0$, where $\boldsymbol{R}'$ and $\boldsymbol{M}_O$ are parallel, the system of forces cannot be further reduced. The force system of this type is called a wrench or screw. If $\boldsymbol{R}'\neq 0$ and $\boldsymbol{M}_O\neq 0$, where $\boldsymbol{R}'$ and $\boldsymbol{M}_O$ are neither perpendicular nor parallel, $\boldsymbol{M}_O$ needs to be resolved into two vector components $\boldsymbol{M}_{O\perp}$ perpendicular to $\boldsymbol{R}'$ and $\boldsymbol{M}_{O\parallel}$ parallel to $\boldsymbol{R}'$. $\boldsymbol{R}'$ and $\boldsymbol{M}_{O\perp}$ can be further reduced to a single force, but $\boldsymbol{R}'$ and $\boldsymbol{M}_{O\parallel}$ can only form a wrench or screw.

Example 2.5 A truss supports the loadings shown in Fig. E2.5(a). Knowing that $F_1=160$ N, $F_2=150$ N, and $F_3=80$ N, determine (a) the equivalent force acting on the truss and (b) the intersection of its line of action with line *AH*.

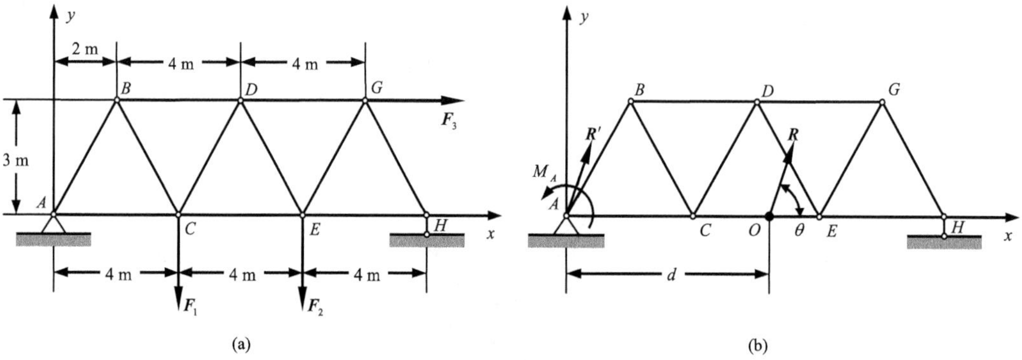

Fig. E2.5

Solution (a) Reduction of the original force system to an equivalent force-couple system at *A*, Fig. E2.5b. Assuming that the unit vectors along *x* and *y* axes are denoted by $\boldsymbol{i}$ and $\boldsymbol{j}$ respectively, and using $\boldsymbol{R}'=\sum \boldsymbol{F}$ and $M_A=\sum M_A(\boldsymbol{F})$, we get

$$\boldsymbol{R}'=\boldsymbol{F}_1+\boldsymbol{F}_2+\boldsymbol{F}_3=(80\boldsymbol{i}-310\boldsymbol{j})\text{ N},\quad M_A=M_A(\boldsymbol{F}_1)+M_A(\boldsymbol{F}_2)+M_A(\boldsymbol{F}_3)=-2080\text{ N}\cdot\text{m}$$

(b) Reduction of the equivalent force-couple system at *A* to an equivalent force at *O*, Fig. E2.5b. Using $\boldsymbol{R}=\boldsymbol{R}'$ and $M_O=M_A+M_O(\boldsymbol{R}')=M_A+(-d)R'_y=0$ yields

$$\boldsymbol{R}=\boldsymbol{R}'=(80\boldsymbol{i}-310\boldsymbol{j})\text{ N},\quad d=\frac{M_A}{R'_y}=6.71\text{ m}$$

or

$$R=\sqrt{R_x^{\,2}+R_y^{\,2}}=320.16\ \text{N},\ \theta=\arctan\frac{R_y}{R_x}=-75.53°,\ \ d=\frac{M_A}{R'_y}=6.71\ \text{m}$$

Problems

2.1 A plate is suspended from two chains *AG* and *BH*. Knowing that the tension in *AG* is 300 N, determine (a) the moment about *B* of the force exerted by the chain *AG*, (b) the smallest force applied at *D* which creates the same moment about *B*.

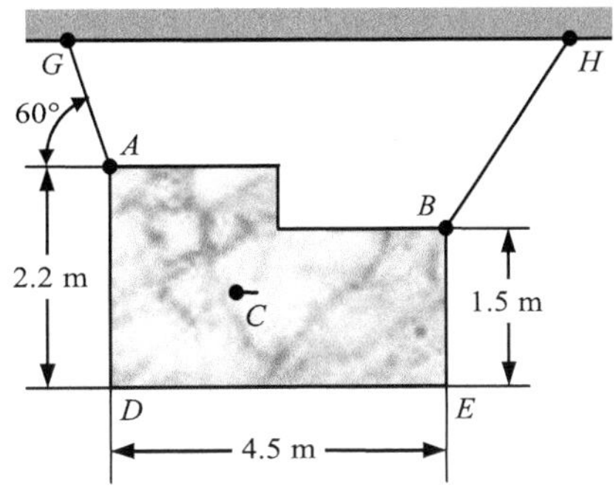

Fig. P2.1

2.2 Cable *AB* is attached to a concrete column. Knowing that the tension in cable *AB* is 500 N, determine the moment about *O* of the force exerted by the cable at *A* on the column.

2.3 Cables *AB*, *AC*, and *AD* are attached to a concrete column. Knowing that the tensions in cables *AB*, *AC*, and *AD* are 800 N, 700 N, and 500 N respectively, determine the moment about *O* of the resultant force exerted by the cables at *A* on the column.

2.4 A block is acted upon by three couples. Knowing that $M_1=10\ \text{N}\cdot\text{m}$, $M_2=15\ \text{N}\cdot\text{m}$, and $M_3=8\ \text{N}\cdot\text{m}$, replace these couples with a single equivalent couple, and specify its magnitude and the direction of its axis.

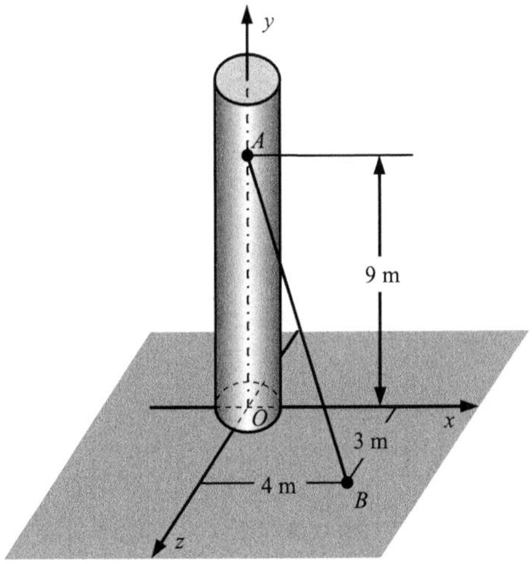

Fig. P2.2

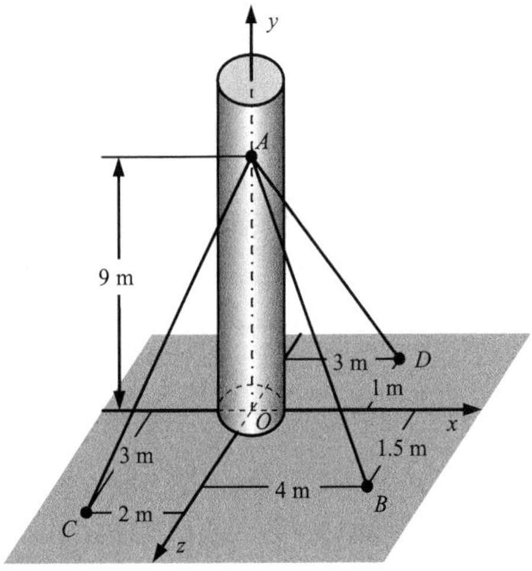

Fig. P2.3

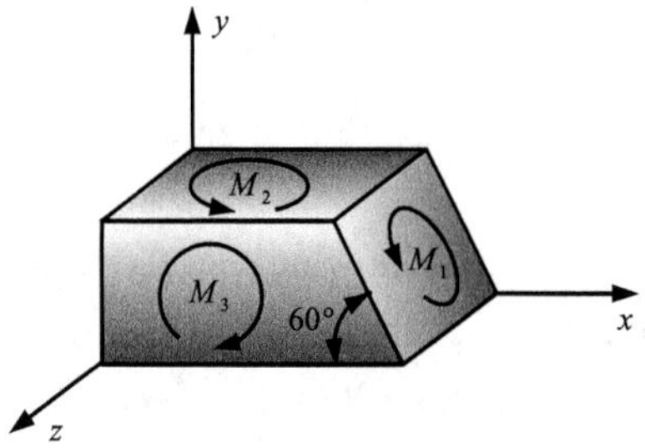

Fig. P2.4

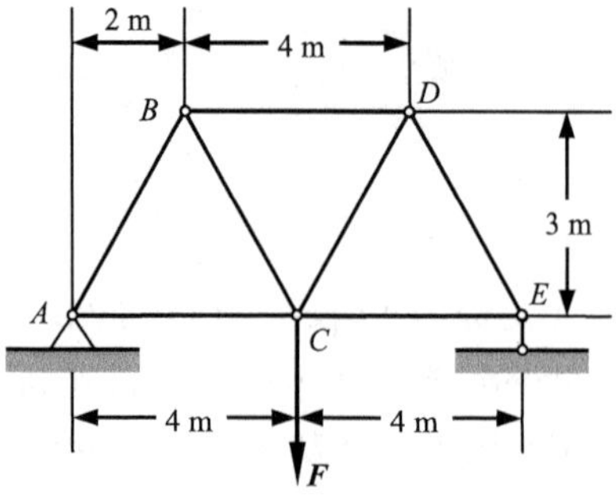

Fig. P2.5

2.5 A vertical force $\boldsymbol{F}$ acts at C of a planar truss. Knowing that $F = 100\ \text{N}$, replace $\boldsymbol{F}$ with an equivalent force-couple system at B.

2.6 A truss supports the loadings as shown. Knowing that $F_1 = F_2 = 100\ \text{N}$, and $F_3 = 90\ \text{N}$, determine the equivalent force acting on the truss and the intersection of its line of action with line AJ.

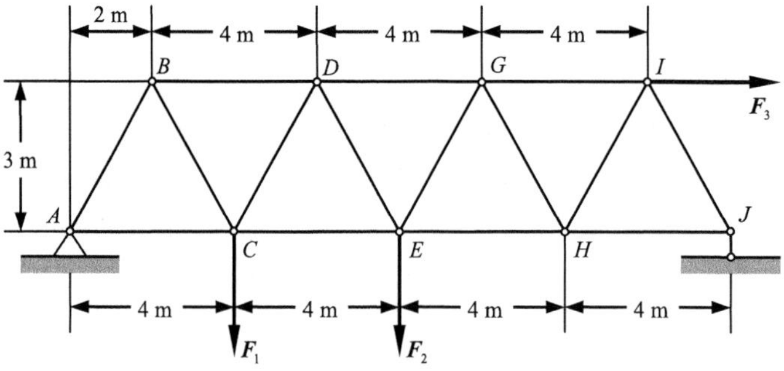

Fig. P2.6

2.7 A cantilever beam AB supports a distributed loading as shown. Knowing that $a = 2\ \text{m}$ and $q = 100\ \text{N/m}$, replace the distributed loading with an equivalent force-couple system at B.

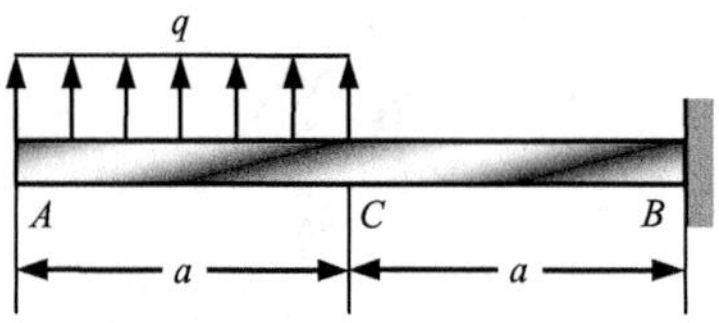

Fig. P2.7

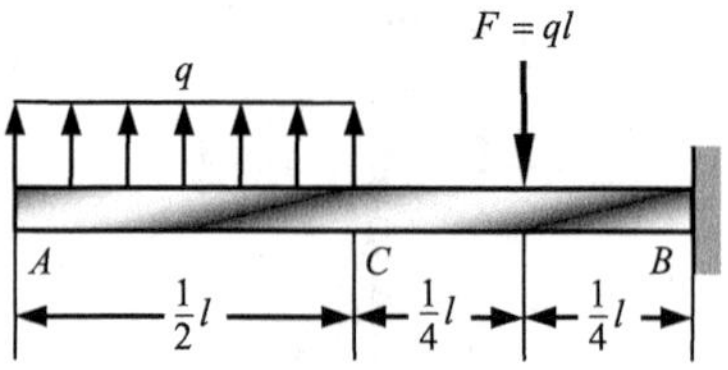

Fig. P2.8

2.8 A cantilever beam AB supports the loadings as shown. Knowing that $l = 8\ \text{m}$ and $q = 80\ \text{N/m}$, determine the equivalent force acting on the beam and the intersection of its line of action with the beam.

Chapter 3 Statics of Rigid Bodies

A rigid body can be considered as a combination of a large number of particles in which all the particles remain at a fixed distance from one another both before and after the action of forces, i.e., a rigid body is defined as one which does not deform when it is subjected to the action of forces.

The forces acting on a rigid body can be reduced to a force-couple system at some arbitrary point. When the force and the couple are both equal to zero, the rigid body is in equilibrium. In solving problems concerning the equilibrium of a rigid body, we must consider all the forces acting on it and exclude any force which is not directly applied to it.

3.1 Equilibrium of Nonconcurrent Coplanar Force System

The equations of equilibrium for a rigid body subjected to nonconcurrent coplanar forces can be expressed as

$$\sum F_x = 0,\ \sum F_y = 0,\ \sum M_A(\boldsymbol{F}) = 0 \tag{3.1}$$

where A is an arbitrary point. The three equations above are independent of each other and can be solved for not more than three unknowns. The following two alternative sets of equilibrium equations can also be often used for solving equilibrium problems.

The first alternative set of equilibrium equations can be written as

$$\sum F_x = 0,\ \sum M_A(\boldsymbol{F}) = 0,\ \sum M_B(\boldsymbol{F}) = 0 \tag{3.2}$$

where the line connecting points A and B is not perpendicular to the x axis.

The second alternative set of equilibrium equations can be expressed as

$$\sum M_A(\boldsymbol{F}) = 0,\ \sum M_B(\boldsymbol{F}) = 0,\ \sum M_C(\boldsymbol{F}) = 0 \tag{3.3}$$

where points A, B, and C do not lie on the same line.

Example 3.1 Member *ABCD* is supported at *C* by a pin and connected by cable *AED* which passes over a fixed pulley at *E*, Fig. E3.1a. Knowing that $F = 150\ \text{N}$ and neglecting friction, determine the tension in the cable and the reaction at *C*.

Solution Taking member *ABCD* as a free body and drawing the free-body diagram, Fig. E3.1b, then we have

$$\sum F_x = 0,\ R_{Cx} + T_1 = 0$$

$$\sum F_y = 0,\ R_{Cy} + T_2 - F = 0$$

$$\sum M_C = 0,\ F \times 1 - T_1 \times 3 + T_2 \times 1.8 = 0$$

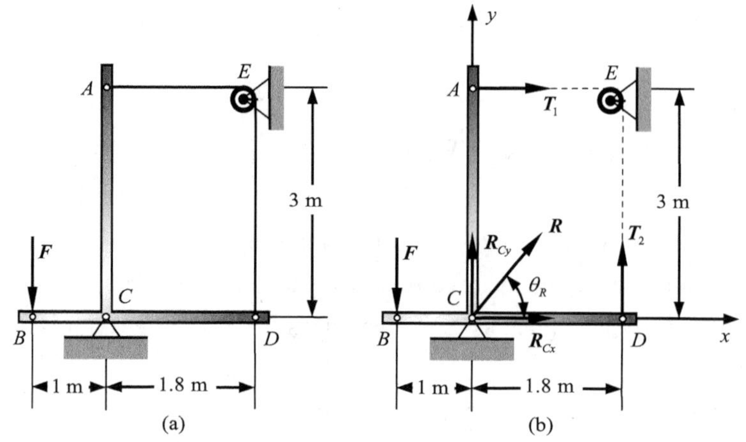

Fig. E3.1

Using $F = 150\text{ N}$ and $T_1 = T_2$, and solving the above equations, we can obtain

$$R_{Cx} = -125\text{ N},\ R_{Cy} = 25\text{ N},\ T_1 = T_2 = 125\text{ N}$$

Thus the tension in the cable is 125 N, and the reaction at C is

$$R = \sqrt{R_{Cx}^{\ 2} + R_{Cy}^{\ 2}} = 127.48\text{ N},\ \theta_R = 180° + \arctan\frac{R_{Cy}}{R_{Cx}} = 168.69°$$

Example 3.2 A homogeneous rod AB of weight W is attached to blocks A and B which move freely along the smooth surfaces, Fig. E3.2a. The stiffness of the spring connected to block A is k, and the spring is unstretched when the rod is horizontal. (a) Neglecting the weight of the blocks, derive an equation in W, k, a, and θ which must be satisfied when the rod is in equilibrium. (b) Determine the value of θ when $W = 45\text{ N}$, $a = 1\text{ m}$, and $k = 50\text{ N/m}$.

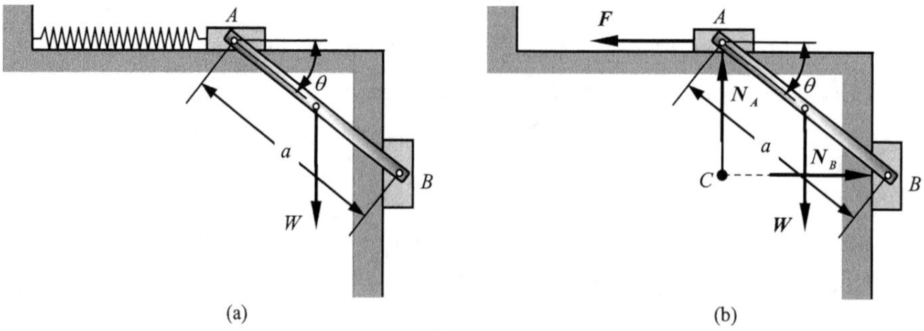

Fig. E3.2

Solution (a) Taking the rod as a free body and drawing its free-body diagram, Fig. E3.2b, we can obtain

$$\sum M_C = 0,\ Fa\sin\theta - \frac{1}{2}Wa\cos\theta = 0$$

Using $F = ka(1-\cos\theta)$ yields

$$\tan\theta - \sin\theta = \frac{W}{2ka}$$

(b) Using $W = 45\ \text{N}$, $a = 1\ \text{m}$, and $k = 50\ \text{N/m}$, thus we get

$$\tan\theta - \sin\theta = 0.45$$

from which it follows that

$$\theta = 50.76°$$

3.2 Two-Force and Three-Force Members

A particular case is the equilibrium of a two-force member subjected to two forces (or two resultants), acting at two different points respectively. It can be shown that if a two-force member is in equilibrium, the two forces must have the same magnitude, opposite directions, and the same line of action.

Another particular case is the equilibrium of a three-force member subjected to three forces (or three resultants), acting at three different points respectively. It can be shown that if a three-force member is in equilibrium, the lines of action of the three forces must be either concurrent or parallel.

Example 3.3 For the frame and loading, Fig. E3.3a, determine the reactions at A and B knowing that $F = 100\ \text{N}$ and $a = 1\ \text{m}$.

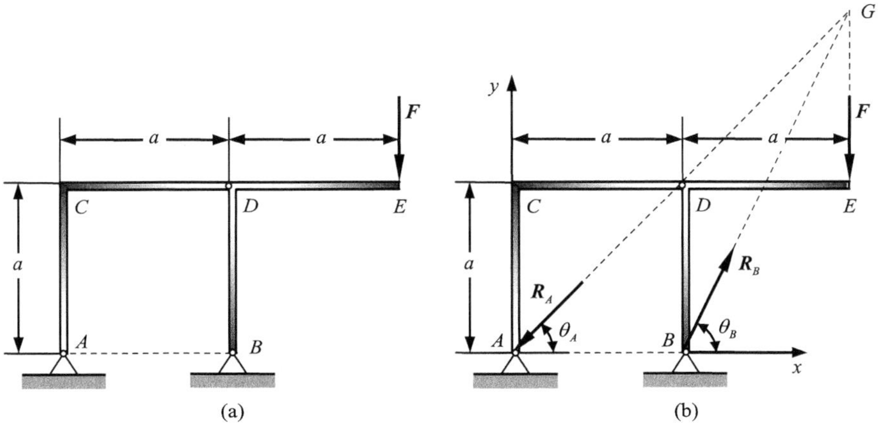

Fig. E3.3

Solution Since ACD is a two-force member in equilibrium, the line of action of the reaction $\boldsymbol{R}_A$ at A must pass through A and D, Fig. E3.3b. Since the entire structure is in equilibrium under the action of the three forces $\boldsymbol{R}_A$ at A, $\boldsymbol{R}_B$ at B, and $\boldsymbol{F}$ at E, the line of action of $\boldsymbol{R}_B$ must pass through the intersection G of the lines of action of $\boldsymbol{R}_A$ and $\boldsymbol{F}$, Fig. E3.3b.

Considering the entire structure in equilibrium, we have

$$\sum F_x = 0, \quad -R_A \cos\theta_A + R_B \cos\theta_B = 0$$

$$\sum F_y = 0, \quad -R_A \sin\theta_A + R_B \sin\theta_B - F = 0$$

where $\tan\theta_A = 1$, and $\tan\theta_B = 2$. Substituting $F = 100\ \text{N}$ yields

$$R_A = 141.4\ \text{N},\ \theta_A = 45°,\ R_B = 223.6\ \text{N},\ \theta_B = 63.43°$$

3.3 Planar Trusses

A planar truss is a two-dimensional structure consisting of straight members connected together at their extremities by pins. Although the members of a planar truss are actually joined together by bolted or welded connections, it is customary to assume that the members are connected together by smooth pins. A planar truss is designed to carry loadings which act in the plane of the structure, it is, however, often to assume that all loadings are applied at the joints of the planar truss and that the weights of the members are also applied to the joints, half of the weight of each member being applied to each of the two joints of the member.

1. Simple Trusses

A truss is designed to carry loadings, thus it must be stable under the action of loadings. The truss shown in Fig. 3.1, which is made of three members connected by three joints at A, B, and C, will be stable under the action of a load applied at C.

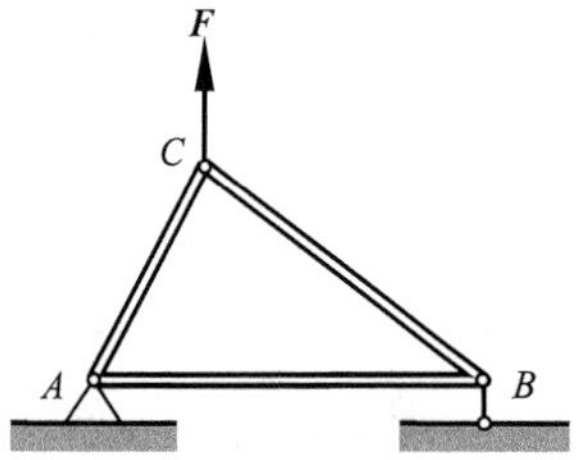

Fig. 3.1

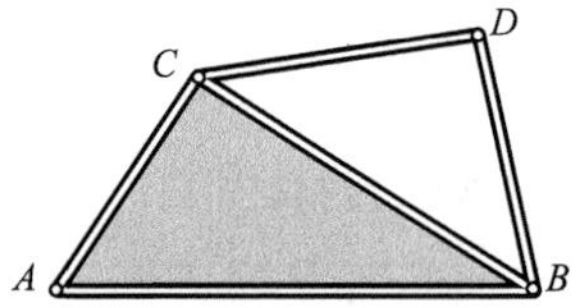

Fig. 3.2

A large truss can be obtained by adding two members BD and CD to the basic triangular truss ABC, Fig. 3.2. This can be repeated as many times as desired, and the resulting truss will be stable if each time two members are attached to two existing joints and connected at a new joint by a joint. A truss which can be constructed from a single triangular truss is called a simple truss.

Assuming that the total number of members and the total number of joints are respectively denoted by m and n in a simple truss, we have

$$m = 2n - 3 \tag{3.4}$$

2. Internal Forces of Trusses

Analysis of a truss requires the determination not only of the external forces acting on it but also of the internal forces which hold together the various parts of the truss. There are two methods, the method of joints and the method of sections, which can be used to determine the internal forces of a truss.

3. Method of Joints

Since the entire truss is in equilibrium, each joint of the truss must also be in equilibrium. The internal forces of the truss can be determined by drawing the free-body diagram of each joint and by solving the equations of equilibrium for each joint.

Example 3.4 Knowing $F = 80$ N and using the method of joints, determine the force in each member of the truss shown and state whether each member is in tension (T) or compression (C), Fig. E3.4a.

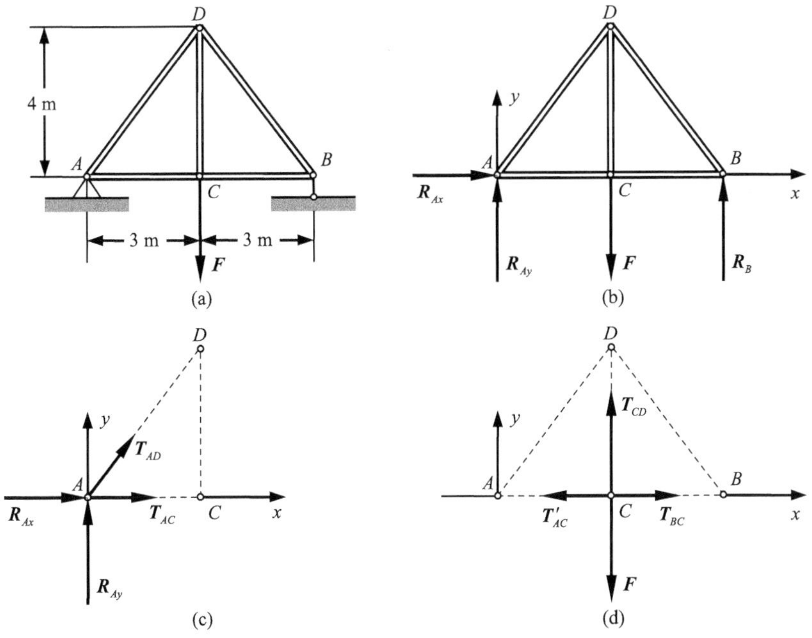

Fig. E3.4

Solution (1) Taking the entire truss as free body and considering its equilibrium, Fig. E3.4b, we have

$$\sum F_x = 0,\ R_{Ax} = 0$$
$$\sum M_A = 0,\ R_B \times 6 - F \times 3 = 0$$
$$\sum M_B = 0,\ -R_{Ay} \times 6 + F \times 3 = 0$$

Solving the above equations gives

$$R_{Ax}=0,\ R_{Ay}=40\ \text{N},\ R_B=40\ \text{N}$$

(2) Taking joint A as free body and considering its equilibrium, Fig. E3.4c, we have

$$\sum F_x=0,\ R_{Ax}+T_{AC}+T_{AD}\times\frac{3}{5}=0$$

$$\sum F_y=0,\ R_{Ay}+T_{AD}\times\frac{4}{5}=0$$

Solving the above equations yields

$$T_{AC}=30\ \text{N (T)},\ T_{AD}=-50\ \text{N (C)}$$

(3) Taking joint C as free body and considering its equilibrium, Fig. E3.4d, we have

$$\sum F_x=0,\ T_{BC}-T'_{AC}=0$$

$$\sum F_y=0,\ T_{CD}-F=0$$

Solving the above equations, we obtain

$$T_{BC}=30\ \text{N (T)},\ T_{CD}=80\ \text{N (T)}$$

(4) Similarly, taking joint B or D as free body and considering its equilibrium, we can obtain

$$T_{BD}=-50\ \text{N (C)}$$

4. Zero-Force Members

If the internal force in a member is zero, this member is a zero-force member. A zero-force member is used to increase the stability of a truss and to provide support if the loading applied to the truss is changed.

Example 3.5 For the truss and loading, determine the zero-force members shown in Fig. E3.5a.

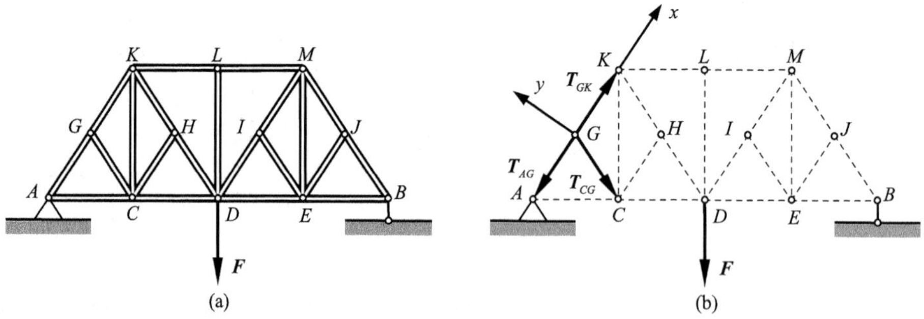

Fig. E3.5

Solution Considering joint G, Fig. E3.5b, and using $\sum F_y=0$, we can find

$$T_{CG}=0$$

Similarly, by inspection of each joint of the truss, we can find

$$T_{CH} = T_{CK} = T_{DL} = T_{EI} = T_{EJ} = T_{EM} = 0$$

5. Method of Sections

The method of joints is most effective when the internal forces in all the members of a truss are to be determined. If, however, the forces in only very few members are desired, the method of sections will be more efficient.

Since the entire truss is in equilibrium, any part of the truss must also be in equilibrium. The internal forces of any part of the truss can be determined by drawing the free-body diagram of this part and by solving equations of equilibrium.

Example 3.6 A truss is loaded as shown in Fig. E3.6a. Knowing that $F_1 = 80$ N and $F_2 = 40$ N and using the method of sections, determine the internal forces in members *CD*, *CH*, and *GH*.

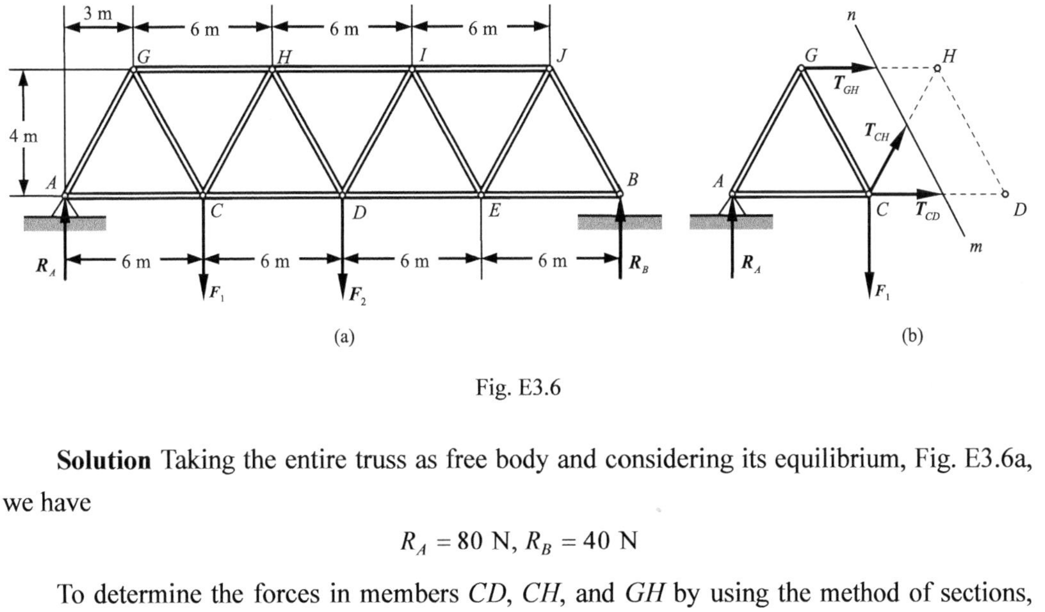

Fig. E3.6

Solution Taking the entire truss as free body and considering its equilibrium, Fig. E3.6a, we have

$$R_A = 80\text{ N},\ R_B = 40\text{ N}$$

To determine the forces in members *CD*, *CH*, and *GH* by using the method of sections, these members need to be cut, Fig. E3.6b. Considering the equilibrium of the left part of the cut trusses, we can obtain

$$\sum M_H = 0,\quad -R_A \times 9 + F_1 \times 3 + T_{CD} \times 4 = 0$$

$$\sum M_C = 0,\quad -R_A \times 6 - T_{GH} \times 4 = 0$$

$$\sum F_y = 0,\ R_A - F_1 + T_{CH} \times \frac{4}{5} = 0$$

Solving the above equations, we obtain

$$T_{CD} = 120\text{ N},\ T_{CH} = 0,\ T_{GH} = -120\text{ N}$$

3.4 Equilibrium of Nonconcurrent Noncoplanar Force System

The equations of equilibrium for a rigid body subjected to nonconcurrent noncoplanar forces can be expressed as

$$\sum F_x = 0,\ \sum F_y = 0,\ \sum F_z = 0,\ \sum M_x(\boldsymbol{F}) = 0,\ \sum M_y(\boldsymbol{F}) = 0,\ \sum M_z(\boldsymbol{F}) = 0 \tag{3.5}$$

The above equations are independent of each other and can be solved for not more than six unknowns.

Example 3.7 A homogeneous square plate has a weight of 150 N and is supported by three vertical cables, Fig. E3.7a. Determine the tension in each cable.

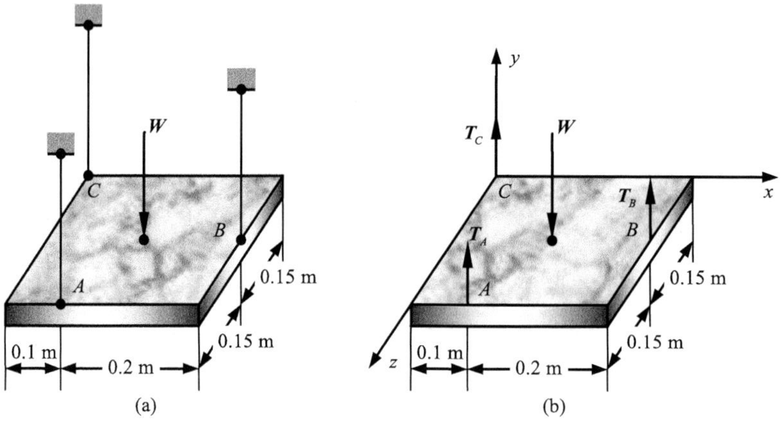

Fig. E3.7

Solution Take the plate as a free body, Fig. E3.7b. From the equilibrium of the plate, we have

$$\sum F_y = 0,\ T_A + T_B + T_C - W = 0$$

$$\sum M_x = 0,\ -T_A \times 0.3 - T_B \times 0.15 + W \times 0.15 = 0$$

$$\sum M_z = 0,\ T_A \times 0.1 + T_B \times 0.3 - W \times 0.15 = 0$$

Solving the equations yields

$$T_A = 45\ \text{N},\ T_B = 60\ \text{N},\ T_C = 45\ \text{N}$$

Problems

3.1 A T-shaped member *ABCD* is supported at *C* by a pin and connected by cable *AED* which passes over a fixed pulley at *E*. Knowing that $q = 250$ N/m and neglecting friction, determine the tension in the cable and the reaction at *C*.

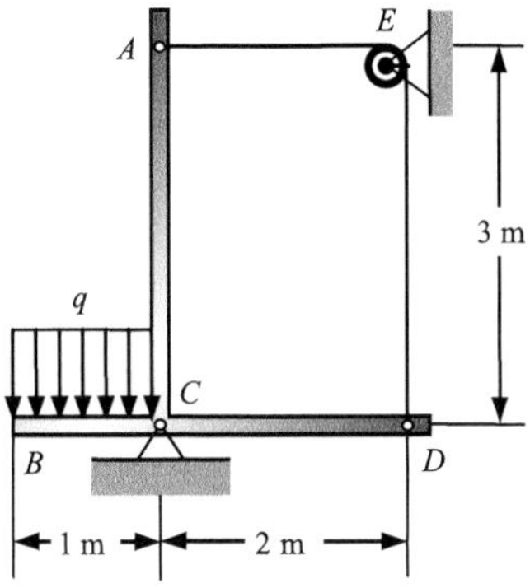

Fig. P3.1

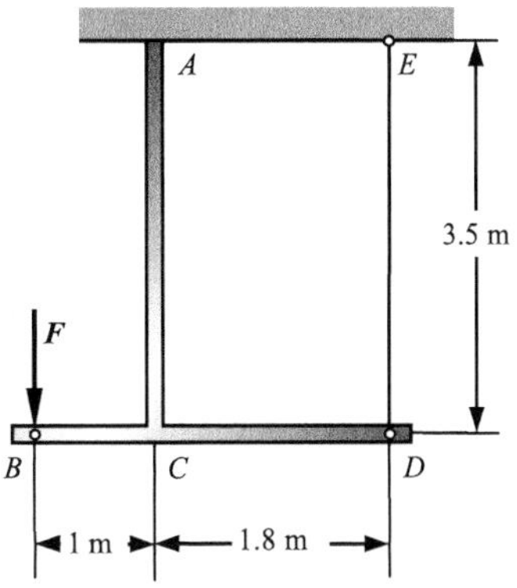

Fig. P3.2

3.2 Member *ABCD* is fixed at *A* and connected by cable *DE*. Knowing that $F = 250$ N and that the tension in the cable is $T = 50$ N, determine the reaction at the fixed support *A*.

3.3 A homogeneous rod *AB* of weight *W* is attached to blocks *A* and *B* which move freely along the smooth surfaces. The stiffness of the spring connected to block *A* is *k*, and the spring is unstretched when the rod is horizontal. (a) Assuming that the weight of each of the blocks is *W*, derive an equation in *W*, *k*, *a*, and θ which must be satisfied when the rod is in equilibrium. (b) Determine the value of θ when $W = 15$ N, $a = 1$ m, and $k = 50$ N/m.

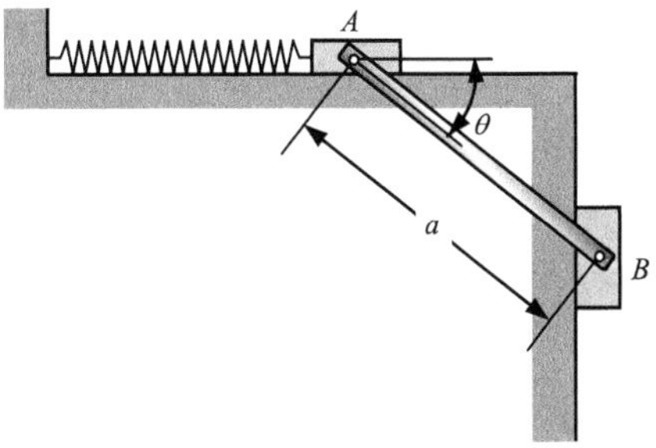

Fig. P3.3

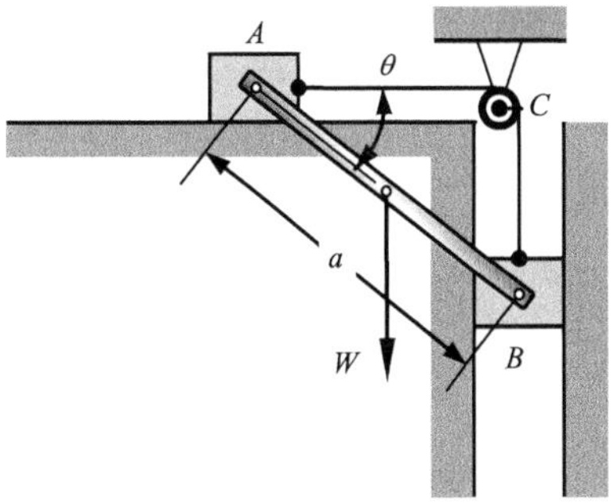

Fig. P3.4

3.4 A homogeneous rod *AB*, of weight *W*, is attached to blocks *A* and *B*, which move freely along the smooth surfaces. The blocks are connected by a cable which passes over a pulley at *C*. (a) Neglecting the weight of the blocks, express the tension in the cable in terms of *W* and θ when the rod is in equilibrium. (b) Determine the value of θ for which the tension in the cable is equal to *W*.

3.5 The maximum allowable value of each of the reactions at *A* and *B* is 450 N. Neglecting the weight of the beam, determine the range of values of the distance *d* for which the beam is safe.

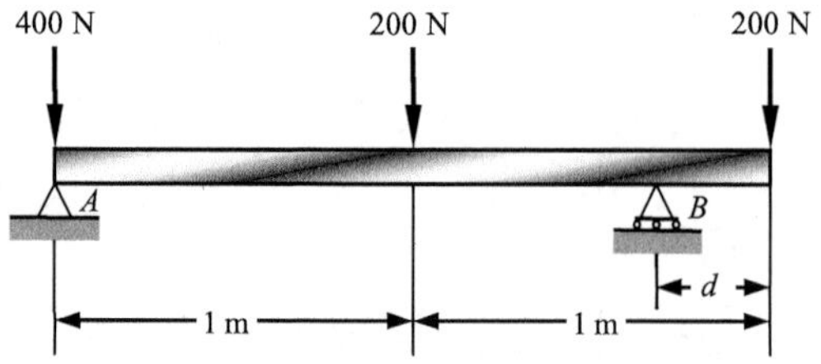

Fig. P3.5

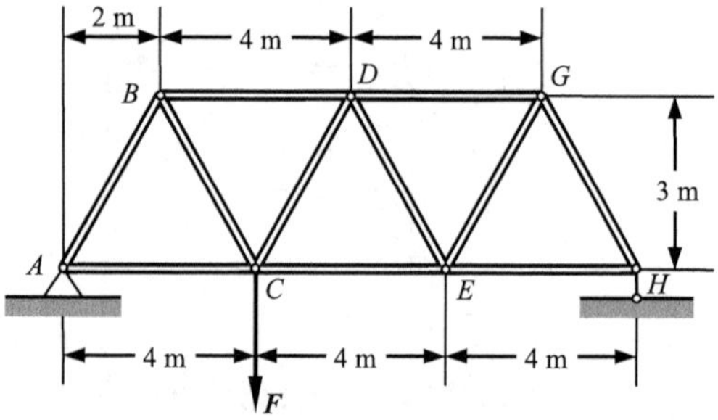

Fig. P3.6

3.6 A vertical force $\boldsymbol{F}$ acts at C of a planar truss. Knowing that $F = 80\ \mathrm{N}$, determine the reactions at A and H.

3.7 A truss supports the loadings shown. Knowing that $F_1 = 160\ \mathrm{N}$, $F_2 = 150\ \mathrm{N}$, and $F_3 = 80\ \mathrm{N}$, determine the reactions at A and H.

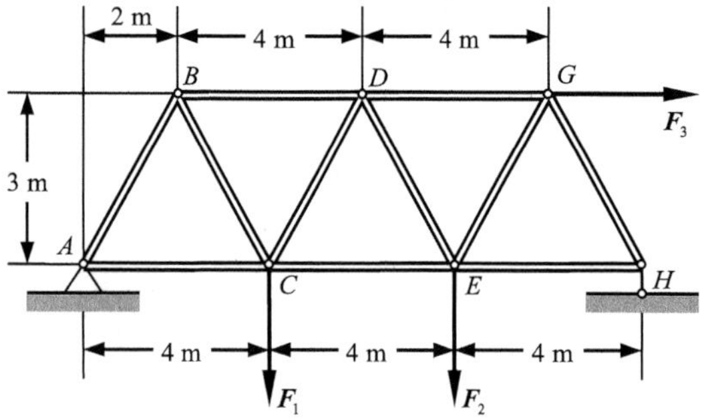

Fig. P3.7

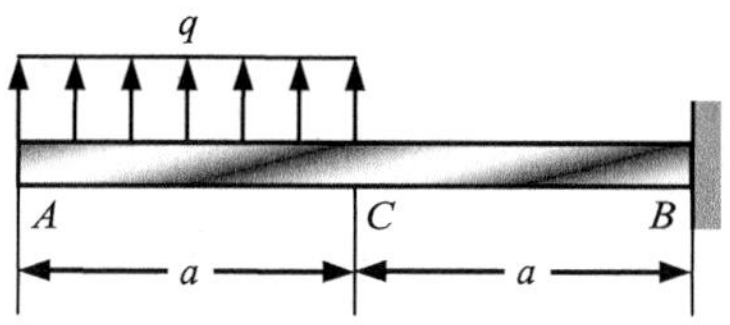

Fig. P3.8

3.8 A cantilever beam AB supports a distributed loading. Knowing that $a = 2\ \mathrm{m}$ and $q = 100\ \mathrm{N/m}$, determine the reactions at B.

3.9 A cantilever beam AB supports the loadings shown. Knowing that $l = 8\ \mathrm{m}$ and $q = 80\ \mathrm{N/m}$, determine the reactions at B.

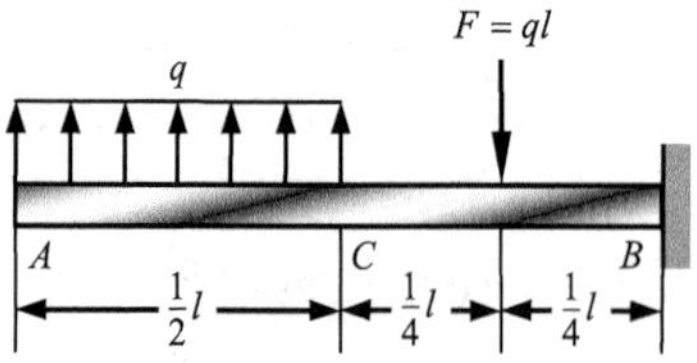

Fig. P3.9

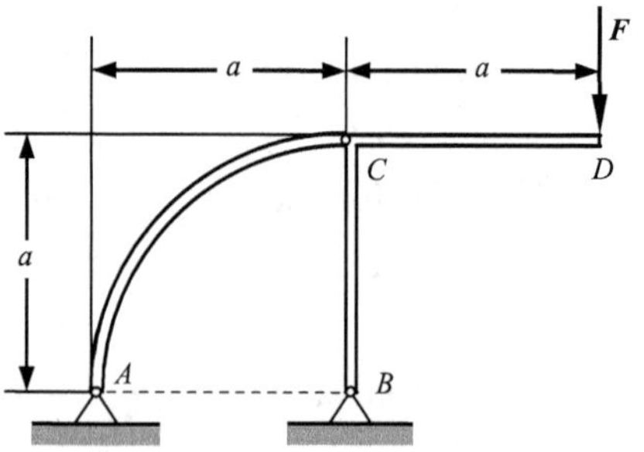

Fig. P3.10

3.10 For the frame and loading shown, determine the reactions at A and B knowing $F = 200\ \text{N}$ and $a = 1\ \text{m}$.

3.11 For the frame and loading shown, knowing that $q = 100\ \text{N/m}$ and $a = 1\ \text{m}$, determine the reactions at A and B.

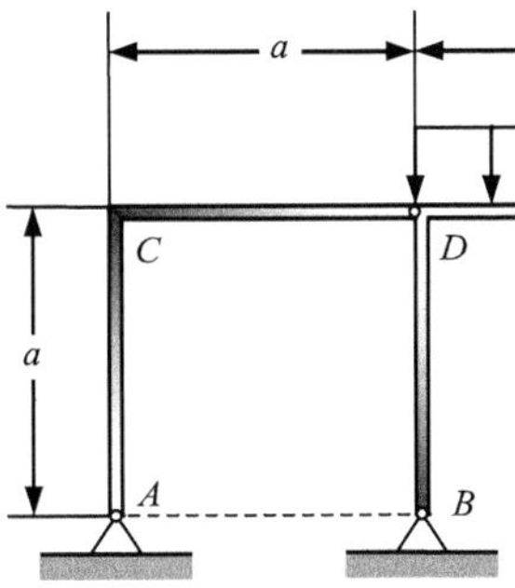

Fig. P3.11

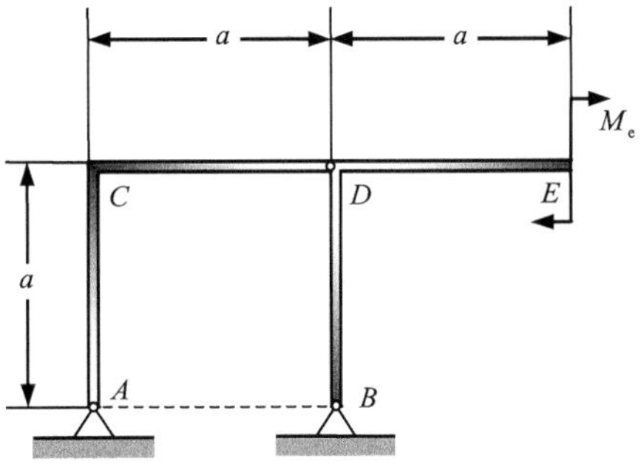

Fig. P3.12

3.12 Knowing $M_e = 100\ \text{N} \cdot \text{m}$ and $a = 1\ \text{m}$, determine the reactions at A and B.

3.13 Using the method of joints determine the force in each member of the truss shown and state whether each member is in tension or compression.

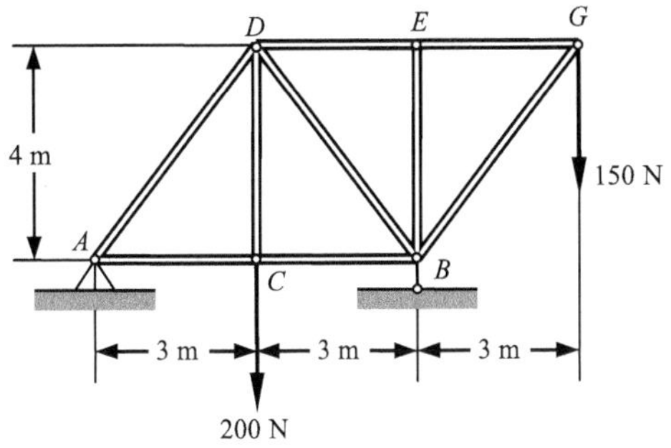

Fig. P3.13

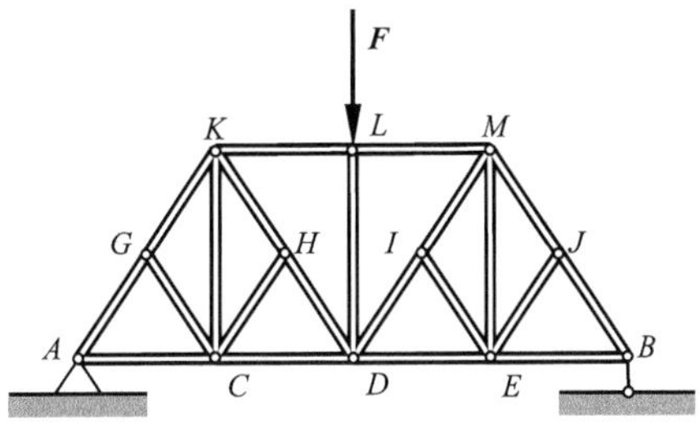

Fig. P3.14

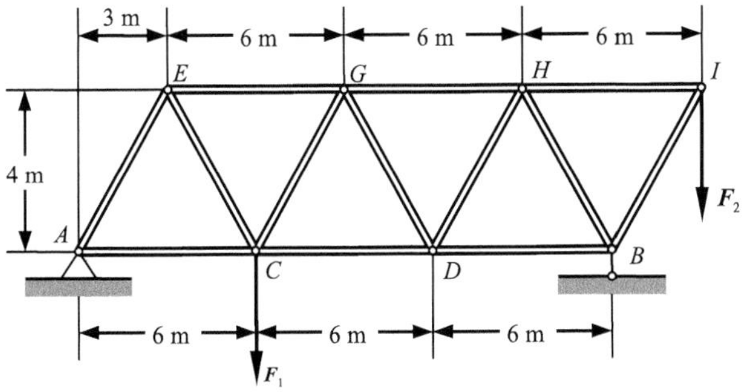

Fig. P3.15

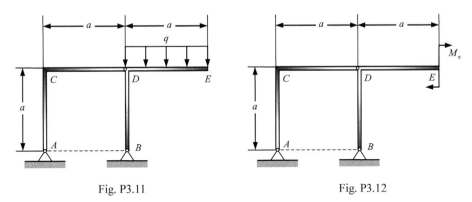

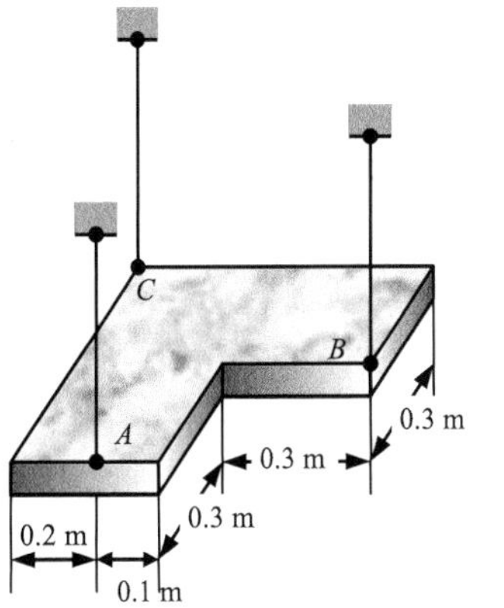

Fig. P3.16

3.14　For the truss and loading shown, determine the zero-force members.

3.15　A truss is loaded as shown. Knowing $F_1 = 90$ N and $F_2 = 60$ N and using the method of sections, determine the forces in members *CD*, *DG*, and *GH*.

3.16　A homogeneous composite plate has a weight of 250 N and is supported by three vertical cables. Determine the tension in each cable.

Chapter 4 Friction

4.1 Laws of Friction

The phenomenon of friction is explained as follows:

(1) A block of weight $\boldsymbol{W}$ is placed on a rough horizontal surface, Fig. 4.1a. The forces acting on the block are its weight $\boldsymbol{W}$ and the normal force N exerted by the rough surface. These vertical forces do not tend to move the block along the rough surface.

(2) A horizontal force F applied to the block, Fig. 4.1b, tends to move the block along the rough surface. However, when F is small, the block does not move. This shows that a tangential force exerted by the rough surface on the block must exist to balance F. This tangential force is called the static friction force denoted by $\boldsymbol{F}_s$, which can be found by solving the equations of equilibrium for the block.

(3) If F is increased, $\boldsymbol{F}_s$ also increases, continuing to balance F, until $\boldsymbol{F}_s$ reaches the maximum static friction force $\boldsymbol{F}_{max}$, Fig. 4.1c. When $\boldsymbol{F}_{max}$ is reached, the block is on the verge of sliding, which is called impending motion.

(4) If F is further increased, $\boldsymbol{F}_s$ cannot balance F any more and the block will begin to slide. As soon as the block starts sliding, Fig. 4.1d, the friction force drops from $\boldsymbol{F}_{max}$ to a lower friction force $\boldsymbol{F}_k$ called the kinetic friction force. When the block is in motion, $\boldsymbol{F}_k$ remains approximately constant.

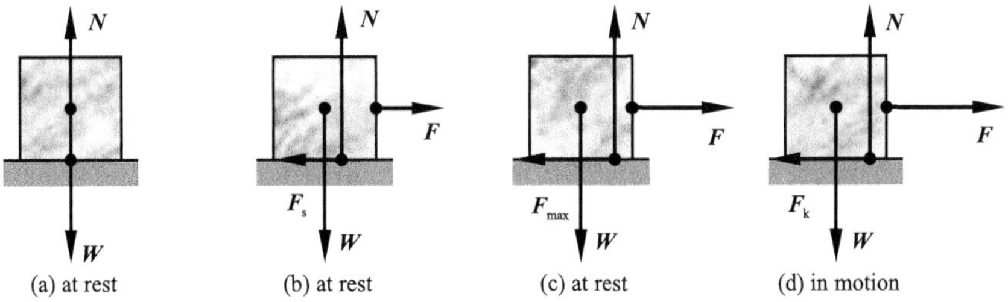

Fig. 4.1

Experimentally, it has been shown that F_{max} is proportional to N. We can thus express the law of static friction as

$$F_{max} = \mu_s N \tag{4.1}$$

where μ_s is the coefficient of static friction.

Similarly, F_k is also proportional to N. Thus the law of kinetic friction can be expressed as

$$F_k = \mu_k N \tag{4.2}$$

where μ_k is the coefficient of kinetic friction.

The coefficients of friction μ_s and μ_k are independent of the contacting area of bodies in contact. They, however, will depend on both the material characteristics and the surface nature of bodies in contact.

4.2 Angle of Static Friction

A block of weight $\boldsymbol{W}$ is placed on a rough horizontal surface and subjected to a horizontal force $\boldsymbol{F}$, Fig. 4.2. When the block is on the verge of sliding, the resultant of the normal force and the maximum static friction force is denoted by $\boldsymbol{R}_s$, and the angle between $\boldsymbol{N}$ and $\boldsymbol{R}_s$ is denoted by φ_s, where φ_s is defined as the angle of static friction. From the geometry of Fig. 4.2, we have

$$\tan\varphi_s = \frac{F_{max}}{N} = \mu_s \tag{4.3}$$

As shown in Fig. 4.3a, if $\varphi > \varphi_s$, then the block will be in motion due to the fact that $\boldsymbol{R}$ and $\boldsymbol{R}_s$ do not have the same line of action. However, if $\varphi = \varphi_s$ (Fig. 4.3b), or $\varphi < \varphi_s$ (Fig. 4.3c), then the block will be at rest, i.e., the block is self-locking.

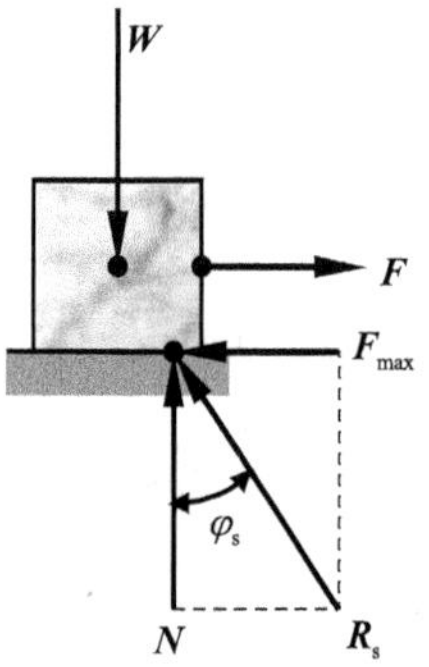

Fig. 4.2

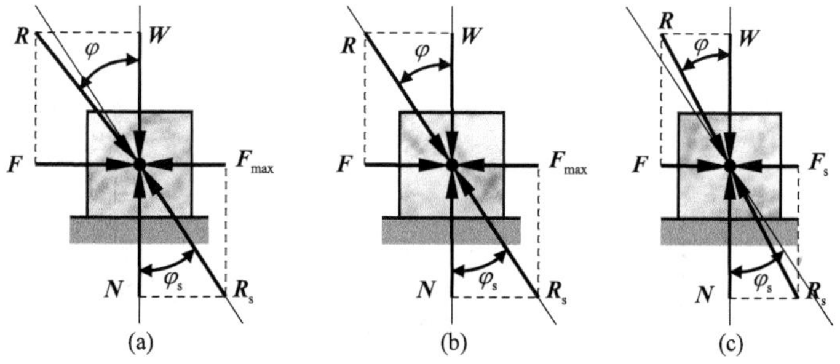

Fig. 4.3

4.3 Problems Involving Friction

There are three types of friction problems, which are described as follows:

(1) Determine whether the body under consideration will be at rest or in motion when all the forces acting on it are given and the coefficient of static friction is known.

(2) Determine the coefficient of static friction when all forces acting on a body are given and the motion of the body is impending.

(3) Determine the force acting on a body when the coefficient of static friction is known

and the motion of the body is impending.

Example 4.1 A force of magnitude $F = 200\ \mathrm{N}$ acts on a block of mass $m = 100\ \mathrm{kg}$ placed on an inclined plane of inclination angle $\theta = 30°$, Fig. E4.1a. Knowing that $\mu_s = 0.3$ and $\mu_k = 0.2$, determine whether the block is in equilibrium, and find the magnitude and direction of the friction force.

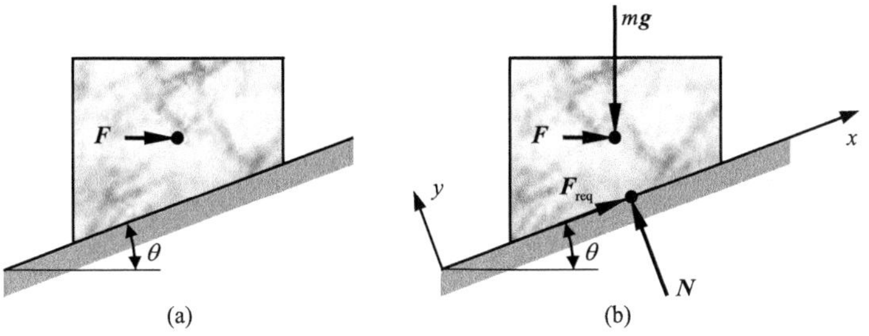

Fig. E4.1

Solution Draw the free-body diagram of the block, Fig. E4.1b. Assuming that the block is in equilibrium and that the required friction force along the plane is $\boldsymbol{F}_{req}$ directed upward and to the right, then the equations of equilibrium for the block can be expressed as

$$\sum F_x = 0,\ F_{req} + F\cos\theta - mg\sin\theta = 0$$

$$\sum F_y = 0,\ N - F\sin\theta - mg\cos\theta = 0$$

Solving the equations above for F_{req} and N, we obtain

$$F_{req} = 317.29\ \mathrm{N}$$

$$N = 949.57\ \mathrm{N}$$

The value of the maximum static friction force is equal to

$$F_{max} = \mu_s N = 284.87\ \mathrm{N}$$

Since $F_{req} > F_{max}$, equilibrium will not be maintained and the block will slide down the plane. The magnitude of friction force can be given, according to the law of kinetic friction force, by

$$F_k = \mu_k N = 189.91\ \mathrm{N}$$

and its direction is up the plane.

Example 4.2 A uniform ladder *AB*, having length l and mass m, leans against a wall, Fig. E4.2a. Assuming that the coefficient of static friction μ_s is the same at *A* and *B*, determine the smallest value of μ_s for which equilibrium is maintained when $\theta = 60°$.

Solution Assume that motion is impending at both *A* and *B*, and draw the free-body diagram of the ladder, Fig. E4.2b. Using the law of static friction, the supplementary equations can be given by

$$F_A = \mu_s N_A,\ F_B = \mu_s N_B$$

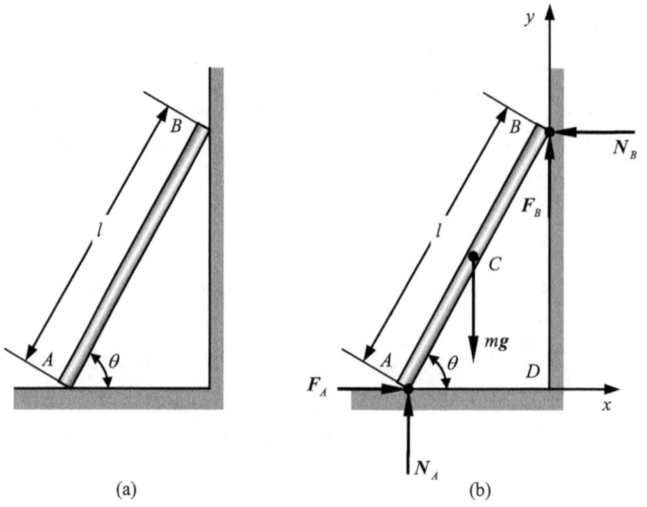

Fig. E4.2

Considering the equilibrium of the ladder, the equilibrium equations can be expressed as

$$\sum F_x = 0,\ F_A - N_B = 0$$

$$\sum F_y = 0,\ F_B + N_A - mg = 0$$

$$\sum M_D = 0,\ N_B l \sin\theta - N_A l \cos\theta + \tfrac{1}{2} mgl\cos\theta = 0$$

Solving the supplementary and equilibrium equations for μ_s, we have

$$\mu_s^2 + 2\mu_s \tan\theta - 1 = 0$$

Using $\theta = 60°$ and solving the above equation, we obtain

$$\mu_s = 0.27 \text{ or } \mu_s = -3.73$$

Physically, the positive root is possible, therefore the smallest value of μ_s is equal to 0.27.

Example 4.3 A collar *B* of weight *W* is attached to the spring *AB* and can move along the rod, Fig. E4.3a. The spring stiffness is $k = 1.8$ kN/m and the spring is unstretched when $\theta = 0$. Knowing that the coefficient of static friction between the collar and the rod is $\mu_s = 0.25$ and that the horizontal distance between points *A* and *B* is $l = 0.6$ m, determine the range of values of *W* for which equilibrium is maintained when $\theta = 30°$.

Solution Assume that the collar is in equilibrium and that the static friction force $\boldsymbol{F}_s$ is directed upward, Fig. E4.3b. From the free-body diagram of the collar, the equations of equilibrium for the collar can be expressed as

$$\sum F_x = 0,\ N - F_{spr}\cos\theta = 0$$

$$\sum F_y = 0,\ F_s + F_{spr}\sin\theta - W = 0$$

Solving these two equations for N and F_s, we obtain

$$N = F_{spr}\cos\theta$$

$$F_s = W - F_{spr}\sin\theta$$

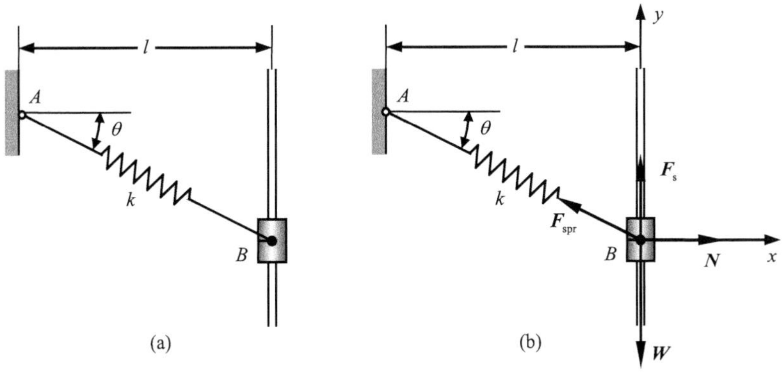

Fig. E4.3

When the collar is in equilibrium, we have

$$|F_s| \leqslant \mu_s N$$

i.e.,

$$-\mu_s F_{spr}\cos\theta \leqslant W - F_{spr}\sin\theta \leqslant \mu_s F_{spr}\cos\theta$$

or

$$F_{spr}(\sin\theta - \mu_s\cos\theta) \leqslant W \leqslant F_{spr}(\sin\theta + \mu_s\cos\theta)$$

Using $F_{spr} = kl(1/\cos\theta - 1)$ and $\theta = 30°$, then we obtain

$$47.37\ \text{N} \leqslant W \leqslant 119.71\ \text{N}$$

Problems

4.1 A force $\boldsymbol{F}$ acts on a block of mass $m = 100$ kg placed on an inclined plane of inclination angle $\theta = 30°$. Knowing that the coefficients of friction between the block and the plane are $\mu_s = 0.3$ and $\mu_k = 0.2$, determine whether the block is in equilibrium, and find the magnitude and direction of the friction force when (a) $F = 500$ N, (b) $F = 1200$ N.

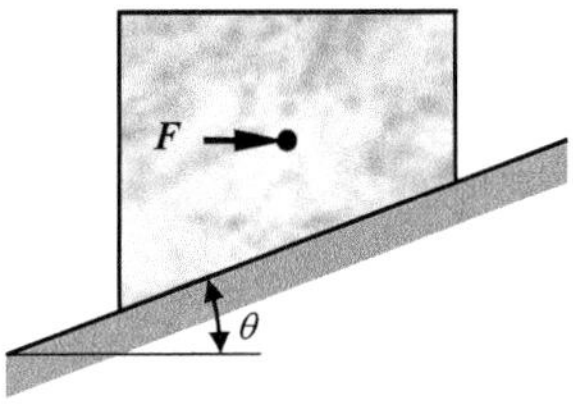

Fig. P4.1

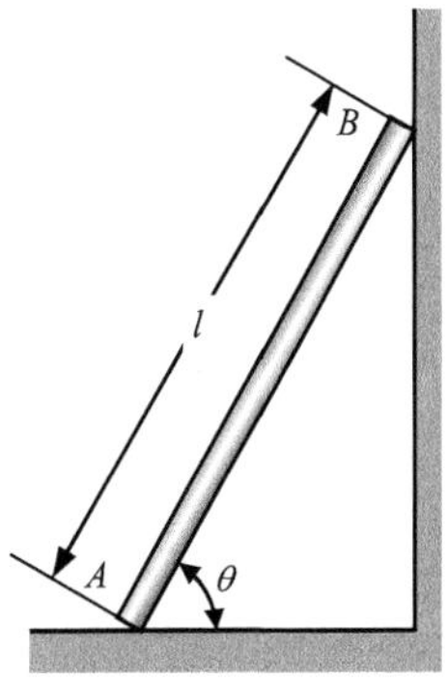

Fig. P4.2

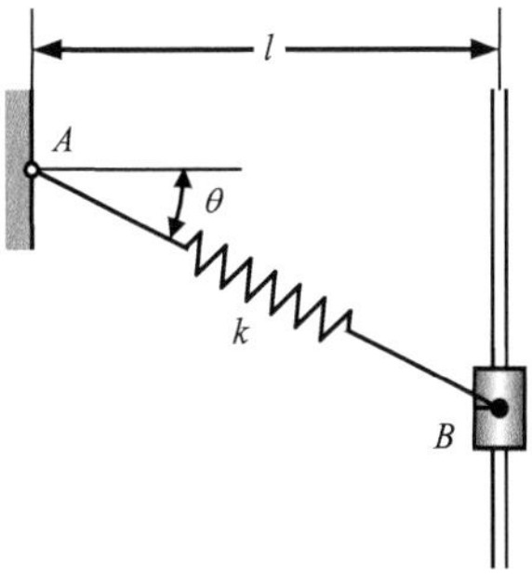

Fig. P4.3

4.2 A uniform ladder AB, having length l and mass m, leans against a wall. Assuming that the coefficient of static friction is μ_s at A and zero at B, determine the smallest value of μ_s for which equilibrium is maintained when $\theta = 60°$.

4.3 A collar B of weight W is attached to the spring AB and can move along the rod. The spring stiffness is $k = 1.8$ kN/m and the spring is unstretched when $\theta = 0$. Knowing that the coefficient of static friction between the collar and the rod is $\mu_s = 0.25$, and that the horizontal distance between points A and B is $l = 0.6$ m, determine the range of values of W for which equilibrium is maintained when $\theta = 15°$.

4.4 A force of magnitude $F = 100$ N acts on a block of mass m placed on a horizontal plane. Knowing that the coefficients of friction between the block and the plane are $\mu_s = 0.3$ and $\mu_k = 0.2$, and considering only values of θ less than or equal to 90°, determine the smallest value of θ for which motion of the block to the right is impending when (a) $m = 15$ kg, (b) $m = 30$ kg.

4.5 A force of magnitude $F = 150$ N acts on a block of mass $m = 10$ kg. Knowing that the coefficients of friction between the block and the plane are $\mu_s = 0.3$ and $\mu_k = 0.2$, determine the range of values of θ for which equilibrium of the block is maintained.

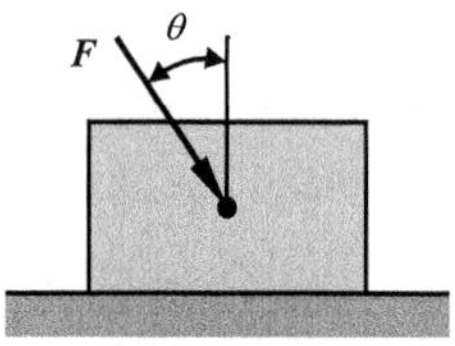

Fig. P4.4

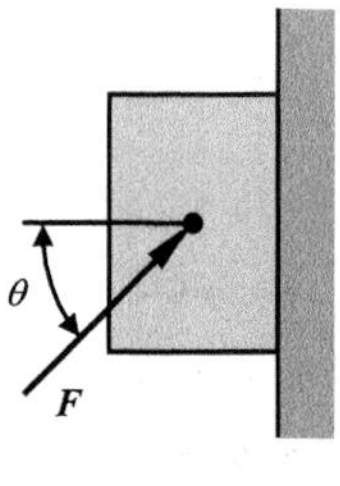

Fig. P4.5

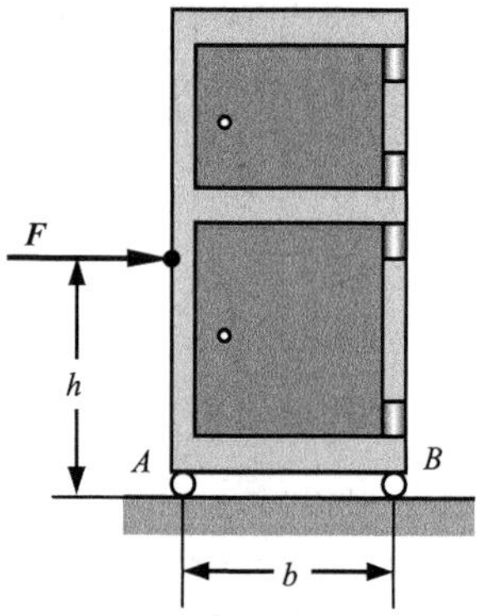

Fig. P4.6

4.6 A cabinet of mass $m = 50$ kg is mounted on casters which can be locked to prevent their rotation. The coefficient of static friction between the floor and each caster is $\mu_s = 0.3$. Knowing that $b = 500$ mm and $h = 650$ mm, determine the magnitude of the force $\boldsymbol{F}$ required for impending motion of the cabinet to the right (a) if all casters are locked, (b) if the caster at B is locked and the caster at A is free to rotate, (c) if the caster at A is locked and the caster at B is free to rotate.

Chapter 5 Kinematics of Particles

A particle moving along a straight line is said to be in rectilinear motion, whereas a particle moving along a curved line is said to be in curvilinear motion.

5.1 Vector Representation

Considering that particle P moves along a curved line, Fig. 5.1, the position of the particle at time t can be represented by position vector $\boldsymbol{r}$ joining O and P, where O is the reference point chosen in space. When the particle is in motion, $\boldsymbol{r}$ is a function of t, that is,

$$\boldsymbol{r} = \boldsymbol{r}(t) \tag{5.1}$$

Assuming that vector $\boldsymbol{r}'$ defines the position of the particle at time $t+\Delta t$, Fig. 5.2, then vector $\Delta\boldsymbol{r} = \boldsymbol{r}' - \boldsymbol{r}$ represents the change of $\boldsymbol{r}$ during the time interval Δt. Thus the velocity of the particle at t can be expressed as

$$\boldsymbol{v} = \lim_{\Delta t \to 0} \frac{\Delta \boldsymbol{r}}{\Delta t} = \dot{\boldsymbol{r}} \tag{5.2}$$

The velocity is a vector, and its direction is always tangent to the path of motion, Fig. 5.3.

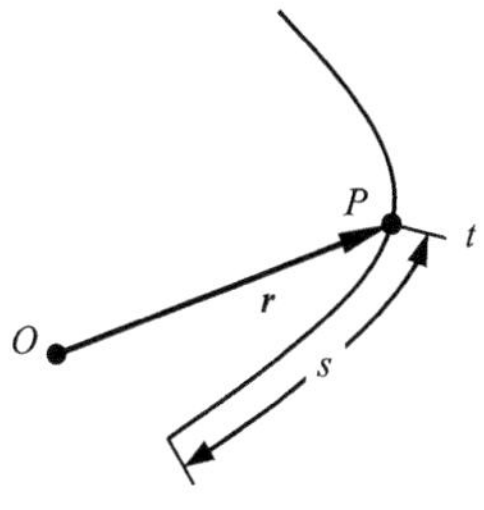

Fig. 5.1

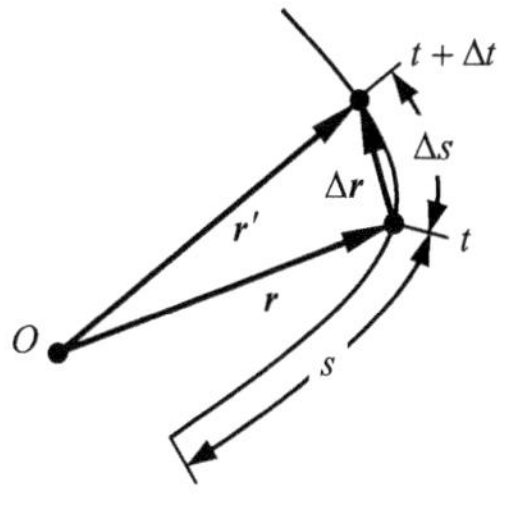

Fig. 5.2

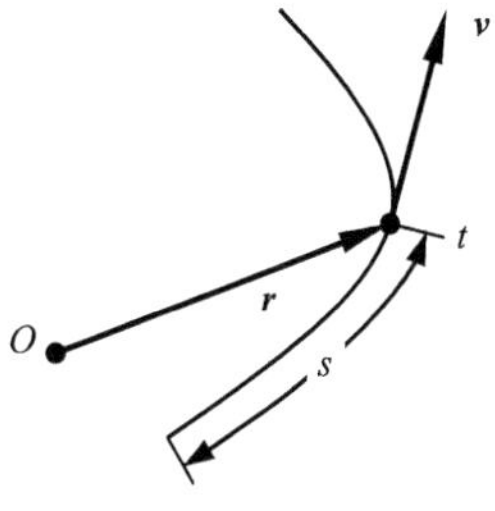

Fig. 5.3

Assuming that vector $\boldsymbol{v}'$ defines the velocity of the particle at time $t+\Delta t$, Fig. 5.4, then vector $\Delta\boldsymbol{v} = \boldsymbol{v}' - \boldsymbol{v}$ represents the change of $\boldsymbol{v}$ during Δt. The acceleration of the particle at t can thus be expressed as

$$\boldsymbol{a} = \lim_{\Delta t \to 0} \frac{\Delta \boldsymbol{v}}{\Delta t} = \dot{\boldsymbol{v}} = \ddot{\boldsymbol{r}} \tag{5.3}$$

The acceleration is a vector, and its direction is usually not tangent to the path of motion, Fig. 5.5.

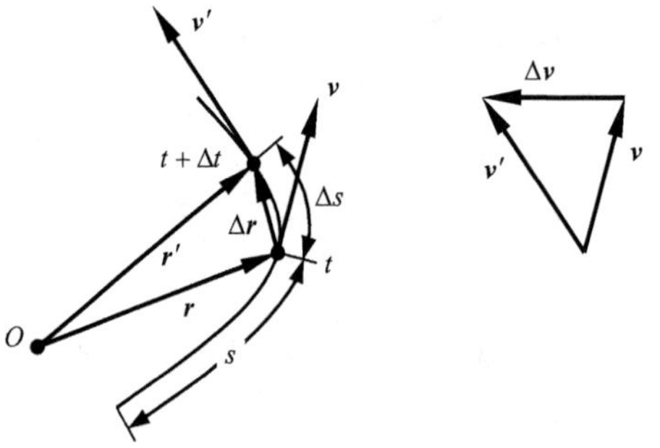

Fig. 5.4

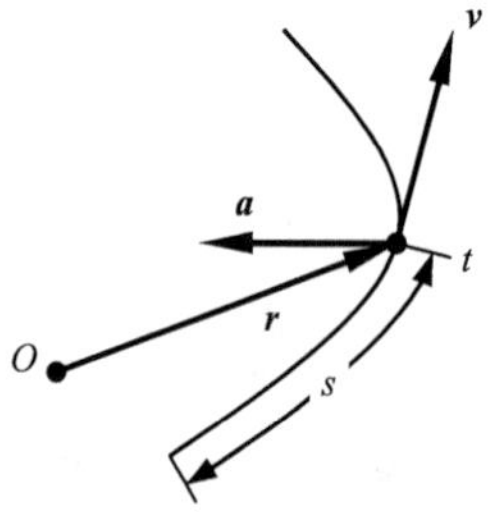

Fig. 5.5

5.2 Rectangular Components

Establishing a rectangular coordinate system $Oxyz$, Fig. 5.6, then position vector $\boldsymbol{r}$ of particle P at time t can be written as

$$\boldsymbol{r} = x\boldsymbol{i} + y\boldsymbol{j} + z\boldsymbol{k} \tag{5.4}$$

where $\boldsymbol{i}$, $\boldsymbol{j}$, and $\boldsymbol{k}$ are the unit vectors along the positive coordinate axes respectively, $x = x(t)$, $y = y(t)$, and $z = z(t)$ are the corresponding scalar components of $\boldsymbol{r}$ on the three coordinate axes.

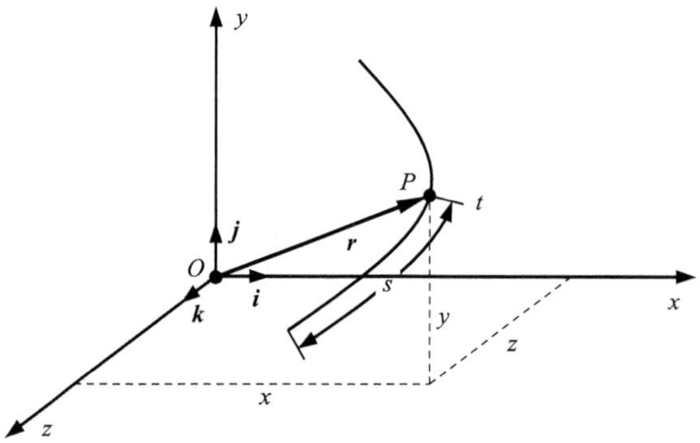

Fig. 5.6

Since the magnitude and direction of $\boldsymbol{i}$, $\boldsymbol{j}$, and $\boldsymbol{k}$ are constants, then the velocity and acceleration of particle P at time t can be expressed, respectively, as

$$\boldsymbol{v} = \dot{\boldsymbol{r}} = \dot{x}\boldsymbol{i} + \dot{y}\boldsymbol{j} + \dot{z}\boldsymbol{k} \tag{5.5}$$

$$\boldsymbol{a} = \ddot{\boldsymbol{r}} = \ddot{x}\boldsymbol{i} + \ddot{y}\boldsymbol{j} + \ddot{z}\boldsymbol{k} \tag{5.6}$$

where $\dot{x}$, $\dot{y}$, $\dot{z}$ and $\ddot{x}$, $\ddot{y}$, $\ddot{z}$ represent, respectively, the scalar components of $\boldsymbol{v}$ and $\boldsymbol{a}$, i.e.,

$$v_x = \dot{x},\ v_y = \dot{y},\ v_z = \dot{z} \tag{5.7}$$

$$a_x = \ddot{x},\ a_y = \ddot{y},\ a_z = \ddot{z} \tag{5.8}$$

Example 5.1 The motion of a particle is defined by the position vector $\boldsymbol{r} = A(\cos t + t\sin t)\boldsymbol{i} + A(\sin t - t\cos t)\boldsymbol{j}$, where t is expressed in seconds. Determine the values of t for which the position vector and the acceleration vector are (a) perpendicular, (b) parallel.

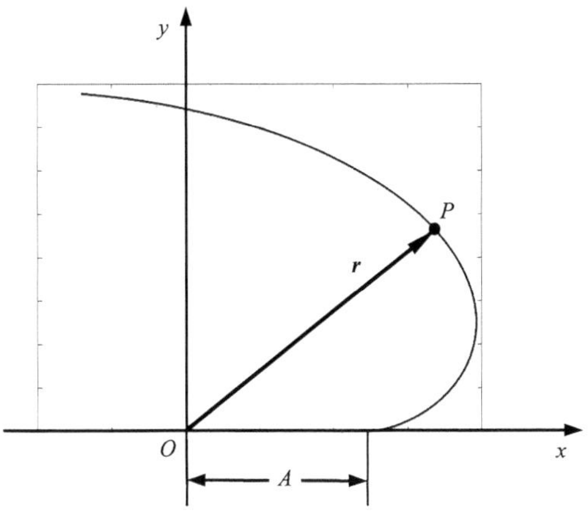

Fig. E5.1

Solution Using $\boldsymbol{r} = A(\cos t + t\sin t)\boldsymbol{i} + A(\sin t - t\cos t)\boldsymbol{j}$, we have

$$\boldsymbol{v} = \dot{\boldsymbol{r}} = A(t\cos t)\boldsymbol{i} + A(t\sin t)\boldsymbol{j}$$

$$\boldsymbol{a} = \dot{\boldsymbol{v}} = A(\cos t - t\sin t)\boldsymbol{i} + A(\sin t + t\cos t)\boldsymbol{j}$$

(a) When the position vector and the acceleration vector are perpendicular, we have

$$\boldsymbol{r} \cdot \boldsymbol{a} = 0$$

from which we obtain

$$t = 1\ \text{s}$$

(b) When the position vector and the acceleration vector are parallel, we have

$$\boldsymbol{r} \times \boldsymbol{a} = 0$$

from which we obtain

$$t = 0$$

5.3 Tangential and Normal Components

Establish a natural coordinate system attached to particle P moving along a curved line, Fig. 5.7, and define three unit vectors $\boldsymbol{e}_\mathrm{t}$, $\boldsymbol{e}_\mathrm{n}$, and $\boldsymbol{e}_\mathrm{b}$, where $\boldsymbol{e}_\mathrm{t}$ is the tangential unit vector tangent to the path of motion pointed toward the direction of motion, $\boldsymbol{e}_\mathrm{n}$ is the principal normal unit vector perpendicular to the path of motion pointed toward the center of curvature of the path of motion, and $\boldsymbol{e}_\mathrm{b}$ is the binormal unit vector perpendicular to the plane containing $\boldsymbol{e}_\mathrm{t}$ and $\boldsymbol{e}_\mathrm{n}$ pointing in the direction of $\boldsymbol{e}_\mathrm{t} \times \boldsymbol{e}_\mathrm{n}$ indicated by the right-hand rule, i.e., $\boldsymbol{e}_\mathrm{b} = \boldsymbol{e}_\mathrm{t} \times \boldsymbol{e}_\mathrm{n}$. The plane containing $\boldsymbol{e}_\mathrm{t}$ and $\boldsymbol{e}_\mathrm{n}$ is called the osculating plane.

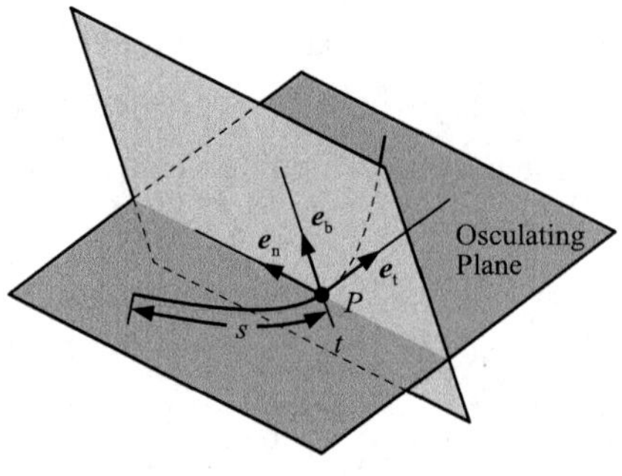

Fig. 5.7

In this natural coordinate system, the velocity vector $\boldsymbol{v}$ of the particle P at time t can be written as

$$\boldsymbol{v} = v\boldsymbol{e}_{\mathrm{t}} \tag{5.9}$$

where $\boldsymbol{e}_{\mathrm{t}}$ is the tangential unit vector. Using the relation above, the acceleration vector $\boldsymbol{a}$ of particle P at time t can be given by

$$\boldsymbol{a} = \dot{\boldsymbol{v}} = \dot{v}\boldsymbol{e}_{\mathrm{t}} + v\dot{\boldsymbol{e}}_{\mathrm{t}} \tag{5.10}$$

Using $\dot{\boldsymbol{e}}_{\mathrm{t}} = \dfrac{v}{\rho}\boldsymbol{e}_{\mathrm{n}}$, where $\boldsymbol{e}_{\mathrm{n}}$ is the principal normal unit vector and ρ is the radius of curvature of the path of motion, then the acceleration vector $\boldsymbol{a}$ of particle P at time t can be rewritten as

$$\boldsymbol{a} = \dot{v}\boldsymbol{e}_{\mathrm{t}} + \frac{v^2}{\rho}\boldsymbol{e}_{\mathrm{n}} \tag{5.11}$$

where $\dot{v}$ and $\dfrac{v^2}{\rho}$ represent, respectively, the tangential and normal components of $\boldsymbol{a}$, i.e.,

$$a_{\mathrm{t}} = \dot{v},\ a_{\mathrm{n}} = \frac{v^2}{\rho} \tag{5.12}$$

The relation above shows that the tangential component a_{t} of the acceleration is equal to the rate of change of the speed of the particle, while the normal component a_{n} is equal to the square of the speed divided by the radius of curvature of the path of motion. If the speed of the particle increases, a_{t} is positive and $\boldsymbol{a}_{\mathrm{t}}$ points in the direction of motion. If the speed of the particle decreases, a_{t} is negative and $\boldsymbol{a}_{\mathrm{t}}$ points against the direction of motion. However, a_{n} is always positive and $\boldsymbol{a}_{\mathrm{n}}$ is always pointed toward the center of curvature of the path of motion.

We thus conclude that the tangential component of the acceleration reflects a change in the magnitude of velocity of the particle, while its normal component reflects a change in the direction of velocity of the particle.

Example 5.2 The motion of a particle is defined by position vector $\boldsymbol{r} = [(2\sin 4t)\boldsymbol{i} + (2\cos 4t)\boldsymbol{j} + (4t)\boldsymbol{k}]\ \mathrm{m}$, where t is expressed in seconds. Determine the radius of curvature of the path along which the particle moves.

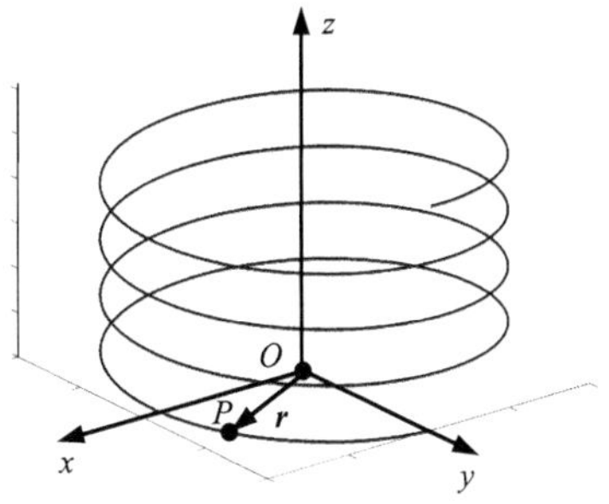

Fig. E5.2

Solution From $\boldsymbol{r}=[(2\sin 4t)\boldsymbol{i}+(2\cos 4t)\boldsymbol{j}+(4t)\boldsymbol{k}]$ m, we have

$$\boldsymbol{v}=\dot{\boldsymbol{r}}=[(8\cos 4t)\boldsymbol{i}+(-8\sin 4t)\boldsymbol{j}+(4)\boldsymbol{k}]\ \text{m/s}$$

$$\boldsymbol{a}=\dot{\boldsymbol{v}}=[(-32\sin 4t)\boldsymbol{i}+(-32\cos 4t)\boldsymbol{j}]\ \text{m/s}^2$$

Thus we can obtain

$$v=\sqrt{(8\cos 4t)^2+(-8\sin 4t)^2+(4)^2}=4\sqrt{5}\ \text{m/s}$$

$$a=\sqrt{(-32\sin 4t)^2+(-32\cos 4t)^2}=32\ \text{m/s}^2$$

Using $a^2=a_\text{t}^2+a_\text{n}^2$, where $a_\text{t}=\dot{v}=0$, $a_\text{n}=\dfrac{v^2}{\rho}$, we can find

$$\rho=\frac{v^2}{a_\text{n}}=\frac{v^2}{a}=2.5\ \text{m}$$

Problems

5.1 The motion of a particle is defined by the position vector $\boldsymbol{r}=(4t^2-3t)\boldsymbol{i}+t^3\boldsymbol{j}$, where r and t are expressed in meters and seconds, respectively. Determine the velocity and acceleration of the particle when (a) $t=0.2$ s, (b) $t=1$ s.

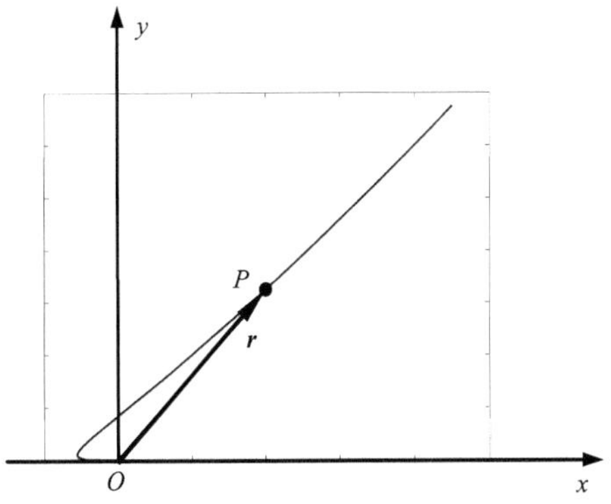

Fig. P5.1

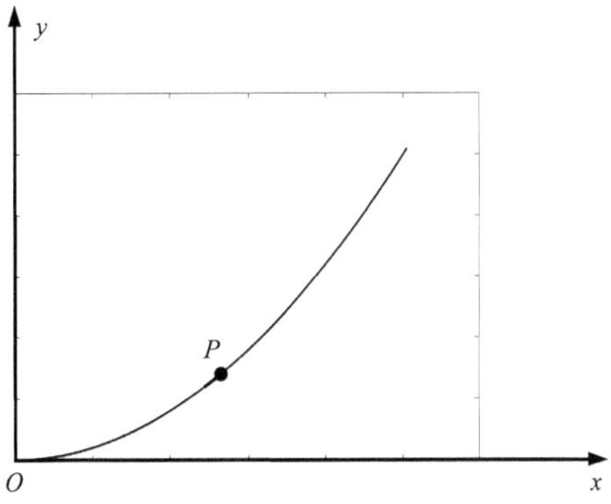

Fig. P5.2

5.2 A particle P moves along a parabolic path $y=\frac{1}{20}x^2$, where x and y are expressed in meters. The particle has a speed of $6\ \text{m/s}$ which is increasing at $2\ \text{m/s}^2$ when $x=6\ \text{m}$. Determine the direction of the velocity, and the magnitude and direction of the acceleration at this instant.

5.3 A particle is defined by the position vector $\boldsymbol{r}=\frac{3}{4}[1-1/(t+1)]\boldsymbol{i}+\frac{1}{2}[\exp(-\pi t/2)\cos 2\pi t]\boldsymbol{j}$, where r and t are expressed in meters and seconds, respectively. Determine the position, the velocity, and the acceleration of the particle when (a) $t=0.5\ \text{s}$, (b) $t=1\ \text{s}$.

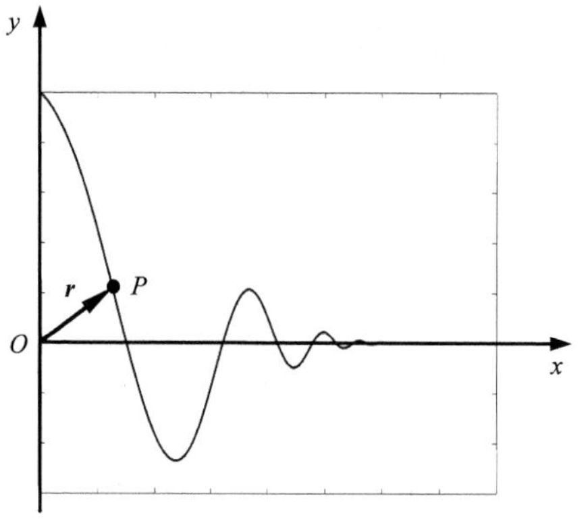

Fig. P5.3

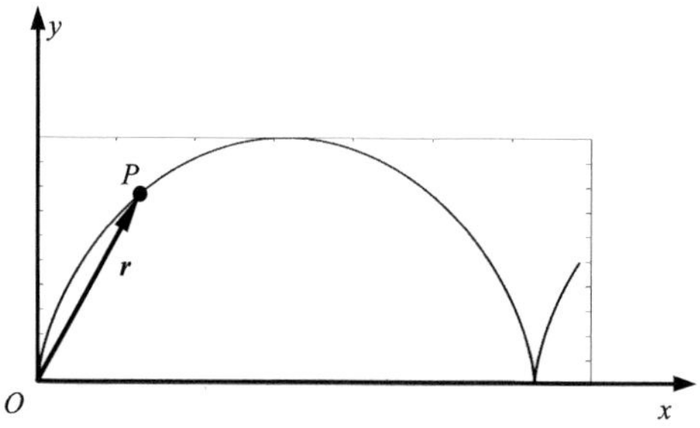

Fig. P5.4

5.4 A particle is defined by the position vector $\boldsymbol{r}=[(\omega t-\sin\omega t)\boldsymbol{i}+(1-\cos\omega t)\boldsymbol{j}]\ \text{m}$, where ω and t are expressed in rad/s and s, respectively. Determine the tangential and normal accelerations of the particle at time t.

Chapter 6 Kinematics of Rigid Bodies in Plane Motion

When all the particles of a rigid body move along paths which are equidistant from a fixed plane, this body is said to be in plane motion. There are three types of plane motion, namely translation, rotation about a fixed axis and general plane motion.

6.1 Translation

A body is said to be in translation if any straight-line segment in the body remains the same direction during the motion. When a body is in translation, all the particles within the body will move along parallel paths of motion. If these paths of motion for a body in translation are straight lines the motion is called rectilinear translation, Fig. 6.1a. However, if the paths of motion are curved lines the motion is called curvilinear translation, Fig. 6.1b.

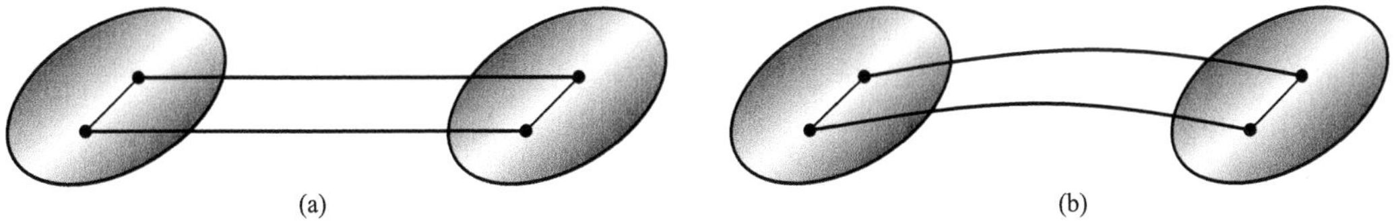

Fig. 6.1

Let A and B be any two particles within a rigid body in translation, Fig. 6.2. Denote, respectively, by $\boldsymbol{r}_A$ and $\boldsymbol{r}_B$ the position vectors of particles A and B with respect to a fixed system of reference $Oxyz$ and by $\boldsymbol{r}_{A/B}$ the position vector of particle A relative to particle B in a translating system of reference $O'x'y'z'$ attached to the body at particle B. From Fig. 6.2, we have

$$\boldsymbol{r}_A = \boldsymbol{r}_B + \boldsymbol{r}_{A/B} \tag{6.1}$$

Differentiating Eq. (6.1) with respect to time t, we obtain

$$\boldsymbol{v}_A = \boldsymbol{v}_B + \boldsymbol{v}_{A/B} \tag{6.2}$$

where $\boldsymbol{v}_A = \dot{\boldsymbol{r}}_A$ and $\boldsymbol{v}_B = \dot{\boldsymbol{r}}_B$ are the velocities of particles A and B respectively, and $\boldsymbol{v}_{A/B} = \dot{\boldsymbol{r}}_{A/B}$ is the velocity of particle A relative to particle B. When the body is in translation, $\boldsymbol{r}_{A/B}$ is a constant vector, i.e., $\dot{\boldsymbol{r}}_{A/B} = 0$, thus we have

$$\boldsymbol{v}_A = \boldsymbol{v}_B \tag{6.3}$$

Differentiating Eq. (6.3), we obtain

$$\boldsymbol{a}_A = \boldsymbol{a}_B \tag{6.4}$$

Thus we conclude that when a rigid body is in translation all the particles of the body have the same velocity and the same acceleration at any given instant, Fig. 6.3.

In the case of curvilinear translation, the velocity and acceleration change in direction as well as in magnitude at every instant. However, in the case of rectilinear translation, the velocity and acceleration remain the same direction during the entire motion.

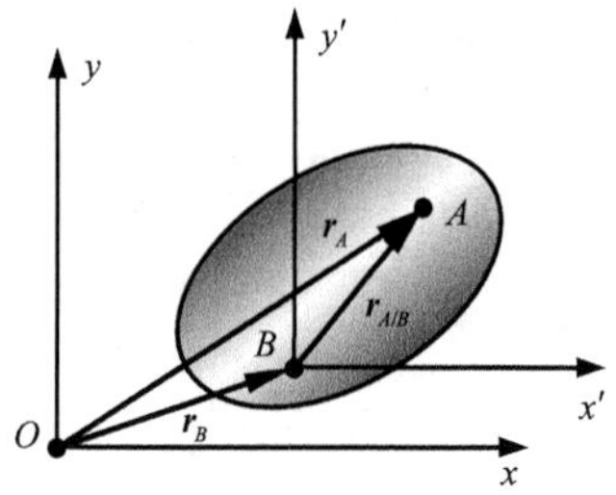

Fig. 6.2

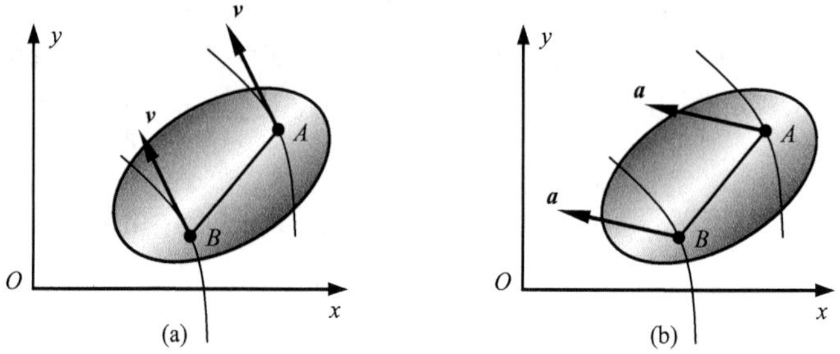

Fig. 6.3

6.2 Rotation about Fixed Axis

When a body rotates about a fixed axis, all the particles forming the body, except those which lie on the axis of rotation, move along circular paths, which are perpendicular to the axis of rotation, Fig. 6.4.

Consider a rigid body rotating about a fixed axis λ, let A be an arbitrary particle on the body, and denote by $\boldsymbol{r}_{A/B}$ the position vector of particle A with respect to an arbitrary point B located on the axis of rotation, as shown in Fig. 6.5. From Fig. 6.5, we have

$$\boldsymbol{v}_A = \boldsymbol{\omega} \times \boldsymbol{r}_{A/B} \tag{6.5}$$

where $\boldsymbol{\omega}$ is the angular velocity of the body, which is directed along the axis of rotation, and equal in magnitude to the rate of change of the angular coordinate.

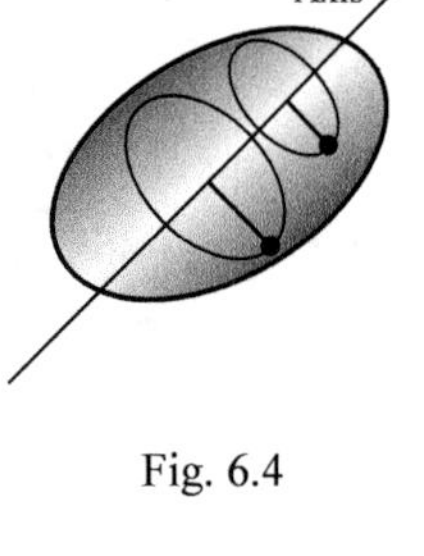

Fig. 6.4

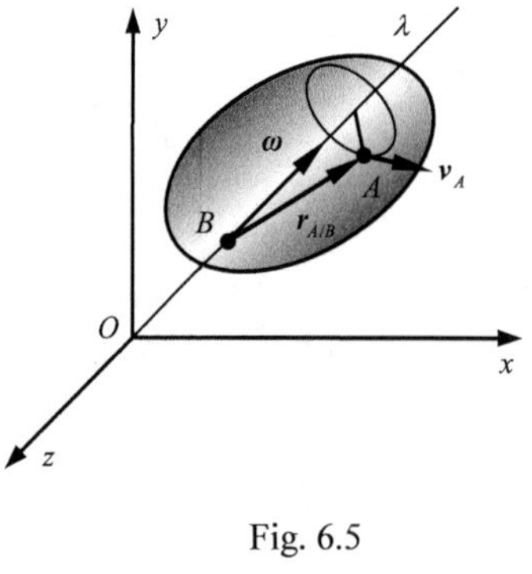

Fig. 6.5

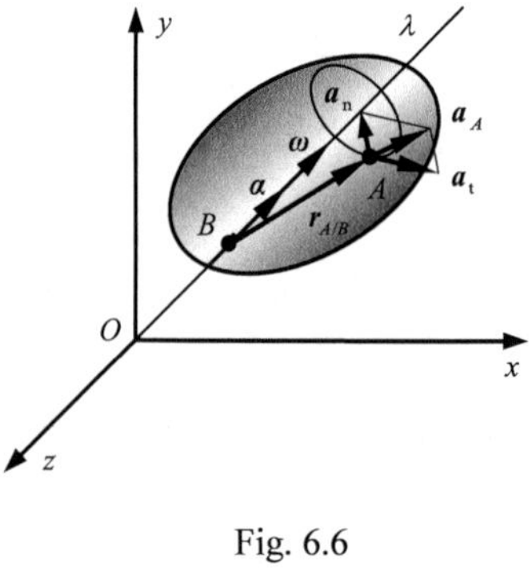

Fig. 6.6

Differentiating Eq. (6.5) with respect to time t, we obtain

$$\boldsymbol{a}_A = \boldsymbol{\alpha} \times \boldsymbol{r}_{A/B} + \boldsymbol{\omega} \times \boldsymbol{v}_A \tag{6.6}$$

where $\boldsymbol{\alpha} = \dot{\boldsymbol{\omega}}$ is the angular acceleration of the body. In the case of a body rotating about a fixed axis, the angular acceleration $\boldsymbol{\alpha}$ is a vector directed along the axis of rotation, and is equal in magnitude to the rate of change of the angular velocity. Using Eq. (6.5), Eq. (6.6) can be rewritten as

$$\boldsymbol{a}_A = \boldsymbol{\alpha} \times \boldsymbol{r}_{A/B} + \boldsymbol{\omega} \times (\boldsymbol{\omega} \times \boldsymbol{r}_{A/B}) \tag{6.7}$$

The equation above can be also expressed as

$$\boldsymbol{a}_A = \boldsymbol{a}_\mathrm{t} + \boldsymbol{a}_\mathrm{n} \tag{6.8}$$

where $\boldsymbol{a}_\mathrm{t} = \boldsymbol{\alpha} \times \boldsymbol{r}_{A/B}$ is the tangential component of the acceleration $\boldsymbol{a}_A$ of particle A tangent to the circle drawn by particle A, and $\boldsymbol{a}_\mathrm{n} = \boldsymbol{\omega} \times \boldsymbol{v}_A = \boldsymbol{\omega} \times (\boldsymbol{\omega} \times \boldsymbol{r}_{A/B})$ is the normal component of the acceleration $\boldsymbol{a}_A$ of particle A directed toward the center of the circle formed by particle A. Eq. (6.8) shows that the acceleration $\boldsymbol{a}_A$ of particle A is equal to the vector sum of the tangential component $\boldsymbol{a}_\mathrm{t}$ and the normal component $\boldsymbol{a}_\mathrm{n}$, Fig. 6.6.

The rotation of a rigid body about a fixed axis can be represented by the in-plane rotation of a plate perpendicular to the rotation axis of the body about the intersection of the plate and the rotation axis of the body, Fig. 6.7.

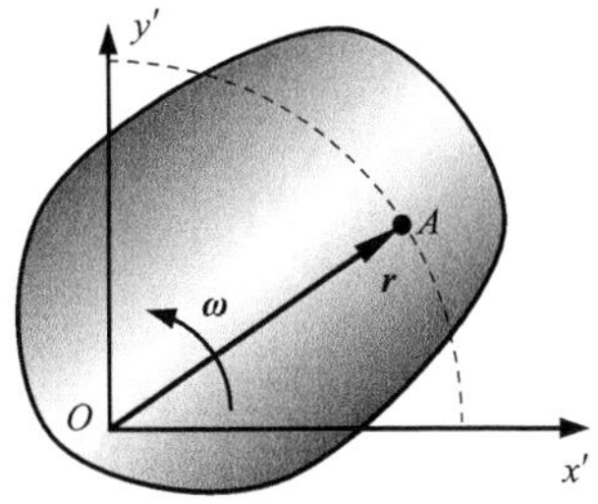

Fig. 6.7

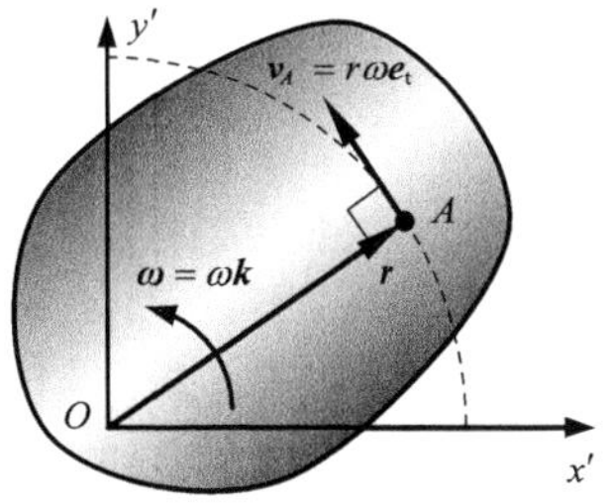

Fig. 6.8

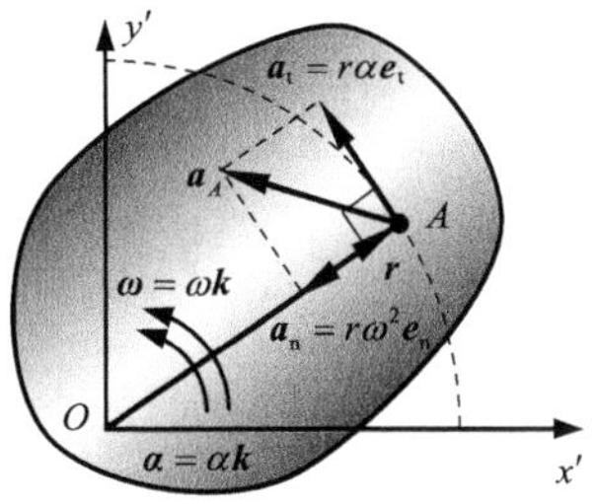

Fig. 6.9

In the case of the in-plane rotation of a plate about a fixed point O, the angular velocity $\boldsymbol{\omega}$ can be expressed as

$$\boldsymbol{\omega} = \omega\boldsymbol{k} \tag{6.9}$$

where $\boldsymbol{k}$ is the unit vector, perpendicular to the plate and pointing out of the plate. ω is the magnitude of the angular velocity, which is positive if the plate rotates counterclockwise, and negative if clockwise.

Using Eq. (6.5), we can express the velocity of particle A on the plate, Fig. 6.8, as

$$\boldsymbol{v}_A = \boldsymbol{\omega} \times \boldsymbol{r} = r\omega\boldsymbol{e}_\mathrm{t} \tag{6.10}$$

where $\boldsymbol{e}_\mathrm{t}$ is the unit vector tangent to the circle described by particle A pointed toward the direction of motion. Thus the magnitude of the velocity is equal to

$$v_A = r\omega \tag{6.11}$$

and its direction can be obtained by rotating $\boldsymbol{r}$ through $90°$ in the direction of rotation of the plate.

Using Eq. (6.7) and (6.8), we can express the acceleration of particle A on the plate, Fig. 6.9, as

$$\boldsymbol{a}_A = \boldsymbol{a}_\mathrm{t} + \boldsymbol{a}_\mathrm{n} = \boldsymbol{\alpha} \times \boldsymbol{r} + \boldsymbol{\omega} \times (\boldsymbol{\omega} \times \boldsymbol{r}) = r\alpha \boldsymbol{e}_\mathrm{t} + r\omega^2 \boldsymbol{e}_\mathrm{n} \tag{6.12}$$

where $\boldsymbol{e}_\mathrm{n}$ is the unit vector, pointed toward the center of the circle described by particle A. The tangential component $\boldsymbol{a}_\mathrm{t} = r\alpha\boldsymbol{e}_\mathrm{t}$ points in the counterclockwise direction if α is positive, and is in the clockwise direction if α is negative. The normal component $\boldsymbol{a}_\mathrm{n} = r\omega^2\boldsymbol{e}_\mathrm{n}$ always points toward the center of the circle.

Example 6.1 The structure shown in Fig. E6.1 consists of two rods *AE* and *CE* and a rectangular plate *ABCD* which are welded together. The structure rotates about the axis *AE* with a constant angular velocity of $\omega = 5$ rad/s. Knowing that the rotation is counterclockwise as viewed from *A*, determine the velocity and acceleration of point *B*.

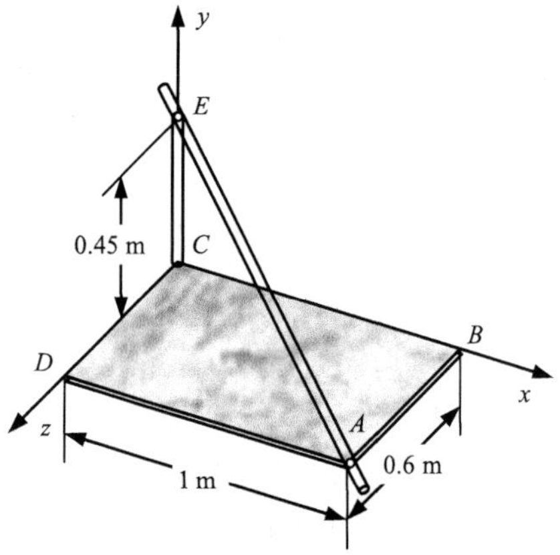

Fig. E6.1

Solution Using $\boldsymbol{r}_{A/E} = (\boldsymbol{i} - 0.45\boldsymbol{j} + 0.6\boldsymbol{k})$ m, we have

$$\boldsymbol{\omega} = \omega \frac{\boldsymbol{r}_{A/E}}{r_{A/E}} = (4\boldsymbol{i} - 1.8\boldsymbol{j} + 2.4\boldsymbol{k}) \text{ rad/s}$$

Using $\boldsymbol{r}_{B/A} = (-0.6\boldsymbol{k})$ m, we have

$$\boldsymbol{v}_B = \boldsymbol{\omega} \times \boldsymbol{r}_{B/A} = \begin{vmatrix} \boldsymbol{i} & \boldsymbol{j} & \boldsymbol{k} \\ 4 & -1.8 & 2.4 \\ 0 & 0 & -0.6 \end{vmatrix} = (1.08\boldsymbol{i} + 2.4\boldsymbol{j}) \text{ m/s}$$

Using $\boldsymbol{\alpha} = 0$, we have

$$\boldsymbol{a}_B = \boldsymbol{\alpha} \times \boldsymbol{r}_{B/A} + \boldsymbol{\omega} \times \boldsymbol{v}_B = \begin{vmatrix} \boldsymbol{i} & \boldsymbol{j} & \boldsymbol{k} \\ 4 & -1.8 & 2.4 \\ 1.08 & 2.4 & 0 \end{vmatrix} = (-5.76\boldsymbol{i} + 2.59\boldsymbol{j} + 11.54\boldsymbol{k}) \text{ m/s}^2$$

6.3 General Plane Motion

When a body is subjected to a general plane motion, it undergoes a combination of translation and rotation, Fig. 6.10. The translation takes place within the plane of reference, and the rotation takes place about an axis perpendicular to the plane of reference.

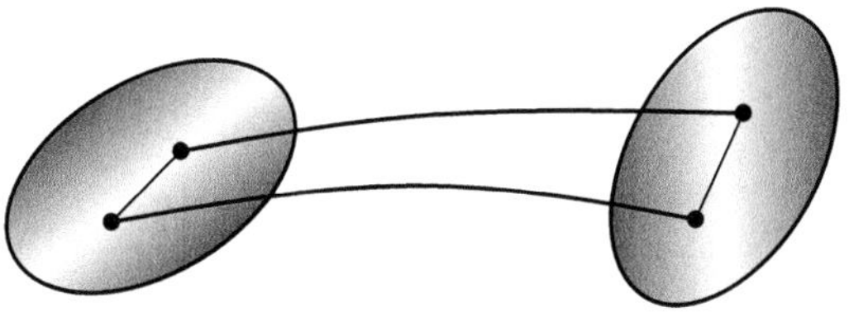

Fig. 6.10

A general plane motion is neither a translation nor a rotation. However, it can always be considered as the combination of a translation and a rotation. Consider, for example, rod AB in general plane motion, Fig. 6.11a, whose two ends slide along a horizontal and a vertical track, respectively. The motion of the rod from A_1B_1 to A_2B_2 can be replaced by a translation of the rod from A_1B_1 to $A_1'B_2$ with the base point B to the right, Fig. 6.11b, and a rotation of the rod from $A_1'B_2$ to A_2B_2 about the base point B counterclockwise, Fig. 6.11c.

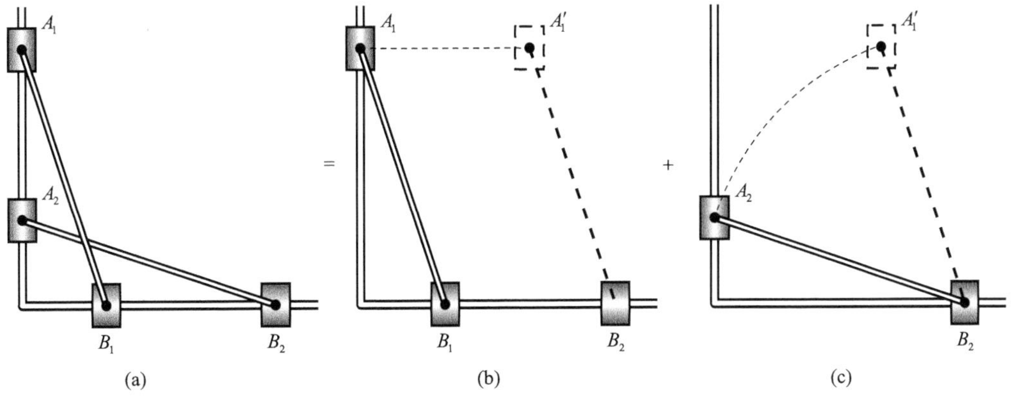

Fig. 6.11

Another example of general plane motion is a wheel rolling without sliding on a horizontal surface, Fig. 6.12a. The motion of the wheel from A_1B_1 to A_2B_2 can be replaced by a translation of the wheel from A_1B_1 to $A_1'B_2$ with the base point B to the right, Fig. 6.12b, and a rotation of the wheel from $A_1'B_2$ to A_2B_2 about the base point B clockwise, Fig. 6.12c.

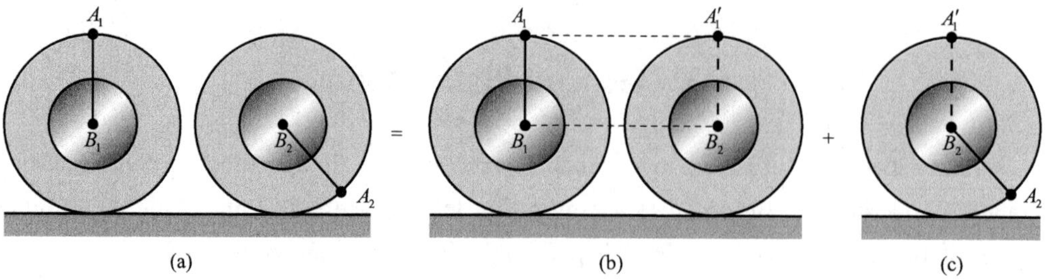

Fig. 6.12

1. Base-Point Method for Determining Velocities

Since any general plane motion of a rigid body can be replaced by a translation with the base point and a simultaneous rotation about the base point, the velocity $\boldsymbol{v}_A$ of A on the rigid body, Fig. 6.13a, can be expressed as

$$\boldsymbol{v}_A = \boldsymbol{v}_B + \boldsymbol{v}_{A/B} \tag{6.13}$$

where $\boldsymbol{v}_B$ is the velocity of the base point B, Fig. 6.13b, and $\boldsymbol{v}_{A/B} = \boldsymbol{\omega} \times \boldsymbol{r}_{A/B}$ is the velocity of A with respect to the base point B, Fig. 6.13c. It should be noted that the choice of a base point is arbitrary, it is, however, convenient to select that point, where the motion is known, as the base point.

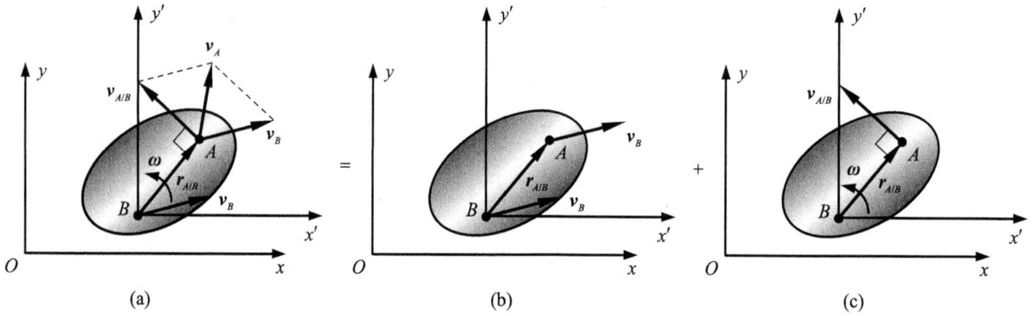

Fig. 6.13

Example 6.2 Collar *A* moves down with a constant velocity of $v_A = 1\ \text{m/s}$, Fig. E6.2a. At the instant $\theta = 30°$, determine (a) the angular velocity of rod *AB*, (b) the velocity of collar *B*.

Solution Choosing point *A* as the base point, then the velocity of point *B* can be expressed as

$$\boldsymbol{v}_B = \boldsymbol{v}_A + \boldsymbol{v}_{B/A}$$

Using the fact that $\boldsymbol{v}_B$ is horizontal, and that $\boldsymbol{v}_{B/A}$ is perpendicular to the line joining *A* and *B*, we can obtain the velocity parallelogram shown in Fig. E6.2b. From this parallelogram, then at the given instant we have

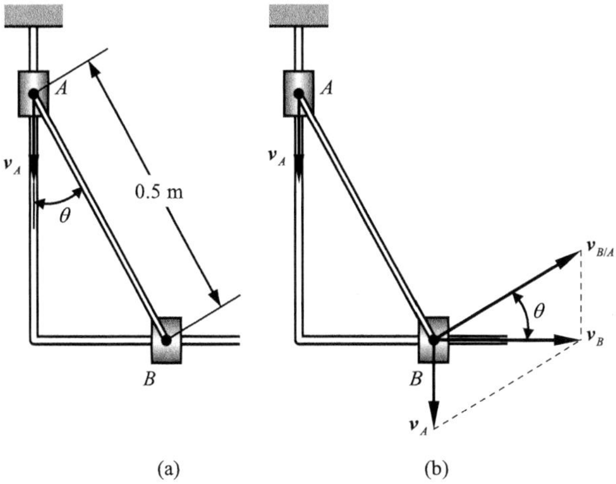

Fig. E6.2

$$v_B = \frac{v_A}{\tan\theta} = 1.73 \text{ m/s}, \ v_{B/A} = \frac{v_A}{\sin\theta} = 2 \text{ m/s}$$

Using $v_{B/A} = AB \cdot \omega$, we have

$$\omega = \frac{v_{B/A}}{AB} = 4 \text{ rad/s (anticlockwise)}$$

2. Instantaneous-Center Method for Determining Velocities

In velocity analysis of a rigid body in general plane motion, we often choose a point, which has zero velocity at the given instant, on the rigid body or outside the rigid body as the base point. Assuming that the velocity at point I is equal to zero at the instant considered, where I is called the instantaneous center of velocity, Fig. 6.14, then the velocity $\boldsymbol{v}_A$ of A at the given instant can be given by

$$\boldsymbol{v}_A = \boldsymbol{\omega} \times \boldsymbol{r} \tag{6.14}$$

where $\boldsymbol{\omega}$ and $\boldsymbol{r}$ are, respectively, the angular velocity of the rigid body and the position vector of A with respect to the instantaneous center of velocity I. It should be noted that, in general, the instantaneous center of velocity does not have zero acceleration.

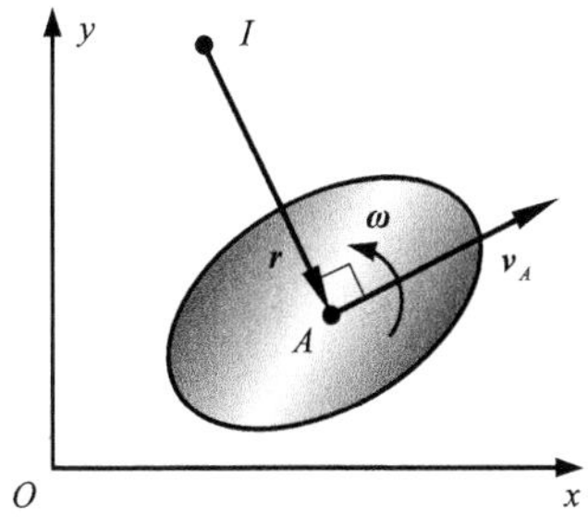

Fig. 6.14

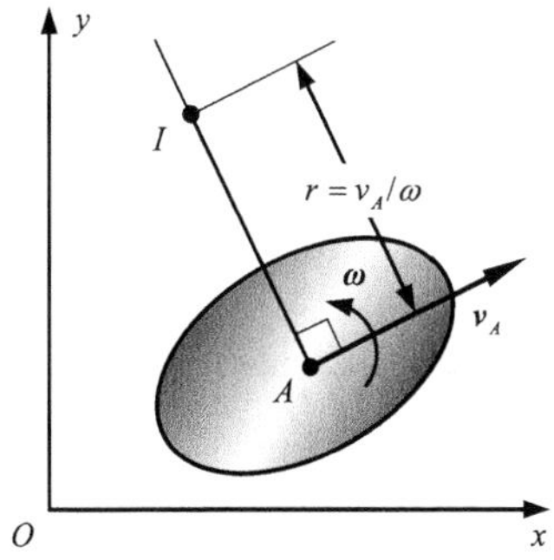

Fig. 6.15

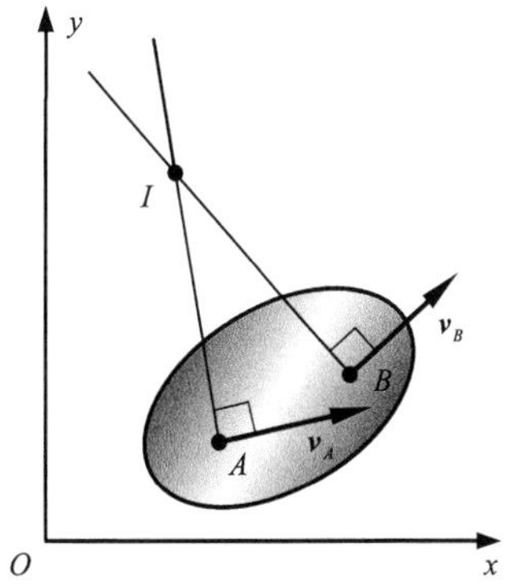

Fig. 6.16

The instantaneous center of velocity at the given instant can be determined by using the following methods:

(1) If $\boldsymbol{v}_A$ and $\boldsymbol{\omega}$ are known, the instantaneous center of velocity can be determined according to Fig. 6.15.

(2) If two nonparallel velocities $\boldsymbol{v}_A$ and $\boldsymbol{v}_B$ are known, the instantaneous center of velocity can be determined according to Fig. 6.16.

(3) If two parallel velocities $\boldsymbol{v}_A$ and $\boldsymbol{v}_B$ are known, the instantaneous center of velocity can be determined according to Fig. 6.17. A special case for Fig. 6.17a should be noted that if $\boldsymbol{v}_A = \boldsymbol{v}_B$ at a given instant, the instantaneous center of velocity will be located at infinity. This special case is called the instantaneous translation.

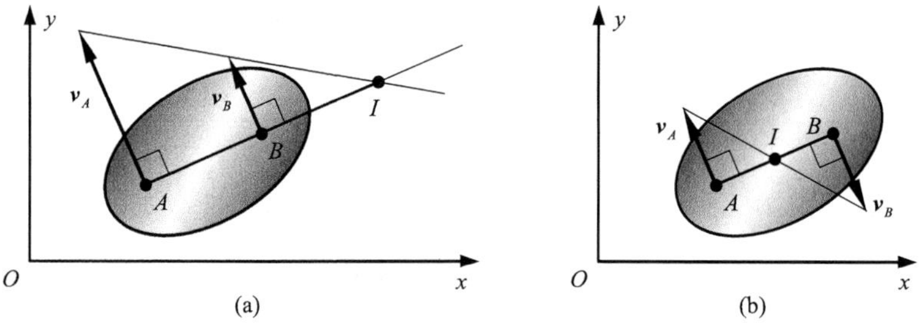

Fig. 6.17

Example 6.3 Collar *A* moves down with a constant velocity of $v_A = 1$ m/s, Fig. E6.3a. At the instant $\theta = 30°$, determine (a) the angular velocity of rod *AB*, (b) the velocity of collar *B*.

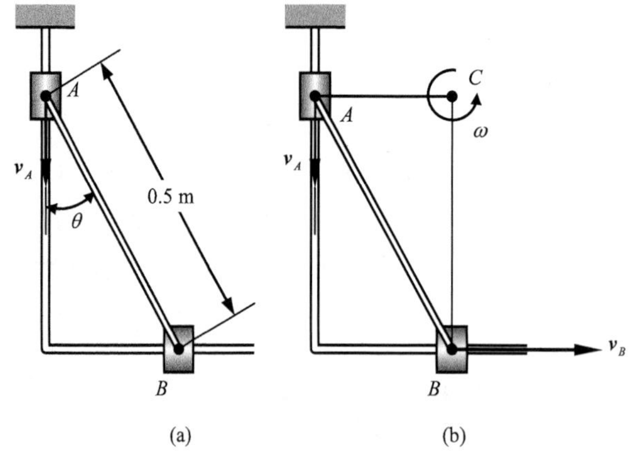

Fig. E6.3

Solution At the given instant the instantaneous center of velocity is located at *C*, Fig. E6.3b, then we have

$$\omega = \frac{v_A}{AB\sin\theta} = 4 \text{ rad/s}; \ v_B = (AB\cos\theta)\omega = 1.73 \text{ m/s}$$

3. Base-Point Method for Determining Accelerations

Since any general plane motion of a rigid body can be replaced by a translation with the base point, and a simultaneous rotation about the base point, the acceleration $\boldsymbol{a}_A$ at A, Fig. 6.18a, can be expressed as

$$\boldsymbol{a}_A = \boldsymbol{a}_B + \boldsymbol{a}_{A/B} = \boldsymbol{a}_B + (\boldsymbol{a}_{A/B})_{\rm t} + (\boldsymbol{a}_{A/B})_{\rm n} \tag{6.15}$$

where $\boldsymbol{a}_B$ is the acceleration of the base point, $(\boldsymbol{a}_{A/B})_{\rm t} = \boldsymbol{\alpha} \times \boldsymbol{r}_{A/B}$ and $(\boldsymbol{a}_{A/B})_{\rm n} = \boldsymbol{\omega} \times (\boldsymbol{\omega} \times \boldsymbol{r}_{A/B})$ are the tangential and normal components of the acceleration of A with respect to the base point B, and $\boldsymbol{a}_{A/B} = (\boldsymbol{a}_{A/B})_{\rm t} + (\boldsymbol{a}_{A/B})_{\rm n}$ is the acceleration of A with respect to the base point B.

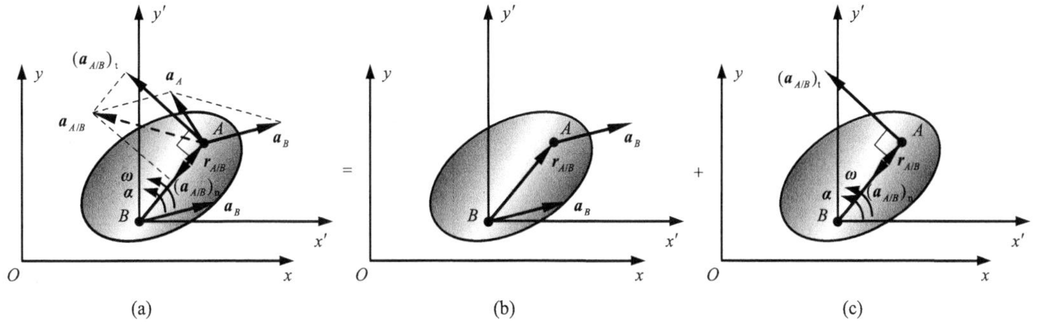

Fig. 6.18

Example 6.4 Collar *A* moves down with a constant velocity of $v_A = 1$ m/s, Fig. E6.4a. At the instant $\theta = 30°$, determine (a) the angular acceleration of rod *AB*, (b) the acceleration of collar *B*.

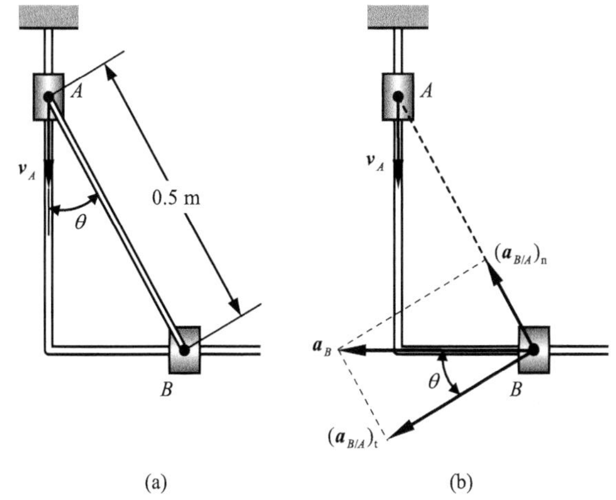

Fig. E6.4

Solution Choosing point A as the base point, the acceleration of point B can be expressed as

$$\boldsymbol{a}_B = \boldsymbol{a}_A + (\boldsymbol{a}_{B/A})_\mathrm{t} + (\boldsymbol{a}_{B/A})_\mathrm{n}$$

Using the fact that $\boldsymbol{a}_B$ is horizontal and that $\boldsymbol{a}_A$ is identically equal to zero, we can obtain the acceleration parallelogram shown in Fig. E6.4b. From this parallelogram, then at the given instant we have

$$a_B = \frac{(a_{B/A})_\mathrm{n}}{\sin\theta},\ (a_{B/A})_\mathrm{t} = \frac{(a_{B/A})_\mathrm{n}}{\tan\theta}$$

Using $\omega = 4$ rad/s (referring to Example 6.2 or 6.3), i.e., $(a_{B/A})_\mathrm{n} = AB\cdot\omega^2 = 8\ \mathrm{m/s^2}$ (directed toward point A), we can obtain

$$a_B = 16\ \mathrm{m/s^2},\ (a_{B/A})_\mathrm{t} = 13.86\ \mathrm{m/s^2}$$

Using $(a_{B/A})_\mathrm{t} = AB\cdot\alpha$, we have

$$\alpha = \frac{(a_{B/A})_\mathrm{t}}{AB} = 27.72\ \mathrm{rad/s^2}\ \text{(clockwise)}$$

Example 6.5 Rod BD is attached to two links AB and CD at B and D, Fig. E6.5a. Knowing that at the instant shown link AB rotates with a constant angular velocity of $\omega = 2$ rad/s clockwise, determine the angular velocity and angular acceleration (a) of rod BD, (b) of link CD.

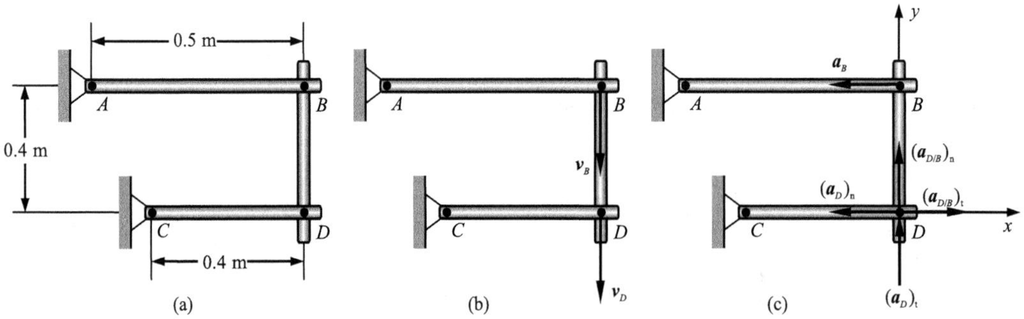

Fig. E6.5

Solution Since link AB rotates about point A clockwise, the velocity of B is vertical down, Fig. E6.5b. Similarly, the velocity of D is also vertical down due to CD rotating about point C clockwise. Therefore, the rod BD is in instantaneous translation due to the same velocity (including magnitude and direction) at B and D at the given instant, from which we have

$$v_D = v_B = AB\cdot\omega_{AB} = 1\ \mathrm{m/s},\ \omega_{BD} = 0,\ \omega_{CD} = \frac{v_D}{CD} = 2.5\ \mathrm{rad/s}\ \text{(clockwise)}$$

Choosing point B of rod BD in instantaneous translation at the given instant as the base point, then the acceleration of point D can be expressed as

$$(\boldsymbol{a}_D)_\mathrm{t} + (\boldsymbol{a}_D)_\mathrm{n} = \boldsymbol{a}_B + (\boldsymbol{a}_{D/B})_\mathrm{t} + (\boldsymbol{a}_{D/B})_\mathrm{n}$$

Establishing the reference system xy, Fig. E6.5c, and denoting by $\boldsymbol{i}$ and $\boldsymbol{j}$ the unit vectors respectively along the positive x and y axes, then we have

$$(a_D)_\mathrm{t}\boldsymbol{j} - (a_D)_\mathrm{n}\boldsymbol{i} = -a_\mathrm{B}\boldsymbol{i} + (a_{D/B})_\mathrm{t}\boldsymbol{i} + (a_{D/B})_\mathrm{n}\boldsymbol{j}$$

or

$$-(a_D)_\mathrm{n} = -a_B + (a_{D/B})_t,\ (a_D)_\mathrm{t} = (a_{D/B})_\mathrm{n}$$

Using $(a_D)_\mathrm{n} = CD \cdot \omega_{CD}{}^2 = 2.5\ \mathrm{m/s^2}$, $a_B = AB \cdot \omega_{AB}{}^2 = 2\ \mathrm{m/s^2}$, and $(a_{D/B})_\mathrm{n} = BD \cdot \omega_{BD}{}^2 = 0$, we can obtain

$$(a_{D/B})_\mathrm{t} = a_B - (a_D)_\mathrm{n} = -0.5\ \mathrm{m/s^2}\ \text{(to the left)},\ (a_D)_\mathrm{t} = 0$$

Using $(a_{D/B})_\mathrm{t} = BD \cdot \alpha_{BD}$ and $(a_D)_\mathrm{t} = CD \cdot \alpha_{CD}$, we have

$$\alpha_{BD} = -1.25\ \mathrm{rad/s^2}\ \text{(clockwise)},\ \alpha_{CD} = 0$$

Problems

6.1 The structure shown consists of two rods AE and CE and a rectangular plate $ABCD$ which are welded together. The structure rotates about the axis AE with an angular velocity of $\omega = 5$ rad/s and an angular acceleration of $\alpha = 10\ \mathrm{rad/s^2}$. Knowing that the angular velocity and angular acceleration are counterclockwise as viewed from A, determine the velocity and acceleration of point P.

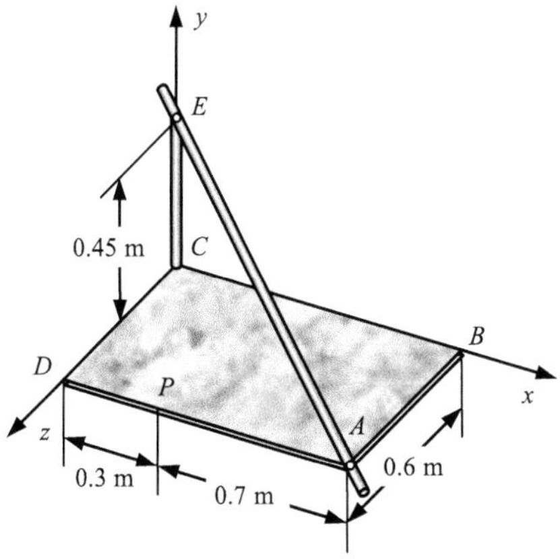

Fig. P6.1

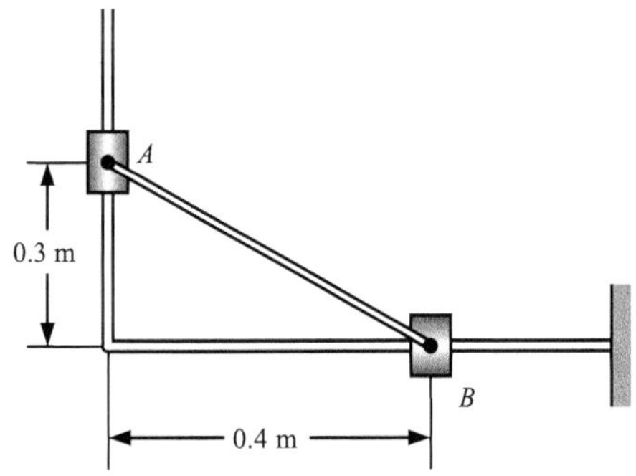

Fig. P6.2

6.2 Collar A moves up with a constant velocity of $v_A = 0.4$ m/s. At the instant shown determine (a) the angular velocity of rod AB and the velocity of collar B, (b) the angular acceleration of rod AB and the acceleration of collar B.

6.3 Rod BDE is attached to two links AB and CD at B and D. Knowing that at the instant shown link AB rotates with a constant angular velocity of $\omega = 2$ rad/s clockwise, determine the velocity and acceleration of point *E*.

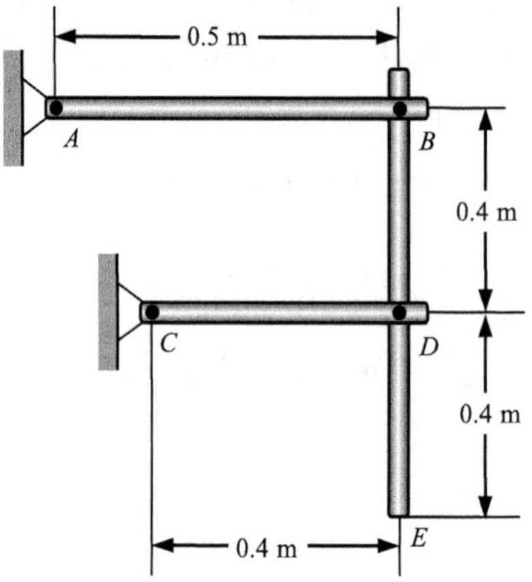

Fig. P6.3

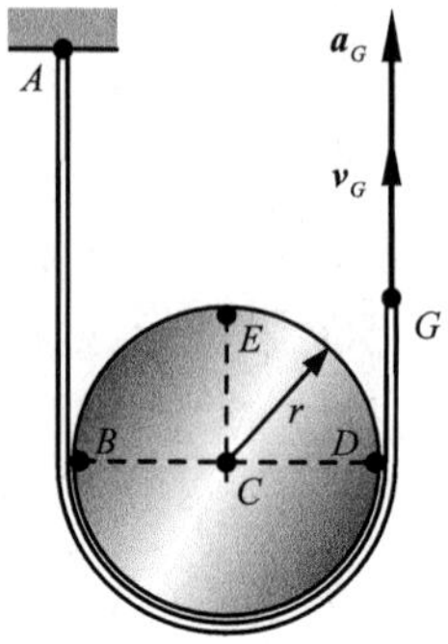

Fig. P6.4

6.4 The motion of a cylinder of radius r is controlled by a cable *AG*. Knowing that end *G* of the cable has a velocity of $v_G = 0.3$ m/s and an acceleration of $a_G = 0.5$ m/s^2, both directed upward, and that $r = 0.15$ m, determine (a) the acceleration of point *B* and of point *D*, (b) the velocity and acceleration of point *E*.

Chapter 7 Resultant Motion of Particle

The motion of a particle is usually analyzed by using a single system of reference. There are many cases, however, where the path of motion for a particle is complicated, so it may be feasible to analyze the motion of a particle by using two or more systems of reference. For example, the motion of a particle located at the tip of an airplane propeller, while the airplane is in flight, is more easily described if one observes first the motion of the airplane relative to the earth and then superposes, using the parallelogram law, the motion of the particle relative to the airplane.

One of the systems attached to the earth is called a fixed system of reference, whereas the other systems moving relative to the earth are called moving systems of reference. The motion of a particle relative to a fixed system of reference is referred to as an absolute motion, whereas the motion relative to a moving system of reference is referred to as a relative motion.

7.1 Time Derivatives of Vector

Assume that $Oxyz$ is a fixed system of reference, and that $O'x'y'z'$ is a moving system of reference, Fig. 7.1.

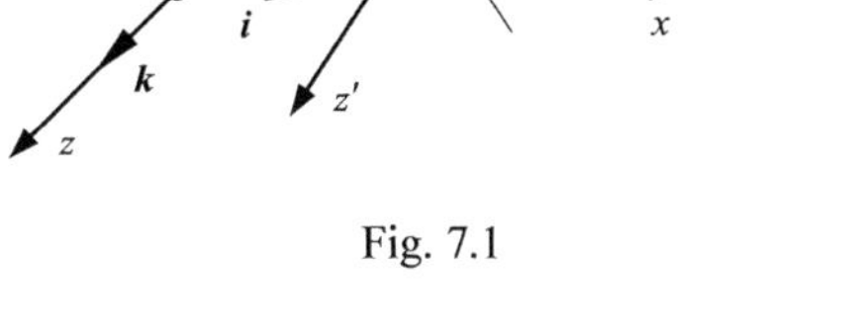

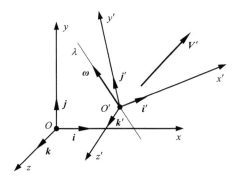

Fig. 7.1

If $O'x'y'z'$ is rotating about axis λ with an angular velocity $\boldsymbol{\omega}$ at a given instant, then vector $\boldsymbol{V}'$ can be expressed as

$$\boldsymbol{V}' = V_x'\boldsymbol{i}' + V_y'\boldsymbol{j}' + V_z'\boldsymbol{k}' \tag{7.1}$$

where V_x', V_y', V_z' and $\boldsymbol{i}'$, $\boldsymbol{j}'$, $\boldsymbol{k}'$ are, respectively, the rectangular components of $\boldsymbol{V}'$ in the rotating system of reference and the unit vectors used in the rotating systems of reference. In the rotating system of reference, $\boldsymbol{i}'$, $\boldsymbol{j}'$, and $\boldsymbol{k}'$ remain unchanged in magnitude and direction, thus we have

$$\dot{\boldsymbol{V}}' = \dot{V}_x'\boldsymbol{i}' + \dot{V}_y'\boldsymbol{j}' + \dot{V}_z'\boldsymbol{k}' \tag{7.2}$$

However, in the fixed system of reference $Oxyz$, $\boldsymbol{i}'$, $\boldsymbol{j}'$ and $\boldsymbol{k}'$ are fixed in magnitude but variable in direction. We thus obtain

$$\{\dot{\boldsymbol{V}}'\}_O = \dot{V}_x'\boldsymbol{i}' + \dot{V}_y'\boldsymbol{j}' + \dot{V}_z'\boldsymbol{k}' + V_x'\dot{\boldsymbol{i}}' + V_y'\dot{\boldsymbol{j}}' + V_z'\dot{\boldsymbol{k}}' \tag{7.3}$$

Using $\dot{\boldsymbol{i}}' = \boldsymbol{\omega}\times\boldsymbol{i}'$, $\dot{\boldsymbol{j}}' = \boldsymbol{\omega}\times\boldsymbol{j}'$, and $\dot{\boldsymbol{k}}' = \boldsymbol{\omega}\times\boldsymbol{k}'$, we have

$$\{\dot{\boldsymbol{V}}'\}_O = \dot{V}_x'\boldsymbol{i}' + \dot{V}_y'\boldsymbol{j}' + \dot{V}_z'\boldsymbol{k}' + \boldsymbol{\omega}\times(V_x'\boldsymbol{i}' + V_y'\boldsymbol{j}' + V_z'\boldsymbol{k}') \tag{7.4}$$

Substituting Eqs. (7.1) and (7.2) into Eq. (7.4) yields

$$\{\dot{\boldsymbol{V}}'\}_O = \dot{\boldsymbol{V}}' + \boldsymbol{\omega}\times\boldsymbol{V}' \tag{7.5}$$

7.2 Resultant of Velocities

Considering that a moving particle P moves along a curved line in space, and assuming that $Oxyz$ is a fixed system of reference, that $O'x'y'z'$ is a moving system of reference, and that the convected point (i.e., that point located on the moving system and coinciding with P at the instant considered) is denoted by M, Fig. 7.2, then we have

$$\boldsymbol{r}_P = \boldsymbol{r}_M + \boldsymbol{r}_{P/M} \tag{7.6}$$

where $\boldsymbol{r}_P$ is the position vector of moving particle P, $\boldsymbol{r}_M$ is the position vector of convected point M, and $\boldsymbol{r}_{P/M}$ is the position vector of the moving particle relative to the convected point.

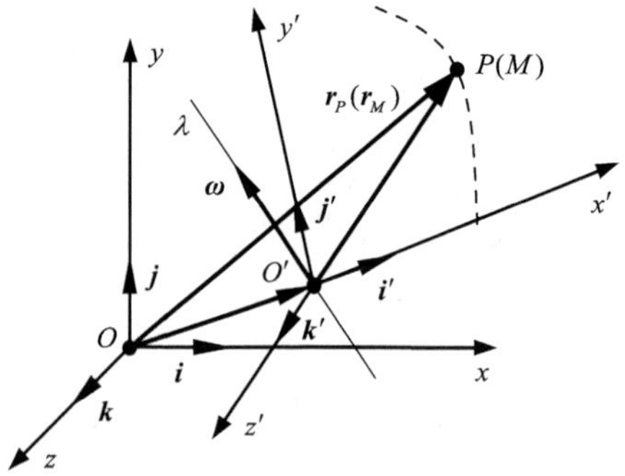

Fig. 7.2

Differentiating Eq. (7.6) with respect to time t within the fixed system $Oxyz$, we obtain

$$\dot{\boldsymbol{r}}_P = \dot{\boldsymbol{r}}_M + \{\dot{\boldsymbol{r}}_{P/M}\}_O \tag{7.7}$$

Assuming that the moving system $O'x'y'z'$ is rotating about the axis λ with an angular velocity $\boldsymbol{\omega}$ at a given instant, and using Eq. (7.5), then we have

$$\{\dot{\boldsymbol{r}}_{P/M}\}_O = \dot{\boldsymbol{r}}_{P/M} + \boldsymbol{\omega} \times \boldsymbol{r}_{P/M} \tag{7.8}$$

Substituting Eq. (7.8) into Eq. (7.7) gives

$$\dot{\boldsymbol{r}}_P = \dot{\boldsymbol{r}}_M + \dot{\boldsymbol{r}}_{P/M} + \boldsymbol{\omega} \times \boldsymbol{r}_{P/M} \tag{7.9}$$

where $\dot{\boldsymbol{r}}_P = \boldsymbol{v}_P$ is the velocity of the moving particle P, $\dot{\boldsymbol{r}}_M = \boldsymbol{v}_M$ is the velocity of the convected point M, $\dot{\boldsymbol{r}}_{P/M} = \boldsymbol{v}_{P/M}$ is the velocity of the moving particle relative to the convected point. Therefore, Eq. (7.9) can be rewritten as

$$\boldsymbol{v}_P = \boldsymbol{v}_M + \boldsymbol{v}_{P/M} + \boldsymbol{\omega} \times \boldsymbol{r}_{P/M} \tag{7.10}$$

Using $\boldsymbol{r}_{P/M} = 0$, Eq. (7.10) can be simplified as

$$\boldsymbol{v}_P = \boldsymbol{v}_M + \boldsymbol{v}_{P/M} \tag{7.11}$$

We thus conclude that, at any given instant, the absolute velocity $\boldsymbol{v}_P$ can be obtained by adding vectorially the convected velocity $\boldsymbol{v}_M$ and the relative velocity $\boldsymbol{v}_{P/M}$.

Example 7.1 Two planes, *A* and *B*, are flying at the same altitude, Fig. E7.1a. Assuming that plane *A* is flying south with a velocity of $v_A = 400$ km/h and that plane *B* is flying 30° north of east with a velocity of $v_B = 500$ km/h, determine the velocity of plane *B* with respect to plane *A*.

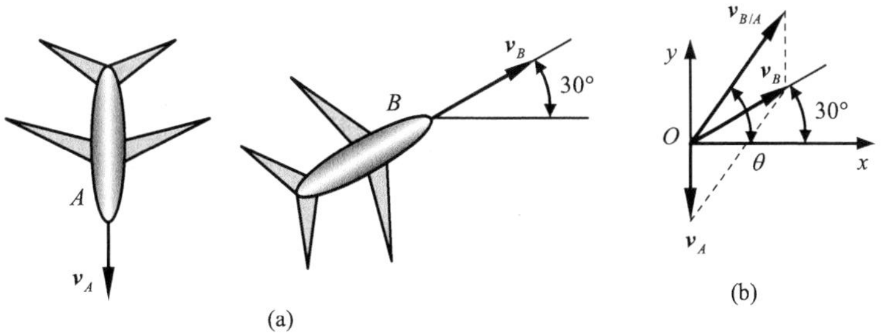

Fig. E7.1

Solution Assume that the earth is chosen as a fixed system, that plane *A* as a moving system, and that plane *B* as a moving particle. From $\boldsymbol{v}_P = \boldsymbol{v}_M + \boldsymbol{v}_{P/M}$, we have

$$\boldsymbol{v}_B = \boldsymbol{v}_A + \boldsymbol{v}_{B/A}$$

Using the system of coordinates shown in Fig. E7.1b, we obtain

$$(v_B \cos 30°)\boldsymbol{i} + (v_B \sin 30°)\boldsymbol{j} = (-v_A)\boldsymbol{j} + (v_{B/A} \cos\theta)\boldsymbol{i} + (v_{B/A} \sin\theta)\boldsymbol{j}$$

where $\boldsymbol{i}$ and $\boldsymbol{j}$ are the unit vectors respectively corresponding to the positive *x* and *y* axes, and θ is the angle between $\boldsymbol{v}_{B/A}$ and the positive *x* axis. Substituting $v_A = 400$ km/h and $v_B = 500$ km/h into the above vector equation, we can obtain

$$v_{B/A} = 781.0 \text{ km/h}, \ \theta = 56.33°$$

Thus the velocity of plane *B* with respect to plane *A* is 781.0 km/h in magnitude and 56.33° north of east in direction.

7.3 Resultant of Accelerations

Differentiating Eq. (7.10) with respect to time t within the fixed system $Oxyz$, Fig. 7.2, we have

$$\dot{\boldsymbol{v}}_P = \dot{\boldsymbol{v}}_M + \{\dot{\boldsymbol{v}}_{P/M}\}_O + \dot{\boldsymbol{\omega}} \times \boldsymbol{r}_{P/M} + \boldsymbol{\omega} \times \{\dot{\boldsymbol{r}}_{P/M}\}_O \tag{7.12}$$

Using Eq. (7.5), we have

$$\{\dot{\boldsymbol{v}}_{P/M}\}_O = \dot{\boldsymbol{v}}_{P/M} + \boldsymbol{\omega} \times \boldsymbol{v}_{P/M} \tag{7.13}$$

and

$$\{\dot{\boldsymbol{r}}_{P/M}\}_O = \dot{\boldsymbol{r}}_{P/M} + \boldsymbol{\omega} \times \boldsymbol{r}_{P/M} \tag{7.14}$$

Substituting Eqs. (7.13) and (7.14) into Eq. (7.12), we have

$$\dot{\boldsymbol{v}}_P = \dot{\boldsymbol{v}}_M + \dot{\boldsymbol{v}}_{P/M} + \boldsymbol{\omega} \times \boldsymbol{v}_{P/M} + \dot{\boldsymbol{\omega}} \times \boldsymbol{r}_{P/M} + \boldsymbol{\omega} \times (\dot{\boldsymbol{r}}_{P/M} + \boldsymbol{\omega} \times \boldsymbol{r}_{P/M}) \tag{7.15}$$

Using $\dot{\boldsymbol{v}}_P = \boldsymbol{a}_P$, $\dot{\boldsymbol{v}}_M = \boldsymbol{a}_M$, $\dot{\boldsymbol{v}}_{P/M} = \boldsymbol{a}_{P/M}$, $\dot{\boldsymbol{\omega}} = \boldsymbol{\alpha}$, and $\dot{\boldsymbol{r}}_{P/M} = \boldsymbol{v}_{P/M}$, and assuming that $\boldsymbol{a}_{\mathrm{C}} = 2\boldsymbol{\omega} \times \boldsymbol{v}_{P/M}$, then we obtain

$$\boldsymbol{a}_P = \boldsymbol{a}_M + \boldsymbol{a}_{P/M} + \boldsymbol{a}_{\mathrm{C}} + \boldsymbol{\alpha} \times \boldsymbol{r}_{P/M} + \boldsymbol{\omega} \times (\boldsymbol{\omega} \times \boldsymbol{r}_{P/M}) \tag{7.16}$$

where $\boldsymbol{a}_P$ is the acceleration of the moving particle P, $\boldsymbol{a}_M$ is the acceleration of the convected point M, $\boldsymbol{a}_{P/M}$ is the acceleration of the moving particle P with respect to the convected point M, $\boldsymbol{a}_{\mathrm{C}}$ is the Coriolis acceleration, and $\boldsymbol{\alpha}$ is the angular acceleration of the moving system of reference. Using $\boldsymbol{r}_{P/M} = \boldsymbol{0}$, Eq. (7.16) can be simplified as

$$\boldsymbol{a}_P = \boldsymbol{a}_M + \boldsymbol{a}_{P/M} + \boldsymbol{a}_{\mathrm{C}} \tag{7.17}$$

We conclude that at any given instant the absolute acceleration $\boldsymbol{a}_P$ is a vector sum of the convected acceleration $\boldsymbol{a}_M$, the relative acceleration $\boldsymbol{a}_{P/M}$, and the Coriolis acceleration $\boldsymbol{a}_{\mathrm{C}}$.

Example 7.2 Knowing that at the instant shown in Fig. E7.2a rod *BD* rotates with a constant counterclockwise angular velocity $\omega_{BD} = 2$ rad/s, determine (a) the angular velocity of rod *AD* and the velocity of collar *D* with respect to rod *BD*, (b) the angular acceleration of rod *AD* and the acceleration of collar *D* with respect to rod *BD*.

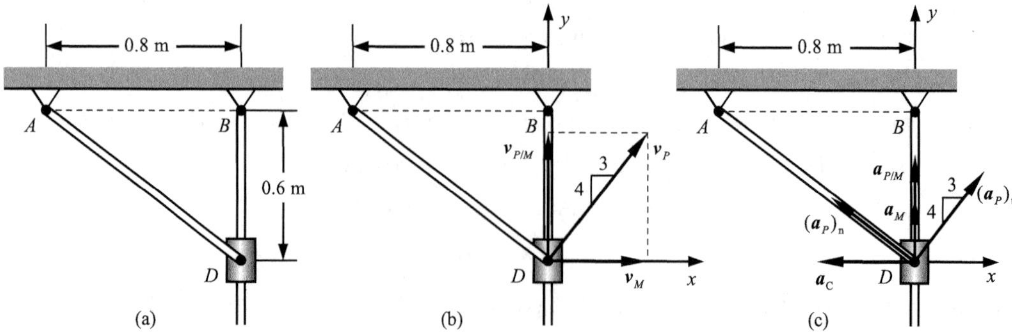

Fig. E7.2

Solution Assume that a fixed system of reference is attached to the earth, that a moving system of reference is attached to rod BD, and that a moving particle is attached to collar D.

(a) Using $\boldsymbol{v}_P = \boldsymbol{v}_M + \boldsymbol{v}_{P/M}$, Fig. E7.2b, we have

$$\frac{3}{5}v_P\boldsymbol{i} + \frac{4}{5}v_P\boldsymbol{j} = v_M\boldsymbol{i} + v_{P/M}\boldsymbol{j}$$

where $\boldsymbol{i}$ and $\boldsymbol{j}$ are the unit vectors respectively corresponding to the positive x and y axes. Solving the above equation and using $v_M = BD \cdot \omega_{BD} = 1.2\ \text{m/s}$, we obtain

$$v_P = \frac{5}{3}v_M = 2\ \text{m/s},\ v_{P/M} = \frac{4}{3}v_M = 1.6\ \text{m/s}$$

Using $v_P = AD \cdot \omega_{AD}$, we can obtain

$$\omega_{AD} = \frac{v_P}{AD} = 2\ \text{rad/s}\ \ \text{(anticlockwise)}$$

(b) Using $(\boldsymbol{a}_P)_\text{t} + (\boldsymbol{a}_P)_\text{n} = \boldsymbol{a}_M + \boldsymbol{a}_{P/M} + \boldsymbol{a}_\text{C}$, Fig. E7.2c, we have

$$[\frac{3}{5}(a_P)_\text{t}\boldsymbol{i} + \frac{4}{5}(a_P)_\text{t}\boldsymbol{j}] + [-\frac{4}{5}(a_P)_\text{n}\boldsymbol{i} + \frac{3}{5}(a_P)_\text{n}\boldsymbol{j}] = a_M\boldsymbol{j} + a_{P/M}\boldsymbol{j} - a_\text{C}\boldsymbol{i}$$

where $(a_P)_\text{n} = AD \cdot \omega_{AD}^2 = 4\ \text{m/s}^2$, $a_M = BD \cdot \omega_{BD}^2 = 2.4\ \text{m/s}^2$ and $a_\text{C} = 2\omega_{BD}v_{P/M} = 6.4\ \text{m/s}^2$. Solving the above equation, we obtain

$$(a_P)_\text{t} = -5.33\ \text{m/s}^2,\ a_{P/M} = -4.27\ \text{m/s}^2$$

Using $(a_P)_\text{t} = AD \cdot \alpha_{AD}$, we can obtain

$$\alpha_{AD} = \frac{(a_P)_\text{t}}{AD} = -5.33\ \text{rad/s}^2\ \ \text{(clockwise)}$$

Example 7.3 As the truck begins to move backward with an acceleration of $2\ \text{m/s}^2$, the outer section D of its boom starts to retract with an acceleration of $1\ \text{m/s}^2$ relative to the truck, Fig. E7.3a. Determine the acceleration of section D.

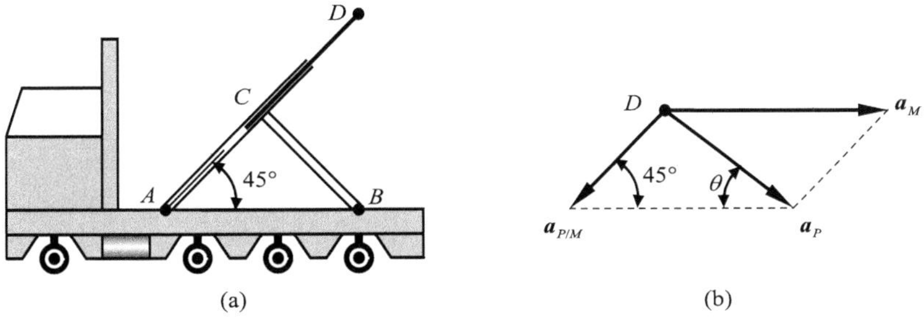

Fig. E7.3

Solution Assume that a fixed system of reference is attached to the earth, that a moving system of reference is attached to the truck, and that a moving particle is attached to section D.

Drawing the acceleration triangle based on $\boldsymbol{a}_P = \boldsymbol{a}_M + \boldsymbol{a}_{P/M}$, Fig. E7.3b, we have

$$a_P = \sqrt{(a_M)^2 + (a_{P/M})^2 - 2a_M a_{P/M}\cos 45°} = 1.47\ \text{m/s}^2,\ \theta = \arcsin(\frac{a_{P/M}}{a_P}\sin 45°) = 28.75°$$

Thus the acceleration of section D is $1.47\ \text{m/s}^2$ in magnitude and $28.75°$ south of east in direction.

Problems

7.1 A collar slides outward at a constant relative speed v along the rod OAB, which rotates counterclockwise with a constant angular velocity of $\omega_B = 2\ \text{rad/s}$. Knowing that the collar is located at A when $\theta = 0$ and that the collar reaches B when $\theta = 90°$, determine the acceleration of the collar when $\theta = 30°$.

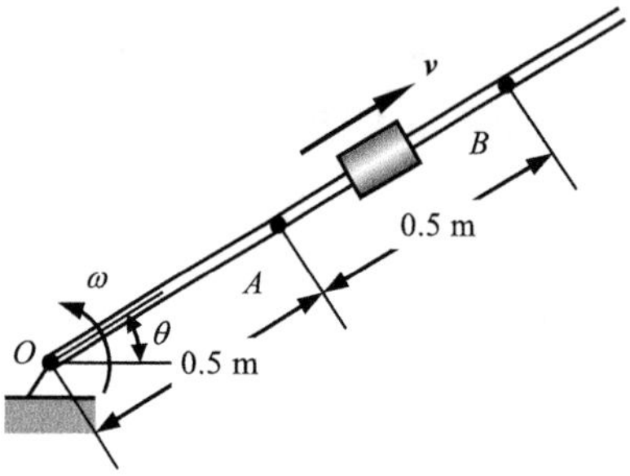

Fig. P7.1

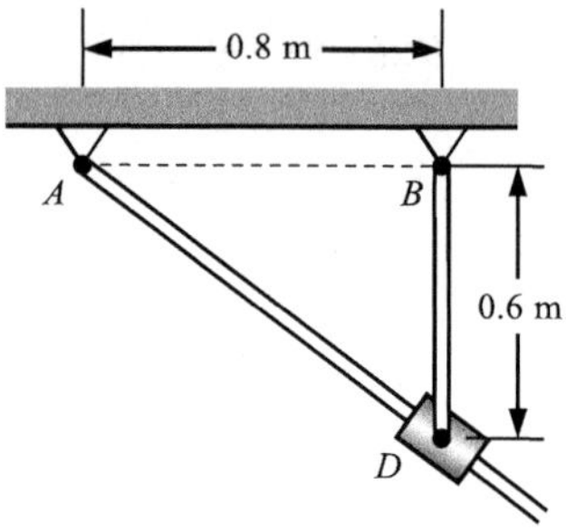

Fig. P7.2

7.2 Knowing that at the instant shown rod BD rotates with a constant counterclockwise angular velocity $\omega_B = 2\ \text{rad/s}$, determine (a) the angular velocity of rod AD and the velocity of collar D with respect to rod AD, (b) the angular acceleration of rod AD and the acceleration of collar D with respect to rod AD.

7.3 As the truck begins to move forward with an acceleration of $3\ \text{m/s}^2$, the outer section D of its boom starts to retract with an acceleration of $2\ \text{m/s}^2$ relative to the truck. Determine the acceleration of section D.

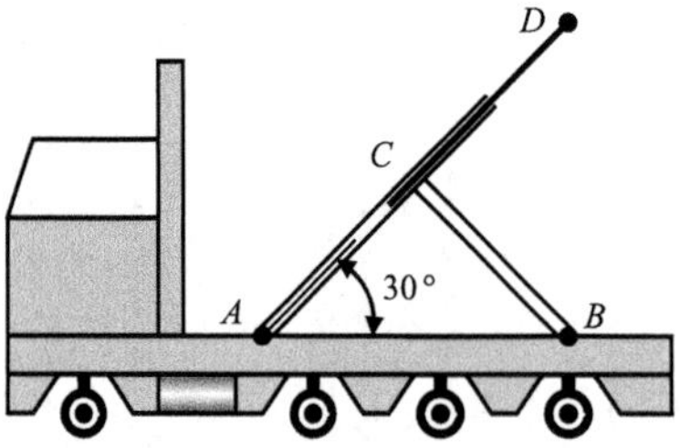

Fig. P7.3

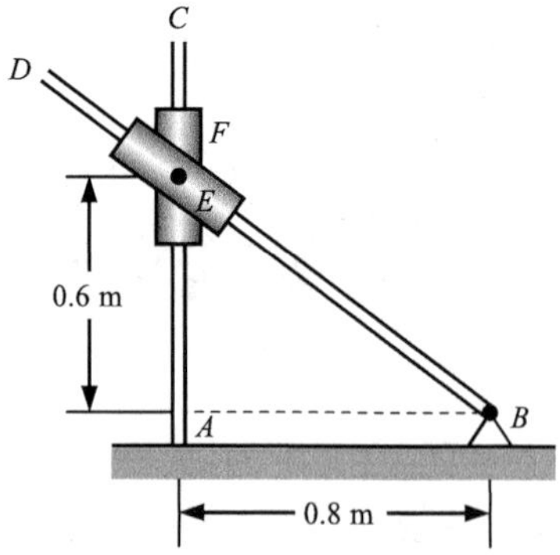

Fig. P7.4

7.4 Collar E slides along rod BD and is attached to collar F that moves along a vertical rod AC. Knowing that the angular velocity and angular acceleration of rod BD are

$\omega = 6$ rad/s and $\alpha = 4$ rad/s^2, both clockwise, determine the velocity and acceleration of collar *E*.

7.5 Plane *A* flying horizontally in a straight line has a velocity of 300 km/h and an acceleration of 5 m/s^2 at the given instant. Plane *B* is flying at the same altitude as plane *A* and is following a circular path of radius 200 m. Knowing that at the given instant the velocity and deceleration of plane *B* are 400 km/h and 3 m/s^2 respectively, determine, at the instant shown, (a) the velocity of *B* relative to *A*, (b) the acceleration of *B* relative to *A*.

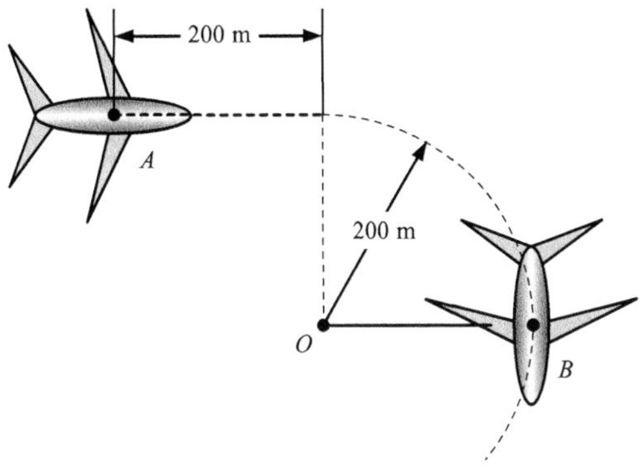

Fig. P7.5

7.6 A rod *OA* is always tangential to the surface of a semicylinder *C* of radius $r = 0.4$ m. Knowing that the velocity and acceleration of the semicylinder are $v = 0.1$ m/s and $a = 0.2$ m/s^2 respectively, both to the right, determine the angular velocity and angular acceleration of the rod when $\theta = 30°$.

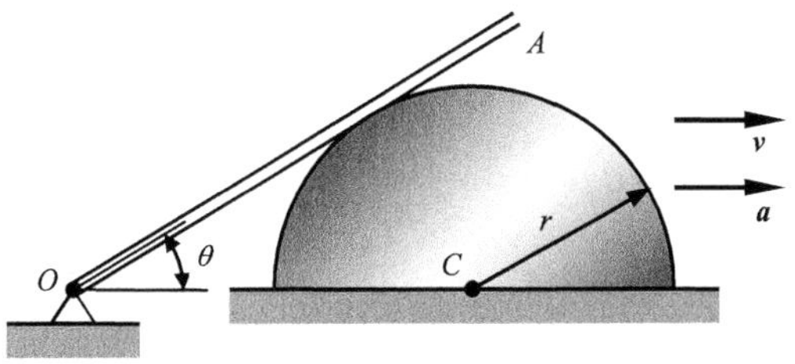

Fig. P7.6

7.7 Rod *AB* moves over a small wheel at *C* while end *A* moves to the right with a constant velocity of 600 mm/s. At the instant shown determine (a) the angular velocity of the rod and the velocity of end *B* of the rod, (b) the angular acceleration of the rod and the acceleration of end *B* of the rod.

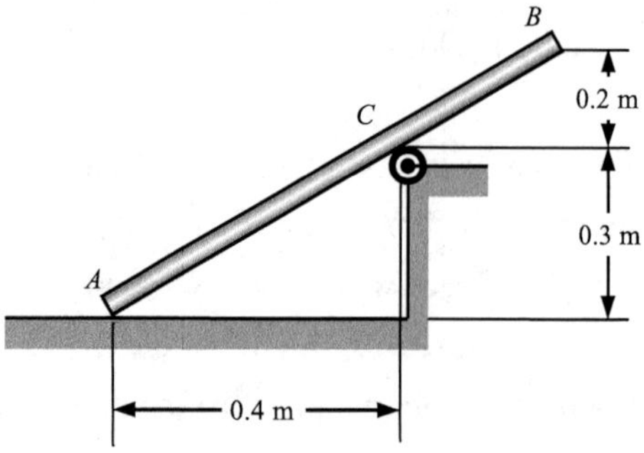

Fig. P7.7

Chapter 8 Kinetics of Particles

8.1 Equations of Motion

From Newton's Second Law, if the resultant force acting on a particle is not zero, the particle will have acceleration proportional to the magnitude of the resultant force and in the direction of the resultant force. This law can be expressed mathematically as

$$m\boldsymbol{a} = \sum \boldsymbol{F} \tag{8.1}$$

where m is the mass of the particle, $\boldsymbol{a}$ is the acceleration of the particle, and $\sum \boldsymbol{F}$ is the resultant force. It should be noted that Newton's Second Law of motion holds only with respect to a Newtonian system of reference or an inertial system of reference. In order to solve problems involving the motion of a particle, it will be more convenient to replace the vector equation above by equivalent scalar equations represented by rectangular or natural coordinates.

Resolving the acceleration and each of the forces into rectangular components, we obtain

$$ma_x = \sum F_x,\ ma_y = \sum F_y,\ ma_z = \sum F_z \tag{8.2}$$

where $a_x = \dot{v}_x = \ddot{x},\ a_y = \dot{v}_y = \ddot{y},\ a_z = \dot{v}_z = \ddot{z}$.

Resolving the acceleration and the forces into components along the tangent to the path (in the direction of motion) and the normal (toward the center of curvature), we obtain

$$ma_{\mathrm{t}} = \sum F_{\mathrm{t}},\ ma_{\mathrm{n}} = \sum F_{\mathrm{n}},\ 0 = \sum F_{\mathrm{b}} \tag{8.3}$$

where $a_{\mathrm{t}} = \dot{v},\ a_{\mathrm{n}} = \dfrac{v^2}{\rho}$.

Example 8.1 A spring AB of stiffness k is attached to a support A and to a collar B of mass m, Fig. E8.1a. The unstretched length of the spring is l. Knowing that the collar is released from rest at $x = \sqrt{3}l$ and neglecting friction between the collar and the horizontal rod, determine the magnitude of the velocity of the collar as it passes through midpoint O.

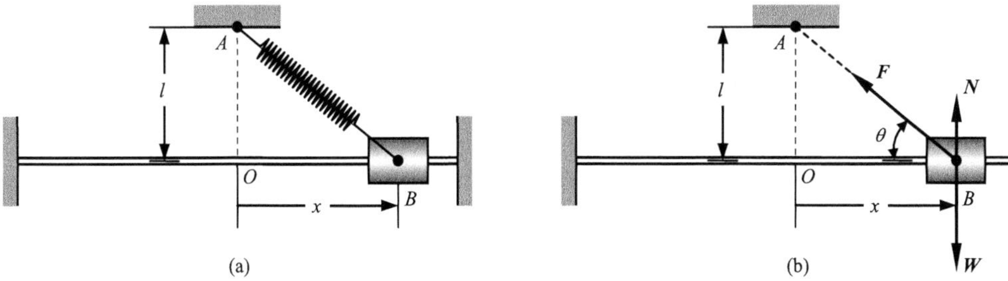

Fig. E8.1

Solution When the collar is located at position x, then the elongation of the spring can be expressed as

$$\delta = \sqrt{l^2 + x^2} - l$$

Thus the corresponding spring force can be given by

$$F = k\delta = k(\sqrt{l^2 + x^2} - l)$$

From the equation of motion, $ma_x = \sum F_x$, we have

$$ma = -F\cos\theta$$

Using $a = \dfrac{\mathrm{d}v}{\mathrm{d}t} = \dfrac{\mathrm{d}x}{\mathrm{d}t}\dfrac{\mathrm{d}v}{\mathrm{d}x} = v\dfrac{\mathrm{d}v}{\mathrm{d}x}$ and $\cos\theta = \dfrac{x}{\sqrt{l^2 + x^2}}$, we have

$$v\mathrm{d}v = -\frac{k}{m}\left(x - \frac{lx}{\sqrt{l^2 + x^2}}\right)\mathrm{d}x$$

Performing integration on the above equation, we obtain

$$\int_0^{v_O} v\mathrm{d}v = -\frac{k}{m}\int_{\sqrt{3}l}^{0}\left(x - \frac{lx}{\sqrt{l^2 + x^2}}\right)\mathrm{d}x$$

i.e.,

$$v_O = l\sqrt{k/m}$$

Example 8.2 The initial velocity of the block in position A is $v_A = 8\ \mathrm{m/s}$, Fig. E8.2a. Knowing that the coefficient of kinetic friction between the block and the plane of inclination is $\mu_\mathrm{k} = 0.25$, determine the distance it moves and the time it takes for the block to reach B with zero velocity if $\theta = 10°$.

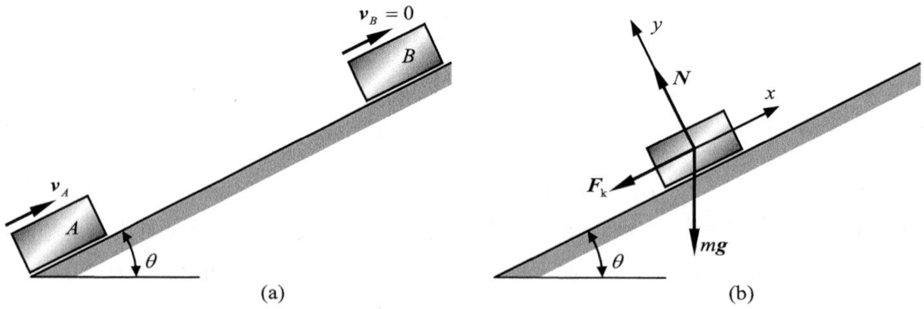

Fig. E8.2

Solution From the equations of motion $ma_x = \sum F_x$ and $ma_y = \sum F_y$, Fig. E8.2b, we have

$$ma_x = \sum F_x,\ \ ma = -mg\sin\theta - F_\mathrm{k}$$

$$ma_y = \sum F_y,\ \ 0 = N - mg\cos\theta$$

Using $F_\mathrm{k} = \mu_\mathrm{k} N$ and solving the above equations, we can obtain

$$a = -(\sin\theta + \mu_k \cos\theta)g$$

(1) Using $a = \frac{dv}{dt} = \frac{dx}{dt}\frac{dv}{dx} = v\frac{dv}{dx}$, we have

$$v dv = -(\sin\theta + \mu_k \cos\theta)g dx$$

Performing integration on the above equation, we obtain

$$\int_{v_A}^{v_B} v dv = \int_0^{x_{AB}} -(\sin\theta + \mu_k \cos\theta)g dx$$

i.e.,

$$x_{AB} = \frac{v_A^2 - v_B^2}{2(\sin\theta + \mu_k \cos\theta)g}$$

Using $v_A = 8$ m/s, $v_B = 0$, $\mu_k = 0.25$, $\theta = 10°$, and $g = 9.81$ m/s^2, we obtain

$$x_{AB} = 7.77 \text{ m}$$

(2) Using $a = \frac{dv}{dt}$, we have

$$dv = -(\sin\theta + \mu_k \cos\theta)g dt$$

Performing integration on the above equation, we obtain

$$\int_{v_A}^{v_B} dv = \int_0^{t_{AB}} -(\sin\theta + \mu_k \cos\theta)g dt$$

i.e.,

$$t_{AB} = \frac{v_A - v_B}{(\sin\theta + \mu_k \cos\theta)g}$$

Using $v_A = 8$ m/s, $v_B = 0$, $\mu_k = 0.25$, $\theta = 10°$, and $g = 9.81$ m/s^2, we obtain

$$t_{AB} = 1.94 \text{ s}$$

8.2 Method of Inertia Force

Returning to Eq. (8.1), and assuming that $\boldsymbol{F}_{\mathrm{I}} = -m\boldsymbol{a}$, we can rewrite Newton's Second Law of motion in the form

$$\sum \boldsymbol{F} + \boldsymbol{F}_{\mathrm{I}} = 0 \tag{8.4}$$

where $\boldsymbol{F}_{\mathrm{I}}$ is called an inertia force, which is equal in magnitude to ma and opposite in direction to $\boldsymbol{a}$. This method proposed by d'Alembert to analyze the motion of a particle is called the method of inertia force or d'Alembert's principle. It should be noted that an inertia force is an imaginary force, but not an actual force.

The method of inertia force can convert a kinetic problem into an equivalent problem in equilibrium, and this kinetic problem can be solved by the methods developed in statics. Although the particle is not in equilibrium, the equations of equilibrium obtained in statics can be applied if an inertia force is applied to the particle in motion.

8.3 Method of Work and Energy

The method of work and energy can be used to solve problems involving force, mass, velocity and displacement.

1. Principle of Work and Energy

Multiplying both sides of Eq. (8.1) by $\mathrm{d}\boldsymbol{r}$ through a dot product, we obtain

$$m\boldsymbol{a}\cdot \mathrm{d}\boldsymbol{r}=\sum \boldsymbol{F}\cdot \mathrm{d}\boldsymbol{r} \tag{8.5}$$

Using $\boldsymbol{a}\cdot \mathrm{d}\boldsymbol{r}=\mathrm{d}\boldsymbol{v}\cdot \boldsymbol{v}=\frac{1}{2}\mathrm{d}(\boldsymbol{v}\cdot \boldsymbol{v})=\frac{1}{2}\mathrm{d}(v^2)$, we have

$$\mathrm{d}T=\mathrm{d}W \tag{8.6}$$

where $\mathrm{d}T=\mathrm{d}(\frac{1}{2}mv^2)$ and $\mathrm{d}W=\sum \boldsymbol{F}\cdot \mathrm{d}\boldsymbol{r}$ are respectively the increment in kinetic energy of the particle and the elementary work done by the forces $\sum \boldsymbol{F}$ during a displacement $\mathrm{d}\boldsymbol{r}$. Integrating from A_1 to A_2, we obtain

$$T_2-T_1=W_{12} \tag{8.7}$$

where $T_1=\frac{1}{2}mv_1^2$ and $T_2=\frac{1}{2}mv_2^2$ are the kinetic energy of the particle respectively at A_1 and A_2, $W_{12}=\sum \int_{A_1}^{A_2} \boldsymbol{F}\cdot \mathrm{d}\boldsymbol{r}$ is the work of the forces acting on the particle when the particle moves from A_1 to A_2.

We thus conclude that, when a particle moves from one point to another, the change in kinetic energy of the particle is equal to the work of the forces acting on the particle. The relation expressed by Eq. (8.6) or (8.7) is called the principle of work and energy. It should be note that the principle of work and energy applies only with respect to a Newtonian's system of reference.

Example 8.3 The initial velocity of the block in position A is $v_A=8$ m/s, Fig. E8.3. Knowing that the coefficient of kinetic friction between the block and the plane of inclination is $\mu_\mathrm{k}=0.25$, determine the distance it moves for the block to reach B with zero velocity if $\theta=10°$.

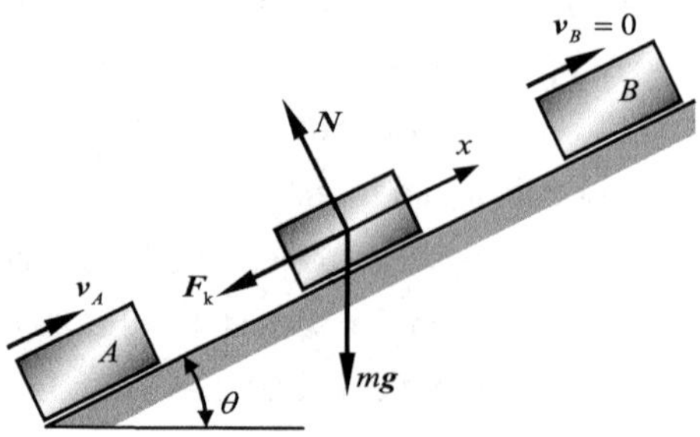

Fig. E8.3

Solution From the principle of work and energy, $T_B - T_A = W_{AB}$, we have

$$\frac{1}{2}mv_B^2 - \frac{1}{2}mv_A^2 = (-F_k - mg\sin\theta)x_{AB}$$

where, $F_k = \mu_k N = \mu_k mg\cos\theta$.

Using $v_A = 8$ m/s, $v_B = 0$, $\mu_k = 0.25$, $\theta = 10°$, and $g = 9.81$ m/s^2, we obtain

$$x_{AB} = \frac{v_A^2 - v_B^2}{2(\sin\theta + \mu_k\cos\theta)g} = 7.77 \text{ m}$$

2. Conservation of Energy

A force acting on a particle is said to be conservative if its work is independent of the path followed by the particle as it moves from one point to another. The gravitational and elastic forces are typical examples of conservative forces.

The work done by a conservative force acting on a particle moving from A_1 to A_2 is defined as the potential energy of the particle at A_1 with respect to A_2, and can be expressed as

$$V_1 - V_2 = W_{12} = \sum\int_{A_1}^{A_2} \boldsymbol{F}\cdot \mathrm{d}\boldsymbol{r} \tag{8.8}$$

where V_1 and V_2 respectively represent the potential energy of the particle at A_1 and A_2.

Using Eqs. (8.7) and (8.8), we can obtain

$$T_1 + V_1 = T_2 + V_2 \tag{8.9}$$

Eq. (8.9) shows that when a particle moves under the action of conservative forces, the sum of the kinetic energy and the potential energy of the particle remains constant. This sum is called mechanical energy.

Example 8.4 A spring *AB* of stiffness *k* is attached to a support *A* and to a collar *B* of mass *m*, Fig. E8.4. The unstretched length of the spring is l. Knowing that the collar is released from rest at $x = \sqrt{3}l$ and neglecting friction between the collar and the horizontal rod, determine the magnitude of the velocity of the collar as it passes through midpoint *O*.

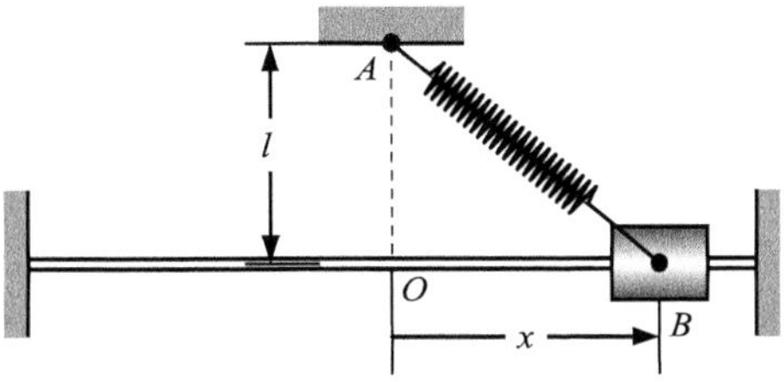

Fig. E8.4

Solution When the collar is located at position $x = \sqrt{3}l$, we have

$$T_1 = 0,\ V_1 = \frac{1}{2}k\delta^2 = \frac{1}{2}k[\sqrt{l^2 + (\sqrt{3}l)^2} - l]^2 = \frac{1}{2}kl^2$$

and when at midpoint O, we have

$$T_2 = \frac{1}{2}mv_O^{\ 2},\ V_2 = 0$$

Using the conservation of mechanical energy, $T_1 + V_1 = T_2 + V_2$, we have

$$0 + \frac{1}{2}kl^2 = \frac{1}{2}mv_O^{\ 2} + 0,\ \text{or}\ \ v_O = l\sqrt{k/m}$$

8.4 Method of Impulse and Momentum

The method of impulse and momentum can be used to solve problems involving force, mass, velocity and time.

1. Principle of Impulse and Momentum

Multiplying both sides of Eq. (8.1) by $\mathrm{d}t$, we have

$$m\boldsymbol{a}\mathrm{d}t = \sum \boldsymbol{F}\mathrm{d}t \tag{8.10}$$

Using $\boldsymbol{a}\mathrm{d}t = \mathrm{d}\boldsymbol{v}$, we obtain

$$\mathrm{d}\boldsymbol{L} = \mathrm{d}\boldsymbol{I} \tag{8.11}$$

where $\mathrm{d}\boldsymbol{L} = \mathrm{d}(m\boldsymbol{v})$ and $\mathrm{d}\boldsymbol{I} = \sum \boldsymbol{F}\mathrm{d}t$ are respectively the increment in momentum of the particle and the elementary impulse of the forces $\sum \boldsymbol{F}$ acting on the particle during a time interval $\mathrm{d}t$. Integrating from t_1 to t_2, we obtain

$$\boldsymbol{L}_2 - \boldsymbol{L}_1 = \boldsymbol{I}_{12} \tag{8.12}$$

where $\boldsymbol{L}_1 = m\boldsymbol{v}_1$ and $\boldsymbol{L}_2 = m\boldsymbol{v}_2$ are respectively the momenta of the particle at t_1 and t_2, $\boldsymbol{I}_{12} = \sum \int_{t_1}^{t_2} \boldsymbol{F}\mathrm{d}t$ is the impulse of the forces acting on the particle during the interval of time from t_1 to t_2.

We thus conclude from Eq. (8.11) or (8.12) that the change in momentum of the particle is equal to the impulse of the forces acting on the particle during the time interval considered. The relation expressed by Eq. (8.11) or (8.12) is called the principle of impulse and momentum. The principle of impulse and momentum is a vector equation, thus Eq. (8.11) or (8.12) is often resolved into rectangular coordinate equations when used to solve problems.

Example 8.5 The initial velocity of the block in position A is $v_A = 8$ m/s, Fig. E8.5. Knowing that the coefficient of kinetic friction between the block and the plane of inclination is $\mu_\mathrm{k} = 0.25$, determine the time it takes for the block to reach B with zero velocity if $\theta = 10°$.

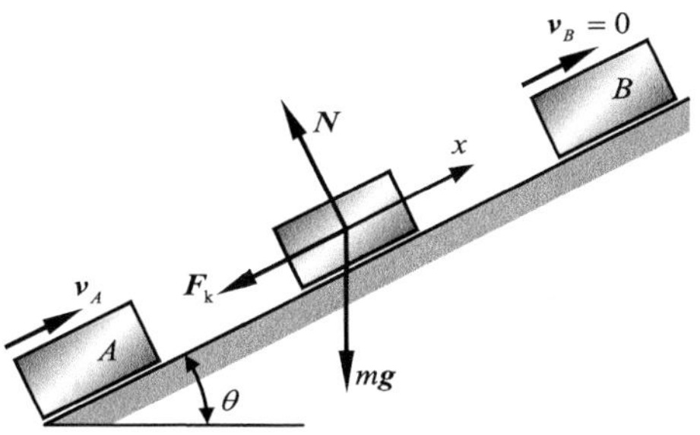

Fig. E8.5

Solution From the principle of impulse and momentum, $\boldsymbol{L}_B - \boldsymbol{L}_A = \boldsymbol{I}_{AB}$, we have

$$mv_B - mv_A = (-F_k - mg\sin\theta)t_{AB}$$

where, $F_k = \mu_k N = \mu_k mg\cos\theta$.

Using $v_A = 8$ m/s, $v_B = 0$, $\mu_k = 0.25$, $\theta = 10°$, and $g = 9.81$ m/s^2, we obtain

$$t_{AB} = \frac{v_A - v_B}{(\sin\theta + \mu_k\cos\theta)g} = 1.94 \text{ s}$$

2. Conservation of Momentum

If the resultant of forces acting on a particle is zero, Eq. (8.12) reduces to

$$\boldsymbol{L}_1 = \boldsymbol{L}_2 \tag{8.13}$$

which expresses that the momentum is conserved.

8.5 Equations of Motion for Particle System

In order to derive the equations of motion for a system of n particles, we apply Newton's Second Law to each individual particle of the system. Consider particle P_i of mass m_i acted upon by the resultant $\boldsymbol{F}_i^{(e)}$ of the external forces acting on particle P_i and by the resultant $\boldsymbol{F}_i^{(i)}$ of the internal forces exerted on particle P_i by all the other particles of the system. Newton's Second Law for particle P_i can be expressed as

$$m_i\boldsymbol{a}_i = \boldsymbol{F}_i^{(e)} + \boldsymbol{F}_i^{(i)} \quad (i = 1, 2, \cdots, n) \tag{8.14}$$

where $\boldsymbol{a}_i$ is the acceleration of particle P_i with respect to a Newtonian system of reference. Taking the moments about a fixed point O of all the terms in Eq. (8.14), we have

$$\boldsymbol{r}_i \times m_i\boldsymbol{a}_i = \boldsymbol{r}_i \times \boldsymbol{F}_i^{(e)} + \boldsymbol{r}_i \times \boldsymbol{F}_i^{(i)} \quad (i = 1, 2, \cdots, n) \tag{8.15}$$

where $\boldsymbol{r}_i$ is the position vector of particle P_i with respect to point O.

Considering all the particles in the system, then from Eqs. (8.14) and (8.15) we can obtain

$$\sum m_i\boldsymbol{a}_i = \sum \boldsymbol{F}_i^{(e)} + \sum \boldsymbol{F}_i^{(i)} \tag{8.16}$$

$$\sum(\boldsymbol{r}_i \times m_i \boldsymbol{a}_i) = \sum(\boldsymbol{r}_i \times \boldsymbol{F}_i^{(e)}) + \sum(\boldsymbol{r}_i \times \boldsymbol{F}_i^{(i)}) \tag{8.17}$$

Using $\sum \boldsymbol{F}_i^{(i)} = 0$ and $\sum(\boldsymbol{r}_i \times \boldsymbol{F}_i^{(i)}) = 0$, Eqs. (8.16) and (8.17) can be simplified as

$$\sum m_i \boldsymbol{a}_i = \sum \boldsymbol{F}_i^{(e)} \tag{8.18}$$

$$\sum(\boldsymbol{r}_i \times m_i \boldsymbol{a}_i) = \sum(\boldsymbol{r}_i \times \boldsymbol{F}_i^{(e)}) \tag{8.19}$$

The linear momentum of a system of particles is defined as the sum of the linear momenta of the various particles of the system, i.e.,

$$\boldsymbol{L} = \sum m_i \boldsymbol{v}_i \tag{8.20}$$

Differentiating Eq. (8.20) with respect to time t, we have

$$\dot{\boldsymbol{L}} = \sum m_i \boldsymbol{a}_i \tag{8.21}$$

The angular momentum about a fixed point O of a system of particles is defined as the sum of the angular momenta about the same point O of the various particles of the system, i.e.,

$$\boldsymbol{H}_O = \sum(\boldsymbol{r}_i \times m_i \boldsymbol{v}_i) \tag{8.22}$$

where $\boldsymbol{r}_i \times m_i \boldsymbol{v}_i$ is the angular momentum about the fixed point O of particle P_i.

Differentiating Eq. (8.22) with respect to time t, we have

$$\dot{\boldsymbol{H}}_O = \sum(\boldsymbol{v}_i \times m_i \boldsymbol{v}_i) + \sum(\boldsymbol{r}_i \times m_i \boldsymbol{a}_i) \tag{8.23}$$

Using $\boldsymbol{v}_i \times \boldsymbol{v}_i = 0$, Eq. (8.23) can be simplified as

$$\dot{\boldsymbol{H}}_O = \sum(\boldsymbol{r}_i \times m_i \boldsymbol{a}_i) \tag{8.24}$$

Substituting Eqs. (8.21) and (8.24) into Eqs. (8.18) and (8.19), we obtain

$$\dot{\boldsymbol{L}} = \sum \boldsymbol{F}_i^{(e)} \tag{8.25}$$

$$\dot{\boldsymbol{H}}_O = \sum(\boldsymbol{r}_i \times \boldsymbol{F}_i^{(e)}) = \sum \boldsymbol{M}_O(\boldsymbol{F}_i^{(e)}) \tag{8.26}$$

We thus conclude that the rates of change of the linear momentum and of the angular momentum about a fixed point O of the system of particles are respectively equal to the resultant force and the resultant moment about O of the external forces acting on the particles of the system.

8.6 Equations of Motion for Mass Center of Particle System

Assuming that $\boldsymbol{r}_C$ represents the position vector of the mass center of the system of particles, then we have

$$m\boldsymbol{r}_C = \sum m_i \boldsymbol{r}_i \tag{8.27}$$

where $m = \sum m_i$ is the total mass of the system of particles.

Differentiating Eq. (8.27) twice with respect to time t, we can obtain

$$m\boldsymbol{a}_C = \sum m_i \boldsymbol{a}_i \tag{8.28}$$

where $\boldsymbol{a}_C$ is the acceleration of the mass center of the system of particles.

Substituting Eq. (8.18) into Eq. (8.28), we obtain

$$m\boldsymbol{a}_C = \sum \boldsymbol{F}_i^{(e)} \tag{8.29}$$

This equation describes the motion of the mass center of a system of particles. We therefore conclude that the mass center of a system of particles moves as if the entire mass of the system and all the external forces were concentrated at that point.

8.7 Equations of Motion for Particle System about Mass Center

It is often convenient to consider the motion of the particles of the system with respect to a centroidal system of reference which translates with respect to the Newtonian system of reference. Denoting, respectively, by $\boldsymbol{r}_i'$ and $\boldsymbol{v}_i'$ the position vector of the particle P_i relative to the mass center C and the velocity of the particle P_i relative to the centroidal system of reference $Cx'y'z'$, we define the angular momentum of the system of particles about the mass center as follows:

$$\boldsymbol{H}_C' = \sum(\boldsymbol{r}_i' \times m_i \boldsymbol{v}_i') \tag{8.30}$$

In a similar manner, we can define

$$\boldsymbol{H}_C = \sum(\boldsymbol{r}_i' \times m_i \boldsymbol{v}_i) \tag{8.31}$$

where $\boldsymbol{v}_i$ is the absolute velocity observed from the Newtonian system of reference $Oxyz$.

Using $\boldsymbol{v}_i - \boldsymbol{v}_i' = \boldsymbol{v}_C$, we have

$$\boldsymbol{H}_C - \boldsymbol{H}_C' = \sum(\boldsymbol{r}_i' \times m_i \boldsymbol{v}_i) - \sum(\boldsymbol{r}_i' \times m_i \boldsymbol{v}_i') = \left(\sum m_i \boldsymbol{r}_i'\right) \times \boldsymbol{v}_C \tag{8.32}$$

Substituting $\sum m_i \boldsymbol{r}_i' = m\boldsymbol{r}_C' = 0$ into Eq. (8.32), we obtain

$$\boldsymbol{H}_C = \boldsymbol{H}_C' \tag{8.33}$$

Using $\boldsymbol{r}_i - \boldsymbol{r}_i' = \boldsymbol{r}_C$, we have

$$\boldsymbol{H}_O - \boldsymbol{H}_C = \sum(\boldsymbol{r}_i \times m_i \boldsymbol{v}_i) - \sum(\boldsymbol{r}_i' \times m_i \boldsymbol{v}_i) = \boldsymbol{r}_C \times \left(\sum m_i \boldsymbol{v}_i\right) \tag{8.34}$$

Substituting $\sum m_i \boldsymbol{v}_i = m\boldsymbol{v}_C$ into Eq. (8.34), we obtain

$$\boldsymbol{H}_O - \boldsymbol{H}_C = \boldsymbol{r}_C \times m\boldsymbol{v}_C \tag{8.35}$$

Differentiating both sides of Eq. (8.35) and using $\boldsymbol{v}_C \times \boldsymbol{v}_C = 0$, we have

$$\dot{\boldsymbol{H}}_O - \dot{\boldsymbol{H}}_C = \boldsymbol{r}_C \times m\boldsymbol{a}_C \tag{8.36}$$

Using Eqs. (8.26) and (8.29), Eq. (8.36) can be written as

$$\dot{\boldsymbol{H}}_C = \dot{\boldsymbol{H}}_O - \boldsymbol{r}_C \times m\boldsymbol{a}_C = \sum(\boldsymbol{r}_i \times \boldsymbol{F}_i^{(e)}) - \boldsymbol{r}_C \times \sum \boldsymbol{F}_i^{(e)} = \sum[(\boldsymbol{r}_i - \boldsymbol{r}_C) \times \boldsymbol{F}_i^{(e)}] \tag{8.37}$$

Using $\boldsymbol{r}_i - \boldsymbol{r}_C = \boldsymbol{r}_i'$, we have

$$\dot{\boldsymbol{H}}_C = \sum(\boldsymbol{r}_i' \times \boldsymbol{F}_i^{(e)}) = \sum \boldsymbol{M}_C(\boldsymbol{F}_i^{(e)}) \tag{8.38}$$

We therefore conclude that the rate of change of the angular momentum about the mass center of a system of particles is equal to the resultant moment about the mass center of the external forces acting on the particles of the system.

Problems

8.1　To transport a block of mass m from the ground A to the roof B, a contractor uses a motor-driven lift consisting of a horizontal platform which can slide along a guide attached to a ladder. The lift starts from rest at A and initially moves with a constant acceleration a_1. The lift then decelerates at a constant rate a_2 and comes to rest at B, near the top of the ladder. Knowing that the coefficient of static friction between the block and the horizontal platform is $\mu_s = 0.3$ and that the inclination of the ladder is $\theta = 60°$, determine the largest allowable acceleration a_1 and the largest allowable deceleration a_2 if the block is not to slide on the platform.

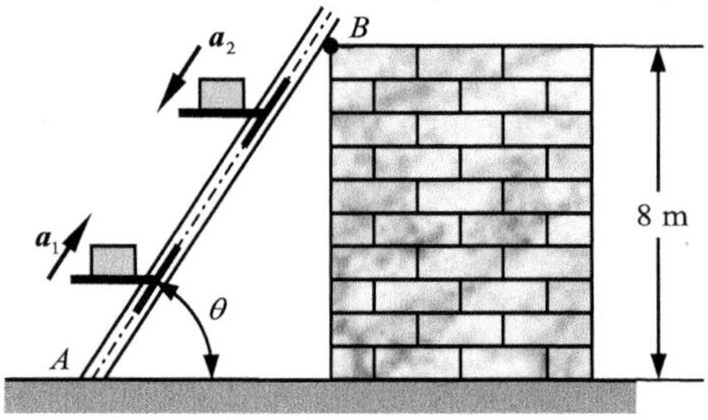

Fig. P8.1

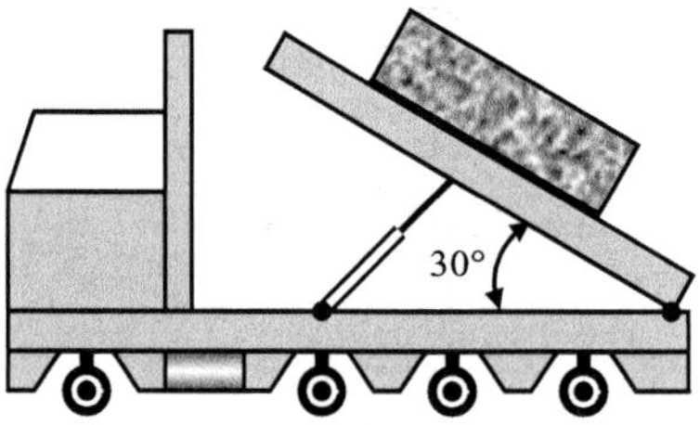

Fig. P8.2

8.2　To unload a stone block from a truck, the driver first tilts the bed of the truck and then accelerates from rest. Knowing that the coefficients of friction between the block and the bed are $\mu_s = 0.4$ and $\mu_k = 0.3$, determine the smallest acceleration of the truck which will cause the block to slide.

8.3　The block B of weight $W_B = 100$ N is supported by the block A of weight $W_A = 300$ N and is attached to a cord to which a horizontal force of magnitude $F = 200$ N is applied. Neglecting friction, determine (a) the acceleration of block A, (b) the acceleration of block B relative to A.

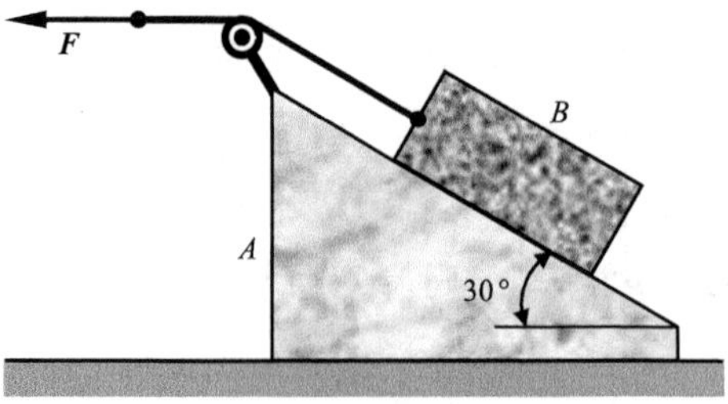

Fig. P8.3

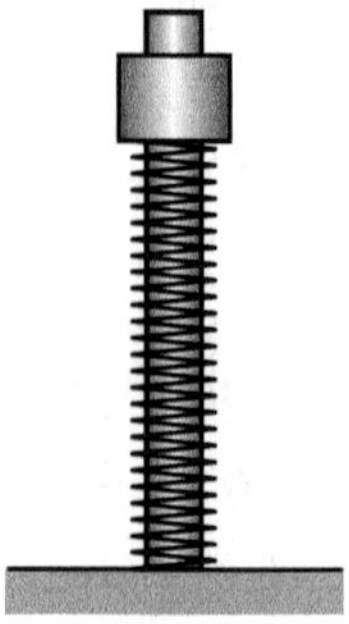

Fig. P8.4

8.4 A collar of weight 30 N can slide without friction on a vertical rod and is held so it just touches an undeformed spring. Knowing that the spring stiffness is $k = 2$ kN/m, determine the maximum deflection of the spring (a) if the collar is slowly released until it reaches an equilibrium position, (b) if the collar is suddenly released.

8.5 A collar of mass 2 kg is attached to a spring and slides without friction along a circular rod in a vertical plane. The spring has an undeformed length of 0.1 m and a stiffness k. The collar is at rest at point A and is given a slight push to get it moving to the right. Knowing that the maximum velocity of the collar is achieved as it passes through point B, determine (a) the spring stiffness k, (b) the maximum velocity of the collar.

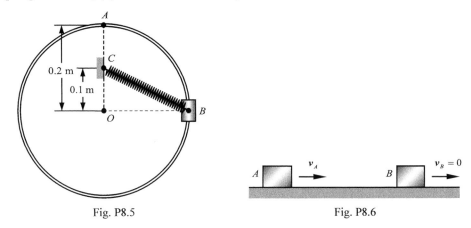

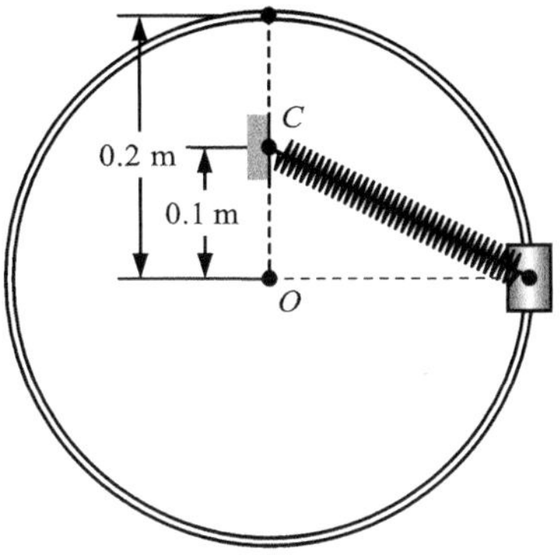

Fig. P8.5

Fig. P8.6

8.6 The initial velocity of the block in position A is $v_A = 2$ m/s. Knowing that the coefficient of kinetic friction between the block and the horizontal plane is $\mu_k = 0.2$, determine the distance it moves and the time it takes for the block to reach B with zero velocity.

8.7 A 2 kg sphere is connected to a fixed point O by an inextensible cord of length 0.5 m. The sphere is resting at A on a frictionless horizontal surface at a distance of 0.3 m from O when it is given a velocity v_0 in a direction perpendicular to line OA. It moves freely until it reaches position B, when the cord becomes taut. Determine the maximum allowable velocity if the impulse of the force exerted on the cord is not to exceed $I = 5$ N·s.

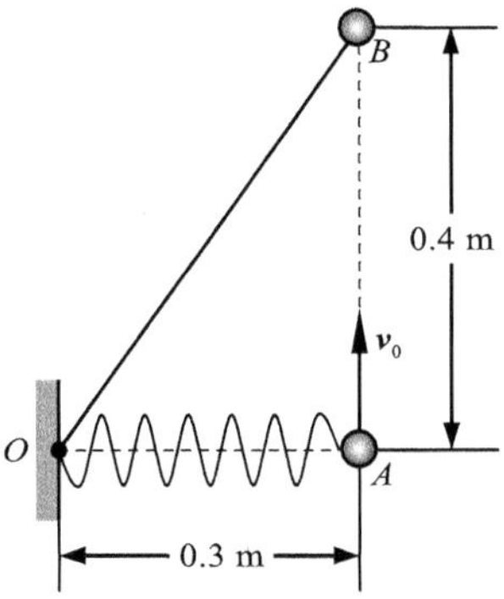

Fig. P8.7

Fig. P8.8

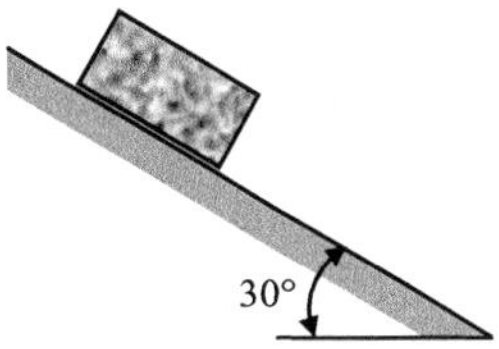

Fig. P8.9

8.8 The triple jump is a track-and-field event. Assuming that an athlete of weight 800 N approaches the takeoff line from the left with a horizontal velocity of 8 m/s, remains in contact with the ground for 0.2 s, and takes off at a 60° angle with a velocity of 10 m/s, determine the average impulsive force exerted by the ground on his foot.

8.9 A block of mass $m = 10$ kg slides from rest down a plane inclined $\theta = 30°$ with the horizontal. Knowing that the coefficient of kinetic friction between the block and the plane is $\mu_k = 0.3$, determine the velocity of the block at the end of $t = 5$ s.

Chapter 9 Kinetics of Rigid Bodies in Plane Motion

9.1 Equations of Motion

For a rigid body in plane motion, the equation of motion for the mass center of the rigid body and the equation of motion for the rigid body about the mass center can be written, respectively, as

$$m\boldsymbol{a}_C = \sum \boldsymbol{F},\ \dot{H}_C = \sum M_C(\boldsymbol{F}) \tag{9.1}$$

where H_C can be expressed as

$$H_C = H'_C = \int r'(\mathrm{d}mv') = (\int r'^2 \mathrm{d}m)\omega = I_C\omega \tag{9.2}$$

where $I_C = \int r'^2 \mathrm{d}m$ is the mass moment of inertia about the mass center of the rigid body and ω is the angular velocity of the rigid body. Differentiating both sides of Eq. (9.2) with respect to time, we obtain

$$\dot{H}_C = I_C\dot{\omega} = I_C\alpha \tag{9.3}$$

where α is the angular acceleration of the rigid body.

Substituting Eq. (9.3) into Eq. (9.1), we obtain the equations of motion for a rigid body in plane motion:

$$m\boldsymbol{a}_C = \sum \boldsymbol{F},\ I_C\alpha = \sum M_C(\boldsymbol{F}) \tag{9.4}$$

Resolving the vector equation above into rectangular coordinate components, Eq. (9.4) can be rewritten as

$$ma_{Cx} = \sum F_x,\ ma_{Cy} = \sum F_y,\ I_C\alpha = \sum M_C(\boldsymbol{F}) \tag{9.5}$$

Example 9.1 A $40\ \mathrm{kg}$ uniform thin panel is placed in a truck with end *A* supported by a smooth vertical surface and end *B* resting on a rough horizontal surface, Fig. E9.1a. Knowing that the deceleration of the truck is $2\ \mathrm{m/s^2}$ and that $l = 2\ \mathrm{m}$ and $\theta = 60°$, determine (a) the reactions at ends *A* and *B*, (b) the minimum required coefficient of static friction at end *B*.

Solution Taking *AB* as a free body and drawing its free-body diagram, Fig. E9.1b, then we have

$$ma_{Cx} = \sum F_x,\quad ma = N_A - F_B$$

$$ma_{Cy} = \sum F_y,\quad 0 = N_B - mg$$

$$I_C\alpha = \sum M_C(\boldsymbol{F}),\ 0 = -N_A(\frac{1}{2}l\sin\theta) + N_B(\frac{1}{2}l\cos\theta) - F_B(\frac{1}{2}l\sin\theta)$$

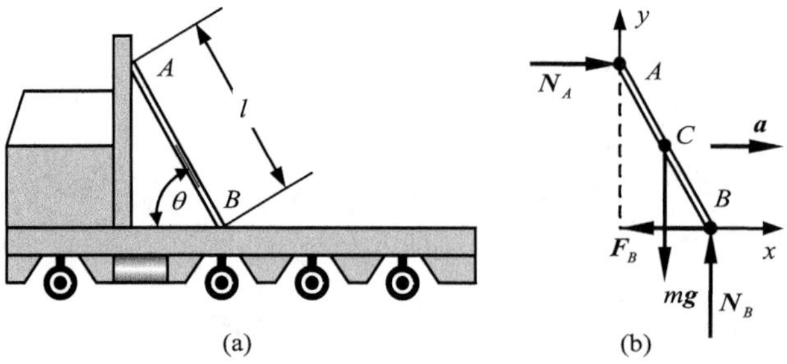

Fig. E9.1

Solving the above equations and using $g = 9.81\ \text{m/s}^2$, we can obtain

$$N_A = 153.28\ \text{N},\ N_B = 392.40\ \text{N},\ F_B = 73.28\ \text{N}$$

Using $F_B \leqslant \mu_s N_B$, we have

$$\mu_s \geqslant 0.19$$

Example 9.2 A uniform rod *AB* of length *l* and mass *m* is supported as shown in Fig. E9.2a. If the cable attached at *B* suddenly breaks, determine (a) the angular acceleration of rod *AB*, (b) the reaction at the pin support *A*.

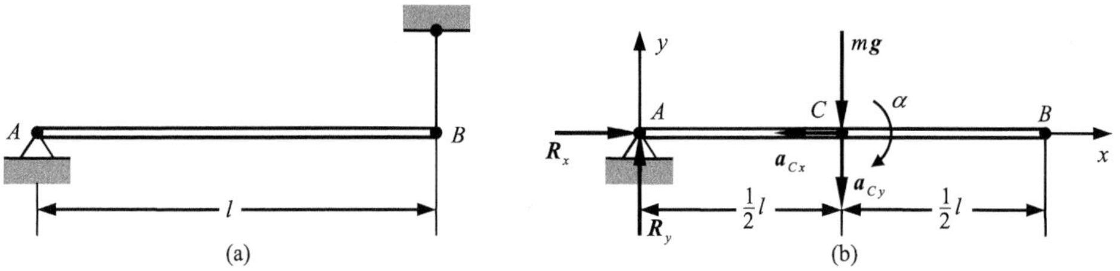

Fig. E9.2

Solution Taking *AB* as a free body and drawing its free-body diagram, Fig. E9.2b, then we have

$$ma_{Cx} = \sum F_x,\quad -ma_{Cx} = R_x$$

$$ma_{Cy} = \sum F_y,\quad -ma_{Cy} = R_y - mg$$

$$I_C\alpha = \sum M_C(\boldsymbol{F}),\quad -I_C\alpha = -R_y\left(\frac{1}{2}l\right)$$

where $a_{Cx} = AC \cdot \omega^2 = 0$, $a_{Cy} = AC \cdot \alpha = \frac{1}{2}l\alpha$, and $I_C = \frac{1}{12}ml^2$.

Solving the above equations, we can obtain

$$R_x = 0,\ R_y = \frac{1}{4}mg,\ \alpha = \frac{3g}{2l}$$

9.2 Method of Inertia Force

The method of inertia force can be used to analyze the motion of a particle. Similarly, this method can also be applied to a rigid body in plane motion. In the case of a rigid body in plane motion, not only must an inertia force equal in magnitude and opposite in direction to $m\boldsymbol{a}_C$ be applied to the rigid body at the mass center but also an inertia couple equal in magnitude and opposite in direction of rotation to $I_C\alpha$ must be applied to the rigid body. Hence, the equations of motion for a rigid body in plane motion can be expressed as

$$\sum \boldsymbol{F} + \boldsymbol{F}_\mathrm{I} = 0,\ \sum M_C(\boldsymbol{F}) + M_{\mathrm{I}C} = 0 \tag{9.6}$$

where $\boldsymbol{F}_\mathrm{I} = -m\boldsymbol{a}_C$ and $M_{\mathrm{I}C} = -I_C\alpha$ are respectively called the inertia force and the inertia couple.

Although the rigid body considered is not in equilibrium, the equations of equilibrium can be used to analyze the motion of the rigid body if an inertia force and an inertia couple are both applied to the rigid body in motion by using the method of inertia force based on d'Alembert's principle. Resolving the above equations into rectangular coordinate components, then we have

$$\sum F_x + F_{\mathrm{I}x} = 0,\ \sum F_y + F_{\mathrm{I}y} = 0,\ \sum M_C(\boldsymbol{F}) + M_{\mathrm{I}C} = 0 \tag{9.7}$$

where $F_{\mathrm{I}x} = -ma_{Cx}$, $F_{\mathrm{I}y} = -ma_{Cy}$, and $M_{\mathrm{I}C} = -I_C\alpha$.

The advantage of the method of inertia force is that it converts the kinetic problem of a rigid body into an equivalent problem in equilibrium and that it allows moments to be conveniently taken about an arbitrary point. Therefore, the above equations can also be expressed as

$$\sum F_x + F_{\mathrm{I}x} = 0,\ \sum F_y + F_{\mathrm{I}y} = 0,\ \sum M_A(\boldsymbol{F}) + M_A(\boldsymbol{F}_\mathrm{I}) + M_{\mathrm{I}C} = 0 \tag{9.8}$$

where A is an arbitrary point. Two alternative forms of the above equations can be given respectively by

$$\sum F_x + F_{\mathrm{I}x} = 0,\ \sum M_A(\boldsymbol{F}) + M_A(\boldsymbol{F}_\mathrm{I}) + M_{\mathrm{I}C} = 0,\ \sum M_B(\boldsymbol{F}) + M_B(\boldsymbol{F}_\mathrm{I}) + M_{\mathrm{I}C} = 0 \tag{9.9}$$

where the line connecting points A and B is not perpendicular to the x axis, and

$$\begin{gathered}\sum M_A(\boldsymbol{F}) + M_A(\boldsymbol{F}_\mathrm{I}) + M_{\mathrm{I}C} = 0,\ \sum M_B(\boldsymbol{F}) + M_B(\boldsymbol{F}_\mathrm{I}) + M_{\mathrm{I}C} = 0,\\ \sum M_C(\boldsymbol{F}) + M_C(\boldsymbol{F}_\mathrm{I}) + M_{\mathrm{I}C} = 0\end{gathered} \tag{9.10}$$

where the points A, B, and C do not lie in a straight line.

Example 9.3 A uniform rod *AB* of length *l* and mass *m* is supported as shown in Fig. E9.3a. If the cable attached at *B* suddenly breaks, determine (a) the angular acceleration of rod *AB*, (b) the reaction at the pin support *A*.

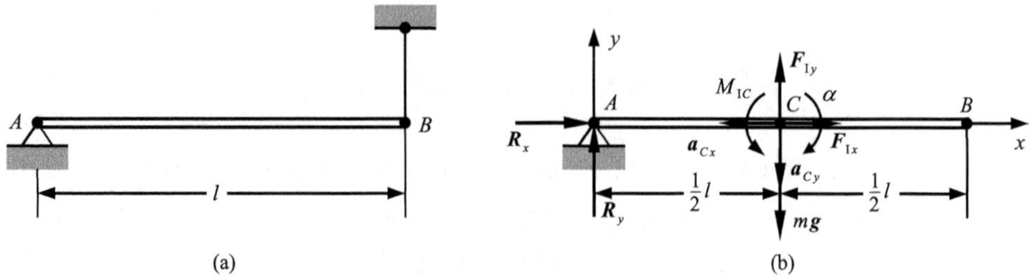

Fig. E9.3

Solution Taking *AB* as a free body and drawing its free-body diagram, Fig. E9.3b, then we have

$$\sum F_x + F_{Ix} = 0, \quad R_x + ma_{Cx} = 0$$

$$\sum F_y + F_{Iy} = 0, \quad R_y - mg + ma_{Cy} = 0$$

$$\sum M_A(\boldsymbol{F}) + M_A(\boldsymbol{F}_I) + M_{IC} = 0, \quad -mg(\frac{1}{2}l) + ma_{Cy}(\frac{1}{2}l) + I_C\alpha = 0$$

Using $a_{Cx} = AC \cdot \omega^2 = 0$, $a_{Cy} = AC \cdot \alpha = \frac{1}{2}l\alpha$, and $I_C = \frac{1}{12}ml^2$, and solving the above equations, we can obtain

$$R_x = 0, \ R_y = \frac{1}{4}mg, \ \alpha = \frac{3g}{2l}$$

9.3 Method of Work and Energy

1. Work of Couple Moment

The elementary work done by a couple of moment M can be expressed as

$$\mathrm{d}W = M\mathrm{d}\theta \tag{9.11}$$

where $\mathrm{d}\theta$ is the angular displacement.

2. Kinetic Energy of Rigid Body

The kinetic energy of a rigid body in plane motion can be defined as

$$T = \int \frac{1}{2}\mathrm{d}mv^2 \tag{9.12}$$

Using $\boldsymbol{v} = \boldsymbol{v}_C + \boldsymbol{v}'$, Eq. (9.12) can be rewritten as

$$T = \int \frac{1}{2}\mathrm{d}m(\boldsymbol{v} \cdot \boldsymbol{v}) = \frac{1}{2}v_C{}^2 \int \mathrm{d}m + \boldsymbol{v}_C \cdot \int \mathrm{d}m\boldsymbol{v}' + \frac{1}{2}\int \mathrm{d}mv'^2 \tag{9.13}$$

Using $\int \mathrm{d}m\boldsymbol{v}' = m\boldsymbol{v}'_C = 0$ and $v' = r'\omega$, we have

$$T = \frac{1}{2}mv_C{}^2 + \frac{1}{2}I_C\omega^2 \tag{9.14}$$

where $I_C = \int r'^2 \mathrm{d}m$ is the mass moment of inertia of the rigid body about the mass center. This equation shows that the kinetic energy of a rigid body in plane motion is equal to the sum of the translational kinetic energy of the rigid body with the mass center and the rotational kinetic energy of the rigid body relative to the translating system attached to the mass center.

3. Principle of Work and Energy

The principle of work and energy for a particle can be applied to each of the particles forming the rigid body. Adding the kinetic energy of the various particles of the rigid body and considering the work done by all the forces acting on the rigid body, then we can obtain the principle of work and energy for a rigid body in plane motion:

$$\mathrm{d}T = \mathrm{d}W \tag{9.15}$$

where $\mathrm{d}T$ and $\mathrm{d}W$ are, respectively, the increment in kinetic energy of the rigid body and the elementary work done by all the forces acting on the rigid body. Integrating the equation above, we have

$$T_2 - T_1 = W_{12} \tag{9.16}$$

where T_1 and T_2 are the initial and final kinetic energies of the rigid body, and W_{12} is the work of all the forces acting on the rigid body.

4. Conservation of Energy

If all the forces that do work are conservative, the principle of work and energy can be replaced by

$$T_1 + V_1 = T_2 + V_2 \tag{9.17}$$

This equation expresses the conservation of mechanical energy.

Example 9.4 A uniform rod of length l and mass m is pivoted about a point O located at a distance d from its center C, Fig. E9.4a. It is released from rest in a horizontal position and swings freely. Determine (a) the distance d for which the angular velocity of the rod as it passes through a vertical position is maximum, (b) the corresponding values of its angular velocity and of the reaction at O.

Solution At position 1, Fig. E9.4a, we have

$$T_1 = V_1 = 0$$

and at position 2, Fig. E9.4b, we have

$$T_2 = \frac{1}{2}I_O\omega^2 = \frac{1}{2}(I_C + md^2)\omega^2 = \frac{1}{2}(\frac{1}{12}ml^2 + md^2)\omega^2,\ V_2 = -mgd$$

Using $T_1 + V_1 = T_2 + V_2$, we can obtain

$$\omega = \sqrt{\frac{2g}{d + \frac{1}{12}l^2 / d}}$$

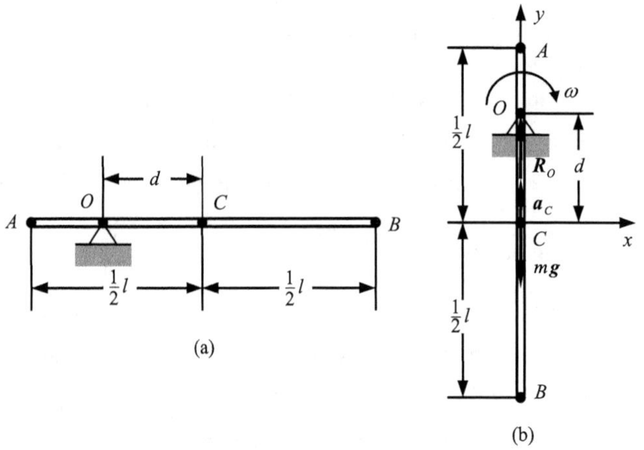

Fig. E9.4

It is obvious that when $d = \frac{1}{12}l^2/d$, i.e., $d = \frac{\sqrt{3}}{6}l$, the angular velocity ω will have a maximum value, i.e.,

$$\omega_{\max} = \sqrt{\frac{2g}{2\sqrt{d(\frac{1}{12}l^2/d)}}} = \sqrt[4]{12}\sqrt{\frac{g}{l}} = 1.86\sqrt{\frac{g}{l}}$$

Using $ma_C = \sum F_y = R_O - mg$ and $a_C = d\omega_{\max}^2 = g$ at position 2, Fig. E9.4b, we have

$$R_O = 2mg$$

9.4 Method of Impulse and Momentum

1. Principle of Impulse and Momentum

For a rigid body in plane motion, we have

$$\mathrm{d}\boldsymbol{L} = \mathrm{d}\boldsymbol{I},\ \mathrm{d}H_C = \mathrm{d}G_C \tag{9.18}$$

where $\mathrm{d}\boldsymbol{L} = \mathrm{d}(m\boldsymbol{v}_C)$ is the increment in linear momentum of the rigid body, $\mathrm{d}\boldsymbol{I} = \sum \boldsymbol{F}\mathrm{d}t$ is the linear impulse of the forces acting on the rigid body during the interval of time $\mathrm{d}t$, $\mathrm{d}H_C = \mathrm{d}(I_C\omega)$ is the increment in angular momentum about the mass center C of the body, and $\mathrm{d}G_C = \sum M_C(\boldsymbol{F})\mathrm{d}t$ is the angular impulse about the mass center C of the forces acting on the body during the interval of time $\mathrm{d}t$. Integrating Eq. (9.18), we write

$$\boldsymbol{L}_2 - \boldsymbol{L}_1 = \boldsymbol{I}_{12},\ H_{C2} - H_{C1} = G_{C12} \tag{9.19}$$

where $\boldsymbol{L}_1 = m\boldsymbol{v}_{C1}$ and $\boldsymbol{L}_2 = m\boldsymbol{v}_{C2}$ are respectively the initial and final linear momenta of the rigid body, $\boldsymbol{I}_{12} = \sum \int_{t_1}^{t_2} \boldsymbol{F}\mathrm{d}t$ is the linear impulse of the forces acting on the rigid body during the interval of time from t_1 to t_2, $H_{C1} = I_C\omega_1$ and $H_{C2} = I_C\omega_2$ are respectively the initial and final angular momenta about the mass center C of the rigid body, and $G_{C12} = \sum\int_{t_1}^{t_2} M_C(\boldsymbol{F})\mathrm{d}t$ is the angular impulse about the mass center C of the forces acting on the body during the interval of time from t_1 to t_2.

Example 9.5 A uniform cylinder of radius r and weight W with an initial anticlockwise angular velocity ω_0 is placed in the corner formed by the floor and a vertical wall, Fig. E9.5a. Denoting by μ_k the coefficient of kinetic friction at A and B, derive an expression for the time required for the cylinder to come to rest.

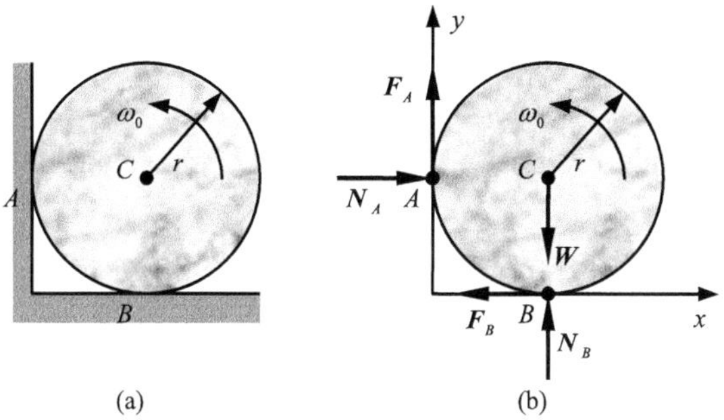

Fig. E9.5

Solution Taking the cylinder as a free body and drawing its free-body diagram, Fig. E9.5b, then we have

$$L_{2x} - L_{1x} = I_{12x}, \quad 0 - 0 = (N_A - F_B)t$$
$$L_{2y} - L_{1y} = I_{12y}, \quad 0 - 0 = (N_B + F_A - W)t$$
$$H_{C2} - H_{C1} = G_{C12}, \quad 0 - I_C\omega_0 = -(F_A + F_B)rt$$

Using $F_A = \mu_k N_A$, $F_B = \mu_k N_B$, and $I_C = \dfrac{1}{2}mr^2$, and solving the equations above, we can obtain

$$t = \frac{(1+\mu_k^{\ 2})I_C\omega_0}{\mu_k(1+\mu_k)Wr} = \frac{(1+\mu_k^{\ 2})r\omega_0}{2\mu_k(1+\mu_k)g}$$

where g is the acceleration of gravity.

2. Conservation of Momentum

If no any external force acts on a rigid body in plane motion, Eq. (9.19) reduces to

$$\boldsymbol{L}_1 = \boldsymbol{L}_2, \ H_{C1} = H_{C2} \tag{9.20}$$

which expresses that the linear momentum and the angular momentum about the mass center C of the rigid body are conserved.

9.5 Impact of Rigid Body in Plane Motion

An instantaneous collision which causes an abrupt change in velocity is called impact. During a very short period of time, a large force is exerted on the body. Since this force leads to time-dependent deformation in the vicinity of the point of contact, a comprehensive treatment of a problem involving impact is rather difficult. However, the following assumptions will allow us to determine the change of the velocity in a relatively simple way. These assumptions mainly include: (1) The impact duration is so small that the change in the position of the body during impact can be neglected; (2) The impulsive force at the point of contact is so large that all the other nonimpulsive forces (e.g., the gravitational and elastic forces) can be neglected during impact.

For a rigid body in plane motion, the principle of impulse and momentum during impact can be written as

$$m\boldsymbol{v}_C' - m\boldsymbol{v}_C = \boldsymbol{I},\ I_C\omega' - I_C\omega = M_C(\boldsymbol{I}) \tag{9.21}$$

where m is the mass of the rigid body, $\boldsymbol{v}_C$ and $\boldsymbol{v}_C'$ are the velocities of the mass center of the rigid body before and after impact, $\boldsymbol{I}$ is the impulse of the impulsive force acting on the rigid body during the impact, I_C is the mass moment of inertia about the mass center of the rigid body, ω and ω' are the angular velocities of the rigid body before and after impact, and $M_C(\boldsymbol{I})$ is the angular impulse about the mass center C of the impulsive force acting on the rigid body during impact.

Example 9.6 A uniform slender rod AB of mass m and length l forms an angle θ with the horizontal direction as it strikes the frictionless surface shown with a vertical velocity $\boldsymbol{v}$ and no angular velocity. Assuming that the impact is perfectly elastic, derive expressions for both the angular velocity of the rod immediately after the impact and the impulse of the impulsive force exerted by the surface on the rod during the impact.

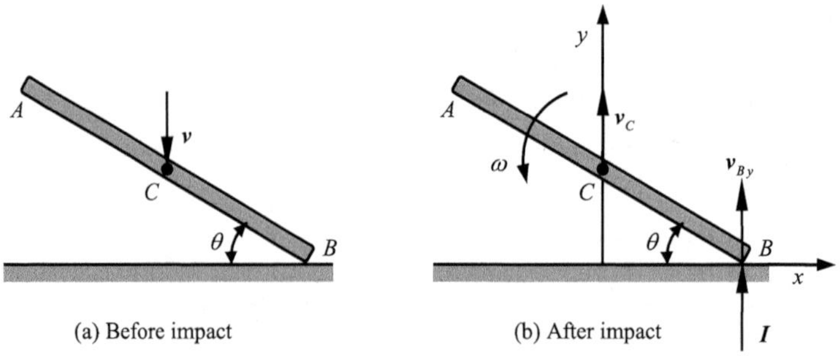

Fig. E9.6

Solution Establishing the coordinate system shown, applying the principle of impulse and momentum to the rod, and noting that the velocity of the mass center of the rod after impact is vertical, and that the impulse of the impulsive force acting on the rod during the impact is also vertical, thus we have

$$mv_C - m(-v) = I,\ I_C\omega - 0 = I(\frac{1}{2}l\cos\theta)$$

where I_C is the mass moment of inertia about the mass center of the rod, and I is the value of the impulse of the impulsive force exerted by the surface on the rod. Using $I_C = \frac{1}{12}ml^2$, $v_{By} = v$, and $v_C = v_{By} - (\frac{1}{2}l\omega)\cos\theta = v - \frac{1}{2}l\omega\cos\theta$, we solve the above equations for ω and I, and obtain

$$\omega = \frac{12v\cos\theta}{(1+3\cos^2\theta)l},\ I = \frac{2mv}{1+3\cos^2\theta}$$

Example 9.7 A bullet of mass m_1 is fired with a horizontal velocity of magnitude v into a uniform slender bar of mass m_2 and length l. Knowing that the bar is initially at rest, determine (a) the angular velocity of the bar immediately after the bullet becomes embedded, (b) the impulse of the impulsive reaction at *A*, and (c) the distance a if the impulsive reaction at *A* is equal to zero.

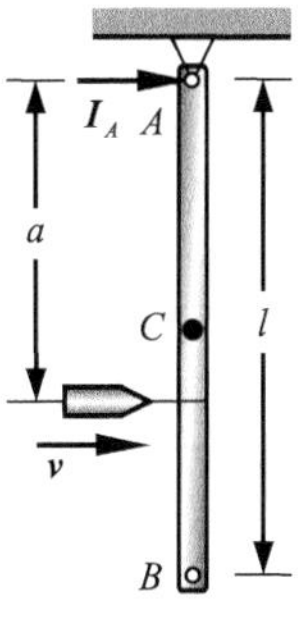

Fig. E9.7

Solution Applying the principle of impulse and momentum to the system consisting of the bullet and the bar, and noting that the impulse of the impulsive reaction acting on the bar during the impact is horizontal, we have

$$m_1(a\omega) + m_2(\frac{1}{2}l\omega) - m_1 v = I_A,\ (m_1a^2)\omega + (\frac{1}{3}m_2l^2)\omega - a(m_1v) = 0$$

where ω is the angular velocity of the bar immediately after the bullet becomes embedded, and I_A is the impulse of the impulsive reaction exerted by the support on the bar at *A*. Solving the above equations for ω and I_A, we obtain

$$\omega = \frac{3m_1av}{3m_1a^2 + m_2l^2},\ I_A = \frac{(3a-2l)lm_1m_2v}{6m_1a^2 + 2m_2l^2}$$

It can be seen from the second expression above that if the impulsive reaction at *A* is equal to zero, then the distance a is equal to

$$a = \frac{2}{3}l$$

The point located on the bar and having a distance $a = 2l/3$ from the pivot point is called the center of percussion. The center of percussion is defined as the point located on an object where a perpendicular impact will not cause impulsive forces at a given pivot point.

Problems

9.1 A uniform cabinet of mass $m = 20$ kg is mounted on casters that allow it to move freely on the floor. Knowing $F = 100$ N, $b = 500$ mm, and $h_C = 800$ mm, determine (a) the acceleration of the cabinet, (b) the range of values of h for which the cabinet will not tip.

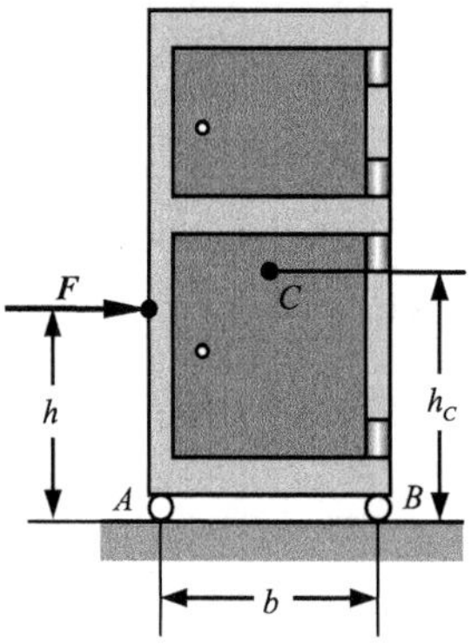

Fig. P9.1

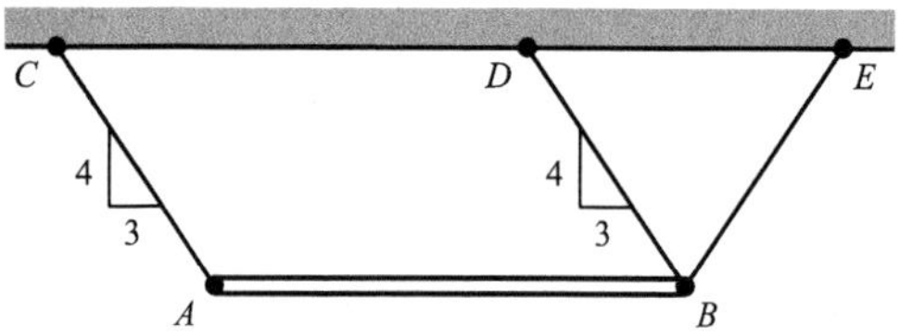

Fig. P9.2

9.2 A uniform slender rod *AB* of mass 2 kg is held in position by three ropes of the same length. Determine, immediately after rope *BE* has been cut, (a) the acceleration of rod *AB*, (b) the tension in each rope.

9.3 A uniform circular plate of mass 5 kg is attached to three cables of the same length. Knowing that the lines joining *C* to *A* and *B* are, respectively, horizontal and vertical, determine, immediately after cable *BF* has been cut, (a) the acceleration of the plate, (b) the tension in each cable.

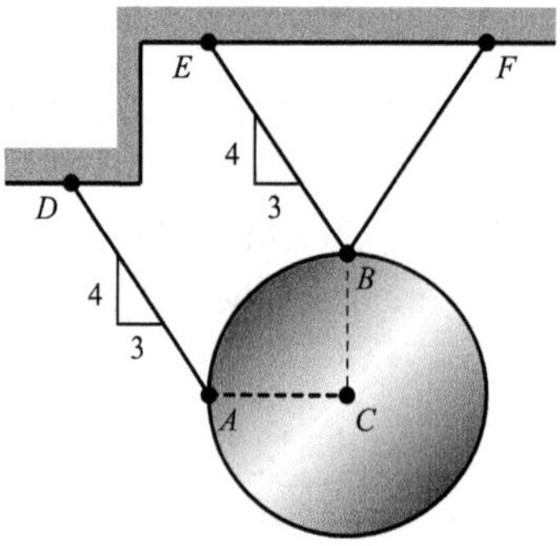

Fig. P9.3

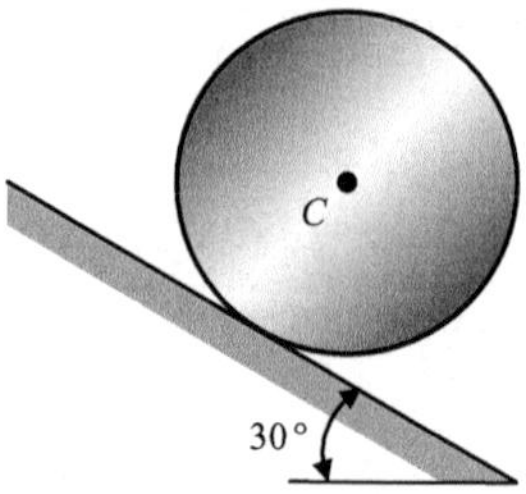

Fig. P9.4

9.4 A uniform cylinder of mass $m = 8$ kg and radius $r = 0.15$ m rolls without sliding down a plane inclined at an angle of $\theta = 30°$ with the horizontal. Determine the friction force and the acceleration of the mass center.

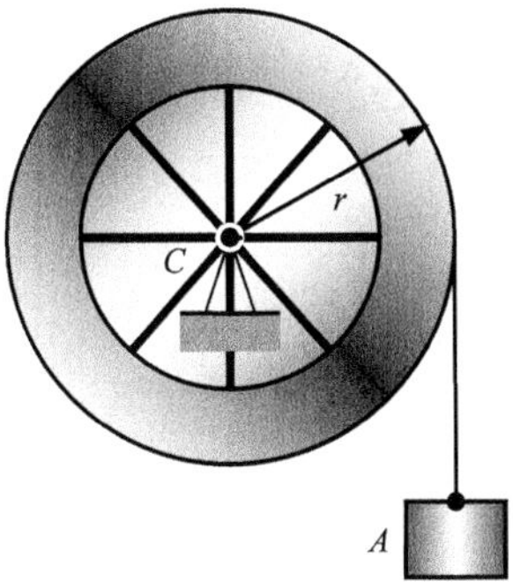

Fig. P9.5

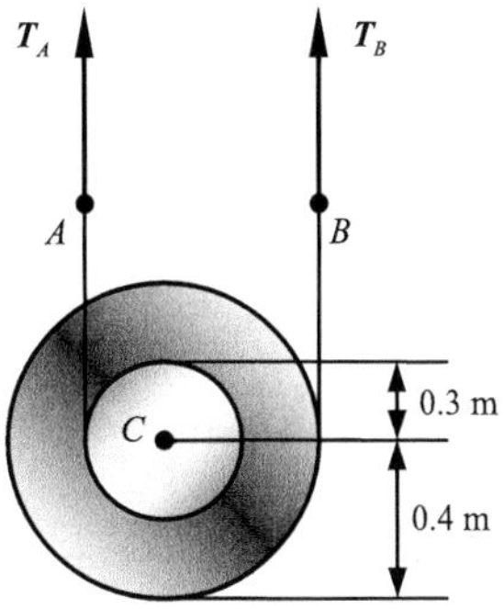

Fig. P9.6

9.5 The flywheel shown has a radius of 500 mm, a mass of 150 kg, and a radius of gyration of 400 mm. A block A of mass 20 kg is attached to a wire that is wrapped around the flywheel, and the system is released from rest. Neglecting the effect of friction, determine (a) the acceleration of block A, (b) the speed of block A after it has moved 2 m.

9.6 A drum of 0.4 m radius is attached to a disk of radius 0.3 m. The disk and drum have a combined mass of 5 kg and a combined radius of gyration of 0.25 m and are suspended by two cords. Knowing that $T_A = 50$ N and $T_B = 30$ N, determine the accelerations of points A and B on the cords.

9.7 A uniform rod AB of length l and mass m is supported from two springs as shown. If the spring attached at B suddenly breaks, determine, at this instant, (a) the angular acceleration of rod AB, (b) the acceleration of point A.

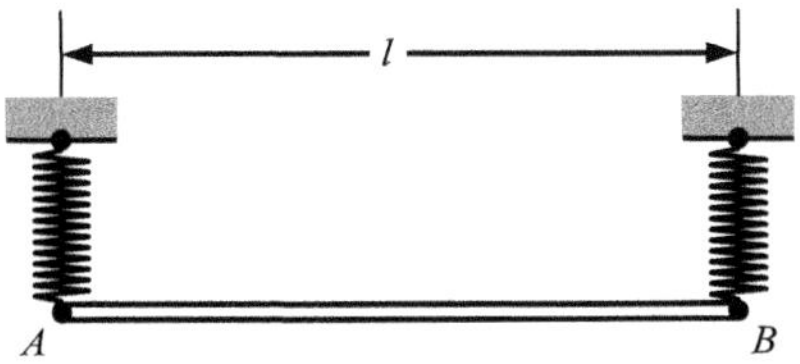

Fig. P9.7

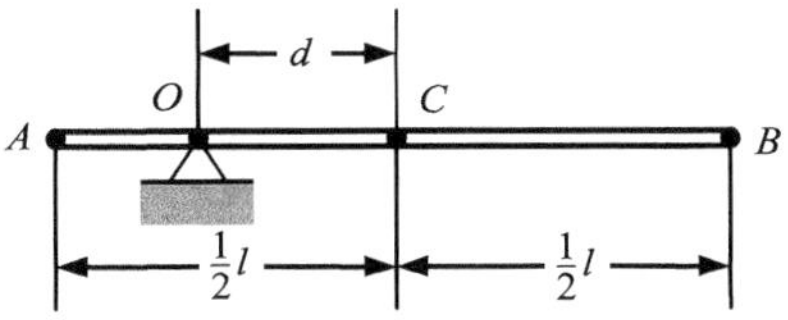

Fig. P9.8

9.8 A uniform rod of length l and mass m is pivoted about a point O located at a distance d from its center C. It is released from rest in a horizontal position and swings freely. Knowing that $d = \frac{3}{8}l$, determine the angular velocity of rod AB and the reaction at point O after the rod has rotated through $90°$.

9.9 A uniform cylinder of radius r and weight W with an initial anticlockwise angular velocity ω_0 is placed in the corner formed by the rough floor and a smooth wall. Denoting by μ_k the coefficient of kinetic friction between the cylinder and the floor, derive an expression for the time required for the cylinder to come to rest.

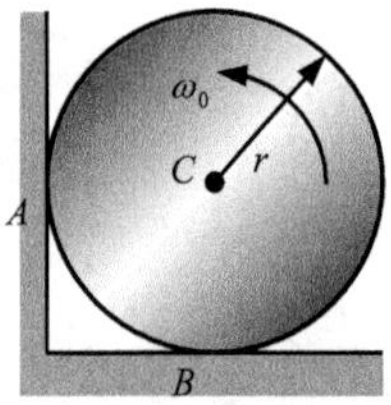

Fig. P9.9

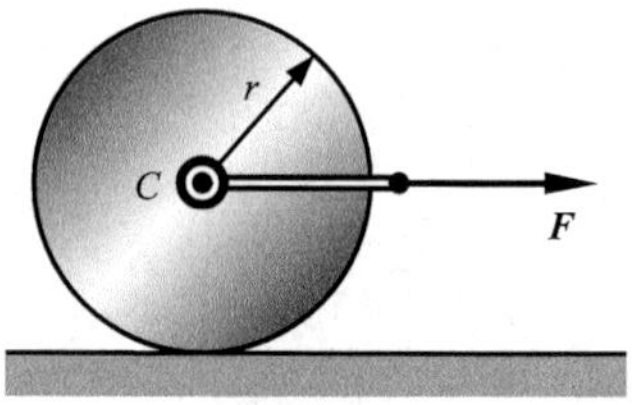

Fig. P9.10

9.10 A uniform cylindrical roller of mass $m = 25$ kg and radius $r = 0.2$ m, initially at rest, is acted upon by a force of magnitude $F = 120$ N. Knowing that the body rolls without slipping, determine (a) the velocity of its center *C* after it has moved 1.5 m, (b) the friction force required to prevent slipping.

9.11 A rope is wrapped around a uniform cylinder of radius *r* and mass *m*. Knowing that the cylinder is released from rest, determine the velocity of the center *C* of the cylinder after it has moved downward a distance *s*.

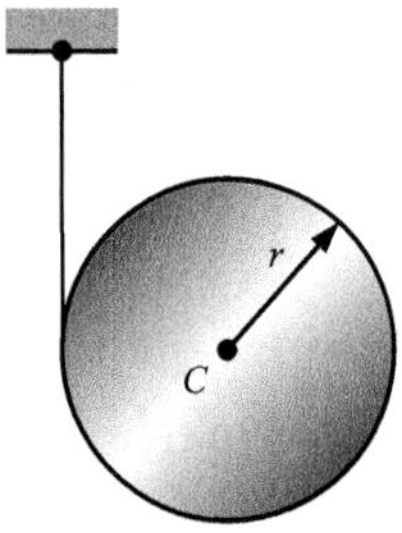

Fig. P9.11

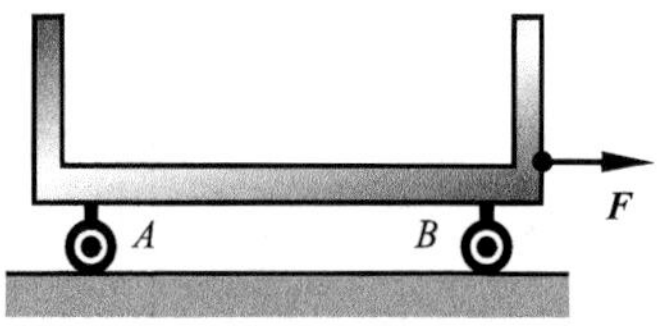

Fig. P9.12

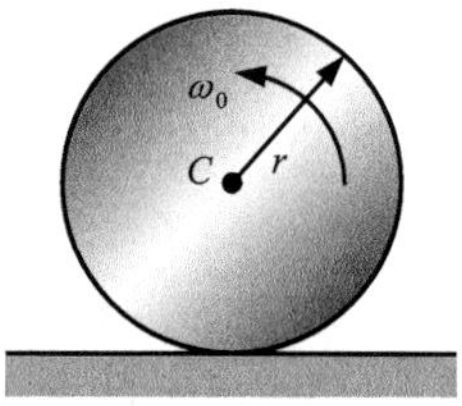

Fig. P9.13

9.12 A 10 kg cradle, subjected to a force of magnitude $F = 30$ N, is supported by two uniform disks that roll without sliding at all surfaces of contact. The mass of each disk is $m = 5$ kg and the radius of each disk is $r = 0.1$ m. Knowing that the system is initially at rest, determine the velocity of the cradle after it has moved 0.5 m.

9.13 A uniform cylinder of radius *r* and mass *m* is placed on a horizontal floor with no linear velocity but with an anticlockwise angular velocity ω_0. Denoting by μ_k the coefficient of kinetic friction between the cylinder and the floor, determine (a) the time at which the cylinder will start rolling without sliding, (b) the linear and angular velocities of the cylinder as the cylinder starts rolling without sliding.

9.14 Two blocks, *A* and *B*, are connected by a cable *AB* which is wrapped over the surface of a disk *O*. Assume that there is no relative sliding between the cable and the disk and that the disk rotates in frictionless bearings. Knowing that at the instant shown block B is moving down $0.4\ \text{m/s}$ and that the spring is compressed $0.2\ \text{m}$, determine the velocity of block B after it has dropped $0.5\ \text{m}$.

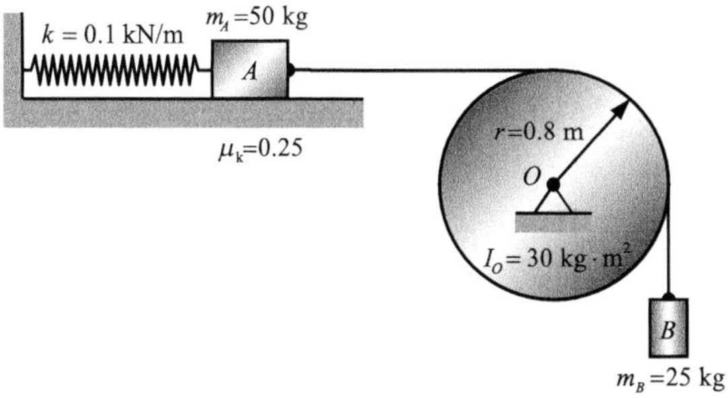

Fig. P9.14

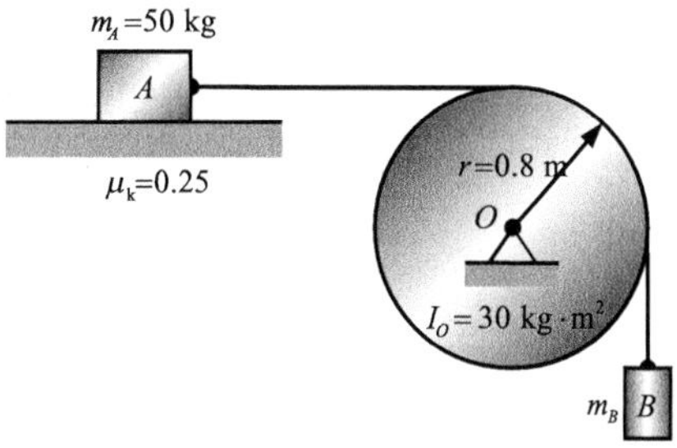

Fig. P9.15

9.15 Two blocks, A and B, are connected by a cable AB which is wrapped over the surface of a disk O. Assuming that there is no relative sliding between the cable and the disk and that the disk rotates in frictionless bearings, determine the mass m_B of block B necessary to cause block A to change its velocity from 4 to $8\ \text{m/s}$ in $6\ \text{s}$.

9.16 A bullet of mass $m_1 = 10\ \text{g}$ is fired with a horizontal velocity of magnitude $v = 500\ \text{m/s}$ into the lower end of a uniform slender bar of mass $m_2 = 5\ \text{kg}$ and length $l = 1\ \text{m}$. Knowing that the bar is initially at rest, determine (a) the angular velocity of the bar immediately after the bullet becomes embedded, (b) the impulse exerted on the bar at point A.

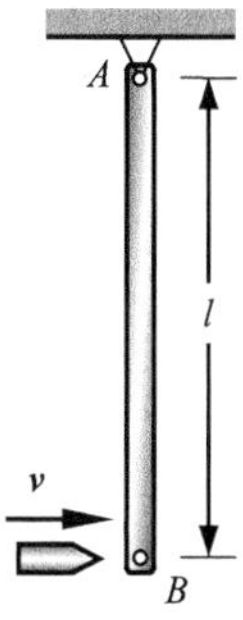

Fig. P9.16

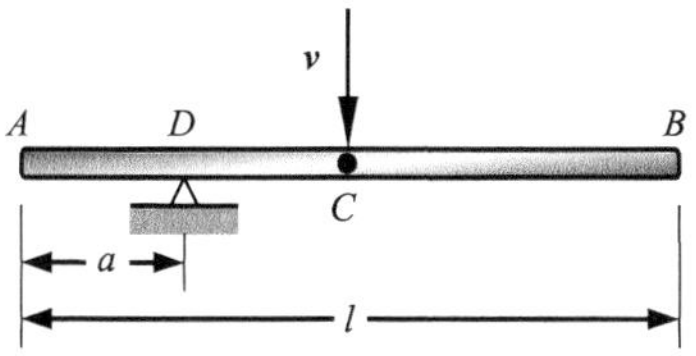

Fig. P9.17

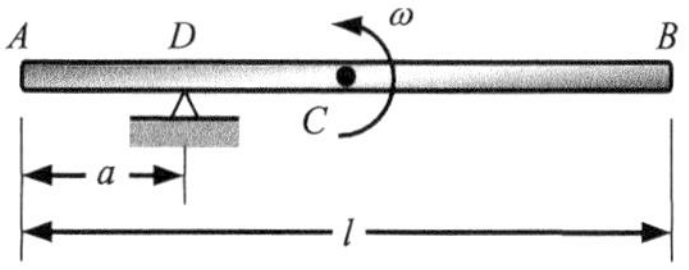

Fig. P9.18

9.17 A uniform slender rod AB of mass m and length l has a vertical velocity of magnitude v and no angular velocity when it strikes a rigid frictionless support at point D. Knowing that $a = l/5$ and assuming that $e = 0$, determine (a) the angular velocity of the rod and the velocity of its mass center immediately after the impact, (b) the impulse exerted on the rod at point D.

9.18 A uniform slender rod AB of mass m and length l has an angular velocity ω counterclockwise and a zero velocity of the mass center when it strikes a rigid frictionless support at point D. Knowing that $a = l/5$ and assuming that $e = 1$, determine (a) the angular velocity of the rod and the velocity of its mass center immediately after the impact, (b) the impulse exerted on the rod at point D.

9.19 A uniform slender rod AB of mass m and length l forms an angle θ with the horizontal direction as it strikes the frictionless surface shown with a vertical velocity $\boldsymbol{v}$ and no angular velocity. Assuming that the coefficient of restitution between the rod and the surface is e, derive an expression for the angular velocity of the rod immediately after the impact.

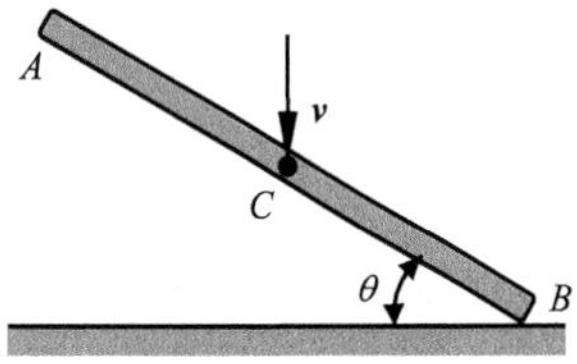

Fig. P9.19

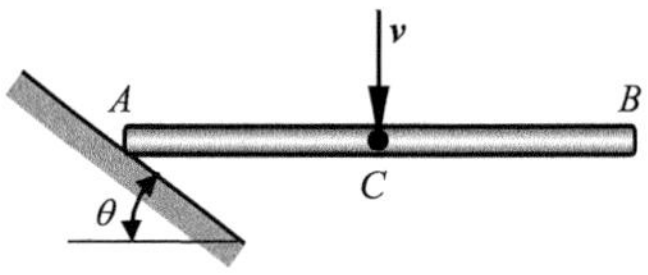

Fig. P9.20

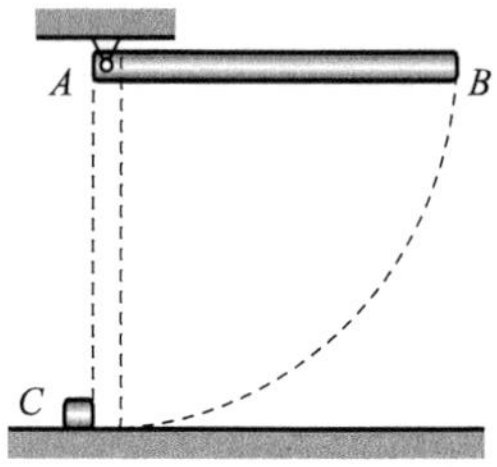

Fig. P9.21

9.20 A uniform slender rod AB of mass m and length l is falling freely with a velocity $\boldsymbol{v}$ when end B strikes a smooth inclined surface with an angle of inclination θ as shown. Assuming that the impact is perfectly elastic, determine the angular velocity of the rod and the velocity of its mass center immediately after the impact.

9.21 A uniform slender rod AB of mass m_1 and length l is released from rest in the horizontal position shown. It swings down to a vertical position and strikes a block C of mass m_2 which is resting on a frictionless surface. Assuming that the coefficient of restitution between the rod and the block is e, determine the velocity of the block immediately after the impact.

Chapter 10 Analytical Mechanics

10.1 Constraints and Virtual Work

A constraint is defined as a geometric or kinematic condition that limits the motion of a body. The number of degrees of freedom is defined as the minimum number of variables required to specify completely the configuration of a given system.

The virtual displacement of a body is defined as a fictitious infinitesimal displacement consistent with the constraints imposed on the body. This displacement can be imagined; i.e., it does not actually take place.

Virtual work is defined to be the work done by a force or a couple undergoing a virtual displacement. The virtual work done by a force $\boldsymbol{F}$ undergoing a virtual displacement $\delta \boldsymbol{r}$ can be expressed as

$$\delta W = \boldsymbol{F} \cdot \delta \boldsymbol{r} \tag{10.1}$$

The virtual work done by a couple of magnitude M during a virtual displacement $\delta\theta$ can be expressed as

$$\delta W = M\delta\theta \tag{10.2}$$

If the virtual work done by all the constraint forces undergoing an arbitrary virtual displacement is equal to zero, this type of constraints is called the ideal constraints.

10.2 Principle of Virtual Work

Consider a system of particles. If $\delta \boldsymbol{r}_i$ is the virtual displacement of particle i, consistent with the constraints, then the virtual work of the forces acting on particle i can be expressed as

$$\delta W_i = (\boldsymbol{F}_i + \boldsymbol{R}_i) \cdot \delta \boldsymbol{r}_i \tag{10.3}$$

where $\boldsymbol{F}_i$ is the resultant of the applied or active forces acting on particle i and $\boldsymbol{R}_i$ is the resultant of the constraint forces acting on the same particle. If the system of particles is in static equilibrium, i.e., $\boldsymbol{F}_i + \boldsymbol{R}_i = 0$, then Eq. (10.3) can be rewritten as

$$\delta W_i = (\boldsymbol{F}_i + \boldsymbol{R}_i) \cdot \delta \boldsymbol{r}_i = 0 \tag{10.4}$$

Summing over i, the total virtual work is equal to

$$\delta W = \sum \delta W_i = \sum (\boldsymbol{F}_i + \boldsymbol{R}_i) \cdot \delta \boldsymbol{r}_i = 0 \tag{10.5}$$

If the system is subject to ideal constraints, i.e., $\sum \boldsymbol{R}_i \cdot \delta \boldsymbol{r}_i = 0$, we then have

$$\delta W = \sum \boldsymbol{F}_i \cdot \delta \boldsymbol{r}_i = 0 \tag{10.6}$$

This relation expresses the principle of virtual work: The necessary and sufficient condition for equilibrium of a system subject to ideal constraints is that the virtual work of the applied forces, for any virtual displacement consistent with the constraints, is equal to zero.

Example 10.1 An unstretched spring of stiffness $k = 800$ N/m is attached to pins at points I and J as shown. The pin at B is attached to member BCD and can slide freely along the vertical slot in the fixed plate. Determine the force in the spring and the horizontal displacement of point H when a horizontal force of magnitude $F = 135$ N, directed to the right, is applied at point G.

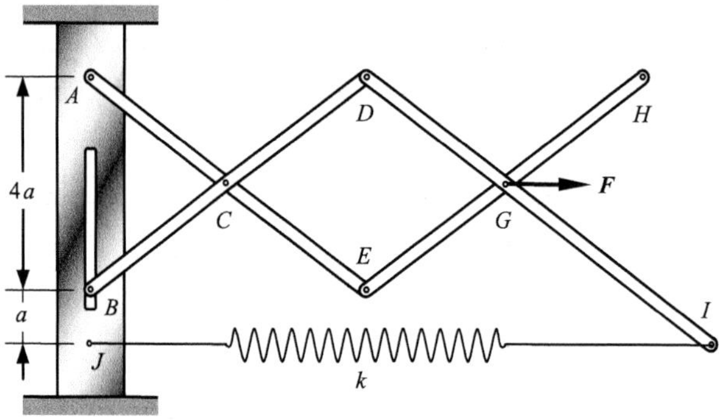

Fig. E10.1

Solution Using $x_G = 3x_C$, $x_H = 4x_C$, $x_I = 4.5x_C$, we have

$$\delta x_G = 3\delta x_C,\ \delta x_H = 4\delta x_C,\ \delta x_I = 4.5\delta x_C$$

From the principle of virtual work, $\delta W = 0$, we have

$$F\delta x_G - F_{\text{spr}}\delta x_I = 0$$

i.e.,

$$F_{\text{spr}} = \frac{\delta x_G}{\delta x_I}F = \frac{3\delta x_C}{4.5\delta x_C}F = 90\ \text{N}$$

Using $F_{\text{spr}} = k\delta x_I$, we obtain

$$x_I = \frac{F_{\text{spr}}}{k} = 112.5\ \text{mm}$$

Using $\delta x_H = 4\delta x_C$ and $\delta x_I = 4.5\delta x_C$

$$\delta x_H = \frac{4}{4.5}\delta x_I = 100\ \text{mm}$$

Example 10.2 The structure is acted upon by the force $\boldsymbol{F}$ at point G, Fig. E10.2a. Neglecting the weight of the structure and knowing that $AB = AC = BC = CD = CE = DG = EG = a$, determine the magnitude of horizontal constraint force at point B.

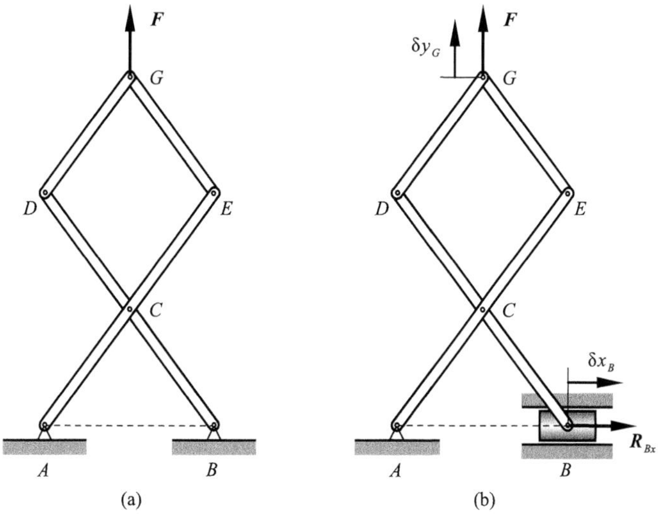

Fig. E10.2

Solution Replace the horizontal constraint at point B with its horizontal constraint force $\boldsymbol{R}_{Bx}$ and assume that the virtual displacement δx_B at point B is positive to the right and the virtual displacement δy_G at point G is positive upward, Fig. E10.2b. Using $x_C{}^2 + y_C{}^2 = a^2$, i.e., $(\frac{x_B}{2})^2 + (\frac{y_G}{3})^2 = a^2$, we have

$$x_B \delta x_B + \frac{4}{9} y_G \delta y_G = 0$$

Using $AB = AC = BC$, i.e. $y_C = \sqrt{3} x_C$, or $y_G = \frac{3\sqrt{3}}{2} x_B$, we have

$$\delta x_B = -\frac{2\sqrt{3}}{3} \delta y_G$$

From the principle of virtual work, $\delta W = 0$, we have

$$R_{Bx} \delta x_B + F \delta y_G = 0$$

i.e.,

$$R_{Bx} = -\frac{\delta y_G}{\delta x_B} F = \frac{\sqrt{3}}{2} F$$

10.3 Generalized Coordinates and Generalized Forces

1. Generalized Coordinates

The independent variables specifying the configuration of a given system are called generalized coordinates. In general, the generalized coordinates can be chosen to be linear and angular coordinates. Sometimes it is suitable to use a larger number of coordinates than the number of degrees of freedom. Then these coordinates must be related via some constraints.

For example, consider a system consisting of n particles. Assuming that the configuration of the system can be described by generalized coordinates $q_1, q_2, \cdots, q_s$, where s is the number of degrees of freedom of the system, then position vector of particle i can be expressed as

$$\boldsymbol{r}_i = \boldsymbol{r}_i(q_1, q_2, \cdots, q_s, t) \quad (i = 1, 2, \cdots, n) \tag{10.7}$$

Assuming that $\delta \boldsymbol{r}_i$ is the virtual displacement of particle i, consistent with the constraints, then $\delta \boldsymbol{r}_i$ can be expressed as

$$\delta \boldsymbol{r}_i = \sum_{k=1}^{s} \frac{\partial \boldsymbol{r}_i}{\partial q_k} \delta q_k \quad (i = 1, 2, \cdots, n) \tag{10.8}$$

2. Generalized Forces

Assuming that particle i undergoes a virtual displacement $\delta \boldsymbol{r}_i$, consistent with the constraints, under the action of an applied force $\boldsymbol{F}_i$, then the virtual work δW done by the applied forces acting on the system can be given by

$$\delta W = \sum_{i=1}^{n} \boldsymbol{F}_i \cdot \delta \boldsymbol{r}_i \tag{10.9}$$

Substituting Eq. (10.8) into Eq. (10.9), we have

$$\delta W = \sum_{i=1}^{n} \boldsymbol{F}_i \cdot \left(\sum_{k=1}^{s} \frac{\partial \boldsymbol{r}_i}{\partial q_k} \delta q_k\right) = \sum_{k=1}^{s} \left[\sum_{i=1}^{n} \left(\boldsymbol{F}_i \cdot \frac{\partial \boldsymbol{r}_i}{\partial q_k}\right)\right] \delta q_k = \sum_{k=1}^{s} Q_k \delta q_k \tag{10.10}$$

where

$$Q_k = \sum_{i=1}^{n} \left(\boldsymbol{F}_i \cdot \frac{\partial \boldsymbol{r}_i}{\partial q_k}\right) \quad (k = 1, 2, \cdots, s) \tag{10.11}$$

is called the generalized force associated with the generalized coordinate q_k.

For a conservative system, the generalized force Q_k can be expressed as

$$Q_k = \sum_{i=1}^{n} \left(\boldsymbol{F}_i \cdot \frac{\partial \boldsymbol{r}_i}{\partial q_k}\right) = -\frac{\partial V}{\partial q_k} \quad (k = 1, 2, \cdots, s) \tag{10.12}$$

where $V = V(q_1, q_2, \cdots, q_s)$ is the potential energy function of the system.

10.4 Lagrange's Equations

Consider that a system consisting of n particles is subjected to holonomic constraints. Assuming that the configuration of the system can be described by generalized coordinates $q_1, q_2, \cdots, q_s$, where s is the number of degrees of freedom of the system, then position vector of particle i can be expressed as

$$\boldsymbol{r}_i = \boldsymbol{r}_i(q_1, q_2, \cdots, q_s, t) \quad (i = 1, 2, \cdots, n) \tag{10.13}$$

Differentiating with respect to t, we have

$$\dot{\boldsymbol{r}}_i = \sum_{k=1}^{s} \frac{\partial \boldsymbol{r}_i}{\partial q_k}\dot{q}_k + \frac{\partial \boldsymbol{r}_i}{\partial t} \quad (i = 1, 2, \cdots, n) \tag{10.14}$$

From this equation, we can obtain

$$\frac{\partial \dot{\boldsymbol{r}}_i}{\partial \dot{q}_k} = \frac{\partial \boldsymbol{r}_i}{\partial q_k} \quad (i = 1, 2, \cdots, n;\ k = 1, 2, \cdots, s) \tag{10.15}$$

Multiplying by $\dot{\boldsymbol{r}}_i$ and differentiating with respect to t, we then have

$$\frac{\mathrm{d}}{\mathrm{d}t}(\dot{\boldsymbol{r}}_i \cdot \frac{\partial \dot{\boldsymbol{r}}_i}{\partial \dot{q}_k}) = \frac{\mathrm{d}}{\mathrm{d}t}(\dot{\boldsymbol{r}}_i \cdot \frac{\partial \boldsymbol{r}_i}{\partial q_k}) = \ddot{\boldsymbol{r}}_i \cdot \frac{\partial \boldsymbol{r}_i}{\partial q_k} + \dot{\boldsymbol{r}}_i \cdot \frac{\partial \dot{\boldsymbol{r}}_i}{\partial q_k} \quad (i = 1, 2, \cdots, n;\ k = 1, 2, \cdots, s) \tag{10.16}$$

i.e.,

$$\frac{\mathrm{d}}{\mathrm{d}t}\frac{\partial}{\partial \dot{q}_k}(\frac{1}{2}m_i\dot{\boldsymbol{r}}_i \cdot \dot{\boldsymbol{r}}_i) = (m_i\ddot{\boldsymbol{r}}_i) \cdot \frac{\partial \boldsymbol{r}_i}{\partial q_k} + \frac{\partial}{\partial q_k}(\frac{1}{2}m_i\dot{\boldsymbol{r}}_i \cdot \dot{\boldsymbol{r}}_i) \quad (i = 1, 2, \cdots, n;\ k = 1, 2, \cdots, s) \tag{10.17}$$

where m_i is the mass of particle i. By summing over i, we obtain

$$\frac{\mathrm{d}}{\mathrm{d}t}\frac{\partial}{\partial \dot{q}_k}(\sum_{i=1}^{n}\frac{1}{2}m_i\dot{\boldsymbol{r}}_i \cdot \dot{\boldsymbol{r}}_i) = \sum_{i=1}^{n}[(m_i\ddot{\boldsymbol{r}}_i) \cdot \frac{\partial \boldsymbol{r}_i}{\partial q_k}] + \frac{\partial}{\partial q_k}(\sum_{i=1}^{n}\frac{1}{2}m_i\dot{\boldsymbol{r}}_i \cdot \dot{\boldsymbol{r}}_i) \quad (k = 1, 2, \cdots, s) \tag{10.18}$$

i.e.,

$$\frac{\mathrm{d}}{\mathrm{d}t}\frac{\partial T}{\partial \dot{q}_k} = \sum_{i=1}^{n}(\boldsymbol{F}_i \cdot \frac{\partial \boldsymbol{r}_i}{\partial q_k}) + \frac{\partial T}{\partial q_k} \quad (k = 1, 2, \cdots, s) \tag{10.19}$$

where $T = \sum_{i=1}^{n}\frac{1}{2}m_i\dot{\boldsymbol{r}}_i \cdot \dot{\boldsymbol{r}}_i$ is the kinetic energy of the system, $\boldsymbol{F}_i = m_i\ddot{\boldsymbol{r}}_i$ is the force acting on particle i. Using the definition of generalized force, the above equation can be rewritten as

$$\frac{\mathrm{d}}{\mathrm{d}t}\frac{\partial T}{\partial \dot{q}_k} - \frac{\partial T}{\partial q_k} = Q_k \quad (k = 1, 2, \cdots, s) \tag{10.20}$$

where $Q_k = \sum_{i=1}^{n}(\boldsymbol{F}_i \cdot \frac{\partial \boldsymbol{r}_i}{\partial q_k})$ is the generalized force associated with the generalized q_k. The above equations are called Lagrange's equations.

For a conservative system, $Q_k = -\frac{\partial V}{\partial q_k}$, then Lagrange's equations can be expressed as

$$\frac{\mathrm{d}}{\mathrm{d}t}\frac{\partial T}{\partial \dot{q}_k} - \frac{\partial T}{\partial q_k} = -\frac{\partial V}{\partial q_k} \quad (k = 1, 2, \cdots, s) \tag{10.21}$$

Defining

$$L = T - V \tag{10.22}$$

where L is called the Lagrangian function. Using $\frac{\partial V}{\partial \dot{q}_k} = 0$, Lagrange's equations for a conservative system can be rewritten as

$$\frac{\mathrm{d}}{\mathrm{d}t}\frac{\partial L}{\partial \dot{q}_k} - \frac{\partial L}{\partial q_k} = 0 \quad (k = 1, 2, \cdots, s) \tag{10.23}$$

If part of the generalized forces are not conservative, say Q_k', and part are derivable from a potential function V, i.e.,

$$Q_k = Q_k' - \frac{\partial V}{\partial q_k} \quad (k = 1, 2, \cdots, s) \tag{10.24}$$

we then obtain Lagrange's equations in the most general form:

$$\frac{\mathrm{d}}{\mathrm{d}t}\frac{\partial L}{\partial \dot{q}_k} - \frac{\partial L}{\partial q_k} = Q_k' \quad (k = 1, 2, \cdots, s) \tag{10.25}$$

Example 10.3 A block of mass m_1 slides on a smooth inclined plane, which itself has a mass m_2 and slides on a smooth horizontal surface. Determine the horizontal acceleration of the block and the acceleration of the inclined plane.

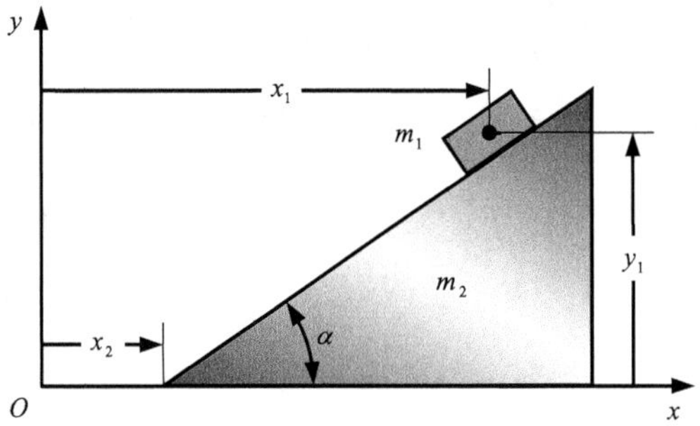

Fig. E10.3

Solution Choosing x_1 and x_2 as the generalized coordinates, then $y_1 = \tan\alpha(x_1 - x_2)$ and $\dot{y}_1 = \tan\alpha(\dot{x}_1 - \dot{x}_2)$. Hence, the kinetic and potential energies of the system can be expressed, respectively, as

$$T = \frac{1}{2}m_1[\dot{x}_1^2 + \tan^2\alpha(\dot{x}_1 - \dot{x}_2)^2] + \frac{1}{2}m_2\dot{x}_2^2, \quad V = m_1 g\tan\alpha(x_1 - x_2) + m_2 gh$$

where h is a constant. The corresponding Lagrangian function is equal to

$$L = T - V = T = \frac{1}{2}m_1[\dot{x}_1^2 + \tan^2\alpha(\dot{x}_1 - \dot{x}_2)^2] + \frac{1}{2}m_2\dot{x}_2^2 - m_1 g\tan\alpha(x_1 - x_2) - m_2 gh$$

For the block of mass m_1, we have

$$\frac{\partial L}{\partial \dot{x}_1} = m_1[\dot{x}_1 + \tan^2\alpha(\dot{x}_1 - \dot{x}_2)], \quad \frac{\partial L}{\partial x_1} = -m_1 g\tan\alpha$$

Substituting into Lagrange's equation, we obtain

$$m_1[\ddot{x}_1 + \tan^2\alpha(\ddot{x}_1 - \ddot{x}_2)] + m_1 g\tan\alpha = 0$$

Similarly, for the inclined plane of mass m_2, we have

$$\frac{\partial L}{\partial \dot{x}_2} = -m_1\tan^2\alpha(\dot{x}_1 - \dot{x}_2) + m_2\dot{x}_2, \quad \frac{\partial L}{\partial x_2} = m_1 g\tan\alpha$$

Substituting into Lagrange's equation, we obtain

$$-m_1\tan^2\alpha(\ddot{x}_1 - \ddot{x}_2) + m_2\ddot{x}_2 - m_1 g\tan\alpha = 0$$

Solving the equations obtained, we have

$$\ddot{x}_1 = -\frac{m_2 g \sin\alpha\cos\alpha}{m_1 \sin^2\alpha + m_2}, \quad \ddot{x}_2 = \frac{m_1 g \sin\alpha\cos\alpha}{m_1 \sin^2\alpha + m_2}$$

Problems

10.1 The mechanism shown is acted upon by the force ***F***. Neglecting the weight of the mechanism and knowing that $AC = BC = CD = CE = DG = EG = a$ and $AB = b$, derive an expression for the magnitude of the force ***R*** required for equilibrium.

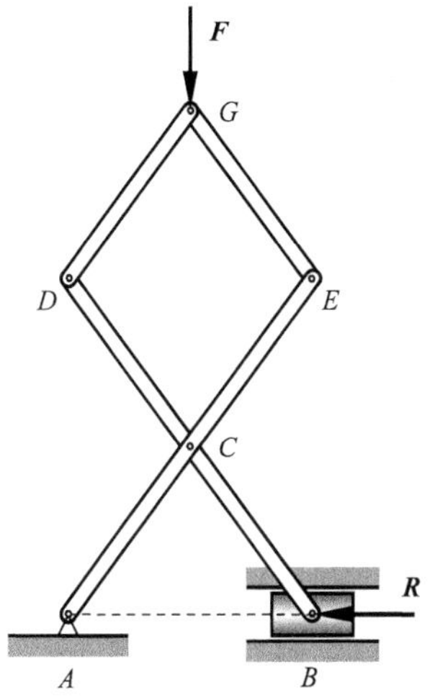

Fig. P10.1

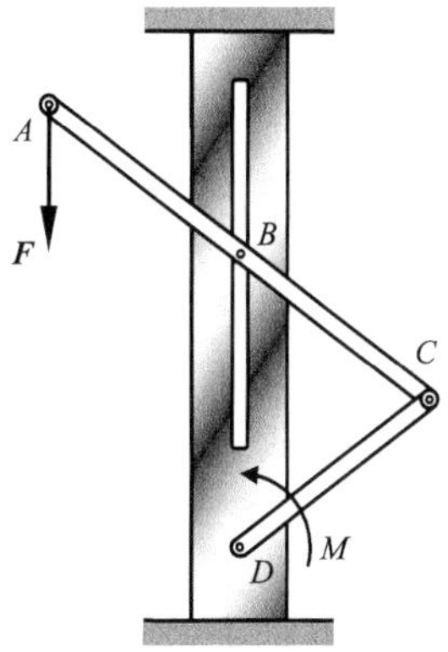

Fig. P10.2

10.2 The pin at B is attached to member ABC and can slide along a slot cut in the fixed plate shown. Neglecting the effect of friction and knowing that $AB = BC = CD = a$ and $BD = b$, derive an expression for the magnitude M of the couple required to maintain equilibrium when the force ***F*** which acts at A is directed (a) vertically downward, (b) horizontally to the left.

10.3 A slender rod AB of length l is attached to a collar at B and rests on a semicircular cylinder of radius r. Neglecting the effect of friction and knowing that the mechanism is acted upon by the force s $\boldsymbol{F}_A$ and $\boldsymbol{F}_B$, derive an expression for θ corresponding to the equilibrium position of the mechanism.

10.4 Two rods ABC and CDE are connected by a pin at C and by a spring AE. The stiffness of the spring is k, and the spring is unstretched when $\theta = 30°$. Knowing that $AB = BC = CD = DE = a$, derive an equation in F, θ, l, and k that must be satisfied when the system is in equilibrium under the action of the loading shown.

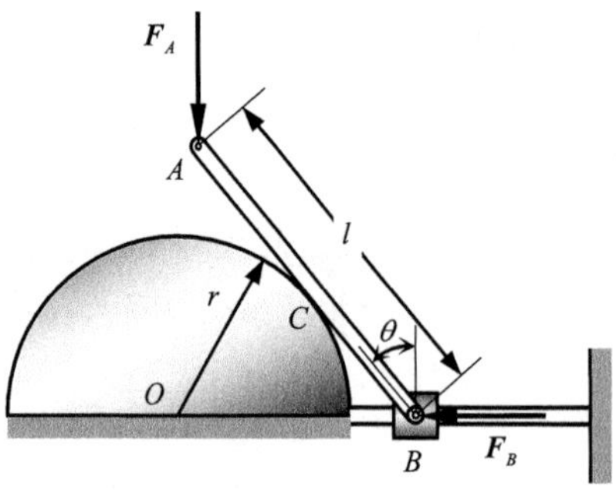

Fig. P10.3

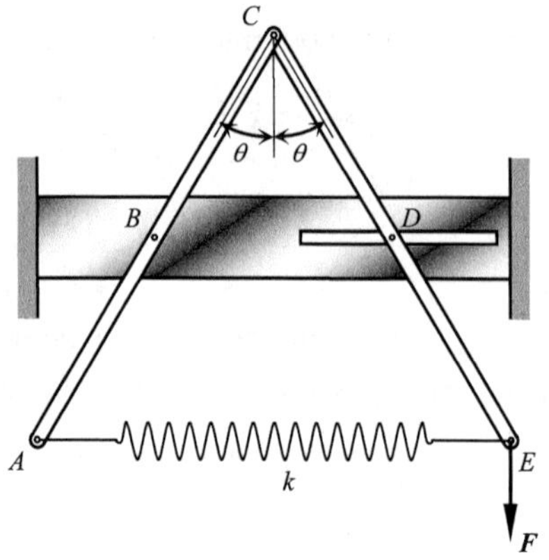

Fig. P10.4

10.5 A horizontal force $\boldsymbol{F}$ of magnitude 150 N is applied to the mechanism at A. The stiffness of the spring is $k = 1.5$ kN/m, and the spring is unstretched when $\theta = 0$. Neglecting the mass of the mechanism and knowing that $a = 250$ mm and $r = 150$ mm, determine the value of θ corresponding to equilibrium.

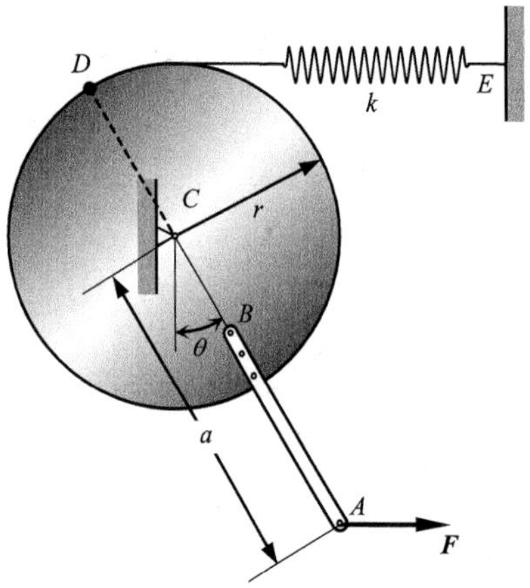

Fig. P10.5

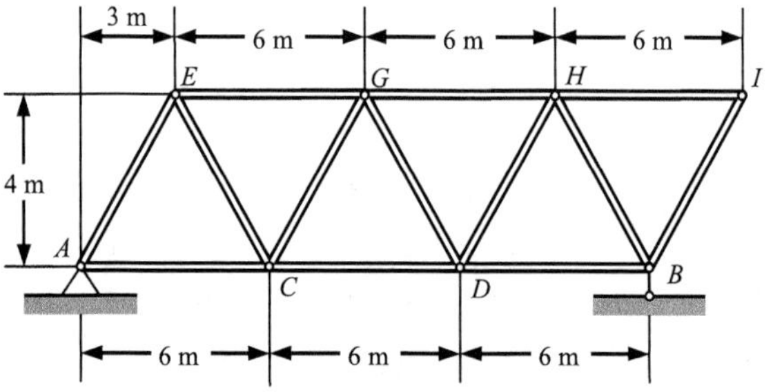

Fig. P10.6

10.6 Determine the vertical displacement of joint I if the length of member AC is increased by 6 mm.

10.7 A small sphere of mass m_1 attached to a string of length l is suspended from the boundary of a uniform disk of mass m_2 and radius r. Assuming that the disk can rotate freely about O, determine the differential equations of motion of the system.

10.8 A block of mass m_1 slides on a smooth inclined plane, which itself has a mass m_2 and slides on a smooth horizontal surface. Determine the acceleration of the block with respect to the inclined plane and the acceleration of the inclined plane.

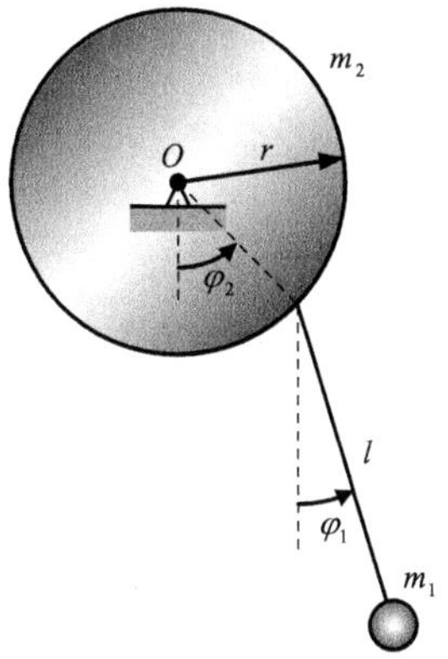

Fig. P10.7

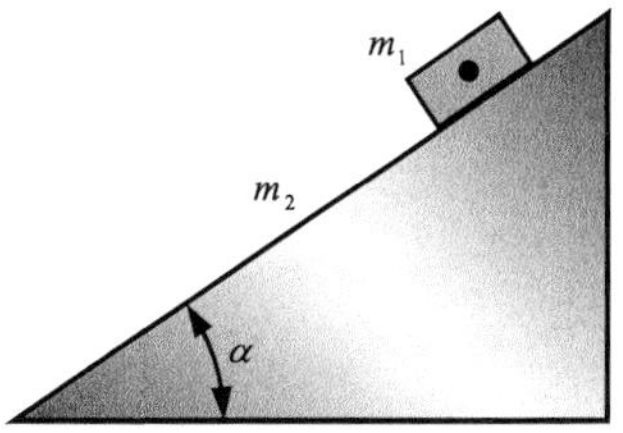

Fig. P10.8

10.9 A two degrees of freedom system of undamped forced vibration, consisting of two block, each of mass m, and four springs, each of stiffness k, is subjected to two periodic forces, each of magnitude $H \sin \omega_f t$, where H and ω_f are the amplitude and the circular frequency of the periodic force, respectively. Determine the differential equations of motion of the system.

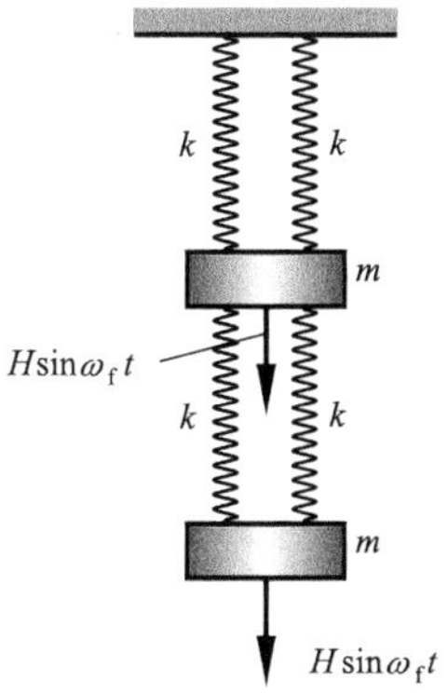

Fig. P10.9

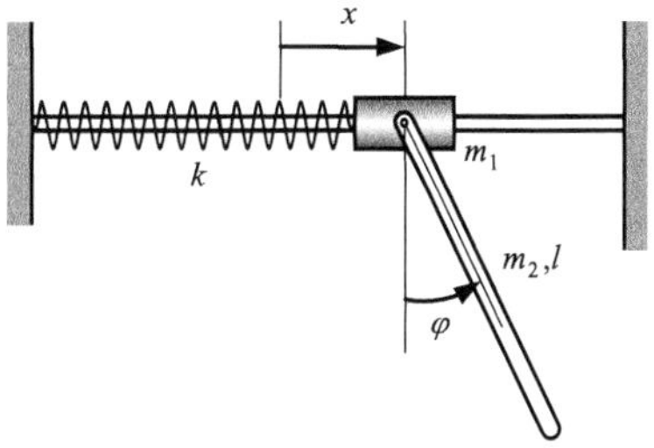

Fig. P10.10

10.10 A collar of mass m_1 can slide on a horizontal beam and is attached to a spring of stiffness k. From the collar hangs a homogeneous rod of length l and mass m_2 which can swing in a vertical plane through the beam. Neglecting friction, find Lagrange's equations of motion for the system.

Chapter 11 Stresses and Strains

11.1 External Forces

An external force that is applied to the surface of a body is called a surface force. If the surface force is distributed over a finite area of the body, it is said to be a distributed load on a surface, Fig. 11.1a. If the surface force is applied along a narrow area, this force is defined as a distributed load along a line, Fig. 11.1b. If the area subjected to a surface force is very small, compared with the surface area of the body, then this surface force can be regarded as a concentrated load, Fig. 11.1c.

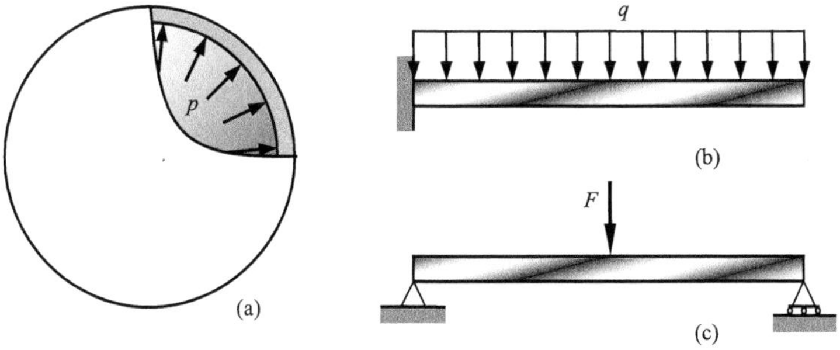

Fig. 11.1

An external force that is applied to every point within a body is called a body force. A gravitational force is an excellent example of the body force since it acts upon each of the particles forming the body.

11.2 Internal Forces

When various external loads are applied to a member, a distributed internal force will be developed at any point within the member. The distributed internal force on any section within the member can be determined by using the method of sections.

We imagine to use a plane, such as plane Π shown in Fig. 11.2a, to cut the member where the distributed internal force need to be determined. The portion of the member to the right of the cut plane is removed, and it is replaced by a distributed internal force acting on the left portion, Fig. 11.2b. For equilibrium of the remaining portion of the member, the distributed internal force can be determined by using the equations of equilibrium. Although the exact distribution of internal force may be unknown, we can use the equations of equilibrium to determine the resultant force $\boldsymbol{R}$ and resultant couple $\boldsymbol{M}_O$ about centroid O on the cut plane, which are caused by the distributed internal force, Fig. 11.3a.

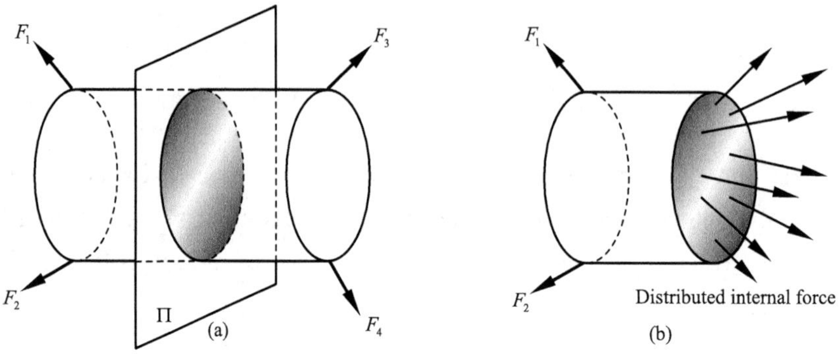

Fig. 11.2

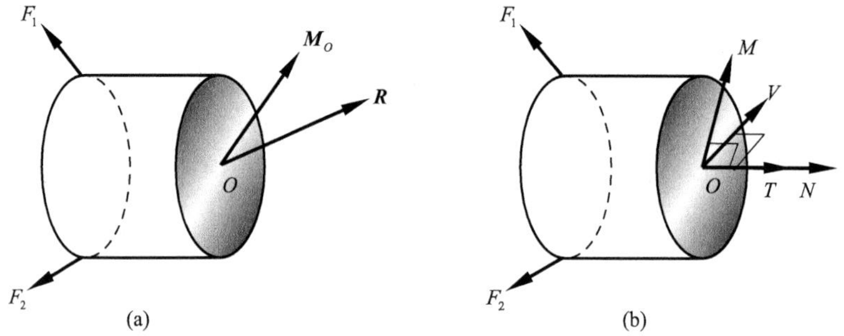

Fig. 11.3

Generally speaking, the resultant force $\boldsymbol{R}$ and resultant couple $\boldsymbol{M}_O$ have arbitrary directions, neither perpendicular nor parallel to the cut plane. However, we can resolve the resultant force and couple into four components, Fig. 11.3b.

(1) **Normal Force** The normal component of the resultant force is called a normal force (an axial force), denoted by N.

(2) **Shearing Force** The tangential component of the resultant force is called a shearing force, denoted by V.

(3) **Torsional Moment** The normal component of the resultant couple is called a torsional moment (twisting moment, or torque), denoted by T.

(4) **Bending Moment** The tangential component of the resultant couple called a bending moment, denoted by M.

11.3 Stresses

A distributed internal force is developed at any point within the member subjected to external loads. To define the stress at a given point P of the section, Fig. 11.4a, we consider a small area ΔA containing P and assume that the resultant force is $\Delta \boldsymbol{F}$ on the area ΔA. In general, the force $\Delta \boldsymbol{F}$ has a unique direction at a given point on the section and can be resolved into components ΔN and ΔV, Fig. 11.4b. ΔN is the normal component perpendicular to the area ΔA, ΔV is the tangential component within the area ΔA.

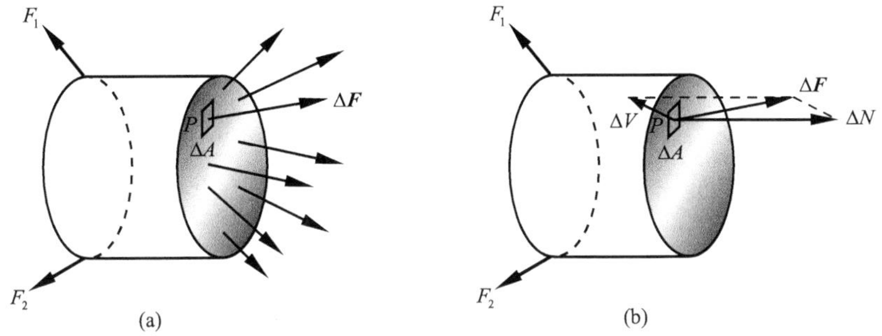

Fig. 11.4

1. Normal Stress

The intensity of the normal force, the normal force per unit area, is defined as a normal stress, denoted by σ. The normal stress at the given point P, Fig. 11.5, can be expressed as

$$\sigma = \lim_{\Delta A \to 0} \frac{\Delta N}{\Delta A} \tag{11.1}$$

where σ is perpendicular to the section. A positive sign is usually used to indicate a tensile stress and a negative sign to indicate a compressive stress. From SI units, with ΔN expressed in N and ΔA in m^2, the normal stress σ is expressed in Pascals (Pa).

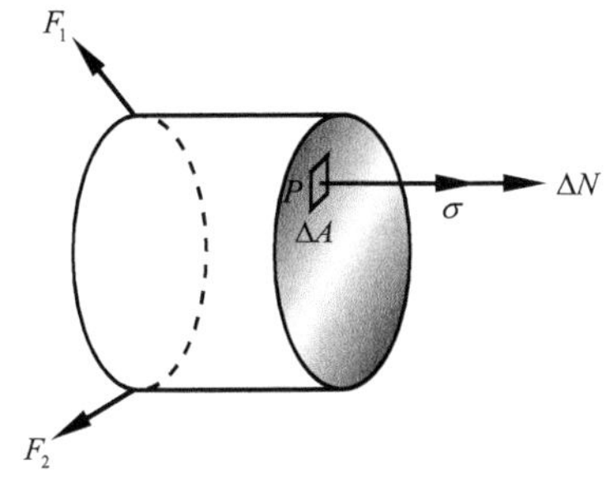

Fig. 11.5

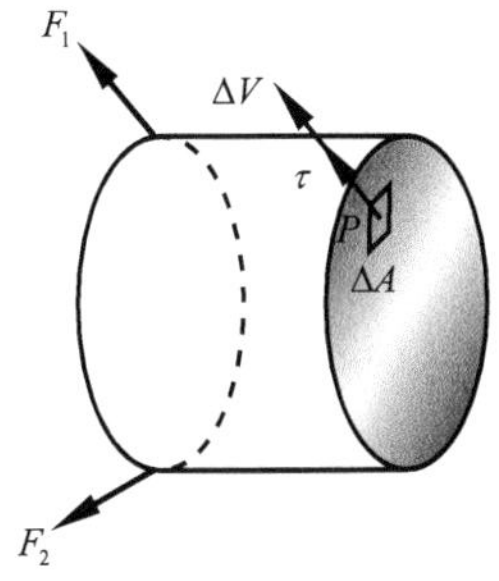

Fig. 11.6

2. Shearing Stress

The intensity of the tangential force, the tangential force per unit area, is called a shearing stress, denoted by τ. The shearing stress at the given point *P*, Fig. 11.6, can be written as

$$\tau = \lim_{\Delta A \to 0} \frac{\Delta V}{\Delta A} \tag{11.2}$$

where τ lies in the section. In SI units, the shearing stress τ is also measured in Pa.

Example 11.1 A load F is applied to a steel rod supported as shown in Fig. E11.1a by a plate into which a 15 mm diameter hole has been drilled. Knowing that the shearing stress must not exceed 120 MPa in the steel, determine the largest load F which may be applied to the rod.

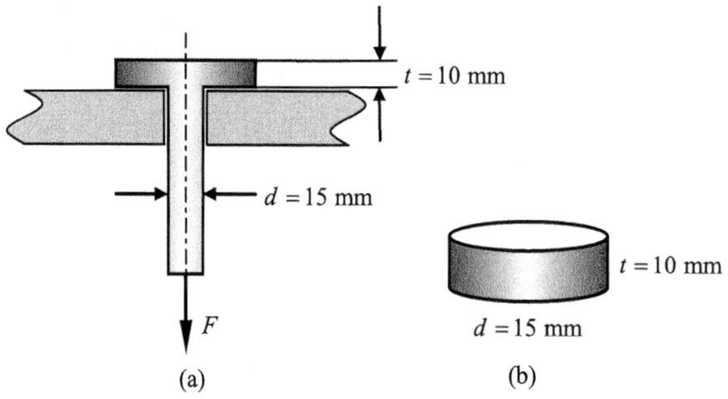

Fig. E11.1

Solution The shearing plane is a cylindrical surface, Fig. E11.1b, and its shearing area is equal to $A = \pi dt$. Since the maximum shearing stress $\tau_{max} = 120$ MPa, then the largest load F_{max} can, from Eq. (11.2), be obtained by

$$F_{max} = \tau_{max} A = 56.5 \text{ kN}$$

3. Bearing Stress

Consider two plates connected by a bolt, Fig. 11.7a. The bolt exerts on the upper plate force F_{bs}, Fig. 11.7b, equal and opposite to force F'_{bs} exerted by the upper plate on the bolt, Fig. 11.7c. F_{bs} represents the resultant of distributed force on the inside surface of a half-cylinder.

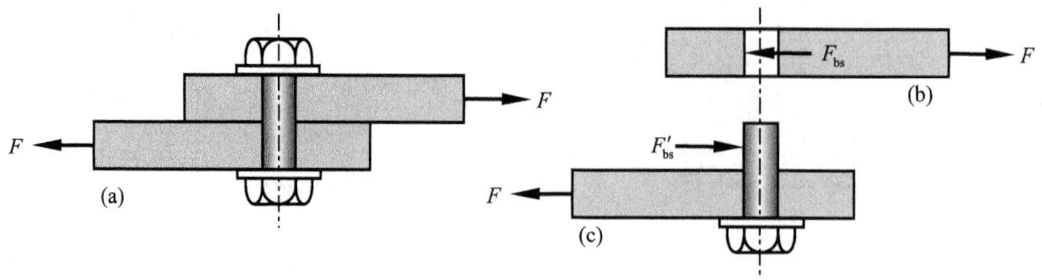

Fig. 11.7

Since the distribution of the force on the contacting surface is quite complicated, an average stress, obtained by dividing F_{bs} by the projected area A_{bs} of the bolt on the plate section, is regarded as the bearing stress σ_{bs}. Since A_{bs} is equal to td, where t is the plate thickness and d is the diameter of the bolt, we have

$$\sigma_{bs} = \frac{F_{bs}}{A_{bs}} = \frac{F_{bs}}{td} \tag{11.3}$$

Example 11.2 A load F is applied to a steel rod supported as shown in Fig. E11.2a by a plate into which a 15 mm diameter hole has been drilled. Knowing that the bearing stress of the steel must not exceed 150 MPa, determine the largest load F which may be applied to the rod.

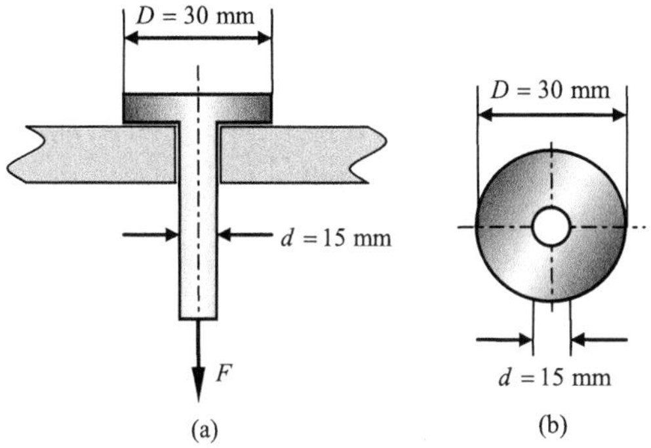

Fig. E11.2

Solution The bearing plane is an annular surface of inner diameter $d = 15$ mm and outer diameter $D = 30$ mm, Fig. E11.2b, and thus the bearing area is equal to $A_{bs} = \frac{1}{4}\pi(D^2 - d^2)$. Since the maximum bearing stress $\sigma_{bs} = 150$ MPa, then the largest load F_{max} can, from Eq. (11.3), be obtained by

$$F_{max} = \sigma_{bs} A_{bs} = 79.5 \text{ kN}$$

11.4 Strains

When external loads are applied to a member, they will tend to change the size and shape of the member.

1. Normal Strain

The elongation or contraction of a line segment per unit of length is referred to as the normal strain denoted by ε. Consider a line segment Δl passing through the given point P within the member, Fig. 11.8a. After the member is deformed under the action of external loads, the line segment Δl has length of $\Delta l'$, Fig. 11.8b. The deformation of the line segment Δl is equal to $\Delta\delta = \Delta l' - \Delta l$.

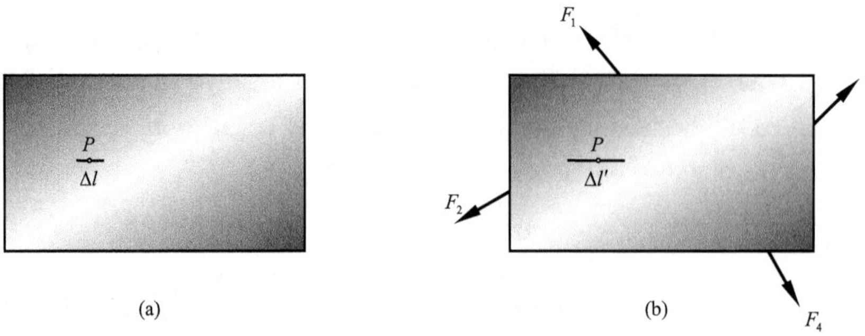

Fig. 11.8

As Δl approaches zero, the quotient of $\Delta\delta$ and Δl will approach a finite limit. This limit is called the normal strain at the given point P along the direction of the line segment, which can be expressed as

$$\varepsilon = \lim_{\Delta l \to 0} \frac{\Delta\delta}{\Delta l} \tag{11.4}$$

where ε is a dimensionless quantity. The normal strain is said to be positive if the line segment elongates and negative if the line segment contracts.

2. Shearing Strain

The change of angle that occurs between two perpendicular line segments is referred to as the shearing strain, denoted by γ. Consider the two perpendicular line segments Δx and Δy intersecting at the given point P, Fig. 11.9a. After the member is deformed under the action of external loads, the included angle between the two line segments is changed from $\frac{1}{2}\pi$ into θ, Fig. 11.9b.

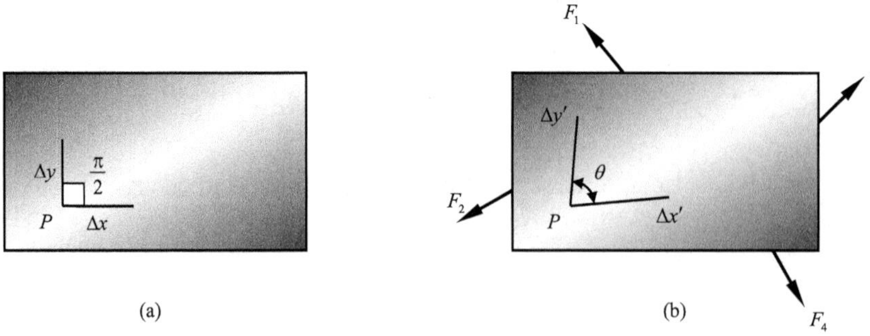

Fig. 11.9

The shearing strain at the given point P in the plane containing the two line segments can be written as

$$\gamma = \frac{1}{2}\pi - \lim_{\substack{\Delta x \to 0 \\ \Delta y \to 0}} \theta \tag{11.5}$$

The shearing strain γ is positive if θ is smaller than $\frac{1}{2}\pi$; otherwise, it is negative. The shearing strain γ is measured in radians (rad).

11.5 Hooke's Law

Most engineering structures are designed to undergo relatively small deformations, involving only the straight-line portion of the stress-strain curve. For the straight-line portion of the stress-strain curve, the stress σ is proportional to the strain ε, and we have

$$\sigma = E\varepsilon \tag{11.6}$$

This relation is known as Hooke's law. E is called the modulus of elasticity, or Young's modulus. Since ε is a dimensionless quantity, E is expressed in the same units as σ.

11.6 Tensile Properties of Low-Carbon Steel

1. Standard Specimen

A standard specimen having certain size and shape must be used in the tensile test of low-carbon steel for purpose of the comparability of experimental results.

One type of standard tensile specimen of low-carbon steel commonly used is shown in Fig. 11.10a. The cross-sectional area A_0 of the central portion of the specimen has been accurately determined and two gauge marks have been inscribed on the portion at a distance l_0 from each other. The distance l_0 is known as the gauge length of the specimen.

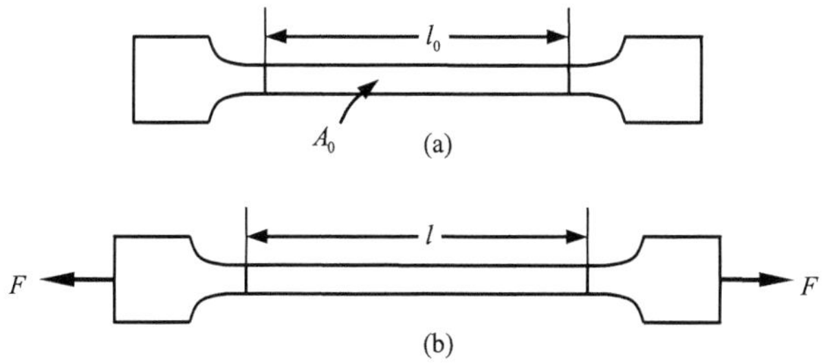

Fig.11.10

2. Tensile Diagram

The specimen is then placed in a universal testing machine, which is used to apply a centric load F. As the load F increases, the distance l between the two gauge marks also increases, Fig. 11.10b. The elongation $\delta = l - l_0$ is recorded for each value of F.

Plotting the magnitude F of the load against the deformation δ, we obtain a load-deformation diagram, Fig. 11.11. This diagram is called the tensile diagram of low-carbon steel.

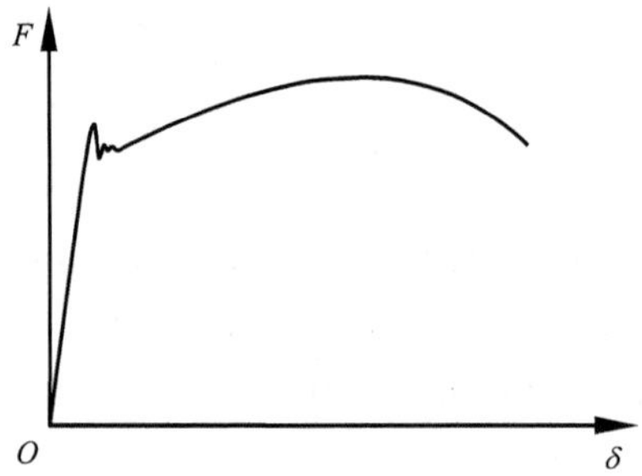

Fig. 11.11

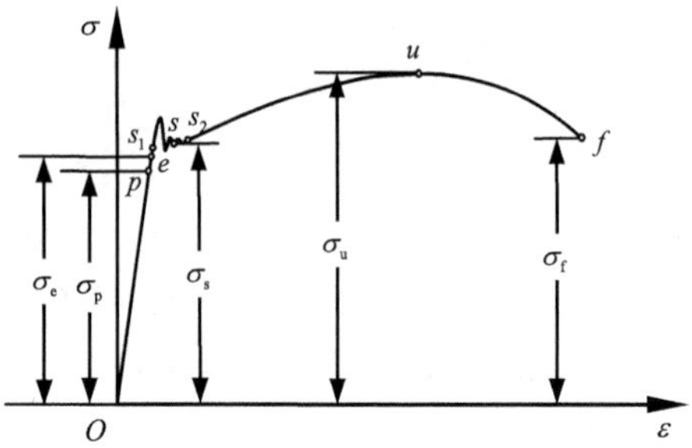

Fig. 11.12

3. Stress-Strain Curve

From each pair of readings F and δ, the stress σ is computed by dividing F by the original cross-sectional area A_0 of the specimen, and the strain ε by dividing the elongation δ by the original distance l_0 between the two gauge marks.

Plotting ε as an abscissa and σ as an ordinate, we obtain a curve that describes the mechanical properties of low-carbon steel and that is independent of the dimensions of the particular specimen used. This curve is called a stress-strain curve of low-carbon steel, Fig. 11.12. From the stress-strain curve of low-carbon steel, we can identify four different ranges in which the material behaves.

(1) Elastic Range (*Oe*). The region *Oe* of the stress-strain curve is called the elastic range. The material behaves elastically in this region. The largest stress corresponding to the elastic range is called the elastic limit, denoted by σ_e.

The straight-line region *Op* within the elastic range *Oe* is called the linearly elastic range. In this range, the material is linearly elastic, i.e., the stress is proportional to the strain. The maximum stress limit to this linear relationship is called the proportional limit, σ_p. If the stress slightly exceeds the proportional limit, the material still respond elastically. This continues until the stress reaches the elastic limit.

In the elastic range, if the load is removed the specimen will return back to its original size and shape. Normally for low-carbon steel, the elastic limit is very close to the proportional limit and therefore rather difficult to determine.

(2) Yielding Range (s_1s_2). A slight increase in stress above the elastic limit will result in a failure of the material and cause it to deform permanently. This phenomenon is called yielding, which is indicated by the region s_1s_2 of the stress-strain curve. The stress at which

yielding is initiated is called the yield stress, yield strength, or yield point, denoted by σ_s, and the deformation that occurs during the yielding range is called plastic deformation. This deformation is caused by slippage of the material along oblique section and is due, therefore, primarily to shearing stresses. Once the yield point is reached, the specimen will continue to elongate without any increase in load.

For low-carbon steels, the yield point is often distinguished by two values. The upper yield point occurs first, followed by a sudden decrease in load-carrying capacity to a lower yield point.

(3) Strain Hardening Range (s_2u). When yielding has ended, a further load can be applied to the specimen, resulting in a curve that rises continuously but becomes flatter until it reaches a maximum stress referred to as the ultimate stress or ultimate strength, σ_u. The rise in the curve in this manner is called strain hardening, and it is identified as the region s_2u.

(4) Necking Range (uf). At the ultimate stress, the cross-sectional area begins to decrease in a local region of the specimen. This phenomenon, which is called necking, is caused by slip planes formed within the material, and the actual strains produced are caused by shearing stress. As a result, a constriction or neck gradually tends to form in this region as the specimen elongates further. Since the cross-sectional area within this region is continuously decreasing, the smaller area can only carry an ever-decreasing load. Hence the stress-strain curve tends to curve downward until the specimen breaks at the fracture stress σ_f.

We note that rupture occurs along a cone-shaped surface which forms an angle of approximately 45° with the cross section of the specimen.

4. Percent Elongation and Percent Reduction in Area

A standard measure of the ductility of a material is its percent elongation δ_1, which is defined as

$$\delta_1 = \frac{l_1 - l_0}{l_0} \times 100\% \tag{11.7}$$

where l_0 and l_1 denote, respectively, the initial gauge length of the tensile test specimen and its final gauge length at rupture.

Another measure of ductility which is sometimes used is the percent reduction in area ψ_1, defined as

$$\psi_1 = \frac{A_0 - A_1}{A_0} \times 100\% \tag{11.8}$$

where A_0 and A_1 denote, respectively, the initial cross-sectional area of the specimen and its minimum cross-sectional area at rupture.

11.7 Stress-Strain Curve of Ductile Materials without Distinct Yield Point

In the case of many ductile materials, the onset of yield is not characterized by a fluctuating or horizontal portion of the stress-strain curve. Instead, the stress keeps increasing until the ultimate strength is reached. Necking then begins, leading eventually to rupture. For such materials, the yield strength can be defined by the offset method. A point s on the stress-strain curve can be obtained by drawing through the point of the horizontal axis of abscissa $\varepsilon = 0.2\%$, an oblique straight-line parallel to the initial straight-line portion of the stress-strain curve, Fig. 11.13. The stress $\sigma_{0.2}$ corresponding to the point s obtained in this fashion is defined as the offset yield strength of the material at 0.2% offset.

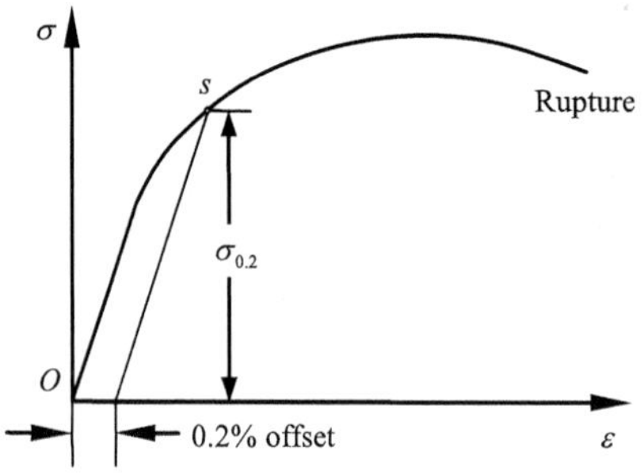

Fig. 11.13

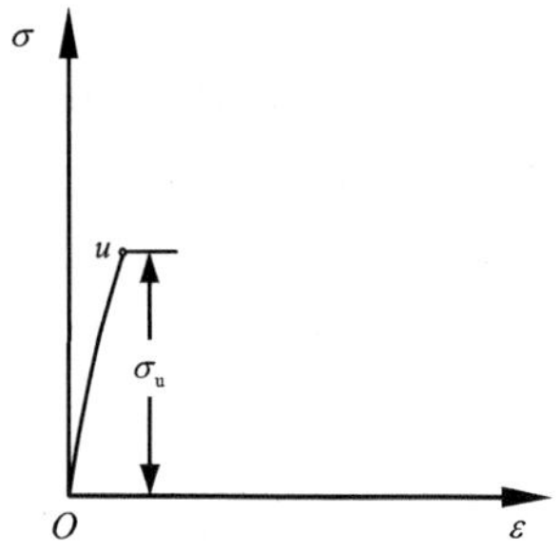

Fig. 11.14

11.8 Ductile and Brittle Materials

Stress-strain curves of various materials vary widely. It is possible, however, to distinguish some common characteristics among the stress-strain curves of various groups of materials and to divide materials into two broad categories on the basis of these characteristics, namely, the ductile materials and brittle materials.

Ductile materials, which comprise structural steel, as well as many alloys of other metals, are characterized by the ability to yield at normal temperatures, Fig. 11.12.

Brittle materials, which comprise cast iron, glass, and stone, are characterized by the fact that rupture occurs without any noticeable prior change in the rate of elongation, Fig. 11.14. We note the absence of any necking of the specimen in the case of a brittle material, and observe that rupture occurs along a surface perpendicular to the load.

11.9 Properties of Materials in Compression

When a specimen made of a ductile material is loaded in compression, the stress-stain curve obtained is essentially the same through its initial straight-line portion and through the beginning of the portion corresponding to yield and stain-hardening. Particularly noteworthy is the fact that for low-carbon steel, the yield strength is the same in both tension and compression. For large value of the strain, the tension and compression stress-strain curve diverge, and it should be noted that necking cannot occur in compression, Fig. 11.15a.

For most brittle materials, say a cast iron, we find that the ultimate strength in compression is much larger than the ultimate strength in tension, Fig. 11.15b.

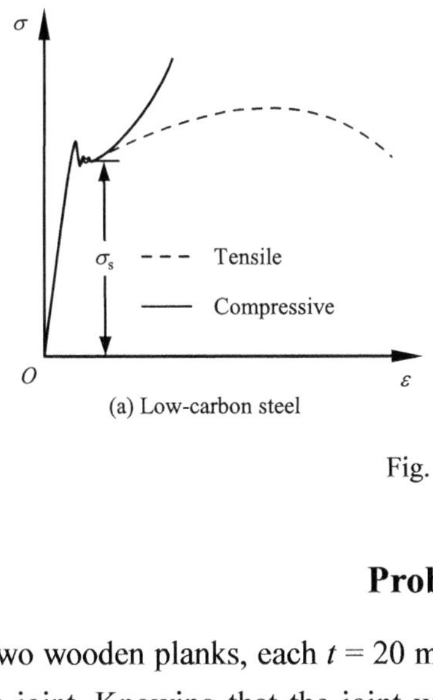

(a) Low-carbon steel

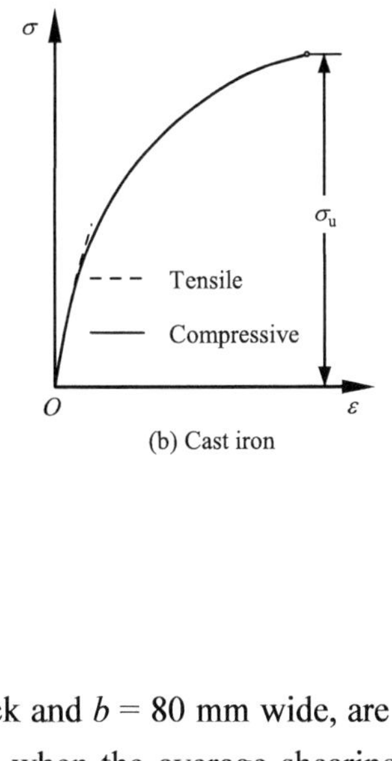

(b) Cast iron

Fig. 11.15

Problems

11.1 Two wooden planks, each $t = 20$ mm thick and $b = 80$ mm wide, are joined by the glued mortise joint. Knowing that the joint will fail when the average shearing stress in the glue reaches $\tau = 1.2$ MPa, determine the largest allowable axial load F_{max} if the length of the cuts is equal to $a = 30$ mm.

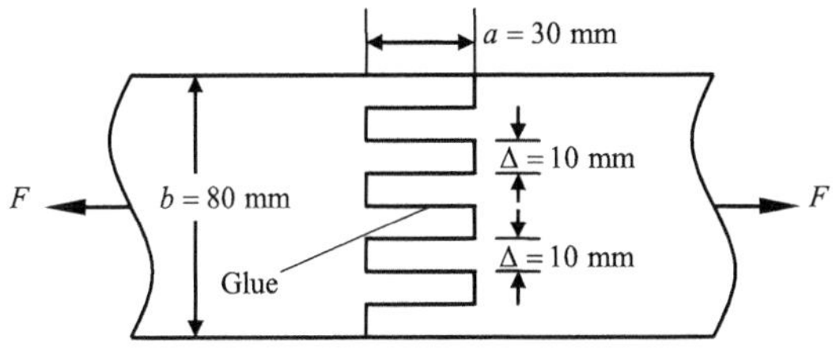

Fig. P11.1

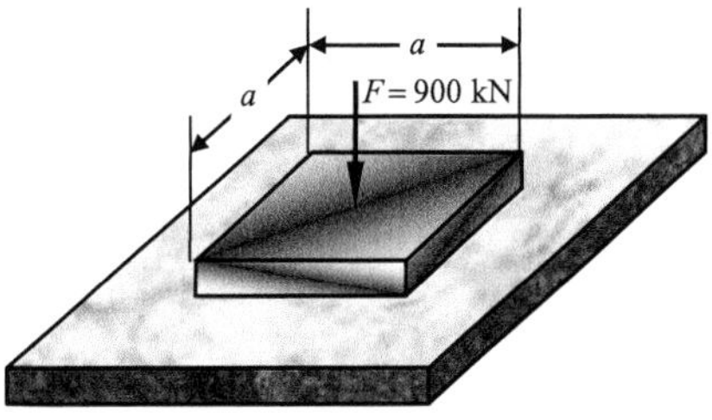

Fig. P11.2

11.2 A load F is uniformly applied to a concrete foundation by a square plate. Knowing that the bearing stress on the concrete foundation must not exceed 12 MPa, determine the side a of the plate which will provide the most economical and safe design.

11.3　The stress–strain diagram for an aluminum alloy that is used for making aircraft parts is shown in Fig. P11.3. If a specimen of this material is stressed to 550 MPa, determine the permanent strain that remains in the specimen when the load is released.

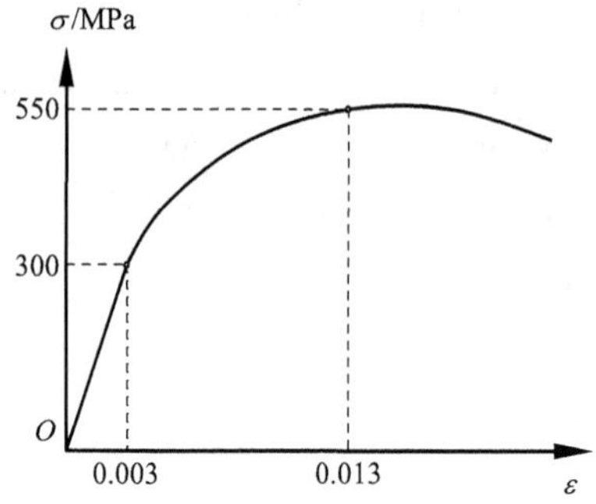

Fig. P11.3

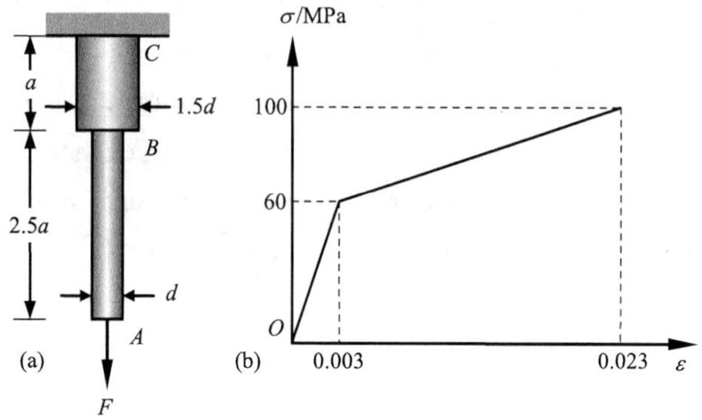

Fig. P11.4

11.4　A circular cross section rod shown in Fig. P11.4a is subjected to an axial load F. If a portion of the stress-strain diagram is shown in Fig. P11.4b, determine the elongation of the rod when the load is applied. Knowing that $a = 50$ mm, $d = 4$ mm, and $F = 1131$ N.

Chapter 12 Tension and Compression

Consider a prismatic bar loaded at its ends by a pair of equal and opposite external forces coinciding with the longitudinal centroidal axis of the bar. If the external forces are directed away from the bar, the bar is said to be in tension, Fig. 12.1a; if the external forces are directed toward the bar, the bar is in compression, Fig. 12.1b.

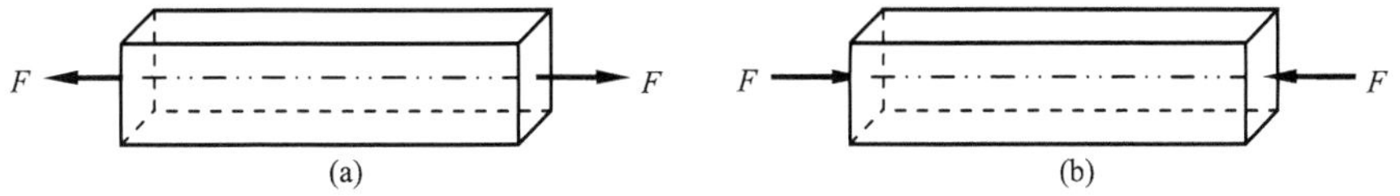

Fig. 12.1

12.1 Axial Force

Under the action of a pair of external forces, Fig. 12.1a, distributed internal forces will appear within the bar. The resultant of these distributed internal forces is called the axial or normal force of the bar denoted by N, Fig. 12.2.

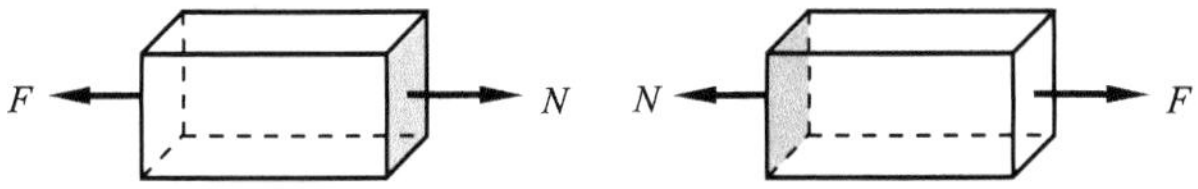

Fig. 12.2

The axial force can be determined by imagining a plane to cut the bar anywhere along its length and oriented perpendicular to the longitudinal axis of the bar. For determination of the axial force on the cut plane, the right (or left) portion of the bar of the cut plane is removed and replaced by the axial force acting on the left (right) portion. By this method of introducing a cut plane, the original axial force now becomes an external force with the remaining portion of the bar. For equilibrium of the remaining portion of the bar, the axial force can be determined by using the equation of static equilibrium along the bar.

12.2 Normal Stress on Cross Section

It is necessary to make some assumption regarding the distribution of the axial force on the cross section of a bar subjected to tension or compression, Fig. 12.3a. Since the external forces act through the centroid of the cross section, it is commonly assumed that the axial force caused by the applied external forces is uniformly distributed across the cross section of the bar, Fig. 12.3b. Therefore, the normal stress on the cross section of the bar can be expressed as

$$\sigma = \frac{N}{A} \tag{12.1}$$

where A is the cross-sectional area, and N is the axial force on the cross section. A positive sign will be used to indicate a tensile stress (bar in tension) and a negative sign to indicate a compressive stress (bar in compression).

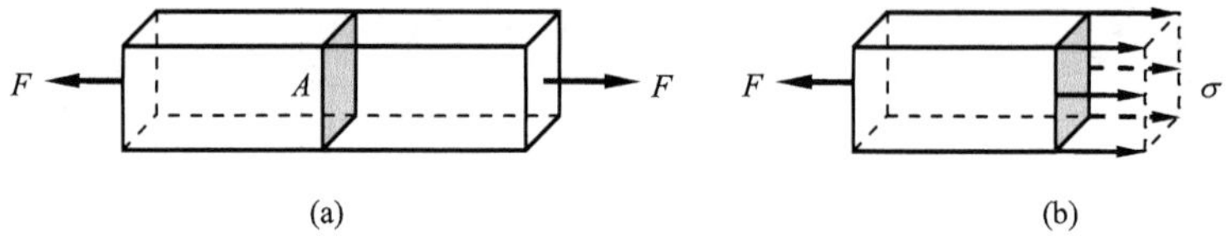

Fig. 12.3

Example 12.1 Each of the two vertical links BD and CE has a 16 mm×26 mm uniform rectangular cross section and each of the three pins has a 6 mm diameter, Fig. E12.1a. Determine the maximum value of the average normal stress in the links (a) BD, and (b) CE.

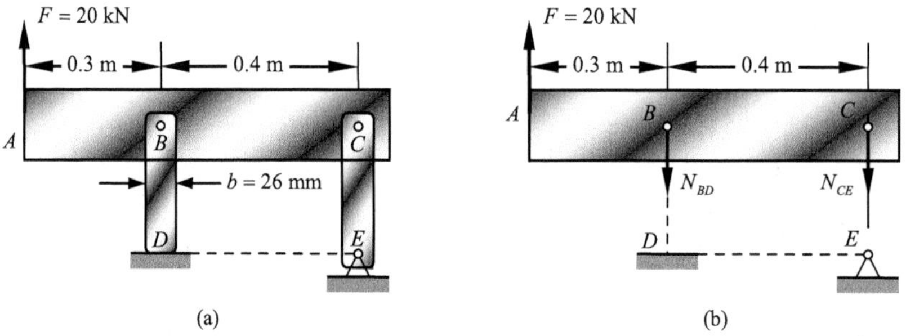

Fig. E12.1

Solution Considering the equilibrium of horizontal beam, Fig. E12.1b, we have

$$N_{BD} = F\frac{l_{AC}}{l_{BC}} = 35\ \text{kN}\text{, and}\ \ N_{CE} = F - N_{BD} = -15\ \text{kN}$$

Since the link BD is in tension, its minimum cross-sectional area is equal to

$$A_{BD} = t(b-d) = 320\ \text{mm}^2$$

Using Eq. (12.1), we obtain the average normal stress in the link BD, which is equal to

$$\sigma_{BD} = \frac{N_{BD}}{A_{BD}} = 109.4\ \text{MPa}$$

Similarly, the average normal stress in the link CE is

$$\sigma_{CE} = \frac{N_{CE}}{A_{CE}} = -36.1 \text{ MPa}$$

12.3 Normal and Shearing Stresses on Oblique Section

Consider a two-force bar of Fig. 12.4a, which is subjected to a pair of tensile forces of magnitude F. If we cut the bar through a plane forming an angle θ with the cross section and draw the free-body diagram of the left portion of the bar, we find from the equilibrium of the free body that the force R acting on the oblique section must be equal to F. Resolving R into components N and V, respectively normal and tangent to the oblique section, Fig. 12.4b, we have

$$N = R\cos\theta,\ V = R\sin\theta \tag{12.2}$$

where the force N represents the resultant of normal forces distributed over the section, and the force V the resultant of tangential distributed forces. The corresponding normal and shearing stresses on the oblique section are obtained by dividing, respectively, N and V by the area A_θ of the oblique section, Fig. 12.4c:

$$\sigma_\theta = N/A_\theta,\ \tau_\theta = V/A_\theta \tag{12.3}$$

Using that $A_\theta = A/\cos\theta$, where A denotes the cross-sectional area of the bar, we obtain

$$\sigma_\theta = \sigma\cos^2\theta,\ \tau_\theta = \frac{1}{2}\sigma\sin 2\theta \tag{12.4}$$

where σ is the normal stress on the cross section. We note from Eq. (12.4) that $\sigma_\theta = \sigma_{\max} = \sigma$ and $\tau_\theta = 0$ when $\theta = 0$, and that $\sigma_\theta = \tau_\theta = \tau_{\max} = \frac{1}{2}\sigma$ as $\theta = 45°$.

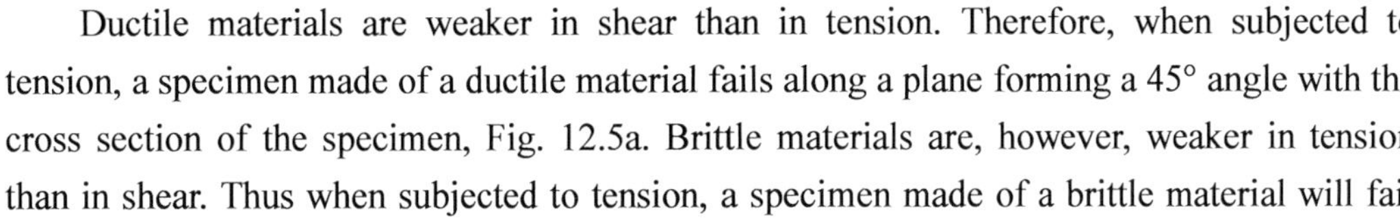

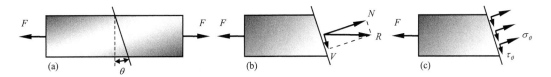

Fig. 12.4

Ductile materials are weaker in shear than in tension. Therefore, when subjected to tension, a specimen made of a ductile material fails along a plane forming a 45° angle with the cross section of the specimen, Fig. 12.5a. Brittle materials are, however, weaker in tension than in shear. Thus when subjected to tension, a specimen made of a brittle material will fail along the cross section of the specimen, Fig. 12.5b.

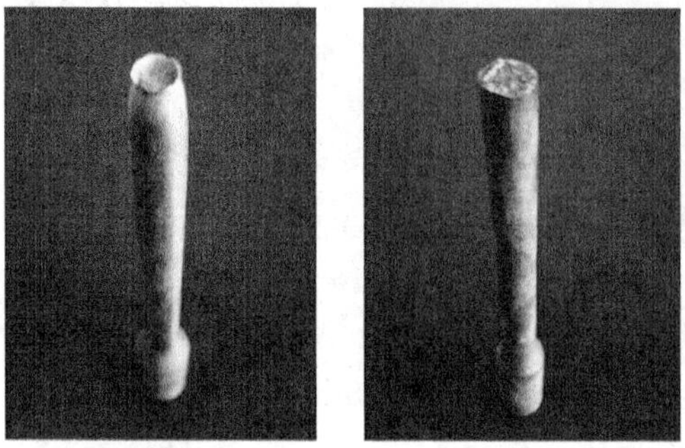

(a)　　(b)

Fig. 12.5

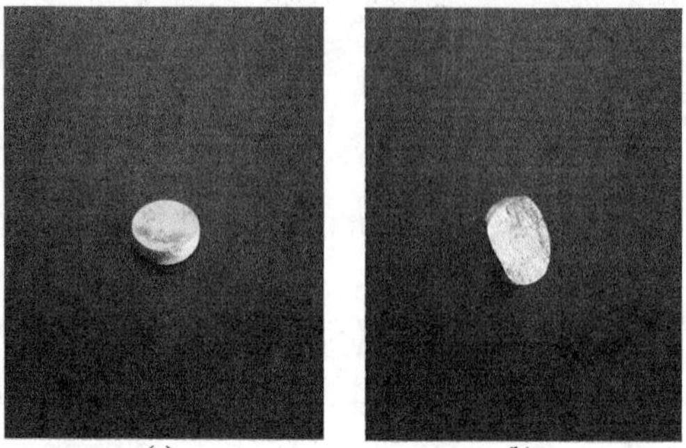

(a)　　(b)

Fig. 12.6

Ductile materials are weaker in shear than in compression. Therefore, when subjected to compression, a specimen made of a ductile material will fail along a plane forming a 45° angle with the cross section of the specimen, Fig. 12.6a. Brittle materials are also weaker in shear than in compression. Thus when subjected to compression, a specimen made of a brittle material will break along a plane forming a 45~55° angle with the cross section of the specimen, Fig. 12.6b.

Example 12.2 Two wooden members having a $90\ \text{mm} \times 120\ \text{mm}$ uniform rectangular cross section are joined by the glued splice shown. Knowing that $F = 100\ \text{kN}$, determine the normal and shearing stresses in the glued splice.

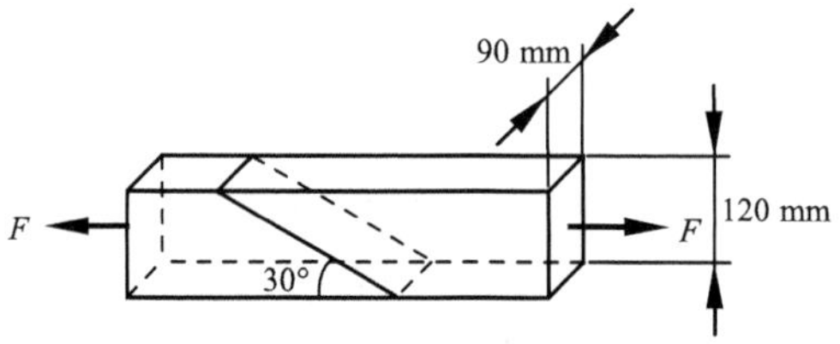

Fig. E12.2

Solution Using Eq. (12.4), we have

$$\sigma_{60^\circ} = \sigma \cos^2 60^\circ = 2.31\ \text{MPa}, \quad \tau_{60^\circ} = \frac{\sigma}{2}\sin(2 \times 60^\circ) = 4.01\ \text{MPa}$$

12.4　Normal Strain

1. Longitudinal Normal Strain

Consider a prismatic bar, of length l and cross-sectional area A, which is suspended from the upper end, Fig. 12.7a. If we apply an external load F to the lower end, the bar will elongate, Fig. 12.7b.

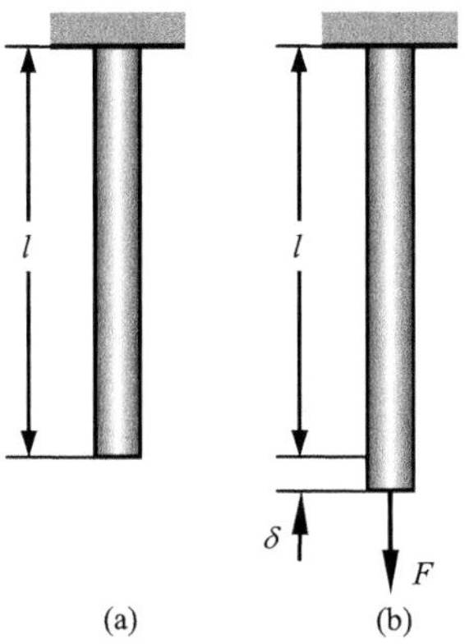

Fig. 12.7

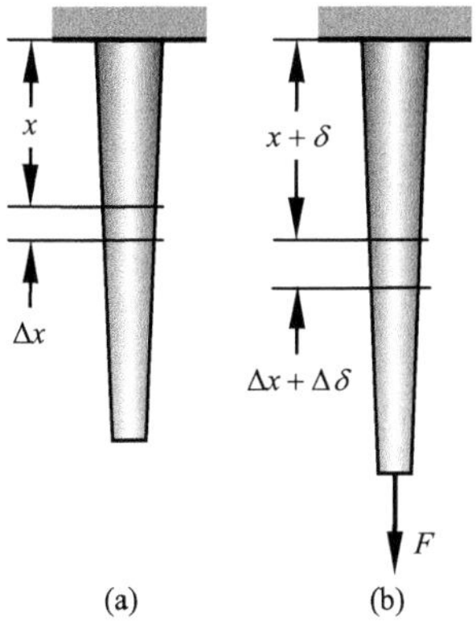

Fig. 12.8

The longitudinal normal strain of the bar under axial loading can be expressed as

$$\varepsilon = \frac{\delta}{l} \tag{12.5}$$

where δ is the longitudinal deformation of the bar, and l is the length of the bar.

In the case of a bar of variable cross-sectional area, the normal stress on the cross section varies along the bar, and it is necessary to define the longitudinal normal strain at a given point by considering a small element of undeformed length Δx, Fig. 12.8a. Denoting by $\Delta\delta$ the deformation of the element under the given loading, Fig. 12.8b, the longitudinal normal strain at the given point located on the x-section can be written as

$$\varepsilon = \lim_{\Delta x \to 0} \frac{\Delta\delta}{\Delta x} = \frac{\mathrm{d}\delta}{\mathrm{d}x} \tag{12.6}$$

2. Lateral Normal Strain

When a bar is axially loaded, the normal stress on the cross section and longitudinal normal strain satisfy Hooke's law, as long as the proportional limit of the material is not exceeded. And all materials considered will be assumed to be both homogeneous and isotropic, i.e., their mechanical properties will be assumed independent of both position and direction. Therefore, for the loading shown in Fig. 12.9 we have $\varepsilon_y = \varepsilon_z$, where ε_y and ε_z are referred to as the lateral strains respectively along the y and z axes.

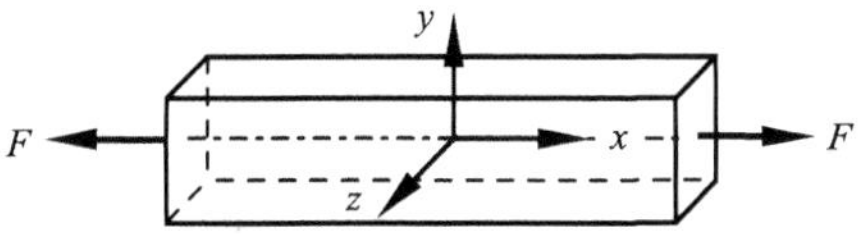

Fig. 12.9

An important constant for given material is its Poisson's ratio, denoted by μ. It is defined as

$$\mu = -\frac{\varepsilon_y}{\varepsilon_x} = -\frac{\varepsilon_z}{\varepsilon_x} \tag{12.7}$$

Therefore, we have the lateral normal strains:

$$\varepsilon_y = \varepsilon_z = -\mu\varepsilon_x \tag{12.8}$$

12.5 Axial Deformation

Consider a prismatic bar of length l and cross-sectional area A subjected to a centric axial load F, Fig. 12.10. If the normal stress on the cross section does not exceed the proportional limit of the material, we can apply Hooke's law, i.e., $\sigma = E\varepsilon$. Using $\sigma = N/A$ and $\varepsilon = \delta/l$, we have

$$\delta = \frac{Nl}{EA} \tag{12.9}$$

where EA is the axial (tensile or compressive) rigidity of the bar. The above formula can be used only if the bar is homogeneous and isotropic, has a uniform cross section, and is subjected to centric axial loads at its ends.

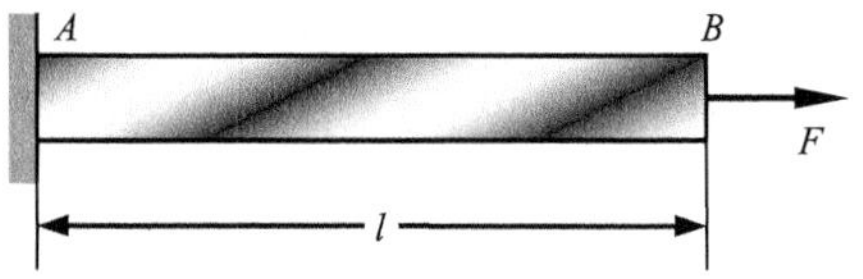

Fig. 12.10

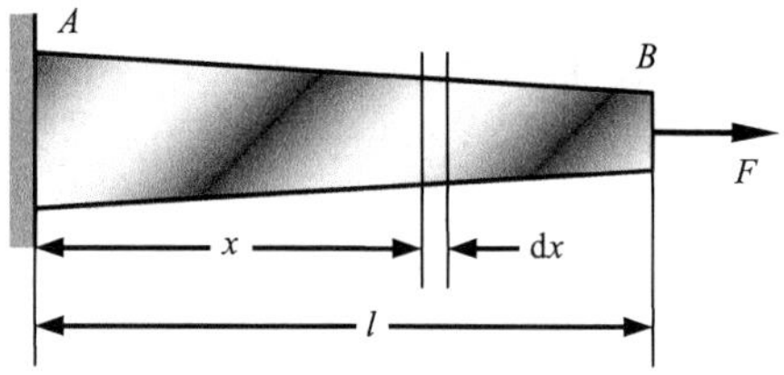

Fig. 12.11

In the case of a bar having variable cross section, Fig. 12.11, we can express the deformation of an element of length $\mathrm{d}x$ as

$$\mathrm{d}\delta = \frac{N\mathrm{d}x}{EA} \tag{12.10}$$

Thus the total deformation δ of the bar is obtained by integrating the above expression over the length l of the bar:

$$\delta = \int \frac{N\mathrm{d}x}{EA} \tag{12.11}$$

Example 12.3 The 4-mm-diameter cable BC is made of a steel with E=200 GPa. Knowing that the maximum stress in the cable must not exceed 190 MPa and that the elongation of the cable must not exceed 6 mm, find the maximum load F that can be applied as shown in Fig. E12.3a.

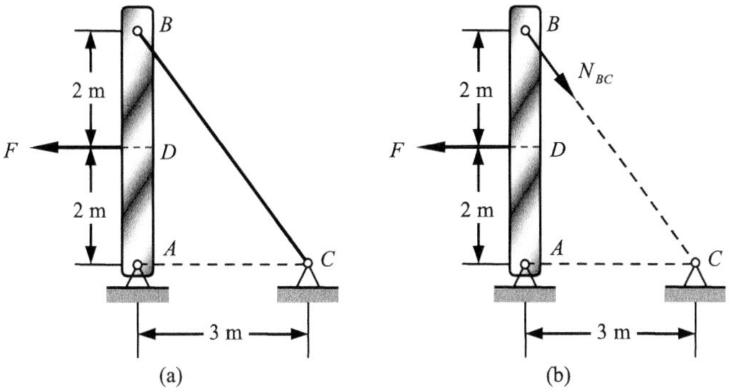

Fig. E12.3

Solution Using AB as a free body and considering its equilibrium, Fig. E12.3b, then we have

$$\sum M_A = 0: \ Fl_{AD} - (N_{BC}\frac{l_{AC}}{l_{BC}})l_{AB} = 0 \text{, i.e., } \ F = 1.2N_{BC}$$

Determining the maximum allowable load based on the stress in the cable, we have

$$F_{\max}^{\sigma} = 1.2N_{BC} = 1.2\sigma_{\max}A_{BC} = 1.2\sigma_{\max}(\frac{1}{4}\pi d_{BC}^2) = 2.87 \text{ kN}$$

Determining the maximum allowable load based on the deformation of the cable, we have

$$F_{\max}^{\delta} = 1.2N_{BC} = 1.2\frac{\delta EA_{BC}}{l_{BC}} = 1.2\frac{\delta E(\frac{1}{4}\pi d_{BC}^2)}{l_{BC}} = 3.62 \text{ kN}$$

The smaller value of the two maximum loads should be chosen, that is, the maximum allowable load $F_{\max}$ that can be applied to the structure is equal to

$$F_{\max} = \min[F_{\max}^{\sigma}, F_{\max}^{\delta}] = 2.87 \text{ kN}$$

12.6 Statically Indeterminate Axially-Loaded Bar

In the problems considered in the preceding sections, we could use the equations of static equilibrium to determine the internal forces produced in the various portions of a bar under given loads. There are some problems encountered in engineering, however, in which the internal forces cannot be determined from the equations of static equilibrium alone. For these problems, the equations of static equilibrium must be complemented with additional equations based on the deformation of the bar. Because the equations of static equilibrium are not sufficient to determine either the reactions or the internal forces, problems of this type are said to be statically indeterminate.

Consider a prismatic bar AB, of length l, cross-sectional area A, and modulus of elasticity

E, which is attached to rigid supports at A and B before being loaded, Fig. 12.12a.

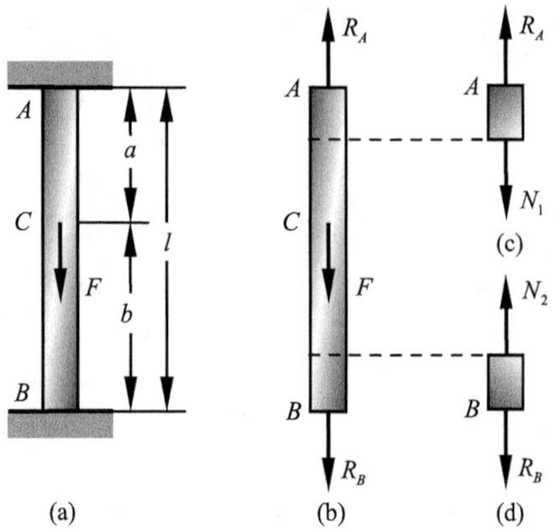

Fig. 12.12

Drawing the free-body diagram of the bar, Fig. 12.12b, we obtain the equilibrium equation

$$\sum F_y = 0: \quad R_A - R_B - F = 0 \tag{12.12}$$

Since this equation is not sufficient to determine the two unknown reactions R_A and R_B, the problem is statically indeterminate to the first degree.

We observe from the geometry that the total deformation δ of the bar must be zero. Denoting by δ_1 and δ_2, respectively, the deformations of the portions AC and BC, we have

$$\delta = \delta_1 + \delta_2 = 0 \tag{12.13}$$

Expressing δ_1 and δ_2 in terms of the corresponding internal forces N_1 and N_2, we have

$$\delta_1 = \frac{N_1 a}{EA}, \quad \delta_2 = \frac{N_2 b}{EA} \tag{12.14}$$

We note from the free-body diagrams shown in Figs. 12.12c and 12.12d that $N_1 = R_A$ and $N_2 = R_B$. Carrying these values into Eq. (12.14), and using Eq. (12.13), we have

$$R_A a + R_B b = 0 \tag{12.15}$$

Eqs. (12.12) and (12.15) can be solved simultaneously for R_A and R_B, we obtain

$$R_A = \frac{Fb}{l}, \quad R_B = -\frac{Fa}{l} \tag{12.16}$$

12.7 Design of Axially-Loaded Bar

In engineering applications, stresses are used in the design of structures that will safely and economically perform a specified function.

The largest force which can be applied to a bar is called the ultimate load for this bar and is denoted by F_u. Since the applied load is centric, we may divide the ultimate load by the original cross-sectional area of the bar to obtain the ultimate stress of the bar. This stress, also known as the ultimate strength, can be written as

$$\sigma_u = \frac{F_u}{A} \tag{12.17}$$

The working load that a bar will be allowed to carry under normal conditions of utilization is smaller than the ultimate load. This maximum working load which can be applied to a bar is referred to as the allowable load, denoted by F_{allow}. Thus, only a fraction of the ultimate-load capacity of the bar is utilized when the allowable load is applied. The remaining portion of the load-carrying capacity of the bar is kept in reserve to assure its safe performance. The ratio of the ultimate load to the allowable load is used to define the factor of safety which can expressed as

$$n = \frac{F_u}{F_{allow}} \tag{12.18}$$

or, an alternative definition of the factor of safety is based on the use of stresses:

$$n = \frac{\sigma_u}{\sigma_{allow}} \tag{12.19}$$

The selection of the factor of safety is one of the most important engineering tasks. On one hand, if a factor of safety is chosen too small, the possibility of failure becomes unacceptably large; on the other hand, if a factor of safety is chosen unacceptably large, the result is an uneconomical design. The choice of the factor of safety that is appropriate for a given design requires engineering judgment based on many considerations.

Problems

12.1 Two solid cylindrical rods *AB* and *BC* are welded together at *B* and loaded as shown. Knowing that d_1=50 mm and d_2=30 mm, find the normal stress in the midsection of (a) rod *AB*, (b) rod *BC*.

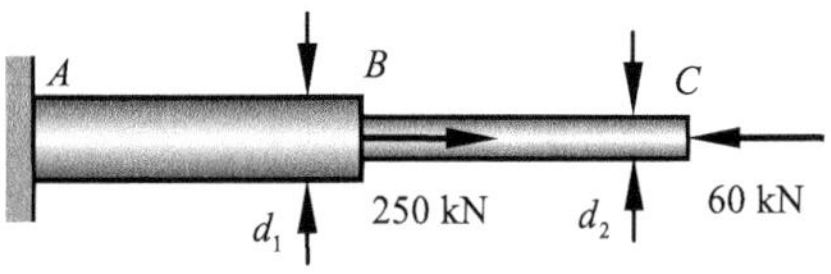

Fig. P12.1

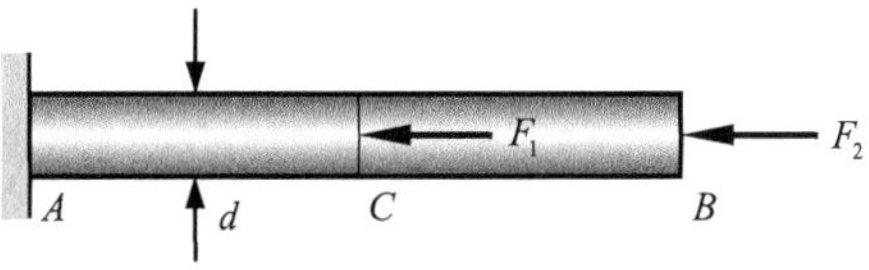

Fig. P12.2

12.2 A solid cylindrical rod *AB* of diameter *d* is subjected to two axial loads, F_1 at *C* and F_2 at *B*. Knowing that *d*=50 mm, F_1=200 kN and F_2=50 kN, determine the normal stresses in portions (a) *AC*, and (b) *BC*.

12.3 Link *AB*, of width *b*=50 mm, and thickness *t*=6 mm, is used to support the end of a horizontal beam. Knowing that the average normal stress in the link is 138 MPa and that the average shearing stress in the pin is 82 MPa, determine (a) the diameter d of the pin, (b) the average bearing stress in the link.

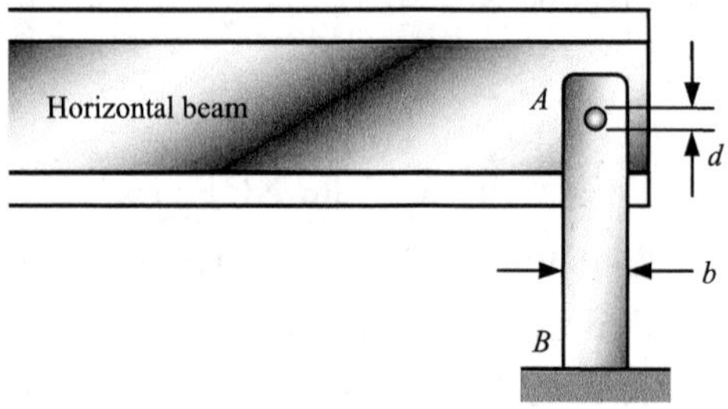

Fig. P12.3

12.4 Each of the two vertical links has a 16 mm×26 mm uniform rectangular cross section and each of the three pins has a 6 mm diameter. Determine (a) the average shearing stress in the pin at B, (b) the average bearing stress at B in link BD, (c) the average bearing stress at B in beam ABC, knowing that this beam has a 10 mm×50 mm uniform rectangular cross section.

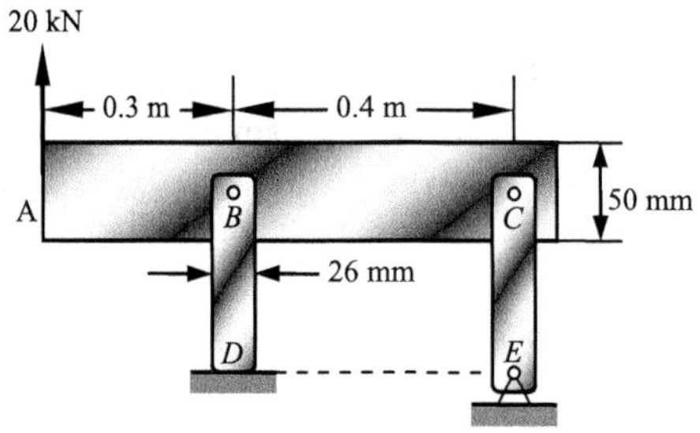

Fig. P12.4

12.5 An axial load is applied at end C of the steel rod ABC. Knowing that E=200 GPa, determine the diameter d_2 of portion BC for which the deflection of point C will be 3 mm.

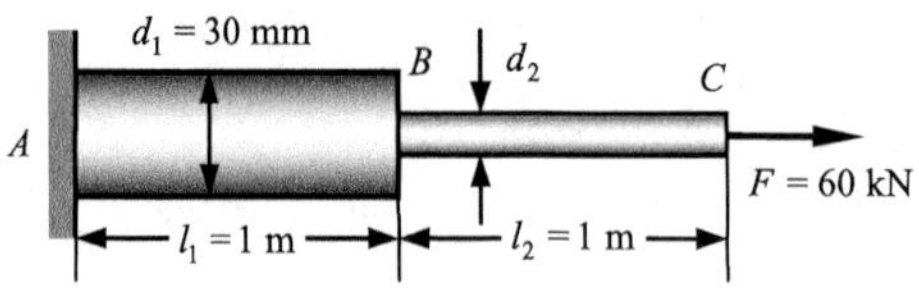

Fig. P12.5

12.6 Considering a rod AB as shown, made of steel (E=200 GPa) and subjected to an axial load F. If the stress in the rod must not exceed 120 MPa and the maximum change in length of the rod must not exceed 0.001 times the length of the rod, determine the smallest diameter of the rod that can be used.

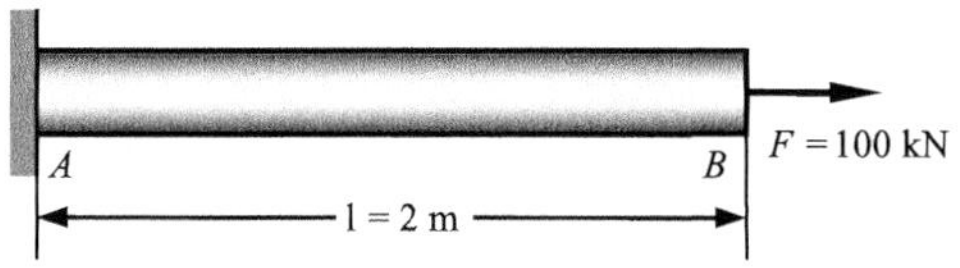

Fig. P12.6

12.7 Both portions of the rod *ABC* are made of a brass for which E=105 GPa. Knowing that the diameter of portion *BC* is d_2=10 mm, determine the largest load F that can be applied if σ_{allow}=100 MPa and the corresponding deflection at point *C* is not to exceed 4 mm.

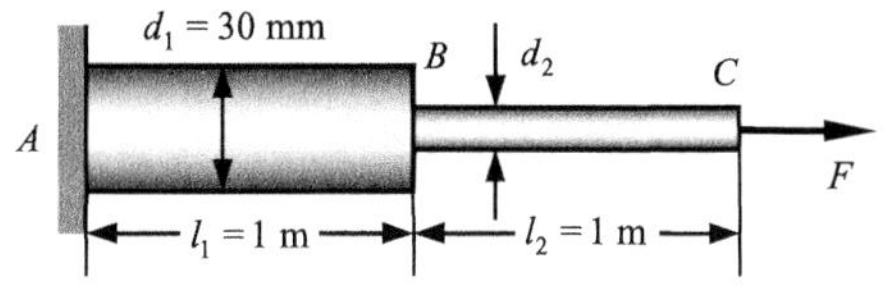

Fig. P12.7

12.8 Each of the links *AB* and *CD* is made of aluminum (E=75 GPa) and has a cross-sectional area of 125 mm^2. Knowing that they are used to support a rigid beam *AC*, determine the deflection of point *E*.

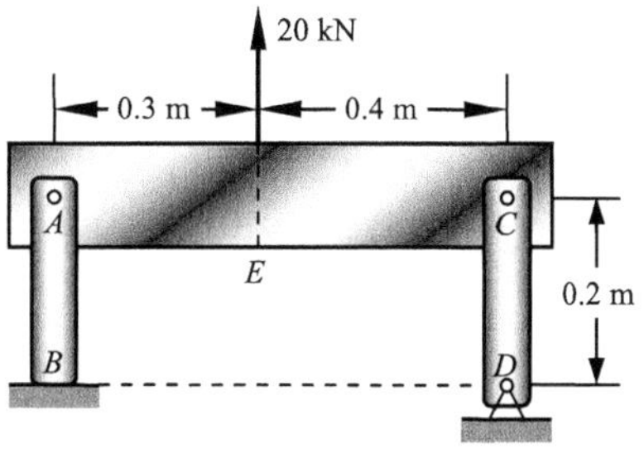

Fig. P12.8

12.9 The rigid bar *AD* is supported by two steel wires of 1.5 mm diameter (E=200 GPa) and a pin support at *A*. Knowing that the wires were initially taut, determine (a) the additional tension in each wire when a 1.0 kN load F is applied at *D*, (b) the corresponding deflection of point *D*.

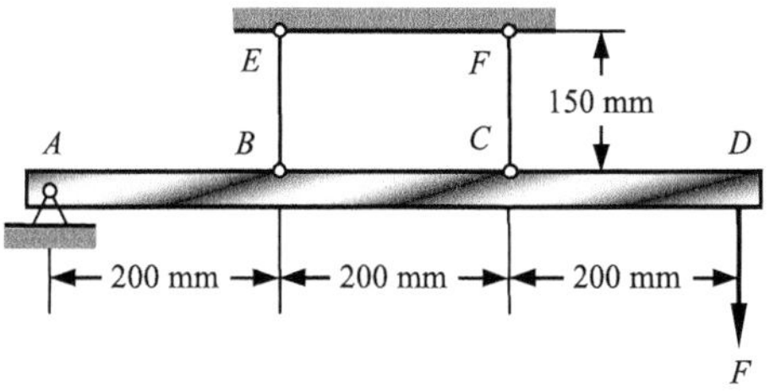

Fig. P12.9

12.10 The rigid bar AD is supported by two steel wires of 1.5 mm diameter (E= 200 GPa) and a pin and bracket at A. Knowing that the wires were initially taut, determine (a) the additional tension in each wire when a 1.0 kN load F is applied at D, (b) the corresponding deflection of point D.

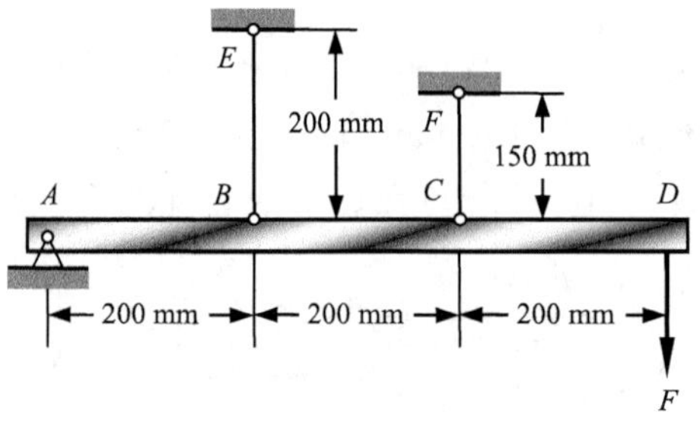

Fig. P12.10

Chapter 13 Torsion

Consider a bar of circular cross section, loaded at its ends by a pair of equal and opposite external couples of magnitude M_e applied in two planes perpendicular to the longitudinal axis of the bar as shown in Fig. 13.1. Such a bar is said to be a shaft in torsion.

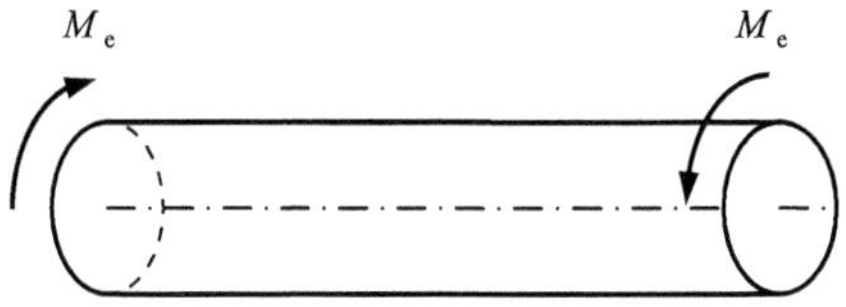

Fig. 13.1

13.1 Torsional Moment

Under the action of a pair of external couples, Fig. 13.1, a distributed internal force will appear on the cross section of the shaft. The resultant couple of this distributed internal force about the longitudinal axis of the shaft is called the torsional moment (twisting moment or torque) of the shaft, denoted by T, Fig. 13.2.

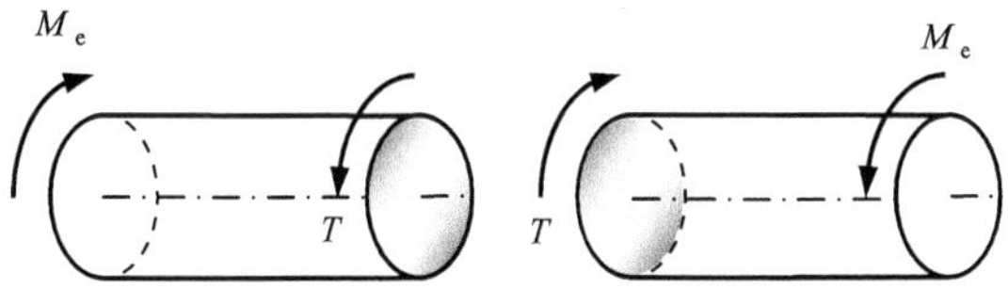

Fig. 13.2

13.2 Hooke's Law in Shear

Consider an element in pure shear, Fig. 13.3a, subjected to shearing stresses τ_{xy} and τ_{yx} applied to four faces of the element respectively perpendicular to the x and y axes. According to the equilibrium of the element, we can obtain

$$\tau_{xy} = \tau_{yx} \tag{13.1}$$

This relation is called the reciprocal theorem of shearing stresses.

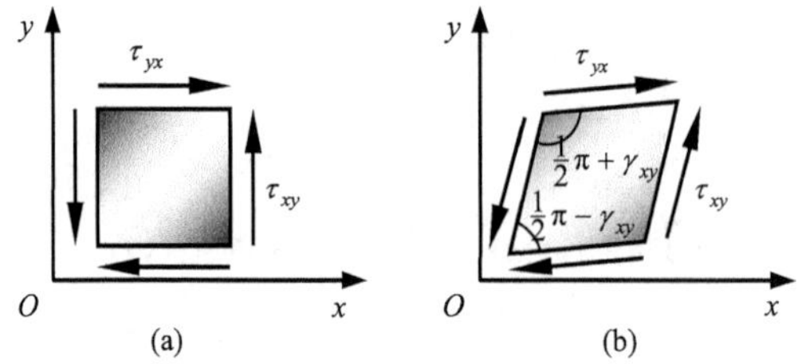

Fig. 13.3

The element is observed to deform into a rhomboid, Fig. 13.3b, under pure shearing stresses. Two of the angles formed by four faces under the shearing stresses are reduced from $\frac{1}{2}\pi$ to $\frac{1}{2}\pi-\gamma_{xy}$, while the other two are increased from $\frac{1}{2}\pi$ to $\frac{1}{2}\pi+\gamma_{xy}$, where γ_{xy} is the shearing strain. For the shearing stress which does not exceed the proportional limit in shear, we have for any homogeneous isotropic material

$$\tau_{xy}=G\gamma_{xy} \tag{13.2}$$

This relation is known as Hooke's law in shear, and *G*, expressed in the same units as τ_{xy}, is called the shear modulus of elasticity of the material.

The moduli of elasticity in tension and shear for a homogeneous isotropic material are related by the following equation:

$$G=\frac{E}{2(1+\mu)} \tag{13.3}$$

where μ is Poisson's ratio.

13.3 Shearing Stress on Cross Section

1. Equilibrium Condition

Considering a circular shaft *AB* subjected to torsion, we cut the shaft through a section perpendicular to the longitudinal axis of the shaft at some arbitrary point *C*, Fig. 13.4a. The section *C* of the shaft must include the elementary shearing forces d*F*, Fig. 13.4b, perpendicular to the radius of the shaft, that portion *BC* exerts on portion *AC* as the shaft is twisted.

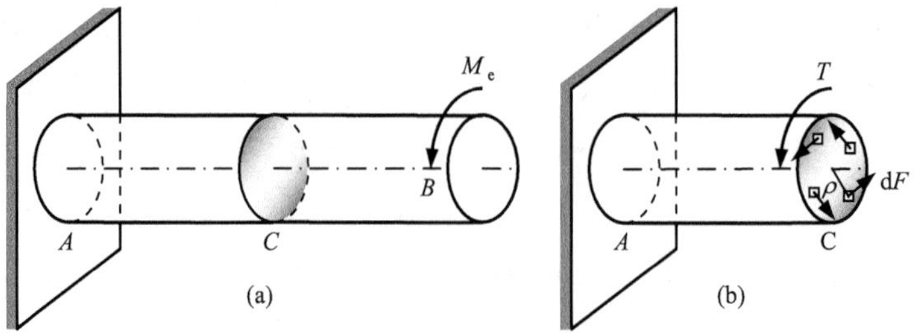

Fig. 13.4

The condition of static equilibrium requires that resultant couple of these elementary forces about the longitudinal centroidal axis of the shaft be equivalent to the torsional moment T on the cross section of the shaft. Denoting by ρ the perpendicular distance from the force $\mathrm{d}F$ to the longitudinal centroidal axis of the shaft, then we have

$$T = \int \rho \mathrm{d}F = \int \rho\tau \mathrm{d}A \tag{13.4}$$

where τ is the shearing stress on the element of area $\mathrm{d}A$.

2. Deformation Condition

When the circular shaft AB is subjected to an external couple M_e, Fig. 13.5a, the shaft will twist and every cross section remains plane and undistorted. Consider a segment CD with length $\mathrm{d}x$, which has been twisted through a relative angle of twist $\mathrm{d}\varphi$, Fig. 13.5b.

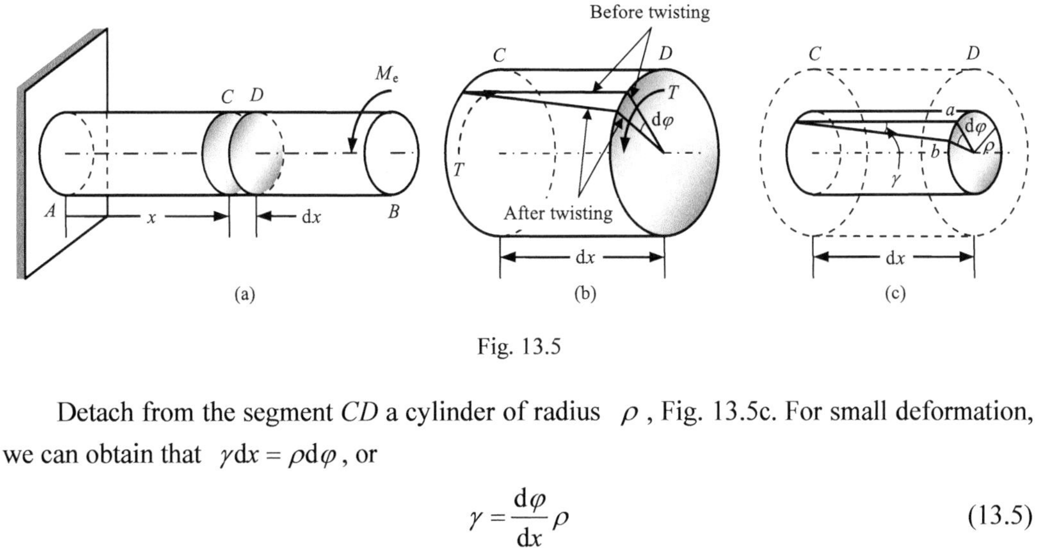

Fig. 13.5

Detach from the segment CD a cylinder of radius ρ, Fig. 13.5c. For small deformation, we can obtain that $\gamma \mathrm{d}x = \rho \mathrm{d}\varphi$, or

$$\gamma = \frac{\mathrm{d}\varphi}{\mathrm{d}x}\rho \tag{13.5}$$

where γ and $\mathrm{d}\varphi$ are expressed in radians. The equation obtained shows that the shearing strain in the circular shaft varies linearly with the distance ρ from the longitudinal centroidal axis of the shaft.

3. Stress-Strain Relation

Assume that the stresses in the shaft remain below the proportional limit. From Hooke's law in shear, $\tau = G\gamma$, we have

$$\tau = G\frac{\mathrm{d}\varphi}{\mathrm{d}x}\rho \tag{13.6}$$

where G is the modulus of elasticity in shear. The equation obtained shows that, as long as the proportional limit is not exceeded in any point of a circular shaft, the shearing stress in the circular shaft varies linearly with the distance ρ from the longitudinal centroidal axis of the shaft. Fig. 13.6a shows the shearing stress distribution in a solid circular shaft and Fig. 13.6b in a hollow circular shaft.

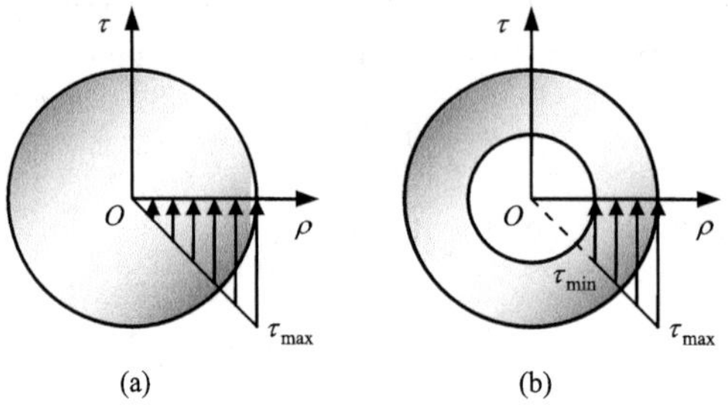

Fig. 13.6

Substituting τ from Eq. (13.6) into Eq. (13.4), we obtain

$$T = \int \rho\tau \mathrm{d}A = G\frac{\mathrm{d}\varphi}{\mathrm{d}x}\int \rho^2 \mathrm{d}A = GI_{\mathrm{p}}\frac{\mathrm{d}\varphi}{\mathrm{d}x} \tag{13.7}$$

where $I_{\mathrm{p}} = \int \rho^2 \mathrm{d}A$ represents the polar moment of inertia of the cross section with respect to its centroid O. For a solid circular shaft of diameter d, $I_{\mathrm{p}} = \frac{1}{32}\pi d^4$; and for a hollow circular shaft of inner diameter d and outer diameter D, $I_{\mathrm{p}} = \frac{1}{32}\pi(D^4 - d^4) = \frac{1}{32}\pi D^4(1-\alpha^4)$, where $\alpha = d/D$.

Using Eqs. (13.6) and (13.7), we have

$$\tau = \frac{T\rho}{I_{\mathrm{p}}} \tag{13.8}$$

It can be seen from the above equation that the maximum shearing stress can be written as

$$\tau_{\max} = \frac{T_{\max}\rho_{\max}}{I_{\mathrm{p}}} \tag{13.9}$$

We note that the ratio $I_{\mathrm{p}}/\rho_{\max}$ depends only upon the geometry of the cross section. This ratio is called the polar modulus of section and is denoted by S_{p}. From Eq. (13.9), we write

$$\tau_{\max} = \frac{T_{\max}}{S_{\mathrm{p}}} \tag{13.10}$$

For a solid circular shaft of diameter d, $S_{\mathrm{p}} = \frac{1}{16}\pi d^3$; and for a hollow circular shaft of inner diameter d and outer diameter D, $S_{\mathrm{p}} = \frac{1}{16}\pi D^3(1-\alpha^4)$, where $\alpha = d/D$.

Example 13.1 Two external couples M_B and M_C are applied to the sections B and C of a variable-section solid shaft ABC, Fig. E13.1. Knowing that D=46 mm, d=30 mm, M_B=400 N·m, and M_C=300 N·m, determine the maximum shearing stress (a) in shaft AB, (b) in shaft *BC*.

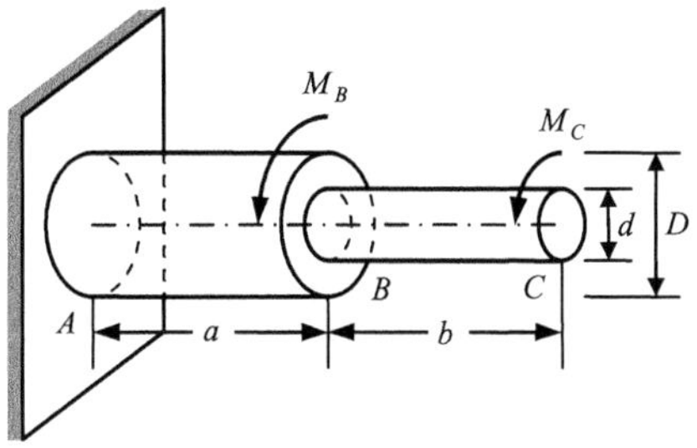

Fig. E13.1

Solution Using the method of sections, we obtain the torsional moments

$$T_{AB} = M_B + M_C = 700\ \text{N}\cdot\text{m}\text{, and } T_{BC} = M_C = 300\ \text{N}\cdot\text{m}$$

For shaft *AB*, from Eq. (13.10), we obtain

$$(\tau_{\max})_{AB} = \frac{T_{AB}}{(S_p)_{AB}} = \frac{T_{AB}}{\frac{1}{16}\pi D^3} = 36.6\ \text{MPa}$$

Similarly, for shaft *BC*, we have

$$(\tau_{\max})_{BC} = \frac{T_{BC}}{(S_p)_{BC}} = \frac{T_{BC}}{\frac{1}{16}\pi d^3} = 56.6\ \text{MPa}$$

13.4 Normal and Shearing Stresses on Oblique Section

Up to this time, the analysis of stresses in a shaft has been limited to shearing stresses on the cross section. We will now consider the stresses on an oblique section. We choose three elements *a*, *b* and *c* located on the front surface of the circular shaft subjected to torsion, Fig. 13.7. Since the faces of element *a* are respectively parallel and perpendicular to the longitudinal axis of the shaft, the only stress on the element will be the shearing stress and the shearing stress is maximum. The faces of element *b*, which form an arbitrary angle with the longitudinal axis of the shaft, will be subjected to a combination of normal and shearing stresses. The element *c* at 45° to the longitudinal axis of the shaft, hence the only stress on element *c* will be the normal stress.

Ductile materials are weak in shear. Therefore, when subjected to torsion, a specimen made of a ductile material breaks along the cross section of the specimen, Fig. 13.8a. Brittle materials are weaker in tension than in shear. Thus when subjected to torsion, a specimen made of a brittle material will break along the plane perpendicular to the direction of the maximum tensile stress, i.e., along the plane forming a 45° angle with the cross section of the specimen, Fig. 13.8b.

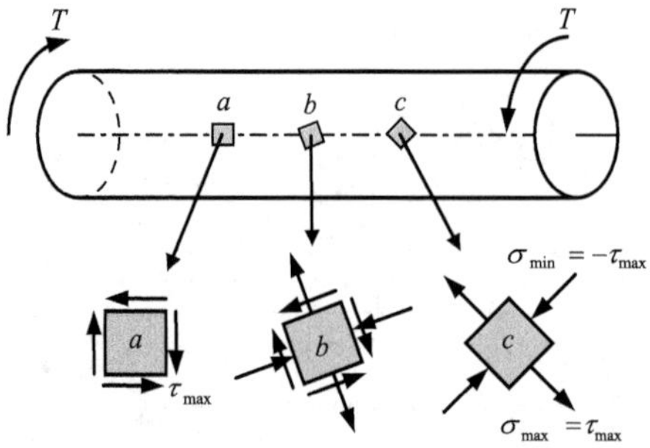

Fig. 13.7

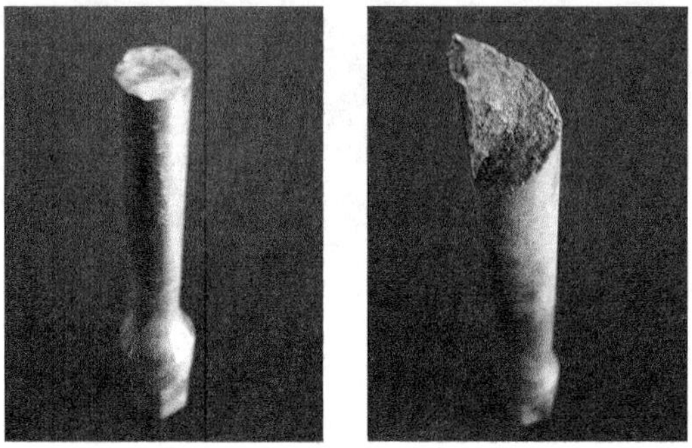

(a) (b)

Fig. 13.8

13.5 Angle of Twist

Considering a circular shaft AB subjected to an external couple M_e at its free end, Fig. 13.9, and assuming that the corresponding torsional moment produced in the shaft is denoted by T and that the entire shaft remains linearly elastic, we recall from Eq. (13.7) that the angle of twist $d\varphi$ of the segment CD can be written as

$$d\varphi = \frac{Tdx}{GI_p} \tag{13.11}$$

where GI_p is the torsional rigidity of the shaft. The above equation shows that, within the linearly elastic range, the angle of twist $d\varphi$ is proportional to the torsional moment T in the shaft. Integrating in x from 0 to l, we obtain the angle of twist of the shaft

$$\varphi = \int \frac{Tdx}{GI_p} \tag{13.12}$$

For a circular shaft of a uniform cross section subjected to a constant torsional moment in the shaft, then Eq. (13.12) can be simplified as

$$\varphi = \frac{Tl}{GI_p} \tag{13.13}$$

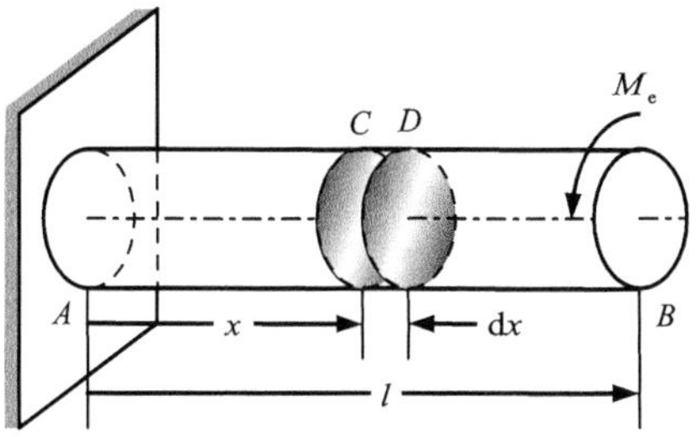

Fig. 13.9

Example 13.2 Two external couples M_B and M_C are applied to the sections B and C of a variable-section solid shaft ABC, made of aluminum (G=77 GPa). Knowing that D=46 mm, d=30 mm, a=750 mm, b=900 mm, M_B=400 N·m, and M_C=300 N·m, determine the angle of twist (a) between the sections B and C, (b) of the section C, Fig. E13.2.

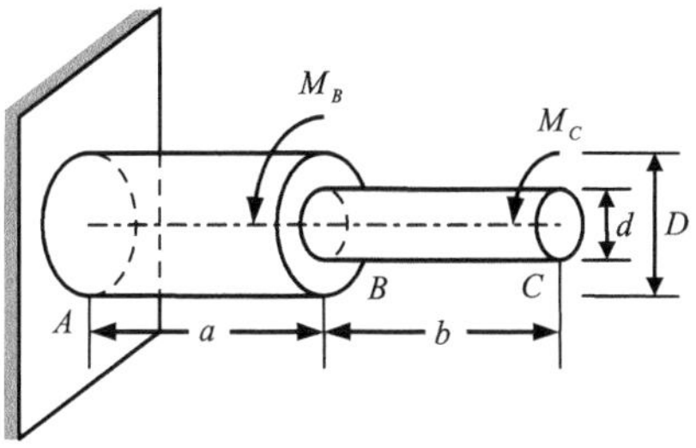

Fig. E13.2

Solution Using Eq. (13.13), we obtain the angle of twist of section C relative to section B

$$\varphi_{C/B}=\frac{T_{BC}l_{BC}}{G(I_{\mathrm{p}})_{BC}}=\frac{T_{BC}l_{BC}}{G\left(\dfrac{1}{32}\pi d^4\right)}=4.41\times10^{-2}\ \mathrm{rad}=2.53^\circ$$

and the absolute angle of twist of section C

$$\varphi_C=\varphi_{C/A}=\varphi_{C/B}+\varphi_{B/A}=\frac{T_{BC}l_{BC}}{G\left(\dfrac{1}{32}\pi d^4\right)}+\frac{T_{AB}l_{AB}}{G\left(\dfrac{1}{32}\pi d^4\right)}=5.96\times10^{-2}\ \mathrm{rad}=3.41^\circ$$

13.6 Statically Indeterminate Shaft

Consider now a circular shaft AB, which is fixed at ends A and B and subjected to an external couple M_e at the section C, Fig. 13.10a. We note that the reactions involve two unknowns, while only one equilibrium condition is available. Therefore, the circular shaft is statically indeterminate to the first degree.

We consider the reaction at the support B as redundant and eliminate the corresponding support B, Fig. 13.10b. The redundant reaction is then treated as an unknown load M_B that, together with the other load M_e, must produce deformation which is compatible with the original support B. The redundant reaction M_B will be determined from the condition that the angle of twist at B of the circular shaft must be zero.

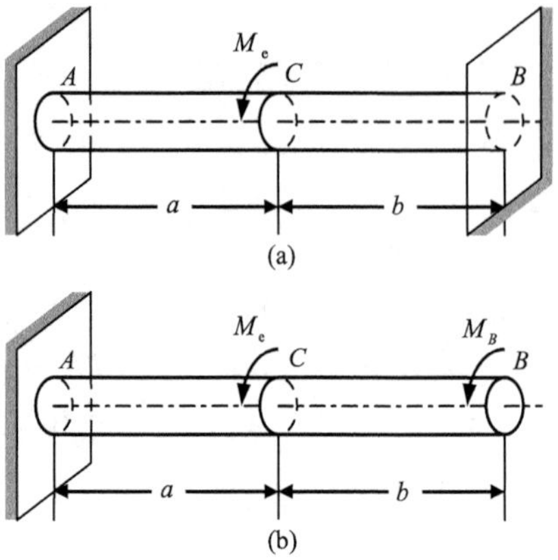

Fig. 13.10

The solution is carried out by considering separately the angle of twist $(\varphi_B)_{M_B}$ caused at B by the redundant reaction M_B, and the angle of twist $(\varphi_B)_{M_e}$ produced at the same section by the external couple M_e. From Eq. (13.13), we find that $(\varphi_B)_{M_B} = \dfrac{M_B(a+b)}{GI_p}$ and $(\varphi_B)_{M_e} = (\varphi_C)_{M_e} = \dfrac{M_e a}{GI_p}$. The angle of twist at B is the sum of these two quantities and must be zero, i.e., $\varphi_B = (\varphi_B)_{M_B} + (\varphi_B)_{M_e} = 0$, or $\dfrac{M_B(a+b)}{GI_p} + \dfrac{M_e a}{GI_p} = 0$. Solving for M_B, we have $M_B = -\dfrac{M_e a}{a+b}$.

13.7 Design of Torsional Shaft

The principal specifications to be met in the design of a shaft are the power to be transmitted and the speed of rotation of the shaft. We need to select the material and the dimensions of the cross section of the shaft, so that the allowable shearing stress will not be exceeded when the shaft is transmitting the required power at the specified speed.

The torsional moment T produced in the rotating shaft to transmit a power P (kW) at a rotating speed ω (rad/s) or n (r/min, revolutions per minute) can be written as

$$T(\text{N}\cdot\text{m}) = \frac{P(\text{W})}{\omega(\text{rad/s})}, \text{ or } T(\text{N}\cdot\text{m}) = 9549\frac{P(\text{kW})}{n(\text{r/min})} \tag{13.14}$$

After having determined the torsional moment T that will be applied to the shaft and having selected the material to be used, we will carry the allowable shearing stress into Eq. (13.9) and the allowable angle of twist into Eq. (13.12). We have

$$\tau_{\max} = \frac{T_{\max}\rho_{\max}}{I_p} \leqslant \tau_{\text{allow}}, \quad \varphi = \int \frac{T\mathrm{d}x}{GI_p} \leqslant \varphi_{\text{allow}} \tag{13.15}$$

where τ_{allow} is the allowable shearing stress, and φ_{allow} is the allowable angle of twist.

Example 13.3 A 1.8 m long tubular shaft of 45 mm outer diameter having a hollow circular section shown is to be made of steel for which τ_{allow}=75 MPa and G=77 GPa. Knowing that the angle of twist of the shaft must not exceed 5° when the shaft is subjected to an external couple of M_e=900 N·m, determine the largest inner diameter d which can be used in the design.

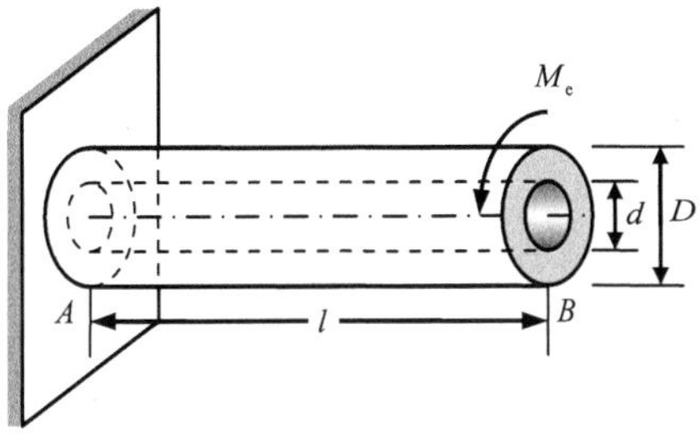

Fig. E13.3

Solution Using $\tau_{\max} = \frac{T\rho_{\max}}{I_p} \leqslant \tau_{\text{allow}}$, we have

$$(I_p)_\tau \geqslant \frac{T\rho_{\max}}{\tau_{\text{allow}}} = 270 \times 10^{-9}\ \text{m}^4$$

Using $\varphi = \frac{Tl}{GI_p} \leqslant \varphi_{\text{allow}}$, we have

$$(I_p)_\varphi \geqslant \frac{Tl}{G\varphi_{\text{allow}}} = 241 \times 10^{-9}\ \text{m}^4$$

Larger value of I_p should be used in the design, thus we have

$$I_p = \max[(I_p)_\tau,\ (I_p)_\varphi] \geqslant 270 \times 10^{-9}\ \text{m}^4$$

For a hollow circular shaft of inner diameter d and outer diameter D, using $I_p = \frac{1}{32}\pi(D^4 - d^4)$, we have

$$d \leqslant \sqrt[4]{D^4 - \frac{32I_p}{\pi}} = 34\ \text{mm}$$

Therefore, the largest inner diameter d which can be used in the design is equal to 34 mm.

Problems

13.1 Knowing that d=30 mm and D=40 mm, determine the external couple M_e which causes a maximum shearing stress of 52 MPa in the hollow shaft.

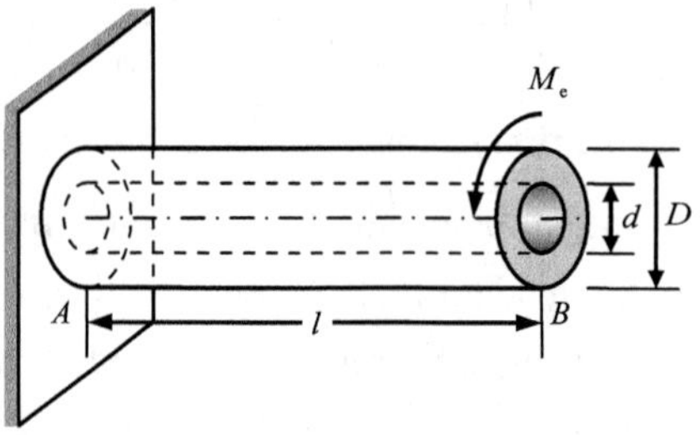

Fig. P13.1

13.2 Two external couples M_B and M_C are applied to the sections B and C of a variable-section solid shaft ABC. Knowing that D=46 mm, d=30 mm, M_B=400 N·m, and M_C=300 N·m, in order to reduce the total mass of the shaft ABC, determine the smallest diameter of shaft AB for which the largest shearing stress in the shaft is not increased.

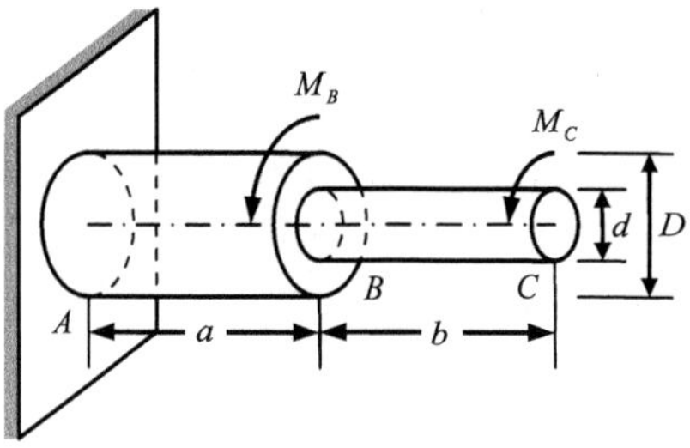

Fig. P13.2

13.3 The pipe AB has an outer diameter of 90 mm and a wall thickness of 6 mm. The solid rod BC has a diameter of 60 mm. Knowing that both the pipe and the rod are made of steel for which the allowable shearing stress is 95 MPa, determine the largest external torque M_e which may be applied at C.

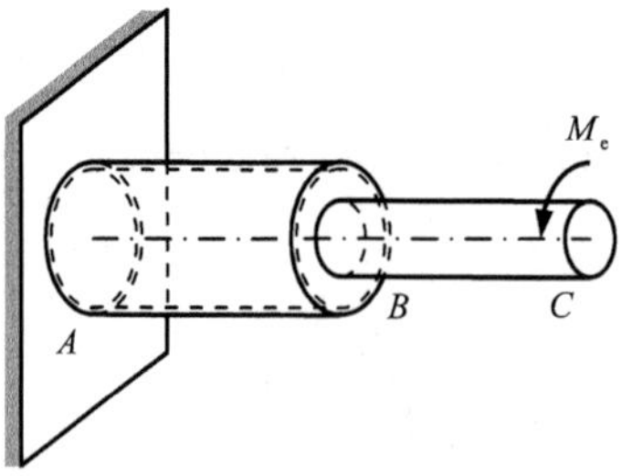

Fig. P13.3

13.4 The allowable stress is 25 MPa in the aluminum rod AB and 50 MPa in the brass rod BC. Knowing that M_e=1.5 kN·m, determine the required diameter of (a) rod AB, and (b) rod *BC*.

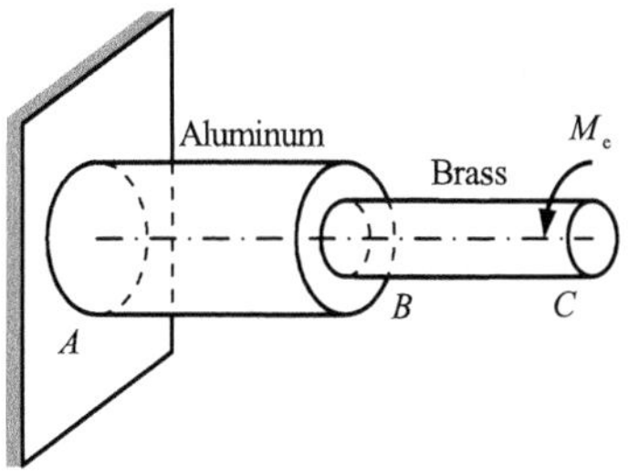

Fig. P13.4

13.5 For the aluminum shaft shown, knowing that G=27 GPa, l=1.1 m, d=10 mm, and D=20 mm, determine (a) the external couple M_e which causes an angle of twist of 5°, (b) the angle of twist caused by the same external couple M_e in a solid cylindrical shaft of the same length and cross-sectional area.

13.6 While a steel shaft of the cross section shown rotates at 120 rpm, a stroboscopic measurement indicates that angle of twist is 2° in a 4-m length. Using G=77 GPa, determine the power being transmitted.

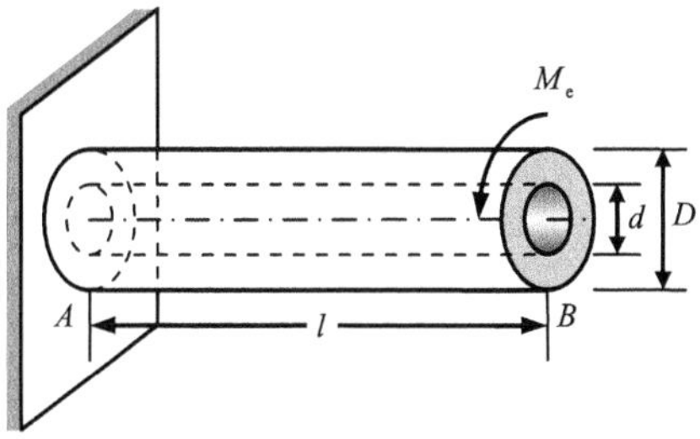

Fig. P13.5

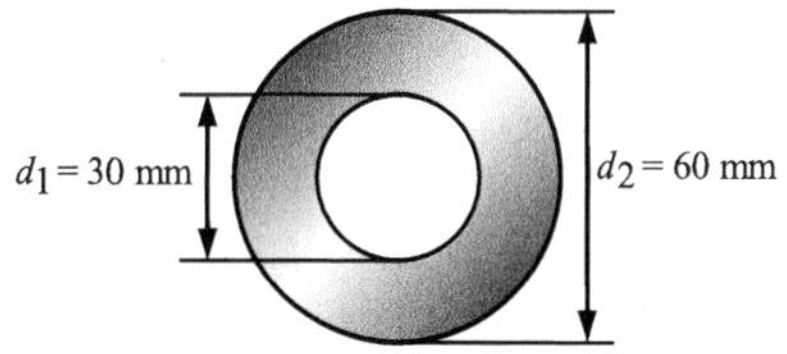

Fig. P13.6

13.7 A 1.8 m long tubular shaft of 45 mm inner diameter having a hollow circular section shown is to be made of steel for which τ_{allow}=75 MPa and G=77 GPa. Knowing that the angle of twist of the shaft must not exceed 5° when the shaft is subjected to an external couple of M_e=900 N·m, determine the smallest outer diameter which can be used in the design.

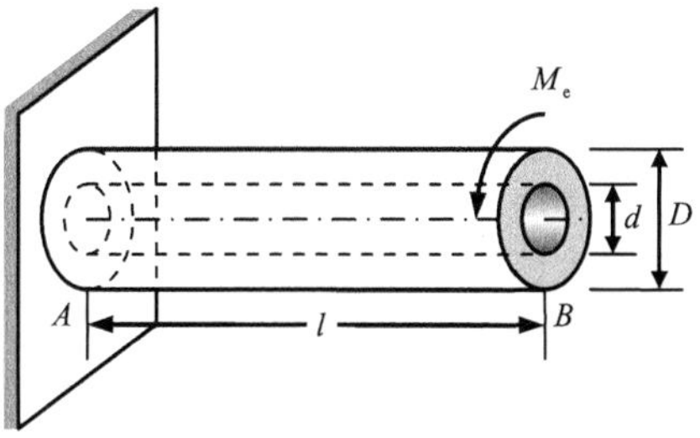

Fig. P13.7

13.8 The design of a machine element requires a D=38 mm outer diameter shaft to transmit 45 kW. (a) If the speed of rotation is 800 rpm, determine the maximum shearing stress in the solid shaft, Fig. P13.8a. (b) If the speed of rotation can be increased 50% to 1200 rpm, determine the largest inner diameter of the hollow shaft, Fig. P13.8b, for which the maximum shearing stress will be the same in each shaft.

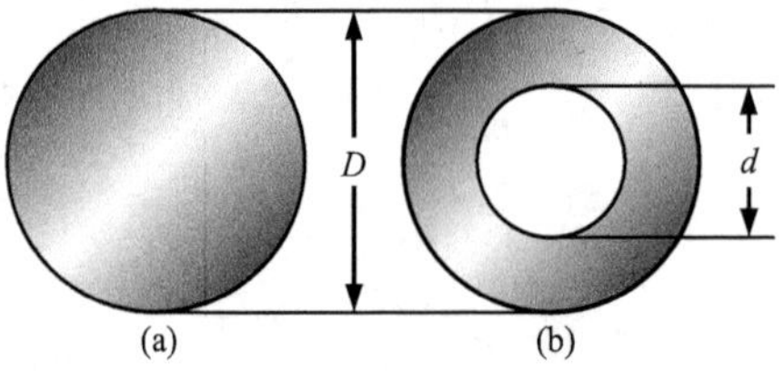

Fig. P13.8

13.9 A 1.5 m long tubular steel shaft (G=77 GPa) of 40 mm outer diameter D and 30 mm inner diameter d is to transmit 120 kW. Knowing that the allowable shearing stress is 65 MPa and that the angle of twist must not exceed 3°, determine the minimum rotating speed at which the shaft may rotate.

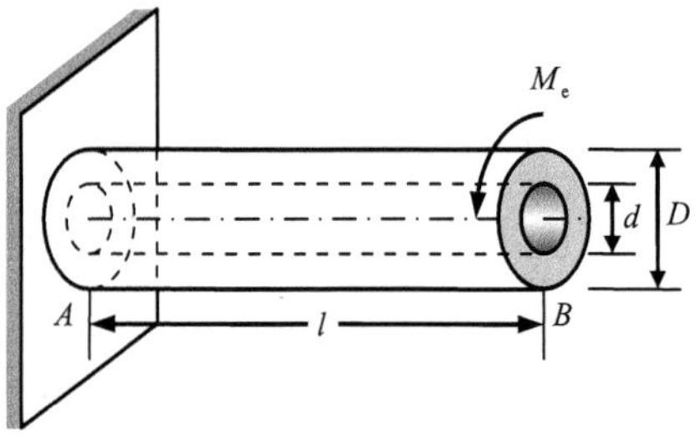

Fig. P13.9

Chapter 14 Bending Internal Forces

A bar subjected to external forces perpendicular to the longitudinal axis or external couples lying in a plane containing the longitudinal axis is called a beam in bending.

Beams are classified according to the way in which they are supported. Several types of frequently-used beams are shown in Fig. 14.1. A simply supported beam is pin-supported at one end and roller-supported at the other, Fig. 14.1a. An overhanging beam is supported at two points and has one or both ends extended beyond the supports, Fig. 14.1b. A cantilevered beam is fixed at one end and free at the other, Fig. 14.1c. External loads commonly applied to beams may consist of concentrated forces, Fig. 14.1a, distributed forces, Fig. 14.1b, and concentrated couples, Fig. 14.1c.

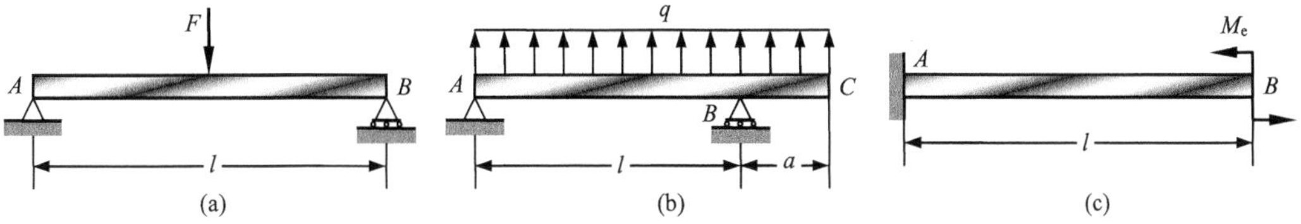

Fig. 14.1

The reactions at the supports of the above beams involve a total of only three unknowns and, therefore, can be determined by the equations of equilibrium. Such beams are said to be statically determinate. If the reactions at the supports of the beams involve more than three unknowns and cannot be determined by the equations of equilibrium alone, such beams are said to be statically indeterminate. The beam shown in Fig. 14.2 is statically indeterminate to the first degree.

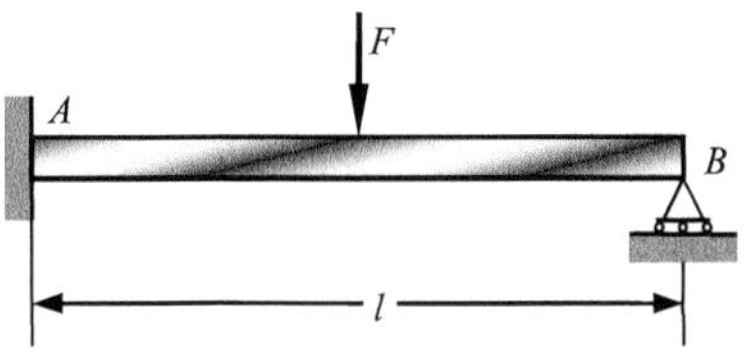

Fig. 14.2

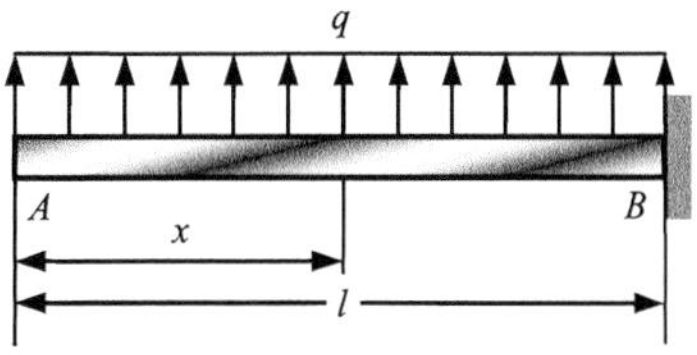

Fig. 14.3

14.1 Shearing-Force and Bending-Moment Diagrams

When external loads are applied to a beam, the beam will develop shearing forces and bending moments that, in general, vary from section to section along the axis of the beam, Fig. 14.3. Therefore, the shearing forces and bending moments are functions of the arbitrary position x along the axis of the beam, and can be expressed as

$$V = V(x),\ M = M(x) \tag{14.1}$$

These two equations are respectively called the shearing-force and bending-moment equations. According to the above equations, the shearing forces and bending moments along the axis of the beam can be plotted and represented by graphs, which are called shearing-force and bending-moment diagrams.

To draw the shearing-force and bending-moment diagrams of the beam, it is first necessary to establish a sign convention for the shearing forces and bending moments. Although the choice of a sign convention is arbitrary, the sign convention often used will be adopted for convenience. The shearing force that causes a clockwise rotation of the beam segment is said to be positive, Fig. 14.4a; the bending moment that causes compression in the upper part of the beam segment is said to be positive, Fig. 14.4b.

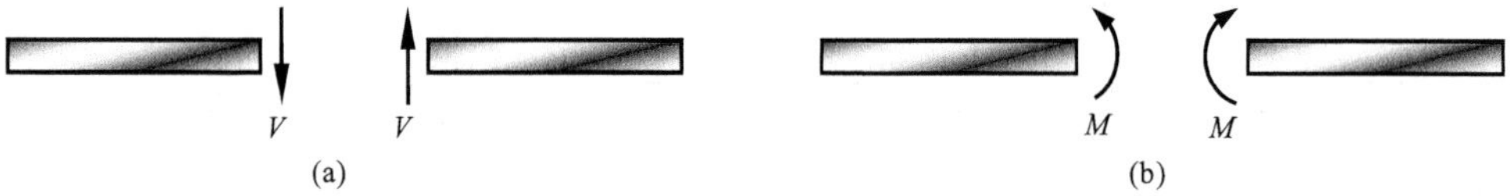

Fig. 14.4

The shearing force and bending moment at any section of the beam can be determined by using the method of sections, i.e., by cutting the beam through the section C where the shearing forces and bending moments are to be determined, Fig. 14.5a, and considering the equilibrium of the beam segment located on the left side of the section, Fig. 14.5b, or the right side of the section, Fig. 14.5c.

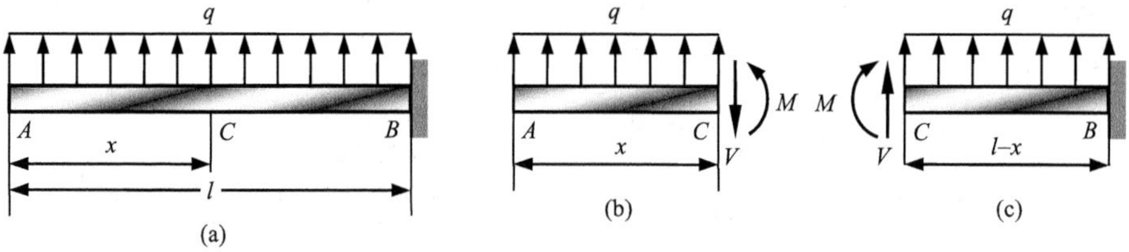

Fig. 14.5

Considering the equilibrium of the left portion AC of the beam, Fig. 14.5b, and using the equations of equilibrium, then we have

$$\begin{aligned} &\sum F_y = 0:\ \ qx - V = 0 \quad (0 \leqslant x < l) \\ &\sum M_C = 0:\ M - \frac{1}{2}qx^2 = 0\ (0 \leqslant x < l) \end{aligned} \tag{14.2}$$

Solving Eq. (14.2) for V and M, we obtain

$$\begin{aligned} V &= qx \quad (0 \leqslant x < l) \\ M &= \frac{1}{2}qx^2 \quad (0 \leqslant x < l) \end{aligned} \tag{14.3}$$

After finding the shearing-force and bending-moment equations for the entire beam, next we can draw the shearing-force and bending-moment diagrams based on the equations obtained. The shearing-force and bending-moment diagrams for the beam shown in Fig. 14.5a are given in Figs. 14.6b and 14.6c.

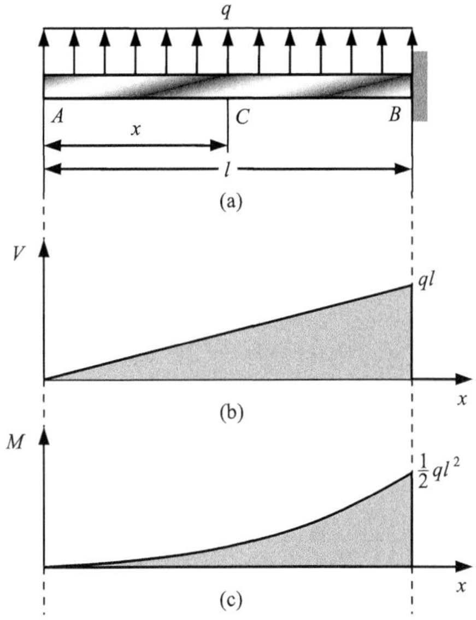

Fig. 14.6

It can be seen from Figs. 14.6b and 14.6c that the section B is a critical section of the beam due to the fact that this section is subjected to the maximum values of shearing forces and bending moments.

Example 14.1 Draw the bending-moment diagram for the structure shown in Fig. E14.1a and determine the maximum bending moment.

Solution Using the method of sections and the condition of equilibrium, we can obtain the bending-moment equations

$$\begin{aligned} M_1 &= F_1 x_1 \quad (0 \leqslant x_1 \leqslant a) \\ M_2 &= F_1 a + F_2 x_2 \quad (0 \leqslant x_2 < b) \end{aligned}$$

It can also be seen from the bending-moment equations obtained that the bending-moment diagram is an oblique straight-line for each portion of the structure. The bending-moment diagram for the entire structure is shown in Fig. E14.1b, and the maximum bending moment, which is located at section A, is equal to

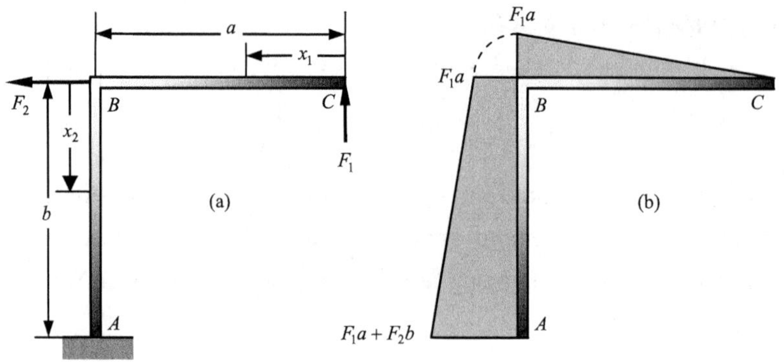

Fig. E14.1

$$M_{max} = F_1a + F_2b$$

Example 14.2 Draw the bending-moment diagram for the structure shown in Fig. E14.2a and determine the maximum bending moment.

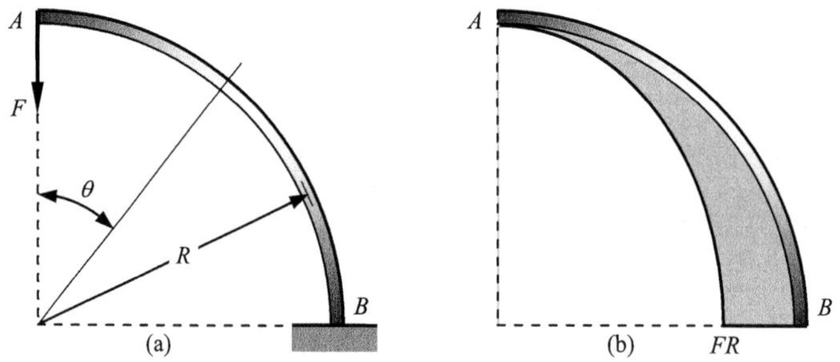

Fig. E14.2

Solution Using the method of sections and the condition of equilibrium, we can obtain the bending-moment equation:

$$M = FR\sin\theta \quad (0 \leqslant \theta < \frac{1}{2}\pi)$$

The bending-moment diagram for the structure is shown in Fig. E14.2b. The maximum bending moment at the fixed end is

$$M_{max} = FR$$

14.2 Relations between Distributed Load, Shearing Force, and Bending Moment

When a beam carries several external loads, the method discussed in the preceding section for drawing shearing force and bending moment diagrams is quite cumbersome. The construction of the shearing-force diagram and, especially, of the bending-moment diagram will be greatly facilitated if certain relations existing between distributed load, shearing force, and bending moment are taken into consideration.

Consider a simply supported beam AB subjected to a distributed load q (q is the intensity of distributed load, i.e., the force per unit length along the axis of the beam), Fig. 14.7a, and let C and D be two sections of the beam at a distance dx from each other. We now detach the portion of the beam CD and draw its free-body diagram, Fig. 14.7b.

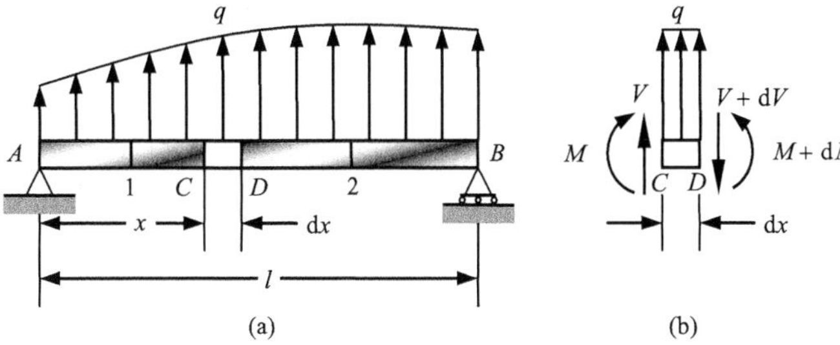

Fig. 14.7

According to the condition of static equilibrium of forces in the vertical direction, we have

$$\sum F_y = 0: \ V-(V+\mathrm{d}V)+q\mathrm{d}x=0, \text{ i.e., } \mathrm{d}V=q\mathrm{d}x \tag{14.4}$$

Dividing both members of Eq. (14.4) by dx, we obtain

$$\frac{\mathrm{d}V}{\mathrm{d}x}=q \tag{14.5}$$

Integrating Eq. (14.5) between sections 1 and 2, we have

$$V_2 = V_1 + \int_{x_1}^{x_2} q\mathrm{d}x \tag{14.6}$$

where $\int_{x_1}^{x_2} q\mathrm{d}x$ is the resultant force of the distributed load between sections 1 and 2.

From the condition of static equilibrium of moments about D, we write

$$\sum M_D = 0: \ (M+\mathrm{d}M)-M-V\mathrm{d}x-\frac{1}{2}q(\mathrm{d}x)^2=0, \text{ i.e., } \mathrm{d}M=V\mathrm{d}x+\frac{1}{2}q(\mathrm{d}x)^2 \tag{14.7}$$

Neglecting the second-order infinitesimal $(\mathrm{d}x)^2$ in Eq. (14.7), and dividing both members of Eq. (14.7) by dx, we obtain

$$\frac{\mathrm{d}M}{\mathrm{d}x}=V \tag{14.8}$$

Integrating Eq. (14.8) between sections 1 and 2, we write

$$M_2 = M_1 + \int_{x_1}^{x_2} V\mathrm{d}x \tag{14.9}$$

where $\int_{x_1}^{x_2} V\mathrm{d}x$ is the area of the shearing-force diagram between sections 1 and 2.

Eqs. (14.5), (14.6), (14.8), and (14.9) provide a convenient means for quickly drawing the shearing-force and bending-moment diagrams of a beam subjected to distributed loads.

Example 14.3 Draw the shearing-force and bending-moment diagrams for the simply supported beam shown in Fig. E14.3a and determine the maximum values of the shearing force and bending moment.

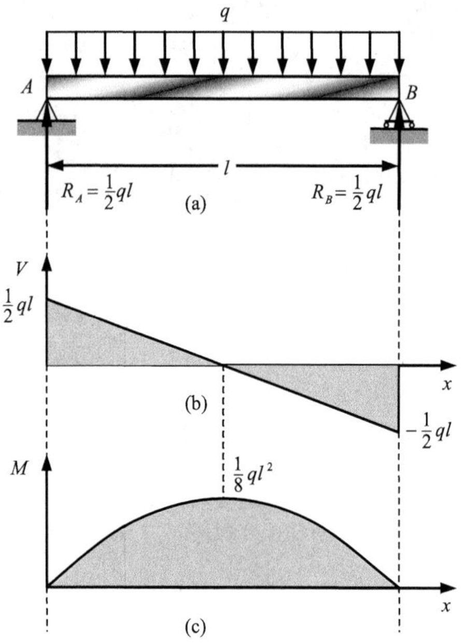

Fig. E14.3

Solution (1) Reactions at A and B are equal to $R_A = R_B = \frac{1}{2}ql$.

(2) Just to the right of the section A of the beam, the shearing force is equal to $V_A^{\mathrm{R}} = R_A = \frac{1}{2}ql$. Using Eq. (14.6), we can obtain the shearing force at any distance x from A

$$V = V_A^{\mathrm{R}} + \int_0^x (-q)\mathrm{d}x = \frac{1}{2}ql - qx \quad (0 < x < l)$$

It can be seen that the shearing-force diagram is an oblique straight line, Fig. E14.3b, and that the end A or B of the beam has a maximum absolute value of shearing forces.

(3) Using $M_A = 0$, the bending moment at any distance x from A can be obtained from Eq. (14.9):

$$M = M_A + \int_0^x V\mathrm{d}x = \frac{1}{2}qlx - \frac{1}{2}qx^2 \quad (0 \leqslant x \leqslant l)$$

It can also be seen that the bending-moment diagram is a parabola, Fig. E14.3c, and that the midsection of the beam is subjected to a maximum positive bending moment.

14.3 Relations between Concentrated Load, Shearing Force, and Bending Moment

Consider a simply supported beam *AB* subjected to a concentrated force *F* and a concentrated couple M_e, Fig. 14.8a, and let *C* and *D* be two sections of the beam at a distance d*x* from each other. Assuming that the concentrated force and couple are located at the midpoint of the segment *CD*, then we detach the segment *CD* and draw its free-body diagram, Fig. 14.8b.

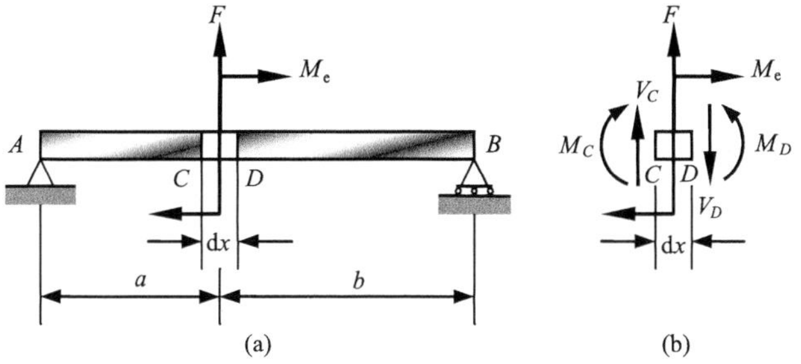

Fig. 14.8

According to the condition of static equilibrium of forces in the vertical direction, we have

$$\sum F_y = 0: \quad V_C - V_D + F = 0 \tag{14.10}$$

i.e.,

$$V_D = V_C + F \tag{14.11}$$

It can be seen that a sudden change of the shearing-force diagram will happen on the section where a concentrated force is applied.

From the condition of static equilibrium of moments about *D*, we write

$$\sum M_D = 0: \quad M_D - M_C - V_C \mathrm{d}x - M_e - F(\frac{1}{2}\mathrm{d}x) = 0 \tag{14.12}$$

Neglecting the infinitesimal d*x* in Eq. (14.12), we obtain

$$M_D = M_C + M_e \tag{14.13}$$

It can also be seen that the bending-moment diagram will have a sudden change at the section subjected to a concentrated couple.

Eqs. (14.11) and (14.13) also provide a convenient means for quickly drawing the shearing-force and bending-moment diagrams of a beam under the action of concentrated loads.

Example 14.4 Draw the shearing-force and bending-moment diagrams for the simply supported beam shown in Fig. E14.4a and determine the maximum values of the shearing force and bending moment.

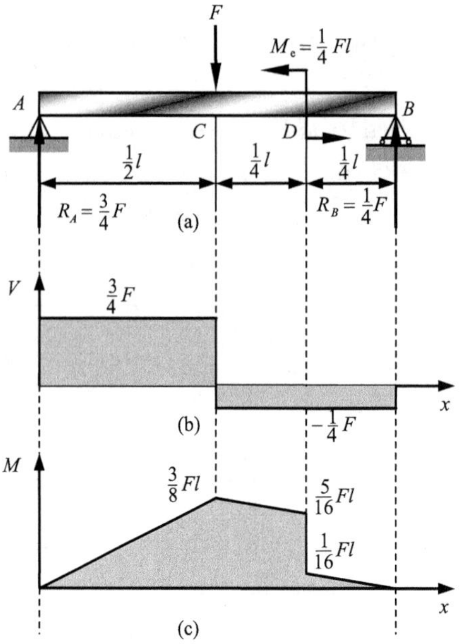

Fig. E14.4

Solution (1) Using the equilibrium condition of the entire beam, we have

$$R_A = \frac{3}{4}F,\ R_B = \frac{1}{4}F$$

(2) The shearing force, close to the right of the section A of the beam, is equal to $V_A^{\mathrm{R}} = R_A = \frac{3}{4}F$. Since no any external load is applied to the segment AC, the shearing force over the segment AC is a constant, and the corresponding shearing-force diagram is a horizontal straight line. Using $V_C^{\mathrm{L}} = V_A^{\mathrm{R}} = \frac{3}{4}F$ and Eq. (14.11), we can obtain the shearing force close to the right of the section C of the beam

$$V_C^{\mathrm{R}} = V_C^{\mathrm{L}} + (-F) = -\frac{1}{4}F$$

Similarly, we can obtain that the shearing force over the segment BC is a constant, and that the shearing-force diagram is also a horizontal straight line.

The shearing-force diagram for the entire beam is shown in Fig. E14.4b, and the maximum shearing forces is equal to $V_{\max} = \frac{3}{4}F$ over the segment AC.

(3) Using $M_A = 0$, the bending moment at the section C can be obtained from Eq. (14.9):

$$M_C = M_A + \int_{x_A}^{x_C} V\mathrm{d}x = \frac{3}{4}F \cdot \frac{1}{2}l = \frac{3}{8}Fl$$

Since the shearing force over the segment AC is constant, the corresponding bending-moment diagram is an oblique straight line. Similarly, the bending-moment diagram over the segment CD is also an oblique straight line, and the bending moment just to the left of the section D is equal to

$$M_D^{\mathrm{L}} = M_C + \int_{x_C}^{x_D} V\mathrm{d}x = \frac{5}{16}Fl$$

There is a concentrated couple on the section D, hence the bending-moment diagram will have a sudden change at that section. Using Eq. (14.13), we have

$$M_D^{\mathrm{R}} = M_D^{\mathrm{L}} + (-M_{\mathrm{e}}) = \frac{1}{16}Fl$$

Likewise, we can obtain the bending-moment diagram over the segment DB and the bending moment on the section B.

$$M_B = M_D^{\mathrm{R}} + \int_{x_D}^{x_B} V\mathrm{d}x = 0$$

The bending-moment diagram for the entire beam is shown in Fig. E14.4c, and the maximum bending moment is located on the section C and equal to $M_{\max} = \frac{3}{8}Fl$.

Problems

14.1-14.18 Draw the shearing-force and bending-moment diagrams for the beam and loading shown.

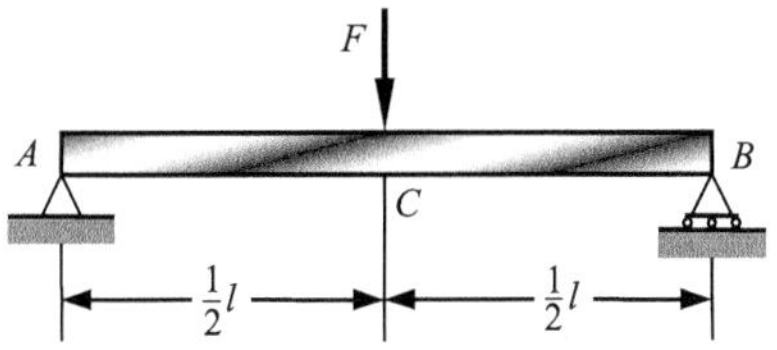

Fig. P14.1

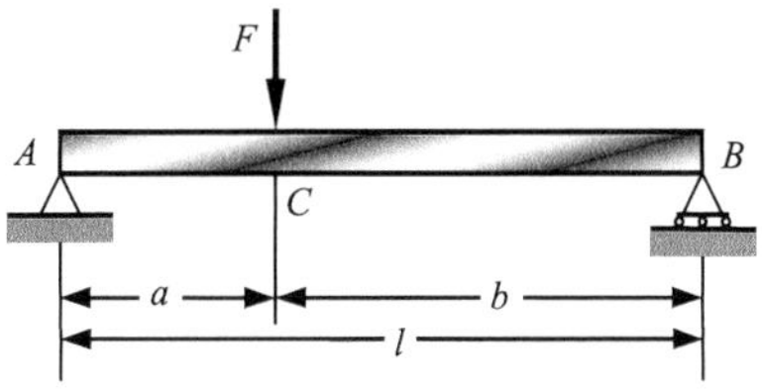

Fig. P14.2

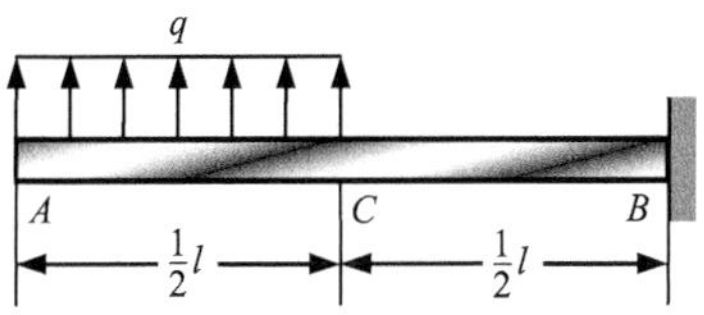

Fig. P14.3

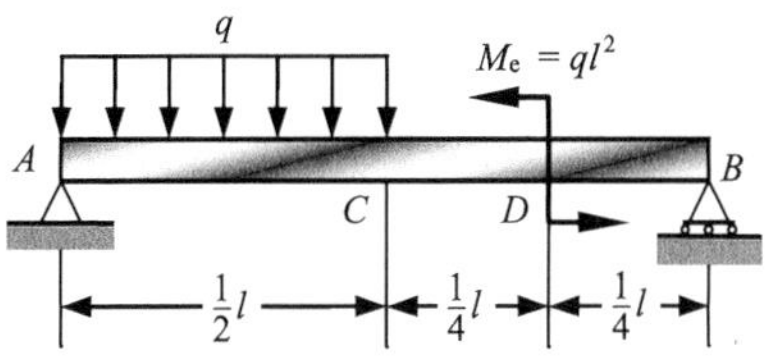

Fig. P14.4

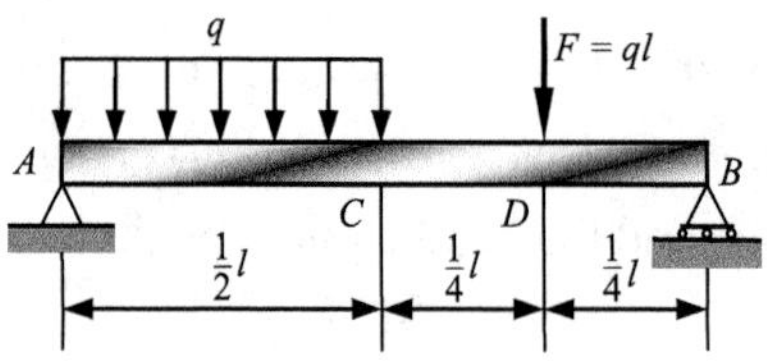

Fig. P14.5

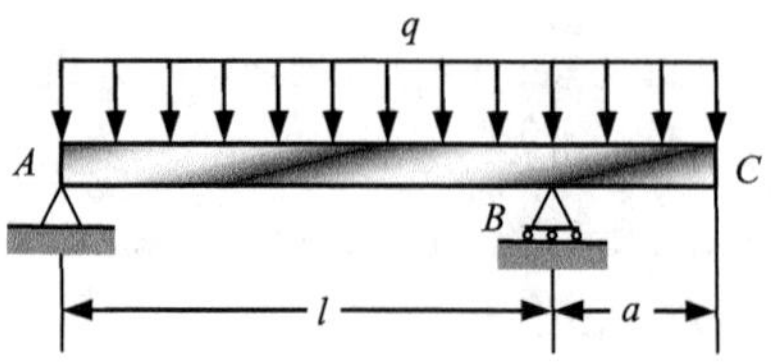

Fig. P14.6

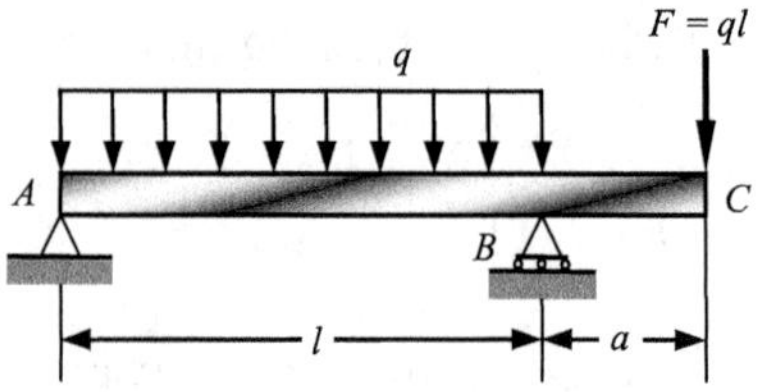

Fig. P14.7

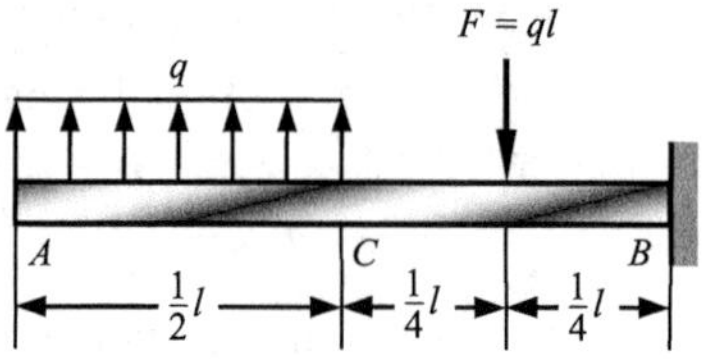

Fig. P14.8

Chapter 15 Bending Stresses

External loads applied to a beam usually include external forces that are perpendicular to the longitudinal axis of the beam and external couples that lie in a plane containing the longitudinal axis of the beam. One of the effects of these external forces and couples acting on a beam is to produce both normal and shearing stresses on the cross section perpendicular to the longitudinal axis of the beam.

If external couples are applied to the beam and no external forces act on the beam, then the bending is called pure bending, Fig. 15.1a. A beam subjected to pure bending has only normal stresses without shearing stresses on the cross section of the beam.

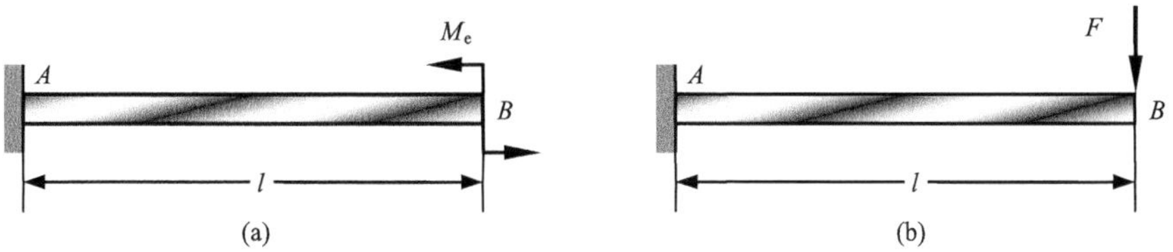

Fig. 15.1

Bending produced by external forces that do not form couples is called transverse-force bending, ordinary bending, or nonuniform bending, Fig. 15.1b. A beam subjected to transverse-force bending has both normal and shearing stresses on the cross section of the beam.

15.1 Normal Stresses on Cross Section in Pure Bending

Consider a prismatic beam AB possessing a longitudinal plane of symmetry and subjected to equal and opposite external couples of magnitude M_e acting in the longitudinal plane of symmetry, Fig. 15.2.

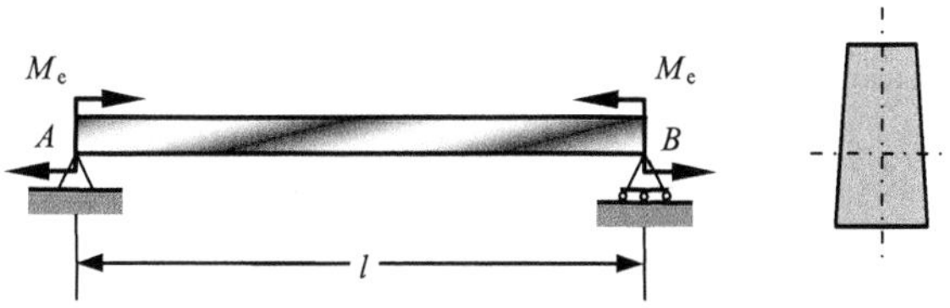

Fig. 15.2

Suppose that the beam is divided into a large number of small elements, Fig. 15.3a. Theses elements will be deformed as shown in Fig. 15.3b when the beam is subjected to pure bending.

The only nonzero stress component exerted on each of the elements is the normal stress on the faces perpendicular to the longitudinal axis of the beam. Thus, each point of a beam in pure bending is in a state of uniaxial stress. For a positive bending moment, the top and bottom surfaces of the beam are observed, respectively, to decrease and increase in length. Therefore, the upper portion of the beam is in compression and the lower portion in tension.

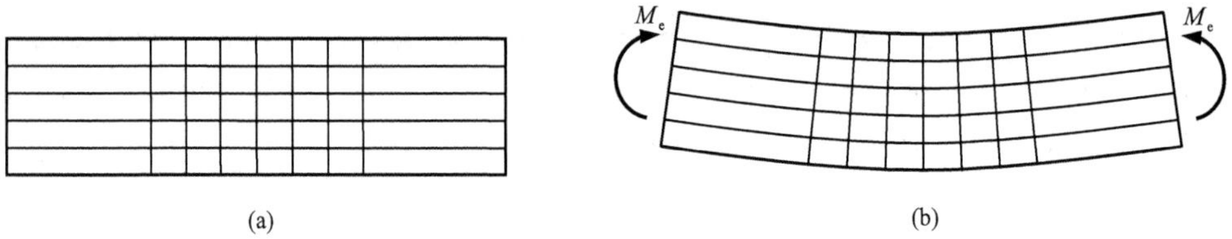

Fig. 15.3

It follows from the above analysis that there must exist a surface parallel to the top and bottom surfaces of the beam, where the normal strain and the normal stress are zero. This surface is called the neutral surface, Fig. 15.4a, and intersects a transverse section along a straight line called the neutral axis of the section, Fig. 15.4b. The origin of coordinates will be selected on the neutral surface so that the distance from any point to the neutral surface will be measured by its coordinate y.

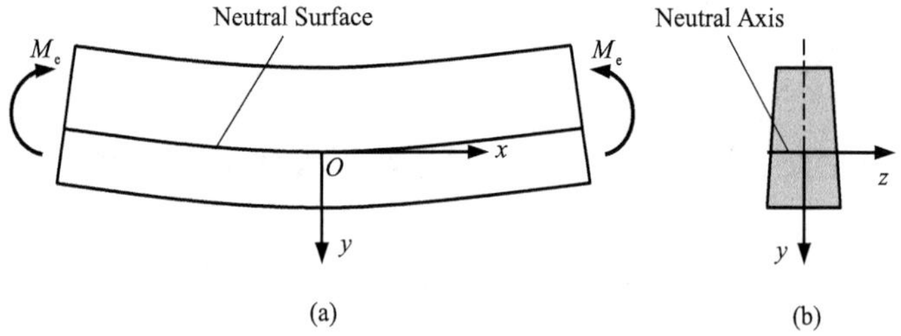

Fig. 15.4

1. Deformation Condition

Denoting by ρ the radius of the neutral surface *aa* and by θ the central angle corresponding to the neutral surface *aa*, Fig. 15.5.

Observing that the length of the neutral surface *aa* after deformation is equal to the length l of the undeformed neutral surface, then we have

$$l = \rho\theta \tag{15.1}$$

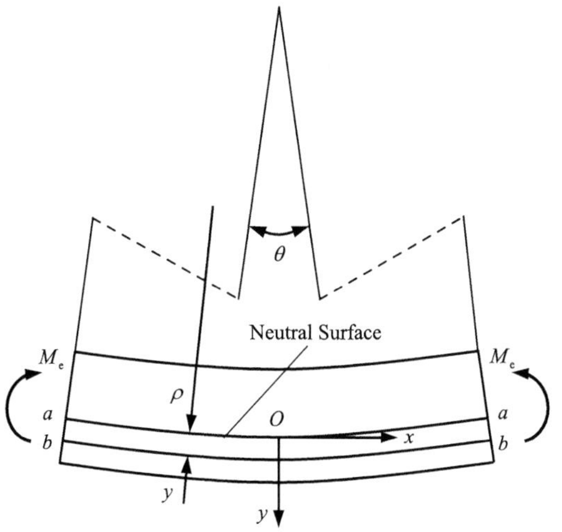

Fig. 15.5

Considering now the arc *bb* located at a distance *y* below the neutral surface, we note that its length after deformation is

$$l'_{bb} = (\rho + y)\theta \tag{15.2}$$

Since the original length of the arc *bb* was equal to $l_{bb} = l$, the deformation of the arc *bb* is

$$\delta = l'_{bb} - l_{bb} = y\theta \tag{15.3}$$

The longitudinal normal strain ε along the arc *bb* can be obtained by dividing δ by the original length l_{bb} of the arc *bb*. Therefore, we obtain

$$\varepsilon = \frac{\delta}{l_{bb}} = \frac{y}{\rho} \tag{15.4}$$

Eq. (15.4) shows that the longitudinal normal strain ε varies linearly with the distance *y* from the neutral surface.

2. Stress-Strain Relation

Assume that the normal stresses in the beam remain below the proportional limit, and Hooke's law for uniaxial stress applies, i.e., $\sigma = E\varepsilon$. From Eq. (15.4), we have

$$\sigma = \frac{E}{\rho} y \tag{15.5}$$

where *E* is the modulus of elasticity of the material. The above equation shows that, in the linearly elastic range, the normal stress varies linearly with the distance from the neutral surface, Fig. 15.6.

3. Equilibrium Condition

According to the condition of static equilibrium in *x* direction, we have

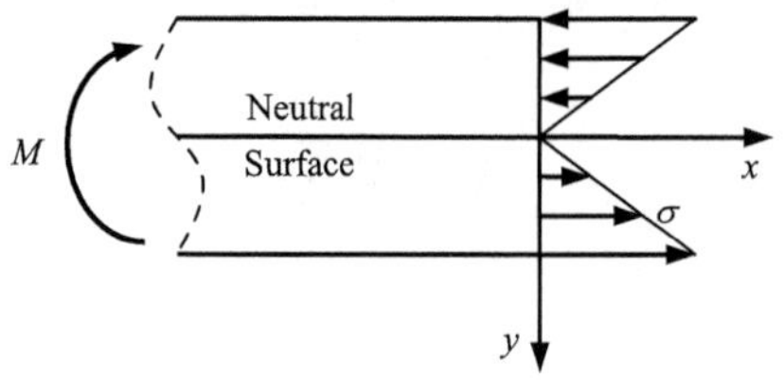

Fig. 15.6

$$\sum F_x = 0: \quad N = \int \sigma \mathrm{d}A = \frac{E}{\rho}\int y \mathrm{d}A = 0 \tag{15.6}$$

from which it follows that

$$\int y \mathrm{d}A = 0 \tag{15.7}$$

This equation shows that, for a beam subjected to pure bending, as long as the stresses remain in the linearly elastic range, the neutral axis passes through the centroid of the section.

Based on the condition of static equilibrium about the neutral axis, we have

$$\sum M_z = 0: \quad M = \int y\sigma \mathrm{d}A = \frac{E}{\rho}\int y^2 \mathrm{d}A \tag{15.8}$$

from which it follows that

$$\frac{1}{\rho} = \frac{M}{E\int y^2 \mathrm{d}A} = \frac{M}{EI} \tag{15.9}$$

where $I = \int y^2 \mathrm{d}A$ is the moment of inertia of the cross section with respect to the centroidal axis. Substituting $1/\rho$ from Eq. (15.9) into Eq (15.5), we obtain the normal stress σ at any distance y from the neutral axis:

$$\sigma = \frac{My}{I} \tag{15.10}$$

The normal stress is compressive ($\sigma < 0$) above the neutral axis ($y < 0$) when the bending moment M is positive, and tensile ($\sigma > 0$) when M is negative.

According to Eq. (15.10), the maximum normal stress is equal to $\sigma_{\max} = \frac{My_{\max}}{I}$. We note that the ratio $I/y_{\max}$ depends only upon the geometry of the cross section. This ratio is called the modulus of section and is denoted by S. From Eq. (15.10), we have

$$\sigma_{\max} = \frac{My_{\max}}{I} = \frac{M}{S} \tag{15.11}$$

For a beam with a rectangular cross section of width b and depth h, we have $S = \frac{1}{6}bh^2$.

Example 15.1 Knowing that the external couple shown acts in a vertical plane, determine the stress at (a) point a, (b) point b on the midsection C of the beam.

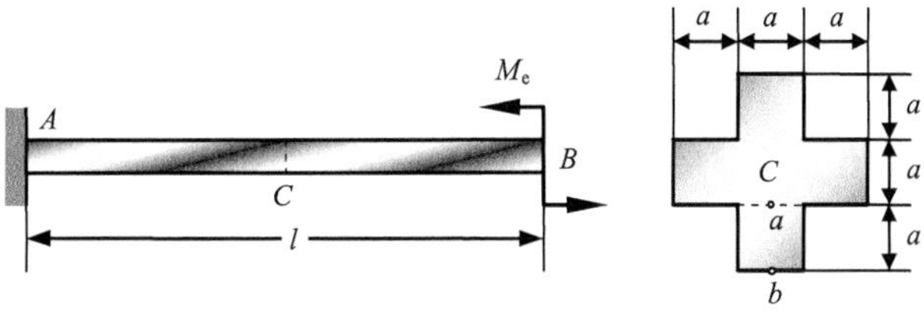

Fig. E15.1

Solution Cutting the beam at the midsection C into two parts and considering the equilibrium of the right part, we obtain the bending moment equal to the external couple applied to the section B, i.e., $M=M_e$.

For the section given in Fig. E15.1, the moment of inertia of the cross section about the neutral axis can be given by

$$I=\frac{1}{12}a^4+\frac{1}{12}a(3a)^3+\frac{1}{12}a^4=\frac{29}{12}a^4$$

Using Eq. (15.10), the stresses at points a and b can be obtained by

$$\sigma_a=\frac{My_a}{I}=\frac{M_e(\frac{1}{2}a)}{\frac{29}{12}a^4}=\frac{6M_e}{29a^3},\quad \sigma_b=\frac{My_b}{I}=\frac{M_e(\frac{3}{2}a)}{\frac{29}{12}a^4}=\frac{18M_e}{29a^3}$$

15.2 Normal and Shearing Stresses on Cross Section in Transverse-Force Bending

A transverse load applied to a beam will result in normal and shearing stresses in the transverse section of the beam. The normal stresses are created by the bending moment and the shearing stresses by the shearing force.

1. Normal Stresses

The formula for determining normal stresses on the cross section of a beam in pure bending is still valid for a beam subjected to transverse loadings as long as the beam under transverse loads is slender, Fig. 15.7, and the normal stresses in the beam subjected to transverse-force bending can be expressed as

$$\sigma=\frac{My}{I} \tag{15.12}$$

where the bending moment M will vary from section to section, i.e., $M=M(x)$ is a function of the position x under the action of transverse loads.

The maximum normal stress for a beam in transverse-force bending can be written as

$$\sigma_{max}=\frac{M_{max}y_{max}}{I}=\frac{M_{max}}{S} \tag{15.13}$$

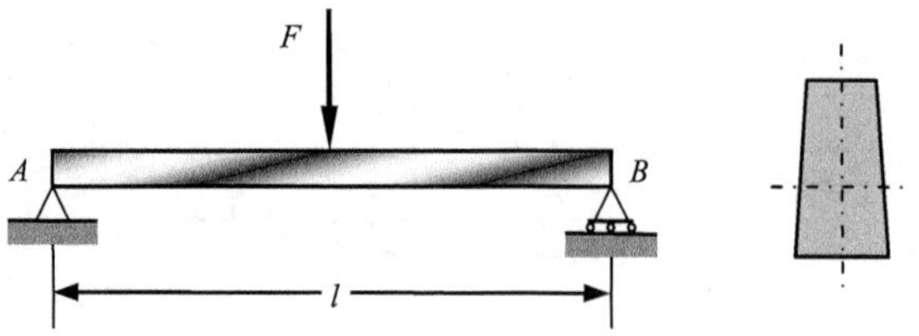

Fig. 15.7

2. Shearing Stresses

Normal stresses are quite important in the design of a beam. Shearing stresses, however, can also be important in the design of a short, stubby beam.

Considering a prismatic beam *AB* with a vertical plane of symmetry that supports external loads (a concentrated force *F*, a concentrated couple M_e, and a distributed load *q*). At a distance *x* from end *A* we detach from the beam an element *abcd* of length dx extending across the width of the beam from the bottom surface of the beam to a horizontal plane located at a distance y from the neutral axis, Fig. 15.8.

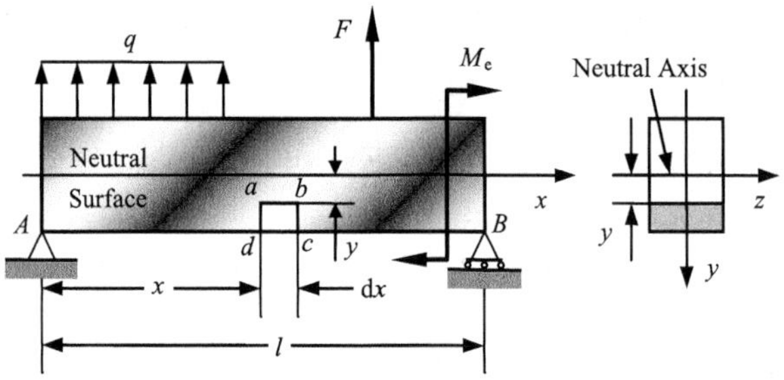

Fig. 15.8

When dx approaches zero, the shearing stresses on the left and right faces of the element are equal in magnitude. As long as the width of the cross section of the beam remains small compared to its depth, the shearing stress can be considered to be a constant in width. Assuming that the shearing stress at the points, located a distance y from the neutral axis of the cross section, on the left or right face of the element is denoted by τ, then the shearing stress on the top face of the element is also equal to τ according to the reciprocal theorem of shearing stresses, Fig. 15.9a.

Considering the equilibrium of the element *abcd* in the horizontal direction, then we obtain

$$\sum F_x = 0: \quad \int [\sigma(x+dx) - \sigma(x)]dA' - \tau(b dx) = 0 \tag{15.14}$$

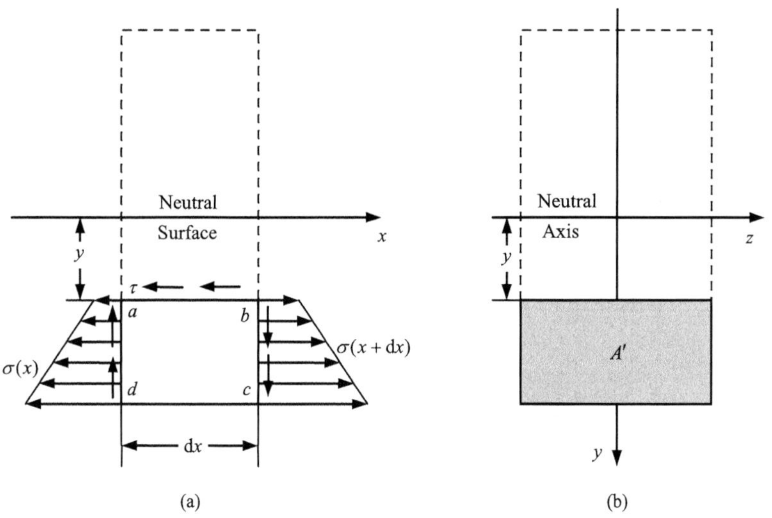

Fig. 15.9

where b is the width of the cross section of the beam and A' is the shaded area of the section located below the line with a distance y from the neutral axis, Fig. 15.9b.

Using $\sigma(x+\mathrm{d}x)-\sigma(x)=\dfrac{M(x+\mathrm{d}x)y'}{I}-\dfrac{M(x)y'}{I}=\dfrac{\mathrm{d}My'}{I}=\dfrac{(V\mathrm{d}x)y'}{I}$, and solving Eq. (15.14) for τ, we have

$$\tau=\frac{V}{Ib}\int y'\mathrm{d}A'=\frac{VQ'}{Ib} \tag{15.15}$$

where $Q'=\int y'\mathrm{d}A'$ represents the first moment with respect to the neutral axis of the area A' that is located below or above the line with a distance y from the neutral axis, I is the centroidal moment of inertia of the entire cross section, and V is the shearing force on the cross section.

For a beam of rectangular section of width b and depth h, Fig. 15.10a, $I=\frac{1}{12}bh^3$, and

$$Q'=A'\overline{y}'=[b(\frac{1}{2}h-y)][\frac{1}{2}(\frac{1}{2}h+y)]=\frac{1}{8}b(h^2-4y^2) \tag{15.16}$$

then we have

$$\tau=\frac{3V}{2bh}[1-(\frac{2y}{h})^2] \tag{15.17}$$

Eq. (15.17) shows that the distribution of shearing stresses in the transverse section of a rectangular beam is parabolic, Fig. 15.10b. The shearing stresses are zero at the top and bottom surfaces of the cross section ($y=\pm\frac{1}{2}h$). Making $y=0$, we obtain the maximum value of the shearing stress on the cross section of a rectangular beam:

$$\tau_{max} = \frac{3V}{2bh} = \frac{3V}{2A} \tag{15.18}$$

where A is the cross-sectional area of the rectangular beam.

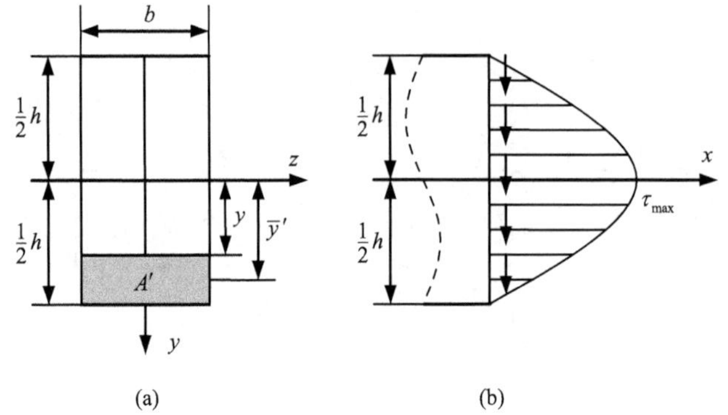

Fig. 15.10

For a beam of I-section (I-beam for short), Fig. 15.11, the shearing stresses on the web vary only very slightly along vertical direction of the section, and the entire shearing force is almost carried by the web. Therefore, the maximum value of the shearing stresses in the cross section can be obtained by dividing the shearing force by the cross-sectional area of the web:

$$\tau_{max} = \frac{V}{A_{web}} = \frac{V}{b_0 h_0} \tag{15.19}$$

where b_0 and h_0 are the width and depth of the web of the I-beam.

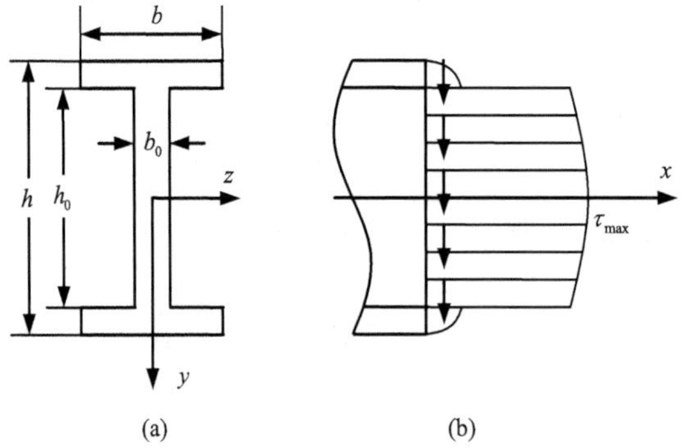

Fig. 15.11

For a beam of solid circular section, Fig. 15.12, the maximum shearing stress is located at the neutral axis of the cross section, and can be expressed as

$$\tau_{max} = \frac{4V}{3A} \tag{15.20}$$

where A is the cross-sectional area of the solid circular beam.

For a hollow circular beam, Fig. 15.13, the maximum shearing stress is also located at the neutral axis of the cross section, and can be expressed as

$$\tau_{\max} = \frac{2V}{A} \tag{15.21}$$

where A is the cross-sectional area of the hollow circular beam.

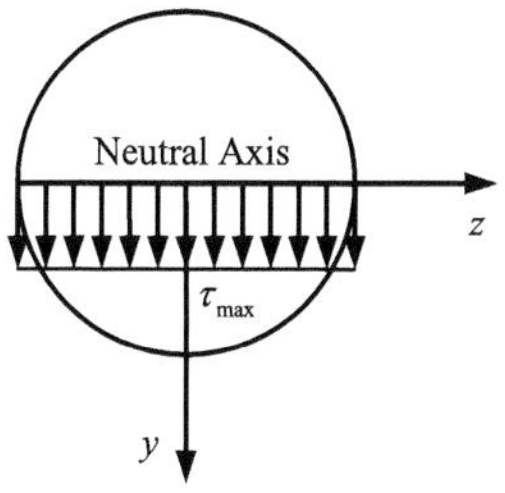

Fig. 15.12

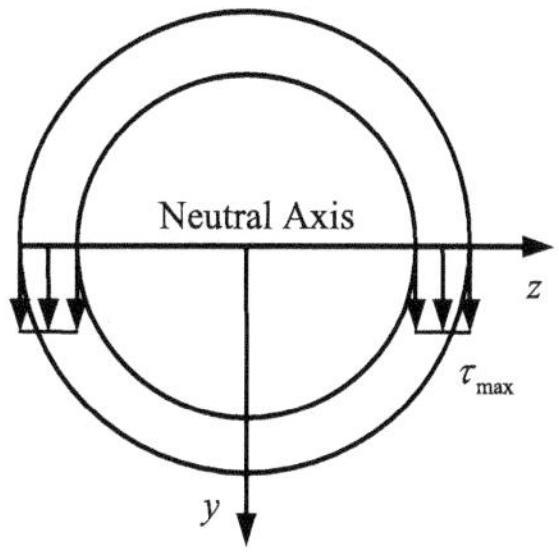

Fig. 15.13

Example 15.2 Knowing that the external force shown acts in a vertical plane, determine the shearing stress at point a on the midsection C of the beam.

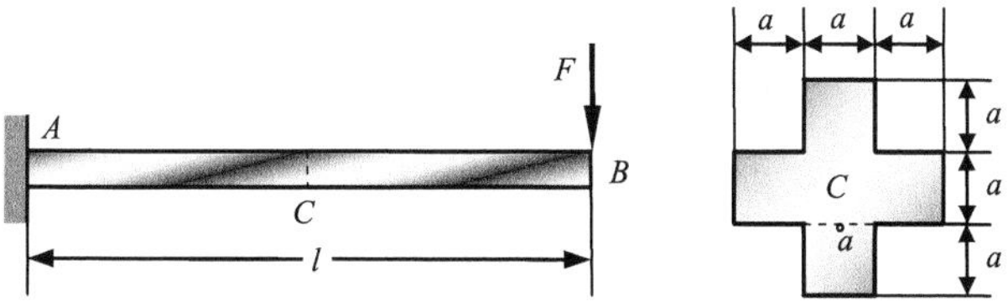

Fig. E15.2

Solution Using the method of sections, then the shearing force is equal to the external force applied to the section B, i.e., V=F. For the given section, the moment of inertia of the entire section about the neutral axis is equal to $I = \frac{29}{12}a^4$, and for the point a, the first moment with respect to the neutral axis of the area that is located below the horizontal line with a distance $\frac{1}{2}a$ from the neutral axis is equal to $Q' = A'\overline{y}' = a^3$. Therefore, the shearing stress at point a on the cross section can be obtained by

$$\tau_a = \frac{VQ'}{Ib} = \frac{Fa^3}{(\frac{29}{12}a^4)a} = \frac{12F}{29a^2}$$

15.3 Design of Bending Beam

The design of a bending beam is usually controlled by the maximum bending moment that will occur in the beam. The largest normal stress in the beam is found at the top or bottom surface of the beam in the critical section and can be obtained by

$$\sigma_{max} = \frac{M_{max} y_{max}}{I} = \frac{M_{max}}{S} \tag{15.22}$$

A safe design requires that $\sigma_{max} \leqslant \sigma_{allow}$, where σ_{allow} is the allowable stress for the material used. Substituting σ_{allow} for σ_{max} in Eq. (15.22), and solving for S, we obtain

$$S \geqslant \frac{M_{max}}{\sigma_{allow}} \tag{15.23}$$

In the design of a beam, a proper procedure should lead to the most economical design. This means that, among beams of the same type and the same material, and other things being equal, the beam with the smallest weight should be selected, since this beam will be the least expensive.

Example 15.3 Knowing that the allowable normal stress for the steel used is 165 MPa, select an I-steel beam to support the external load as shown in Fig. E15.3.

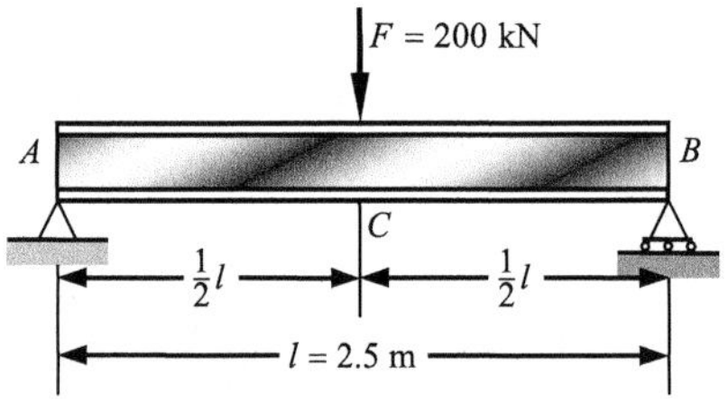

Fig. E15.3

Solution The bending moment is maximum at section C and equal to $M_{max} = \frac{1}{4}Fl = 125\ \text{kN} \cdot \text{m}$. Thus the minimum modulus of section is

$$S_{min} = \frac{M_{max}}{\sigma_{allow}} = 757.58\ \text{cm}^3$$

Referring to the table of Shape Steels in Appendix IV, we choose a group of I-steel beams having a modulus of section at least as large as S_{min}, Tab. 15.1.

Tab. 15.1

Designation	Weight/ (kg/m)	Section Modulus/cm^3
32c	62.7	760
36a	60.0	875
36b	65.7	919

From Tab. 15.1, it can be seen that both 32c and 36a I-steel beams can be chosen for a safe design. However, the most economical design should choose 36a I-steel beam since its weight is only 60.0 kg/m even though it has a large modulus of section, compared to 32c I-steel beam.

The weight of the beam chosen is 1.47 kN. This weight is small compared to the 200 kN external load and can be neglected in the design.

Problems

15.1 Knowing that the external couple shown acts in a vertical plane, determine the stress at (a) point P_1, (b) point P_2 on the midsection C of the beam.

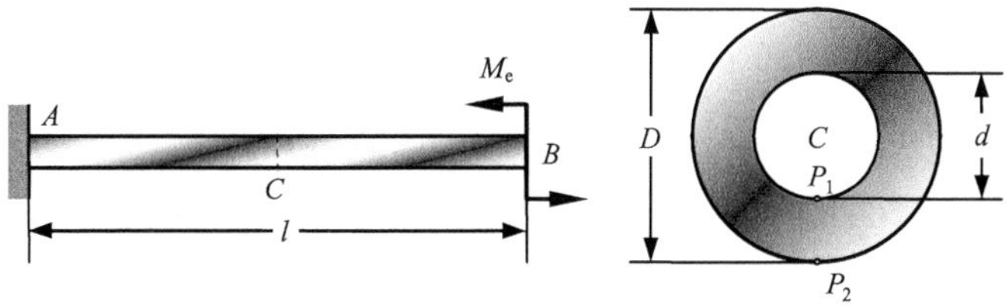

Fig. P15.1

15.2 Knowing that the I-section beam shown is made of steel for which the allowable stress σ_{allow}=150 MPa, determine the largest couple that can be applied to the beam when it is bent about the neutral axis. Neglect the effect of fillets.

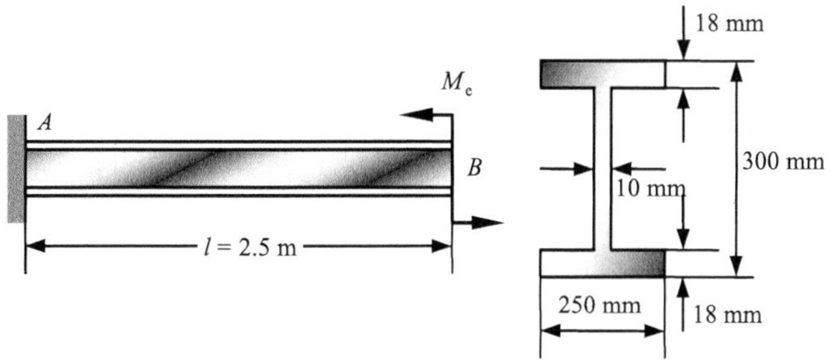

Fig. P15.2

15.3 An external couple is applied to a T-section beam shown. Determine the maximum tensile and compressive stresses in the beam when it is bent about the neutral axis. Neglect the effect of fillets.

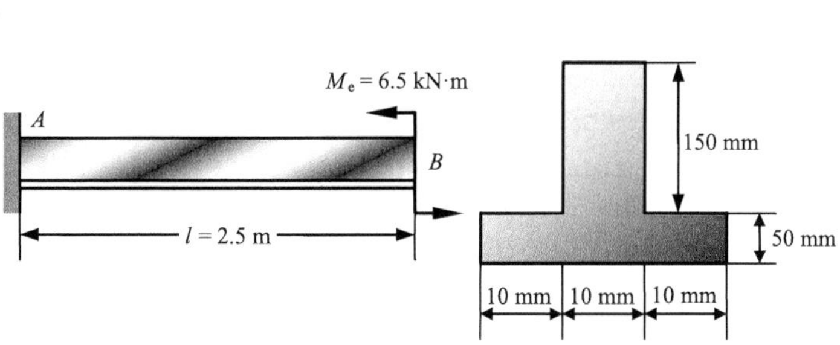

Fig. P15.3

15.4-15.6　Knowing that the external force shown acts in a vertical plane, determine the shearing stress at point P on the midsection C of the beam.

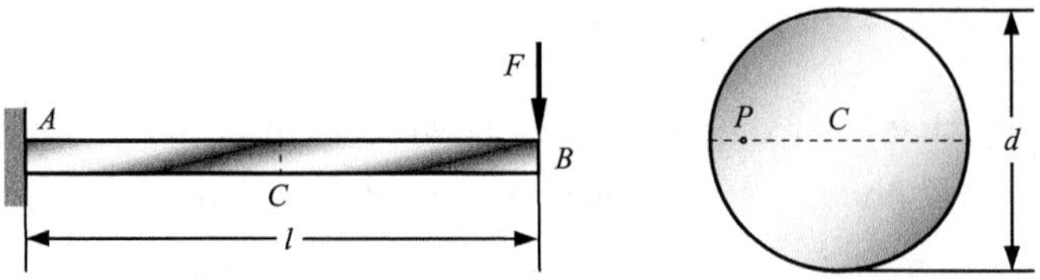

Fig. P15.4

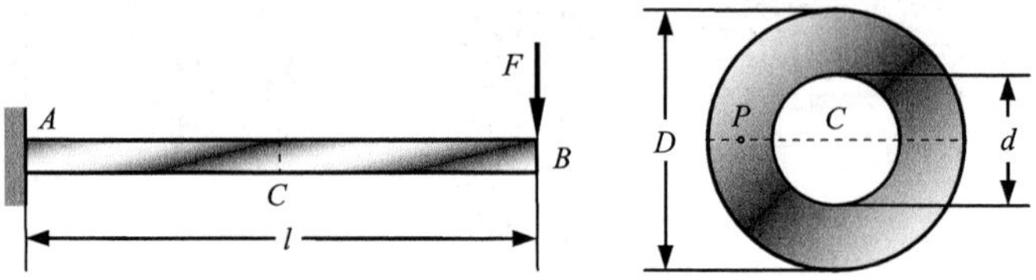

Fig. P15.5

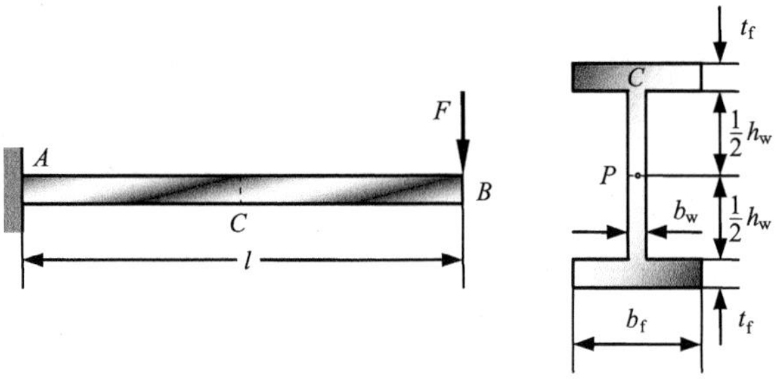

Fig. P15.6

15.7　Select an I-section beam to support the external load as shown. The allowable normal stress for the steel used is 165 MPa.

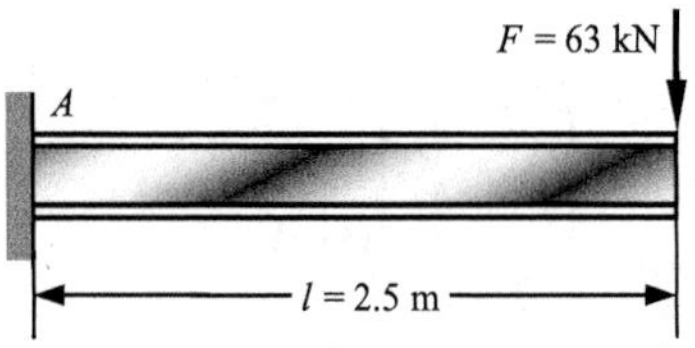

Fig. P15.7

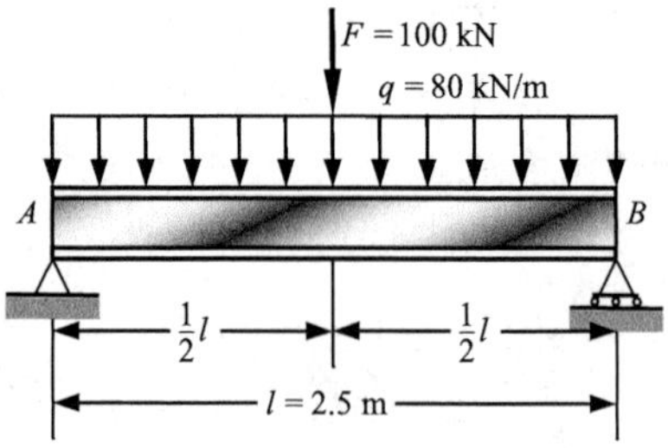

Fig. P15.8

15.8　Knowing that the allowable normal stress for the steel used is 165 MPa, select the most economical I-section beam to support the external loads as shown.

Chapter 16 Bending Deformation

A prismatic beam subjected to pure bending, Fig. 16.1, is bent into an arc of circle and that, within the linearly elastic range, the curvature of the neutral surface can be expressed as

$$\frac{1}{\rho}=\frac{M}{EI} \tag{16.1}$$

where the bending moment M is a constant, and EI is the flexural/bending rigidity of the beam.

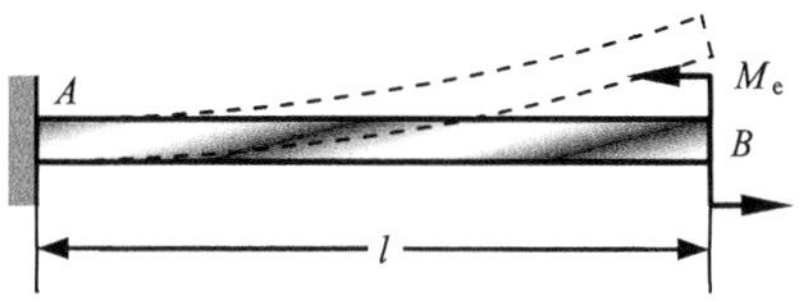

Fig. 16.1

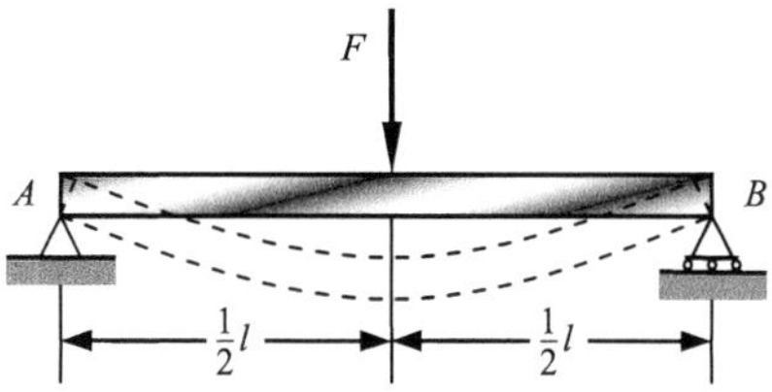

Fig. 16.2

When a slender beam is subjected to a transverse loading, Fig. 16.2, Eq. (16.1) remains valid for any given transverse section. However, both the bending moment and the curvature of the neutral surface will vary from section to section. Denoting by x the distance of the section from the left end of the beam, we have

$$\frac{1}{\rho}=\frac{M(x)}{EI} \tag{16.2}$$

where the bending moment $M(x)$ is a function of x.

The curvature of the neutral surface of the beam after deformation can be expressed mathematically as

$$\frac{1}{\rho}=\frac{w''}{[1+(w')^2]^{3/2}} \tag{16.3}$$

where $w=w(x)$ is the deflection function, Fig. 16.3, and w' and w'' are the first and second derivatives of the function $w=w(x)$. For a small deformation beam, the slope w' is very small, and its square is negligible compared to unity. We have, therefore,

$$\frac{1}{\rho}=w'' \tag{16.4}$$

Substituting ρ from Eq. (16.4) into Eq. (16.2), we obtain

$$w''=\frac{M(x)}{EI} \tag{16.5}$$

The above relation is called the differential equation of the deflection curve for a beam in bending.

16.1 Method of Integration

Integrating Eq. (16.5) in x, we have

$$w' = \int \frac{M(x)}{EI} \mathrm{d}x + C_1 \tag{16.6}$$

where C_1 is a constant of integration. Denoting by $\theta(x)$ the angle, measured in radians, that the tangent to the neutral surface after deformation forms with the horizontal, Fig. 16.3, and recalling that this angle is very small, we have

$$w' = \tan\theta \approx \theta \tag{16.7}$$

Thus, we write Eq. (16.6) in the alternative form

$$\theta = \int \frac{M(x)}{EI} \mathrm{d}x + C_1 \tag{16.8}$$

Integrating Eq. (16.8) in x, we have

$$w = \int \left(\int \frac{M(x)}{EI} \mathrm{d}x \right) \mathrm{d}x + C_1 x + C_2 \tag{16.9}$$

where C_2 is a second constant.

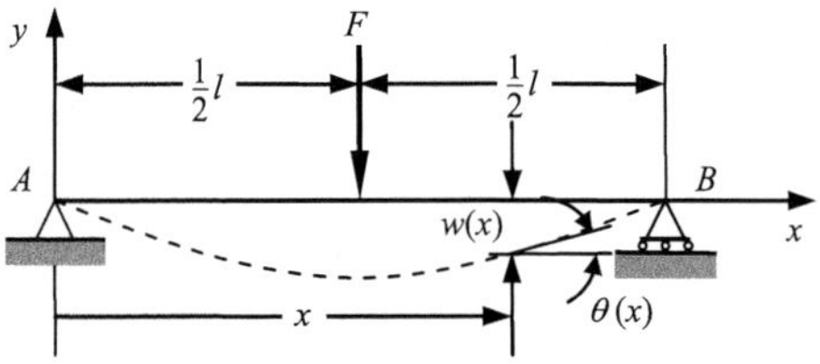

Fig. 16.3

The constants C_1 and C_2 are determined from the boundary conditions. Three types of supports often used and their boundary conditions are given in Fig. 16.4.

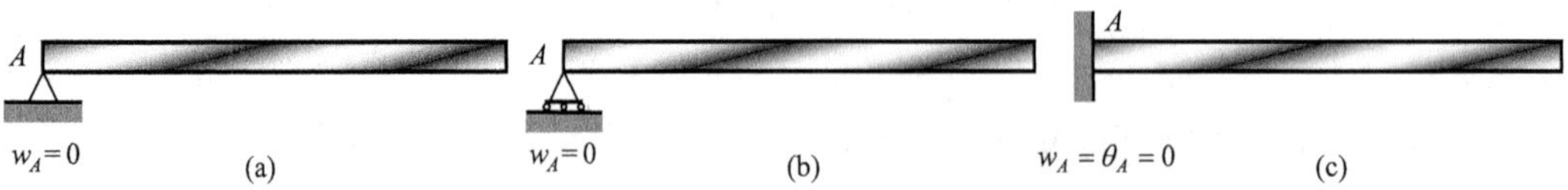

Fig. 16.4

Example 16.1 For the loading shown, determine (a) the equation of the deflection curve for the cantilevered beam AB, (b) the deflection and slope at the free end.

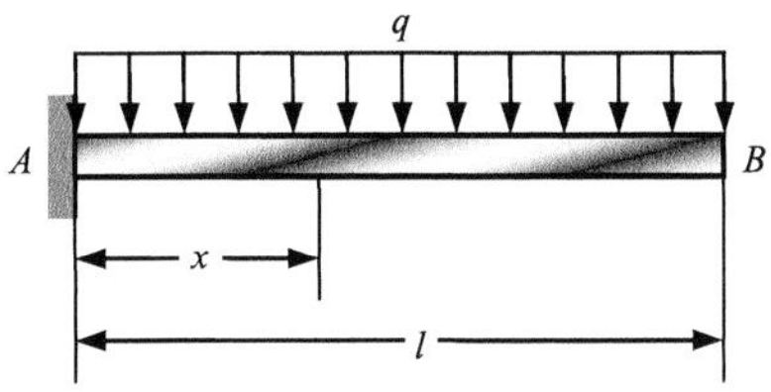

Fig. E16.1

Solution Considering the x section of the beam, Fig. E16.1, then we have bending moment $M(x) = -\frac{1}{2}q(l-x)^2$ and the differential equation of the deflection curve $w'' = -\frac{q(l-x)^2}{2EI}$. Using the method of integration, we obtain

$$\theta = \int [-\frac{q(l-x)^2}{2EI}]\mathrm{d}x + C_1 = -\frac{q}{6EI}(3l^2x - 3lx^2 + x^3) + C_1$$

$$w = \int \{\int [-\frac{q(l-x)^2}{2EI}]\mathrm{d}x\}\mathrm{d}x + C_1x + C_2 = -\frac{q}{24EI}(6l^2x^2 - 4lx^3 + x^4) + C_1x + C_2$$

Using the boundary conditions $\theta = w = 0$ at $x = 0$, we obtain $C_1 = 0$ and $C_2 = 0$. Thus the equation of the deflection curve can be given by

$$w = -\frac{q}{24EI}(6l^2x^2 - 4lx^3 + x^4)$$

From the above equation, we can obtain the deflection and slope at the free end of the beam

$$w_B = -\frac{ql^4}{8EI}, \quad \theta_B = -\frac{ql^3}{6EI}$$

16.2 Method of Superposition

When a beam is subjected to several loads, it is often convenient to compute separately the deflection and slope caused by each of the given loads. The deflection and slope due to the combined loads are then obtained by applying the principle of superposition and adding the values of the deflection and slope corresponding to the various loads. The deflections and slopes of often used beams for various loadings and supports are shown in Tab. 16.1.

Consider a simple supported beam subjected to two loads, Fig. 16.5a. Using the principle of superposition, the deflection and slope at any section of the beam can be obtained by superposing the deflections and slopes caused respectively by the concentrated load, Fig. 16.5b, and by the distributed load, Fig. 16.5c. Therefore, the deflection and slope at any section of the beam subjected to combined loads can be expressed as

$$w = w_F + w_q, \quad \theta = \theta_F + \theta_q \tag{16.10}$$

Tab. 16.1 Deflection Curves

Beam and Load	Deflection Curves	Critical Deflection	Critical Slope
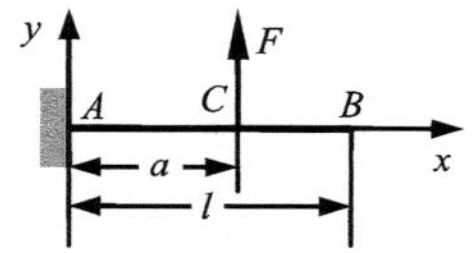	$w=\frac{Fx^2}{6EI}(3a-x)$ $(0\leqslant x\leqslant a)$ $w=\frac{Fa^2}{6EI}(3x-a)$ $(a\leqslant x\leqslant l)$	$w_{\max}=w_B=\frac{Fa^2}{6EI}(3l-a)$ $w_C=\frac{Fa^3}{3EI}$	$\theta_{\max}=\theta_B=\theta_C=\frac{Fa^2}{2EI}$
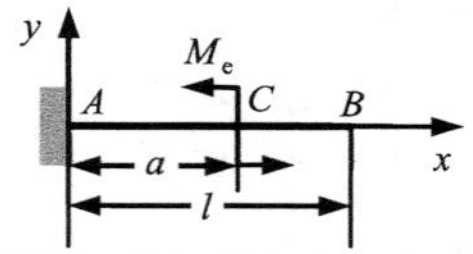	$w=\frac{M_e x^2}{2EI}$ $(0\leqslant x\leqslant a)$ $w=\frac{M_e a}{2EI}(2x-a)$ $(a\leqslant x\leqslant l)$	$w_{\max}=w_B=\frac{M_e a}{2EI}(2l-a)$ $w_C=\frac{M_e a^2}{2EI}$	$\theta_{\max}=\theta_B=\theta_C=\frac{M_e a}{EI}$
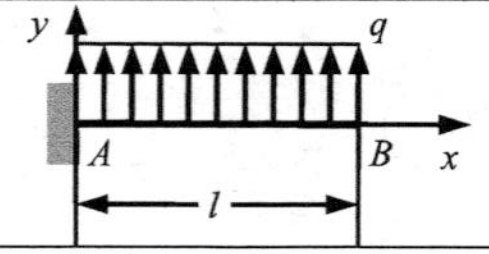	$w=\frac{qx^2}{24EI}(6l^2-4lx+x^2)$	$w_{\max}=w_B=\frac{ql^4}{8EI}$	$\theta_{\max}=\theta_B=\frac{ql^3}{6EI}$
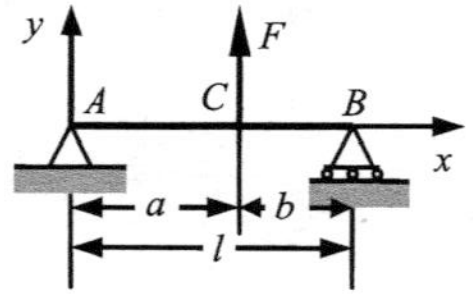	$w=\frac{Fbx}{6EIl}(l^2-b^2-x^2)$ $(0\leqslant x\leqslant a)$ $w=\frac{F}{6EIl}[l(x-a)^3+b(l^2-b^2)x-bx^3]$ $(a\leqslant x\leqslant l)$	For $a\geqslant b$: $w_{\max}=w[\sqrt{\frac{1}{3}(l^2-b^2)}]=\frac{Fb\sqrt{(l^2-b^2)^3}}{9\sqrt{3}EIl}$ $w(\frac{1}{2}l)=\frac{Fb(3l^2-4b^2)}{48EI}$	For $a\geqslant b$: $\theta_A=\frac{Fab(l+b)}{6EIl}$ $\theta_{\max}=-\theta_B=\frac{Fab(l+a)}{6EIl}$
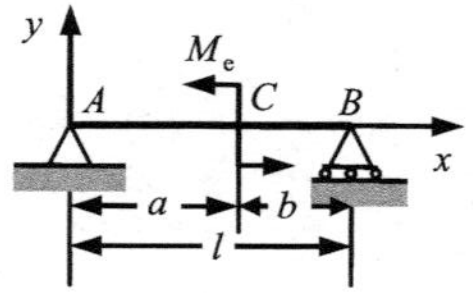	$w=\frac{M_e x}{6EIl}(x^2+3b^2-l^2)$ $(0\leqslant x\leqslant a)$ $w=\frac{M_e}{6EIl}[x^3-3l(x-a)^2-(l^2-3b^2)x]$ $(a\leqslant x\leqslant l)$	For $a\geqslant b$: $w_{\max}=-w[\sqrt{\frac{1}{3}(l^2-3b^2)}]=\frac{M_e\sqrt{(l^2-3b^2)^3}}{9\sqrt{3}EIl}$ $w(\frac{1}{2}l)=-\frac{M_e(l^2-4b^2)}{16EI}$	$\theta_A=-\frac{M_e}{6EIl}(l^2-3b^2)$ $\theta_B=-\frac{M_e}{6EIl}(l^2-3a^2)$ $\theta_{\max}=\theta_C=\frac{M_e}{6EIl}[3(a^2+b^2)-l^2]$
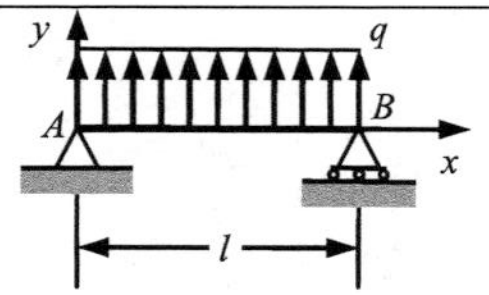	$w=\frac{qx}{24EI}(l^3-2lx^2+x^3)$	$w_{\max}=w(\frac{1}{2}l)=\frac{5ql^4}{384EI}$	$\theta_{\max}=\theta_A=-\theta_B=\frac{ql^3}{24EI}$

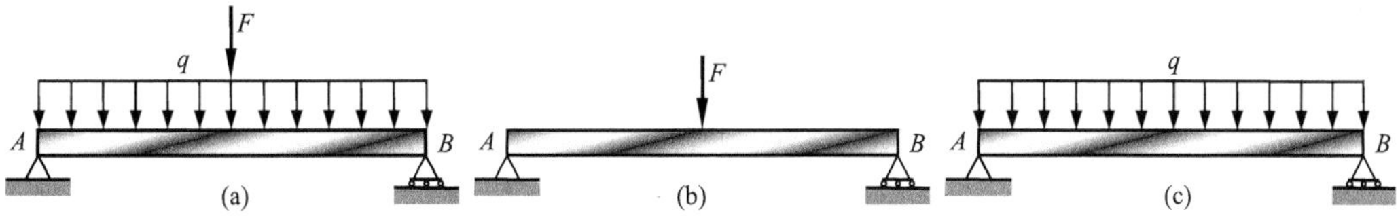

Fig. 16.5

Example 16.2 For the beam and loading shown, Fig. E16.2, determine the deflection at the midsection C, and the slope at the end B.

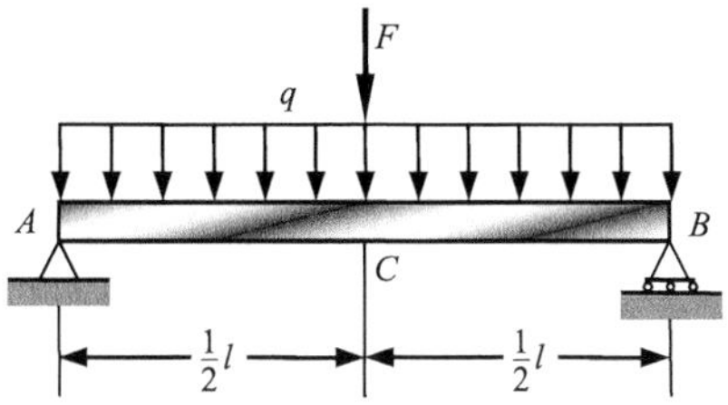

Fig. E16.2

Solution For the concentrated load F and from Tab. 16.1, we have

$$(w_C)_F = -\frac{Fl^3}{48EI}, \quad (\theta_B)_F = \frac{Fl^2}{16EI}$$

Similarly, for the distributed load q and from Tab. 16.1, we have

$$(w_C)_q = -\frac{5ql^4}{384EI}, \quad (\theta_B)_q = \frac{ql^3}{24EI}$$

Using the principle of superposition, then we obtain the deflection at the midsection C and the slope at the end B:

$$w_C = (w_C)_F + (w_C)_q = -\left(\frac{Fl^3}{48EI} + \frac{5ql^4}{384EI}\right), \quad \theta_B = (\theta_B)_F + (\theta_B)_q = \frac{Fl^2}{16EI} + \frac{ql^3}{24EI}$$

16.3 Statically Indeterminate Beam

Consider a prismatic beam AB, Fig. 16.6, which has a fixed end at A and is supported by a roller at B. We note that the reactions involve four unknowns, while only three equilibrium conditions are available. Therefore, the beam is statically indeterminate to the first degree.

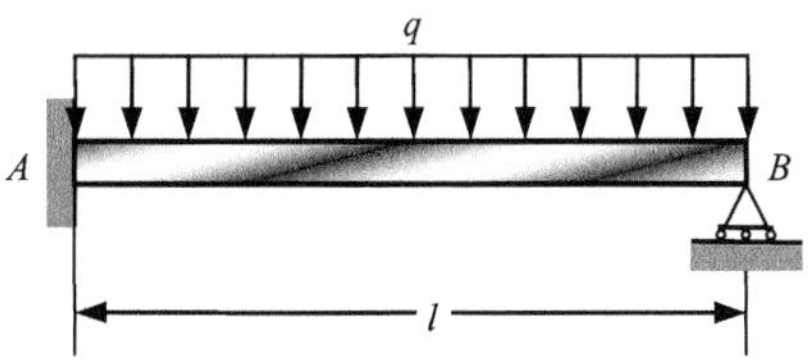

Fig. 16.6

For the statically indeterminate beam to the first degree shown in Fig. 16.6, we designate one of the reactions, say the reaction at *B*, as redundant and eliminate the corresponding support. The redundant reaction is then treated as an unknown load that, together with the other loads, must produce deformation which is compatible with the original support.

Example 16.3 For the beam and loading shown in Fig. 16.6, determine the reaction at the roller support.

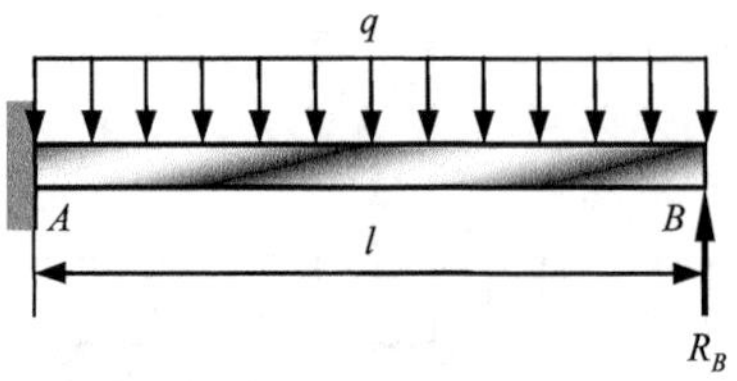

Fig. E16.3

Solution Considering the reaction at *B* as redundant and releasing the beam from support *B*, Fig. E16.3, then the reaction R_B is considered as a redundant reaction and will be determined from the condition that the deflection of the beam at *B* must be zero.

From Tab. 16.1, the deflection caused at *B* by the redundant reaction R_B can be given by

$$(w_B)_{R_B} = \frac{R_B l^3}{3EI}$$

Similarly, the deflection produced at *B* by the uniformly distributed load q can be expressed as

$$(w_B)_q = -\frac{ql^4}{8EI}$$

Using the principle of superposition, the deflection at *B* can be written as

$$w_B = (w_B)_{R_B} + (w_B)_q = \frac{R_B l^3}{3EI} - \frac{ql^4}{8EI}$$

Since the deflection of the beam at *B* must be zero, then we have

$$\frac{R_B l^3}{3EI} - \frac{ql^4}{8EI} = 0$$

Solving for R_B, we obtain the reaction at the roller support *B*:

$$R_B = \frac{3}{8}ql$$

Problems

16.1-16.2 For the loading shown, determine (a) the equation of the deflection curve for the cantilevered beam AB, (b) the deflection and slope at the free end.

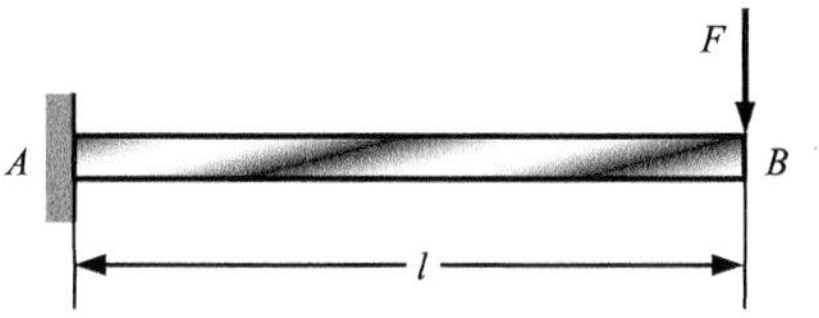

Fig. P16.1

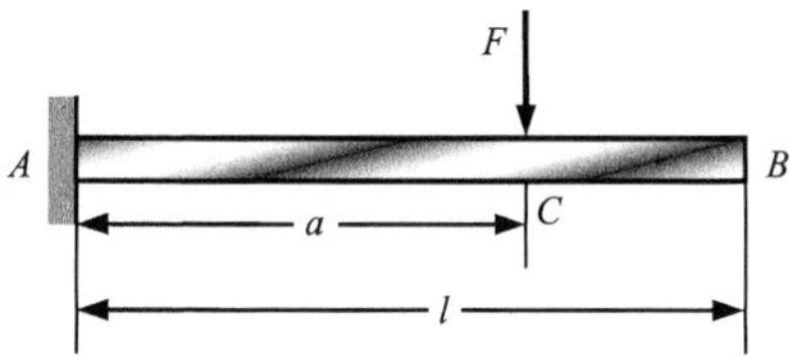

Fig. P16.2

16.3-16.4 For the loading shown, determine (a) the equation of the deflection curve for the simple supported beam AB, (b) the deflection at the midsection C and slope at the section B.

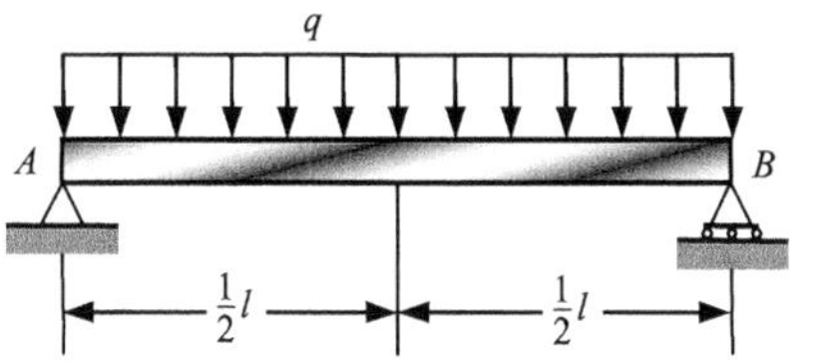

Fig. P16.3

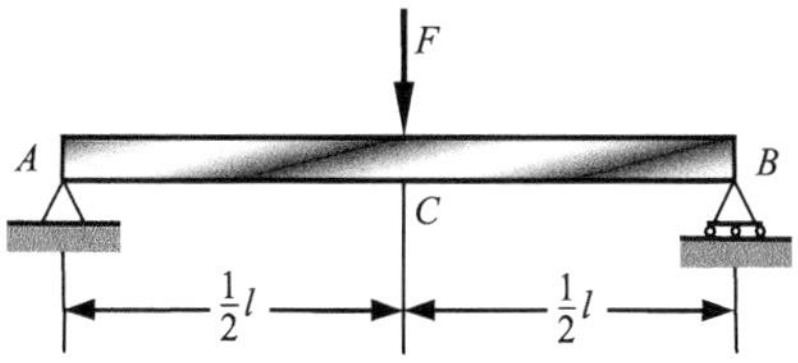

Fig. P16.4

16.5 For the beam and loading shown, determine (a) the deflection at the midsection C, (b) the slope at the end B.

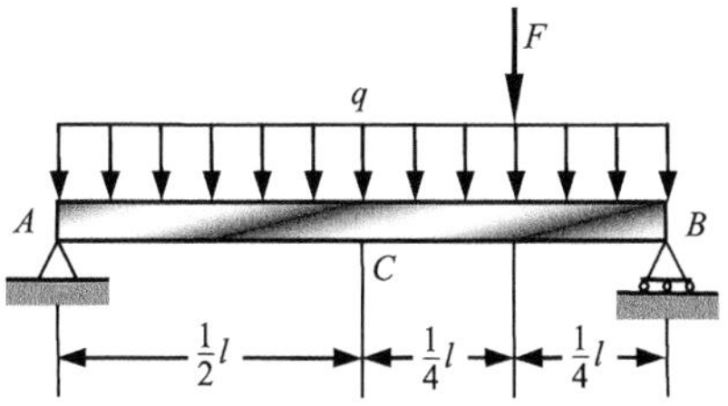

Fig. P16.5

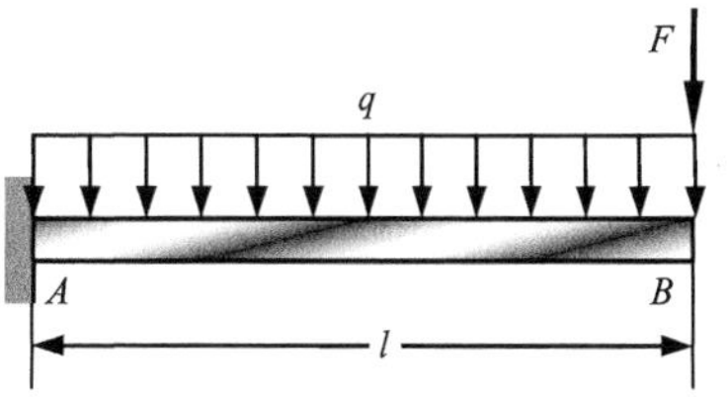

Fig. P16.6

16.6-16.9 For the cantilever beam and loading shown, determine the deflection and slope at point B.

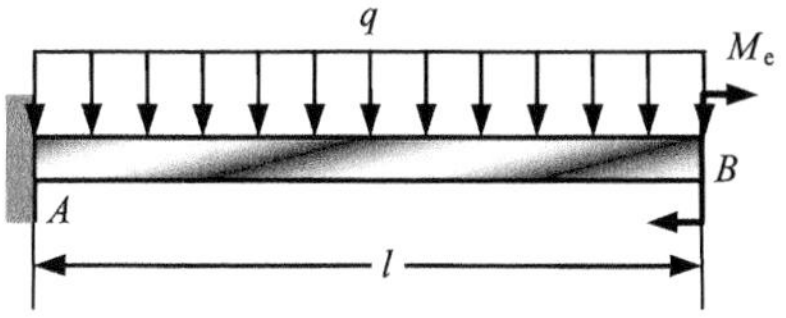

Fig. P16.7

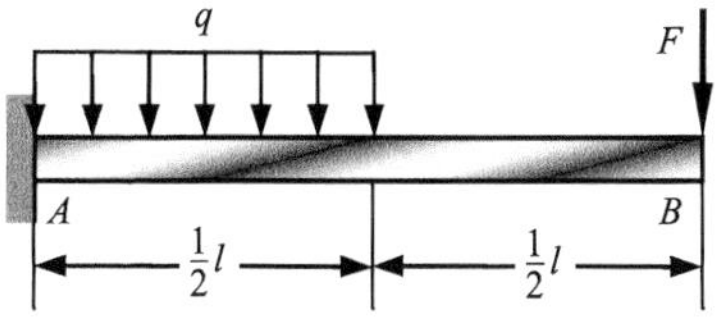

Fig. P16.8

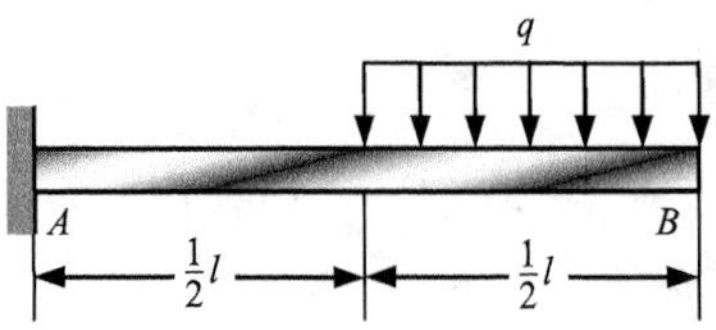

Fig. P16.9

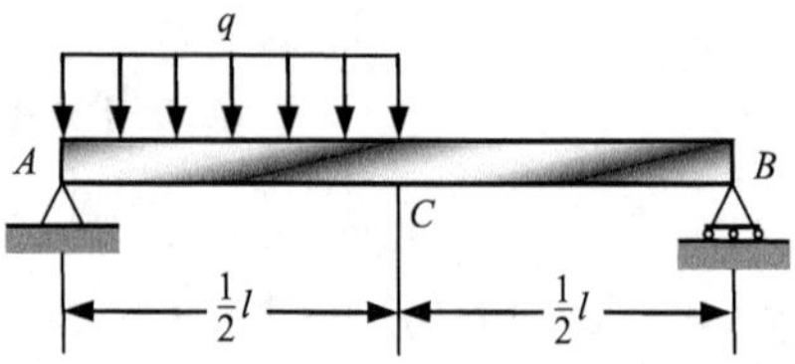

Fig. P16.10

16.10 For the loading shown, determine the deflection at the midsection C and slope at the section B.

16.11-16.12 For the beam and loading shown, determine the reaction at the roller support B.

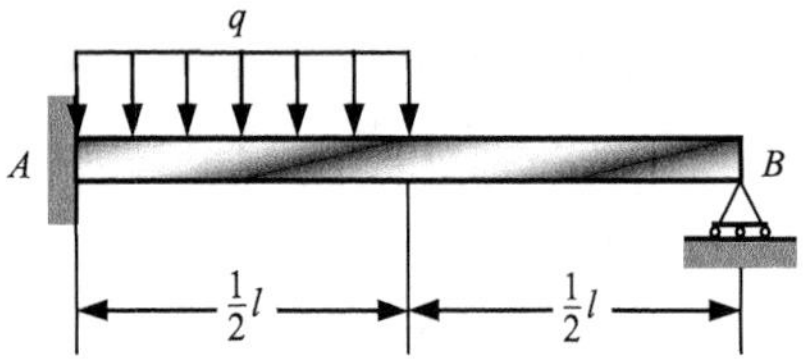

Fig. P16.11

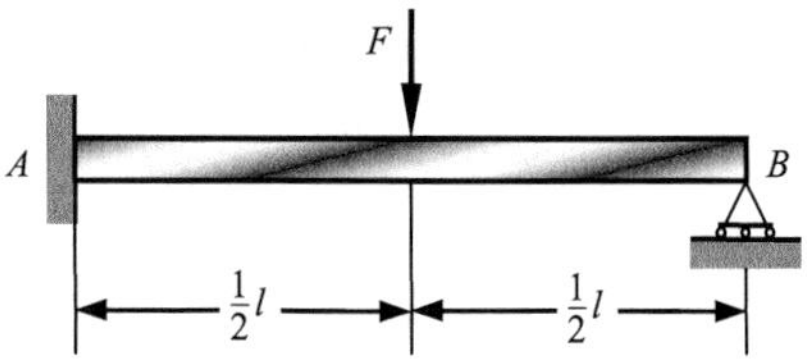

Fig. P16.12

Chapter 17 Stress Analysis and Strength Theories

The most general state of stress at a given point P may be represented by six components. Three of these components, σ_x, σ_y, and σ_z, define the normal stresses acting on the faces of a small cubic element centered at P and having the same orientation as the coordinate axes, Fig. 17.1, and the other three, $\tau_{xy}(\tau_{yx}=\tau_{xy})$, $\tau_{yz}(\tau_{zy}=\tau_{yz})$, and $\tau_{zx}(\tau_{xz}=\tau_{zx})$, define the shearing stresses on the same element. The same state of stress can be represented by a different set of components if the coordinate axes are rotated. The discussion of the stress state will mainly deal with plane stress, i.e., with a situation in which two parallel faces of the cubic element are free of any stress, Fig. 17.2.

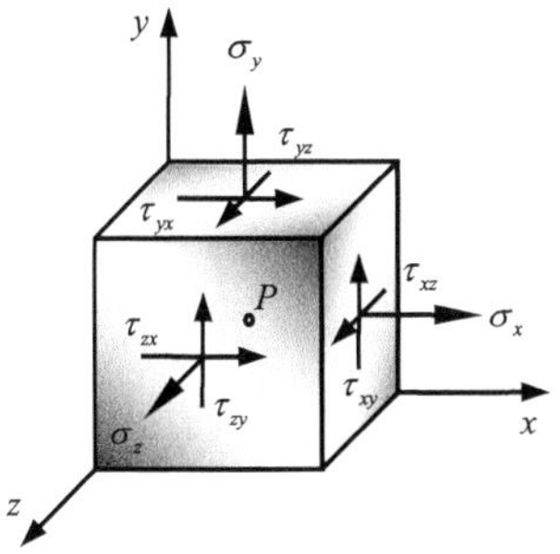

Fig. 17.1

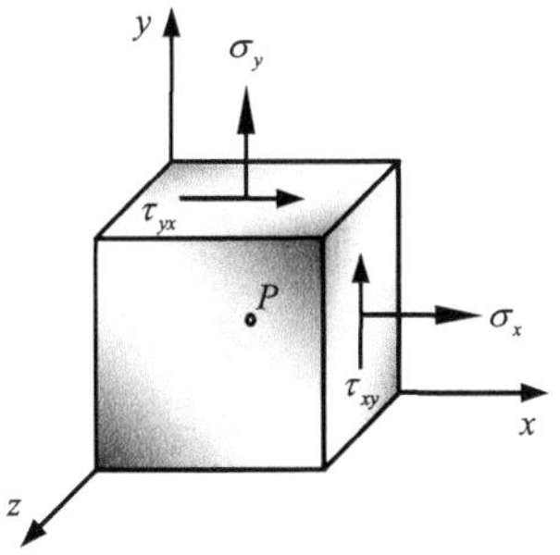

Fig. 17.2

17.1 Stress Transformation

Consider that a state of plane stress exists at point P and is characterized by the stress components σ_x, σ_y, and $\tau_x(\tau_y=\tau_x)$, Fig. 17.3a. We now determine the stress components $\sigma_{x'}$, $\sigma_{y'}$, and $\tau_{x'}(\tau_{y'}=\tau_{x'})$ after it has been rotated through an angle θ counterclockwise about the z axis, Fig. 17.3b.

Consider an element with faces respectively perpendicular to the x, y, and x' axes, Fig. 17.4. We observe that, if the area of the oblique face is denoted by ΔA, the areas of the vertical and horizontal faces are respectively equal to $\Delta A\cos\theta$ and $\Delta A\sin\theta$.

Using the equations of static equilibrium along the x' and y' axes, respectively, we have

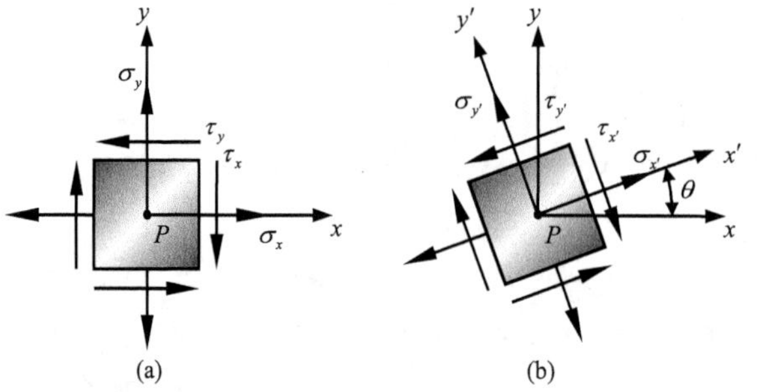

Fig. 17.3

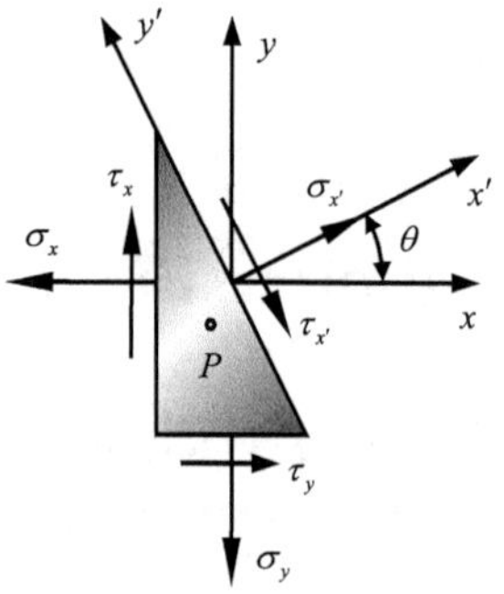

Fig. 17.4

$$
\begin{aligned}
\sum F_{x'} = 0:\quad & \sigma_{x'}\Delta A - \sigma_x(\Delta A\cos\theta)\cos\theta + \tau_x(\Delta A\cos\theta)\sin\theta - \\
& \sigma_y(\Delta A\sin\theta)\sin\theta + \tau_y(\Delta A\sin\theta)\cos\theta = 0 \\
\sum F_{y'} = 0:\quad & -\tau_{x'}\Delta A + \sigma_x(\Delta A\cos\theta)\sin\theta + \tau_x(\Delta A\cos\theta)\cos\theta - \\
& \sigma_y(\Delta A\sin\theta)\cos\theta - \tau_y(\Delta A\sin\theta)\sin\theta = 0
\end{aligned}
\tag{17.1}
$$

Solving for $\sigma_{x'}$ and $\tau_{x'}$, and using $\tau_y = \tau_x$, $\cos^2\theta = \frac{1}{2}(1+\cos 2\theta)$, $\sin^2\theta = \frac{1}{2}(1-\cos 2\theta)$, and $2\sin\theta\cos\theta = \sin 2\theta$, we obtain

$$
\begin{aligned}
\sigma_{x'} &= \frac{1}{2}(\sigma_x + \sigma_y) + \frac{1}{2}(\sigma_x - \sigma_y)\cos 2\theta - \tau_x \sin 2\theta \\
\tau_{x'} &= \frac{1}{2}(\sigma_x - \sigma_y)\sin 2\theta + \tau_x \cos 2\theta
\end{aligned}
\tag{17.2}
$$

If $\sigma_{y'}$ is needed, it can be obtained by substituting $\theta + \frac{1}{2}\pi$ for θ into Eq. (17.2). This yields

$$
\sigma_{y'} = \frac{1}{2}(\sigma_x + \sigma_y) - \frac{1}{2}(\sigma_x - \sigma_y)\cos 2\theta + \tau_x \sin 2\theta \tag{17.3}
$$

Example 17.1 For the given stress state shown in Fig. E17.1a, determine the normal and shearing stresses after the element has been rotated through 20° clockwise.

Solution According to the reference system given in Fig. E17.1a, then we have

$$\sigma_x = 60\ \text{MPa},\ \sigma_y = -40\ \text{MPa},\ \tau_x = -50\ \text{MPa},\ \theta = -20°$$

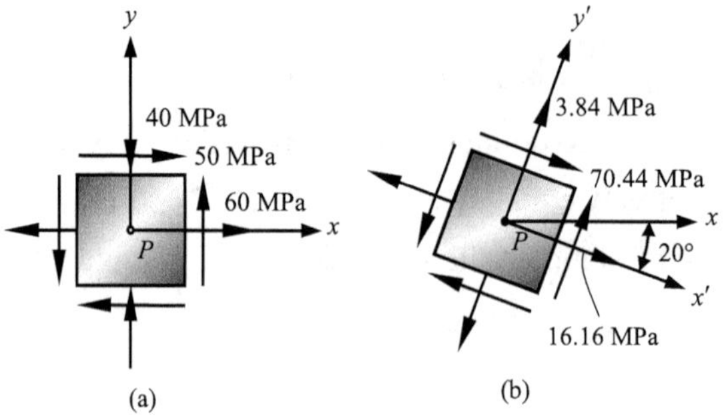

Fig. E17.1

Using Eqs. (17.2) and (17.3), we obtain

$$\sigma_{x'} = \frac{1}{2}(\sigma_x + \sigma_y) + \frac{1}{2}(\sigma_x - \sigma_y)\cos 2\theta - \tau_x \sin 2\theta = 16.16 \text{ MPa}$$

$$\sigma_{y'} = \frac{1}{2}(\sigma_x + \sigma_y) - \frac{1}{2}(\sigma_x - \sigma_y)\cos 2\theta + \tau_x \sin 2\theta = 3.84 \text{ MPa}$$

$$\tau_{x'} = \frac{1}{2}(\sigma_x - \sigma_y)\sin 2\theta + \tau_x \cos 2\theta = -70.44 \text{ MPa}$$

The normal and shearing stresses are shown in Fig. E17.1b after the element has been rotated through 20° clockwise.

17.2 Principal Stresses

Setting $\sigma_{\text{avg}} = \frac{1}{2}(\sigma_x + \sigma_y)$, $R = \sqrt{[\frac{1}{2}(\sigma_x - \sigma_y)]^2 + \tau_x^2}$, and using Eq. (17.2), we have

$$(\sigma_{x'} - \sigma_{\text{avg}})^2 + \tau_{x'}^2 = R^2 \tag{17.4}$$

which is the equation of a circle of radius R centered at point C of abscissa σ_{avg} and ordinate 0, Fig. 17.5. This circle is called Mohr's circle or the stress circle.

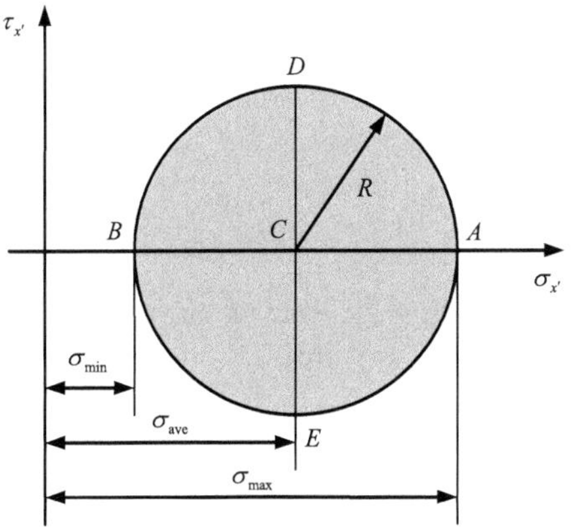

Fig. 17.5

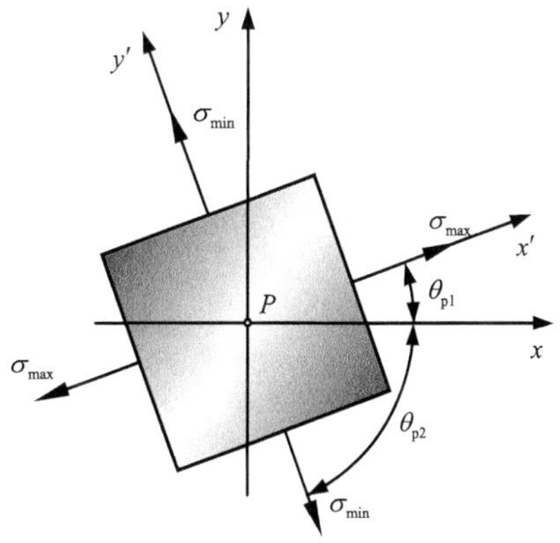

Fig. 17.6

The two points A and B where the circle of Fig. 17.5 intersects the horizontal axis are of special interest: Point A corresponds to the maximum normal stress σ_{max}, while point B corresponds to the minimum normal stress σ_{min}. Besides, these both points correspond to zero value of the shearing stress. The directions of the maximum and minimum normal stresses corresponding to A and B can be obtained by setting $\tau_{x'} = 0$ in Eq. (17.2), i.e.,

$$\tan 2\theta_{\text{p}} = \frac{-2\tau_x}{\sigma_x - \sigma_y} \tag{17.5}$$

This equation defines two values θ_{p1} and θ_{p2} that are 90° apart. Either of these values can be used to determine the orientation of the corresponding element, Fig. 17.6.

The plane containing a maximum or minimum normal stress is called the principal plane at point *P*, and the corresponding maximum or minimum normal stress is called the in-plane principal stress at point *P*. It is clear that no shearing stress is exerted on a principal plane. According to Eq. (17.4), we have

$$\sigma_{\max,\min} = \sigma_{\text{avg}} \pm R = \frac{1}{2}(\sigma_x + \sigma_y) \pm \sqrt{[\frac{1}{2}(\sigma_x - \sigma_y)]^2 + \tau_x^2} \tag{17.6}$$

There exist three perpendicular principal stresses at any point within a homogenous isotropic material subjected to plane stress. If $\sigma_{\min} \geqslant 0$, three principal stresses can be given by $\sigma_1 = \sigma_{\max}$, $\sigma_2 = \sigma_{\min}$ and $\sigma_3 = 0$; if $\sigma_{\max} \geqslant 0 \geqslant \sigma_{\min}$, then $\sigma_1 = \sigma_{\max}$, $\sigma_2 = 0$ and $\sigma_3 = \sigma_{\min}$; if $\sigma_{\max} \leqslant 0$, then $\sigma_1 = 0$, $\sigma_2 = \sigma_{\max}$ and $\sigma_3 = \sigma_{\min}$.

Example 17.2 For the given stress state, Fig. E17.2a, determine the principal stresses and the corresponding principal planes.

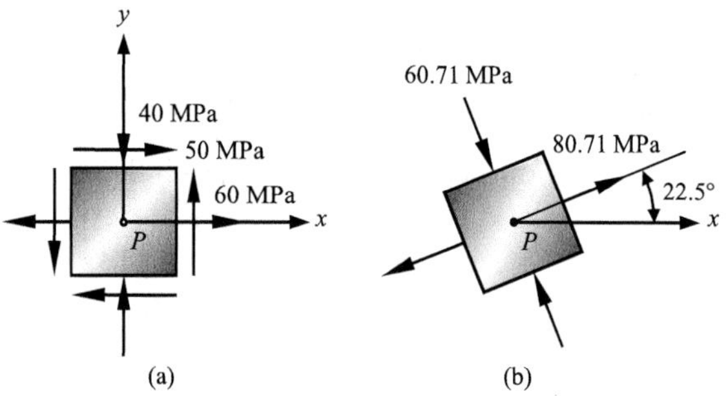

Fig. E17.2

Solution Referring to the coordinate system given in Fig. E17.2a, we have

$$\sigma_x = 60 \text{ MPa}, \sigma_y = -40 \text{ MPa}, \tau_x = -50 \text{ MPa}$$

Using Eqs. (17.6), we obtain the in-plane principal stresses as follows:

$$\sigma_{\max} = \frac{1}{2}(\sigma_x + \sigma_y) + \sqrt{[\frac{1}{2}(\sigma_x - \sigma_y)]^2 + \tau_x^2} = 80.71 \text{ MPa}$$

$$\sigma_{\min} = \frac{1}{2}(\sigma_x + \sigma_y) - \sqrt{[\frac{1}{2}(\sigma_x - \sigma_y)]^2 + \tau_x^2} = -60.71 \text{ MPa}$$

Using Eq. (17.5), we can obtain the orientations of the in-plane principal stresses

$$\theta_p = \frac{1}{2}\arctan\frac{-2\tau_x}{\sigma_x - \sigma_y} = \begin{matrix} 22.5^\circ & (\text{corresponding to } \sigma_{\max}) \\ -67.5^\circ & (\text{corresponding to } \sigma_{\min}) \end{matrix}$$

The principal stresses and the corresponding principal planes are shown in Fig. E17.2b. Therefore, three principal stresses are $\sigma_1 = 80.71 \text{ MPa}$, $\sigma_2 = 0$, and $\sigma_3 = -60.71 \text{ MPa}$.

17.3 Maximum Shearing Stress

Referring to the stress circle shown in Fig. 17.5, points *D* and *E* located on the vertical diameter correspond to the largest and smallest shearing stresses. Since the abscissa of D and E is $\sigma_{\text{avg}}=\frac{1}{2}(\sigma_x+\sigma_y)$, the directions of the maximum and minimum shearing stresses corresponding to these points can be obtained by setting $\sigma_{x'}=\frac{1}{2}(\sigma_x+\sigma_y)$ in Eq. (17.2). Thus, we have

$$\tan 2\theta_{\text{s}}=\frac{\sigma_x-\sigma_y}{2\tau_x} \tag{17.7}$$

This equation defines two values θ_{s1} and θ_{s2} which are 90° apart. Either of these values can be used to determine the orientation of the element corresponding to the maximum in-plane shearing stress, Fig. 17.7.

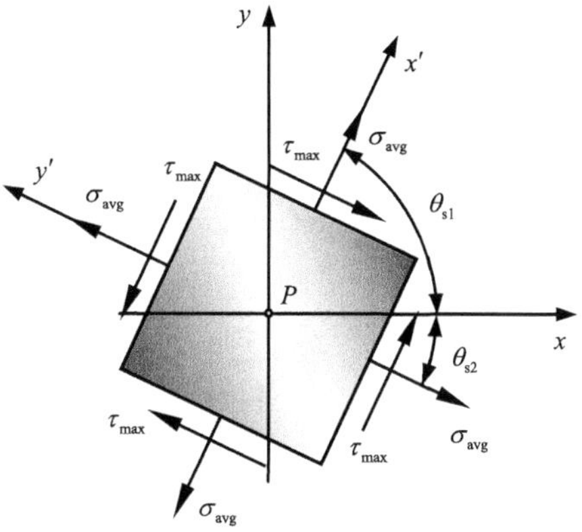

Fig. 17.7

From Eq. (17.4), we have

$$(\tau_{\max})_{\text{in-plane}}=R=\sqrt{[\frac{1}{2}(\sigma_x-\sigma_y)]^2+\tau_x^2} \tag{17.8}$$

and the normal stress on the plane containing the maximum in-plane shearing stress is

$$\sigma_{\text{avg}}=\frac{1}{2}(\sigma_x+\sigma_y) \tag{17.9}$$

We note from Eqs. (17.5) and (17.7) that $\tan 2\theta_{\text{s}}\cdot\tan 2\theta_{\text{p}}=-1$ and thus conclude that the plane containing a maximum in-plane shearing stress is at 45° to the principal plane.

We should note that the shearing stresses $\tau_{\max}=\frac{1}{2}(\sigma_1-\sigma_3)$ may be larger than the shearing stress $(\tau_{\max})_{\text{in-plane}}$ defined by Eq. (17.8). This occurs when the principal stresses defined by Eq. (17.6) have the same sign, i.e., when they are either both tensile or both compressive.

Example 17.3 For the given stress state, determine (a), the maximum in-plane shearing stress, (b) the orientation of the planes of maximum in-plane shearing stress, (c) the normal stress on the planes of maximum in-plane shearing stress.

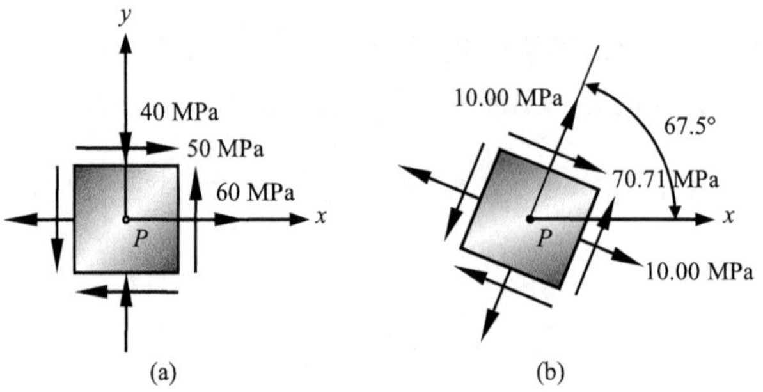

Fig. E17.3

Solution For the coordinate system given in Fig. E17.3a, we have

$$\sigma_x = 60 \text{ MPa}, \ \sigma_y = -40 \text{ MPa}, \ \tau_x = -50 \text{ MPa}$$

(a) Using Eq. (17.8), we obtain the maximum in-plane shearing stresses:

$$(\tau_{\max})_{\text{in-plane}} = \sqrt{[\frac{1}{2}(\sigma_x - \sigma_y)]^2 + \tau_x^2} = 70.71 \text{ MPa}$$

(b) Using Eq. (17.7), we obtain the orientation of the planes of maximum in-plane shearing stress:

$$\theta_s = \frac{1}{2}\arctan\frac{\sigma_x - \sigma_y}{2\tau_x} = \begin{matrix} 67.5° \\ -22.5° \end{matrix}$$

(c) Using Eq. (17.9), we obtain the normal stress on the planes of maximum in-plane shearing stress:

$$\sigma_{\text{avg}} = \frac{1}{2}(\sigma_x + \sigma_y) = 10.00 \text{ MPa}$$

The maximum in-plane shearing stress and the normal stress on the planes of maximum in-plane shearing stress, as well as the directions of these stresses, are shown in Fig. E17.3b.

17.4 Pressure Vessels

Thin-walled pressure vessels provide an important application of plane stress. The analysis of stresses in thin-walled pressure vessels will be limited to the two types of vessels most frequently encountered: cylindrical pressure vessel and spherical pressure vessel.

1. Cylindrical Pressure Vessel

Consider a cylindrical vessel of inner diameter d and wall thickness t subjected to pressure p, Fig. 17.8. We now determine the stresses exerted on a small element of wall with sides respectively parallel and perpendicular to the longitudinal axis of the cylinder.

It is clear that no shearing stress is exerted on the element. The normal stresses σ_1 and σ_2 shown in Fig. 17.8 are therefore principal stresses. The stress σ_1 is known as the hoop or circumferential stress, and the stress σ_2 is called the longitudinal or axial stress.

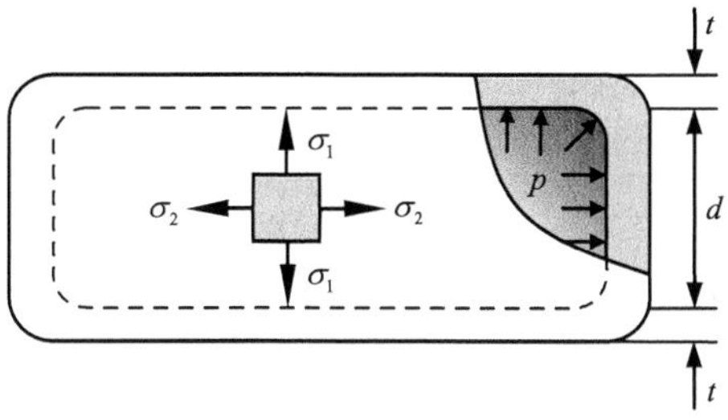

Fig. 17.8

To determine the longitudinal stress σ_2, considering the equilibrium of the free body shown in Fig. 17.9, we have

$$\sum F_x = 0: \quad \sigma_2(\pi dt) - p(\frac{1}{4}\pi d^2) = 0 \tag{17.10}$$

Solving this equation for σ_2, then we obtain

$$\sigma_2 = \frac{pd}{4t} \tag{17.11}$$

In order to determine the hoop stress σ_1, we detach a portion of the vessel, Fig. 17.10. Using the condition of equilibrium, we have

$$\sum F_z = 0: \quad \sigma_1(2t\Delta x) - p(d\Delta x) = 0 \tag{17.12}$$

Solving for σ_1, then we obtain the hoop stress:

$$\sigma_1 = \frac{pd}{2t} \tag{17.13}$$

Using Eqs. (17.11) and (17.13), we have the maximum shearing stresses as follows:

$$(\tau_{\max})_{\text{in-plane}} = \frac{1}{2}(\sigma_1 - \sigma_2) = \frac{pd}{8t}, \quad \tau_{\max} = \frac{1}{2}(\sigma_1 - \sigma_3) = \frac{pd}{4t} \tag{17.14}$$

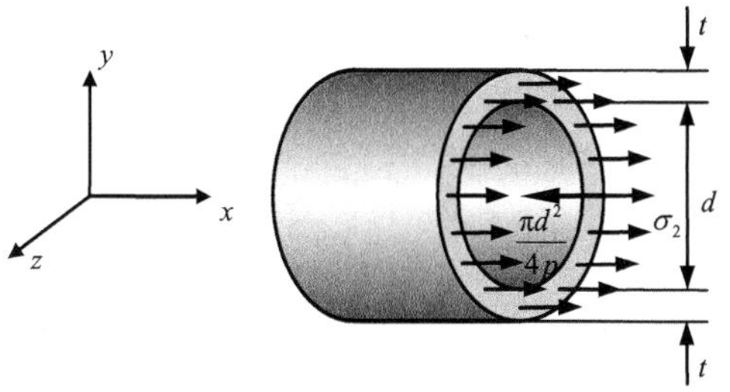

Fig. 17.9

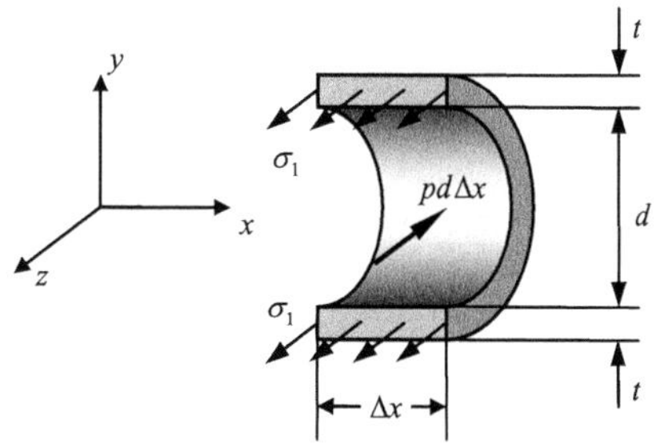

Fig. 17.10

2. Spherical Pressure Vessel

We now consider a spherical vessel of inner diameter d and wall thickness t under pressure p, Fig. 17.11a. Using the symmetry of vessel, the two in-plane principal stresses must be equal, i.e., $\sigma_1 = \sigma_2$.

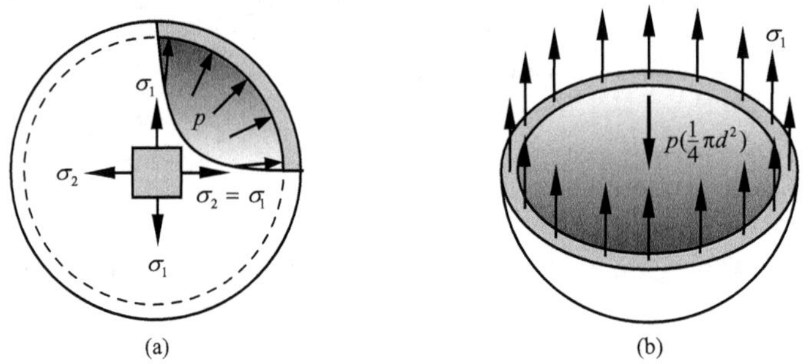

Fig. 17.11

To determine the stress σ_1 (or σ_2), we consider the free body shown in Fig. 17.11b. From the condition of static equilibrium of the free body in the vertical direction, we have

$$\sigma_1(\pi dt) - p\left(\frac{1}{4}\pi d^2\right) = 0 \tag{17.15}$$

Solving the above equation for σ_1, we have

$$\sigma_1 = \sigma_2 = \frac{pd}{4t} \tag{17.16}$$

Using Eq. (17.16), we obtain the maximum shearing stresses

$$(\tau_{\max})_{\text{in-plane}} = \frac{1}{2}(\sigma_1 - \sigma_2) = 0, \quad \tau_{\max} = \frac{1}{2}(\sigma_1 - \sigma_3) = \frac{pd}{8t} \tag{17.17}$$

17.5 Generalized Hooke's Law

Consider an element with side lengths dx, dy and dz, Fig. 17.12a. When this element is subjected to principal stresses σ_1, σ_2 and σ_3, it will deform into a new element having side lengths $(1+\varepsilon_1)dx$, $(1+\varepsilon_2)dy$ and $(1+\varepsilon_3)dz$, Fig. 17.12b, where ε_1, ε_2 and ε_3 are the normal strains corresponding to the principal stresses σ_1, σ_2 and σ_3, respectively.

Assuming that the material is homogenous and isotropic and behaves in a linear-elastic manner, then we can determine the relations between the stresses and strains by using the principle of superposition.

We first consider the normal strain in the x direction. When σ_1 is applied, the element elongates in the x direction and the normal strain ε_1' in this direction, caused by σ_1, is

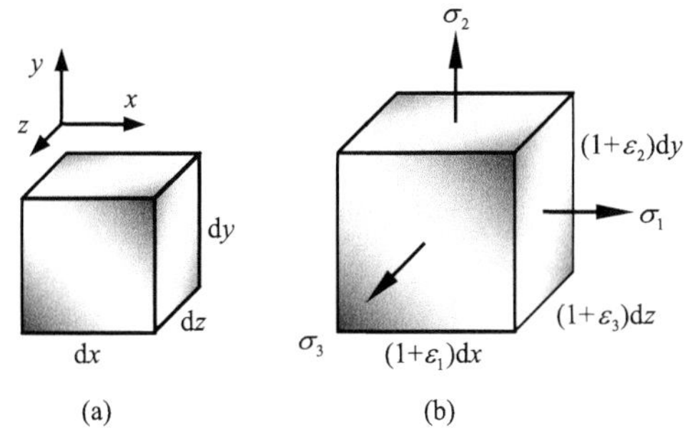

Fig. 17.12

$$\varepsilon_1' = \frac{\sigma_1}{E} \tag{17.18}$$

Application of σ_2 causes the element to contract with a strain ε_1'' in the x direction and ε_1'' can be expressed as

$$\varepsilon_1'' = -\mu\frac{\sigma_2}{E} \tag{17.19}$$

Similarly, σ_3 will cause a contraction of the element in the x direction and the normal strain ε_1''' along this direction can be written as

$$\varepsilon_1''' = -\mu\frac{\sigma_3}{E} \tag{17.20}$$

Using the principle of superposition, the normal strain ε_1 in the x direction, caused by σ_1, σ_2 and σ_3, can be obtained by

$$\varepsilon_1 = \varepsilon_1' + \varepsilon_1'' + \varepsilon_1''' = \frac{1}{E}[\sigma_1 - \mu(\sigma_2 + \sigma_3)] \tag{17.21}$$

Likewise, the normal strains in the y and z directions can also be obtained by using the principle of superposition. Thus the relations between the stresses and strains for a stress state subjected to three principal stresses can expressed as

$$\varepsilon_1 = \frac{1}{E}[\sigma_1 - \mu(\sigma_2 + \sigma_3)],\ \varepsilon_2 = \frac{1}{E}[\sigma_2 - \mu(\sigma_3 + \sigma_1)],\ \varepsilon_3 = \frac{1}{E}[\sigma_3 - \mu(\sigma_1 + \sigma_2)] \tag{17.22}$$

The above relations are referred to as the generalized Hooke's law for the state of principal stresses of a homogeneous isotropic material. The results obtained are valid only if the stresses do not exceed the proportional limit.

Example 17.4 A 40 mm square is scribed on the side of a large steel pressure vessel. After pressurization the biaxial stress condition at the square is as shown in Fig. E17.4. Knowing that E=200 GPa and μ=0.30, determine the change in length of side AB, side BC, and diagonal AC.

Solution Knowing that $\sigma_1 = 60\text{ MPa}, \sigma_2 = 30\text{ MPa}, \sigma_3 = 0$, then from generalized Hooke's law, we have

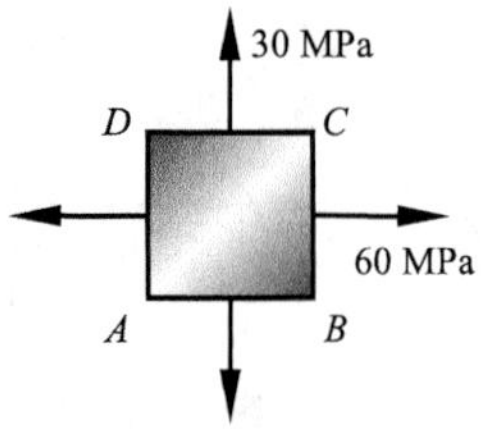

Fig. E17.4

$$\varepsilon_1 = \frac{1}{E}[\sigma_1 - \mu(\sigma_2 + \sigma_3)] = 255 \times 10^{-6},\ \varepsilon_2 = \frac{1}{E}[\sigma_2 - \mu(\sigma_3 + \sigma_1)] = 60 \times 10^{-6}$$

Using $\varepsilon_{AB} = \delta_{AB}/l_{AB} = \varepsilon_1$, we have

$$\delta_{AB} = \varepsilon_1 l_{AB} = 10.2 \times 10^{-3}\ \text{mm}$$

Similarly, we can obtain

$$\delta_{BC} = \varepsilon_2 l_{BC} = 2.4 \times 10^{-3}\ \text{mm}$$

Using $\delta_{AC} = \sqrt{(l_{AB} + \delta_{AB})^2 + (l_{BC} + \delta_{BC})^2} - \sqrt{l_{AB}{}^2 + l_{BC}{}^2}$, we have

$$\delta_{AC} = 8.9 \times 10^{-3}\ \text{mm}$$

Applying the principle of superposition to the stress state shown in Fig. 17.13, we can obtain the following the generalized Hooke's law for a homogeneous isotropic material under the most general stress state if the stresses do not exceed the proportional limit:

$$\begin{aligned} &\varepsilon_x = \frac{1}{E}[\sigma_x - \mu(\sigma_y + \sigma_z)],\ \varepsilon_y = \frac{1}{E}[\sigma_y - \mu(\sigma_z + \sigma_x)],\ \varepsilon_z = \frac{1}{E}[\sigma_z - \mu(\sigma_x + \sigma_y)] \\ &\gamma_{xy} = \frac{\tau_{xy}}{G},\ \gamma_{yz} = \frac{\tau_{yz}}{G},\ \gamma_{zx} = \frac{\tau_{zx}}{G} \end{aligned} \tag{17.23}$$

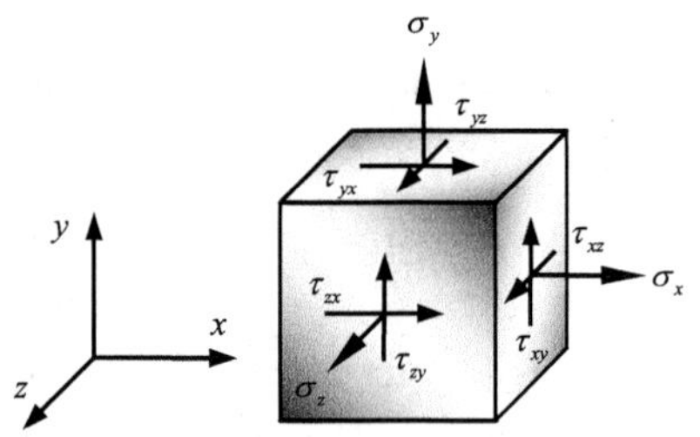

Fig. 17.13

17.6 Strength Theories

A structural member made of a ductile or brittle material is usually designed so that that material will not fail under expected loadings. If the material is ductile, the failure is usually specified by yielding, whereas if the material is brittle, it is specified by fracture.

A member subjected to a uniaxial stress is safe as long as $\sigma < \sigma_s$ for a ductile material, where σ_s is the yield strength of the material, or $\sigma < \sigma_u$ for a brittle material, where σ_u is the ultimate strength of the material, Fig. 17.14.

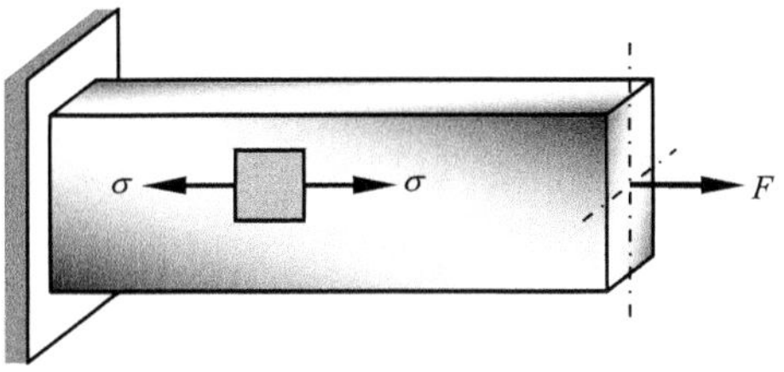

Fig. 17.14

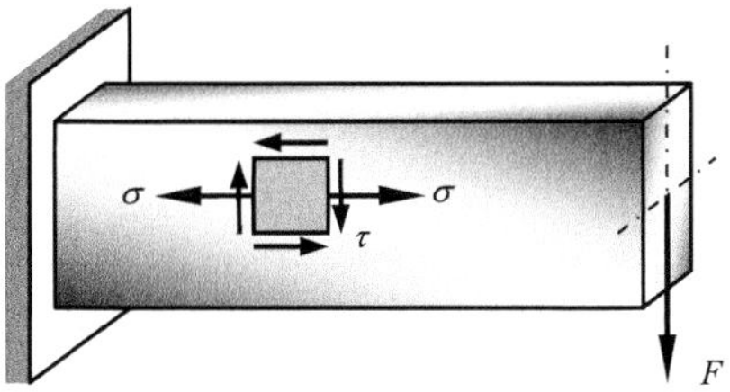

Fig. 17.15

When a member is in a state of plane stress or biaxial stress, it is impossible to predict whether or not the member will fail by using the test, Fig. 17.15. Thus, for a complex stress state, some theory regarding the failure of materials must be established. Such theory is called strength theory or failure theory.

Two types of strength theory are available: one applicable to brittle fracture and the other suitable for ductile yielding.

1. Maximum-Normal-Stress Theory

This theory states that the failure of materials subjected to a biaxial or triaxial stress occurs when the maximum normal stress reaches the limit value at which the failure occurs in a tension test on the same material. This theory is based on the fact that the brittle fracture is caused by the maximum normal stress.

Assume that three principal stresses are represented by σ_1, σ_2 and σ_3, and sorted from large to small, i.e., $\sigma_1 \geqslant \sigma_2 \geqslant \sigma_3$. According to the maximum-normal-stress theory, a given member fails when the maximum normal stress in the member reaches the ultimate strength obtained from the tensile test on the same material, thus the failure criterion can be expressed as

$$\sigma_{eq1} = \sigma_1 = \sigma_u \tag{17.24}$$

where σ_{eq1} is the equivalent stress corresponding to the maximum-normal-stress theory, and σ_u is the ultimate strength of the material obtained from the tensile test.

Assuming that the factor of safety is denoted by n, then the strength condition for the material subjected to a biaxial or triaxial stress can expressed as

$$\sigma_{eq1} = \sigma_1 \leqslant \sigma_{allow} = \frac{\sigma_u}{n} \tag{17.25}$$

where σ_{allow} is the allowable stress, and n is the factor of safety. The maximum-normal-stress theory is in good agreement with experimental evidence on brittle fracture.

Example 17.5 The state of plane stress shown occurs in a member made of cast-iron with σ_{ut}=160 MPa and σ_{uc}=320 MPa. Using the maximum-normal-stress theory, determine whether fracture occurs. If fracture does not occur, determine the corresponding factor of safety.

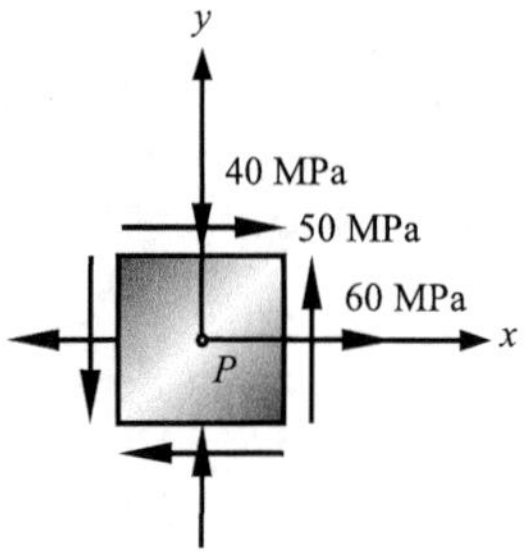

Fig. E17.5

Solution Referring to Example 17.2, we have

$$\sigma_1 = 80.71\ \text{MPa},\ \sigma_2 = 0,\ \text{and}\ \sigma_3 = -60.71\ \text{MPa}$$

Since $\sigma_{eq1} = \sigma_1 = 80.71\ \text{MPa} < \sigma_{ut} = 160\ \text{MPa}$, then fracture does not occur. The factor of safety is equal to

$$n = \frac{\sigma_{ut}}{\sigma_{eq1}} = 1.98$$

2. Maximum-Normal-Strain Theory

This theory states that the failure of materials subjected to a biaxial or triaxial stress occurs when the maximum normal strain reaches the limit value at which the failure occurs in a tension test on the same material. This theory is based on the fact that the brittle fracture is caused by the maximum normal strain.

According to the maximum-normal-strain theory, a given member fails when the maximum normal strain in the member reaches the ultimate tensile strain obtained from the tensile test on the same material, thus the failure criterion can be expressed as

$$\varepsilon_1 = \frac{1}{E}[\sigma_1 - \mu(\sigma_2 - \sigma_3)] = \frac{1}{E}\sigma_b \tag{17.26}$$

where ε_1 is the largest normal strain, and μ is the Poisson's ratio. Alternatively, this criterion can also expressed as

$$\sigma_{eq2} = \sigma_1 - \mu(\sigma_2 - \sigma_3) = \sigma_b \tag{17.27}$$

where σ_{eq2} is the equivalent stress corresponding to the maximum-normal-strain theory. Therefore, the strength condition for the material subjected to a biaxial or triaxial stress can expressed as

$$\sigma_{eq2} = \sigma_1 - \mu(\sigma_2 - \sigma_3) \leqslant \sigma_{allow} = \frac{\sigma_b}{n} \tag{17.28}$$

where σ_{allow} is the allowable stress, and n is the factor of safety. At present, the maximum-normal-strain theory has been seldom used.

3. Maximum-Shearing-Stress Theory

This theory states that the failure of materials subjected to a biaxial or triaxial stress occurs when the maximum shearing stress reaches the limit value at which the failure occurs in a tension test on the same material. This theory is based on the fact that ductile yielding is caused by the maximum shearing stress.

According to the maximum-shearing-stress theory, a given member fails when the maximum shearing stress in the member reaches the yield strength obtained from the tensile test on the same material, thus the failure criterion can be expressed as

$$\tau_{max} = \frac{1}{2}(\sigma_1 - \sigma_3) = \frac{1}{2}\sigma_s \tag{17.29}$$

where τ_{max} is the largest shearing stress, and σ_s is the yield strength of the material obtained from the tensile test. Alternatively, this criterion can also expressed as

$$\sigma_{eq3} = \sigma_1 - \sigma_3 = \sigma_s \tag{17.30}$$

where σ_{eq3} is the equivalent stress corresponding to the maximum-shearing-stress theory. Therefore, the strength condition for the material subjected to a biaxial or triaxial stress can expressed as

$$\sigma_{eq3} = \sigma_1 - \sigma_3 \leqslant \sigma_{allow} = \frac{\sigma_s}{n} \tag{17.31}$$

where σ_{allow} is the allowable stress, and n is the factor of safety. The maximum-shearing-stress theory is widely used for ductile yielding.

Example 17.6 The state of plane stress shown occurs in a member made of steel with σ_s=310 MPa. Using the maximum-shearing-stress theory, determine whether yielding occurs. If yielding does not occur, determine the corresponding factor of safety.

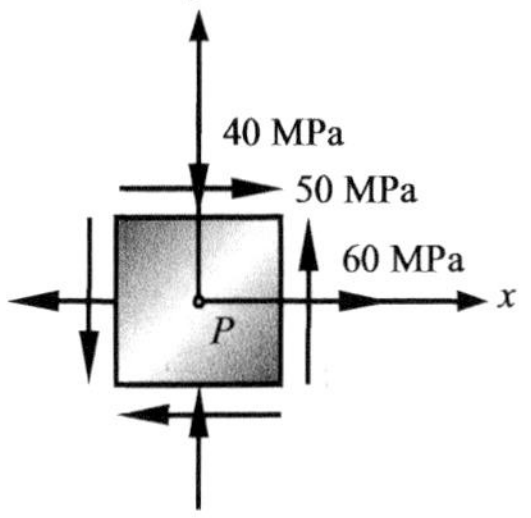

Fig. E17.6

Solution Referring to Example 17.2, we have

$$\sigma_1 = 80.71 \text{ MPa},\ \sigma_2 = 0,\ \text{and}\ \sigma_3 = -60.71 \text{ MPa}$$

Since $\sigma_{\text{eq3}} = \sigma_1 - \sigma_3 = 141.42\ \text{MPa} < \sigma_s = 310\ \text{MPa}$, then yielding does not occur. The factor of safety is equal to

$$n = \frac{\sigma_s}{\sigma_{\text{eq3}}} = 2.19$$

4. Maximum-Distortion-Energy Theory

This theory states that the yielding of materials subjected to a biaxial or triaxial stress occurs when the distortion-energy density (distortion energy per unit volume) reaches the limit value at which the failure occurs in a tension test on the same material. This theory is based on the fact that ductile yielding is caused by the maximum distortion energy associated with change in shape of the material.

According to the maximum-distortion-energy theory, a given member fails when the maximum distortion energy in the member reaches the ultimate distortion energy obtained from the tensile test on the same material, thus the failure criterion can be expressed as

$$u_d = \frac{(1+\mu)}{6E}[(\sigma_1 - \sigma_2)^2 + (\sigma_2 - \sigma_3)^2 + (\sigma_3 - \sigma_1)^2] = \frac{(1+\mu)}{6E} 2\sigma_s^{\ 2} \tag{17.32}$$

where u_d is the distortion-energy density corresponding to a biaxial or triaxial stress, and E is the modulus of elasticity. Alternatively, this criterion can also expressed as

$$\sigma_{\text{eq4}} = \sqrt{\frac{1}{2}[(\sigma_1 - \sigma_2)^2 + (\sigma_2 - \sigma_3)^2 + (\sigma_3 - \sigma_1)^2]} = \sigma_s \tag{17.33}$$

where σ_{eq4} is the equivalent stress corresponding to the maximum-distortion-energy theory. Therefore, the strength condition for the material subjected to a biaxial or triaxial stress can expressed as

$$\sigma_{\text{eq4}} = \sqrt{\frac{1}{2}[(\sigma_1 - \sigma_2)^2 + (\sigma_2 - \sigma_3)^2 + (\sigma_3 - \sigma_1)^2]} \leqslant \sigma_{\text{allow}} = \frac{\sigma_s}{n} \tag{17.34}$$

where σ_{allow} is the allowable stress, and n is the factor of safety. The maximum-distortion-energy theory is excellent agreement with experiments on ductile yielding.

Example 17.7 The state of plane stress shown occurs in a member made of steel with σ_s=310 MPa. Using the maximum-distortion-energy theory, determine whether yielding occurs. If yielding does not occur, determine the corresponding factor of safety.

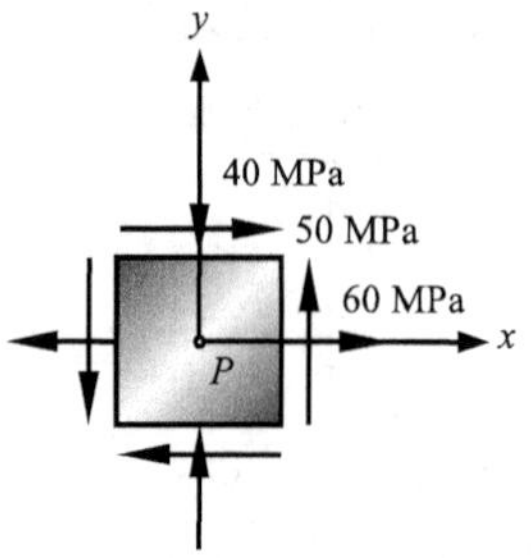

Fig. E17.7

Solution Referring to Example 17.2, we have

$$\sigma_1 = 80.71 \text{ MPa}, \ \sigma_2 = 0, \ \text{and} \ \sigma_3 = -60.71 \text{ MPa}$$

Since $\sigma_{\text{eq4}} = \sqrt{\frac{1}{2}[(\sigma_1 - \sigma_2)^2 + (\sigma_2 - \sigma_3)^2 + (\sigma_3 - \sigma_1)^2]} = 122.88 \text{ MPa} < \sigma_s = 310 \text{ MPa}$, then yielding does not occur. The factor of safety is equal to

$$n = \frac{\sigma_s}{\sigma_{\text{eq4}}} = 2.52$$

Problems

17.1 For the given stress state, determine the normal and shearing stresses after the element shown has been rotated through 30° anticlockwise.

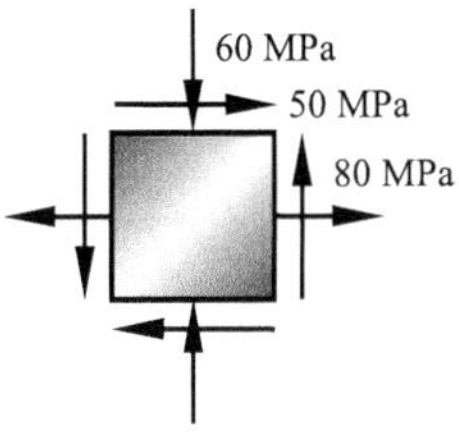

Fig. P17.1

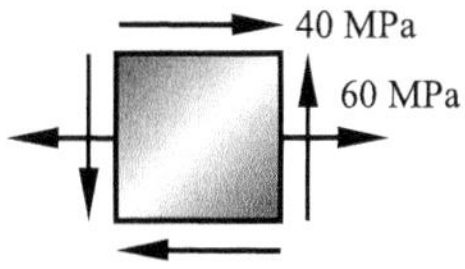

Fig. P17.2

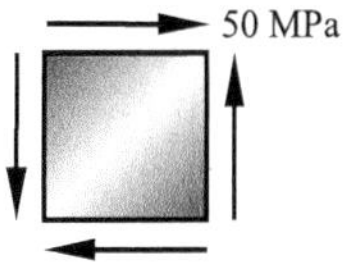

Fig. P17.3

17.2 For the given stress state, determine the normal and shearing stresses after the element shown has been rotated through 15° clockwise.

17.3 For the given stress state, determine the normal and shearing stresses after the element shown has been rotated through 55° clockwise.

17.4 For the given stress state, determine the normal and shearing stresses after the element shown has been rotated through 45° anticlockwise.

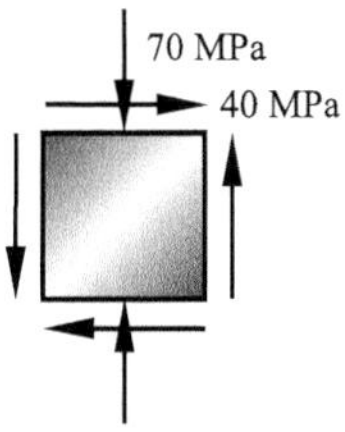

Fig. P17.4

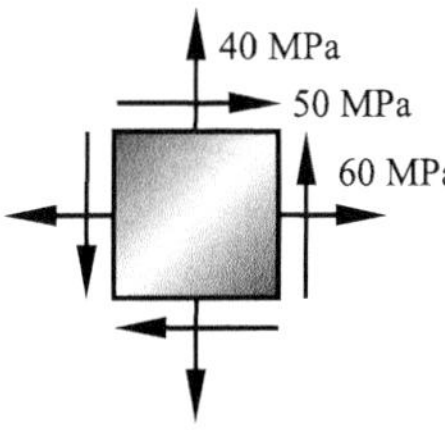

Fig. P17.5

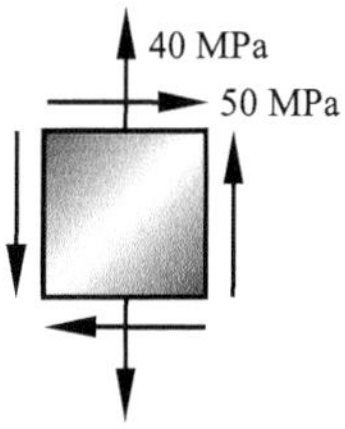

Fig. P17.6

17.5-17.8 For the given stress state, determine the principal stresses and the corresponding principal planes.

17.9 For the given stress state, determine (a) the maximum in-plane shearing stress, (b) the orientation of the planes of maximum in-plane shearing stress, (c) the normal stress on the planes of maximum in-plane shearing stress.

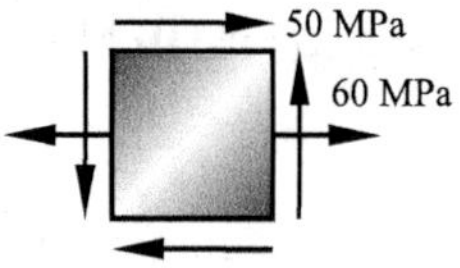

Fig. P17.7

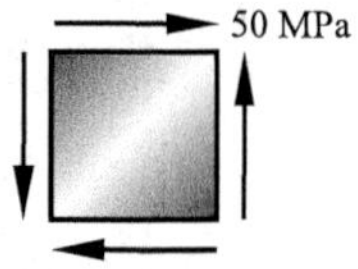

Fig. P17.8

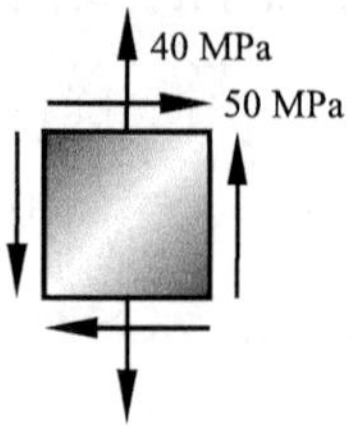

Fig. P17.9

17.10　A 40 mm square plate is subjected to plane stress shown. Knowing that E = 200 GPa and μ=0.30, determine the change in length of side AB, side BC, and diagonal AC.

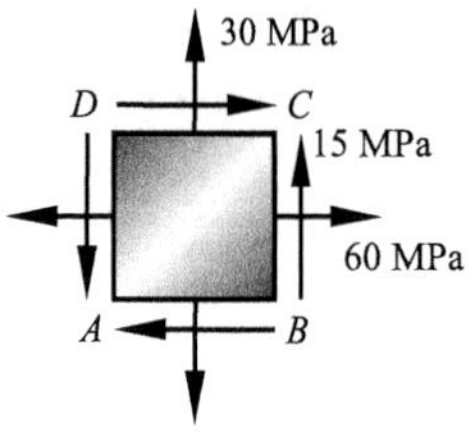

Fig. P1710

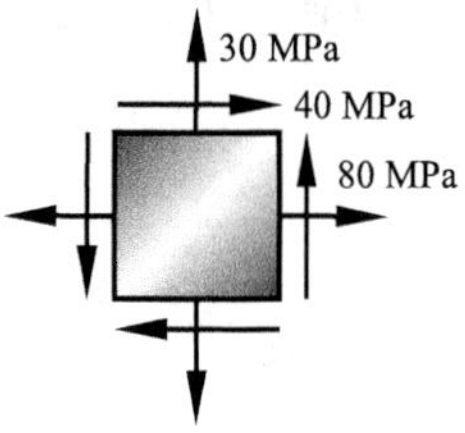

Fig. P17.11

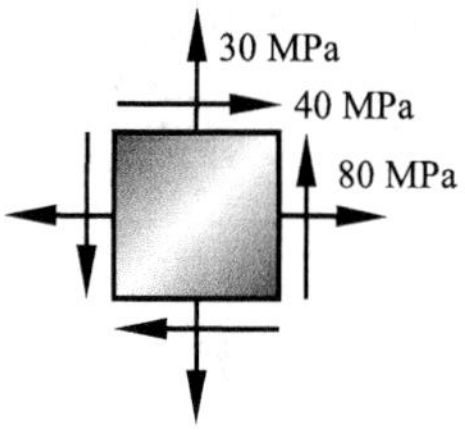

Fig. P17.12

17.11　The state of plane stress shown occurs in a member made of steel with σ_s = 310 MPa. Using the maximum-shearing-stress theory, determine whether yielding occurs. If yielding does not occur, determine the corresponding factor of safety.

17.12　The state of plane stress shown occurs in a member made of steel with σ_s = 310 MPa. Using the maximum-distortion-energy theory, determine whether yielding occurs. If yielding does not occur, determine the corresponding factor of safety.

Chapter 18 Combined Loadings

In preceding chapters, we discussed members subjected to a single internal force. For instance, we analyzed bars in tension or compression, shafts in torsion, and beams in bending. However, in many structures a member is often subjected to two or more internal forces. These internal forces applied instantaneously to a member are called combined loadings.

The stress distribution within a member subjected to combined loadings can be determined by using the principle of superposition. The stress distribution due to each internal force is first determined, and then these distributed stresses are superimposed to determine the resultant distributed stress. The principle of superposition can be used provided a linear relation exists between the stresses and the applied loads. Also, the geometry of the member should not undergo significant change when the loads are applied. This is to ensure that the stress produced by one load is not related to the stress produced by any other load. Most ordinary structures satisfy these two conditions, and therefore the principle of superposition is extensively used in stress and strain analysis of engineering structures under combined loadings.

18.1 Eccentric Tension or Compression

The distribution of stress in the cross section of a bar under an axial loading can be assumed uniform only if the line of action of the external load passes through the centroid of the cross section. Such a loading is said to be centric.

Consider now the distribution of stress when the line of action of the external load does not pass through the centroid of the cross section, i.e., when the loading is eccentric, Fig. 18.1.

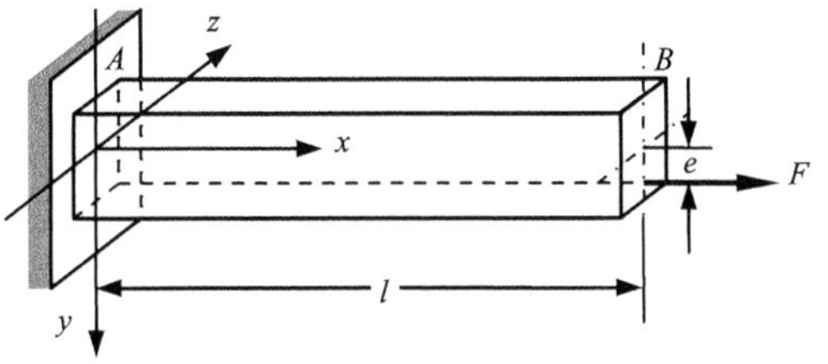

Fig. 18.1

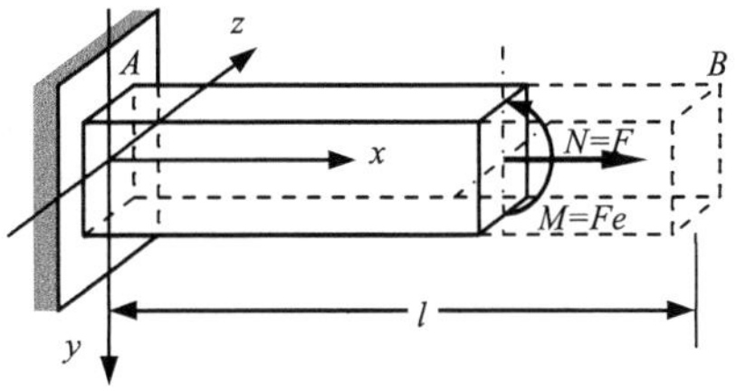

Fig. 18.2

Assuming that the external load is applied in a plane of symmetry of the bar and that the eccentric distance is e, then the internal forces acting on any cross section include an axial force $N = F$ applied at the centroid of the cross section and a bending moment $M = Fe$

acting in the plane of symmetry of the bar, Fig. 18.2.

If the material remains linearly elastic and is subjected to a small deformation, then the stress distribution due to the eccentric loading can be obtained by superposing the uniform stress distribution σ_N corresponding to the axial force N and the linear stress distribution σ_M corresponding to the bending moment M. Therefore, we have

$$\sigma = \sigma_N + \sigma_M = \frac{N}{A} + \frac{My}{I} \tag{18.1}$$

where A is the cross-sectional area, I is the moment of inertia of the cross section about the centroidal axis, and y is measured from the centroidal axis of the cross section.

Depending upon the geometry of the cross section and the eccentricity of the external load, the resultant stress may be positive and negative, or may have the same sign. From Eq. (18.1), we know that there is a line in the cross section along which $\sigma = 0$. This line represents the neutral axis of the cross section. We note that the neutral axis does not coincide with the centroidal axis of the cross section.

The resultant stress has a maximum value at the top or bottom surface, which can be expressed as

$$\sigma_{\max} = \frac{N}{A} + \frac{My_{\max}}{I} = \frac{N}{A} + \frac{M}{S} \tag{18.2}$$

where $y_{\max}$ is the perpendicular distance from the points at the top or bottom to the centroidal axis, and $S = I/y_{\max}$ is the modulus of section about the centroidal axis.

From the analysis given above, the points located at the top or bottom of the cross section will have a maximum stress. Assuming that the allowable stress of the material used is denoted by σ_{allow}, then the strength condition for the bar subjected to an eccentric tension or compression can be expressed as

$$\sigma_{\max} \leqslant \sigma_{\text{allow}} \tag{18.3}$$

Example 18.1 Two external loads of magnitude F=50 kN are applied to the end of a 20a I-steel beam. Knowing that e=80 mm, determine the stress on the bottom surface of the beam.

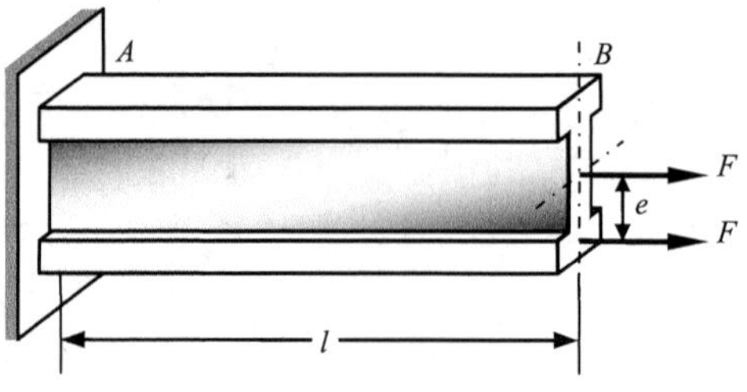

Fig. E18.1

Solution The axial force and bending moment at any section of the beam can be expressed as

$$N = 2F = 100 \text{ kN}, \text{ and } M = Fe = 4 \text{ kN} \cdot \text{m}$$

Referring to the table of Shape Steels in Appendix IV, we have the cross-sectional area and section modulus of the 20a I-steel beam as follows:

$$A = 35.56 \text{ cm}^2, \text{ and } S = 237 \text{ cm}^3$$

Using Eq. (18.2), we have

$$\sigma = \sigma_N + \sigma_M = \frac{N}{A} + \frac{M}{S} = 45.00 \text{ MPa}$$

18.2 Transverse-Force Bending of I-Beam

Depending on the shape of the cross section of a beam and the value of the shearing force at the critical section where the bending moment has a maximum value, it may happen that the largest stress will not occur at the top or bottom of the section, but at some other point within the section. A combination of large values of the normal and shearing stresses near the junction of the web and the flanges of an I-section beam can result in a value of the stress that is larger than the value on the surfaces of the I-beam.

Consider an I-section beam AB subjected to a transverse loading, Fig. 18.3. The shearing force remains constant over the entire beam, while the bending moment will reach a maximum value at the fixed end. Therefore, the critical section is located at the fixed end and the internal forces at the critical section can be expressed as

$$V = F, \ M_{\max} = Fl \tag{18.4}$$

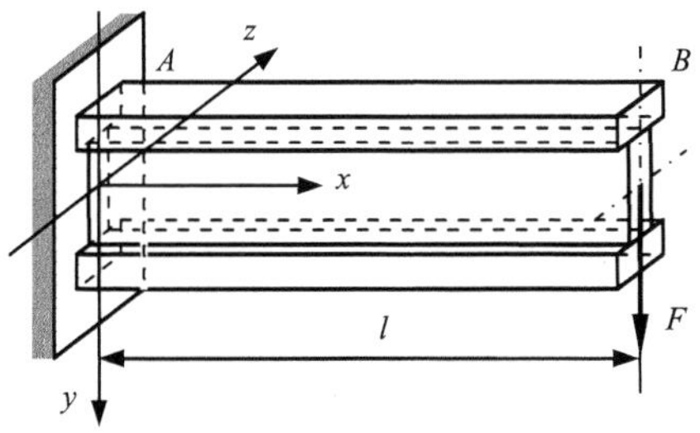

Fig. 18.3

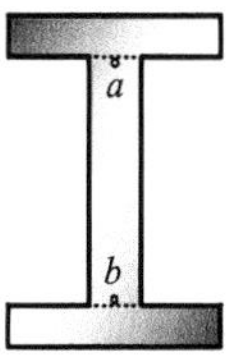

Fig. 18.4

The shearing stress at the critical section caused by the shearing force will reach a maximum value on the neutral axis of the I-section beam, and the maximum shearing stress can be expressed as

$$\tau_{\max} = \frac{VQ'_{\max}}{Ib} = \frac{V}{A_{\mathrm{w}}} \tag{18.5}$$

where I is the moment of inertia of the I-shaped section about the neutral axis, $Q'_{\max}$ is the first moment of the area above the neutral axis about the neutral axis, b is the width of the web, and A_w is the cross-sectional area of the web.

The normal stress at the critical section caused by the bending moment has a maximum stress on the top or bottom surface, which can be written as

$$\sigma_{max} = \frac{M_{max} y_{max}}{I} = \frac{M_{max}}{S} \tag{18.6}$$

where y_{max} is the perpendicular distance from the points at the top or bottom surface to the neutral axis, and $S = I/y_{max}$ is the modulus of section about the neutral axis.

At any other point at the critical section, the material is subjected simultaneously to the normal stress

$$\sigma = \frac{M_{max} y}{I} \tag{18.7}$$

where y is the distance from the neutral axis, and to the shearing stress

$$\tau = \frac{VQ'}{Ib} \tag{18.8}$$

where Q' is the first moment about the neutral axis of the portion of the cross-sectional area located below or above the point where the stress is computed, and b is the width of the cross section at that point.

Using stress transformation, we can obtain the principal stresses at any point of the critical section. For a beam of I-section, the large shearing stresses will occur at the junctions of the web with the flanges of the beam, where the normal stresses are also large, thus the principal stresses at such points are larger than the stresses at the top or bottom surface and on the neutral axis. We should be particularly aware of this possibility when selecting I-beams, and calculate the principal stresses at the junctions a and b of the web with the flanges of the beam, Fig. 18.4.

The principal stresses at point a or point b on the critical section can, by using stress transformation, be expressed as, respectively

$$\sigma_{1,3} = \frac{\sigma}{2} \pm \sqrt{\left(\frac{\sigma}{2}\right)^2 + \tau^2} \tag{18.9}$$

If the material is known to be ductile, then the maximum-shearing-stress theory or the maximum-distortion-energy theory should be used for design of the I-section beam.

If the maximum-shearing-stress theory is applied, then the equivalent stress at point a or b can be written as

$$\sigma_{eq3} = \sigma_1 - \sigma_3 = \sqrt{\sigma^2 + 4\tau^2} \tag{18.10}$$

where σ_{eq3} is the equivalent stress corresponding to the maximum-shearing-stress theory. Assuming that the allowable stress of the material used is σ_{allow}, then the maximum-shearing-stress strength condition at the junction points of the I-section beam subjected to transverse-force bending can be expressed as

$$\sigma_{\text{eq3}} \leqslant \sigma_{\text{allow}} \tag{18.11}$$

If the maximum-distortion-energy theory is applied, then the strength condition at the junction points of the I-section beam subjected to transverse-force bending can be given by

$$\sigma_{\text{eq4}} = \sqrt{\frac{1}{2}[(\sigma_1-\sigma_2)^2+(\sigma_2-\sigma_3)^2+(\sigma_3-\sigma_1)^2]} = \sqrt{\sigma^2+3\tau^2} \leqslant \sigma_{\text{allow}} \tag{18.12}$$

where σ_{eq4} is the equivalent stress corresponding to the maximum-distortion-energy theory.

Example 18.2 An external load, F=150 kN, is applied to the free end of an I-section beam made of steel for which the yield stress σ_s=235 MPa. Using the maximum-shearing-stress theory, determine whether yielding occurs. If yielding does not occur, determine the factor of safety. Neglect the effect of fillets.

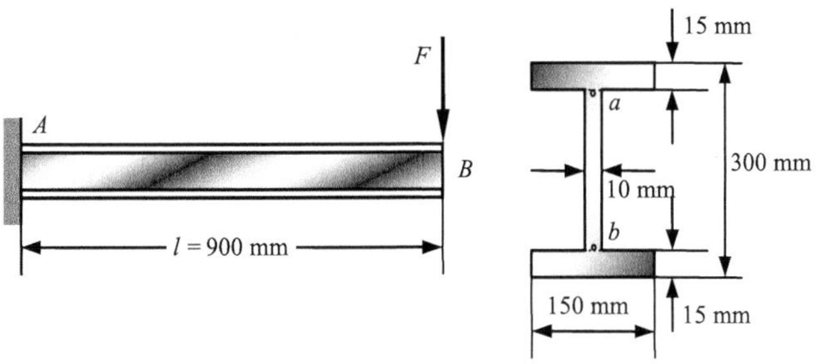

Fig. E18.2

Solution The critical section is located at the fixed end A of the beam, where the bending moment and shearing force can be expressed as

$$M_{\max} = Fl = 135 \text{ kN}\cdot\text{m}, \quad V = F = 150 \text{ kN}$$

The moment of inertia of the cross section is equal to

$$I = \frac{1}{12}(150)(300)^3 - \frac{1}{12}(140)(270)^3 = 1.07865\times10^8 \text{ mm}^4$$

The maximum normal and shearing stresses on the cross section at the fixed end are

$$\sigma_{\max} = \frac{M_{\max}y_{\max}}{I} = 187.73 \text{ MPa}, \quad \tau_{\max} = \frac{V}{A_w} = 55.56 \text{ MPa}$$

The normal and shearing stresses at point a (or b) on the cross section at the fixed end are

$$\sigma_a = \frac{y_a}{y_{\max}}\sigma_{\max} = 168.96 \text{ MPa}, \quad \tau_a \approx \tau_{\max} = \frac{V}{A_w} = 55.56 \text{ MPa}$$

The equivalent stress corresponding to the maximum-shearing-stress theory can be expressed as

$$\sigma_{\text{eq3}} = \sqrt{\sigma_a^2 + 4\tau_a^2} = 202.23 \text{ MPa}$$

The fact that $\sigma_{\text{eq3}} > \sigma_{\max}$ has shown that the critical point is located at point a (or b). Since $\sigma_{\text{eq3}} < \sigma_s = 235 \text{ MPa}$, no yielding occurs. The factor of safety is equal to

$$n = \frac{\sigma_s}{\sigma_{eq3}} = 1.16$$

18.3 Axial Loading and Bending

Consider a rectangular beam AB, which is fixed at the left end and subjected to two external loads, F_1 and F_2, at the free end, Fig. 18.5. Provided the material used remains linearly elastic, and is only subjected to a small deformation, then we can use the principle of superposition to determine the resultant stress distribution in the beam due to combined loadings.

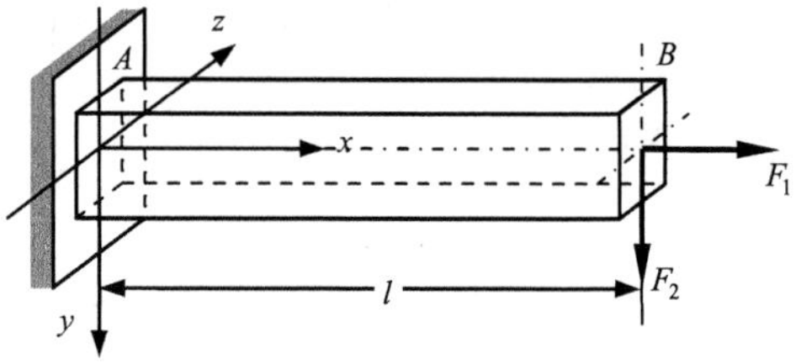

Fig. 18.5

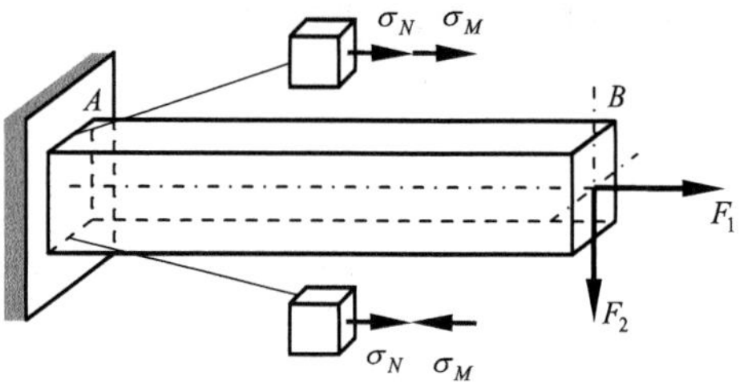

Fig. 18.6

The critical section is located at the fixed end and the internal forces at the critical section can be expressed as

$$V = F_2,\ M_{max} = F_2 l,\ N = F_1 \tag{18.13}$$

The shearing stress at the critical section caused by the shearing force will reach a maximum value on the centroidal axis of the rectangular cross section, and the maximum shearing stress can be expressed as

$$\tau_{max} = \frac{VQ'_{max}}{Ib} = \frac{3V}{2A} \tag{18.14}$$

where I is the moment of inertia of the cross section about the centroidal axis, Q'_{max} is the first moment of the cross section above the centroidal axis about the centroidal axis, b is the width of the rectangular beam, and A is the cross-sectional area of the rectangular beam. The stress produced by the shearing force generally has a much smaller contribution compared with that produced by the axial force or bending moment, and hence it can be neglected.

The normal stress at the critical section caused by the bending moment has a maximum stress at the top or bottom surface, which can be written as

$$\sigma_M = \frac{M_{max} y_{max}}{I} = \frac{M_{max}}{S} \tag{18.15}$$

where y_{max} is the perpendicular distance from the points at the top or bottom surface to the centroidal axis, and $S = I/y_{max}$ is the modulus of section about the centroidal axis.

The normal stress at the critical section caused by the axial force has a uniform distribution, and can be expressed as

$$\sigma_N = \frac{N}{A} \tag{18.16}$$

where A is the cross-sectional area of the rectangular beam.

The points located at the top or bottom of the critical section will have a maximum stress, Fig. 18.6. The maximum value can be given by

$$\sigma_{\max} = \sigma_N + \sigma_M = \frac{N}{A} + \frac{M_{\max}}{S} \tag{18.17}$$

Assuming that the allowable stress of the material used is σ_{allow}, then the strength condition for the beam subjected to both axial loading and bending is

$$\sigma_{\max} \leqslant \sigma_{\text{allow}} \tag{18.18}$$

Example 18.3 Two external loads, F_1=100 kN and F_2=4 kN, are applied to the free end of a 20a I-steel beam. Knowing that l=1 m, determine the maximum tensile stress in the beam.

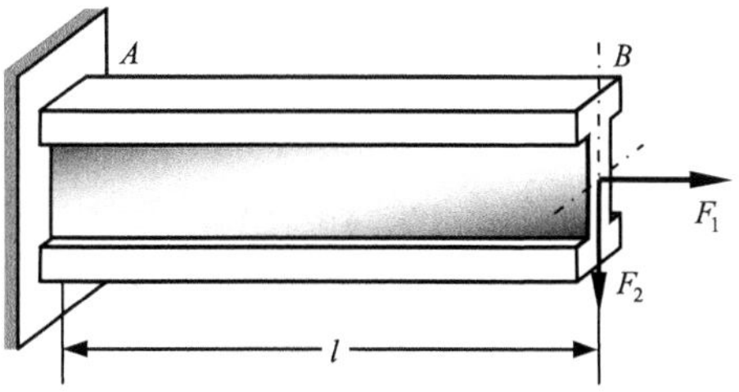

Fig. E18.3

Solution The maximum tensile stress will develop on the top-surface at the fixed end of the beam. The axial force and bending moment at the fixed end can be expressed as

$$N = F_1 = 100 \text{ kN} \text{ , and } M_{\max} = F_2 l = 4 \text{ kN} \cdot \text{m}$$

For a 20a I-steel beam, we have

$$A = 35.56 \text{ cm}^2 \text{, and } S = 237 \text{ cm}^3$$

Using Eq. (18.17), we obtain

$$\sigma_{\max}^{\text{t}} = \sigma_N + \sigma_M = \frac{N}{A} + \frac{M_{\max}}{S} = 45.00 \text{ MPa}$$

18.4 Torsion and Bending

Consider a circular shaft AB, which is fixed at the left end and subjected to two external loads, F and M_e, at the free end, Fig. 18.7. Assuming that the material remains linearly elastic, and that the deformation is small, then we can use the principle of superposition to determine the resultant stress distribution in the shaft due to combined loadings.

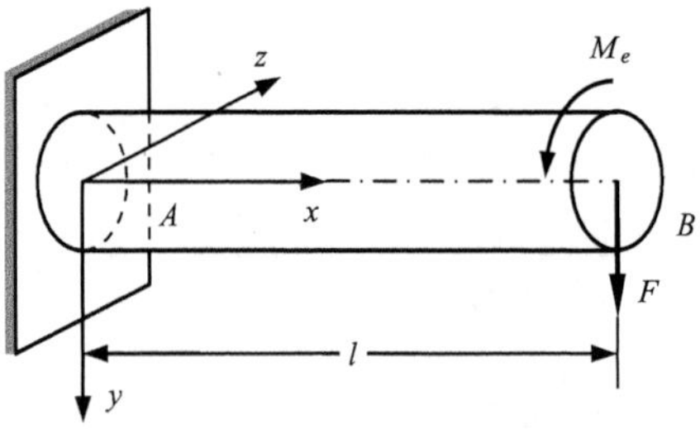

Fig. 18.7

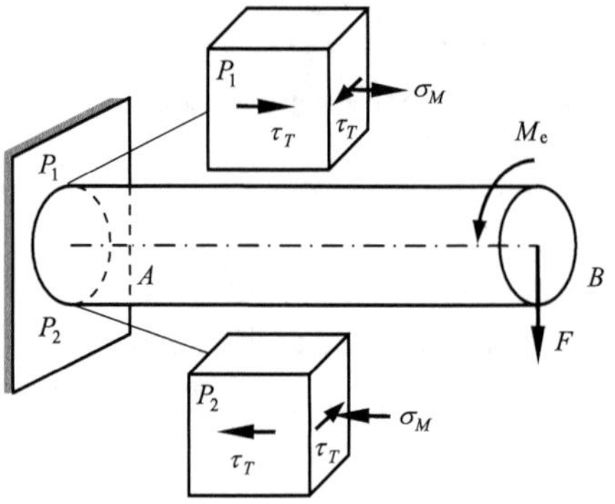

Fig. 18.8

The critical section is located at the section A and the internal forces at the critical section can be expressed as

$$V = F, \ M_{\max} = Fl, \ T = M_e \tag{18.19}$$

The shearing stress at the critical section caused by the shearing force has a maximum value on the neutral axis of the circular cross section, that is

$$\tau_V = \frac{VQ'_{\max}}{Ib} = \frac{4V}{3A} \tag{18.20}$$

where I is the moment of inertia of the circular cross section about the neutral axis, $Q'_{\max}$ is the first moment of the cross-sectional area above the neutral axis with respect to the neutral axis, b is the diameter of the circular shaft, and A is the cross-sectional area of the circular shaft. This stress is small, and can be neglected. Therefore, we only need to consider the resultant stress caused the torsional and bending moments.

The normal stress on the critical section caused by the bending moment has a maximum value at the top or bottom points of the critical section, that is

$$\sigma_M = \frac{M_{\max} y_{\max}}{I} = \frac{M_{\max}}{S} \tag{18.21}$$

where $y_{\max}$ is the perpendicular distance from the points at the top or bottom surface to the neutral axis, and $S = I/y_{\max}$ is the modulus of section.

The shearing stress at the critical section caused by the torsional moment reaches a maximum value on the surface of the circular shaft, and the maximum value can be expressed as

$$\tau_T = \frac{T_{\max}\rho_{\max}}{I_p} = \frac{T_{\max}}{S_p} \tag{18.22}$$

where I_p is the polar moment of inertia of the cross section with respect to the centroid of the cross section, $\rho_{\max}$ is the radius of the cross section, and $S_p = I_p/\rho_{\max}$ is the polar modulus of section.

It has been shown that the critical point has a state of plane stress and is located at the top or bottom point of the critical section, Fig. 18.8.

The principal stresses at the top or bottom point on the critical section can, by using stress transformation, be expressed as

$$\sigma_{1,3} = \frac{\sigma_M}{2} \pm \sqrt{(\frac{\sigma_M}{2})^2 + \tau_T^{\ 2}} \tag{18.23}$$

For a ductile material, if the maximum-shearing-stress theory is applied, then the equivalent stress can be written as

$$\sigma_{\text{eq3}} = \sigma_1 - \sigma_3 = \sqrt{\sigma_M^{\ 2} + 4\tau_T^{\ 2}} \tag{18.24}$$

where σ_{eq3} is the equivalent stress corresponding to the maximum-shearing-stress theory.

Using Eqs. (18.21) and (18.22), and $S_{\text{p}} = 2S$, Eq. (18.24) can be rewritten as

$$\sigma_{\text{eq3}} = \frac{\sqrt{M_{\max}^{\ 2} + T_{\max}^{\ 2}}}{S} \tag{18.25}$$

where $S = \frac{1}{32}\pi d^3$ for a solid shaft with diameter d, and $S = \frac{1}{32}\pi D^3(1-\alpha^4)$, where $\alpha = d/D$, for hollow shaft with outer diameter D and inner diameter d.

Assuming that the allowable stress of the material used is σ_{allow}, then the maximum-shearing-stress strength condition for the shaft subjected to both torsion and bending can be expressed as

$$\sigma_{\text{eq3}} \leqslant \sigma_{\text{allow}} \tag{18.26}$$

For a ductile material, if the maximum-distortion-energy theory is applied, then the strength condition for the shaft subjected to both torsion and bending can be given by

$$\sigma_{\text{eq4}} = \sqrt{\sigma_M^{\ 2} + 3\tau_T^{\ 2}} = \frac{\sqrt{M_{\max}^{\ 2} + \frac{3}{4}T_{\max}^{\ 2}}}{S} \leqslant \sigma_{\text{allow}} \tag{18.27}$$

where σ_{eq4} is the equivalent stress corresponding to the maximum-distortion-energy theory.

Example 18.4 The steel pipe *AB*, having σ_s=325 MPa yield strength, l=200 mm length, D=80 mm outer diameter, and t=5 mm wall thickness, is subjected to two external loads, F= 6 kN and M_e=1.5 kN·m, at the free end. Using the maximum-shearing-stress theory, determine whether yielding occurs. If yielding does not occur, determine the factor of safety.

Solution The critical point is located at the top or bottom surface at the fixed end of the shaft. The bending and torsional moments at the fixed end can be expressed as

$$M_{\max} = Fl = 1.2\ \text{kN}\cdot\text{m}\ \text{, and}\ \ T = M_{\text{e}} = 1.5\ \text{kN}\cdot\text{m}$$

Using Eq. (18.26), we have

$$\sigma_{\text{eq3}} = \frac{\sqrt{M_{\max}^{\ 2} + T^2}}{S} = \frac{\sqrt{M_{\max}^{\ 2} + T^2}}{\frac{1}{32}\pi D^3\{1-[(D-2t)/D]^4\}} = 92.35\ \text{MPa}$$

Since $\sigma_{\text{eq3}} < \sigma_{\text{s}}$, no yielding occurs. Thus the factor of safety is equal to

$$n = \frac{\sigma_s}{\sigma_{eq3}} = 3.52$$

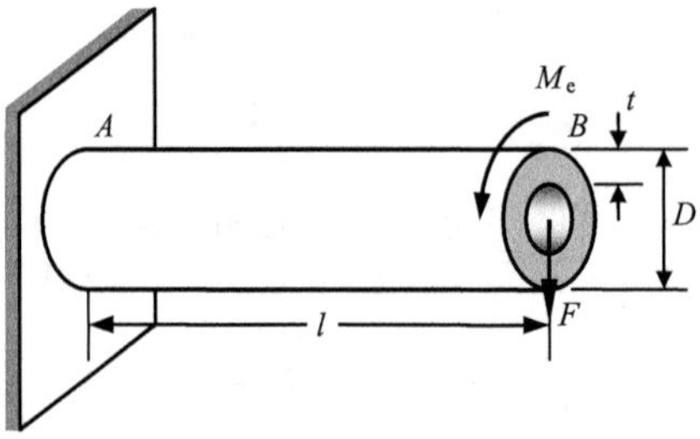

Fig. E18.4

Problems

18.1　An external load of magnitude F=80 kN is applied to the end of a 22a I-steel beam. Knowing that e=80 mm, determine the stresses on the top and bottom surfaces of the beam.

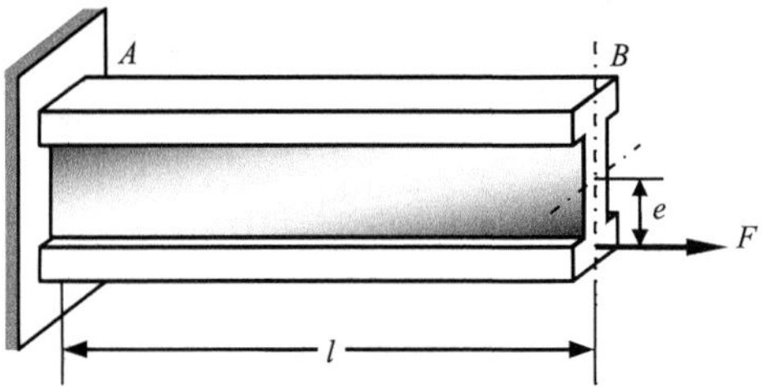

Fig. P18.1

18.2　Three external loads of magnitude F=80 kN are applied to the end of a 20a I-steel beam. Knowing that e=80 mm, determine the maximum tensile and compressive stresses in the beam.

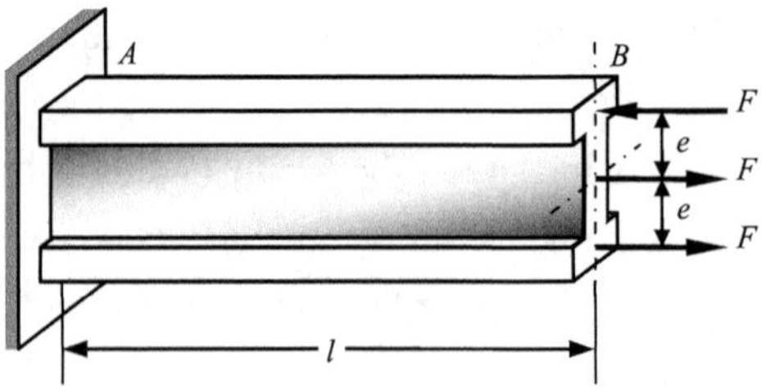

Fig. P18.2

18.3　An external load, F=180 kN, is applied to the free end of an H-steel beam for which the yield stress σ_s=235 MPa. Using the maximum-distortion-energy theory, determine whether yielding occurs. If yielding does not occur, determine the factor of safety. Neglect the effect of fillets.

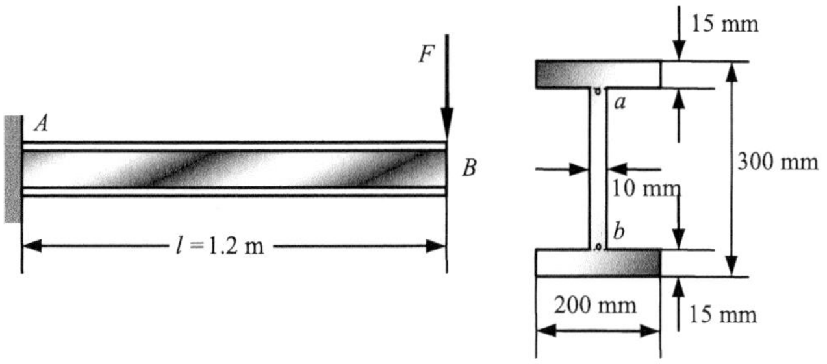

Fig. P18.3

18.4 Two external loads, F_1=100 kN and F_2=20 kN, are applied to the free end of a 36a I-steel beam for which the yield stress σ_s=235 MPa. Using the maximum-shearing-stress theory, determine whether yielding occurs. If yielding does not occur, determine the factor of safety.

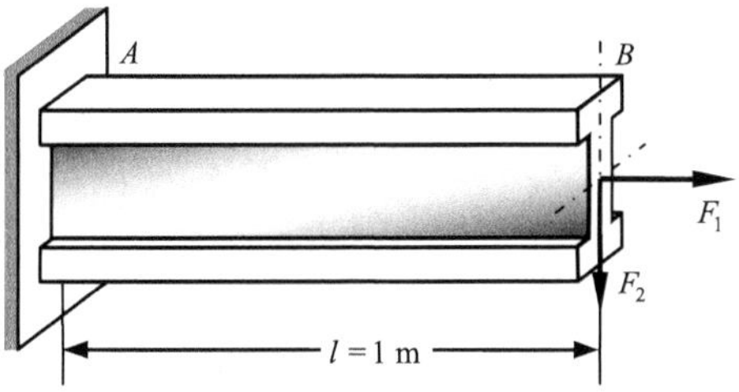

Fig. P18.4

18.5 The steel pipe AB, having σ_s=325 MPa yield strength, l=200 mm length, D= 80 mm outer diameter, and t=5 mm wall thickness, is subjected to two external loads, F=6 kN and M_e=1.5 kN·m, at the free end. Using the maximum-distortion-energy theory, determine whether yielding occurs. If yielding does not occur, determine the factor of safety.

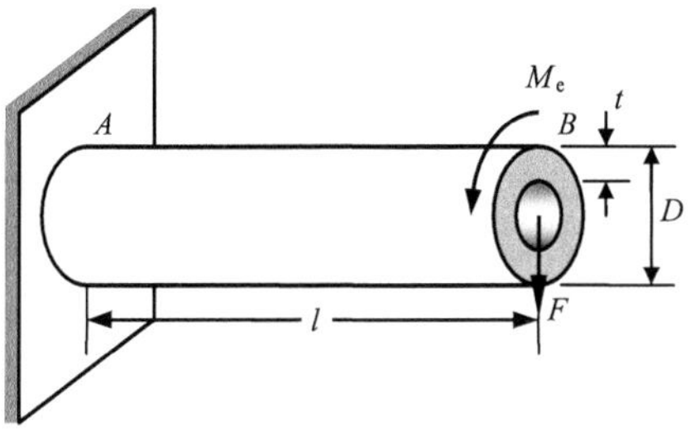

Fig. P18.5

18.6 A cantilevered I-steel beam supports an external load as shown. Knowing that F=200 kN and σ_{allow}=180 MPa, determine (a) the maximum value of the normal stress in the beam, (b) the maximum value of the principal stresses at the junction of a flange and the web, (c) whether the specified I-steel beam is acceptable.

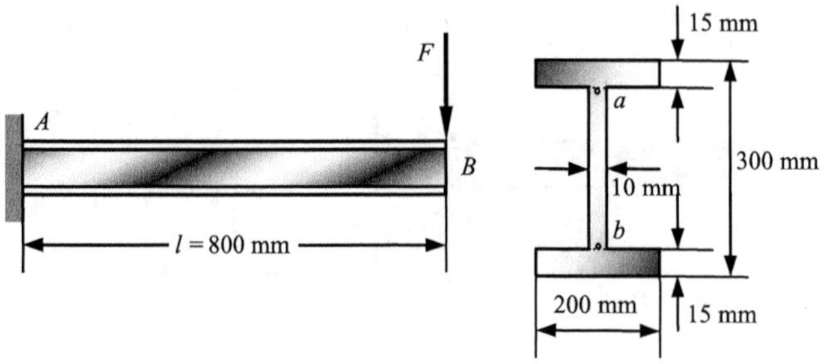

Fig. P18.6

Chapter 19 Stability of Columns

A long slender member subjected to axial compression is called a column. The failure of a column occurs by buckling, i.e., by a sudden lateral deflection. It should be noted that the failure of a short column occurs by yielding or fracturing of the material. The buckling of a column may occur even though the maximum stress in the column is less than the yield point and ultimate strength of the material.

Quite often the buckling of a column can lead to a sudden failure of a structure, and as a result, special attention must be given to the design of a column so that it can safely support a given load without buckling.

19.1 Critical Load of Long Column with Pin Supports

The maximum axial compressive force that a column can support when it is on the verge of buckling is called the critical load, denoted by F_{cr}. Consider a column AB of length l to support a given load F, Fig. 19.1a. The column is pin-supported at both ends and F is a centric axial load. If the cross-sectional area of the column is selected so that the stress on the cross section is less than the allowable stress for the material used, and if the deformation falls within the given specifications, we might conclude that the column has been properly designed. However, it may happen that, as the load is applied, the column will buckle. Instead of remaining straight, it will suddenly become sharply curved, Fig. 19.1b. We now determine the critical value F_{cr} of the load for the column shown in Fig. 19.1a.

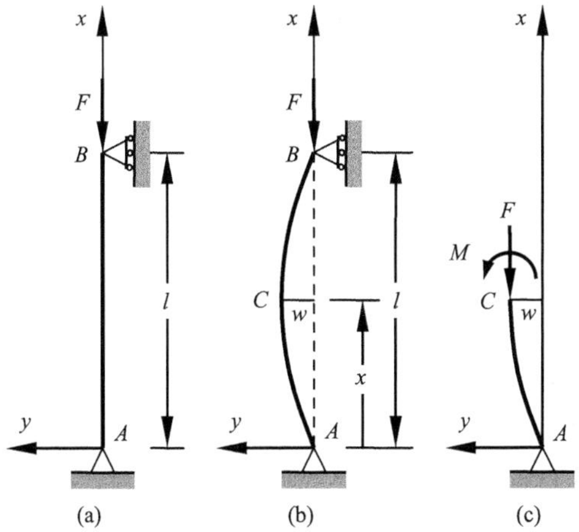

Fig. 19.1

Considering the equilibrium of the free body *AC*, Fig. 19.1c, we find that the bending moment at the cross section *C* is $M = -Fw$. Substituting *M* into $w'' = M/EI$, we have

$$w'' + \frac{F}{EI}w = 0 \tag{19.1}$$

This is a linear homogeneous differential equation of the second order with constant coefficients. Setting

$$k^2 = \frac{F}{EI} \tag{19.2}$$

we write Eq. (19.1) in the form

$$w'' + k^2 w = 0 \tag{19.3}$$

The general solution of Eq. (19.3) is

$$w = A\sin kx + B\cos kx \tag{19.4}$$

Recalling the boundary conditions that must be satisfied at ends *A* and *B* of the column, we first make $x = 0,\ w = 0$ in Eq. (19.4) and find that $B = 0$. Substituting next $x = l,\ w = 0$, we obtain

$$A\sin kl = 0 \tag{19.5}$$

This equation is satisfied either if $A = 0$, or if $\sin kx = 0$. If the first of these conditions is satisfied, Eq. (19.5) reduces to $w \equiv 0$ and the column is always straight. For the second condition to be satisfied, we must have $kl = n\pi$ or, using Eq. (19.2) and solving for *F*,

$$F = \frac{n^2\pi^2 EI}{l^2} \tag{19.6}$$

The smallest value of *F* defined by Eq. (19.6) corresponds to $n = 1$. We thus have

$$F_{cr} = \frac{\pi^2 EI}{l^2} \tag{19.7}$$

This expression obtained is known as Euler's formula. In the case of a column with a circular or square cross section, the moment of inertia *I* of the cross section is the same about any centroidal axis, and the column is likely to buckle about any centroidal axis. For other shape of cross section, the critical load should be computed by making $I = I_{min}$ in Eq. (19.7).

19.2 Critical Load of Long Column with Other Supports

Euler's formula (19.7) was derived for a column pin-supported at both ends. Now the critical load F_{cr} will be determined for columns with different end conditions. In the case of a column with one fixed end *A* and one free end *B* supporting a load *F*, Fig. 19.2a. The column will behave as the upper half of a pin-supported column, Fig. 19.2b. The critical load for the column of Fig. 19.2a is thus the same as for the pin-ended column of Fig. 19.2b and can be obtained from Euler's formula (19.7) by using a column length equal to twice the actual length

l of the given column. Therefore, the critical load is expressed as

$$F_{cr} = \frac{\pi^2 EI}{(2l)^2} = \frac{\pi^2 EI}{(\mu l)^2} = \frac{\pi^2 EI}{l_e^2} \tag{19.8}$$

where μ is the effective-length factor, and $l_e = \mu l$ is the effective length. The above expression is an extension of Euler's formula, and thus is also called Euler's formula.

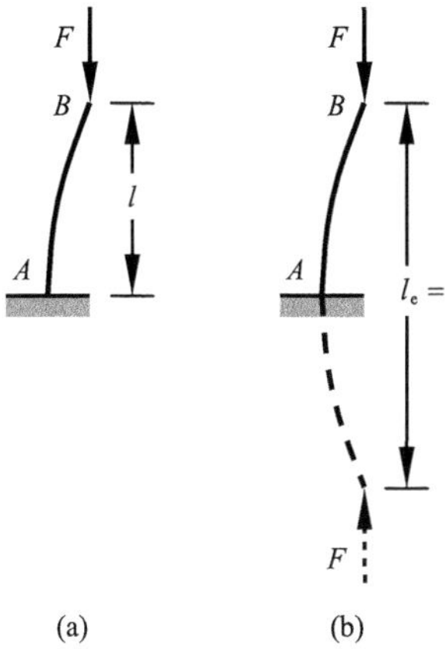

Fig. 19.2

The effective-length factor and effective length corresponding to various end conditions are shown in Fig. 19.3.

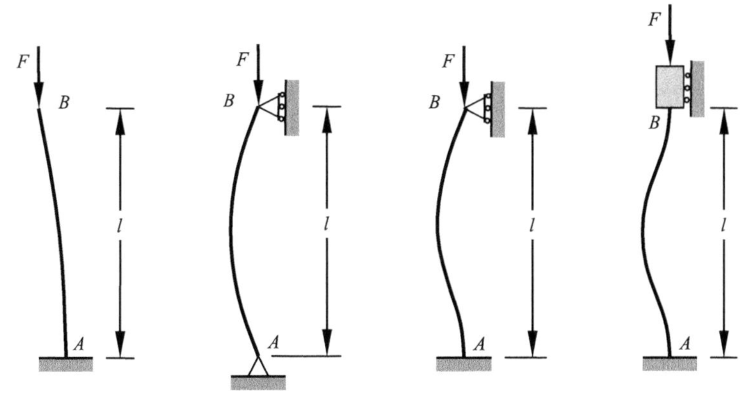

$\mu = 2$; $l_e = 2l$ (a) one end fixed the orther free

$\mu = 1$; $l_e = l$ (b) both ends pinned

$\mu = 0.7$; $l_e = 0.7l$ (c) one end fixed the other pinned

$\mu = 0.5$; $l_e = 0.5l$ (d) both ends fixed

Fig. 19.3

19.3 Critical Stress of Long Column

The value of the stress corresponding to the critical load is called the critical stress and is denoted by σ_{cr}. Recalling Eq. (19.8) and setting $i^2 = I/A$, where A is the cross-sectional area and i is the radius of gyration, we have

$$\sigma_{cr} = \frac{F_{cr}}{A} = \frac{\pi^2 E}{(\mu l/i)^2} = \frac{\pi^2 E}{\lambda^2} \tag{19.9}$$

where $\lambda = \mu l/i$ is called the slenderness ratio of the column. The above equation is also called Euler's formula. Eq. (19.9) shows that the critical stress is proportional to the modulus of elasticity of the material, and inversely proportional to the square of the slenderness ratio of the column.

Euler's formula was derived based on the fact that the critical stress is below the proportional limit of the material, that is,

$$\sigma_{cr} = \frac{\pi^2 E}{\lambda^2} \leqslant \sigma_p \tag{19.10}$$

Defining $\lambda_p = \sqrt{\pi^2 E/\sigma_p}$, then the above equation can be rewritten as

$$\lambda \geqslant \lambda_p \tag{19.11}$$

where λ_p is the critical slenderness ratio. The above equation shows that Euler's formula is valid only when $\lambda \geqslant \lambda_p$. For a low-carbon steel (Q235), E=206 GPa, and σ_p=200 MPa, then λ_p=100.8. For an aluminum alloy, E=70 GPa and σ_p=175 MPa, then λ_p=62.8.

Example 19.1 A long column consists of a steel tube that has a 30 mm outer diameter and a 20 mm inner diameter. Knowing that l=1.5 m and E=200 GPa, determine the critical load and critical stress for the support condition shown.

Solution The critical load for the support condition shown can be determined by using Euler's formula. For the column given, it can be cut at the midsection C, Fig. E19.1, and divided into two portions, BC and AC, each equivalent to a column that is fixed at one end and

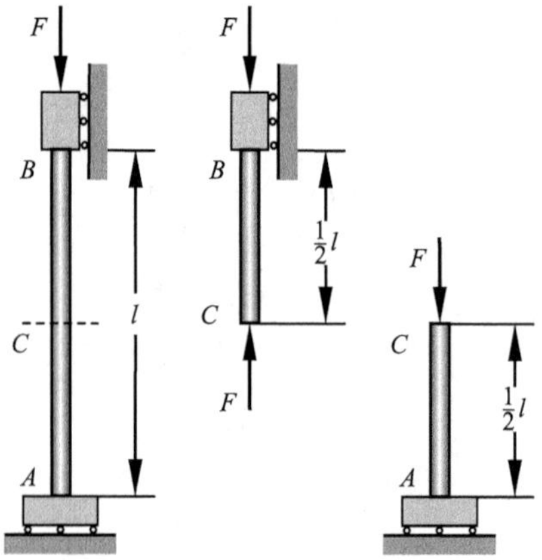

Fig. E19.1

free at the other. Therefore, the critical load for the original column is equal to that for each of the two portions. Using $\mu_{BC}=\mu_{AC}=2$ and $l_{BC}=l_{AC}=\frac{1}{2}l$, we obtain the critical load for the support condition shown in Fig. 19.1 as follows:

$$F_{cr}=(F_{cr})_{BC}=(F_{cr})_{AC}=\frac{\pi^2 EI}{(\mu_{AC}l_{AC})^2}=\frac{\pi^2 E[\frac{1}{64}\pi(D^4-d^4)]}{l^2}=27.99\ \text{kN}$$

Using Eq. (19.9), the critical stress is given by

$$\sigma_{cr}=\frac{F_{cr}}{A}=\frac{F_{cr}}{\frac{1}{4}\pi(D^2-d^2)}=71.28\ \text{MPa}$$

19.4 Critical Stress of Intermediate Column

For a long column, where the slenderness ratio λ is large, its failure is closely predicted by Euler's formula, and the value of σ_{cr} depends on the modulus of elasticity E of the used material by Euler's formula $\sigma_{cr}=\pi^2 E/\lambda^2$, but not on its yield strength σ_s or ultimate strength σ_u.

For a short column, its failure occurs essentially as a result of yielding or fracture, and we have $\sigma_{cr}=\sigma_s$ for a ductile material or $\sigma_{cr}=\sigma_u$ for a brittle material.

For an intermediate column, its failure is dependent on both E and σ_s (or σ_u). The failure of such a column is an extremely complex phenomenon and hence empirical formulas are often used for the design of an intermediate column under centric loading. The simplest empirical formula is a linear formula, which can be written as

$$\sigma_{cr}=a-b\lambda \tag{19.12}$$

where a and b are two constants depending on the material used. For a low carbon steel, $a=304$ MPa and $b=1.12$ MPa. For a cast iron, $a=332.2$ MPa and $b=1.454$ MPa.

Example 19.2 A 20a I-steel column is subjected to a centric compressive load F at the free end. Knowing that σ_p=200 MPa, E=206 GPa, a=304 MPa, and b=1.12 MPa, determine the critical load of the column if the length of the column is (a) l=1.2 m, (b) l=1.0 m.

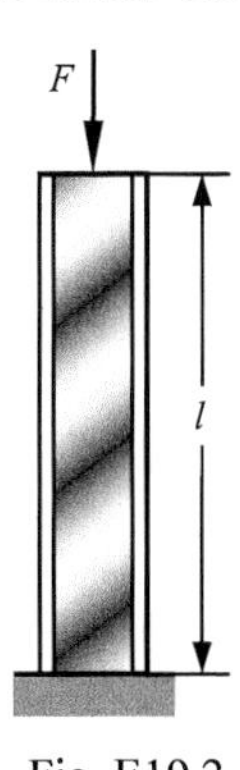

Fig. E19.2

Solution For the column given, the effective-length factor is $\mu = 2.0$, and the critical slenderness ratios equals

$$\lambda_p = \sqrt{\frac{\pi^2 E}{\sigma_p}} = 101$$

Referring to the table of Shape Steels in Appendix IV, we have the cross-sectional area and minimum moment of inertia for the 20a I-steel column as follows:

$$A = 35.56 \text{ cm}^2, \text{ and } I_{min} = 158 \text{ cm}^4$$

(a) The slenderness ratio is

$$\lambda = \frac{\mu l}{i} = \frac{\mu l}{\sqrt{I_{min}/A}} = 114$$

Since $\lambda > \lambda_p$, then Euler's formula should be used to determine the critical load, which can be given by

$$F_{cr} = A\sigma_{cr} = A\frac{\pi^2 E}{\lambda^2} = 556 \text{ kN}$$

(b) The slenderness ratio is

$$\lambda = \frac{\mu l}{\sqrt{I_{min}/A}} = 95$$

Since $\lambda < \lambda_p$, then the empirical formula should be used to determine the critical load, which can be given by

$$F_{cr} = A\sigma_{cr} = A(a - b\lambda) = 703 \text{ kN}$$

19.5 Design of Column

A straight line empirical formula is often used for an intermediate column, and Euler's formula is used for a long column. Therefore, three formulas of the critical stress used for design of long, intermediate, and short columns can be expressed as

$$\begin{aligned} \sigma_{cr} &= \frac{\pi^2 E}{\lambda^2} \qquad && (\lambda \geqslant \lambda_p) \\ \sigma_{cr} &= a - b\lambda && (\lambda_s \leqslant \lambda < \lambda_p) \\ \sigma_{cr} &= \sigma_s && (\lambda < \lambda_s) \end{aligned} \tag{19.13}$$

where $\lambda_s = (a - \sigma_s)/b$ and $\lambda_p = \sqrt{\pi^2 E/\sigma_p}$.

The ratio of the critical stress σ_{cr} to the factor of safety n is used to define the allowable stress σ_{allow}. The working stress σ within a column under centric load should satisfy the following condition if the column is stable:

$$\sigma \leqslant \sigma_{allow} = \frac{\sigma_{cr}}{n} \tag{19.14}$$

Example 19.3 A hollow circular steel column having a D=120 mm outer diameter and a t=15 mm wall thickness is subjected to a centric compressive load F at the free end. Determine the critical load if the length of the column is (a) 1.8 m, (b) 1.2 m, and (c) 0.6 m. Use σ_p= 280 MPa, σ_s=350 MPa, E=210 GPa, a=461 MPa, and b=2.568 MPa.

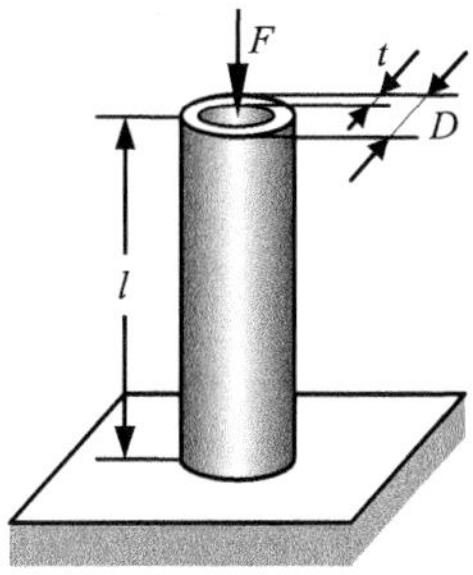

Fig. E19.3

Solution For the column given in Fig. E19.3, the effective-length factor $\mu = 2.0$. And the critical slenderness ratios for the proportional limit and yielding strength are equal to

$$\lambda_p = \sqrt{\frac{\pi^2 E}{\sigma_p}} = 86, \text{ and } \lambda_s = \frac{a-\sigma_s}{b} = 43$$

(a) The slenderness ratio is

$$\lambda = \frac{\mu l}{i} = \frac{\mu l}{\sqrt{I/A}} = \frac{\mu l}{\sqrt{[\frac{1}{64}\pi(D^4 - d^4)]/[\frac{1}{4}\pi(D^2 - d^2)]}} = \frac{4\mu l}{\sqrt{D^2 + d^2}} = 96$$

Since $\lambda > \lambda_p$, then Euler's formula should be chosen to determine the critical load. The corresponding critical load can be computed by

$$F_{cr} = A\sigma_{cr} = \frac{1}{4}\pi(D^2 - d^2)\frac{\pi^2 E}{\lambda^2} = 1113 \text{ kN}$$

(b) The slenderness ratio is

$$\lambda = \frac{4\mu l}{\sqrt{D^2 + d^2}} = 64$$

Since $\lambda_s < \lambda < \lambda_p$, then the empirical formula should be chosen to determine the critical load. The corresponding critical load can be given by

$$F_{cr} = A\sigma_{cr} = \frac{1}{4}\pi(D^2 - d^2)(a - b\lambda) = 1468 \text{ kN}$$

(c) The slenderness ratio is

$$\lambda = \frac{4\mu l}{\sqrt{D^2 + d^2}} = 32$$

Since $\lambda < \lambda_s$, then the strength formula should be chosen to determine the critical load. The corresponding critical load can be expressed as

$$F_{cr} = A\sigma_{cr} = \frac{1}{4}\pi(D^2 - d^2)\sigma_s = 1732 \text{ kN}$$

Example 19.4 A plane structure *ABC* is composed of three bars, *AB*, *AC* and *BC*, and connected by three pins at *A*, *B* and *C*. The bars *AB*, *AC* and *BC*, each having diameter of 32 mm, have lengths of 1 m, 0.6 m and 0.8 m, respectively. The joint *C* is subjected to a concentrated load *F*. Knowing that $\sigma_p = 280$ MPa, $\sigma_s = 350$ MPa, $E = 210$ GPa, $a = 461$ MPa, $b = 2.568$ MPa and $0 < \theta < 90°$, determine the critical load F_{cr} which can be applied to the structure and the corresponding critical angle θ_{cr}. (Note: Only in-plane buckling needs to be considered.)

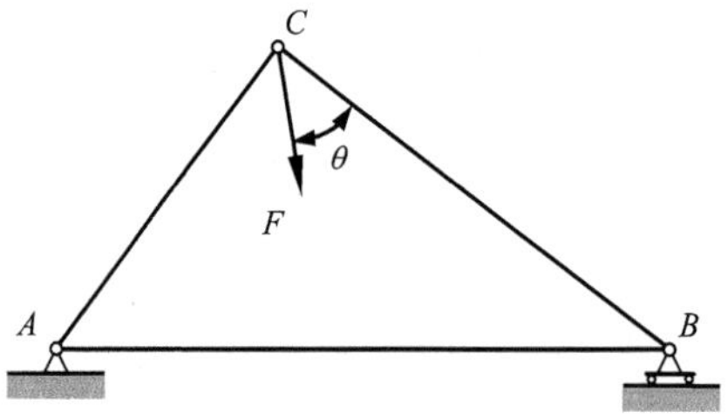

Fig. E19.4

Solution $\lambda_p = \sqrt{\frac{\pi^2 E}{\sigma_p}} = 86$, $\lambda_s = \frac{a - \sigma_s}{b} = 43$, $\lambda_{AC} = \frac{4\mu l_{AC}}{d} = 75$, $\lambda_{BC} = \frac{4\mu l_{BC}}{d} = 100$

Since $\lambda_s < \lambda_{AC} < \lambda_p < \lambda_{BC}$, i.e. *AC* and *BC* are respectively intermediate and long columns, then the empirical and Euler's formulas should be applied to calculating the critical loads of *AC* and *BC* respectively, i.e.

$$(F_{AC})_{cr} = (\frac{1}{4}\pi d^2)(a - b\lambda_{AC}) = 215.86 \text{ kN}, \quad (F_{BC})_{cr} = (\frac{1}{4}\pi d^2)\frac{\pi^2 E}{\lambda_{BC}^2} = 166.69 \text{ kN}$$

Thus the critical load and corresponding critical angle can be expressed as

$$F_{cr} = \sqrt{(F_{AC})_{cr}^2 + (F_{BC})_{cr}^2} = 272.73 \text{ kN}, \quad \theta_{cr} = \arctan\frac{(F_{AC})_{cr}}{(F_{BC})_{cr}} = 52.32°$$

Problems

19.1 Each of the four long columns consists of a steel rod that has a 30 mm diameter. Knowing that l=1.5 m and E=200 GPa, determine the critical load F_{cr} for each support condition shown.

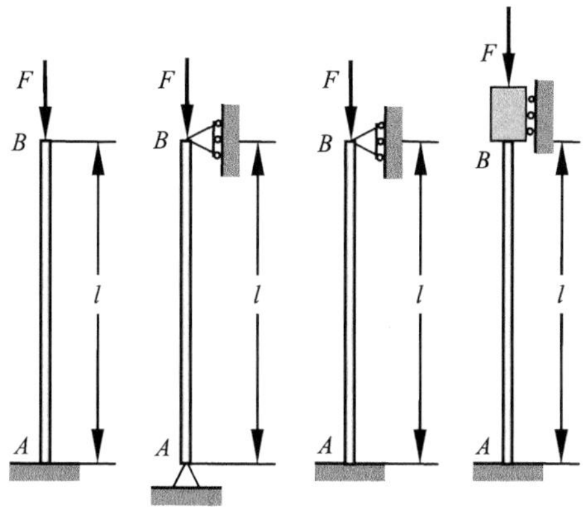

Fig. P19.1

19.2 Each of the four long columns consists of an aluminum tube that has a 30 mm outer diameter and a 5 mm wall thickness. Using l=1.2 m, E=70 GPa and a factor of safety of n=2.3, determine the allowable load F_{allow} for each support condition shown.

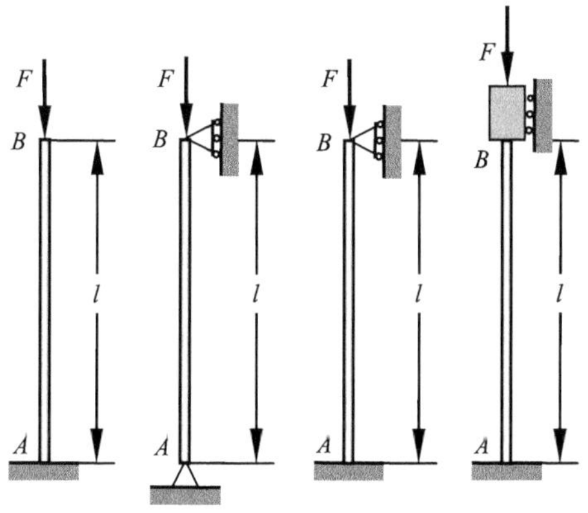

Fig. P19.2

19.3 A solid circular steel column having a d=150 mm diameter is subjected to a centric compressive load F at the free end. Determine the critical load if the length of the column is (a) 2.0 m, (b) 1.2 m, and (c) 0.5 m. Use σ_p=280 MPa, σ_s=350 MPa, E=210 GPa, a=461 MPa, and b=2.568 MPa.

19.4 A 25a I-steel column is subjected to a centric compressive load F at the free end. Knowing that σ_p=200 MPa, E=206 GPa, a=304 MPa, and b=1.12 MPa, determine the critical load of the column if the length of the column is (a) l=1.5 m, (b) l=1.0 m.

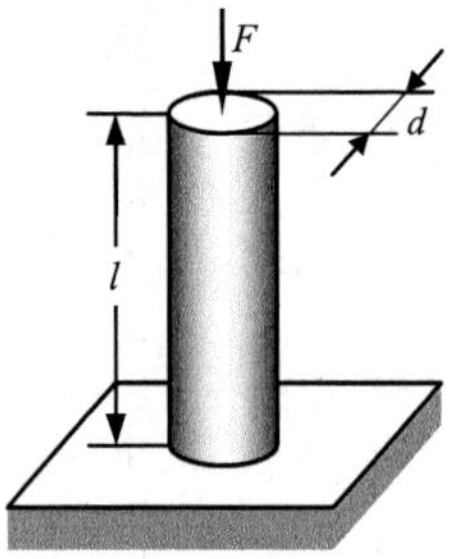

Fig. P19.3

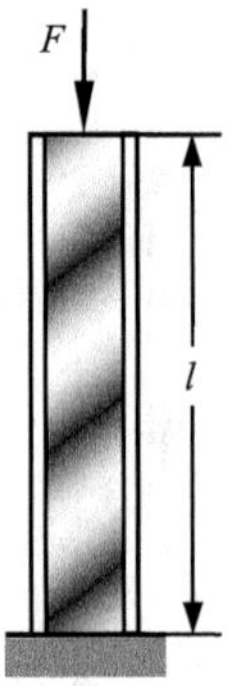

Fig. P19.4

Chapter 20 Energy Methods

20.1 External Work

1. Work of External Force

Consider a deformable body supported at A and B shown in Fig. 20.1. When an external force is applied to the body gradually from zero to a final value F, the corresponding displacement of the force application point in the force direction also slowly increased from zero to a final value δ. If the material is within a linearly elastic range, then the work done by the external force applied to the body can be expressed as

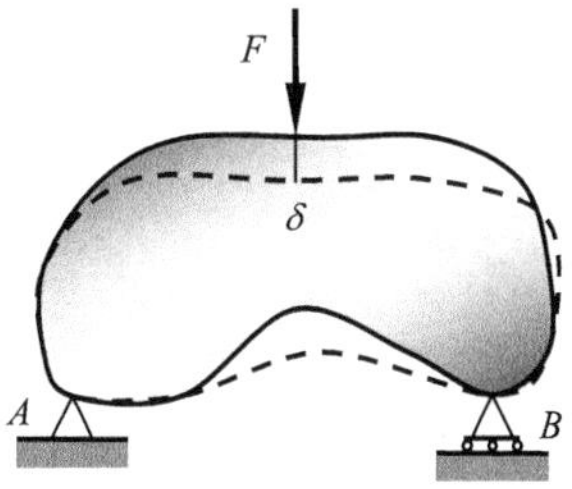

Fig. 20.1

$$W_e = \frac{1}{2} F\delta \tag{20.1}$$

2. Work of External Couple

When an external couple is applied to the body gradually from zero to a final value M_e, the corresponding angular displacement of the couple is slowly increased from zero to a final value θ. If the material is within a linearly elastic range, then the work done by the external couple can be expressed, Fig. 20.2, as

$$W_e = \frac{1}{2} M_e\theta \tag{20.2}$$

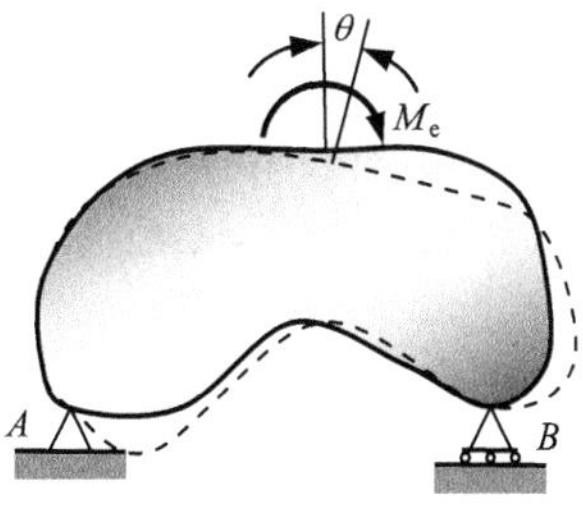

Fig. 20.2

3. Work of Multiple External Loads

Consider a deformable body supported at A and B shown in Fig. 20.3, which is subjected to a series of external forces $F_1, F_2, \cdots, F_i, \cdots, F_n$. When these external forces are applied to the body gradually from zero to their final values, the corresponding displacements of the force application points in the force directions are also slowly increased from zero to their final values. If the material is within a linearly elastic range, then the work done by the external forces applied to the body can be expressed as

$$W_e = \sum \frac{1}{2} F_i \Delta_i \tag{20.3}$$

where Δ_i is a function of the forces $F_1, F_2, \cdots, F_i, \cdots, F_n$, that is, $\Delta_i = \Delta_i(F_1, F_2, \cdots, F_i, \cdots, F_n)$ $(i = 1, 2, \cdots, n)$. It should be noted that F_i is a generalized force and Δ_i is the corresponding generalized displacement. If F_i is a couple, then Δ_i is an angle of rotation corresponding to the couple.

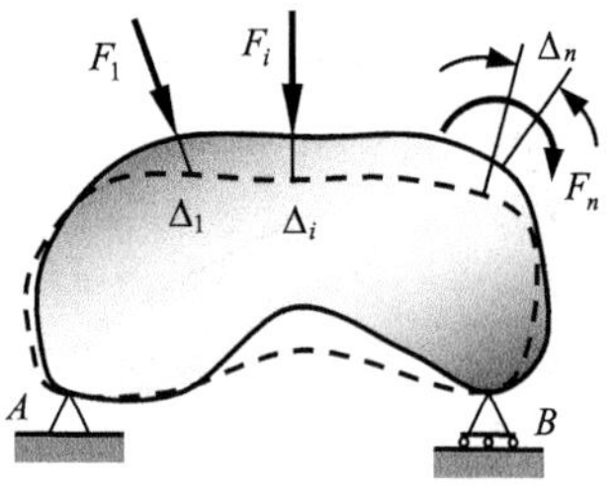

Fig. 20.3

4. Work of Combined Loadings

Consider a slender circular member subjected to combined loadings, Fig. 20.4. If the deformation is small, the external work done by the combined loadings can be written as

$$W_e = \sum \frac{1}{2} F_i \delta_i \tag{20.4}$$

where δ_i is a function of the force F_i, that is, $\delta_i = \delta_i(F_i)$ $(i = 1, 2, 3)$.

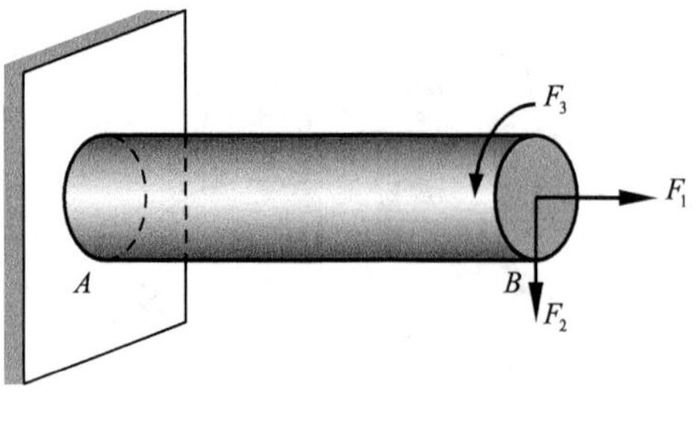

Fig. 20.4

20.2 Stain-Energy Density

1. Strain-Energy Density for Uniaxial Stress

When a member is subjected to uniaxial stress, Fig. 20.5, the strain-energy density is equal to the area under the stress-strain curve and can be expressed as

$$u = \int \sigma \mathrm{d}\varepsilon \tag{20.5}$$

where ε is the normal strain corresponding to σ. For values of σ within the linear elastic range, we have $\sigma = E\varepsilon$, where E is the modulus of elasticity of the material. Substituting σ into Eq. (20.5) and performing the integration, we have

$$u = \frac{1}{2}E\varepsilon^2 = \frac{1}{2}\sigma\varepsilon = \frac{\sigma^2}{2E} \tag{20.6}$$

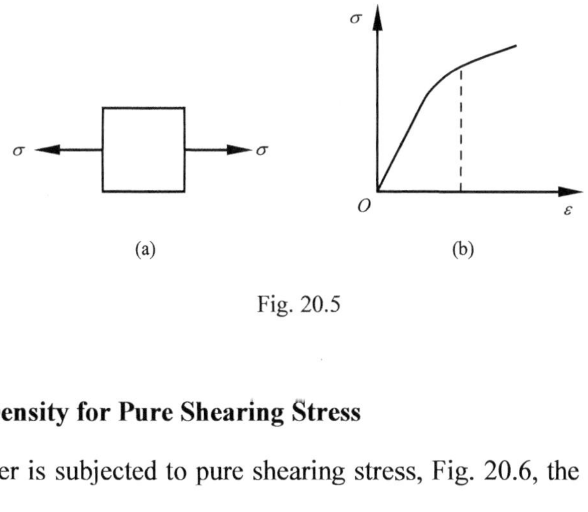

Fig. 20.5

2. Strain-Energy Density for Pure Shearing Stress

When a member is subjected to pure shearing stress, Fig. 20.6, the strain-energy density can be expressed as

$$u = \int \tau \mathrm{d}\gamma \tag{20.7}$$

where γ is the shearing strain corresponding to τ. From Eq. (20.7), we can see that the strain-energy density u is equal to the area under the stress-strain curve. For values of τ within the proportional limit, i.e. $\tau = G\gamma$, where G is the modulus of elasticity in shear, we obtain

$$u = \frac{1}{2}G\gamma^2 = \frac{1}{2}\tau\gamma = \frac{\tau^2}{2G} \tag{20.8}$$

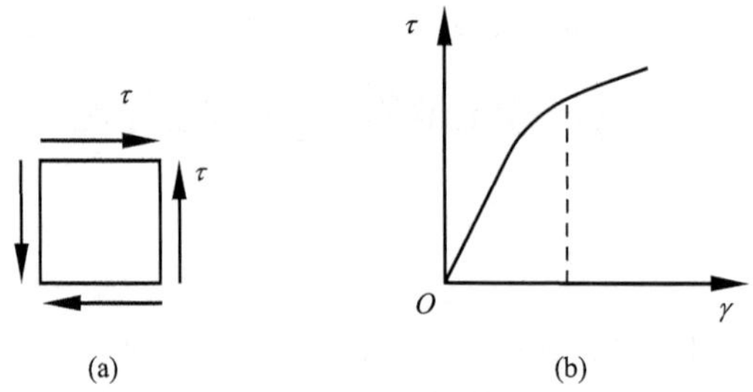

Fig. 20.6

20.3 Strain Energy

1. Axially Tensile or Compressive Strain Energy

Consider a prismatic bar in axial tension or compression, which is fixed at the left end and subjected at the right end to a gradually increasing external force F coinciding with the axial line of the bar, Fig. 20.7.

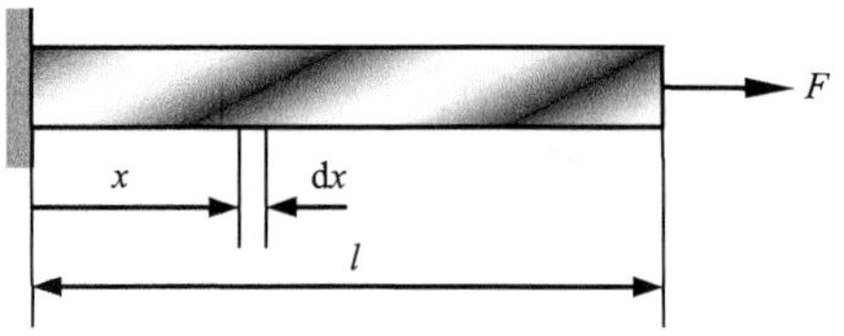

Fig. 20.7

According to Eq. (20.6), the strain energy of the bar can be written as

$$V_\varepsilon = \int \frac{\sigma^2}{2E} A\mathrm{d}x \tag{20.9}$$

Using $\sigma = \dfrac{N}{A}$, where N is the axial force, and A is the cross-sectional area, we have

$$V_\varepsilon = \int \frac{N^2}{2EA} \mathrm{d}x \tag{20.10}$$

In the case of a bar having a uniform cross-sectional area A and a constant axial force N, Eq. (20.10) can be rewritten as

$$V_\varepsilon = \frac{N^2 l}{2EA} \tag{20.11}$$

2. Torsional Strain Energy

Consider a circular shaft in torsion, which is fixed at the left end and subjected at the right end to a gradually increasing external couple M_e applied in a plane perpendicular to the longitudinal axis of the shaft, Fig. 20.8.

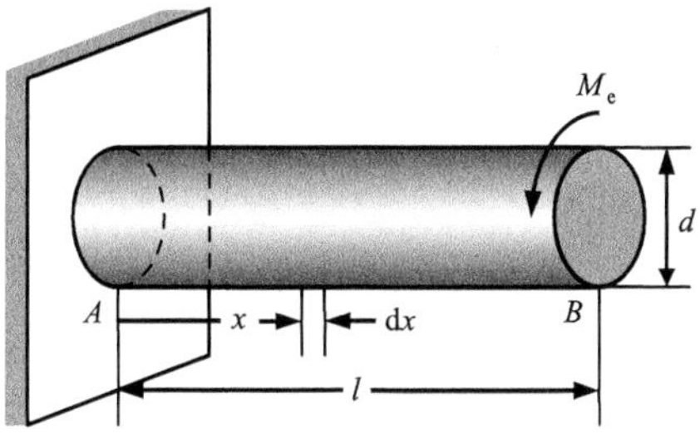

Fig. 20.8

From Eq. (20.8), the strain energy for the shaft can be given by

$$V_{\varepsilon} = \int (\int \frac{\tau^2}{2G} \mathrm{d}A) \mathrm{d}x \tag{20.12}$$

Since $\tau = \dfrac{T\rho}{I_p}$, where T is the torsional moment, I_p is the polar moment of inertia of the cross section, and ρ is the distance from the centroid of the cross section, then we have

$$V_{\varepsilon} = \int (\frac{T^2}{2G{I_p}^2} \int \rho^2 \mathrm{d}A) \mathrm{d}x \tag{20.13}$$

Using $I_p = \int \rho^2 \mathrm{d}A$, Eq. (20.13) can be rewritten as

$$V_{\varepsilon} = \int \frac{T^2}{2GI_p} \mathrm{d}x \tag{20.14}$$

If the shaft has a uniform polar moment of inertia I_p and a constant torsional moment T, then Eq. (20.14) can be expressed as

$$V_{\varepsilon} = \frac{T^2 l}{2GI_p} \tag{20.15}$$

3. Bending Strain Energy

Consider a prismatic beam in bending, which is fixed at the left end and subjected at the right end to a gradually increasing vertical external force F applied in a symmetric plane containing the longitudinal axis of the beam, Fig. 20.9. Since the strain energy caused by the shearing stress is usually small, compared to that caused by the normal stress, and can be neglected for a slender beam. Therefore, the strain energy for the beam in bending can be expressed as

$$V_{\varepsilon} = \int (\int \frac{\sigma^2}{2E} \mathrm{d}A) \mathrm{d}x \tag{20.16}$$

Using $\sigma = \frac{My}{I}$, where M is the bending moment, I is the moment of inertia of the cross section, and y is the perpendicular distance from the neutral axis, we have

$$V_\varepsilon = \int \left(\frac{M^2}{2EI^2}\int y^2 \mathrm{d}A\right)\mathrm{d}x \tag{20.17}$$

Substituting $I = \int y^2 \mathrm{d}A$ into the above equation, then Eq. (20.17) can be rewritten as

$$V_\varepsilon = \int_0^l \frac{M^2}{2EI}\mathrm{d}x \tag{20.18}$$

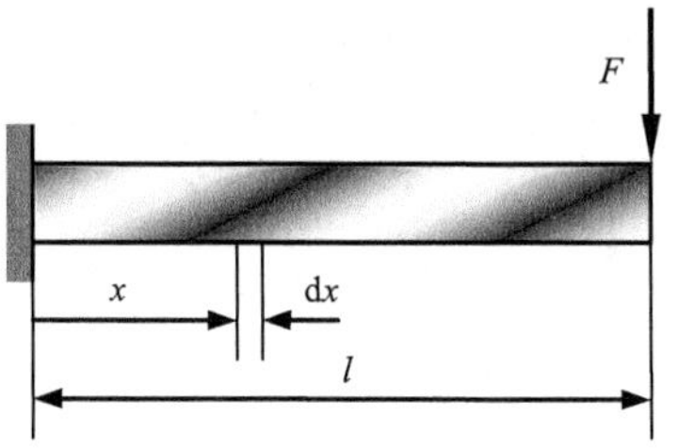

Fig. 20.9

For a beam in pure bending, Fig. 20.10, the bending moment M is a constant. If the beam has a uniform moment of inertia I, then the above equation can be expressed as

$$V_\varepsilon = \frac{M^2 l}{2EI} \tag{20.19}$$

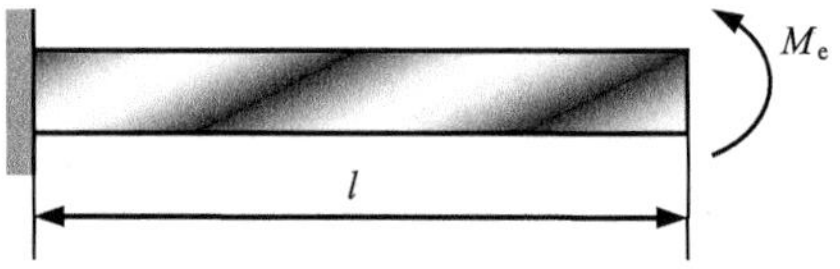

Fig. 20.10

4. Combined Loading Strain Energy

Consider a circular member subjected to combined loadings, Fig. 20.11. For a slender member, the strain energy corresponding to the shearing stress caused the shearing force can be neglected, thus the strain energy for the member subjected to combined loadings can be expressed as

$$V_\varepsilon = \int \frac{N^2}{2EA}\mathrm{d}x + \int \frac{T^2}{2GI_\mathrm{p}}\mathrm{d}x + \int \frac{M^2}{2EI}\mathrm{d}x \tag{20.20}$$

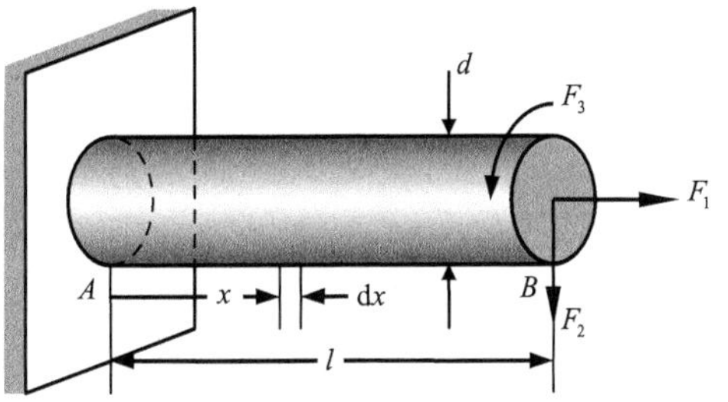

Fig. 20.11

20.4 Principle of Work and Energy

If we neglect the loss of energy, then the conservation of energy requires that the work done by the external forces applied to a body be transformed entirely into the strain energy stored in the deformable body. This relation can be stated mathematically as

$$V_\varepsilon = W_e \tag{20.21}$$

where V_ε is the strain energy and W_e is the external work. The equation above is called the principle of work and energy, and can be used to determine the deformation of a structure.

Example 20.1 Knowing that the flexural rigidity is *EI*, determine the deflection at the free end *B* of the prismatic cantilever *AB* using the principle of work and energy.

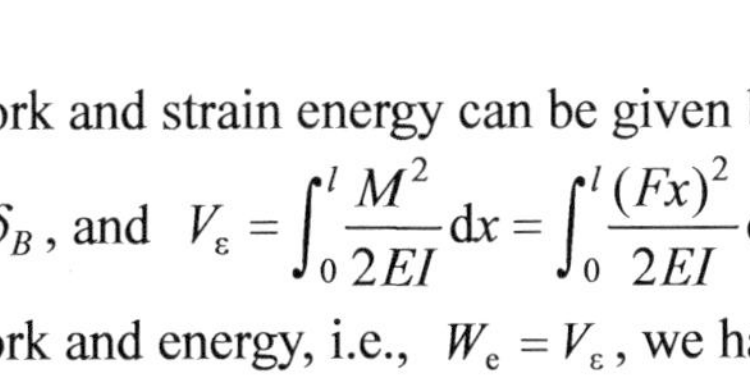

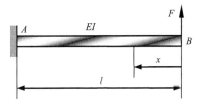

Fig. E20.1

Solution The external work and strain energy can be given by

$$W_e = \frac{1}{2} F\delta_B \text{, and } V_\varepsilon = \int_0^l \frac{M^2}{2EI}\mathrm{d}x = \int_0^l \frac{(Fx)^2}{2EI}\mathrm{d}x = \frac{F^2l^3}{6EI}$$

Using the principle of work and energy, i.e., $W_e = V_\varepsilon$, we have

$$\frac{1}{2} F\delta_B = \frac{F^2l^3}{6EI}\text{, or } \delta_B = \frac{Fl^3}{3EI}$$

Example 20.2 Using the principle of work and energy, determine the slope at the free end *B* of the prismatic cantilever *AB*. Know that the flexural rigidity is *EI*.

Solution $W_e = \frac{1}{2} M_e\theta_B$, $V_\varepsilon = \int_0^l \frac{M^2}{2EI}\mathrm{d}x = \int_0^l \frac{M_e^2}{2EI}\mathrm{d}x = \frac{M_e^2 l}{2EI}$

Using $W_e = V_\varepsilon$, we have

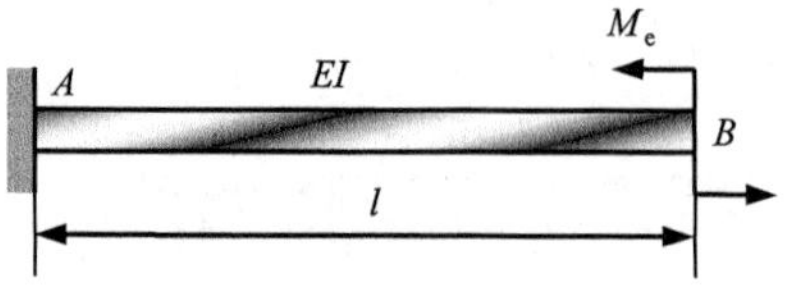

Fig. E20.2

$$\frac{1}{2}M_e\theta_B = \frac{M_e^2 l}{2EI}, \text{ or } \theta_B = \frac{M_e l}{EI}$$

20.5 Reciprocal Theorem

Consider that a linearly elastic body, supported at A and B, is subjected to generalized forces, as shown in Fig. 20.12. Assuming that the generalized forces, F_1 and F_3, are first applied to the body and then the generalized forces, F_2 and F_4, are applied to the body again, the strain energy stored in the body can be expressed as

$$V_{\varepsilon 1} = \frac{1}{2}F_1\Delta_1 + \frac{1}{2}F_3\Delta_3 + \frac{1}{2}F_2\Delta_2 + \frac{1}{2}F_4\Delta_4 + F_1\Delta_1' + F_3\Delta_3' \tag{20.22}$$

where Δ_1 (or Δ_3) is the generalized displacement at the application point of F_1 (or F_3) in the direction of F_1 (or F_3), caused by both F_1 and F_3, Δ_2 (or Δ_4) is the generalized displacement at the application point of F_2 (or F_4) in the direction of F_2 (or F_4), caused by both F_2 and F_4, and Δ_1' (or Δ_3') is the generalized displacement at the application point of F_1 (or F_3) in the direction of F_1 (or F_3), caused by both F_2 and F_4.

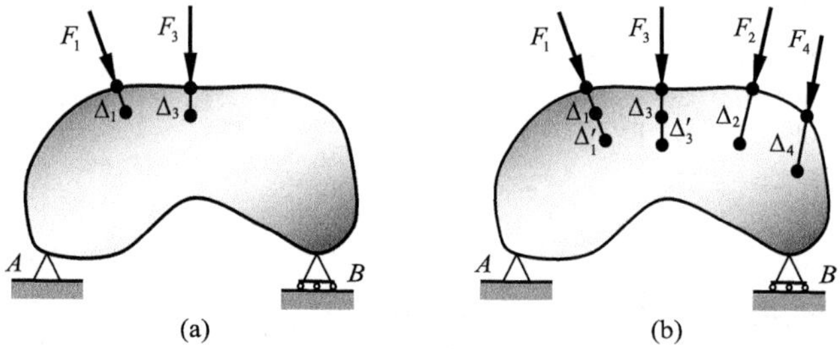

Fig. 20.12

When the generalized forces, F_2 and F_4, are first applied to the body and then the generalized forces, F_1 and F_3, are applied to the body again, the strain energy stored in the body can be given by

$$V_{\varepsilon 2} = \frac{1}{2}F_2\Delta_2 + \frac{1}{2}F_4\Delta_4 + \frac{1}{2}F_1\Delta_1 + \frac{1}{2}F_3\Delta_3 + F_2\Delta_2' + F_4\Delta_4' \tag{20.23}$$

where Δ_2' (or Δ_4') is the generalized displacement at the application point of F_2 (or F_4) in the direction of F_2 (or F_4), caused by both F_1 and F_3.

By comparison of Eqs. (20.22) and (20.23) and using $V_{\varepsilon 1} = V_{\varepsilon 2}$, we have

$$F_1\Delta_1' + F_3\Delta_3' = F_2\Delta_2' + F_4\Delta_4' \tag{20.24}$$

We thus conclude that the work done by the first group of generalized forces on the generalized displacement caused by the second group of generalized forces is equal to the work done by the second group of generalized forces on the generalized displacement caused by the first group of generalized forces. The relation above is called the reciprocal theorem of work.

Assuming that $F_3 = F_4 = 0$ and $F_1 = F_2$, then Eq. (20.24) can be simplified as

$$\Delta_1' = \Delta_2' \tag{20.25}$$

We thus conclude that, when two generalized forces are equal in magnitude, the generalized displacement at the application point of the first force in the direction of the first force caused by the second force is equal to the generalized displacement at the application point of the second force in the direction of the second force caused by the first force. This relation is called the reciprocal theorem of displacement.

Example 20.3 A linearly elastic sphere is subjected to two radial tensile loads, equal in magnitude and opposite in direction, as shown in Fig. E20.3a. Knowing that the sphere has modulus of elasticity E, Poisson's ratio μ and diameter d, determine the change in volume of the sphere.

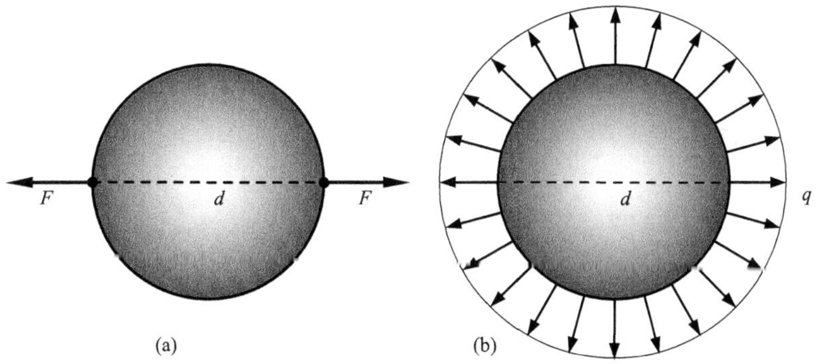

Fig. E20.3

Solution Assuming that the sphere is subjected to triaxial stress in uniform tension, as shown in Fig. E20.3b, then we can obtain

$$\varepsilon_1 = \frac{1}{E}[\sigma_1 - \mu(\sigma_2 + \sigma_3)] = \frac{1-2\mu}{E}q$$

Using the reciprocal theorem of work, we have

$$F(\varepsilon_1 d) = q\Delta V$$

i.e.,

$$\Delta V = \frac{F(\varepsilon_1 d)}{q} = \frac{1-2\mu}{E}Fd$$

20.6 Castigliano's Theorem

Consider a deformable body supported at A and B shown in Fig. 20.13, which is subjected to a series of external forces $F_1, F_2, \cdots, F_i, \cdots, F_n$. When these external forces are applied to the body gradually from zero to their final values, the corresponding displacements of the force application points in the force directions are also slowly increased from zero to their final values. If the material is within a linearly elastic range, then the work done by the external forces applied to the body can be expressed as

$$W_e = \sum \frac{1}{2} F_i \Delta_i \tag{20.26}$$

where Δ_i is a linear homogeneous function of the external forces $F_1, F_2, \cdots, F_i, \cdots, F_n$. It can be seen from Eq. (20.26) that the external work W_e is a quadratic homogeneous function of the external forces $F_1, F_2, \cdots, F_i, \cdots, F_n$. Using the principle of work and energy, i.e., $W_e = V_\varepsilon$, then the strain energy is also a quadratic homogeneous function of the external forces $F_1, F_2, \cdots, F_i, \cdots, F_n$ and can be expressed as

$$V_\varepsilon = V_\varepsilon(F_1, F_2, \cdots, F_i, \cdots, F_n) \tag{20.27}$$

If any one of the external forces, say the ith external force F_i, is increased by $\mathrm{d}F_i$ while the other external forces remain constant, then the strain energy will also be increased by $\mathrm{d}V_\varepsilon$, which can be denoted by

$$\mathrm{d}V_\varepsilon = \frac{\partial V_\varepsilon}{\partial F_i} \mathrm{d}F_i \tag{20.28}$$

Therefore, the total strain energy can be written as

$$V_{\varepsilon 1} = V_\varepsilon + \mathrm{d}V_\varepsilon = \sum \frac{1}{2} F_i \Delta_i + \frac{\partial V_\varepsilon}{\partial F_i} \mathrm{d}F_i \tag{20.29}$$

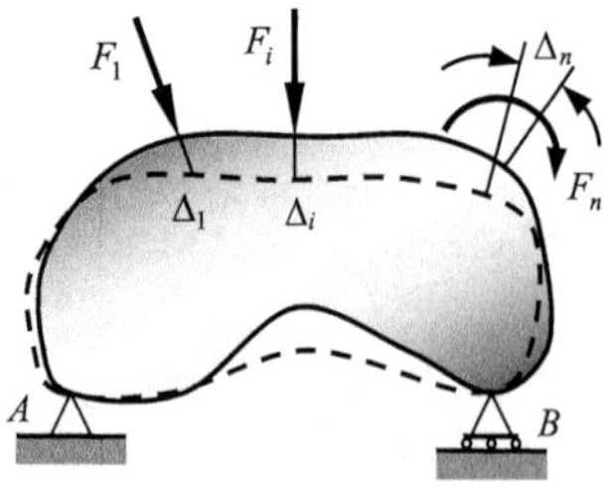

Fig. 20.13

The total strain energy $V_{\varepsilon 1}$ should be independent of the order in which the external forces are applied to the body. For example, we can apply the external force $\mathrm{d}F_i$ to the body first, then apply the external forces $F_1, F_2, \cdots, F_i, \cdots, F_n$. The total strain energy should be the same. When the external force $\mathrm{d}F_i$ is applied first, the external work $\mathrm{d}W_e'$ done by the external force $\mathrm{d}F_i$ can be written as

$$\mathrm{d}W'_{\mathrm{e}} = \tfrac{1}{2}\mathrm{d}F_i \mathrm{d}\Delta_i \tag{20.30}$$

where $\mathrm{d}\Delta_i$ is the displacement corresponding to the external force $\mathrm{d}F_i$ .When the external forces $F_1, F_2, \cdots, F_i, \cdots, F_n$ are applied, the external work done by the external force $F_1, F_2, \cdots, F_i, \cdots, F_n$ and the external force $\mathrm{d}F_i$ can be expressed as

$$W'_{\mathrm{e}} = \sum \frac{1}{2} F_i \Delta_i + \mathrm{d}F_i \Delta_i \tag{20.31}$$

Therefore, the total work can be written as

$$W_{\mathrm{e}2} = W'_{\mathrm{e}} + \mathrm{d}W'_{\mathrm{e}} = \sum \frac{1}{2} F_i \Delta_i + \mathrm{d}F_i \Delta_i + \frac{1}{2}\mathrm{d}F_i \mathrm{d}\Delta_i \tag{20.32}$$

Using the principle of work and energy, the total strain energy can be given by

$$V_{\varepsilon 2} = W_{\mathrm{e}2} = \sum \frac{1}{2} F_i \Delta_i + \mathrm{d}F_i \Delta_i + \frac{1}{2}\mathrm{d}F_i \mathrm{d}\Delta_i \tag{20.33}$$

Using $V_{\varepsilon 1} = V_{\varepsilon 2}$, we have

$$\sum \frac{1}{2} F_i \Delta_i + \frac{\partial V_\varepsilon}{\partial F_i}\mathrm{d}F_i = \sum \frac{1}{2} F_i \Delta_i + \mathrm{d}F_i \Delta_i + \frac{1}{2}\mathrm{d}F_i \mathrm{d}\Delta_i \tag{20.34}$$

Neglecting the second-order differential, we obtain

$$\Delta_i = \frac{\partial V_\varepsilon}{\partial F_i} \tag{20.35}$$

where F_i is a generalized force, and Δ_i is the corresponding generalized displacement. If F_i is a couple, then Δ_i is an angle of rotation corresponding to the couple. Eq. (20.35) is called Castigliano's second theorem or Castigliano's theorem, which can be stated as follows: the generalized displacement Δ_i at the application point of F_i in the direction of F_i is equal to the first partial derivative of the strain energy with respect to the generalized force F_i .

When Castigliano's theorem is used for a member subjected to combined loadings it can be expressed as

$$\Delta_i = \frac{\partial V_\varepsilon}{\partial F_i} = \int \frac{N}{EA}\frac{\partial N}{\partial F_i}\mathrm{d}x + \int \frac{T}{GI_{\mathrm{p}}}\frac{\partial T}{\partial F_i}\mathrm{d}x + \int \frac{M}{EI}\frac{\partial M}{\partial F_i}\mathrm{d}x \tag{20.36}$$

Example 20.4 Knowing that the flexural rigidity is *EI*, determine the deflection at the free end *B* of the prismatic cantilever *AB* using Castigliano's theorem.

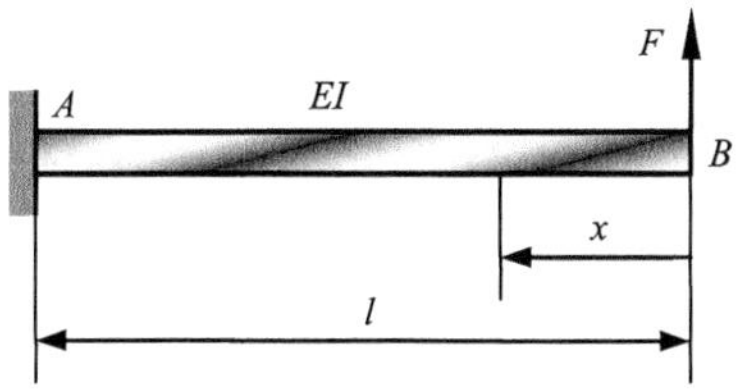

Fig. E20.4

Solution The strain energy is equal to

$$V_\varepsilon = \int_0^l \frac{M^2}{2EI}\mathrm{d}x = \int_0^l \frac{(Fx)^2}{2EI}\mathrm{d}x = \frac{F^2l^3}{6EI}$$

Using Castigliano's second theorem, we have

$$\delta_B = \frac{\partial V_\varepsilon}{\partial F} = \frac{\partial}{\partial F}\left(\frac{F^2l^3}{6EI}\right) = \frac{Fl^3}{3EI}$$

Example 20.5 Using Castigliano's theorem, determine the slope at the free end B of the prismatic cantilever AB. Know that the flexural rigidity is EI.

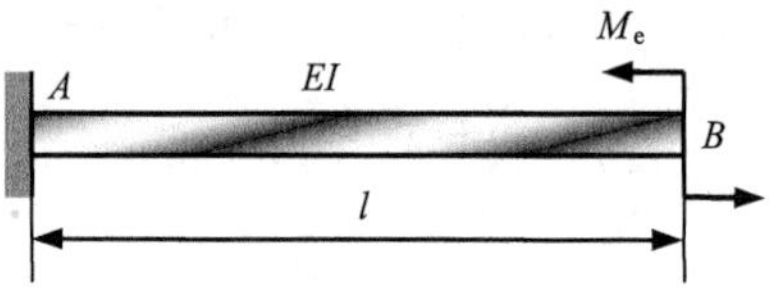

Fig. E20.5

Solution The strain energy is

$$V_\varepsilon = \int_0^l \frac{M^2}{2EI}\mathrm{d}x = \int_0^l \frac{M_e^2}{2EI}\mathrm{d}x = \frac{M_e^2 l}{2EI}$$

Using $\Delta_i = \dfrac{\partial V_\varepsilon}{\partial F_i}$, we have

$$\theta_B = \frac{\partial V_\varepsilon}{\partial M_e} = \frac{\partial}{\partial M_e}\left(\frac{M_e^2 l}{2EI}\right) = \frac{M_e l}{EI}$$

The only displacements that can be determined from Castigliano's second theorem are those that correspond to external forces applied to the body. If we want to find a displacement at a point on the body where there is no external force, then a fictitious (or dummy) force corresponding to the displacement to be determined must be applied to the body. We can then determine the displacement by calculating the strain energy and taking the partial derivative with respect to the fictitious force. The result obtained is that displacement produced simultaneously by the actual forces and the fictitious force. By setting the fictitious force equal to zero, we can then obtain the displacement produced only by the actual forces.

Example 20.6 Knowing that the flexural rigidity is EI, determine the slope at the free end B of the prismatic cantilever AB, Fig. E20.6a, using Castigliano's theorem.

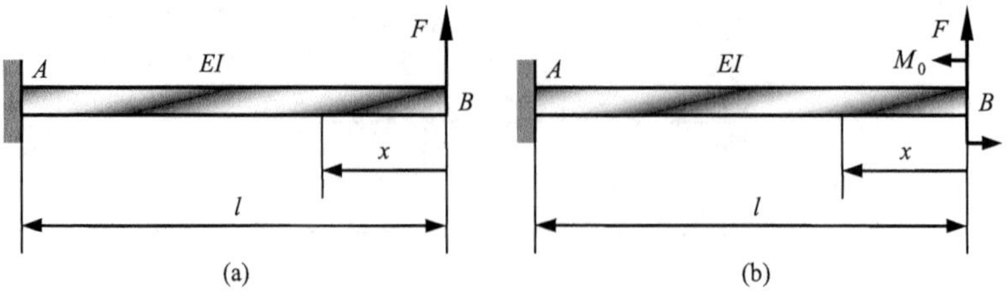

Fig. E20.6

Solution Adding a couple M_0 at the free end B of the beam, Fig. E20.6b, then the bending moment at the x-section is equal to

$$M = Fx + M_0 \quad (0 \leqslant x < l)$$

Therefore, the corresponding strain energy can be expressed as

$$V_\varepsilon = \int_0^l \frac{M^2}{2EI}\mathrm{d}x = \int_0^l \frac{(Fx+M_0)^2}{2EI}\mathrm{d}x = \frac{\frac{1}{3}F^2l^3 + FM_0l^2 + M_0{}^2 l}{2EI}$$

Using Castigliano's second theorem, i.e., $\Delta_i = \dfrac{\partial V_\varepsilon}{\partial F_i}$, we have

$$\theta_B = \left(\frac{\partial V_\varepsilon}{\partial M_0}\right)_{M_0=0} = \left(\frac{Fl^2 + 2M_0 l}{2EI}\right)_{M_0=0} = \frac{Fl^2}{2EI}$$

20.7 Principle of Virtual Work

Consider that a deformable body supported at A and B is balanced under the action of the loads F, q and M_e, as shown in Fig. 20.14.

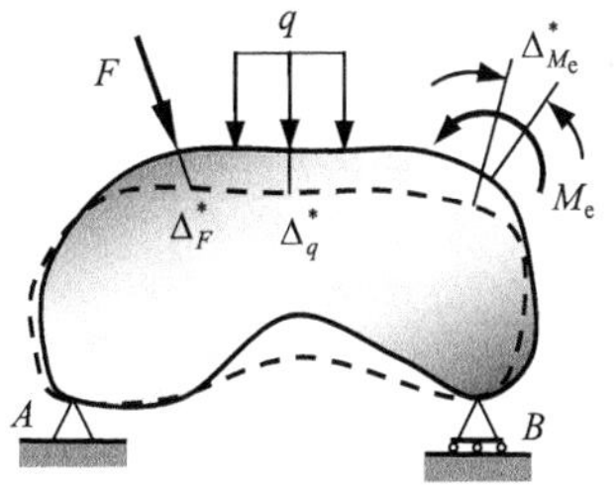

Fig. 20.14

When a virtual deformation is produced in the body due to other factors, such as load change or temperature fluctuation, then the virtual work done by the loads F, q and M_e on the virtual deformation can be expressed as

$$W_e = F\Delta^*_F + \int q\Delta^*_q \mathrm{d}x + M_e\Delta^*_{M_e} \tag{20.37}$$

where Δ^*_F, Δ^*_q and $\Delta^*_{M_e}$ are respectively the corresponding virtual displacements at the application points of F, q and M_e in the directions of F, q and M_e.

When the body is subjected to the virtual deformation, the internal forces caused by the loads F, q and M_e applied to the body will do virtual work, which can be given by

$$W_i = \int N\mathrm{d}\delta^* + \int M\mathrm{d}\theta^* + \int V\mathrm{d}\lambda^* + \int T\mathrm{d}\varphi^* \tag{20.38}$$

where N, M, V and T are respectively the axial force, bending moment, shearing force, and torsional moment caused by the loads F, q and M_e, and δ^*, θ^*, λ^* and φ^* are respectively the virtual deformations corresponding to N, M, V and T.

Using $W_i = W_e$, we have

$$\int N\mathrm{d}\delta^* + \int M\mathrm{d}\theta^* + \int V\mathrm{d}\lambda^* + \int T\mathrm{d}\varphi^* = F\Delta_F^* + \int q\Delta_q^*\mathrm{d}x + M_e\Delta_{M_e}^* \tag{20.39}$$

This is the principle of virtual work, which can be stated as follows: The virtual work done by the internal forces on the virtual deformation is equal to the virtual work done by the external loads on the virtual displacements. The principle of virtual work can be used for both linearly and nonlinearly elastic bodies.

Example 20.7 As shown in Fig. E20.7a, a truss ABC consisting of two rods AC and BC, each having a constant cross-sectional area A, is subjected to a concentrated force F at joint C. The stress-strain relation is $\sigma = E\varepsilon$ for AC and $\sigma^2 = k\varepsilon$ for BC, where E and k are constants. Using the principle of virtual work, determine the horizontal and vertical components of displacement at C. (Note: The stability of structure does not need to be considered.)

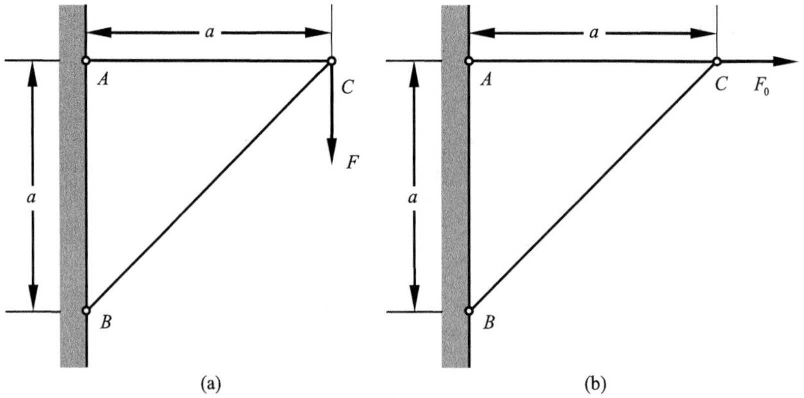

Fig. E20.7

Solution When a horizontal force F_0, in the direction to the right, is applied to the truss at C, as shown in Fig. E20.7b, the internal forces produced in AC and BC can be given by

$$N_{AC} = F_0,\ N_{BC} = 0$$

When the original load F is applied to the structure, the deformations produced in AC and BC can be regarded as virtual deformations, which can be expressed as

$$\delta_{AC}^* = \frac{Fa}{EA},\ \delta_{BC}^* = [-\frac{(-\sqrt{2}F)^2}{kA^2}](\sqrt{2}a) = -\frac{2\sqrt{2}F^2a}{kA^2}$$

and the displacement produced at C can be regarded as virtual displacement, which is equal to

$$\delta_C^* = \Delta_H$$

where Δ_H is the horizontal components of displacement at C to be determined.

Using the principle of virtual work, we have

$$F_0\Delta_H = N_{AC}\frac{Fa}{EA} + N_{BC}(-\frac{2\sqrt{2}F^2a}{kA^2}), \text{ i.e. } \Delta_H = \frac{Fa}{EA}$$

If a vertical force downward is applied to the truss at C, in the same way, we can find the vertical component of displacement at C, equal to

$$\Delta_V = \frac{Fa}{EA} + (-\sqrt{2})(-\frac{2\sqrt{2}F^2a}{kA^2}) = \frac{Fa}{EA} + \frac{4F^2a}{kA^2}$$

20.8 Unit Load Method

Consider that a deformable body supported at A and B is balanced under the action of the loads F, q and M_e, as shown in Fig. 20.15a.

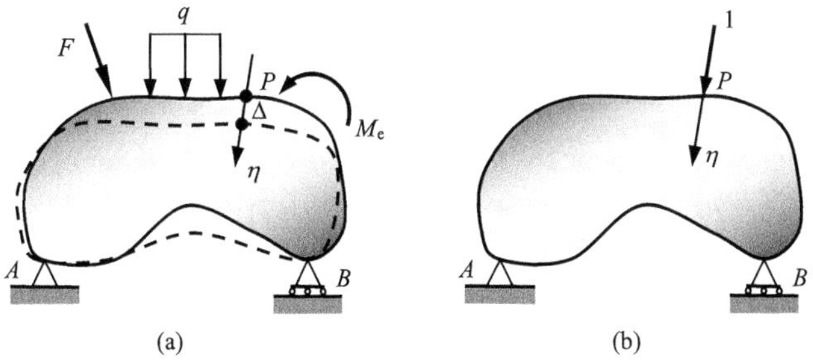

Fig. 20.15

In order to determine the displacement Δ at P in the direction of η, a unit load is first applied to the body at P in the direction of η, as shown in Fig. 20.15b, and then the original loads are also applied to the body. If the deformations produced by the original loads are regarded as virtual deformations, then from the principle of virtual work we can obtain

$$\Delta = \int \bar{N}\mathrm{d}\delta + \int \bar{M}\mathrm{d}\theta + \int \bar{V}\mathrm{d}\lambda + \int \bar{T}\mathrm{d}\varphi \tag{20.40}$$

where $\bar{N}$, $\bar{M}$, $\bar{V}$, and $\bar{T}$ are respectively the axial force, bending moment, shearing force, and torsional moment produced by the unit load, and δ, θ, λ, and φ are the deformations, respectively corresponding to the axial force, bending moment, shearing force, and torsional moment produced by the loads F, q and M_e applied to the body. The above relation, which can be used to determine the displacement at any point in any direction of a structure, is called the unit load method. If an angle of rotation is needed to be determined, then a unit couple is required to be applied at the point where the angle of rotation is to be determined. The unit load method can be used for both linearly and nonlinearly elastic bodies.

For a linearly elastic member subjected to combined loadings, Eq. (20.40) can be rewritten as

$$\Delta = \int \frac{N(x)\bar{N}(x)}{EA}\mathrm{d}x + \int \frac{M(x)\bar{M}(x)}{EI}\mathrm{d}x + \int \frac{T(x)\bar{T}(x)}{GI_{\mathrm{p}}}\mathrm{d}x \tag{20.41}$$

where N, M, and T are respectively the axial force, bending moment, and torsional moment produced by the combined loadings, and EA, EI, and GI_{p} are respectively the axially tensile or compressive, bending, and torsional rigidities of the member subjected to combined loadings.

Example 20.8 Knowing that the flexural rigidity is *EI*, determine the slope at the free end *B* of the prismatic cantilever *AB*, Fig. E20.8a, using the unit load method.

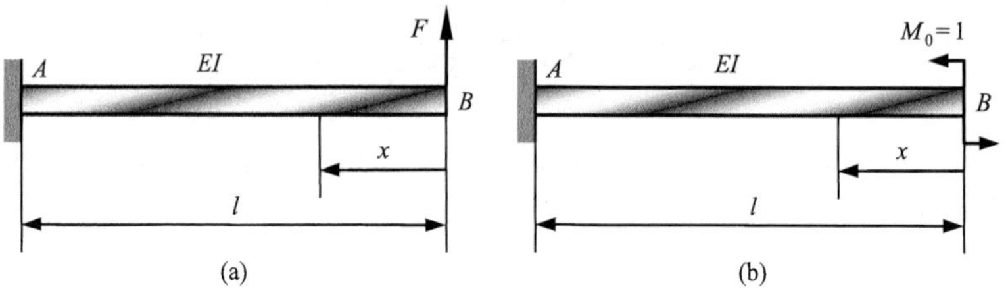

Fig. E20.8

Solution The bending moment for the original load, Fig. E20.8a, can be expressed as

$$M(x) = Fx \quad (0 \leqslant x < l)$$

The bending moment for the unit couple, Fig. E20.8b, can be expressed as

$$\bar{M}(x) = M_0 = 1 \quad (0 < x < l)$$

From the unit load method, we have

$$\theta_B = \int_0^l \frac{M(x)\bar{M}(x)}{EI}\mathrm{d}x = \frac{Fl^2}{2EI}$$

Example 20.9 Two rods *AB* and *BC* of the same flexural rigidity *EI* are welded together at *B*, Fig. E20.9a. For the loading shown and using the unit load method, determine the vertical deflection of point *C*, and the slope of section *C*.

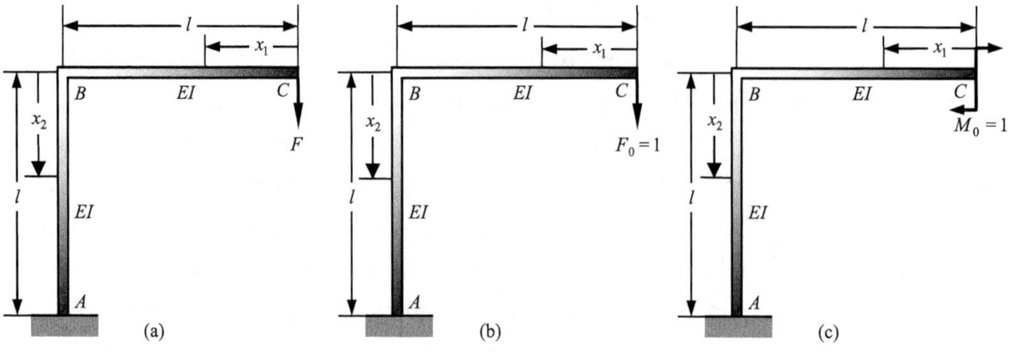

Fig. E20.9

Solution The bending moment for the original load, Fig. E20.9a, can be expressed as

$$M(x_1) = -Fx_1 \quad (0 \leqslant x_1 \leqslant l), \quad M(x_2) = -Fl \quad (0 \leqslant x_2 < l)$$

In order to determine the vertical deflection of point *C*, add a downward unit load $F_0 = 1$ at point *C*, Fig. E20.9b. The bending moment corresponding to the unit load can be expressed as

$$\bar{M}(x_1) = -x_1 \quad (0 \leqslant x_1 \leqslant l), \quad \bar{M}(x_2) = -l \quad (0 \leqslant x_2 < l)$$

Using the unit load method, we have

$$w_{Cy} = \int_0^l \frac{M(x_1)\bar{M}(x_1)}{EI} \mathrm{d}x_1 + \int_0^l \frac{M(x_2)\bar{M}(x_2)}{EI} \mathrm{d}x_2 = \frac{4Fl^3}{3EI}$$

In order to determine the slope of point *C*, add a clockwise unit couple $M_0 = 1$ at point *C*, Fig. E20.9c. The bending moment corresponding to the unit couple can be expressed as

$$\bar{M}(x_1) = -1 \quad (0 \leqslant x_1 \leqslant l), \quad \bar{M}(x_2) = -1 \quad (0 \leqslant x_2 < l)$$

Using the unit load method, we have

$$\theta_C = \int_0^l \frac{M(x_1)\bar{M}(x_1)}{EI} \mathrm{d}x_1 + \int_0^l \frac{M(x_2)\bar{M}(x_2)}{EI} \mathrm{d}x_2 = \frac{3Fl^2}{2EI}$$

Example 20.10 Two rods *AB* and *BC* of the same flexural rigidity *EI* are welded together at *B*, Fig. E20.10a, which has a fixed end at *A* and is supported by a roller at *C*. For the loading and supports shown and using the unit load method, determine the reaction at point *C*.

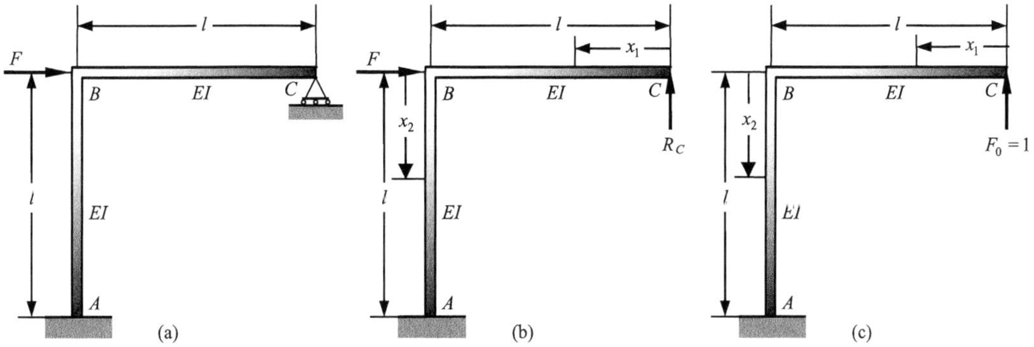

Fig. E20.10

Solution The structure shown in Fig. E20.10a is statically indeterminate to the first degree. We designate the reaction R_C at point *C* as redundant and eliminate the corresponding support, Fig. E20.10b.

Referring to E20.10b, the bending moment can be expressed as

$$M(x_1) = R_C x_1 \quad (0 \leqslant x_1 \leqslant l), \quad M(x_2) = R_C l - Fx_2 \quad (0 \leqslant x_2 < l)$$

In order to determine the vertical deflection of point *C*, we add a unit load $F_0 = 1$ at point *C* in the vertical direction, Fig. E20.10c. The bending moment corresponding to the unit load can be expressed as

$$\bar{M}(x_1)=x_1\ \ (0\leqslant x_1\leqslant l),\ \ \bar{M}(x_2)=l\ \ (0\leqslant x_2<l)$$

Using the unit load method, we have

$$w_{Cy}=\int_0^l \frac{M(x_1)\bar{M}(x_1)}{EI}\mathrm{d}x_1+\int_0^l \frac{M(x_2)\bar{M}(x_2)}{EI}\mathrm{d}x_2=\frac{(8R_C-3F)l^3}{6EI}$$

Since $w_{Cy}=0$, then we obtain

$$R_C=\frac{3}{8}F$$

20.9 Impact Loading

In the preceding discussion, all loadings are static; that is, they are applied gradually to a member and remain constant when they reach a final value. Some loadings, however, are dynamic; that is, they are applied suddenly to a member. These dynamic loadings are called impact loadings. When an object strikes a member, a very large stress will be developed within the member during impact.

1. Vertical Impact

Consider a block-and-spring system shown in Fig. 20.16. When a block of weight P and kinetic energy T strikes a spring without consideration of mass, it will attach to the spring and move downward. Assume that the block compresses the spring a distance Δ_d before coming to rest. If there is no dissipation of energy during the impact, then the conservation of energy requires that the kinetic and potential energy of the block be transformed entirely into the strain energy stored in the spring; or in other words, the sum of kinetic and potential energy of the block is equal to the work needed to displace the free end of the spring by an amount Δ_d.

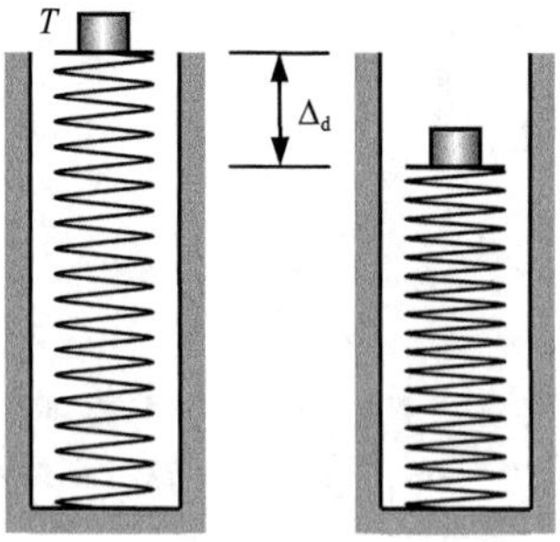

Fig. 20.16

From the conservation of energy, we have

$$T+P\Delta_d=\frac{1}{2}F_d\Delta_d \tag{20.42}$$

where F_d is the force between the block and spring when the spring contraction reaches Δ_d .

Assuming that the spring is within linearly elastic range, i.e., $\frac{F_d}{\Delta_d}=\frac{P}{\Delta_{st}}$ where Δ_{st} is the static displacement of the free end of the spring when the block is applied statically to the free end of the spring, then we obtain

$$\frac{F_d}{P}=\frac{\Delta_d}{\Delta_{st}}=K_d \tag{20.43}$$

where K_d is the impact factor. Substituting Eq. (20.43) into Eq. (20.42), we can obtain

$$\frac{1}{2}K_d^{\ 2}-K_d-\frac{T}{P\Delta_{st}}=0 \tag{20.44}$$

Solving the above equation for K_d, we obtain the following positive root:

$$K_d=1+\sqrt{1+\frac{2T}{P\Delta_{st}}} \tag{20.45}$$

Assuming that the block is released from rest, falls a distance h and strikes the spring, then the impact factor can be given by

$$K_d=1+\sqrt{1+\frac{2h}{\Delta_{st}}} \tag{20.46}$$

Once K_d is determined, the dynamic stress in the spring and the dynamic deformation of the spring can be expressed as

$$\sigma_d=K_d\sigma_{st},\quad \delta_d=K_d\delta_{st} \tag{20.47}$$

where σ_{st} and δ_{st} are respectively the static stress and static deformation when the weight is applied statically to the spring.

If the block is held just above the spring, h=0, and released from rest, then, from Eq. (20.45) or Eq. (20.46), the impact factor is K_d=2, and the resulting dynamic stress in the spring is $\sigma_d=2\sigma_{st}$; i.e., when the block is dropped suddenly with no initial velocity from the top of the spring (suddenly applied loading), the dynamic stress is twice as large as the static stress caused by the block placed statically on the spring (statically applied load).

Example 20.11 A block of weight P, initially at rest, is dropped from a height h onto the midpoint C of the simple beam AB. Knowing that the modulus of elasticity of the beam is E=105 GPa, determine (a) the maximum deflection of the beam, and (b) the maximum normal stress in the beam.

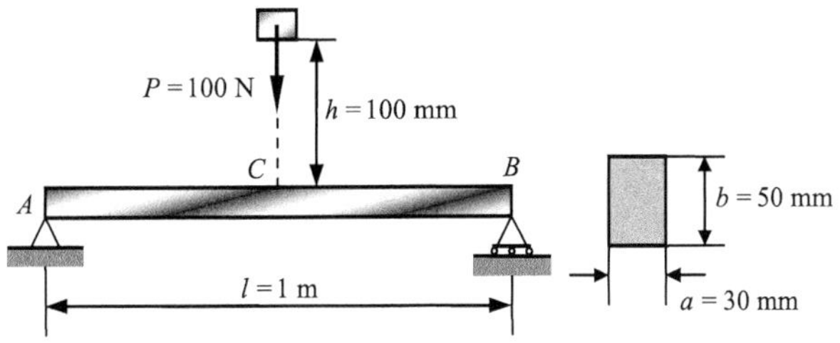

Fig. E20.11

Solution (a) Referring to Tab. 16.1, we have

$$\Delta_{st} = \frac{Pl^3}{48EI} = \frac{Pl^3}{48E(\frac{1}{12}ab^3)} = 6.35 \times 10^{-2} \text{ mm}$$

Using Eq. (20.46), we obtain

$$K_d = 1 + \sqrt{1 + \frac{2h}{\Delta_{st}}} = 57.13$$

Therefore, the maximum dynamic deflection is equal to

$$\Delta_d = K_d \Delta_{st} = 3.63 \text{ mm}$$

(b) Referring to Eq. (15.11), we have the maximum static stress

$$\sigma_{st} = \frac{M}{S} = \frac{\frac{1}{4}Pl^2}{\frac{1}{6}ab^2} = 2.00 \text{ MPa}$$

Using Eq. (20.47), we obtain the maximum dynamic normal stress in the beam

$$\sigma_d = K_d \sigma_{st} = 114.26 \text{ MPa}$$

Example 20.12 A collar of mass *m* is released from rest in the position shown and is stopped by a plate attached at the free end *B* of the vertical rod *AB* of diameter *d*. Knowing that *E*=70 GPa, determine (a) the maximum elongation of the rod, and (b) the maximum normal stress in the rod. ($g = 9.81 \text{ m/s}^2$)

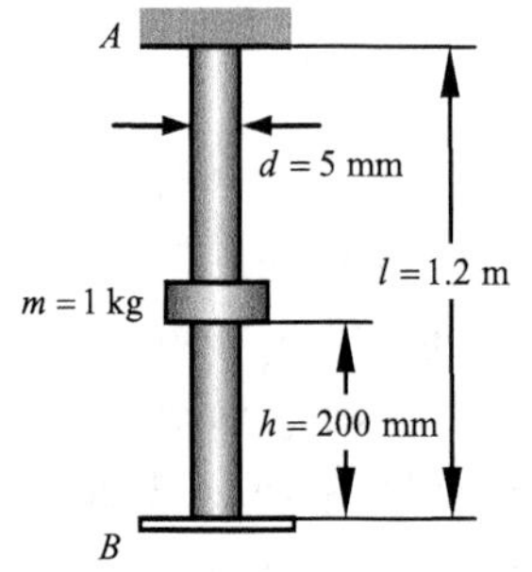

Fig. E20.12

Solution (a) Maximum elongation of the rod

$$\Delta_{st} = \frac{mgl}{EA} = \frac{mgl}{E(\frac{1}{4}\pi d^2)} = 8.56 \times 10^{-3} \text{ mm}, \quad K_d = 1 + \sqrt{1 + \frac{2h}{\Delta_{st}}} = 217.17, \quad \Delta_d = K_d \Delta_{st} = 1.86 \text{ mm}$$

(b) Maximum normal stress in the rod

$$\sigma_{st} = \frac{mg}{A} = \frac{mg}{\frac{1}{4}\pi d^2} = 0.50 \text{ MPa}, \quad \sigma_d = K_d \sigma_{st} = 108.58 \text{ MPa}$$

Example 20.13 A plane mechanism, *ABCD*, is composed of two blocks, *A* and *D*, two smooth fixed pulleys, *B* and *C*, and an elastic cable, *ABCD*. The cable has length of 0.6 m and cross-sectional area of 50 mm^2. The blocks *A* and *B*, each having weigh of 120 N, are moving downward and upward with a constant speed of 56 mm/s. Knowing that $E = 10$ GPa, and assuming that the block *D* is suddenly seized up and stops moving, determine the maximum stress in the cable.

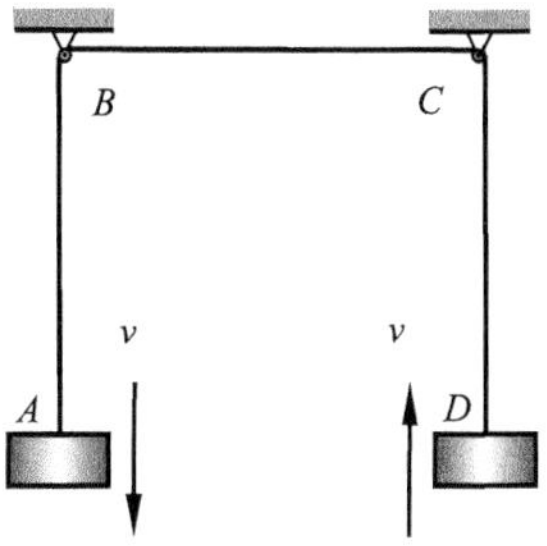

Fig. E20.13

Solution $\Delta_{st} = \dfrac{P_A l}{EA} = 144\ \mu m$, $K_d = 1 + \sqrt{\dfrac{2T_A}{P_A \Delta_{st}}} = 2.49$, $\sigma_{st} = \dfrac{P_A}{A} = 2.4$ MPa, $\sigma_d = K_d \sigma_{st}$ $= 5.98$ MPa

2. Horizontal Impact

Consider a block-and-spring system placed on a smooth horizontal surface, Fig. 20.17. The block of weight *P* sliding on the surface with kinetic energy *T* collides with the spring and compresses it a distance Δ_d before coming to a maximum displacement. If we neglect the loss of energy during impact and the mass of the spring and assume that the spring is within linearly elastic range and that the block sticks to the spring and moves with it, then kinetic energy of the block is transferred entirely to the strain energy of the spring when the block comes to a rest.

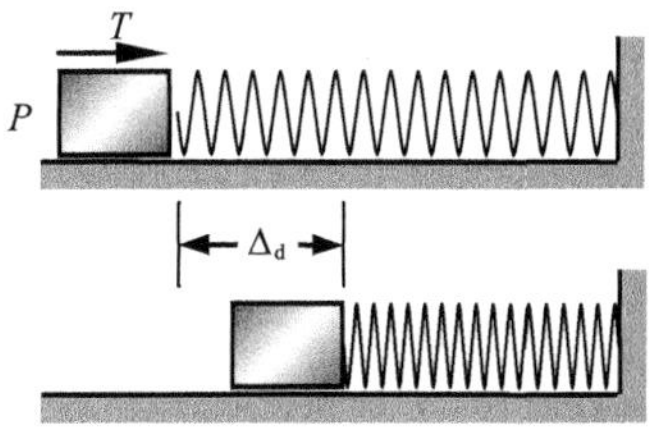

Fig. 20.17

From the conservation of energy, we have

$$T = \frac{1}{2} F_d \Delta_d = \frac{1}{2} P \Delta_{st} K_d^2 \tag{20.48}$$

The positive root of Eq. (20.48) can be expressed as

$$K_d = \sqrt{\frac{2T}{P\Delta_{st}}} \tag{20.49}$$

where Δ_{st} is the static displacement of the free end of the spring when an equivalent force, equal to the weight P of the block, is applied statically to the free end of the spring. Using $T = \frac{1}{2}\frac{P}{g}v^2$, Eq. (20.49) can also be rewritten as

$$K_d = \sqrt{\frac{v^2}{g\Delta_{st}}} \tag{20.50}$$

From Eq. (20.49) or (20.50), the dynamic stress in the spring and the dynamic deformation of the spring can be expressed, respectively, as

$$\sigma_d = K_d\sigma_{st}, \quad \delta_d = K_d\delta_{st} \tag{20.51}$$

where σ_{st} and δ_{st} are the static stress in the spring and the static deformation of the spring when an equivalent force, equal to the weight P of the block, is applied statically to the spring.

Example 20.14 A sphere of mass m is moving with a velocity v to the right and hits squarely the free end A of the post AB of diameter d. Using E=200 GPa, determine (a) the maximum deflection of the post, and (b) the maximum normal stress in the post.

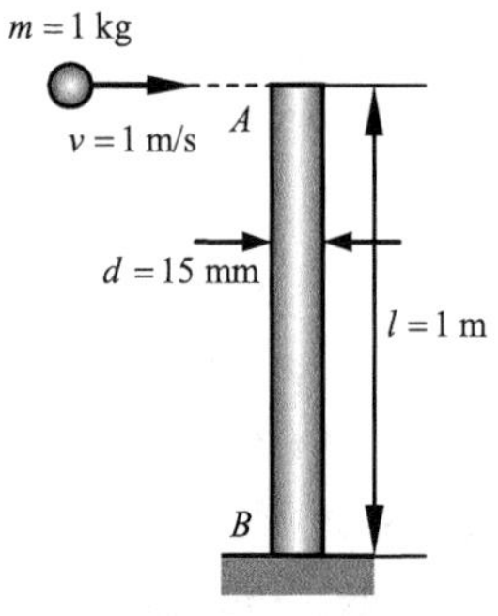

Fig. E20.14

Solution (a) Maximum deflection of the post

$$\Delta_{st} = \frac{mgl^3}{3EI} = \frac{mgl^3}{3E(\frac{1}{64}\pi d^4)} = 6.58\ \text{mm}, \quad K_d = \sqrt{\frac{v^2}{g\Delta_{st}}} = 3.94, \quad \Delta_d = K_d\Delta_{st} = 25.9\ \text{mm}$$

(b) Maximum normal stress in the post

$$\sigma_{st} = \frac{M}{W} = \frac{mgl}{\frac{1}{32}\pi d^3} = 29.6\ \text{MPa}, \quad \sigma_d = K_d\sigma_{st} = 117\ \text{MPa}$$

Example 20.15 A sphere of mass m moving with a velocity v hits squarely the free end A of the rod AB. Knowing that E=100 GPa, determine (a) the displacement of point A, and (b) the maximum normal stress in the rod.

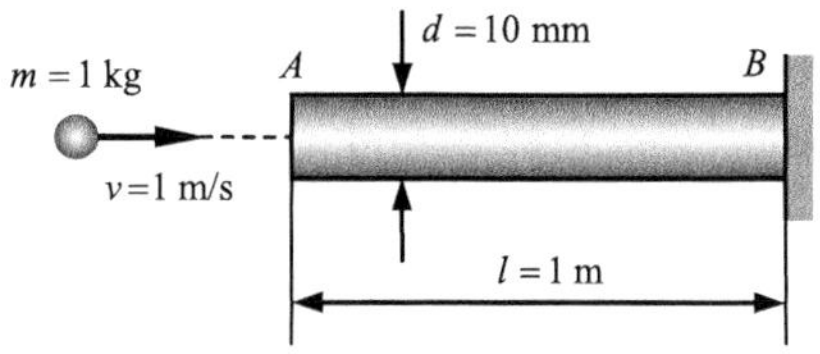

Fig. E20.15

Solution (a) Maximum displacement of point *A*

$$\Delta_{st}=\frac{mgl}{EA}=\frac{mgl}{E(\frac{1}{4}\pi d^2)}=1.25\times10^{-6}\ \text{m},\quad K_d=\sqrt{\frac{v^2}{g\Delta_{st}}}=285.57,\quad \Delta_d=K_d\Delta_{st}=0.357\ \text{mm}$$

(b) Maximum normal stress in the rod

$$\sigma_{st}=\frac{mg}{A}=\frac{mg}{\frac{1}{4}\pi d^2}=0.125\ \text{MPa},\quad \sigma_d=K_d\sigma_{st}=35.7\ \text{MPa}$$

Problems

20.1 Using the principle of work and energy, determine the deflection at the free end *C* of the cantilever *ABC*. Know that the flexural rigidity is 2*EI* for portion *AB* and *EI* for portion *BC*.

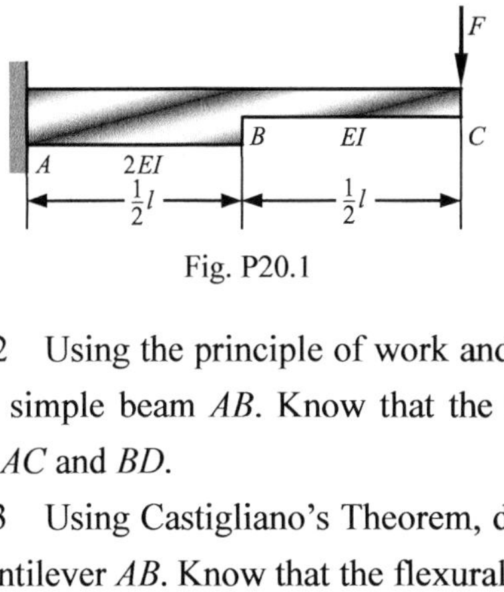

Fig. P20.1

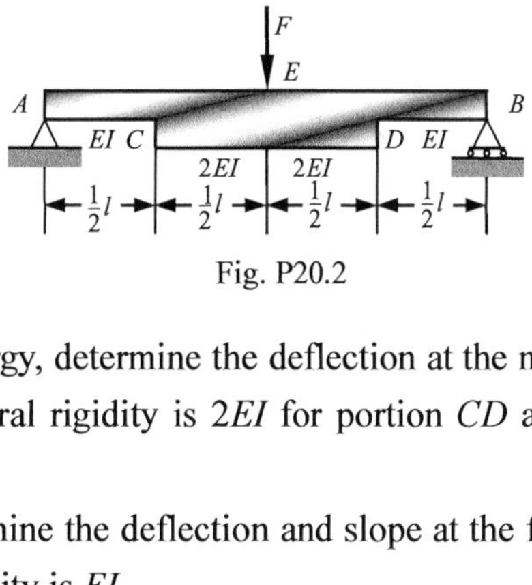

Fig. P20.2

20.2 Using the principle of work and energy, determine the deflection at the midsection *E* of the simple beam *AB*. Know that the flexural rigidity is 2*EI* for portion *CD* and *EI* for portions *AC* and *BD*.

20.3 Using Castigliano's Theorem, determine the deflection and slope at the free end *B* of the cantilever *AB*. Know that the flexural rigidity is *EI*.

20.4 Using Castigliano's Theorem, determine the deflection at the free end *C* of the cantilever *ABC*. Know that the flexural rigidity is 2*EI* for portion *A*B and *EI* for portion *BC*.

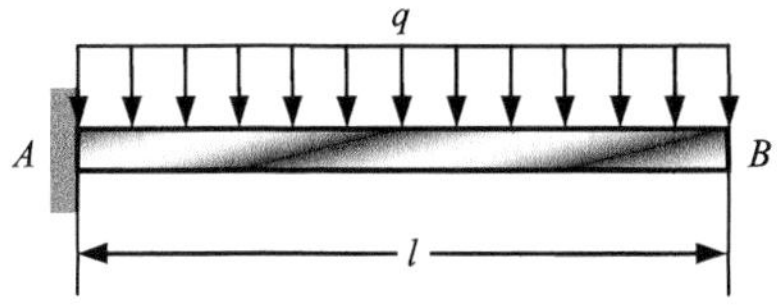

Fig. P20.3

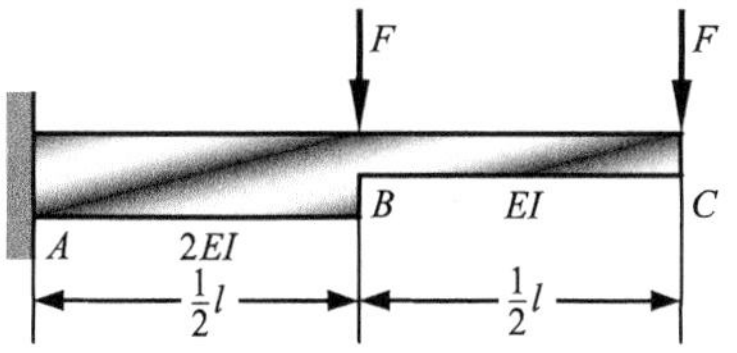

Fig. P20.4

20.5 Using Castigliano's Theorem, determine the deflection at the midsection *C* of the simple-supported beam *AB*. Know that the flexural rigidity is *EI*.

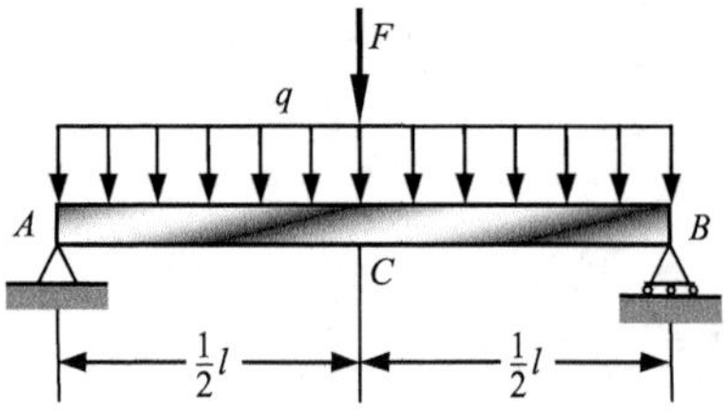

Fig. P20.5

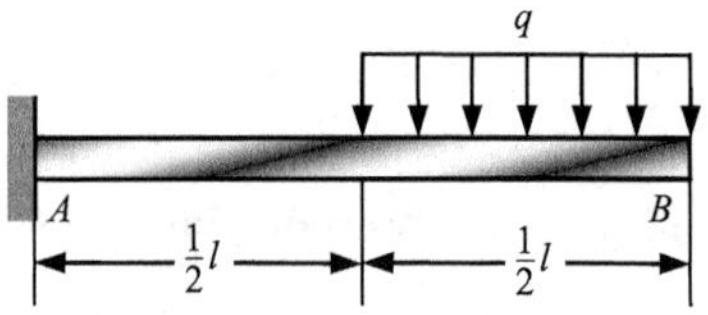

Fig. P20.6

20.6 Using Castigliano's Theorem, determine the deflection and slope at point *B*. Know that the flexural rigidity is *EI*.

20.7 Using the principle of virtual work, determine the deflection and slope at the free end *B* of the cantilever *AB* subjected to a uniform load q. Know that the stress-strain relation is $\sigma^2 = k\varepsilon$, where k is a constant.

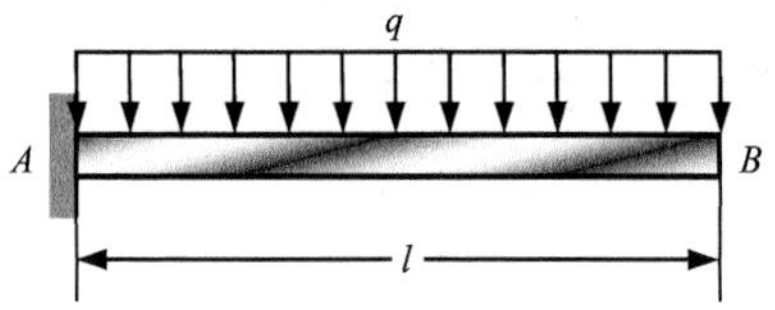

Fig. P20.7

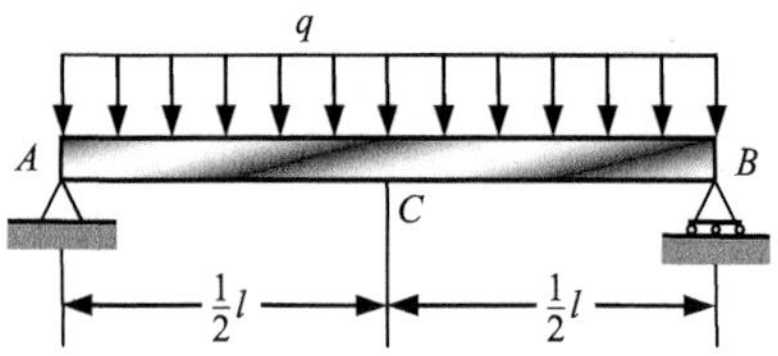

Fig. P20.8

20.8 Using the principle of virtual work, determine the deflection at the midsection *C* of the simple beam *AB* subjected to a uniform load q. Know that the stress-strain relation is $\sigma^2 = k\varepsilon$, where k is a constant.

20.9 Using the unit load method, determine the deflection and slope at the free end *B* of the cantilever *AB*. Know that the flexural rigidity is *EI*.

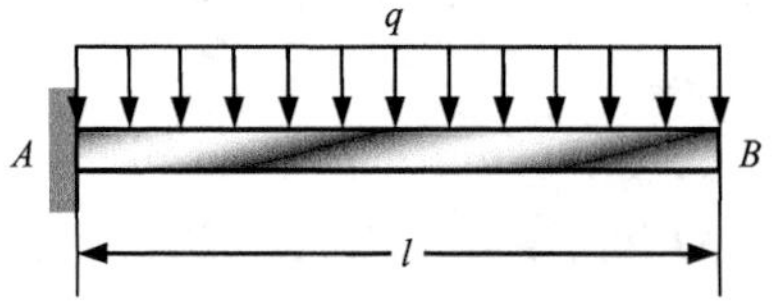

Fig. P20.9

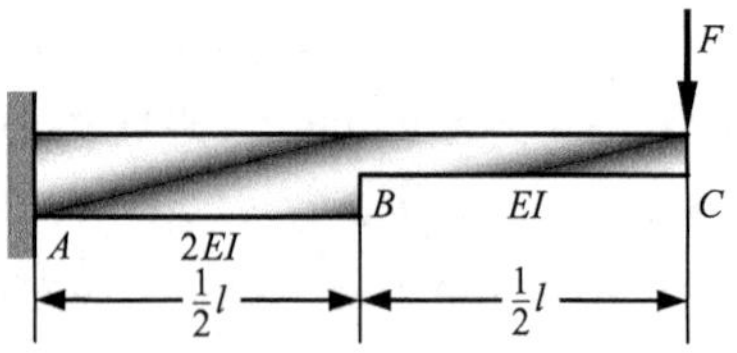

Fig. P20.10

20.10 Using the unit load method, determine the deflection and slope at the midsection *B* of the cantilever *ABC*. Know that the flexural rigidity is 2*EI* for portion *A*B and *EI* for portion *BC*.

20.11-20.12 Using the unit load method, determine the deflection at the midsection *C* and slope at the section B. Know that the flexural rigidity is EI.

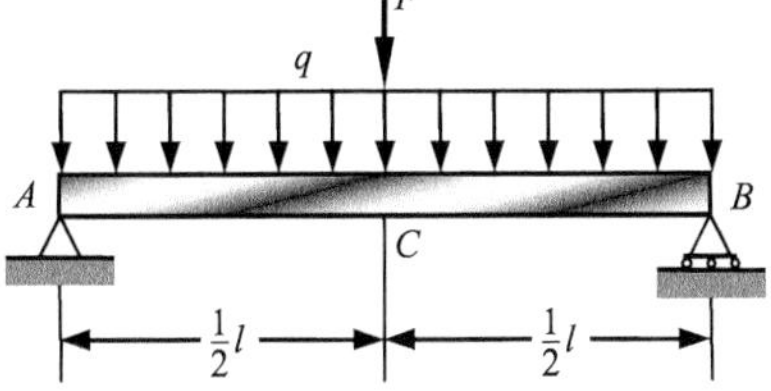

Fig. P20.11

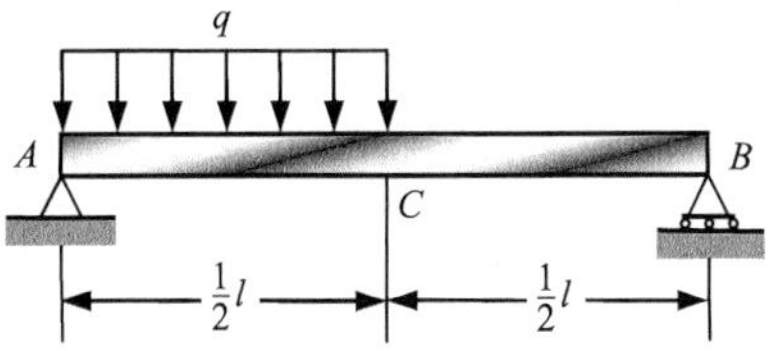

Fig. P20.12

20.13 Using the unit load method, determine the horizontal and vertical deflections of point A. Know that the flexural rigidity is EI.

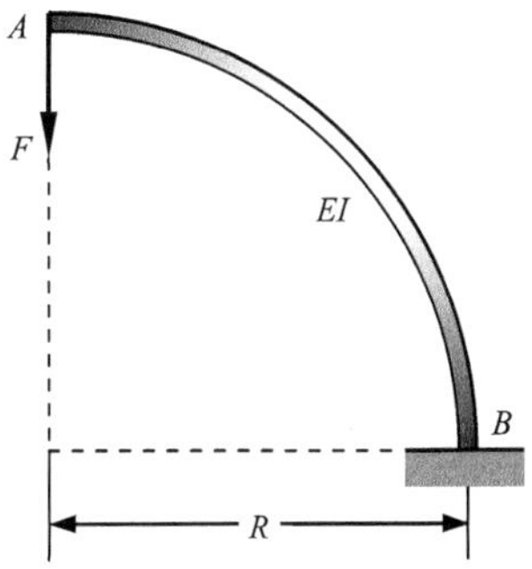

Fig. P20.13

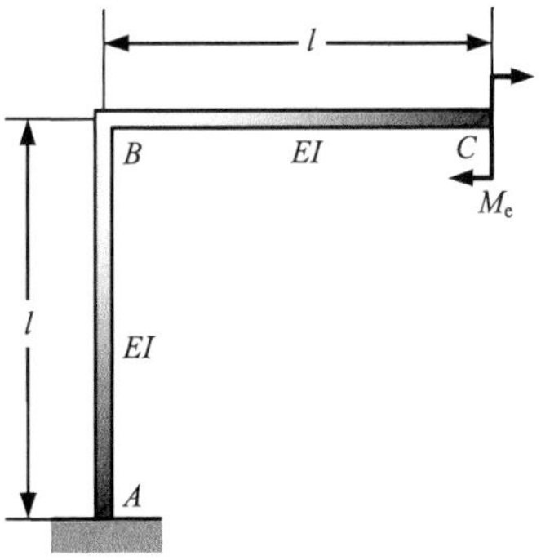

Fig. P20.14

20.14-20.15 Two rods AB and BC of the same flexural rigidity EI are welded together at B. For the loading shown and using the unit load method, determine the horizontal and vertical deflection of point C, and the slope of section C.

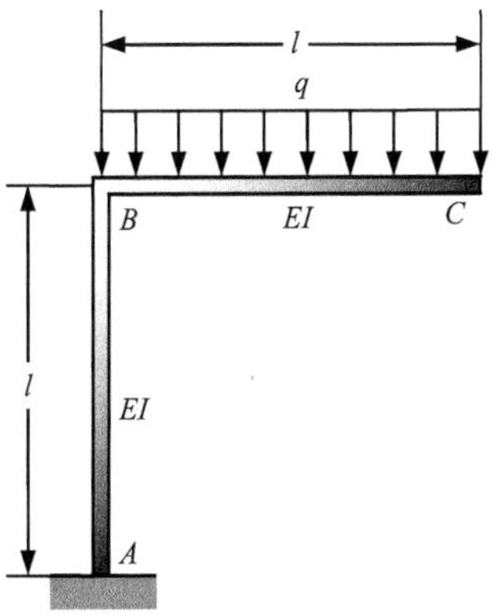

Fig. P20.15

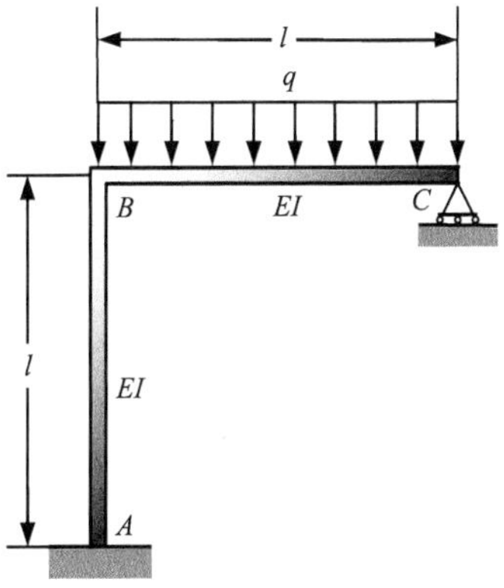

Fig. P20.16

20.16 Two rods *AB* and *BC* of the same flexural rigidity *EI* are welded together at *B*, which has a fixed end at *A* and is supported by a roller at *C*. For the loading and supports shown and using the unit load method, determine the reaction at point *C*.

20.17 A block of mass *m*, initially at rest, falls from a height *h* onto the free end *B* of the cantilever beam *AB*. Knowing that *E*=206 GPa, determine (a) the maximum deflection of the beam, and (b) the maximum normal stress in the beam.

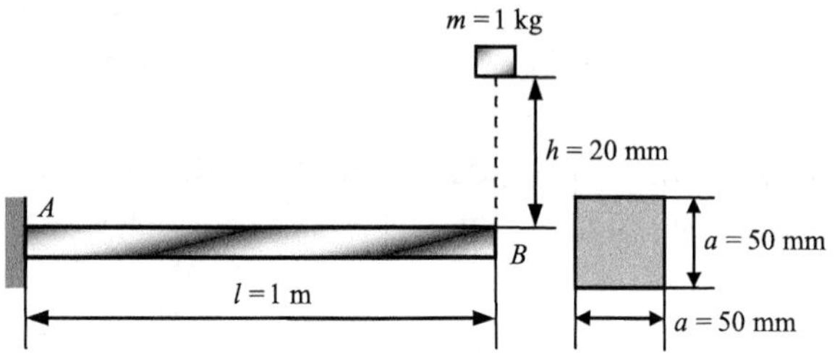

Fig. P20.17

20.18 A block of mass *m*, initially at rest, falls a distance *h* before it strikes the midsection *C* of the cantilever beam *AB*. Knowing that *E*=206 GPa, determine (a) the maximum deflection of the beam, and (b) the maximum normal stress in the beam.

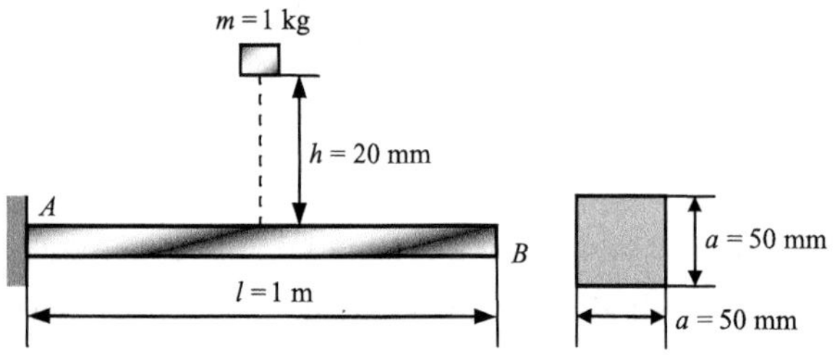

Fig. P20.18

20.19 A block of mass *m* falls from a height *h* onto the free end *C* of the overhanging beam *ABC* of diameter *d*. Knowing that *E*=73 GPa, determine (a) the deflection of point *C*, and (b) the maximum normal stress in the beam.

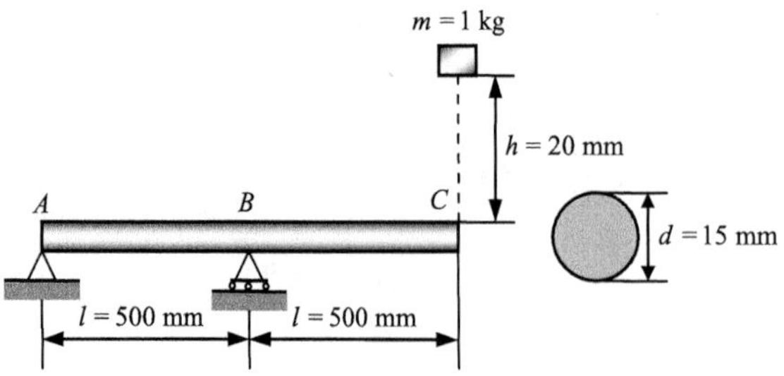

Fig. P20.19

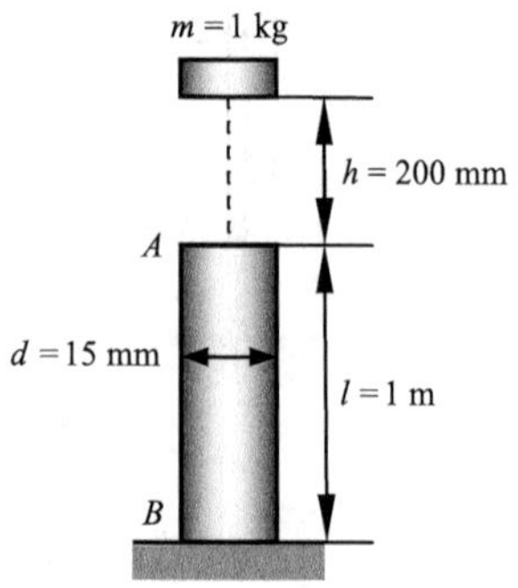

Fig. P20.20

20.20 A cylinder of mass m is released from rest in the position shown and strikes the free end A of the vertical rod AB of diameter d. Knowing that E=200 GPa, determine the maximum normal stress in the rod.

20.21 A sphere of mass m moving with a velocity v strikes the post AB squarely at the midpoint C. Using E=200 GPa, determine (a) the displacement of point A, and (b) the maximum normal stress in the post.

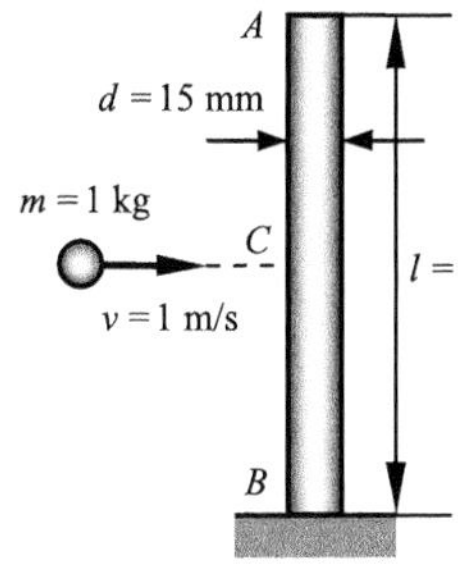

Fig. P20.21

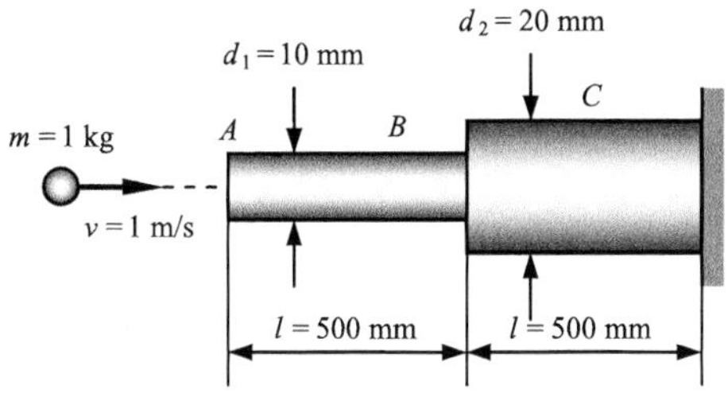

Fig. P20.22

20.22 A block of mass m moving with a velocity v strikes squarely the free end A of the nonuniform rod ABC. Determine (a) the displacement of point A, and (b) the maximum normal stress in the rod. Knowing that E=100 GPa, d_1=10 mm, and d_2=20 mm.

References

Beer F P, et al. Mechanics of Materials[M]. 8th Ed. New York: McGraw-Hill Education, 2020.

Beer F P, et al. Vector Mechanics for Engineers: Dynamics[M]. 12th Ed. New York: McGraw-Hill Education, 2019.

Beer F P, et al. Vector Mechanics for Engineers: Statics[M]. 12th Ed. New York: McGraw-Hill Education, 2019.

Hibbeler R C. Engineering Mechanics: Dynamics[M]. 14th Ed. Hoboken: Pearson Prentice Hall, 2016.

Hibbeler R C. Engineering Mechanics: Statics[M]. 13th Ed. Upper Saddle River: Pearson Prentice Hall, 2013.

Hibbeler R C. Mechanics of Materials[M]. 8th Ed. Upper Saddle River: Pearson Prentice Hall, 2011.

Appendix I Center of Gravity

I.1 2D Body

Considering a two-dimensional body shown in Fig. I.1, the center of gravity of this 2D body can be defined as

$$\bar{x} = \frac{\int x\mathrm{d}W}{W}, \bar{y} = \frac{\int y\mathrm{d}W}{W} \tag{I.1}$$

where $W = \int \mathrm{d}W$ is the weight of the 2D body.

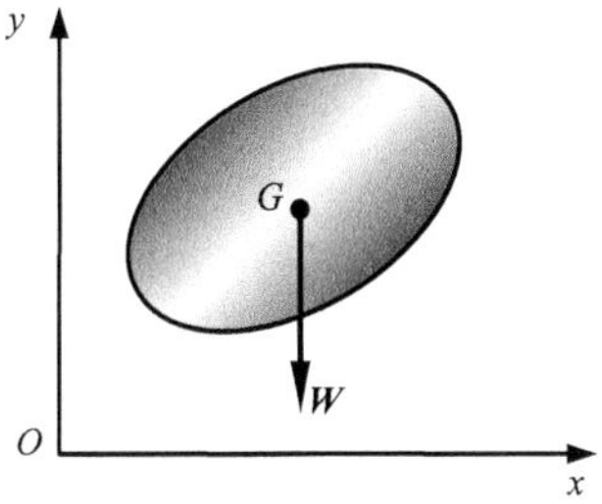

Fig. I.1

If a 2D body has both homogeneous density ρ and uniform thickness t, the center of gravity G of the 2D body will coincide with its centroid C, Fig. I.2. Using $W = \rho g A t$, the centroid of the 2D body can be expressed as

$$\bar{x} = \frac{\int x\mathrm{d}A}{A}, \bar{y} = \frac{\int y\mathrm{d}A}{A} \tag{I.2}$$

where $A = \int \mathrm{d}A$ is the area of the 2D body.

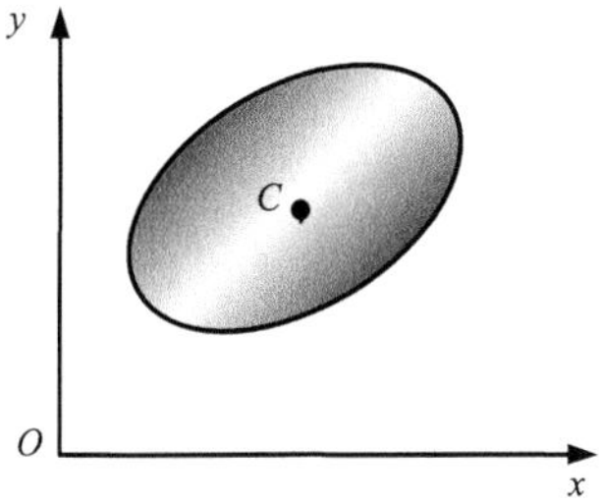

Fig. I.2

I.2 2D Composite Body

Considering a two-dimensional composite body shown in Fig. I.3, the center of gravity of this 2D composite body can be determined by

$$\bar{X}=\frac{\sum \bar{x}_i W_i}{W},\ \bar{Y}=\frac{\sum \bar{y}_i W_i}{W} \tag{I.3}$$

where $W=\sum W_i$ is the weight of the 2D composite body.

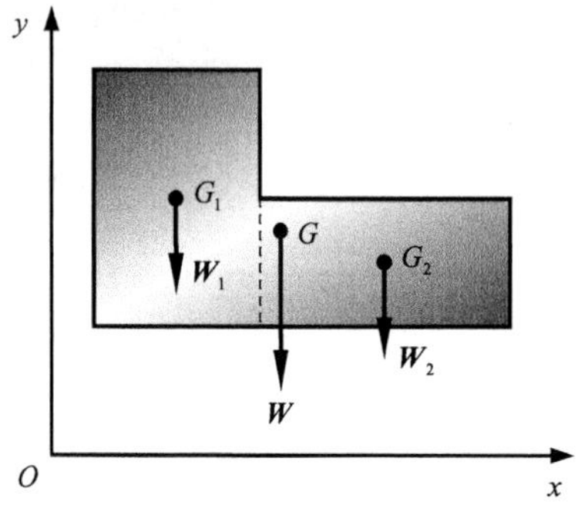

Fig. I.3

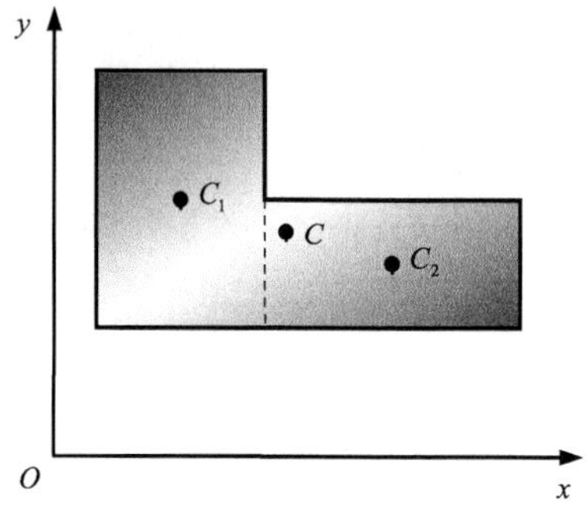

Fig. I.4

If a 2D composite body has both homogeneous density ρ and uniform thickness t, the center of gravity G of the 2D composite body will coincide with its centroid C, Fig. I.4. Using $W_i=\rho g A_i t$, the centroid of the 2D composite body can be given by

$$\bar{X}=\frac{\sum \bar{x}_i A_i}{A},\ \bar{Y}=\frac{\sum \bar{y}_i A_i}{A} \tag{I.4}$$

where $A=\sum A_i$ is the area of the 2D composite body.

I.3 3D Body

Considering a three-dimensional body shown in Fig. I.5, the center of gravity of this 3D body can be defined as

$$\bar{x}=\frac{\int x\mathrm{d}W}{W},\ \bar{y}=\frac{\int y\mathrm{d}W}{W},\ \bar{z}=\frac{\int z\mathrm{d}W}{W} \tag{I.5}$$

where $W=\int \mathrm{d}W$ is the weight of the 3D body.

If a 3D body is made of a homogeneous material of density ρ, the center of gravity G of the 3D body will coincide with its centroid C, Fig. I.6. Using $W=\rho g V$, the centroid of the 3D body can be expressed as

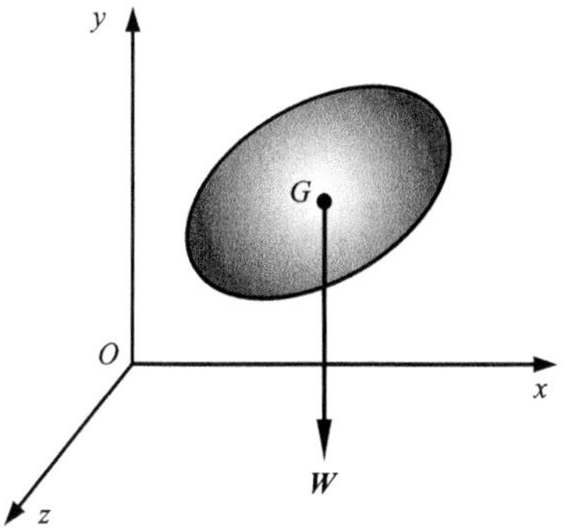

Fig. I.5

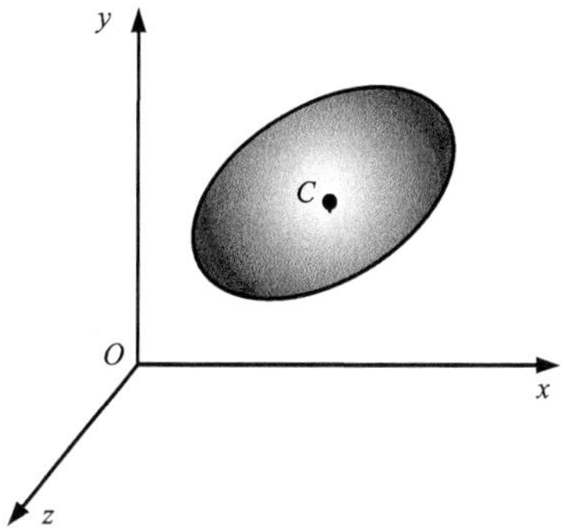

Fig. I.6

$$\bar{x} = \frac{\int x \mathrm{d}V}{V}, \quad \bar{y} = \frac{\int y \mathrm{d}V}{V}, \quad \bar{z} = \frac{\int z \mathrm{d}V}{V} \tag{I.6}$$

where $V = \int \mathrm{d}V$ is the volume of the 3D body.

I.4 3D Composite Body

Considering a three-dimensional composite body shown in Fig. I.7, the center of gravity of this 3D composite body can be defined as

$$\bar{X} = \frac{\sum \bar{x}_i W_i}{W}, \quad \bar{Y} = \frac{\sum \bar{y}_i W_i}{W}, \quad \bar{Z} = \frac{\sum \bar{z}_i W_i}{W} \tag{I.7}$$

where $W = \sum W_i$ is the weight of the 3D composite body.

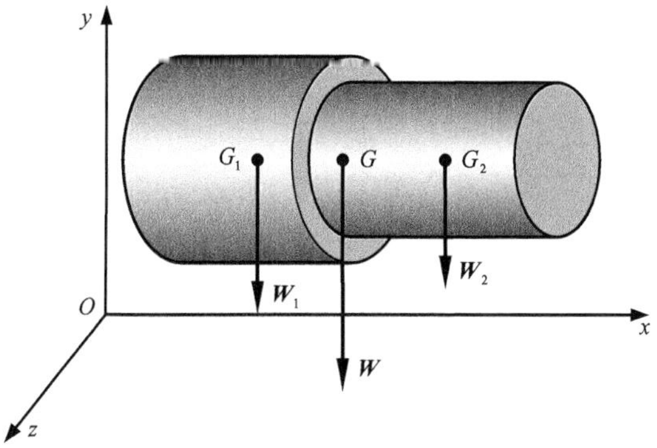

Fig. I.7

If a 3D composite body is made of a homogeneous material of density ρ, the center of gravity G of the 3D composite body will coincide with its centroid C, Fig. I.8. Using $W_i = \rho g V_i$, the centroid of the 3D composite body can be expressed as

$$\bar{X} = \frac{\sum \bar{x}_i V_i}{V}, \quad \bar{Y} = \frac{\sum \bar{y}_i V_i}{V}, \quad \bar{Z} = \frac{\sum \bar{z}_i V_i}{V} \tag{I.8}$$

where $V = \sum V_i$ is the volume of the 3D composite body.

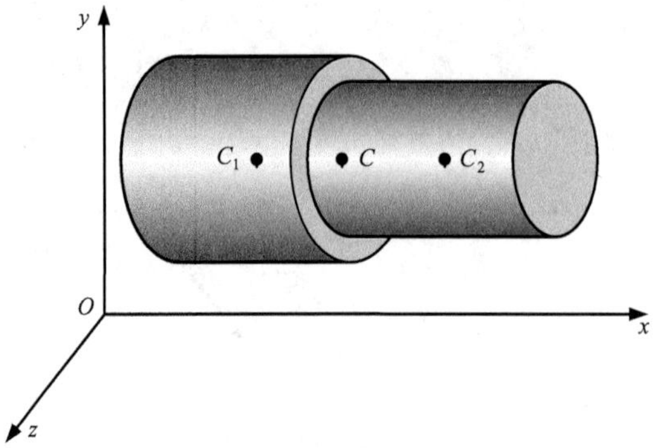

Fig. I.8

Appendix II Mass Moment of Inertia

II.1 Moment of Inertia and Radius of Gyration

Considering a three-dimensional body of mass m, Fig. II.1, the mass moments of inertia of the body with respect to the x, y and z axes can be defined as

$$I_x = \int (y^2 + z^2)\mathrm{d}m,\ I_y = \int (z^2 + x^2)\mathrm{d}m,\ I_z = \int (x^2 + y^2)\mathrm{d}m \tag{II.1}$$

The mass radius of gyration of the body with respect to the x, y and z axes can be defined, respectively, as

$$i_x = \sqrt{\frac{I_x}{m}},\ i_y = \sqrt{\frac{I_y}{m}},\ i_z = \sqrt{\frac{I_z}{m}} \tag{II.2}$$

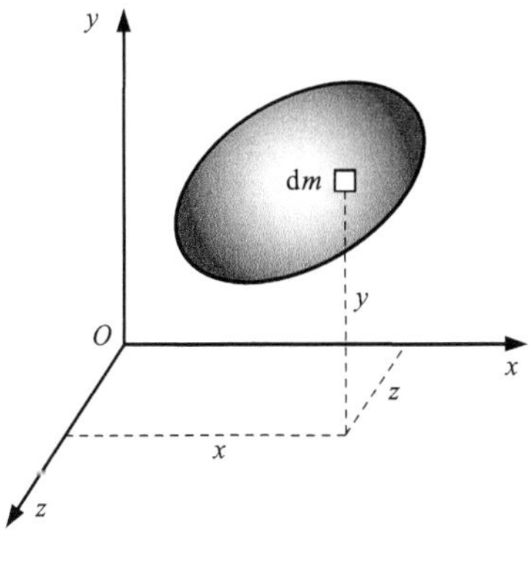

Fig. II.1

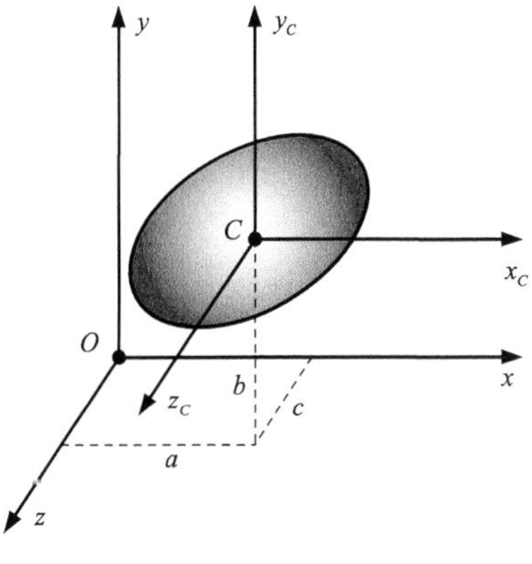

Fig. II.2

II.2 Parallel-Axis Theorem

Considering a three-dimensional body of mass m, Fig. II.2, the mass moments of inertia of the body with respect to the x and x_C can be defined, respectively, as

$$I_x = \int (y^2 + z^2)\mathrm{d}m,\ I_{x_C} = \int ({y_C}^2 + {z_C}^2)\mathrm{d}m \tag{II.3}$$

Using $y = y_C + b$ and $z = z_C + c$, we have

$$I_x = \int (y^2 + z^2)\mathrm{d}m = \int ({y_C}^2 + {z_C}^2)\mathrm{d}m + 2\int (by_C + cz_C)\mathrm{d}m + (b^2 + c^2)\int \mathrm{d}m \tag{II.4}$$

Using $\int (by_C + cz_C)\mathrm{d}m = 0$, the equation above can be simplified as

$$I_x = I_{x_C} + m(b^2 + c^2) \tag{II.5}$$

Similarly, we can obtain

$$I_y = I_{y_C} + m(c^2 + a^2),\ I_z = I_{z_C} + m(a^2 + b^2) \tag{II.6}$$

The relations expressed by Eqs. (II.5) and (II.6) are called the parallel-axis theorem. These relations are often used to determine the mass moment of inertia of a body with respect to an arbitrary axis if the mass moment of inertia with respect to the centroidal axis is known.

Appendix III　Properties of Area

III.1　First Moment (Static Moment)

Considering a shaded area A shown in Fig. III.1, the first moments of the area A with respect to the z and y axes can be defined, respectively, as

$$Q_z = \int y \mathrm{d}A , \quad Q_y = \int z \mathrm{d}A \tag{III.1}$$

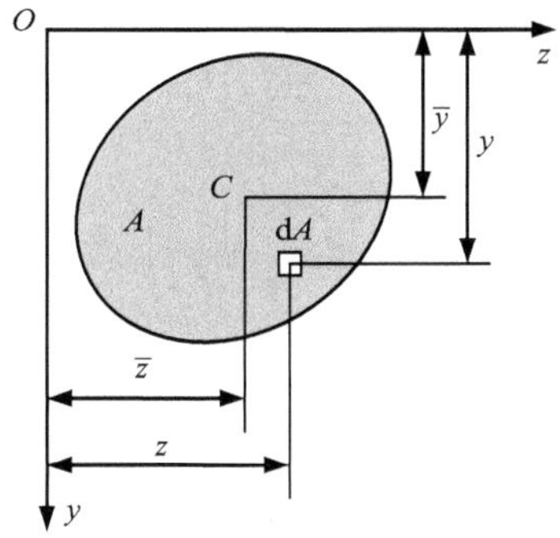

Fig. III.1

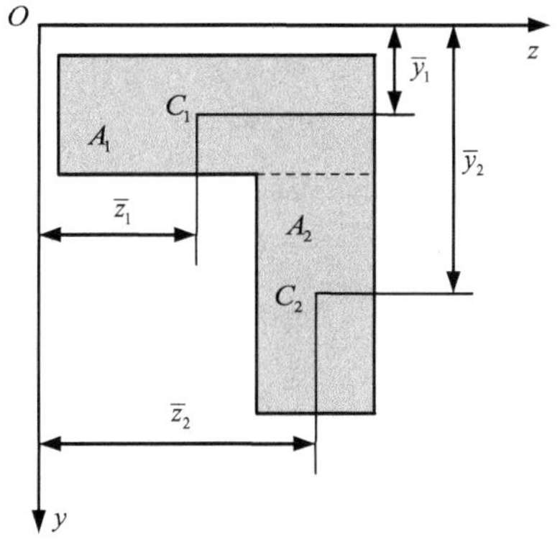

Fig. III.2

The centroid of the area A can be defined as

$$\overline{y} = \frac{\int y \mathrm{d}A}{A} , \quad \overline{z} = \frac{\int z \mathrm{d}A}{A} \tag{III.2}$$

Comparing Eq. (III.1) with Eq. (III.2), we have

$$Q_z = A\overline{y} , \quad Q_y = A\overline{z} \tag{III.3}$$

For a composite area, consisting of areas A_1 and A_2, shown in Fig. III.2, the first moments of the composite area with respect to the z and y axes can, respectively, be expressed as

$$Q_z = \sum (Q_z)_i = \sum A_i \overline{y}_i , \quad Q_y = \sum (Q_y)_i = \sum A_i \overline{z}_i \tag{III.4}$$

The centroid of the composite area can be written as

$$\overline{y} = \frac{\sum A_i \overline{y}_i}{\sum A_i} , \quad \overline{z} = \frac{\sum A_i \overline{z}_i}{\sum A_i} \tag{III.5}$$

III.2　Moment of Inertia and Polar Moment of Inertia

1. Moment of Inertia

Considering a shaded area A shown in Fig. III.3, the moments of inertia of the area A with respect to the z and y axes can be defined as

$$I_z = \int y^2 \mathrm{d}A, \quad I_y = \int z^2 \mathrm{d}A \tag{III.6}$$

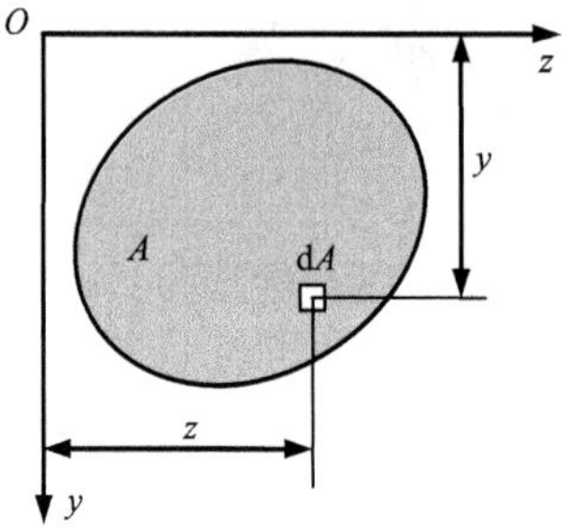

Fig. III.3

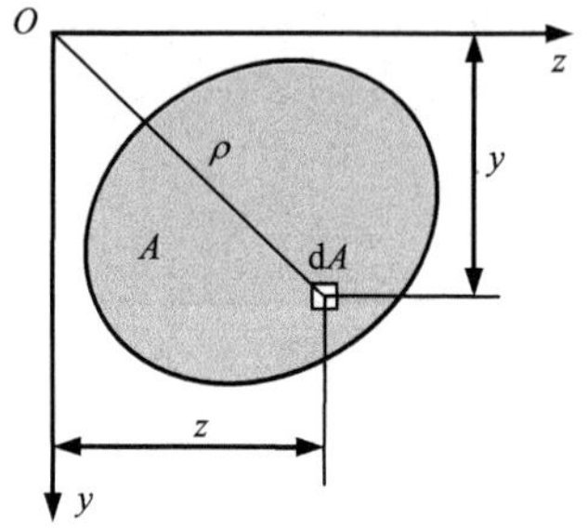

Fig. III.4

2. Polar Moment of Inertia

Considering a shaded area A shown in Fig. III.4, the polar moment of inertia of the area A with respect to the origin O can be defined as

$$I_{\mathrm{p}} = \int \rho^2 \mathrm{d}A \tag{III.7}$$

Using $\rho^2 = y^2 + z^2$, we have

$$I_{\mathrm{p}} = \int \rho^2 \mathrm{d}A = \int y^2 \mathrm{d}A + \int z^2 \mathrm{d}A = I_z + I_y \tag{III.8}$$

III.3　Radius of Gyration and Polar Radius of Gyration

1. Radius of Gyration

Considering the shaded area A shown in Fig. III.3, the radius of gyration of the area A with respect to the z and y axes can be defined, respectively, as

$$i_z = \sqrt{\frac{I_z}{A}}, \quad i_y = \sqrt{\frac{I_y}{A}} \tag{III.9}$$

2. Polar Radius of Gyration

Considering the shaded area A shown in Fig. III.4, the polar radius of gyration of the area A with respect to the origin O can be defined as

$$i_{\mathrm{p}} = \sqrt{\frac{I_{\mathrm{p}}}{A}} \tag{III.10}$$

III.4 Product of Inertia

Considering a shaded area A shown in Fig. III.3, the product of inertia of the area A can be defined as

$$I_{zy} = \int zy\mathrm{d}A \tag{III.11}$$

III.5 Parallel-Axis Theorem

Considering a shaded area A shown in Fig. III.5, the moments of inertia of the area A with respect to the z and z_C axes can be defined, respectively, as

$$I_z = \int y^2\mathrm{d}A, \quad I_{z_C} = \int {y_C}^2\mathrm{d}A \tag{III.12}$$

Using $y = y_C + b$, we have

$$I_z = \int y^2\mathrm{d}A = \int (y_C + b)^2\mathrm{d}A = \int {y_C}^2\mathrm{d}A + 2b\int y_C\mathrm{d}A + b^2\int \mathrm{d}A \tag{III.13}$$

Using $\int y_C\mathrm{d}A = 0$, the equation above can be simplified as

$$I_z = I_{z_C} + Ab^2 \tag{III.14}$$

Similarly, we can obtain

$$I_y = I_{y_C} + Aa^2, \quad I_{zy} = I_{z_C y_C} + Aab, \quad I_{\mathrm{p}} = (I_{\mathrm{p}})_C + A(a^2 + b^2) \tag{III.15}$$

The relations expressed by Eqs. (III.14) and (III.15) are called the parallel-axis theorem. These relations are often used to determine the moment of inertia of an area with respect to an arbitrary axis and the polar moment of inertia of an area with respect to an arbitrary point if the moment of inertia with respect to the centroidal axis and the polar moment of inertia with respect to the centroid are known.

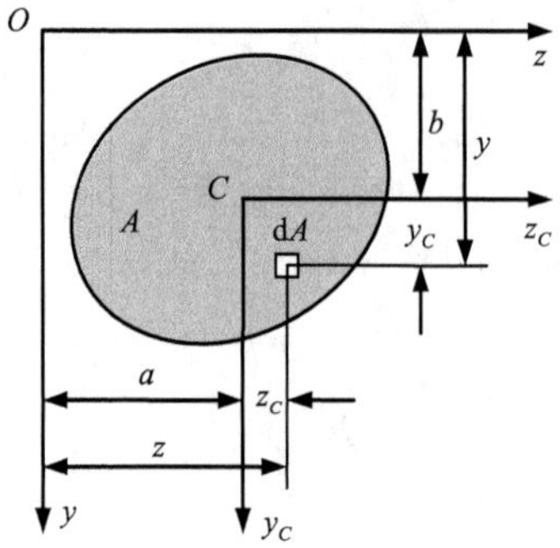

Fig. III.5

Appendix IV Shape Steels

IV.1 I Steel

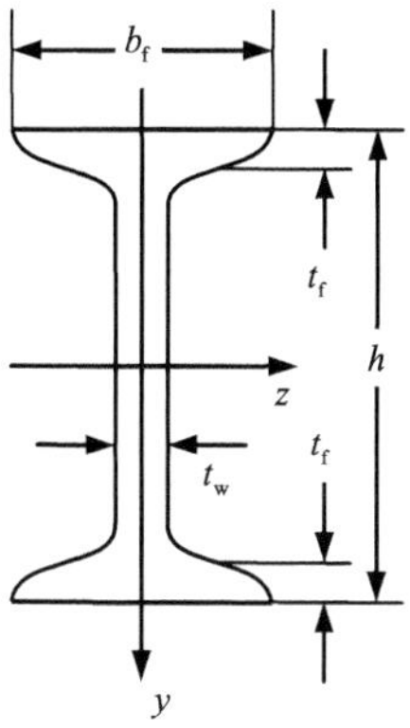

Designation	Weight /(kg/m)	Depth h/mm	Flange		Web	Area A/cm^2	Moment of Inertia		Section Modulus	
			Width b_f/mm	Thickness t_f/mm	Thickness t_w/mm		I_z/cm^4	I_y/cm^4	S_z/cm^3	S_y/cm^3
10	11.3	100	68	7.6	4.5	14.33	245	33.0	49.0	9.72
12	14.0	120	74	8.4	5.0	17.80	436	46.9	72.7	12.7
12.6	14.2	126	74	8.4	5.0	18.10	488	46.9	77.5	12.7
14	16.9	140	80	9.1	5.5	21.50	712	64.4	102	16.1
16	20.5	160	88	9.9	6.0	26.11	1130	93.1	141	21.2
18	24.1	180	94	10.7	6.5	30.74	1660	122	185	26.0
20a	27.9	200	100	11.4	7.0	35.56	2370	158	237	31.5
20b	31.1		102		9.0	39.55	2500	169	250	33.1
22a	33.1	220	110	12.3	7.5	42.10	3400	225	309	40.9
22b	36.5		112		9.5	46.50	3570	239	325	42.7
24a	37.5	240	116	13.0	8.0	47.71	4570	280	381	48.4
24b	41.2		118		10.0	52.51	4800	297	400	50.4
25a	38.1	250	116		8.0	48.51	5020	280	402	48.3
25b	42.0		118		10.0	53.51	52800	309	423	52.4
27a	42.8	270	122	13.7	8.5	54.52	6550	345	485	56.6
27b	47.0		124		10.5	59.92	6870	366	509	58.9
28a	43.5	280	122		8.5	55.39	7110	345	508	56.6
28b	47.9		124		10.5	60.97	7480	379	534	61.2
30a	48.1	300	126	14.4	9.0	61.22	8950	400	597	63.5
30b	52.8		128		11.0	67.22	9400	422	627	65.9
30c	57.5		130		13.0	73.22	9850	445	657	68.5

Continued

Designation	Weight /(kg/m)	Depth h/mm	Flange		Web	Area A/cm^2	Moment of Inertia		Section Modulus	
			Width b_f/mm	Thickness t_f/mm	Thickness t_w/mm		I_z/cm^4	I_y/cm^4	S_z/cm^3	S_y/cm^3
32a	52.7	320	130	15.0	9.5	67.12	11100	460	692	70.8
32b	57.7		132		11.5	73.52	11600	502	723	76.0
32c	62.7		134		13.5	79.92	12200	544	760	81.2
36a	60.0	360	136	15.8	10.0	76.42	15800	552	875	81.2
36b	65.7		138		12.0	83.64	16500	582	919	84.3
36c	71.3		140		14.0	90.84	17300	612	962	87.4
40a	67.6	400	142	16.5	10.5	86.07	21700	660	1090	93.2
40b	73.8		144		12.5	94.07	22800	692	1140	96.2
40c	80.1		146		14.5	102.1	23900	727	1190	99.6
45a	80.4	450	150	18.0	11.5	102.4	32200	855	1430	114
45b	87.4		152		13.5	111.4	33800	894	1500	118
45c	94.5		154		15.5	120.4	35300	938	1570	122
50a	93.6	500	158	20.0	12.0	119.2	46500	1120	1860	142
50b	101		160		14.0	129.2	48600	1170	1940	146
50c	109		162		16.0	139.2	50600	1220	2080	151
55a	105	550	166	21.0	12.5	134.1	62900	1370	2290	164
55b	114		168		14.5	145.1	65000	1420	2390	170
55c	123		170		16.5	156.1	68400	1480	2490	175
56a	106	560	166		12.5	135.4	65600	1370	2340	165
56b	115		168		14.5	146.6	68500	1490	2450	174
56c	124		170		16.5	157.8	71400	1560	2550	183
63a	121	630	176	22.0	13.0	154.6	93900	1700	2980	193
63b	131		178		15.0	167.2	98100	1810	3160	204
63c	141		180		17.0	179.8	102000	1920	3300	214

IV.2 Channel Steel

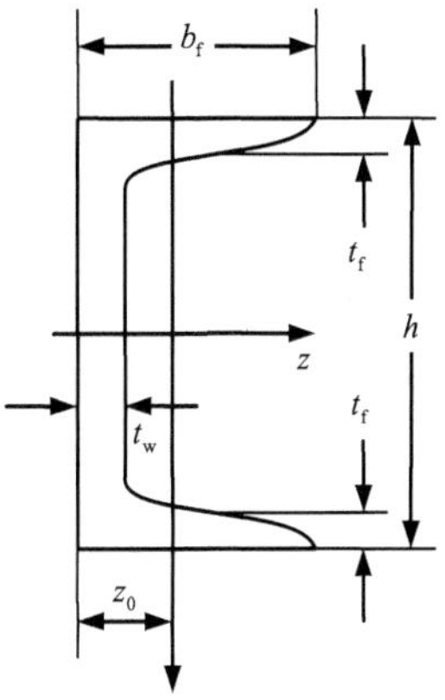

Designation	Weight /(kg/m)	Depth h/mm	Flange		Web	Area A/cm^2	Moment of Inertia		Section Modulus		z_0/cm
			Width b_f/mm	Thickness t_f/mm	Thickness t_w/mm		I_z/cm^4	I_y/cm^4	S_z/cm^3	S_y/cm^3	
5	5.44	50	37	7.0	4.5	6.925	26.0	8.3	10.4	3.55	1.35
6.3	6.63	63	40	7.5	4.8	8.446	50.8	11.9	16.1	4.50	1.36
6.5	6.51	65	40	7.5	4.3	8.292	55.2	12.0	17.0	4.59	1.38
8	8.04	80	43	8.0	5.0	10.24	101	16.6	25.3	5.79	1.43
10	10.0	100	48	8.5	5.3	42.74	198	25.6	39.7	7.8	1.52
12	12.1	120	53	9.0	5.5	15.36	346	37.4	57.7	10.2	1.62
12.6	12.3	126	53	9.0	5.5	15.69	391	38.0	62.1	10.2	1.59
14a	14.5	140	58	9.5	6.0	18.51	564	53.2	80.5	13.0	1.71
14b	16.7		60		8.0	21.31	609	61.1	87.1	14.1	1.67
16a	17.2	160	63	10.0	6.5	21.95	866	73.3	108	16.3	1.80
16b	19.8		65		8.5	25.15	935	83.4	117	17.6	1.75
18a	20.2	180	68	10.5	7.0	25.69	1270	98.6	141	20.0	1.88
18b	23.0		70		9.0	29.29	1370	111	152	21.5	1.84
20a	22.6	200	73	11.0	7.0	28.83	1780	128	178	24.2	2.01
20b	25.8		75		9.0	32.83	1910	144	191	25.9	1.95
22a	25.0	220	77	11.5	7.0	31.83	2390	158	218	28.2	2.10
22b	28.5		79		9.0	36.23	2570	176	234	30.1	2.03
24a	26.9	240	78	12.0	7.0	34.21	3050	174	254	30.5	2.1
24b	30.6		80		9.0	39.01	3280	194	274	32.5	2.03
24c	34.4		82		11.0	43.81	3510	213	293	34.4	2.00
25a	27.4	250	78	12.0	7.0	34.91	3370	176	270	30.6	2.07
25b	31.3		80		9.0	39.91	3530	196	282	32.7	1.98
25c	35.3		82		11.0	44.91	3690	218	295	35.9	1.92

Continued

Designation	Weight /(kg/m)	Depth h/mm	Flange		Web	Area A/cm^2	Moment of Inertia		Section Modulus		z_0/cm
			Width b_f/mm	Thickness t_f/mm	Thickness t_w/mm		I_z/cm^4	I_y/cm^4	S_z/cm^3	S_y/cm^3	
27a	30.8		82		7.5	39.27	4360	216	323	35.5	2.13
27b	35.1	270	84		9.5	44.67	4690	239	347	37.7	2.06
27c	39.3		86	12.5	11.5	50.07	5020	261	372	39.8	2.03
28a	31.4		82		7.5	40.02	4760	218	340	35.7	2.10
28b	35.8	280	84		9.5	45.62	5130	242	366	37.9	2.02
28c	40.2		86		11.5	51.22	5500	268	393	40.3	1.95
30a	34.5		85		7.5	43.89	6050	260	4.3	41.1	2.17
30b	39.2	300	87	13.5	9.5	49.89	6500	289	433	44.0	2.13
30c	43.9		89		11.5	55.89	6950	316	463	46.4	2.09
32a	38.1		88		8.0	48.50	7600	305	475	46.5	2.24
32b	43.1	320	90	14.0	10.0	54.90	8140	336	509	49.2	2.16
32c	48.1		92		12.0	61.30	8690	374	543	52.6	2.09
36a	47.8		96		9.0	60.89	11900	455	660	63.5	2.44
36b	53.5	360	98	16.0	11.0	68.09	12700	497	703	66.9	2.37
36c	59.1		100		13.0	75.29	13400	536	746	70.0	2.34
40a	58.9		100		10.5	75.04	17600	592	879	78.8	2.49
40b	65.2	400	102	18.0	12.5	83.04	18600	640	932	82.5	2.44
40c	71.5		104		14.5	91.04	19700	688	986	86.2	2.42

IV.3 Equal Angle Steel

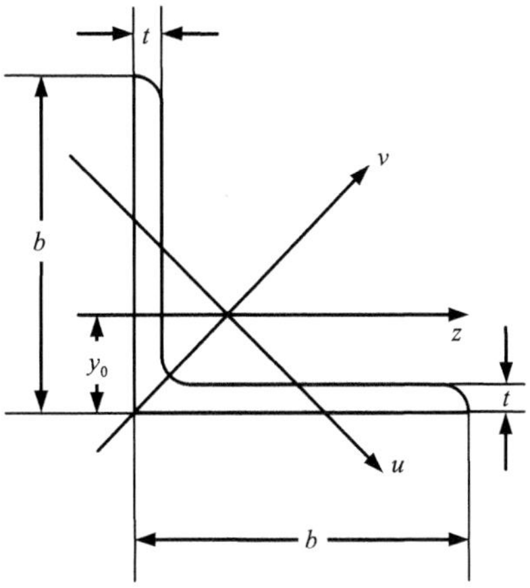

Designation	Weight /(kg/m)	Width b/mm	Thickness t/mm	Area A/cm^2	Moment of Inertia			y_0/cm
					I_z/cm^4	I_v/cm^4	I_u/cm^4	
2	0.89	20	3	1.132	0.40	0.63	0.17	0.60
	1.15		4	1.459	0.50	0.78	0.22	0.64
2.5	1.12	25	3	1.432	0.82	1.29	0.34	0.73
	1.46		4	1.859	1.03	1.62	0.43	0.76
3.0	1.37	30	3	1.749	1.46	2.31	0.61	0.85
	1.79		4	2.276	1.84	2.92	0.77	0.89
3.6	1.66	36	3	2.109	2.58	4.09	1.07	1.00
	2.16		4	2.756	3.29	5.22	1.37	1.04
	2.65		5	3.382	3.95	6.24	1.65	1.07
4.0	1.85	40	3	2.359	3.59	5.69	1.49	1.09
	2.42		4	3.086	4.60	7.29	1.91	1.13
	2.98		5	3.792	5.53	8.76	2.30	1.17
4.5	2.09	45	3	2.659	5.17	8.20	2.14	1.22
	2.74		4	3.486	6.65	10.6	2.75	1.26
	3.37		5	4.292	8.04	12.7	3.33	1.30
	3.99		6	5.077	9.33	14.8	3.89	1.33
5	2.33	50	3	2.971	7.18	11.4	2.98	1.34
	3.06		4	3.897	9.26	14.7	3.82	1.38
	3.77		5	4.803	11.2	17.8	4.64	1.42
	4.46		6	5.688	13.1	20.7	5.42	1.46
5.6	2.62	56	3	3.343	10.2	16.1	4.24	1.48
	3.45		4	4.390	13.2	20.9	5.46	1.53
	4.25		5	5.415	16.0	25.4	6.61	1.57
	5.04		6	6.420	18.7	29.7	7.73	1.61
	5.81		7	7.404	21.2	33.6	8.82	1.64
	6.57		8	8.367	23.6	37.4	9.89	1.68

Continued

Designation	Weight /(kg/m)	Width b/mm	Thickness t/mm	Area A/cm^2	Moment of Inertia			y_0/cm
					I_z/cm^4	I_y/cm^4	I_u/cm^4	
6	4.58	60	5	5.829	19.9	31.6	8.21	1.67
	5.43		6	6.914	23.4	36.9	9.60	1.70
	6.26		7	7.977	26.4	41.9	11.0	1.74
	7.08		8	90.20	29.5	46.7	12.3	1.78
6.3	3.91	63	4	4.978	19.0	30.2	7.89	1.70
	4.82		5	6.143	23.2	36.8	9.57	1.74
	5.72		6	7.288	27.1	43.0	11.2	1.78
	6.60		7	8.412	30.9	49.0	12.8	1.82
	7.47		8	9.515	34.5	54.6	14.3	1.85
	9.15		10	11.66	41.1	64.9	17.3	1.93
7	4.37	70	4	5.570	26.4	41.8	11.0	1.86
	5.40		5	6.876	32.2	51.1	13.3	1.91
	6.41		6	8.160	37.8	59.9	15.6	1.95
	7.40		7	9.424	43.1	68.4	17.8	1.99
	8.37		8	10.67	48.2	76.4	20.0	2.03
7.5	5.82	75	5	7.412	40.0	63.3	16.6	2.04
	6.91		6	8.797	47.0	74.4	19.5	2.07
	7.98		7	10.16	53.6	85.0	22.2	2.11
	9.03		8	11.50	60.0	95.1	24.8	2.15
	10.1		9	12.83	66.1	105	27.5	2.18
	11.1		10	14.13	72.0	114	30.1	2.22
8	6.21	80	5	7.912	48.8	77.3	20.3	2.15
	7.38		6	9.397	57.4	91.0	23.7	2.19
	8.53		7	10.86	65.6	104	27.1	2.23
	9.66		8	12.30	73.5	117	30.4	2.27
	10.8		9	13.73	81.1	129	33.6	2.31
	11.9		10	15.13	88.4	140	36.8	2.35
9	8.35	90	6	10.64	82.8	131	34.3	2.44
	9.66		7	12.30	94.8	150	39.2	2.48
	10.9		8	13.94	106	169	44.0	2.52
	12.2		9	15.57	118	187	48.7	2.56
	13.5		10	17.17	129	204	53.3	2.59
	15.9		12	20.31	149	236	62.2	2.67
10	9.37	100	6	11.93	115	182	47.9	2.67
	10.8		7	13.80	132	209	54.7	2.71
	12.3		8	15.64	148	235	61.4	2.76
	13.7		9	17.46	164	260	68.0	2.80
	15.1		10	19.26	180	285	74.4	2.84
	17.9		12	22.80	209	331	86.8	2.91
	20.6		14	26.26	237	374	99.0	2.99
	23.3		16	29.63	263	414	111	3.06
11	11.9	110	7	15.20	177	281	73.4	2.96
	13.5		8	17.24	199	316	82.4	3.01
	16.7		10	21.26	242	384	100	3.09
	19.8		12	25.20	283	448	117	3.16
	22.8		14	29.06	321	508	133	3.24

Continued

Designation	Weight /(kg/m)	Width b/mm	Thickness t/mm	Area A/cm^2	Moment of Inertia			y_0/cm
					I_z/cm^4	I_y/cm^4	I_u/cm^4	
12.5	15.5	125	8	19.75	297	471	123	3.37
	19.1		10	24.37	362	574	149	3.45
	22.7		12	28.91	423	671	175	3.53
	26.2		14	33.37	482	764	200	3.61
			16	37.74	537	851	224	3.68
14	21.5	140	10	27.37	515	817	212	3.82
	25.5		12	32.51	604	959	249	3.90
	29.5		14	37.57	689	1090	284	3.98
	33.4		16	42.54	770	1220	319	4.06
15	18.6	150	8	23.75	521	827	215	3.99
	23.1		10	29.37	638	1010	262	4.08
	27.4		12	34.91	749	1190	308	4.15
	31.7		14	40.37	856	1360	352	4.23
	33.8		15	43.06	907	1440	374	4.27
	35.9		16	45.74	958	1520	395	4.31
16	24.7	160	10	31.50	780	1240	322	4.31
	29.4		12	37.44	917	1460	377	4.39
	34.0		14	43.30	1050	1670	432	4.47
	38.5		16	49.07	1180	1870	485	4.55
18	33.2	180	12	42.24	1320	2100	543	4.89
	38.4		14	48.90	1510	2410	622	4.97
	43.5		16	55.47	1700	2700	699	5.05
	48.6		18	61.96	1880	2990	762	5.13
20	42.9	200	14	54.64	2100	3340	864	5.46
	48.7		16	62.01	2370	3760	971	5.54
	54.4		18	69.30	2620	4160	1080	5.62
	60.1		20	76.51	2870	4550	1180	5.69
	71.2		24	90.66	3340	5290	1380	5.87
22	53.9	220	16	68.67	3190	5060	1310	6.03
	60.3		18	76.75	3540	5620	1450	6.11
	66.5		20	84.76	3870	6150	1590	6.18
	72.8		22	92.68	4200	6670	1730	6.26
	78.9		24	100.5	4520	7170	1870	6.33
	85.0		26	108.3	4830	7690	2000	6.41
25	69.0	250	18	87.84	5270	8370	2170	6.84
	76.2		20	97.05	5780	9180	2380	6.92
	83.3		22	106.2	6280	9970	2580	7.00
	90.4		24	115.2	6770	10700	2790	7.07
	97.5		26	124.2	7240	11500	2980	7.15
	104		28	133.0	7700	12200	3180	7.22
	111		30	141.8	8160	12900	3380	7.30
	118		32	150.5	8600	13600	3570	7.37
	128		35	163.4	9240	14600	3850	7.48

IV.4 Unequal Angle Steel

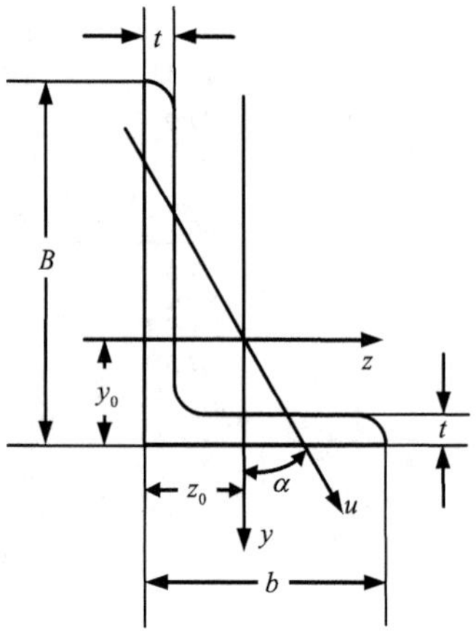

Designation	Weight /(kg/m)	Width		Thickness t/mm	Area A/cm^2	Moment of Inertia			z_0/cm	y_0/cm	tanα
		B/mm	b/mm			I_z/cm^4	I_y/cm^4	I_u/cm^4			
2.5/1.6	0.91	25	16	3	1.162	0.70	0.22	0.14	0.42	0.86	0.392
	1.18			4	1.499	0.88	0.27	0.17	0.46	0.90	0.381
3.2/2	1.17	32	20	3	1.492	1.53	0.46	0.28	0.49	1.08	0.382
	1.52			4	1.939	1.93	0.57	0.35	0.53	1.12	0.374
4/2.5	1.48	40	25	3	1.890	3.08	0.93	0.56	0.59	1.32	0.385
	1.94			4	2.467	3.93	1.18	0.71	0.63	1.37	0.381
4.5/2.8	1.69	45	28	3	2.149	4.45	1.34	0.80	0.64	1.47	0.383
	2.20			4	2.806	5.69	1.70	1.02	0.68	1.51	0.380
5/3.2	1.91	50	32	3	2.431	6.24	2.02	1.20	0.73	1.60	0.404
	2.49			4	3.177	8.02	2.58	1.53	0.77	1.65	0.402
5.6/3.6	2.15	56	36	3	2.743	8.88	2.92	1.73	0.80	1.78	0.408
	2.82			4	3.590	11.5	3.76	2.23	0.85	1.82	0.408
	3.47			5	4.415	13.9	4.49	2.67	0.88	1.87	0.404
6.3/4	3.19	63	40	4	4.058	16.5	5.23	3.12	0.92	2.04	0.398
	3.92			5	4.993	20.0	6.31	3.76	0.95	2.08	0.396
	4.64			6	5.908	23.4	7.29	4.34	0.99	2.12	0.393
	5.34			7	6.802	26.5	8.24	4.97	1.03	2.15	0.389
7/4.5	3.57	70	45	4	4.553	23.2	7.55	4.40	1.02	2.24	0.410
	4.40			5	5.609	28.0	9.13	5.40	1.06	2.28	0.407
	5.22			8	6.644	32.5	10.6	6.35	1.09	2.32	0.404
	6.01			7	7.658	37.2	12.0	7.16	1.13	2.36	0.402
7.5/5	4.81	75	50	5	6.126	34.9	12.6	7.41	1.17	2.40	0.435
	5.70			6	7.260	41.1	14.7	8.54	1.21	2.44	0.435
	7.43			8	9.467	52.4	18.5	10.9	1.29	2.52	0.429
	9.10			10	11.59	62.7	22.0	13.1	1.36	2.60	0.423

Continued

Designation	Weight /(kg/m)	Width		Thickness t/mm	Area A/cm^2	Moment of Inertia			z_0/cm	y_0/cm	tanα
		B/mm	b/mm			I_z/cm^4	I_y/cm^4	I_u/cm^4			
8/5	5.00	80	50	5	6.376	42.0	12.8	7.66	1.14	2.60	0.388
	5.93			6	7.560	49.5	15.0	8.85	1.18	2.65	0.387
	6.85			7	8.724	56.2	17.0	10.2	1.21	2.69	0.384
	7.75			8	9.867	62.8	18.9	11.4	1.25	2.73	0.381
9/5.6	5.66	90	56	5	7.212	60.5	18.3	11.0	1.25	2.91	0.385
	6.72			6	8.557	71.0	21.4	12.9	1.29	2.95	0.384
	7.76			7	9.881	81.0	24.4	14.7	1.33	3.00	0.382
	8.78			8	11.18	91.0	27.2	16.3	1.36	3.04	0.380
10/6.3	7.55	100	63	6	9.618	99.1	30.9	18.4	1.43	3.24	0.394
	8.72			7	11.11	113	35.3	21.0	1.47	3.28	0.394
	9.88			8	12.58	127	39.4	23.5	1.50	3.32	0.391
	12.1			10	15.47	154	47.1	28.3	1.58	3.40	0.387
10/8	8.35	100	80	6	10.64	107	61.2	31.7	1.97	2.95	0.627
	9.66			7	12.30	123	70.1	36.2	2.01	3.00	0.626
	10.9			8	13.94	138	78.6	40.6	2.05	3.04	0.625
	13.5			10	17.17	167	94.7	49.1	2.13	3.12	0.622
11/7	8.35	110	70	6	10.64	133	42.9	25.4	1.57	3.53	0.403
	9.66			7	12.30	153	49.0	29.0	1.61	3.57	0.402
	10.9			8	13.94	172	54.9	32.5	1.65	3.62	0.401
	13.5			10	17.17	208	65.9	39.2	1.72	3.07	0.397
12.5/8	11.1	125	80	7	14.10	228	74.4	43.8	1.80	4.01	0.408
	12.6			8	15.99	257	83.5	49.2	1.84	4.06	0.407
	15.5			10	19.71	312	101	59.5	1.92	4.14	0.404
	18.3			12	23.35	364	117	69.4	2.00	4.22	0.400
14/9	14.2	140	90	8	18.04	366	121	70.8	2.04	4.50	0.411
	17.5			10	22.26	446	140	85.8	2.12	4.58	0.409
	20.7			12	26.40	522	170	100	2.19	4.66	0.406
	23.9			14	30.46	594	192	114	2.27	4.74	0.403
15/9	14.8	150	90	8	18.84	442	123	74.1	1.97	4.92	0.364
	18.3			10	23.26	539	14	89.9	2.05	5.01	0.362
	21.7			12	27.60	532	173	105	2.12	5.09	0.359
	25.0			14	31.86	721	196	120	2.20	5.17	0.356
	26.7			15	33.95	764	207	127	2.24	5.21	0.354
	28.3			16	36.03	806	217	134	2.27	5.25	0.352
16/10	19.9	160	100	10	25.32	669	205	122	2.28	5.24	0.390
	23.6			12	30.05	785	239	142	2.36	5.32	0.388
	27.2			14	34.71	896	271	162	2.43	5.40	0.385
	30.8			16	39.28	1000	302	183	2.51	5.48	0.382
18/11	22.3	180	110	10	28.37	956	278	167	2.44	5.89	0.376
	26.5			12	33.71	1120	325	195	2.52	5.98	0.374
	30.5			14	38.97	1290	370	222	2.59	6.06	0.372
	34.6			16	44.14	1440	412	249	2.67	6.14	0.369
20/12.5	29.8	200	125	12	37.91	1570	483	286	2.83	6.54	0.392
	34.4			14	43.87	1800	551	327	2.91	6.62	0.390
	39.0			16	49.74	2020	615	366	2.99	6.70	0.388
	43.6			18	55.53	2240	677	405	3.06	6.78	0.385

第 1 章　质点静力学

如果物体尺寸并不影响所考虑问题的解，则该物体可以理想化为一个质点。质点具有有限质量，但尺寸可以忽略不计。所有作用于质点的力都可假定施加于同一点，因而形成汇交力系。

1.1　平面汇交力系合成

平面汇交力系是指所有力的作用线位于同一平面并交于同一点。

1. 几何法

作用于质点的两个力可以由一个合力代替，以给定力为边绘制平行四边形，通过汇交点的对角线即为合力。例如，图 1.1a 所示作用于质点 O 的力 $\boldsymbol{F}_1$ 和 $\boldsymbol{F}_2$ 可以通过图 1.1b 所示力 $\boldsymbol{R}$ 代替。力 $\boldsymbol{R}$ 称为力 $\boldsymbol{F}_1$ 和 $\boldsymbol{F}_2$ 的合力。合力 $\boldsymbol{R}$ 通过以力 $\boldsymbol{F}_1$ 和 $\boldsymbol{F}_2$ 为邻边所画平行四边形确定。通过 O 点的对角线表示合力 $\boldsymbol{R}$，即 $\boldsymbol{R}=\boldsymbol{F}_1+\boldsymbol{F}_2$。这就是平行四边形定律。

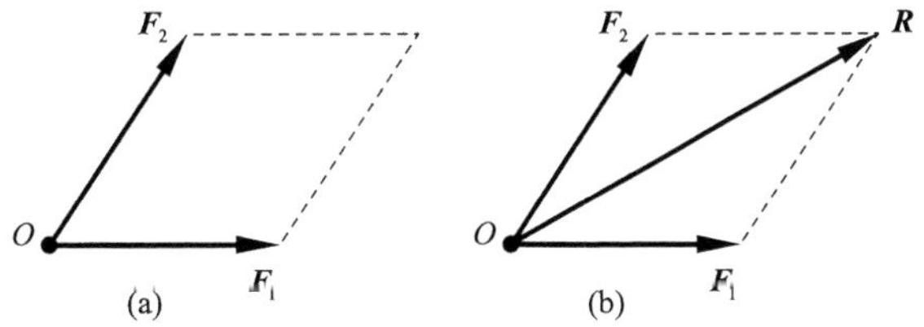

图 1.1

仅考虑半个平行四边形时，通过绘制图 1.2b 所示三角形可以得到替代方法。把力 $\boldsymbol{F}_1$ 和 $\boldsymbol{F}_2$ 进行首尾相连，并把力 $\boldsymbol{F}_1$ 的尾与力 $\boldsymbol{F}_2$ 的头相连，即可得到合力 $\boldsymbol{R}$，即 $\boldsymbol{R}=\boldsymbol{F}_1+\boldsymbol{F}_2$。这就是三角形法则。

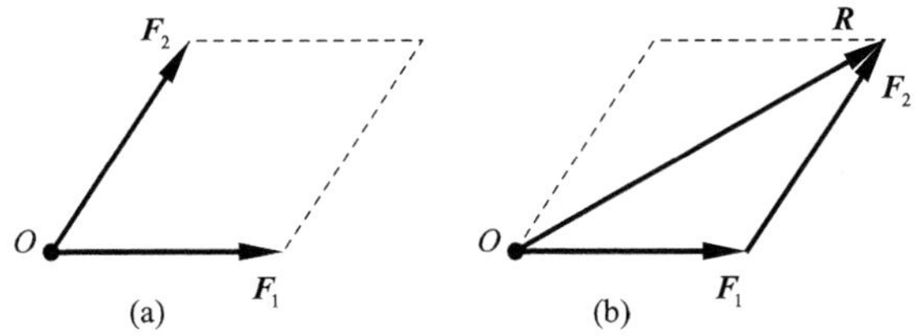

图 1.2

如果有三个或三个以上的平面汇交力作用于质点，则通过重复应用三角形法则即可获得合力。考虑图 1.3a 所示的平面汇交力 $\boldsymbol{F}_1$ 、$\boldsymbol{F}_2$ 和 $\boldsymbol{F}_3$ 作用于质点 O，把作用于质点的所有的力进行首尾相连，由第一个力的尾指向最后一个力的头的矢量即表示合力 $\boldsymbol{R}$ ，如图 1.3b 所示，即 $\boldsymbol{R}=\boldsymbol{F}_1+\boldsymbol{F}_2+\boldsymbol{F}_3$ 。这就是多边形法则。

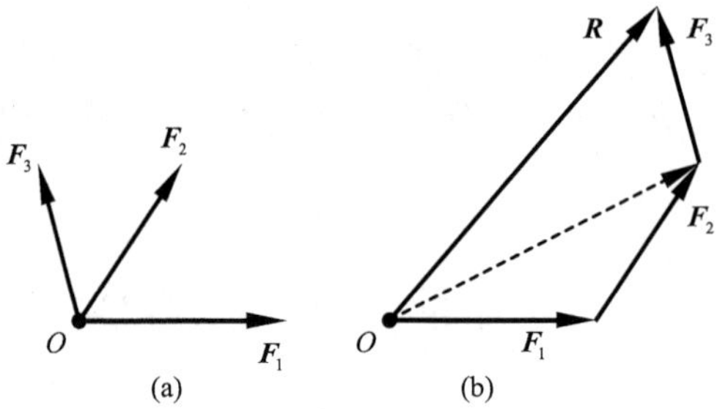

图 1.3

由此得出结论，作用于质点的平面汇交力系可由通过汇交点的合力代替，合力等于平面汇交力的矢量和，即

$$\boldsymbol{R}=\sum\boldsymbol{F} \tag{1.1}$$

例 1.1 如图 E1.1a 所示，两杆 AC 和 AD 连接到柱 AB 上点 A。已知 $F_1=150\ \mathrm{N}$ 、$\theta_1=30°$ 和 $\theta_2=15°$ ，求：(a) F_2 ，如果合力竖直向上；(b)合力大小。

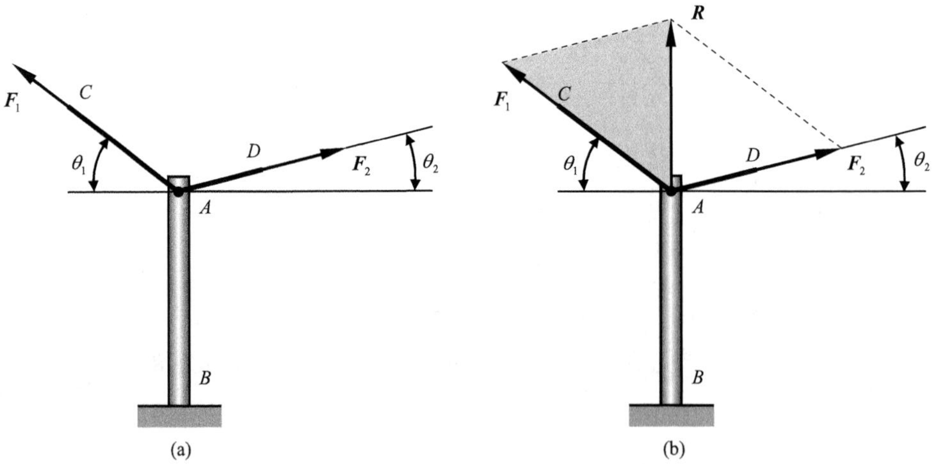

图 E1.1

解 根据平行四边形定律，$\boldsymbol{F}_1$ 和 $\boldsymbol{F}_2$ 可由 $\boldsymbol{R}$ 代替，如图 E1.1b 所示。考虑图 E1.1b 所示阴影三角形，并利用正弦定理，有

$$\frac{F_1}{\sin(90°-\theta_2)}=\frac{F_2}{\sin(90°-\theta_1)}=\frac{R}{\sin(\theta_1+\theta_2)}$$

代入 $F_1=150\ \mathrm{N}$ 、 $\theta_1=30°$ 和 $\theta_2=15°$ ，得

$$F_2 = \frac{\sin(90° - \theta_1)}{\sin(90° - \theta_2)} F_1 = 134.49\ \text{N},\ R = \frac{\sin(\theta_1 + \theta_2)}{\sin(90° - \theta_2)} F_1 = 109.81\ \text{N}$$

例 1.2　如图 E1.2a 所示，两杆 AC 和 AD 连接到柱 AB 上点 A。已知 $F_1 = 120$ N、$F_2 = 100$ N、$\theta_1 = 35°$ 和 $\theta_2 = 20°$，求合力。

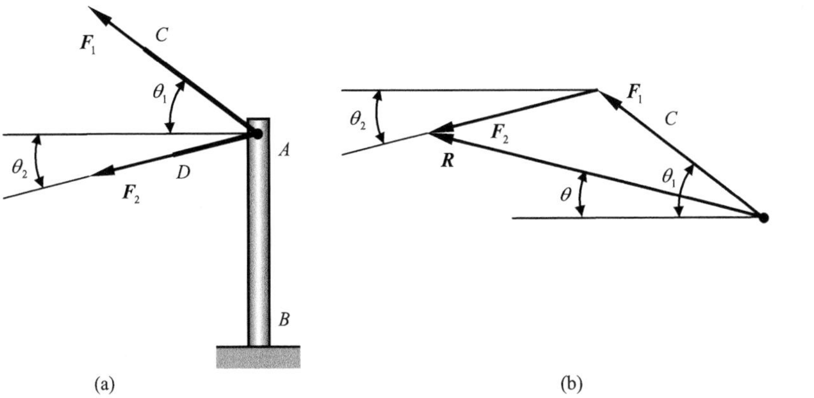

图 E1.2

解　力三角形如图 E1.2b 所示。利用余弦和正弦定理，有

$$R^2 = F_1^2 + F_2^2 - 2F_1F_2\cos[180° - (\theta_1 + \theta_2)],\ \frac{F_2}{\sin(\theta_1 - \theta)} = \frac{R}{\sin[180° - (\theta_1 + \theta_2)]}$$

代入 $F_1 = 120$ N、$F_2 = 100$ N、$\theta_1 = 35°$ 和 $\theta_2 = 20°$，得

$$R = \sqrt{F_1^2 + F_2^2 + 2F_1F_2\cos(\theta_1 + \theta_2)} = 195.36\ \text{N},\ \theta = \theta_1 - \arcsin[\frac{F_2}{R}\sin(\theta_1 + \theta_2)] = 10.21°$$

2. 解析法

作用于质点的两个或两个以上的力能够由合力代替。反之，作用于质点的力也能够由两个或两个以上的分力代替。例如，图 1.4a 所示 $\boldsymbol{F}$ 能够由 $\boldsymbol{F}_1$ 和 $\boldsymbol{F}_2$ 代替，其中 $\boldsymbol{F}_1$ 和 $\boldsymbol{F}_2$ 为 $\boldsymbol{F}$ 的矢量分量。用 $\boldsymbol{F}_1$ 和 $\boldsymbol{F}_2$ 代替 $\boldsymbol{F}$ 称为力的分解。显然，对于 $\boldsymbol{F}$ 存在无穷多组矢量分量，如图 1.4b 所示。

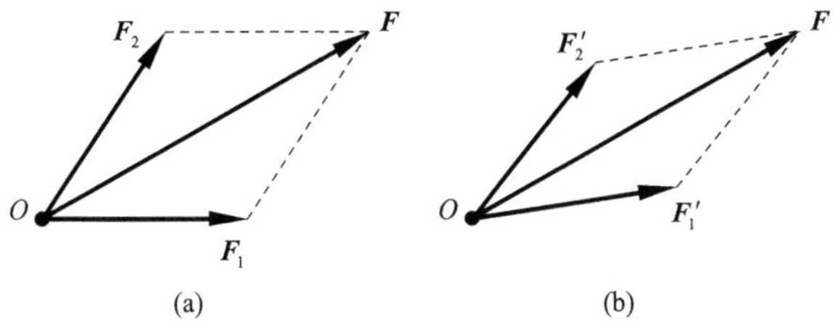

图 1.4

把力分解为相互垂直的分量通常很方便。例如，如图 1.5 所示 $\boldsymbol{F}$ 可以分解为两个矢量分量 $\boldsymbol{F}_x$ 和 $\boldsymbol{F}_y$。

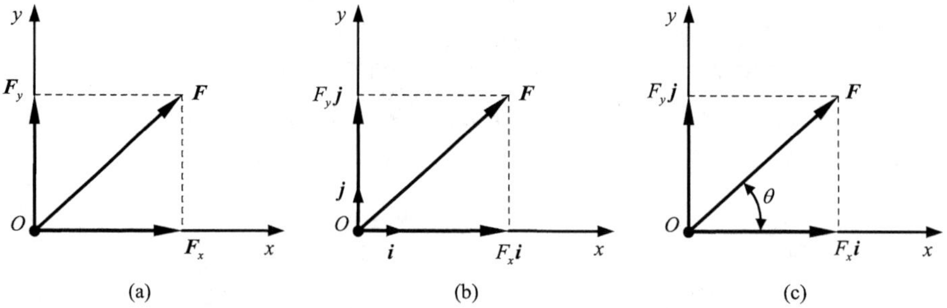

图 1.5

如图 1.5b 所示，通过引进两个单位矢量 $\boldsymbol{i}$ 和 $\boldsymbol{j}$ ，则 $\boldsymbol{F}$ 也可表示为 $\boldsymbol{F}=F_x\boldsymbol{i}+F_y\boldsymbol{j}$ ，其中 F_x 和 F_y 称为 $\boldsymbol{F}$ 的标量分量。

由 F 表示 $\boldsymbol{F}$ 的大小，θ 表示 $\boldsymbol{F}$ 与正 x 轴的夹角，如图 1.5c 所示，则标量分量 F_x 和 F_y 可表示为 $F_x=F\cos\theta,\ F_y=F\sin\theta$ 。

采用几何法求平面汇交力合力常常需要进行大量的几何或三角计算，尤其是求三个或三个以上力的合力更是如此。相反，通过使用解析法，这类问题就很容易得到求解。

如图 1.6 所示，考虑作用于质点 O 的 $\boldsymbol{F}_1$ 、 $\boldsymbol{F}_2$ 和 $\boldsymbol{F}_3$ ，则利用几何法，合力 $\boldsymbol{R}$ 可表示为 $\boldsymbol{R}=\boldsymbol{F}_1+\boldsymbol{F}_2+\boldsymbol{F}_3$ 。包括合力在内，把每个力分解为直角分量，有

$$R_x\boldsymbol{i}+R_y\boldsymbol{j}=(F_{1x}+F_{2x}+F_{3x})\boldsymbol{i}+(F_{1y}+F_{2y}+F_{3y})\boldsymbol{j} \tag{1.2}$$

由此得

$$R_x=F_{1x}+F_{2x}+F_{3x},\ R_y=F_{1y}+F_{2y}+F_{3y} \tag{1.3}$$

因此得出结论，作用于质点的合力沿任意轴的标量分量等于给定力沿相同轴的标量分量的代数和，即

$$R_x=\sum F_x,\ R_y=\sum F_y \tag{1.4}$$

因此，合力大小 R 和合力与正 x 轴夹角 θ 可分别写为

$$R=\sqrt{R_x^2+R_y^2},\ \theta=\arctan\frac{R_y}{R_x} \tag{1.5}$$

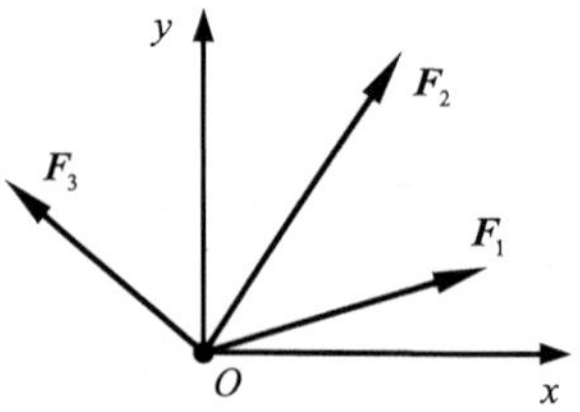

图 1.6

例 1.3 如图E1.3a所示，两杆 AC 和 AD 连接到柱 AB 上点 A。已知 $F_1=150\ \text{N}$ 、$\theta_1=30°$ 和 $\theta_2=15°$ ，求：(a) F_2 ，如果合力竖直向上；(b)合力大小。

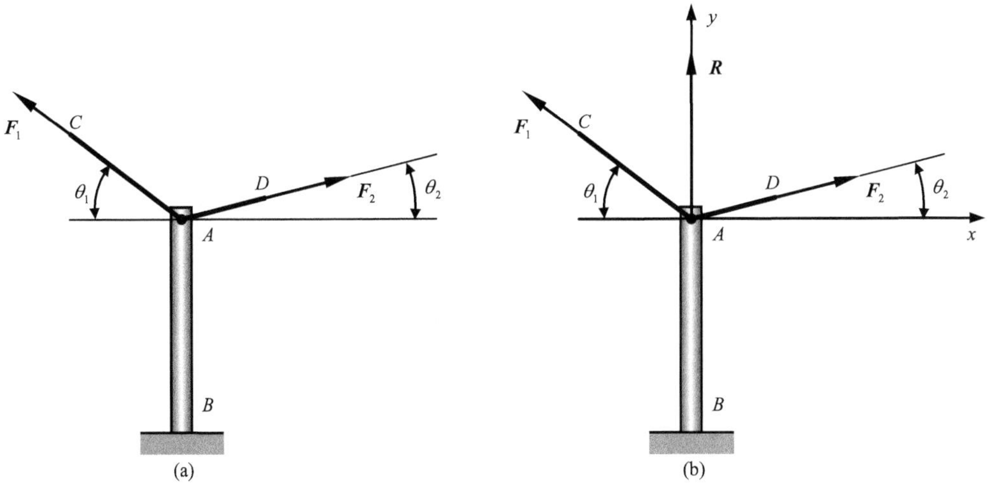

图 E1.3

解　建立图 E1.3b 所示坐标系，则合力的标量分量可表示为

$$R_x = -F_1\cos\theta_1 + F_2\cos\theta_2,\ R_y = F_1\sin\theta_1 + F_2\sin\theta_2$$

因合力在竖直方向，即 $R_x = 0$ ，则有

$$F_2 = \frac{\cos\theta_1}{\cos\theta_2}F_1 = 134.49\ \text{N},\ R=\sqrt{R_x^2+R_y^2} = F_2\sin\theta_2 + F_1\sin\theta_1 = 109.81\ \text{N}$$

例 1.4　受三个力作用的物块放在倾角 $\alpha = 25°$ 的斜面上，如图 E1.4a 所示。假设 $\theta = 40°$ 、 $F_1 = 150$ N 、 $F_2 = 250$ N 和 $F_3 = 200$ N ，求作用于物块的力的合力。

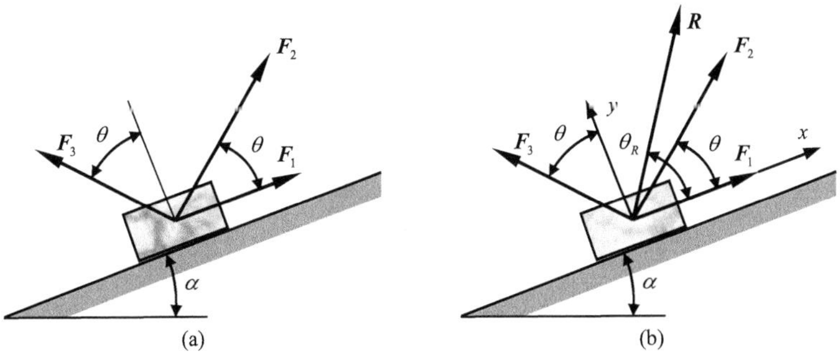

图 E1.4

解　建立图 E1.4b 所示坐标系，则合力的标量分量可表示为

$$R_x = F_1 + F_2\cos\theta - F_3\sin\theta = 212.95\ \text{N},\ R_y = F_2\sin\theta + F_3\cos\theta = 313.91\ \text{N}$$

利用上述标量分量，可得合力的大小和方向

$$R = \sqrt{R_x^2 + R_y^2} = 379.32\ \text{N},\ \theta_R = \arctan\frac{R_y}{R_x} = 55.85°$$

由此得，合力大小为 $R = 379.32$ N ，倾角为 $\alpha + \theta_R = 80.85°$ 。

1.2 平面汇交力系平衡

在求解质点平衡问题时，必须考虑作用于质点的所有力。为了做到这，选择质点为自由体(脱离体)、绘制单独的自由体图、显示作用于质点上的所有力。

1. 几何法

如果作用于质点的力的合力等于零，则质点保持平衡。因此，质点在平面汇交力系作用下的平衡条件可表示为

$$\sum \boldsymbol{F} = \boldsymbol{0} \tag{1.6}$$

从式(1.6)可看出，如果作用于质点的力形成封闭多边形，则质点保持平衡。如图 1.7a 所示，考虑质点 O 受力 $\boldsymbol{F}_1$、$\boldsymbol{F}_2$ 和 $\boldsymbol{F}_3$ 作用，利用多边形法则可得给定力的合力。从点 O 处由 $\boldsymbol{F}_1$ 开始，把力安排为首尾相连，即可发现 $\boldsymbol{F}_3$ 的头与起点 O 重合，如图 1.7b 所示。因此，给定力的合力为零，质点保持平衡。

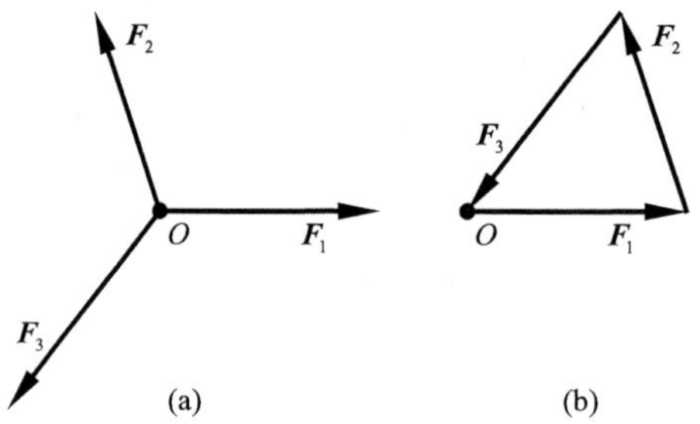

图 1.7

例 1.5 三根绳索在 A 处连接，按图 E1.5a 加载。求绳索 AB 和 AC 中的拉力。

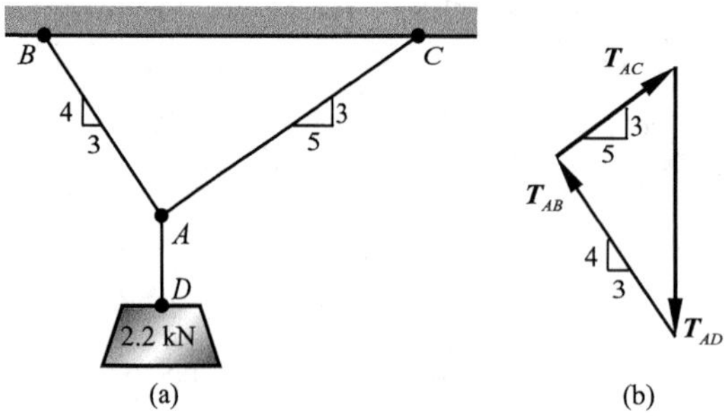

图 E1.5

解 因节点 A 平衡，故作用于节点 A 的所有的力将形成封闭三角形，如图 E1.5b 所示。根据正弦定理，有

$$\frac{T_{AB}}{\sin[\arctan(5/3)]}=\frac{T_{AC}}{\sin[\arctan(3/4)]}=\frac{T_{AD}}{\sin[\arctan(3/5)+\arctan(4/3)]}$$

代入 $T_{AD}=2.2\text{ kN}$ ，得

$$T_{AB}=1.90\text{ kN},\ T_{AC}=1.33\text{ kN}$$

2. 解析法

受平面汇交力作用质点的平衡条件可表示为 $\sum \boldsymbol{F}=\boldsymbol{0}$ 。每个力都分解为直角分量，有

$$(\sum F_x)\boldsymbol{i}+(\sum F_y)\boldsymbol{j}=\boldsymbol{0} \tag{1.7}$$

由此得出受平面汇交力作用质点的平衡条件可表示为

$$\sum F_x=0,\ \sum F_y=0 \tag{1.8}$$

上式称为平衡方程。

例 1.6　三根绳索在 A 处连接，按图 E1.6a 加载。求绳索 AB 和 AC 中的拉力。

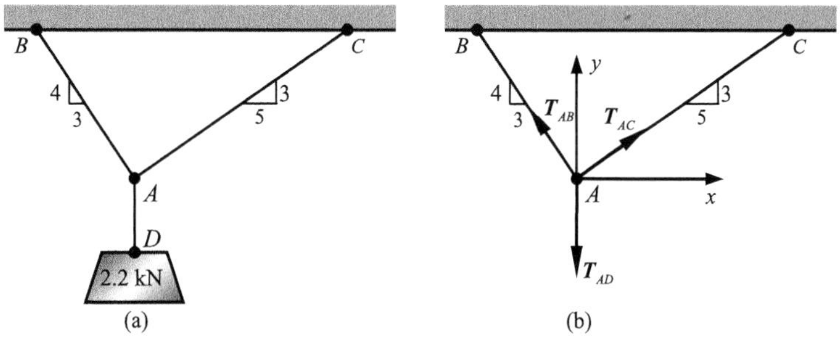

图 E1.6

解　如图 E1.6b 所示，建立坐标系，并考虑质点 A 的平衡，则有

$$\sum F_x=0,\ -T_{AB}\frac{3}{\sqrt{3^2+4^2}}+T_{AC}\frac{5}{\sqrt{5^2+3^2}}=0$$

$$\sum F_y=0,\ T_{AB}\frac{4}{\sqrt{3^2+4^2}}+T_{AC}\frac{3}{\sqrt{5^2+3^2}}-T_{AD}=0$$

解上述方程，得

$$T_{AB}=1.90\text{ kN},\ T_{AC}=1.33\text{ kN}$$

1.3　空间汇交力系合成

如图 1.8a 所示，空间力 $\boldsymbol{F}$ 可以分解为三个矢量分量 $\boldsymbol{F}_x$ 、 $\boldsymbol{F}_y$ 和 $\boldsymbol{F}_z$ ，即 $\boldsymbol{F}=\boldsymbol{F}_x+\boldsymbol{F}_y+\boldsymbol{F}_z$ 。

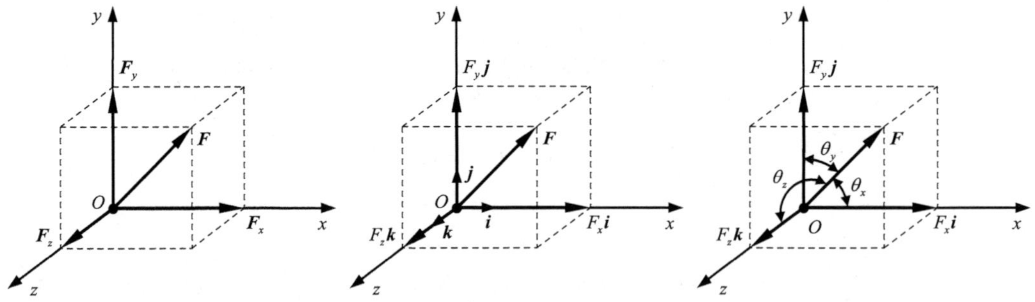

(a) (b) (c)

图 1.8

如图 1.8b 所示，通过引入三个单位矢量 $\boldsymbol{i}$、$\boldsymbol{j}$ 和 $\boldsymbol{k}$，则 $\boldsymbol{F}$ 可表示为 $\boldsymbol{F}=F_x\boldsymbol{i}+F_y\boldsymbol{j}+F_z\boldsymbol{k}$，其中 F_x、F_y 和 F_z 分别为力 $\boldsymbol{F}$ 的标量分量。

由 F 表示 $\boldsymbol{F}$ 的大小，θ_x、θ_y 和 θ_z 表示 $\boldsymbol{F}$ 与正 x、y 和 z 轴的夹角，如图 1.8c 所示，则标量分量可表示为

$$F_x=F\cos\theta_x,\ F_y=F\cos\theta_y,\ F_z=F\cos\theta_z \tag{1.9}$$

式中，$\cos\theta_x$、$\cos\theta_y$ 和 $\cos\theta_z$ 为 $\boldsymbol{F}$ 的方向余弦。这些方向余弦满足如下关系：

$$\cos^2\theta_x+\cos^2\theta_y+\cos^2\theta_z=1 \tag{1.10}$$

如图 1.9 所示，如果已知 $\boldsymbol{F}$ 与正 y 轴之间的夹角为 γ，正 z 轴与由 $\boldsymbol{F}$ 和 y 轴构成的平面之间的夹角为 φ，则相应标量分量可表示为

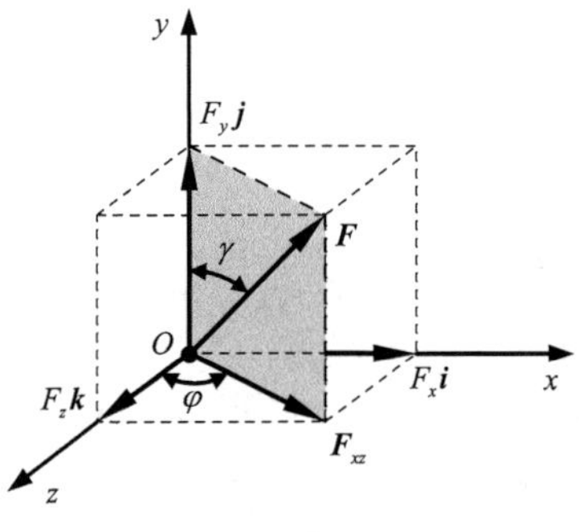

图 1.9

$$F_x=F_{xz}\sin\varphi=F\sin\gamma\sin\varphi,\ F_y=F\cos\gamma,\ F_z=F_{xz}\cos\varphi=F\sin\gamma\cos\varphi \tag{1.11}$$

如图 1.10 所示，考虑作用于质点 O 的三个力 $\boldsymbol{F}_1$、$\boldsymbol{F}_2$ 和 $\boldsymbol{F}_3$，采用几何法，则这些力的合力 $\boldsymbol{R}$ 可表示为 $\boldsymbol{R}=\boldsymbol{F}_1+\boldsymbol{F}_2+\boldsymbol{F}_3$。包括合力在内，把每个力分解为直角分量，有

$$R_x\boldsymbol{i}+R_y\boldsymbol{j}+R_z\boldsymbol{k}=(F_{1x}+F_{2x}+F_{3x})\boldsymbol{i}+(F_{1y}+F_{2y}+F_{3y})\boldsymbol{j}+(F_{1z}+F_{2z}+F_{3z})\boldsymbol{k} \tag{1.12}$$

由此得出

$$R_x=F_{1x}+F_{2x}+F_{3x},\ R_y=F_{1y}+F_{2y}+F_{3y},\ R_z=F_{1z}+F_{2z}+F_{3z} \tag{1.13}$$

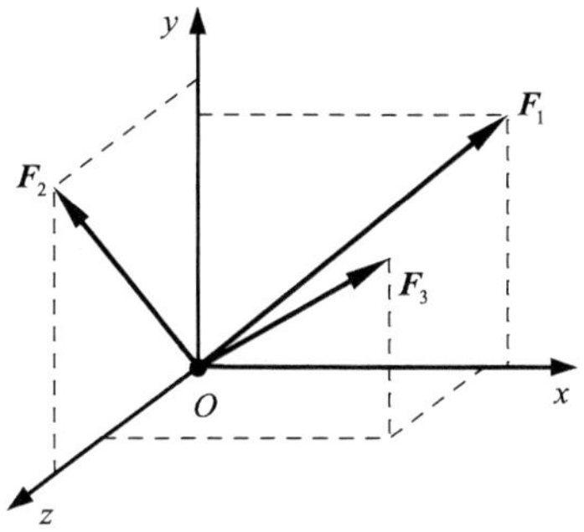

图 1.10

因此得出结论，作用于质点的力的合力沿任意轴的标量分量等于给定力沿相同轴的标量分量的代数和，即

$$R_x = \sum F_x,\ R_y = \sum F_y,\ R_z = \sum F_z \tag{1.14}$$

因此，合力大小 R 和合力与正 x、y 和 z 轴的夹角 θ_x、θ_y 和 θ_z 可分别写为

$$R = \sqrt{R_x^2 + R_y^2 + R_z^2},\ \theta_x = \arccos\frac{R_x}{R},\ \theta_y = \arccos\frac{R_y}{R},\ \theta_z = \arccos\frac{R_z}{R} \tag{1.15}$$

例 1.7　如图 E1.7 所示，已知 $F_1 = 300\ \text{N}$、$F_2 = 200\ \text{N}$ 和 $F_3 = 100\ \text{N}$，求合力大小和方向。

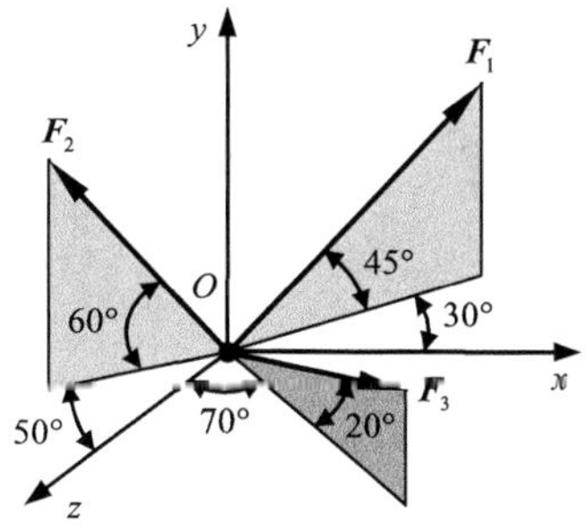

图 E1.7

解　合力的三个标量分量分别为

$$R_x = F_1\cos 45°\cos 30° - F_2\cos 60°\sin 50° + F_3\cos 20°\sin 70° = 195.41\ \text{N}$$

$$R_y = F_1\sin 45° + F_2\sin 60° + F_3\sin 20° = 419.54\ \text{N}$$

$$R_z = -F_1\cos 45°\sin 30° + F_2\cos 60°\cos 50° + F_3\cos 20°\cos 70° = -9.65\ \text{N}$$

因此，合力大小和方向为

$$R = \sqrt{R_x^2 + R_y^2 + R_z^2} = 462.92\ \text{N}$$

$$\theta_x = \arccos\frac{R_x}{R} = 65.0°,\ \theta_y = \arccos\frac{R_y}{R} = 25.0°,\ \theta_z = \arccos\frac{R_z}{R} = 91.2°$$

1.4　空间汇交力系平衡

受空间汇交力作用质点的平衡条件为$\sum \boldsymbol{F}=\boldsymbol{0}$。每个力都分解为直角分量，有

$$(\sum F_x)\boldsymbol{i}+(\sum F_y)\boldsymbol{j}+(\sum F_z)\boldsymbol{k}=\boldsymbol{0} \tag{1.16}$$

由此得出，受空间汇交力作用质点的平衡条件可表示为

$$\sum F_x=0,\ \sum F_y=0,\ \sum F_z=0 \tag{1.17}$$

上式称为平衡方程。

例 1.8　如图 E1.8a 所示，重量$W=300\ \text{N}$的水平均质圆板通过三根金属丝悬挂，金属丝与竖直方向形成30°夹角，求每根金属丝的拉力。

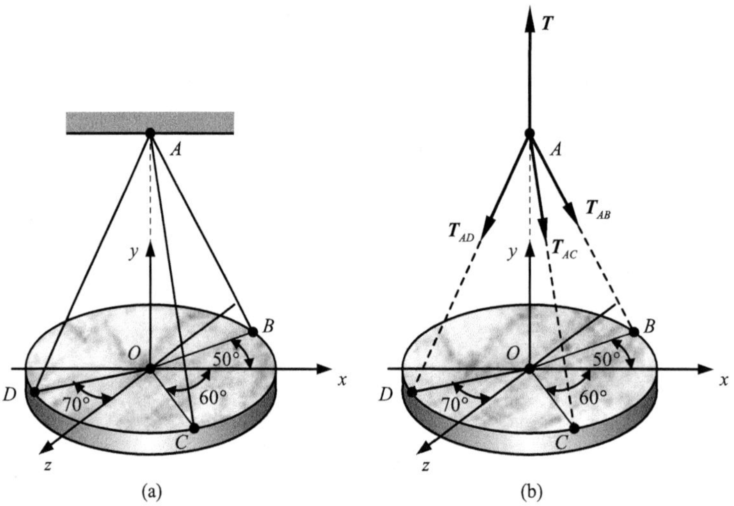

图 E1.8

解　如图 E1.8b 所示，因整个系统平衡，故有

$$T=W=300\ \text{N}$$

取点 A 为自由体，绘制受力图，有

$$\sum F_x=0,\ T_{AB}\sin 30°\cos 50°+T_{AC}\sin 30°\cos 60°-T_{AD}\sin 30°\sin 70°=0$$

$$\sum F_y=0,\ T-T_{AB}\cos 30°-T_{AC}\cos 30°-T_{AD}\cos 30°=0$$

$$\sum F_z=0,\ -T_{AB}\sin 30°\sin 50°+T_{AC}\sin 30°\sin 60°+T_{AD}\sin 30°\cos 70°=0$$

解方程，得

$$T_{AB}=140.71\ \text{N},\ T_{AC}=71.44\ \text{N},\ T_{AD}=134.26\ \text{N}$$

习　　题

1.1　两杆 AC 和 AD 连接于柱 AB 上 A 点。已知$F_2=100\ \text{N}$、$\theta_1=20°$和$\theta_2=10°$，

求：(a) F_1，如果合力竖直向上；(b)合力大小。

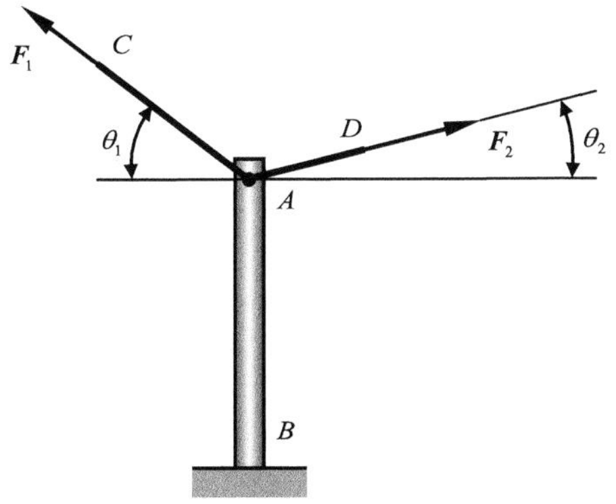

图 P1.1

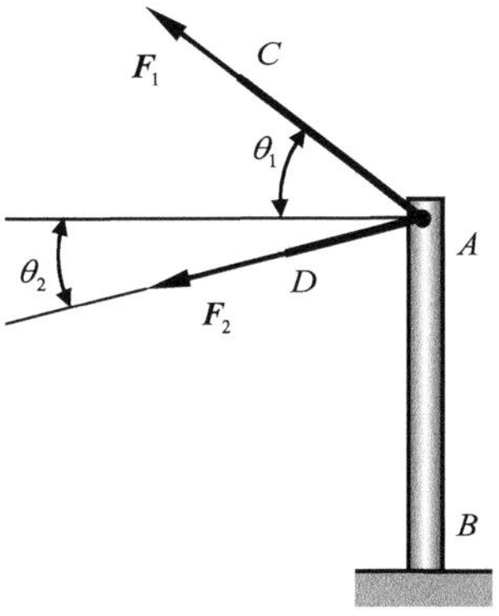

图 P1.2

1.2　两杆 AC 和 AD 连接于柱 AB 上 A 点。已知 $F_1 = 150$ N、$\theta_1 = 30°$ 和 $\theta_2 = 15°$，求：(a) F_2，如果合力水平向左；(b)合力大小。

1.3　已知 $F_1 = 110$ N、$F_2 = 90$ N、$\theta_1 = 40°$ 和 $\theta_2 = 25°$，求合力。

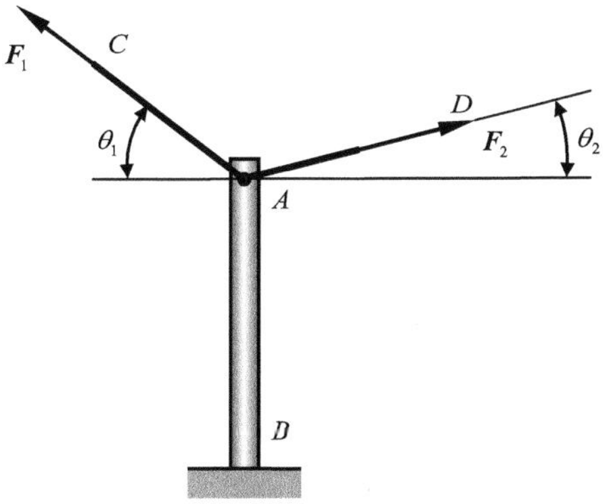

图 P1.3

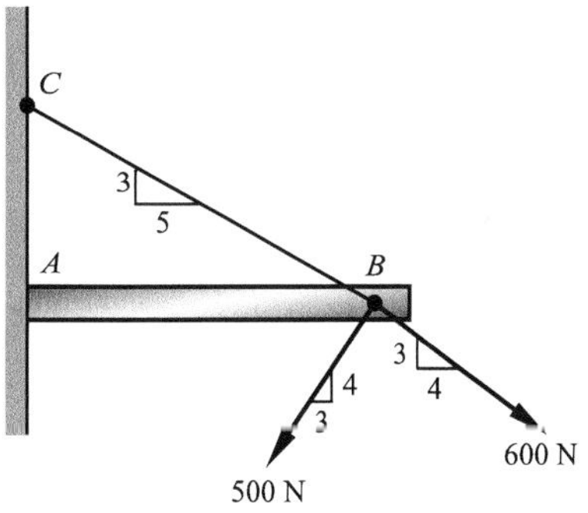

图 P1.4

1.4　已知 BC 绳拉力为 650 N，求作用于梁 AB 上点 B 处三个力的合力。

1.5　三根绳索在 A 处连接，加载如图所示，求(a) AB 和(b) AC 中的拉力。

1.6　假设杠杆在图示位置保持平衡，并已知 $F_1 = 200$ N、$F_2 = 175$ N 和 $\theta_1 = 30°$，求(a) θ_2；(b)两杆作用于杠杆的力的合力。

1.7　轴环 A 可沿无摩擦竖直杆滑动，并与穿过无摩擦定滑轮 B 的弹簧 C 相连。假设当 $h = 0.3$ m 时弹簧未伸长，当 $h = 0.4$ m 时轴环平衡，并已知 $a = 0.4$ m 和弹簧刚度 $k = 500$ N/m，求轴环重量和杆给轴环的作用力。

1.8　重量为 W 的轴环 B 可沿竖直杆自由运动。假设弹簧刚度为 k，当 $\theta = 0$ 时弹簧未伸长，并已知 $W = 13.5$ N、$l = 150$ mm 和 $k = 120$ N/m，求与平衡相对应的 θ 值。

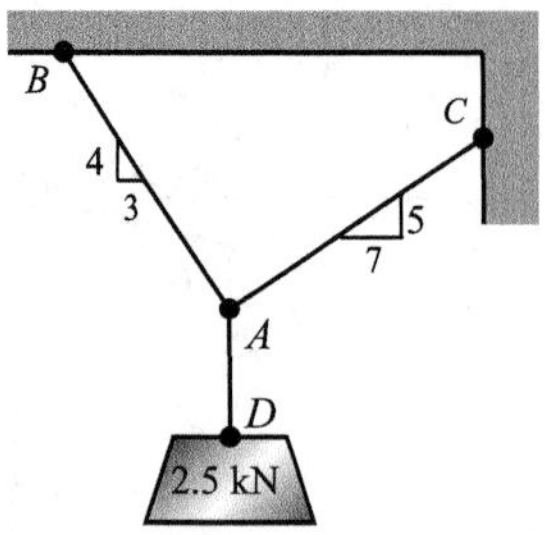

图 P1.5

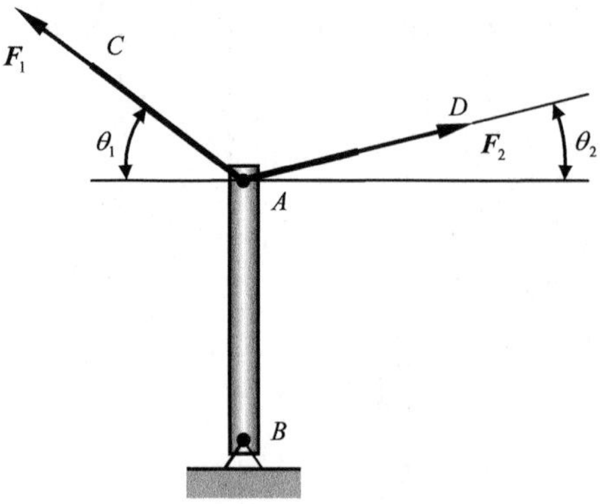

图 P1.6

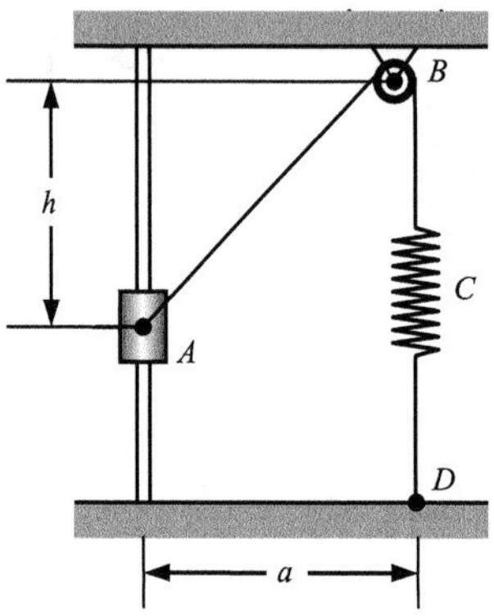

图 P1.7

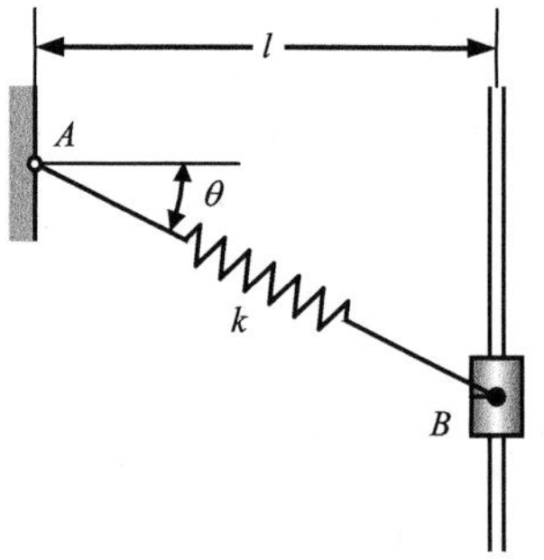

图 P1.8

1.9　水平圆板通过三根金属丝悬挂，金属丝与竖直方向夹角均为30°。已知金属丝 AB 作用于板的力的 x 分量为 50 N，求(a)金属丝 AB 的拉力和(b)作用于 B 处的力与坐标轴的夹角 θ_x、θ_y 和 θ_z。

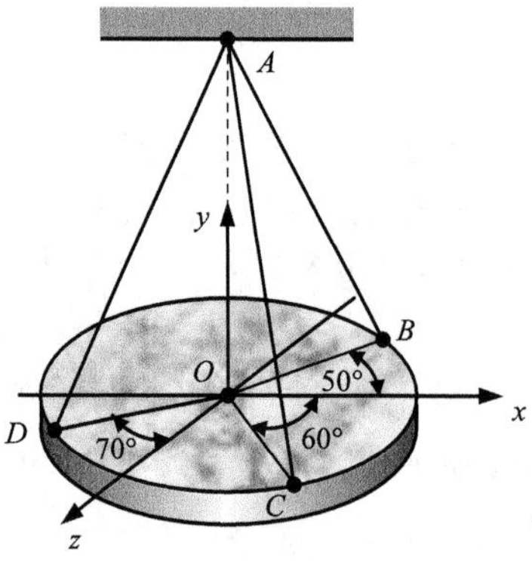

图 P1.9

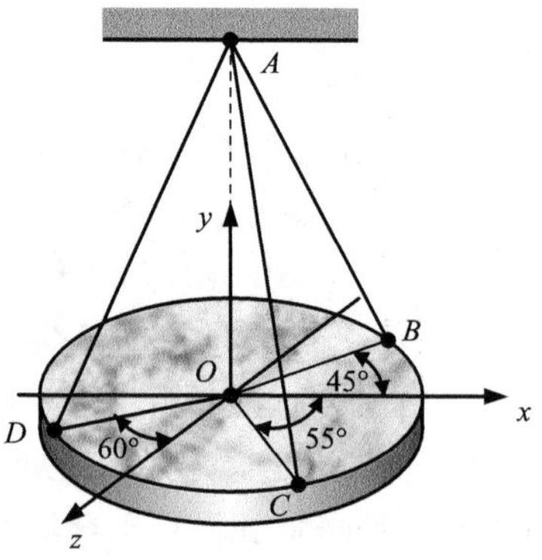

图 P1.10

1.10　重量 $W=200$ N 的水平均质圆板通过三根金属丝悬挂，金属丝与竖直方向的夹角均为35°，求每根金属丝的拉力。

第2章 力系简化

2.1 对点力矩

如图2.1所示，假设点O为参考点，A为力$\boldsymbol{F}$作用线上任意点，则$\boldsymbol{F}$对O力矩$\boldsymbol{M}_O(\boldsymbol{F})$定义为$\boldsymbol{r}$与$\boldsymbol{F}$的矢积，即

$$\boldsymbol{M}_O(\boldsymbol{F})=\boldsymbol{r}\times\boldsymbol{F} \tag{2.1}$$

式中，$\boldsymbol{r}$为由O指向A的位置矢量，$\boldsymbol{M}_O(\boldsymbol{F})$满足：

(1) 力矩大小等于$rF\sin\theta=Fd$，其中θ为$\boldsymbol{r}$和$\boldsymbol{F}$之间夹角，d为力矩臂(即O到$\boldsymbol{F}$作用线的垂直距离)。

(2) 力矩方向通过右手法则确定。右手法则表述如下：如果右手四指由$\boldsymbol{r}$向$\boldsymbol{F}$弯曲，那么右手拇指则指向力矩方向。

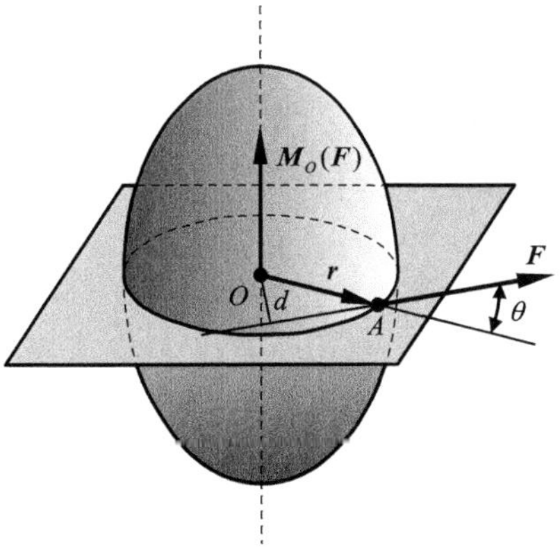

图 2.1

如图2.2所示，建立直角坐标系$Oxyz$，$\boldsymbol{i}$，$\boldsymbol{j}$，$\boldsymbol{k}$分别为沿x，y，z轴的单位矢量，则$\boldsymbol{F}$对O力矩$\boldsymbol{M}_O(\boldsymbol{F})$可展开为

$$\boldsymbol{M}_O(\boldsymbol{F})=\begin{vmatrix}\boldsymbol{i} & \boldsymbol{j} & \boldsymbol{k}\\ x & y & z\\ F_x & F_y & F_z\end{vmatrix}=(yF_z-zF_y)\boldsymbol{i}+(zF_x-xF_z)\boldsymbol{j}+(xF_y-yF_x)\boldsymbol{k} \tag{2.2}$$

式中，x，y，z为$\boldsymbol{r}$的坐标，F_x，F_y，F_z为$\boldsymbol{F}$的标量分量。

例 2.1 如图E2.1a，缆绳AB和AC与混凝土立柱相连。已知缆绳AB和AC中的拉力分别为800 N和500 N，求缆绳作用于立柱点A的合力对点O力矩。

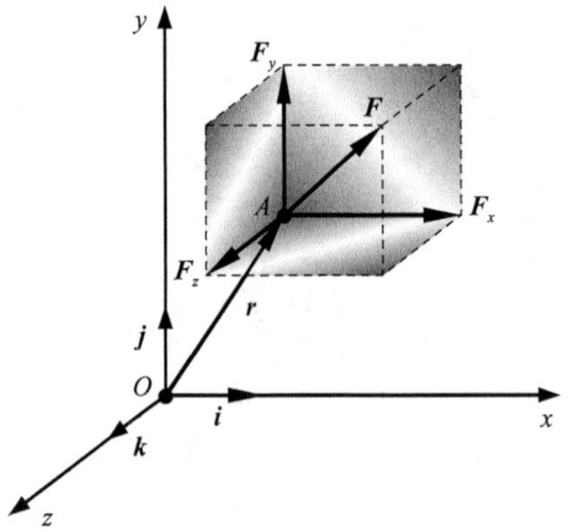

图 2.2

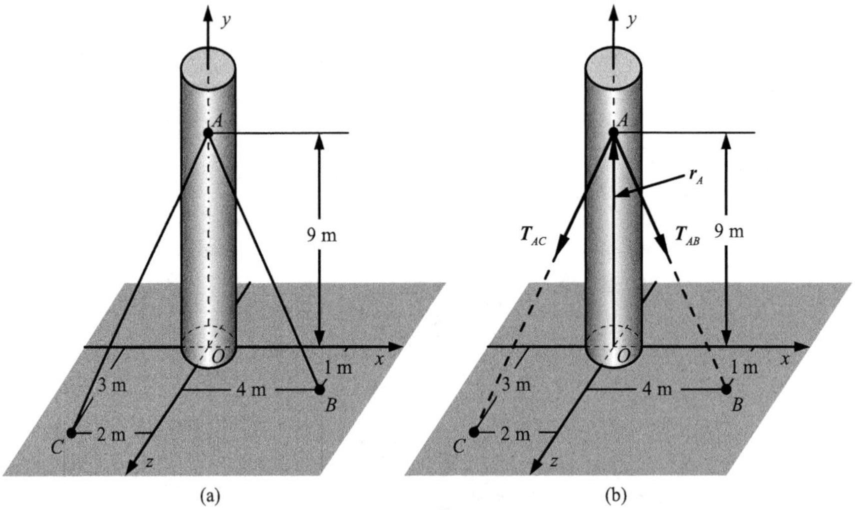

图 E2.1

解　设 $\boldsymbol{i}$、$\boldsymbol{j}$ 和 $\boldsymbol{k}$ 分别为沿 x、y 和 z 轴的单位矢量，则有

$$\boldsymbol{r}_A=(9\boldsymbol{j})\ \text{m},\ \boldsymbol{r}_{AB}=(4\boldsymbol{i}-9\boldsymbol{j}+\boldsymbol{k})\ \text{m},\ \boldsymbol{r}_{AC}=(-2\boldsymbol{i}-9\boldsymbol{j}+3\boldsymbol{k})\ \text{m}$$

$$\boldsymbol{T}_{AB}=T_{AB}\frac{\boldsymbol{r}_{AB}}{r_{AB}}=(800\ \text{N})\frac{4\boldsymbol{i}-9\boldsymbol{j}+\boldsymbol{k}}{\sqrt{4^2+(-9)^2+1^2}}=80.81(4\boldsymbol{i}-9\boldsymbol{j}+\boldsymbol{k})\ \text{N}$$

$$\boldsymbol{T}_{AC}=T_{AC}\frac{\boldsymbol{r}_{AC}}{r_{AC}}=(500\ \text{N})\frac{-2\boldsymbol{i}-9\boldsymbol{j}+3\boldsymbol{k}}{\sqrt{(-2)^2+(-9)^2+3^2}}=51.57(-2\boldsymbol{i}-9\boldsymbol{j}+3\boldsymbol{k})\ \text{N}$$

利用 $\boldsymbol{R}=\boldsymbol{T}_{AB}+\boldsymbol{T}_{AC}$，得

$$\boldsymbol{R}=(220.1\boldsymbol{i}-1191.4\boldsymbol{j}+235.5\boldsymbol{k})\ \text{N}$$

因此，合力对点 O 力矩等于

$$\boldsymbol{M}_O(\boldsymbol{R})=\boldsymbol{r}_A\times\boldsymbol{R}=\begin{vmatrix}\boldsymbol{i} & \boldsymbol{j} & \boldsymbol{k}\\ 0 & 9 & 0\\ 220.1 & -1191.4 & 235.5\end{vmatrix}\ \text{N}\cdot\text{m}=(2119.5\boldsymbol{i}-1980.9\boldsymbol{k})\ \text{N}\cdot\text{m}$$

2.2 对轴力矩

如图2.3所示，假设OL为过点O之轴，则$\boldsymbol{F}$对OL力矩M_{OL}定义为$\boldsymbol{M}_O(\boldsymbol{F})$在$OL$上的投影，即

$$M_{OL}=\boldsymbol{\lambda}\cdot\boldsymbol{M}_O(\boldsymbol{F})=\boldsymbol{\lambda}\cdot(\boldsymbol{r}\times\boldsymbol{F}) \tag{2.3}$$

式中，$\boldsymbol{\lambda}$为沿OL方向的单位矢量。

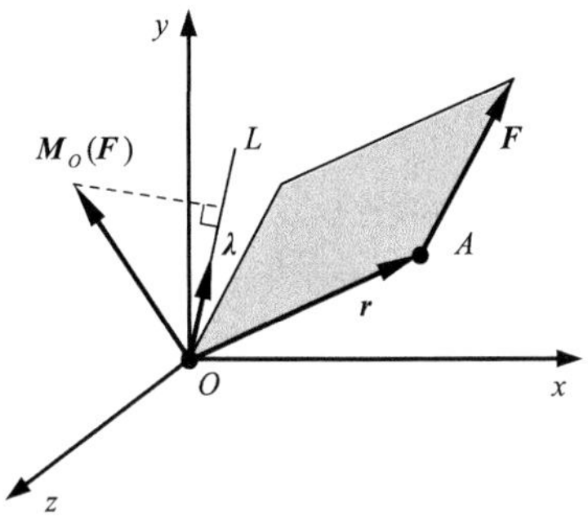

图2.3

2.3 力矩定理

假设力$\boldsymbol{F}_1$，$\boldsymbol{F}_2$，$\boldsymbol{F}_3$作用于同一点A，这些力的合力由$\boldsymbol{R}$表示，如图2.4所示，则有

$$\boldsymbol{M}_O(\boldsymbol{R})=\sum\boldsymbol{M}_O(\boldsymbol{F}) \tag{2.4}$$

因此得到结论：汇交力的合力对一点的力矩等于各汇交力对同一点的力矩的矢量和。这种关系称为力矩原理或伐里农定理。

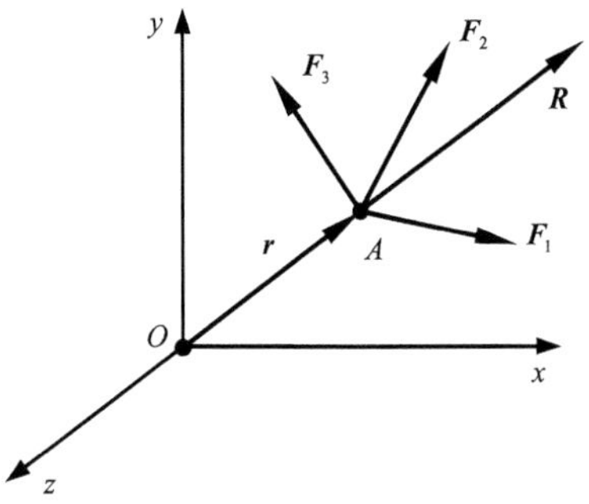

图2.4

例2.2 如图E2.2a所示，板由两链条悬挂。已知BH中拉力为200 N，求：(a)链条BH施加到板上的力对点A力矩；(b)如对点A产生相同力矩，则作用于点E的最小力。

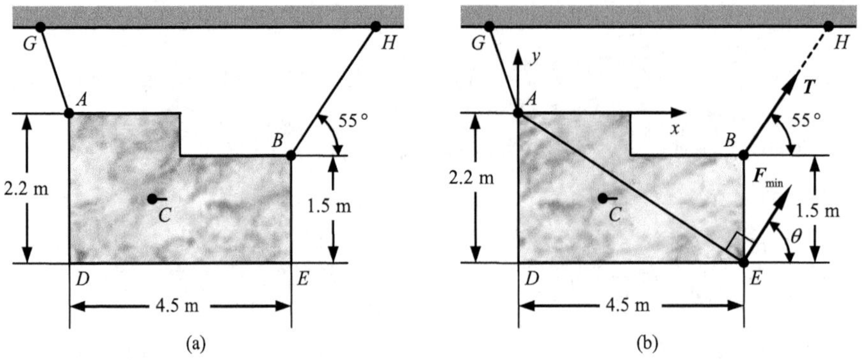

图 E2.2

解　(a) 建立如图 E2.2b 所示坐标系 Axy，则有

$$x_B = 4.5\ \text{m}, y_B = -0.7\ \text{m};\ T_x = T\cos 55° = 114.72\ \text{N},\ T_y = T\sin 55° = 163.83\ \text{N}$$

利用力矩原理，得

$$M_A(\boldsymbol{T}) = M_A(\boldsymbol{T}_x) + M_A(\boldsymbol{T}_y) = -y_B T_x + x_B T_y = 817.54\ \text{N}\cdot\text{m}$$

(b) 利用 $M_A(\boldsymbol{F}_{\min}) = r_E F_{\min}\sin 90° = M_A(\boldsymbol{T})$，得

$$F_{\min} = \frac{M_A(\boldsymbol{T})}{r_E \sin 90°} = 163.21\ \text{N},\ \theta = \arctan\frac{DE}{AD} = 63.95°$$

2.4　力　偶　矩

大小相同、作用线平行、方向相反的两个力 $\boldsymbol{F}$ 和 $\boldsymbol{F}'$ 将形成力偶，如图 2.5 所示。显然，两个力在任何方向的分量之和等于零。然而，两个力对任意点的矩之和并不等于零。因此，两个力不会使物体发生平移，但会使物体具有转动趋势。

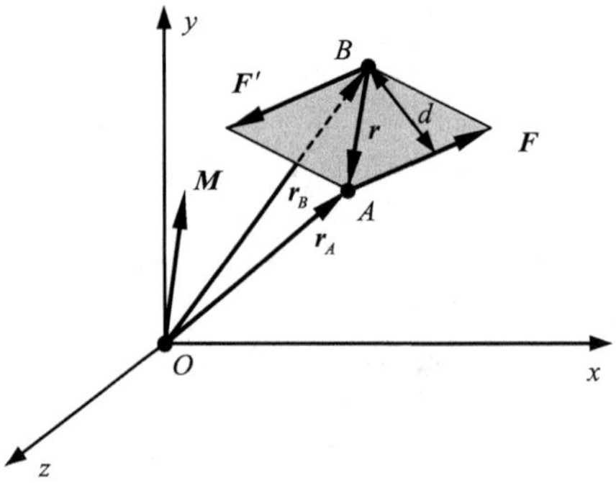

图 2.5

假设 $\boldsymbol{r}_A$ 和 $\boldsymbol{r}_B$ 分别为 $\boldsymbol{F}$ 和 $\boldsymbol{F}'$ 作用点的位置矢量，则 $\boldsymbol{F}$ 和 $\boldsymbol{F}'$ 对 O 的矩之和为 $\boldsymbol{M}_O(\boldsymbol{F},\boldsymbol{F}') = \boldsymbol{r}_A\times\boldsymbol{F} + \boldsymbol{r}_B\times\boldsymbol{F}' = \boldsymbol{r}_A\times\boldsymbol{F} + \boldsymbol{r}_B\times(-\boldsymbol{F}) = (\boldsymbol{r}_A - \boldsymbol{r}_B)\times\boldsymbol{F} = \boldsymbol{r}\times\boldsymbol{F}$。$\boldsymbol{M}_O(\boldsymbol{F},\boldsymbol{F}')$ 与 O 选择无关，因此可重写为

$$\boldsymbol{M}=\boldsymbol{r}\times\boldsymbol{F} \tag{2.5}$$

式中，$\boldsymbol{M}$ 为力偶矩。

因力偶矩与参考点选择无关，故它是自由矢量，即可作用于刚体上任何位置，只要方向保持不变。

例 2.3 如图 E2.3 所示，三个力偶作用于物块。已知 $M_1=10\ \text{N}\cdot\text{m}$、$M_2=15\ \text{N}\cdot\text{m}$ 和 $M_3=8\ \text{N}\cdot\text{m}$，用单一等效力偶代替这三个力偶，并求等效力偶的大小和方向。

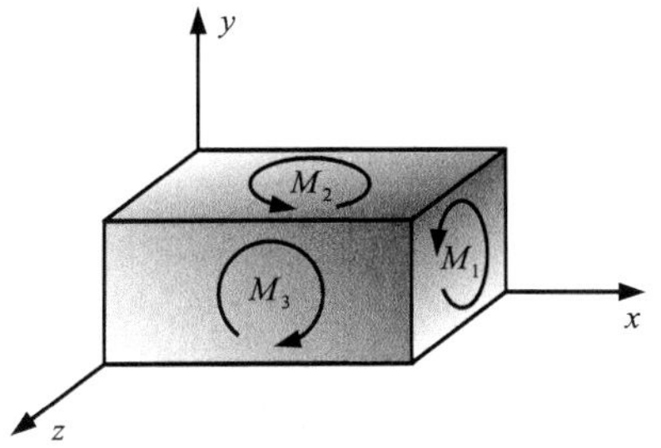

图 E2.3

解 设沿 x、y 和 z 轴的单位矢量分别为 $\boldsymbol{i}$、$\boldsymbol{j}$ 和 $\boldsymbol{k}$，则三个力偶矩可写为

$$\boldsymbol{M}_1=(10\boldsymbol{i})\ \text{N}\cdot\text{m},\ \boldsymbol{M}_2=(15\boldsymbol{j})\ \text{N}\cdot\text{m},\ \boldsymbol{M}_3=(-8\boldsymbol{k})\ \text{N}\cdot\text{m}$$

因此，等效力偶矩可表示为

$$\boldsymbol{M}=\boldsymbol{M}_1+\boldsymbol{M}_2+\boldsymbol{M}_3=(10\boldsymbol{i}+15\boldsymbol{j}-8\boldsymbol{k})\ \text{N}\cdot\text{m}$$

等效力偶矩的大小和方向分别为

$$M=\sqrt{10^2+15^2+(-8)^2}=19.72\ \text{N}\cdot\text{m}$$

$$\theta_x=\arccos(\frac{10}{19.72})=59.53°,\ \theta_y=\arccos(\frac{15}{19.72})=40.48°,\ \theta_z=\arccos(\frac{-8}{19.72})=113.93°$$

2.5 力 的 等 效

力可作用于其作用线上任何点而不改变其对刚体的作用效果。例如，图 2.6a 所示作用于点 O 的力 $\boldsymbol{F}$ 可以通过图 2.6b 所示具有相同大小、相同方向但作用于相同作用线上不同点 O' 的力 $\boldsymbol{F}'$ 代替。力 $\boldsymbol{F}$ 和 $\boldsymbol{F}'$ 对刚体具有相同效果，因而称为等效。该原理称为可传性原理，表明只要作用于刚体上的力沿其作用线移动，则力对刚体的作用效果保持不变。因此，作用于刚体上的力是滑移矢量。

作用于刚体 A 点的力 $\boldsymbol{F}$，A 点的位置矢量为 $\boldsymbol{r}$，如图 2.7a 所示。利用可传性原理，若不改变力对刚体的作用效果，则作用于刚体的力 $\boldsymbol{F}$ 不能移到不在作用线上的点 O。然而，我们可以在点 O 附加两个力 $\boldsymbol{F}'$ 和 $\boldsymbol{F}''$，使 $\boldsymbol{F}'=\boldsymbol{F}$ 和 $\boldsymbol{F}''=-\boldsymbol{F}$，而不改变原力对刚体的作用效果，如图 2.7b 所示。通过这种变换，力 $\boldsymbol{F}'$ 作用于点 O，其余两个力 $\boldsymbol{F}$ 和 $\boldsymbol{F}''$ 将形成矩为 $\boldsymbol{M}_O=\boldsymbol{r}\times\boldsymbol{F}$ 的力偶，如图 2.7c 所示。

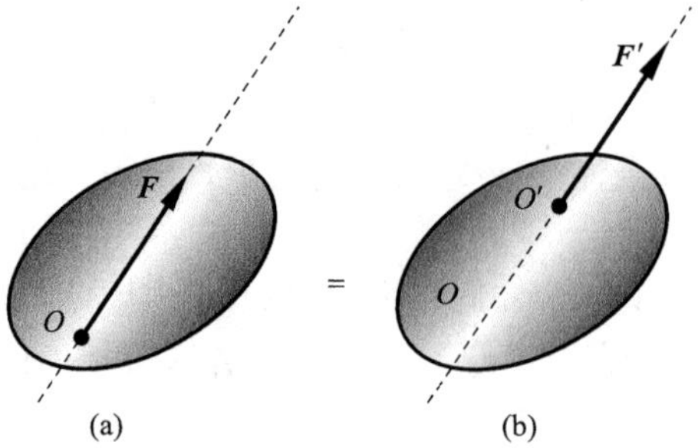

图 2.6

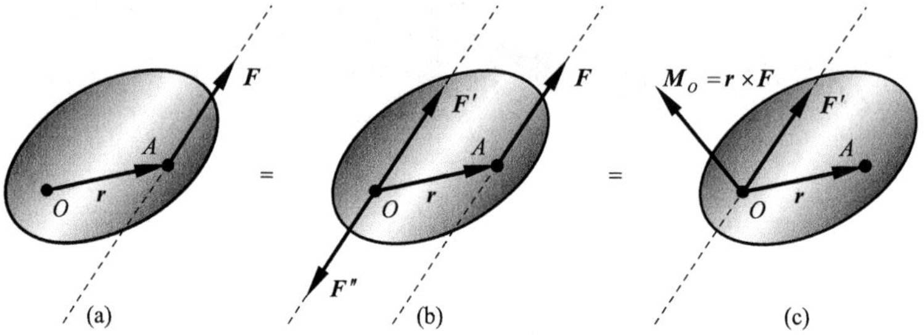

图 2.7

因此，我们得出结论：只要附加力偶，则作用于刚体上的力 $\boldsymbol{F}$ 可以移到任意点 O，附加力偶矩等于力 $\boldsymbol{F}$ 对点 O 力矩 $\boldsymbol{M}_O$。因 $\boldsymbol{M}_O$ 是自由矢量，故 $\boldsymbol{M}_O$ 可以附加在任意位置，但为方便起见，$\boldsymbol{M}_O$ 通常附加在点 O。

例 2.4　如图 E2.4a 所示，垂直力 $\boldsymbol{F}$ 作用于平面桁架的点 C。已知 $F=80\ \text{N}$，试用作用于点 G 的力–力偶系代替力 $\boldsymbol{F}$。

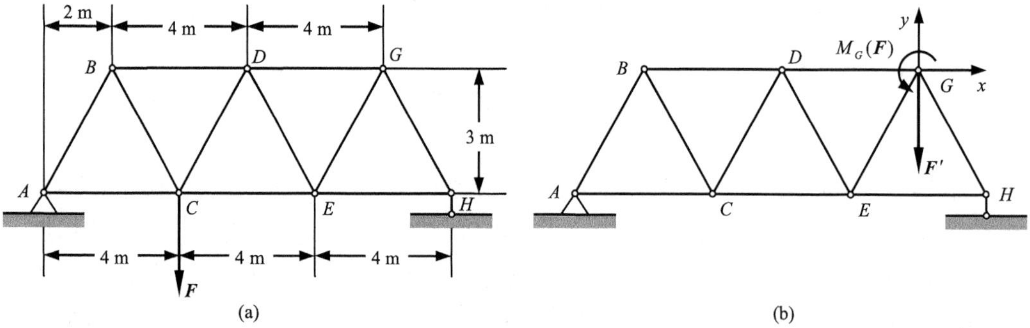

图 E2.4

解　如图 E2.4b，等效力–力偶系为

$$\boldsymbol{F}' = \boldsymbol{F} = (-80\boldsymbol{j})\ \text{N}，\ M_G(\boldsymbol{F}) = 480\ \text{N} \cdot \text{m}$$

式中，$\boldsymbol{j}$ 为沿正 y 轴的单位矢量。

2.6　力系简化为力-力偶系

如图 2.8a 所示，力 $\boldsymbol{F}_1$，$\boldsymbol{F}_2$，$\boldsymbol{F}_3$ 分别作用于刚体上由位置矢量 $\boldsymbol{r}_1$，$\boldsymbol{r}_2$，$\boldsymbol{r}_3$ 定义的点 A_1，A_2，A_3。$\boldsymbol{F}_1$ 可从 A_1 移到 O，只要力偶 $\boldsymbol{M}_O(\boldsymbol{F}_1)=\boldsymbol{r}_1\times\boldsymbol{F}_1$ 添加到力系。对 $\boldsymbol{F}_2$ 和 $\boldsymbol{F}_3$ 重复该过程，将得到由 $\boldsymbol{F}_1'$，$\boldsymbol{F}_2'$，$\boldsymbol{F}_3'$，$\boldsymbol{M}_O(\boldsymbol{F}_1)=\boldsymbol{r}_1\times\boldsymbol{F}_1$，$\boldsymbol{M}_O(\boldsymbol{F}_2)=\boldsymbol{r}_2\times\boldsymbol{F}_2$，$\boldsymbol{M}_O(\boldsymbol{F}_3)=\boldsymbol{r}_3\times\boldsymbol{F}_3$ 构成的新力系，如图 2.8b。在新力系中，$\boldsymbol{F}_1'$，$\boldsymbol{F}_2'$，$\boldsymbol{F}_3'$ 可由力 $\boldsymbol{R}'$ 代替，同理力偶 $\boldsymbol{M}_O(\boldsymbol{F}_1)=\boldsymbol{r}_1\times\boldsymbol{F}_1$，$\boldsymbol{M}_O(\boldsymbol{F}_2)=\boldsymbol{r}_2\times\boldsymbol{F}_2$，$\boldsymbol{M}_O(\boldsymbol{F}_3)=\boldsymbol{r}_3\times\boldsymbol{F}_3$ 也可由力偶 $\boldsymbol{M}_O$ 代替，如图 2.8c。其中 $\boldsymbol{R}'$ 和 $\boldsymbol{M}_O$ 可分别表示为

$$\boldsymbol{R}'=\sum\boldsymbol{F}'=\sum\boldsymbol{F},\ \boldsymbol{M}_O=\sum\boldsymbol{M}_O(\boldsymbol{F}) \tag{2.6}$$

上式表明，力 $\boldsymbol{R}'$ 可通过相加所有力而得到，力偶 $\boldsymbol{M}_O$ 可通过相加所有力对 O 力矩而得到。

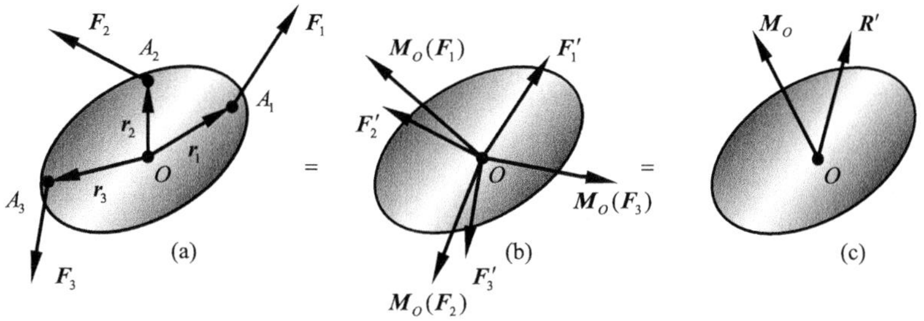

图 2.8

2.7　力-力偶系简化

作用于刚体上的任何力系都可简化为由力 $\boldsymbol{R}'$ 和力偶 $\boldsymbol{M}_O$ 构成的作用于 O 的等效力-力偶系。力系的简化结果分析如下：

(1) 如果 $\boldsymbol{R}'=0$ 和 $\boldsymbol{M}_O=0$，则力系平衡。

(2) 如果 $\boldsymbol{R}'=0$，但 $\boldsymbol{M}_O\neq0$，则力系简化为单个力偶 $\boldsymbol{M}=\boldsymbol{M}_O$，称为合力偶。

(3) 如果 $\boldsymbol{R}'\neq0$，但 $\boldsymbol{M}_O=0$，则力系简化为单个力 $\boldsymbol{R}=\boldsymbol{R}'$，称为合力。如果 $\boldsymbol{R}'\neq0$ 和 $\boldsymbol{M}_O\neq0$，$\boldsymbol{R}'$ 和 $\boldsymbol{M}_O$ 互相垂直，则力系可进一步简化为合力 $\boldsymbol{R}=\boldsymbol{R}'$。对汇交力系、平面力系或平行力系，$\boldsymbol{R}'$ 和 $\boldsymbol{M}_O$ 总是相互垂直。

(4) 如果 $\boldsymbol{R}'\neq0$ 和 $\boldsymbol{M}_O\neq0$，$\boldsymbol{R}'$ 和 $\boldsymbol{M}_O$ 互相平行，则力系不能进一步简化。这类力系称为力螺旋。如果 $\boldsymbol{R}'\neq0$ 和 $\boldsymbol{M}_O\neq0$，$\boldsymbol{R}'$ 和 $\boldsymbol{M}_O$ 既不垂直也不平行，$\boldsymbol{M}_O$ 需要分解为垂直 $\boldsymbol{R}'$ 的矢量分量 $\boldsymbol{M}_{O\perp}$ 和平行 $\boldsymbol{R}'$ 的矢量分量 $\boldsymbol{M}_{O\parallel}$。$\boldsymbol{R}'$ 和 $\boldsymbol{M}_{O\perp}$ 可进一步简化为单个力，然而 $\boldsymbol{R}'$ 和 $\boldsymbol{M}_{O\parallel}$ 只能形成力螺旋。

例 2.5　桁架承受图 E2.5a 所示载荷作用。已知 $F_1=160\ \text{N}$，$F_2=150\ \text{N}$ 和 $F_3=80\ \text{N}$，求(a)等效力以及(b)等效力与直线 AH 的交点。

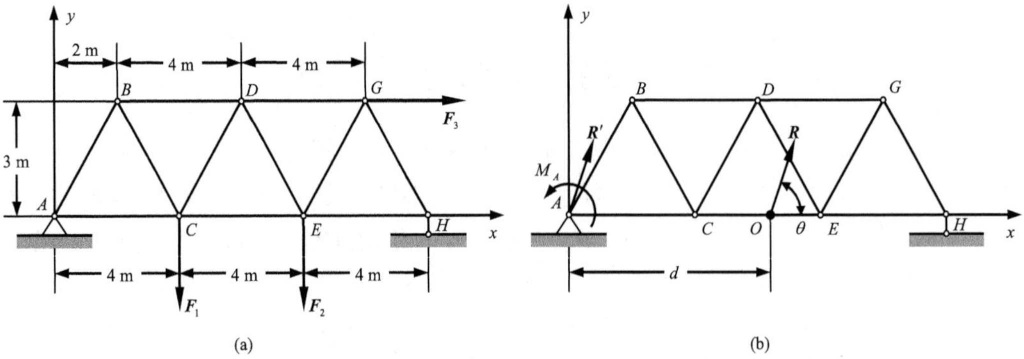

图 E2.5

解　(a)原始力系可以简化为作用于 A 处的等效力–力偶系，如图 E2.5b。设沿 x 和 y 轴的单位矢量分别为 $\boldsymbol{i}$ 和 $\boldsymbol{j}$，利用 $\boldsymbol{R}'=\sum\boldsymbol{F}$ 和 $M_A=\sum M_A(\boldsymbol{F})$，得

$$\boldsymbol{R}'=\boldsymbol{F}_1+\boldsymbol{F}_2+\boldsymbol{F}_3=(80\boldsymbol{i}-310\boldsymbol{j})\ \text{N}\text{，}\quad M_A=M_A(\boldsymbol{F}_1)+M_A(\boldsymbol{F}_2)+M_A(\boldsymbol{F}_3)=-2080\ \text{N}\cdot\text{m}$$

(b)上述作用于 A 处的等效力–力偶系可简化为作用于 O 处的等效力，如图 E2.5b。利用 $\boldsymbol{R}=\boldsymbol{R}'$ 和 $M_O=M_A+M_O(\boldsymbol{R}')=M_A+(-d)R'_y=0$，得

$$\boldsymbol{R}=\boldsymbol{R}'=(80\boldsymbol{i}-310\boldsymbol{j})\ \text{N}\text{，}\quad d=\frac{M_A}{R'_y}=6.71\ \text{m}$$

或

$$R=\sqrt{{R_x}^2+{R_y}^2}=320.16\ \text{N}\text{，}\quad \theta=\arctan\frac{R_y}{R_x}=-75.53^\circ\text{，}\quad d=\frac{M_A}{R'_y}=6.71\ \text{m}$$

习　　题

2.1　平板由两根链条 AG 和 BH 悬挂。已知 AG 中的拉力为 300 N，求：(a)链条 AG 施加到平板上的力对点 B 力矩；(b)如果对点 B 产生相同力矩，则作用于点 D 的最小力。

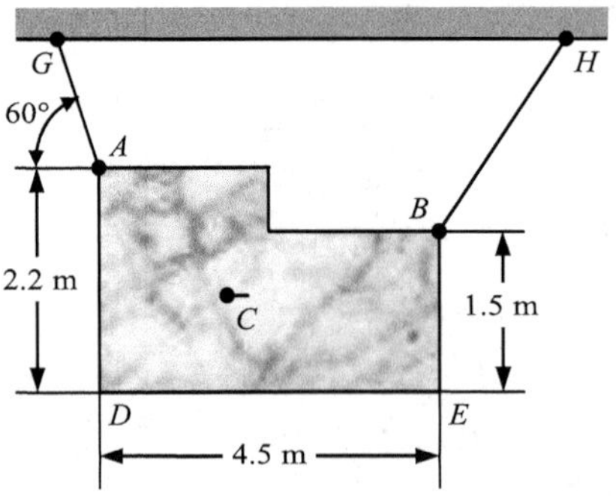

图 P2.1

2.2　缆绳 AB 连在混凝土立柱上。已知缆绳 AB 中的拉力为 500 N，求缆绳作用于立柱上点 A 的力对点 O 的力矩。

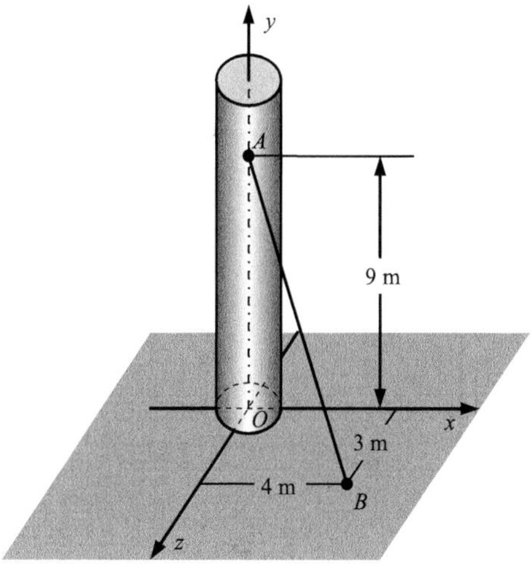

图 P2.2

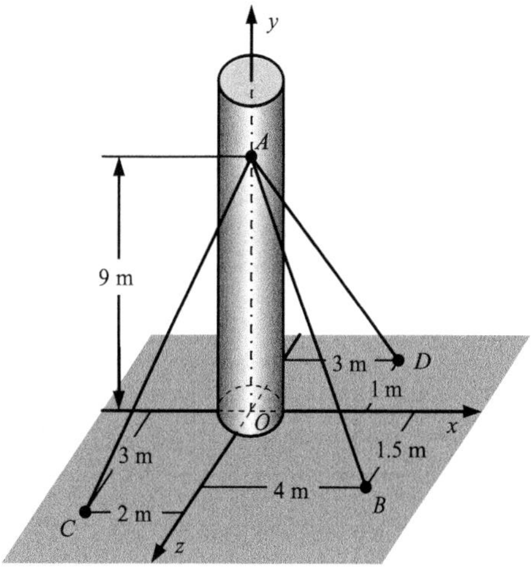

图 P2.3

2.3　缆绳 *AB*、*AC* 和 *AD* 连在混凝土立柱上。已知缆绳 *AB*、*AC* 和 *AD* 中的拉力分别为 800 N、700 N 和 500 N，求缆绳作用于立柱上点 *A* 的合力对点 *O* 的力矩。

2.4　三个力偶作用于物块。已知 $M_1 = 10\ \mathrm{N \cdot m}$，$M_2 = 15\ \mathrm{N \cdot m}$ 和 $M_3 = 8\ \mathrm{N \cdot m}$，试用一个等效力偶代替这三个力偶，并求等效力偶的大小和方向。

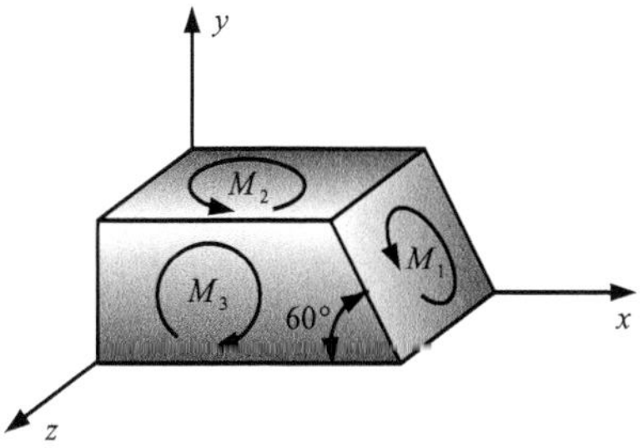

图 P2.4

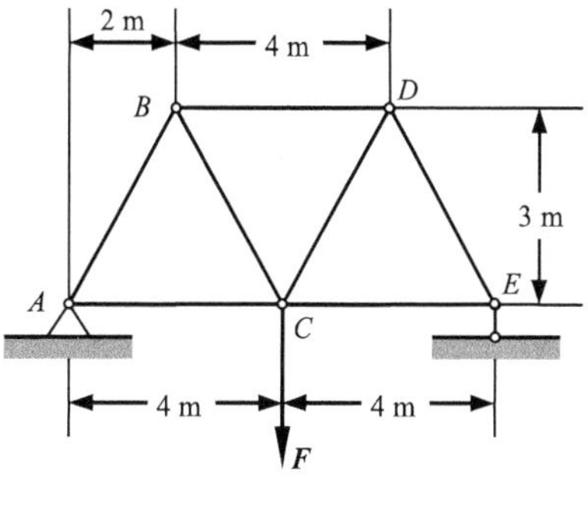

图 P2.5

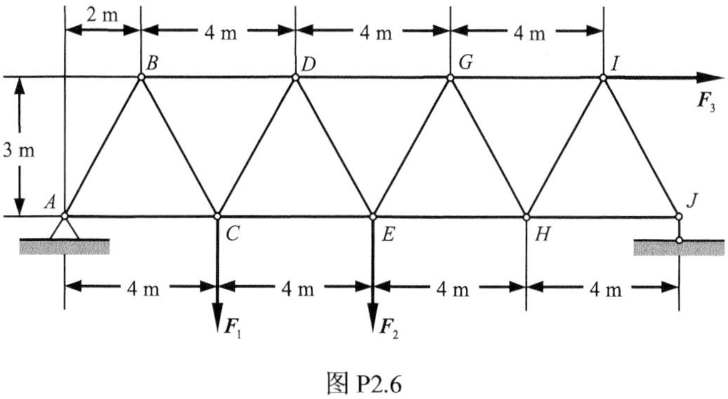

图 P2.6

2.5　垂直力 $\boldsymbol{F}$ 作用于平面桁架的点 C。已知 $F=100\ \text{N}$，试用作用于点 B 的力–力偶系代替力 $\boldsymbol{F}$ 。

2.6　桁架受图示载荷作用。已知 $F_1=F_2=100\ \text{N}$ 和 $F_3=90\ \text{N}$，求等效力以及等效力与直线 AJ 的交点。

2.7　悬臂梁 AB 承受图示分布载荷。已知 $a=2\ \text{m}$ 和 $q=100\ \text{N/m}$，试用 B 处等效力–力偶系代替分布载荷。

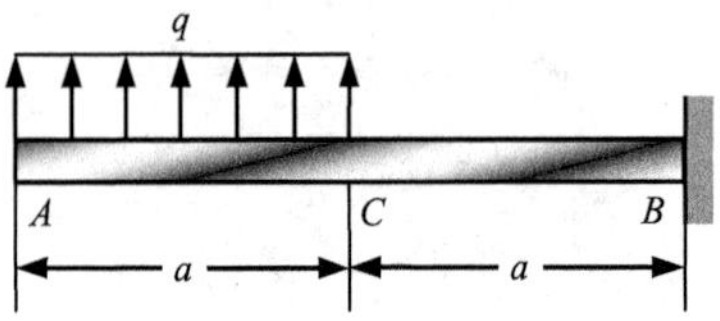

图 P2.7

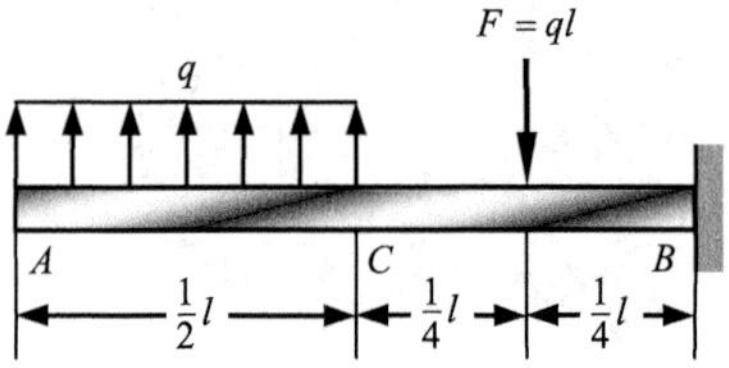

图 P2.8

2.8　悬臂梁 AB 承受图示分布载荷。已知 $l=8\ \text{m}$ 和 $q=100\ \text{N/m}$，试求等效力以及等效力与梁的交点。

第 3 章　刚体静力学

刚体可视为大量质点的集合，其中所有质点在力作用前后都保持固定距离，即刚体在力作用下并不发生变形。

作用于刚体上的力可简化为作用于任意点的力-力偶系。当力和力偶都等于零，则刚体处于平衡状态。求解刚体平衡问题，需要考虑作用于刚体的所有力，并排除没有直接作用于刚体的力。

3.1　平面一般力系平衡

刚体受平面一般力系作用的平衡方程可表示为

$$\sum F_x = 0,\ \sum F_y = 0,\ \sum M_A(\boldsymbol{F}) = 0 \tag{3.1}$$

式中，A 为任意点。上述三个方程相互独立，因此最多可以求解三个未知量。下面两组替代平衡方程也常用于求解平衡问题。

第一组替代平衡方程可写成

$$\sum F_x = 0,\ \sum M_A(\boldsymbol{F}) = 0,\ \sum M_B(\boldsymbol{F}) = 0 \tag{3.2}$$

式中，A 和 B 连线不能垂直于 x 轴。

第二组替代平衡方程可表示为

$$\sum M_A(\boldsymbol{F}) = 0,\ \sum M_B(\boldsymbol{F}) = 0,\ \sum M_C(\boldsymbol{F}) = 0 \tag{3.3}$$

式中，A 、B 和 C 不能共线。

例 3.1　构件 $ABCD$ 由 C 处销钉支撑，并与穿过 E 处定滑轮的缆绳 AED 连接。已知 $F = 150\ \text{N}$，不计摩擦，试求缆绳拉力和 C 处反力。

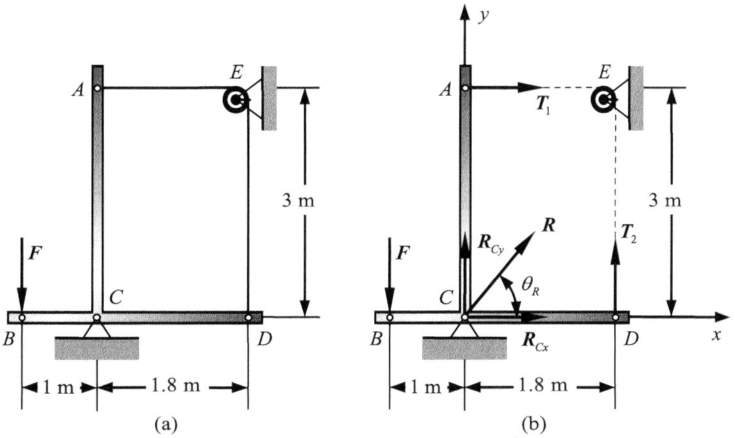

图 E3.1

解　如图 E3.1b，取 T 形构件为自由体，画受力图，得

$$\sum F_x = 0,\ R_{Cx} + T_1 = 0$$

$$\sum F_y = 0,\ R_{Cy} + T_2 - F = 0$$

$$\sum M_C = 0,\ F \times 1 - T_1 \times 3 + T_2 \times 1.8 = 0$$

利用 $F = 150\text{ N}$ 和 $T_1 = T_2$，解上述方程，得

$$R_{Cx} = -125\text{ N},\ R_{Cy} = 25\text{ N},\ T_1 = T_2 = 125\text{ N}$$

因此，缆绳拉力为 125 N，C 处反力为

$$R = \sqrt{R_{Cx}{}^2 + R_{Cy}{}^2} = 127.48\text{ N},\ \theta_R = 180° + \arctan\frac{R_{Cy}}{R_{Cx}} = 168.69°$$

例 3.2　如图 E3.2a 所示，重量为 W 的均质杆 AB 连接到可沿光滑表面自由运动的物块 A 和 B。与物块 A 相连弹簧的刚度为 k，当杆水平时弹簧未伸长。(a) 忽略物块重量，推导当杆平衡时 W、k、a 和 θ 需要满足的方程；(b) 当 $W = 45\text{ N}$、$a = 1\text{ m}$ 和 $k = 50\text{ N/m}$ 时，求 θ 值。

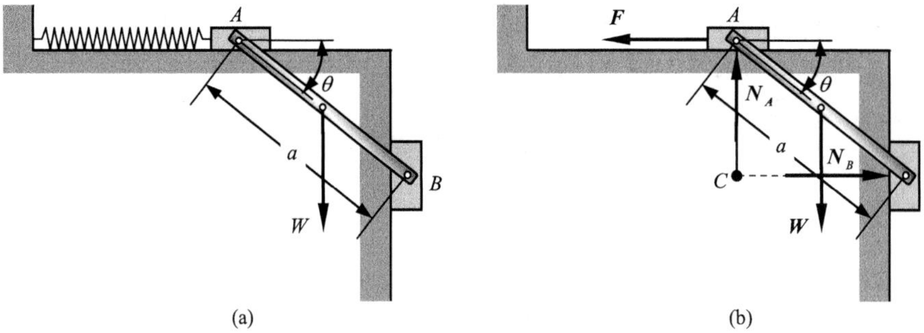

图 E3.2

解　(a) 如图 E3.2b，取杆为自由体，画受力图，得

$$\sum M_C = 0,\ Fa\sin\theta - \frac{1}{2}Wa\cos\theta = 0$$

利用 $F = ka(1 - \cos\theta)$，得

$$\tan\theta - \sin\theta = \frac{W}{2ka}$$

(b) 利用 $W = 45\text{ N}$、$a = 1\text{ m}$ 和 $k = 50\text{ N/m}$，得

$$\tan\theta - \sin\theta = 0.45$$

解上式，得

$$\theta = 50.76°$$

3.2　二力和三力构件

一种特例是二力构件的平衡，构件受二力(或二合力)作用，二力分别作用于二不同点。可以证明，如果二力构件平衡，那么二力必须大小相同、方向相反和作用线相同。

另一特例是三力构件的平衡，构件受三力(或三合力)作用，三力分别作用于三不同点。可以证明，如果三力构件平衡，那么三力作用线要么汇交，要么平行。

例 3.3　图 E3.3a 所示刚架与载荷，已知 $F=100\ \text{N}$ 和 $a=1\ \text{m}$，试求 A 和 B 处反力。

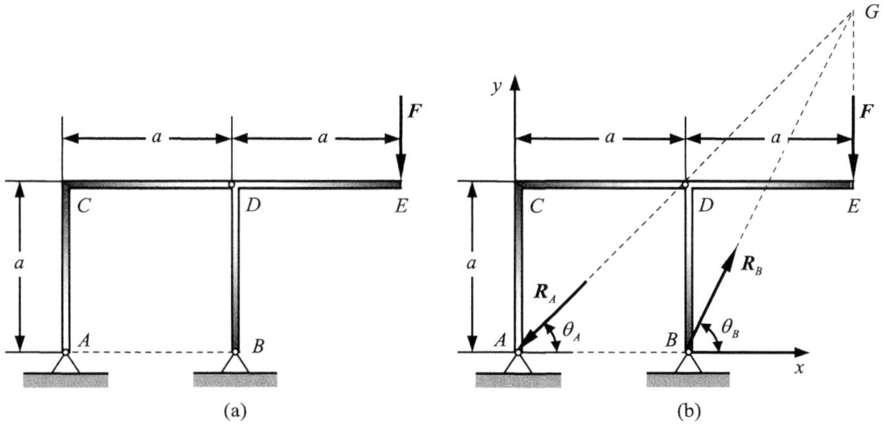

图 E3.3

解　如图 E3.3b，因 ACD 是平衡二力构件，故 A 处反力 $\boldsymbol{R}_A$ 作用线必过 A 和 D。因整个结构在 $\boldsymbol{R}_A$、$\boldsymbol{R}_B$ 和 $\boldsymbol{F}$ 作用下处于平衡状态，故 $\boldsymbol{R}_B$ 作用线必过 $\boldsymbol{R}_A$ 和 $\boldsymbol{F}$ 作用线交点。

考虑整个结构平衡，得

$$\sum F_x=0,\ \ -R_A\cos\theta_A+R_B\cos\theta_B=0$$

$$\sum F_y=0,\ \ -R_A\sin\theta_A+R_B\sin\theta_B-F=0$$

式中，$\tan\theta_A=1$、$\tan\theta_B=2$。代入 $F=100\ \text{N}$，得

$$R_A=141.4\ \text{N},\ \theta_A=45°,\ R_B=223.6\ \text{N},\ \theta_B=63.43°$$

3.3　平 面 桁 架

平面桁架是由直杆通过铰接而成的二维结构。在平面桁架中，虽然杆件实际上是通过螺栓或焊接进行连接，但是通常都假定杆件是通过光滑销钉进行连接。平面桁架是用于承受作用于结构平面内的载荷，然而常常假设所有载荷都是作用于平面桁架的节点，并且每根杆件的重量也平分到节点。

1. 简单桁架

桁架都是设计成能承受载荷，因此桁架在载荷作用下必须保持稳定。图 3.1 所示桁

架由三根杆件通过三个节点连接而成，在载荷作用下是稳定的。

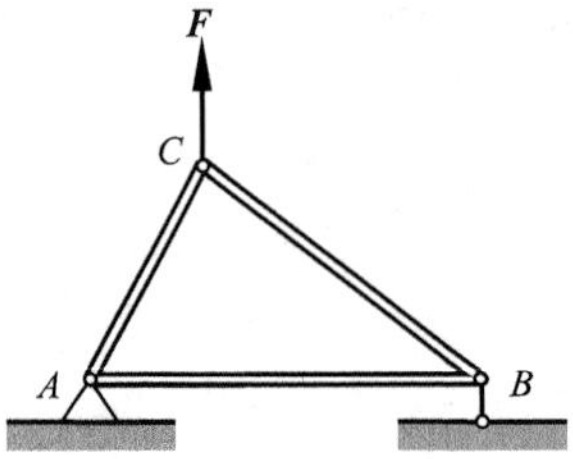

图 3.1

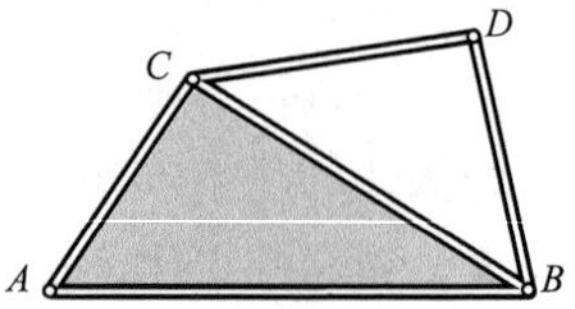

图 3.2

如图 3.2 所示，在最基本的三角形桁架 *ABC* 上添加杆件 *BD* 和 *CD* 即可得到较大桁架。这可进行多次重复，最终的桁架是稳定的，只要每次增加两根杆件和一个节点。由一个三角形桁架通过上述方法构成的桁架称为简单桁架。

假设简单桁架的杆件总数和节点总数分别为 m 和 n，则有

$$m = 2n - 3 \tag{3.4}$$

2. 桁架内力

桁架分析不仅需要确定外力，而且还需要确定内力。确定桁架内力的方法主要有两种：即节点法和截面法。

3. 节点法

因为整个桁架平衡，因此桁架的每个节点也必须平衡。通过画每个节点的受力图，并解每个节点的平衡方程，则可确定桁架内力。

例 3.4　已知 $F = 80$ N，采用节点法求图 E3.4a 所示桁架每根杆件的内力，并说明每根杆件是受拉还是受压。

解　(1) 如图 E3.4b，取整个桁架为自由体，得

$$\sum F_x = 0,\ R_{Ax} = 0$$

$$\sum M_A = 0,\ R_B \times 6 - F \times 3 = 0$$

$$\sum M_B = 0,\ -R_{Ay} \times 6 + F \times 3 = 0$$

解上述方程，得

$$R_{Ax} = 0\,,\ R_{Ay} = 40\ \text{N}\,,\ R_B = 40\ \text{N}$$

(2) 取节点 *A* 为自由体，如图 E3.4c，有

$$\sum F_x = 0,\ R_{Ax} + T_{AC} + T_{AD} \times \frac{3}{5} = 0$$

$$\sum F_y = 0,\ R_{Ay} + T_{AD} \times \frac{4}{5} = 0$$

求解，得

$$T_{AC} = 30\ \text{N (拉)},\ T_{AD} = -50\ \text{N (压)}$$

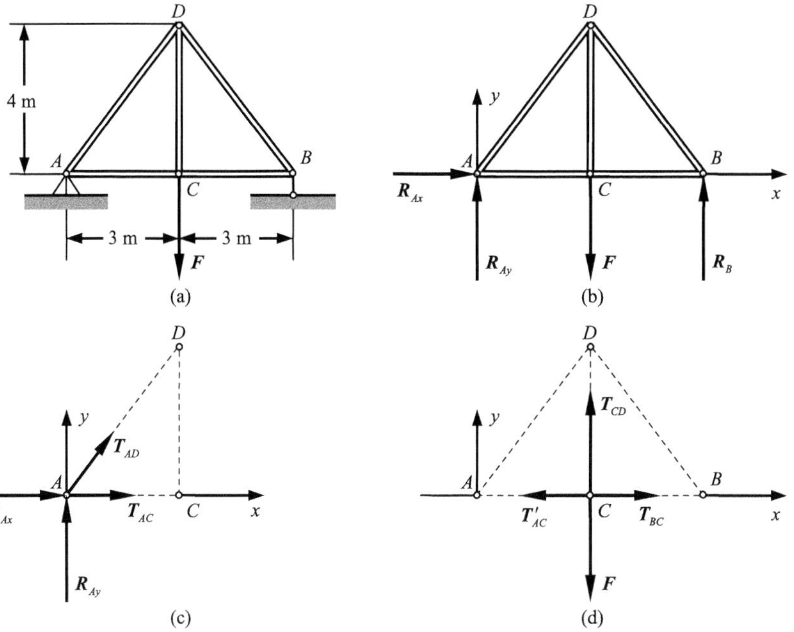

图 E3.4

(3)取节点 C 为自由体，如图 E3.4d，有

$$\sum F_x = 0,\ T_{BC} - T'_{AC} = 0$$

$$\sum F_y = 0,\ T_{CD} - F = 0$$

求解，得

$$T_{BC} = 30\ \text{N}\ (\text{拉}),\ T_{CD} = 80\ \text{N}\ (\text{拉})$$

(4)同理，取节点 B 或 D 为自由体，得

$$T_{BD} = -50\ \text{N}\ (\text{压})$$

4. 零力杆

如果杆件中的内力等于零，则该杆为零力杆。零力杆一方面用于增加桁架稳定性，另一方面当作用于桁架的载荷发生变化时用于支撑载荷。

例 3.5　对图 E3.5a 所示桁架和载荷，求零力杆。

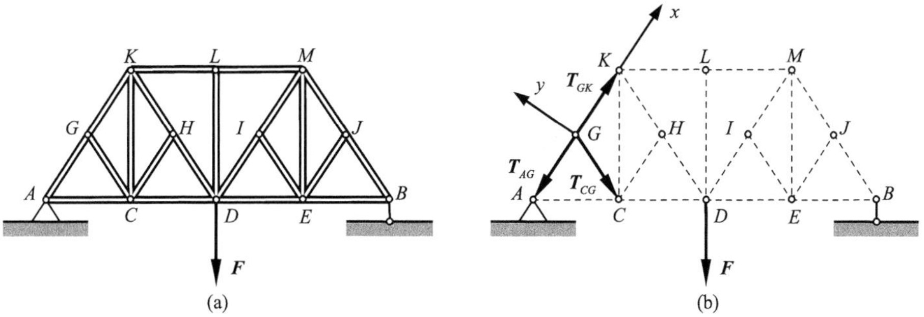

图 E3.5

解　考虑图 E3.5b 所示节点 G，利用 $\sum F_y = 0$，得

$$T_{CG} = 0$$

同理，通过检查每个节点，得

$$T_{CH} = T_{CK} = T_{DL} = T_{EI} = T_{EJ} = T_{EM} = 0$$

5. 截面法

若需求桁架中所有杆件内力，则节点法是最有效的方法。然而，如果仅需求很少几根杆件内力，那么截面法将更为有效。

因为整个桁架平衡，因此桁架的任何部分也必须平衡。桁架任何部分的内力都可通过取该部分为自由体并解平衡方程而得到。

例 3.6　桁架加载如图 E3.6a 所示。已知 $F_1 = 80\ \text{N}$ 和 $F_2 = 40\ \text{N}$，试用截面法求杆件 CD、CH 和 GH 的内力。

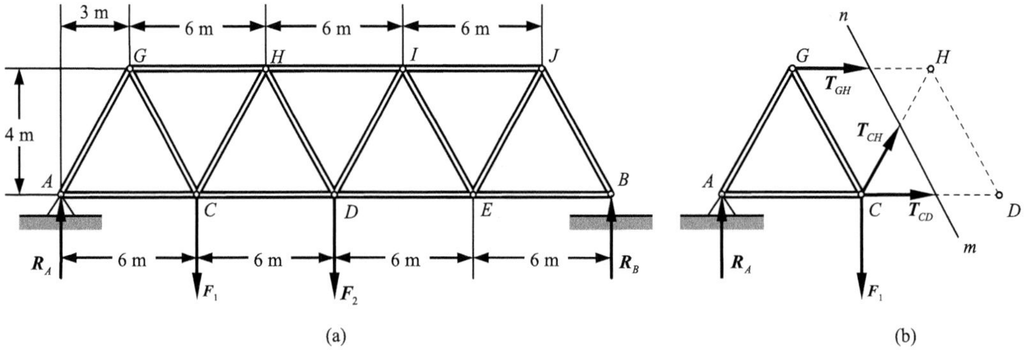

图 E3.6

解　取整个桁架为自由体，得

$$R_A = 80\ \text{N}, R_B = 40\ \text{N}$$

为了求杆件 CD、CH 和 GH 的内力，这些杆件应该被截开，如图 E3.6b。考虑桁架截开后左部的平衡，得

$$\sum M_H = 0,\ -R_A \times 9 + F_1 \times 3 + T_{CD} \times 4 = 0$$

$$\sum M_C = 0,\ -R_A \times 6 - T_{GH} \times 4 = 0$$

$$\sum F_y = 0,\ R_A - F_1 + T_{CH} \times \frac{4}{5} = 0$$

解上述方程，得

$$T_{CD} = 120\ \text{N},\ T_{CH} = 0,\ T_{GH} = -120\ \text{N}$$

3.4　空间一般力系平衡

空间一般力系的平衡方程为

$$\sum F_x = 0,\ \sum F_y = 0,\ \sum F_z = 0,\ \sum M_x(\boldsymbol{F}) = 0,\ \sum M_y(\boldsymbol{F}) = 0,\ \sum M_z(\boldsymbol{F}) = 0 \tag{3.5}$$

上述六个方程相互独立，因此最多可以求解六个未知量。

例 3.7　重量为 150 N 的均质方板由三根垂直缆绳支撑，如图 E3.7a 所示。试求每根缆绳的拉力。

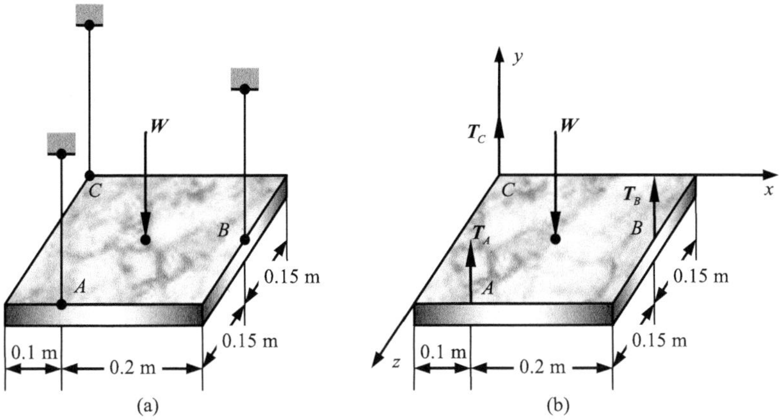

图 E3.7

解　取板为自由体，如图 E3.7b。根据板的平衡，得

$$\sum F_y = 0,\ T_A + T_B + T_C - W = 0$$
$$\sum M_x = 0,\ -T_A \times 0.3 - T_B \times 0.15 + W \times 0.15 = 0$$
$$\sum M_z = 0,\ T_A \times 0.1 + T_B \times 0.3 - W \times 0.15 = 0$$

求解，得

$$T_A = 45\ \text{N},\ T_B = 60\ \text{N},\ T_C = 45\ \text{N}$$

习　　题

3.1　T 形构件 $ABCD$ 由 C 处销钉支撑，并与穿过 E 处定滑轮的缆绳 AED 连接。已知 $q = 250\ \text{N/m}$，不计摩擦，试求缆绳拉力和 C 处反力。

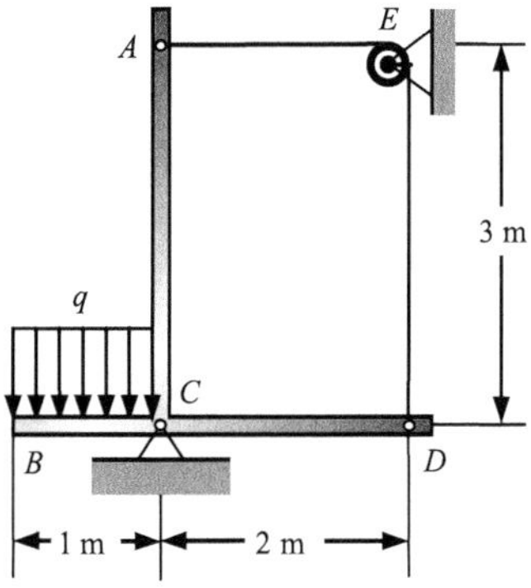

图 P3.1

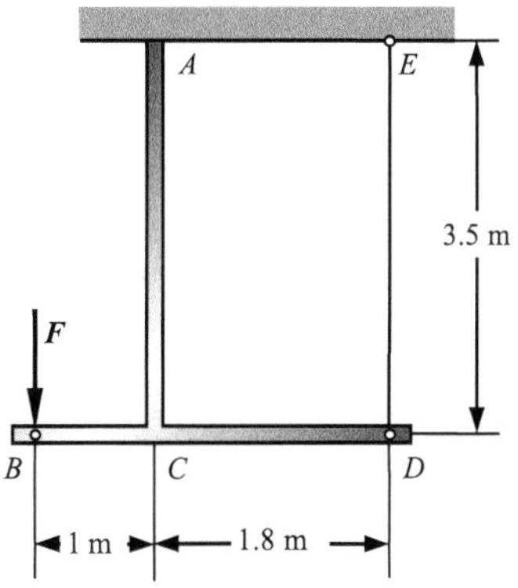

图 P3.2

3.2 构件 $ABCD$ 在 A 处固定，并由缆绳 DE 连接。已知 $F=250$ N 和缆绳拉力 $T=50$ N，试求 A 处反力。

3.3 重量为 W 的均质杆 AB 连接到可沿光滑表面自由运动的物块 A 和 B 上。与物块 A 相连弹簧的刚度为 k，当杆水平时弹簧未伸长。(a)假设每个物块的重量均为 W，推导当杆平衡时 W、k、a 和 θ 需要满足的方程；(b)当 $W=15$ N 、$a=1$ m 和 $k=50$ N/m 时，求 θ 值。

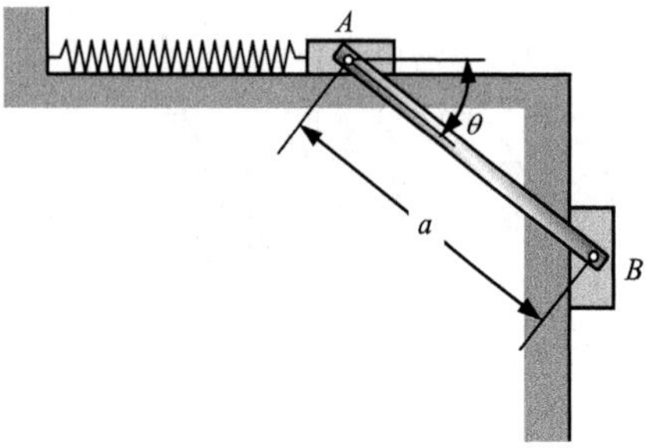

图 P3.3

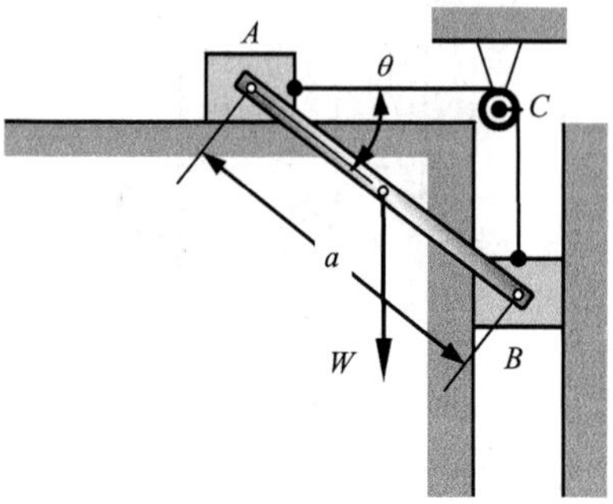

图 P3.4

3.4 重量为 W 的均质杆 AB 连接到可沿光滑表面自由运动的物块 A 和 B 上。物块由穿过定滑轮 C 的缆绳连接。(a)不计物块重量，试用 W 和 θ 表示当杆平衡时的缆绳拉力；(b)当缆绳拉力等于 W 时，求 θ 值。

3.5 A 和 B 处反力的最大允许值为 450 N。不计梁重，求在保证梁安全的情况下距离 d 的取值范围。

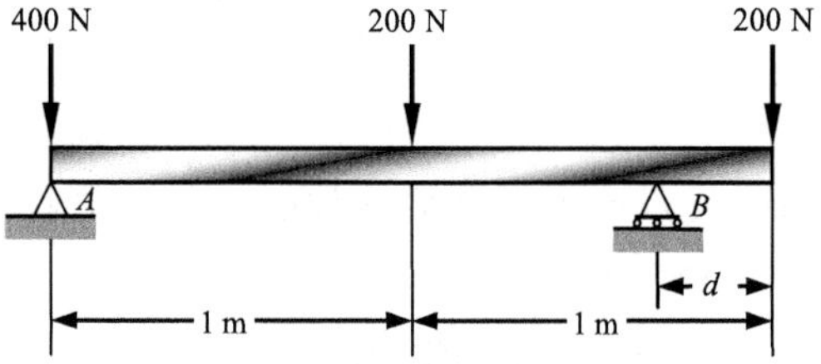

图 P3.5

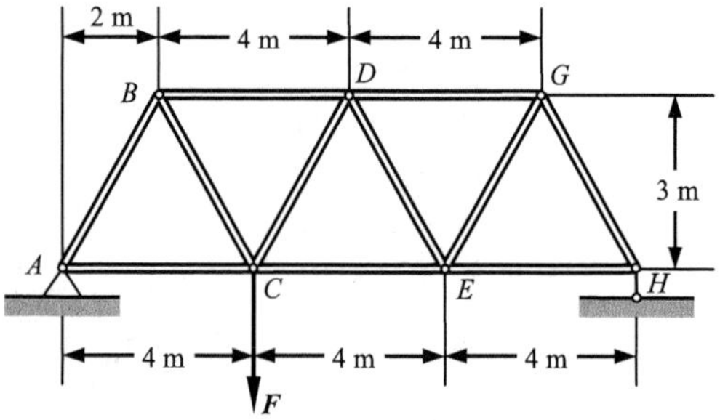

图 P3.6

3.6 垂直力 $\boldsymbol{F}$ 作用于平面桁架的点 C。已知 $F=80$ N，求 A 和 H 处反力。

3.7 桁架承受图示载荷。已知 $F_1=160$ N 、 $F_2=150$ N 和 $F_3=80$ N，求 A 和 H 处反力。

3.8 悬臂梁 AB 支撑分布载荷。已知 $a=2$ m 和 $q=100$ N/m，求 B 处反力。

3.9 悬臂梁 AB 支撑图示载荷。已知 $l=8$ m 和 $q=80$ N/m，求 B 处反力。

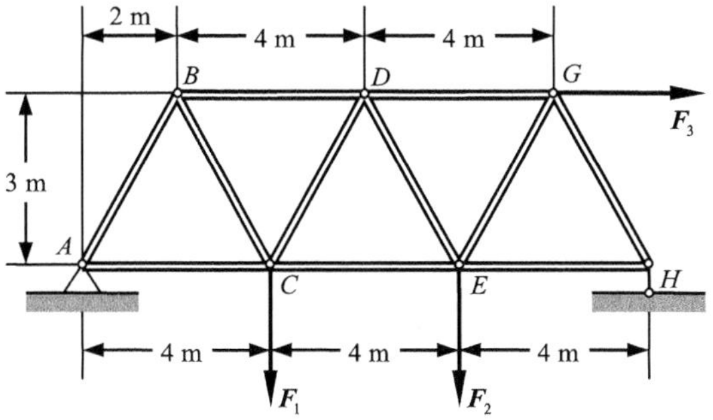

图 P3.7

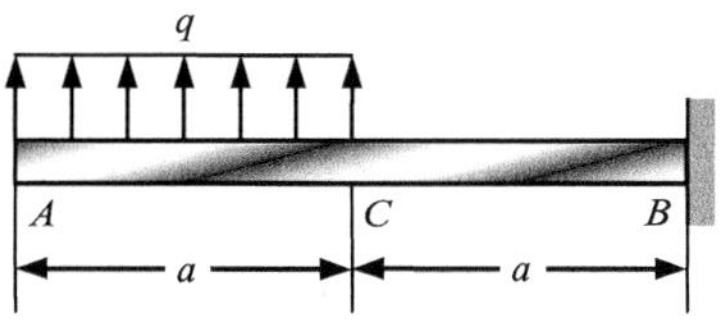

图 P3.8

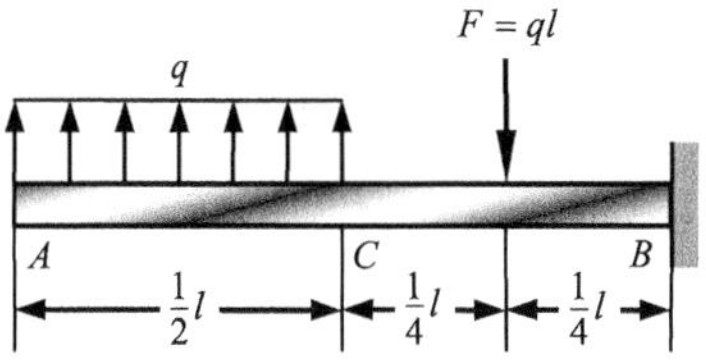

图 P3.9

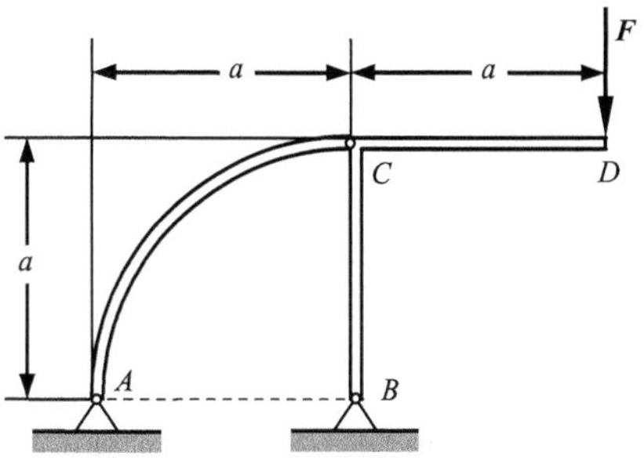

图 P3.10

3.10　刚架和载荷如图，已知 $F=200\ \text{N}$ 和 $a=1\ \text{m}$，求 A 和 B 处反力。

3.11　刚架和载荷如图，已知 $q=100\ \text{N/m}$ 和 $a=1\ \text{m}$，求 A 和 B 处反力。

3.12　已知 $M_e=100\ \text{N}\cdot\text{m}$ 和 $a=1\ \text{m}$，求 A 和 B 处反力。

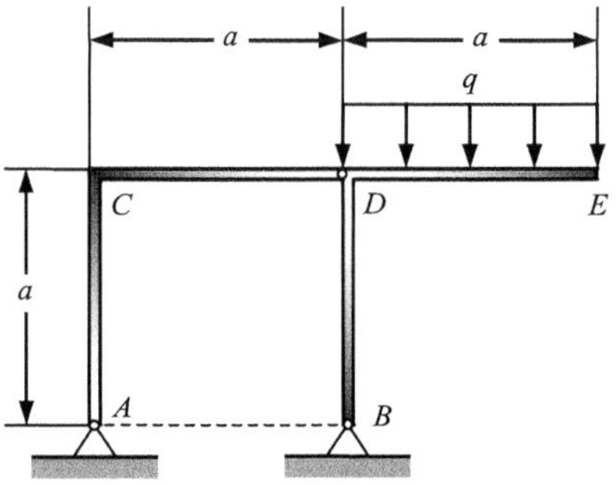

图 P3.11

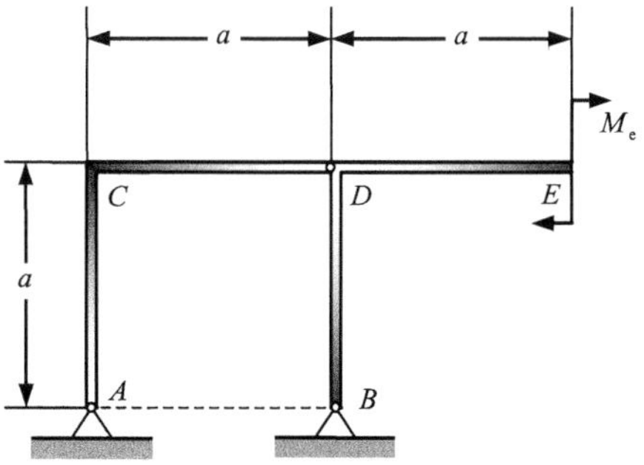

图 P3.12

3.13　用节点法求图示桁架每根杆件的内力，并说明每根杆件是受拉还是受压。

3.14　对图示桁架和载荷，求零力杆。

3.15　桁架加载如图。已知 $F_1=90\ \text{N}$ 和 $F_2=60\ \text{N}$，试用截面法求杆件 CD、DG 和 GH 的内力。

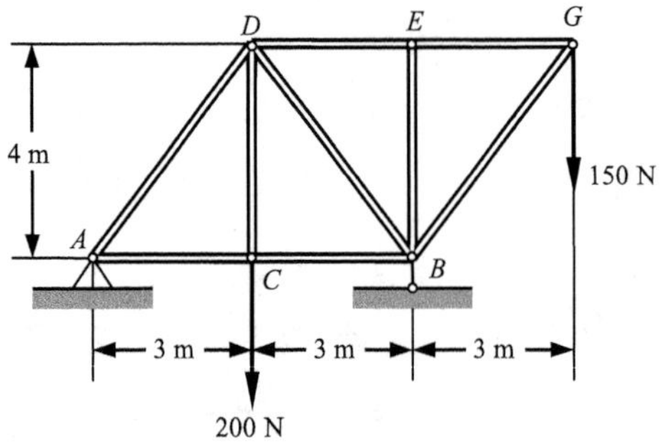

图 P3.13

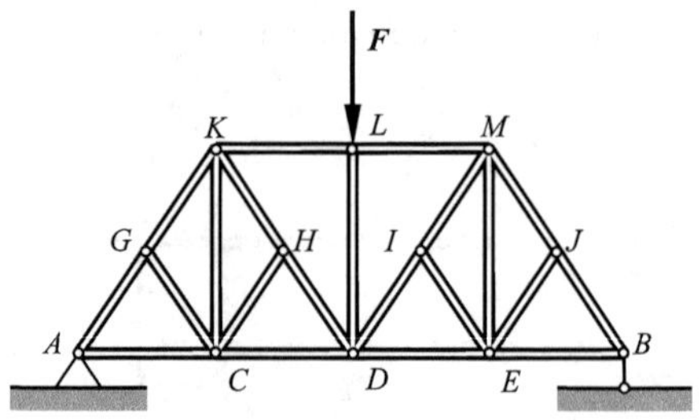

图 P3.14

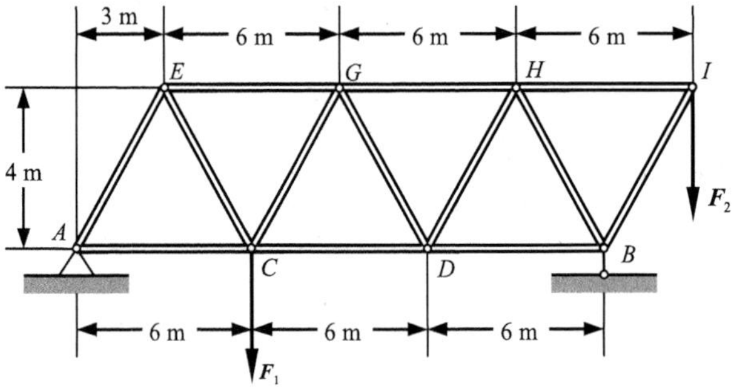

图 P3.15

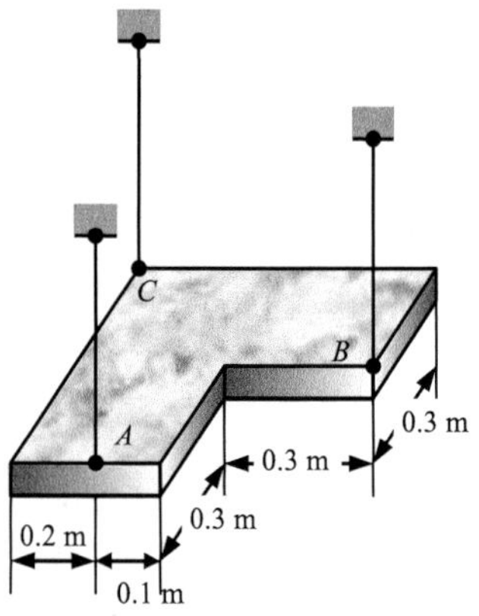

图 P3.16

3.16　重量为 250 N 的均质组合板由三根垂直缆绳支撑，试求每根缆绳的拉力。

第4章 摩　擦

4.1 摩擦定律

摩擦现象解释如下：

(1) 如图 4.1a 所示，重量为 $\boldsymbol{W}$ 的物块放在粗糙水平面上。作用于物块上的力是物块重量 $\boldsymbol{W}$ 和由粗糙面施加的法向力 $\boldsymbol{N}$ 。这些垂直力没有沿粗糙面移动物块的趋势。

(2) 如图 4.1b 所示，水平力 $\boldsymbol{F}$ 作用于物块，该力有沿粗糙面移动物块的趋势。然而，当 $\boldsymbol{F}$ 不大时，物块并不运动。这就表明，必定存在由粗糙面施加到物块的切向力与 $\boldsymbol{F}$ 平衡。该切向力称为静摩擦力 $\boldsymbol{F}_{\mathrm{s}}$ ，可通过求解物块平衡方程而得到。

(3) 如图 4.1c 所示，如果 $\boldsymbol{F}$ 增大，则 $\boldsymbol{F}_{\mathrm{s}}$ 也相应增大，继续与 $\boldsymbol{F}$ 平衡，直到 $\boldsymbol{F}_{\mathrm{s}}$ 达到最大静摩擦力 $\boldsymbol{F}_{\max}$ 。当达到 $\boldsymbol{F}_{\max}$ ，则物块处于滑动临界状态。

(4) 如图 4.1d 所示，如果 $\boldsymbol{F}$ 进一步增大， $\boldsymbol{F}_{\mathrm{s}}$ 不再与 $\boldsymbol{F}$ 平衡，则物块开始滑动。物块一旦滑动，摩擦力就从 $\boldsymbol{F}_{\max}$ 降为动摩擦力 $\boldsymbol{F}_{\mathrm{k}}$ 。物块运动后， $\boldsymbol{F}_{\mathrm{k}}$ 近似保持常数。

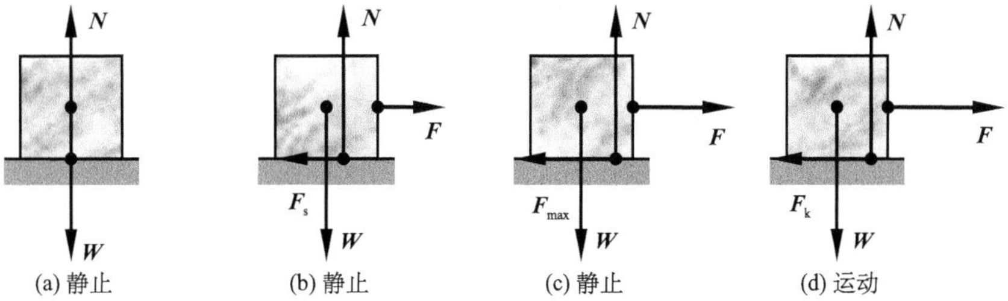

图 4.1

实验已经表明，最大静摩擦力 $F_{\max}$ 与粗糙面施加于物块的法向力 N 成正比。因此，有

$$F_{\max} = \mu_{\mathrm{s}} N \tag{4.1}$$

式中， μ_{s} 为静摩擦系数。上述关系称为静摩擦定律。

同样，动摩擦力 F_{k} 也与作用于物块的法向力 N 成正比。因此，动摩擦定律可表示为

$$F_{\mathrm{k}} = \mu_{\mathrm{k}} N \tag{4.2}$$

式中， μ_{k} 为动摩擦系数。

摩擦系数 μ_{s} 和 μ_{k} 与接触面积无关，但与接触物体的材料特性和表面性质有关。

4.2 静摩擦角

重为 $\boldsymbol{W}$ 的物块放在粗糙水平面上，受水平力 $\boldsymbol{F}$ 作用，如图 4.2 所示。当物块处于滑

动临界状态，由法向力与最大静摩擦力形成的合力由 $\boldsymbol{R}_s$ 表示，$\boldsymbol{N}$ 和 $\boldsymbol{R}_s$ 之间的夹角由 φ_s 表示，称为静摩擦角。根据图 4.2，有

$$\tan\varphi_s = \frac{F_{max}}{N} = \mu_s \tag{4.3}$$

如图 4.3a 所示，如果 $\varphi > \varphi_s$，那么物块会由于 $\boldsymbol{R}$ 和 $\boldsymbol{R}_s$ 不可能共线而运动。然而，如果 $\varphi = \varphi_s$（图 4.3b），或 $\varphi < \varphi_s$（图 4.3c），那么物块静止，即物块自锁。

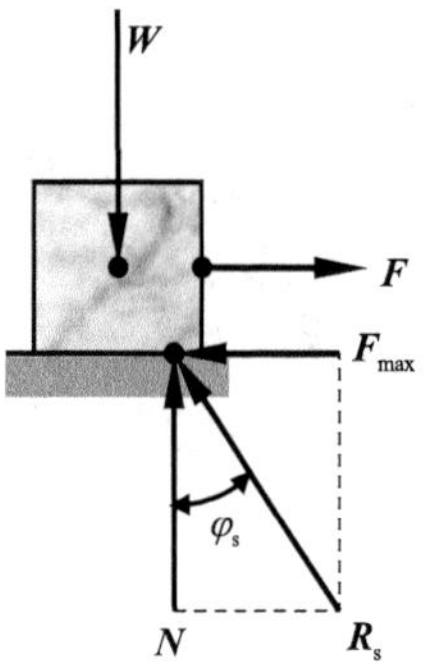

图 4.2

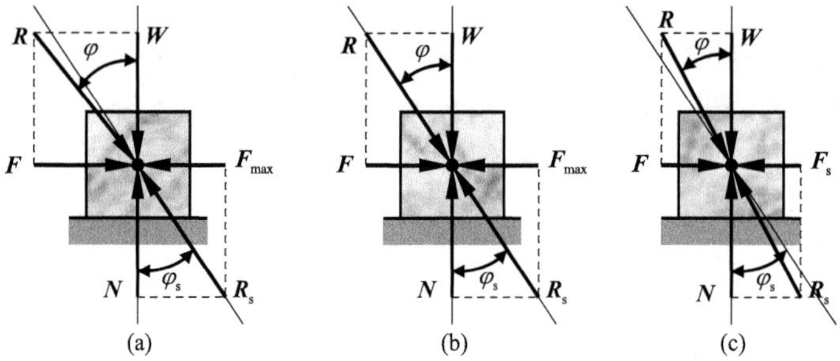

图 4.3

4.3 摩 擦 问 题

存在三类摩擦问题，说明如下：

(1) 所有作用力给定，且静摩擦系数已知，确定物体是静止还是运动。

(2) 所有作用力给定，且物体运动处于临界状态，确定静摩擦系数。

(3) 静摩擦系数已知，且物体运动处于临界状态，确定作用力。

例 4.1　$F = 200$ N 的力作用于质量 $m = 100$ kg 的物块上，物块放在倾角 $\theta = 30°$ 的斜面上，如图 E4.1a 所示。已知 $\mu_s = 0.3$ 和 $\mu_k = 0.2$，确定物块是否平衡，并求摩擦力的大小和方向。

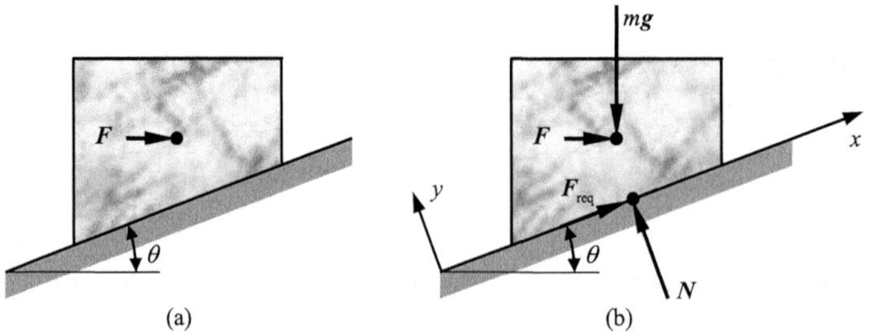

图 E4.1

解　画物块受力图，如图 E4.1b。假设物块平衡和所需摩擦力为 $\boldsymbol{F}_{req}$（方向沿斜面向上），则物块平衡方程为

$$\sum F_x = 0,\ F_{req} + F\cos\theta - mg\sin\theta = 0$$
$$\sum F_y = 0,\ N - F\sin\theta - mg\cos\theta = 0$$

解方程，得

$$F_{req} = 317.29\ \text{N}$$
$$N = 949.57\ \text{N}$$

最大静摩擦力等于

$$F_{max} = \mu_s N = 284.87\ \text{N}$$

因 $F_{req} > F_{max}$，则物块不能平衡，即物块将沿斜面下滑。根据动摩擦定律，摩擦力大小为

$$F_k = \mu_k N = 189.91\ \text{N}$$

方向沿斜面向上。

例 4.2 如图 E4.2a，长度为l、质量为m的均质梯子 AB 靠在墙上。假设 A 和 B 处具有相同静摩擦系数 μ_s，求梯子在$\theta = 60°$保持平衡时 μ_s 的最小值。

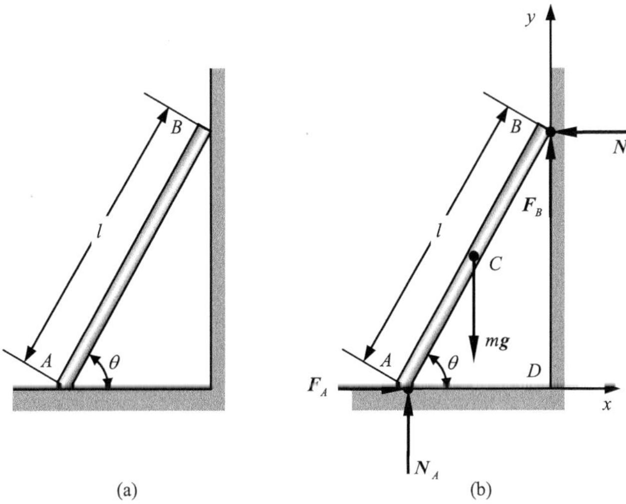

图 E4.2

解 假设 A 和 B 处都处于临界运动状态，梯子受力图如图 E4.2b 所示。利用静摩擦定律，补充方程为

$$F_A = \mu_s N_A,\ F_B = \mu_s N_B$$

考虑梯子平衡，平衡方程可写为

$$\sum F_x = 0,\ F_A - N_B = 0$$
$$\sum F_y = 0,\ F_B + N_A - mg = 0$$
$$\sum M_D = 0,\ N_B l\sin\theta - N_A l\cos\theta + \tfrac{1}{2}mgl\cos\theta = 0$$

求解补充和平衡方程，得

$$\mu_s^2 + 2\mu_s \tan\theta - 1 = 0$$

利用$\theta = 60°$并解方程，得

$$\mu_s = 0.27 \text{ 或 } \mu_s = -3.73$$

物理上，正根才是合理的，因此μ_s的最小值等于0.27。

例 4.3　如图 E4.3a，重量为 W 的环 B 连到弹簧 AB 并可沿杆运动。弹簧刚度$k = 1.8\ \text{kN/m}$，且当$\theta = 0$时弹簧未伸长。已知环与杆之间的静摩擦系数$\mu_s = 0.25$，点 A 和 B 之间的水平距离$l = 0.6\ \text{m}$，求当$\theta = 30°$环维持平衡时 W 值的范围。

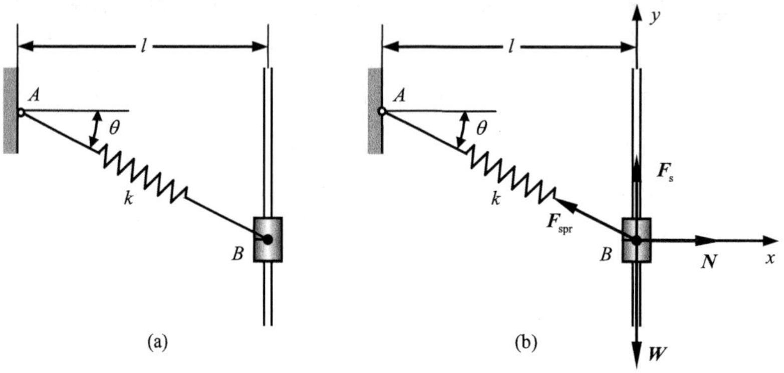

图 E4.3

解　如图 E4.3b，假设环平衡，并且静摩擦力 $\boldsymbol{F}_s$ 向上，则根据环受力图，得环平衡方程为

$$\sum F_x = 0,\ N - F_{spr}\cos\theta = 0$$

$$\sum F_y = 0,\ F_s + F_{spr}\sin\theta - W = 0$$

解方程，得

$$N = F_{spr}\cos\theta$$

$$F_s = W - F_{spr}\sin\theta$$

当环平衡，则

$$|F_s| \leqslant \mu_s N$$

即

$$-\mu_s F_{spr}\cos\theta \leqslant W - F_{spr}\sin\theta \leqslant \mu_s F_{spr}\cos\theta$$

或

$$F_{spr}(\sin\theta - \mu_s\cos\theta) \leqslant W \leqslant F_{spr}(\sin\theta + \mu_s\cos\theta)$$

利用$F_{spr} = kl(1/\cos\theta - 1)$和$\theta = 30°$，则得

$$47.37\ \text{N} \leqslant W \leqslant 119.71\ \text{N}$$

习 题

4.1 力 $\boldsymbol{F}$ 作用于质量 $m=100\ \text{kg}$ 的物块上，物块放在倾角 $\theta=30°$ 的斜面上。已知物块与平面之间的摩擦系数 $\mu_s=0.3$ 和 $\mu_k=0.2$，确定物块是否平衡，并求当(a) $F=500\ \text{N}$ 和(b) $F=1200\ \text{N}$ 时的摩擦力的大小和方向。

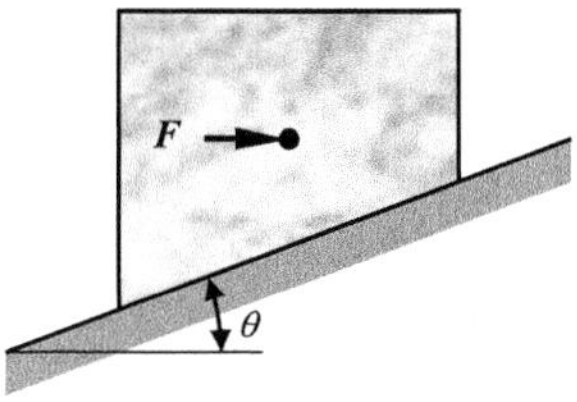

图 P4.1

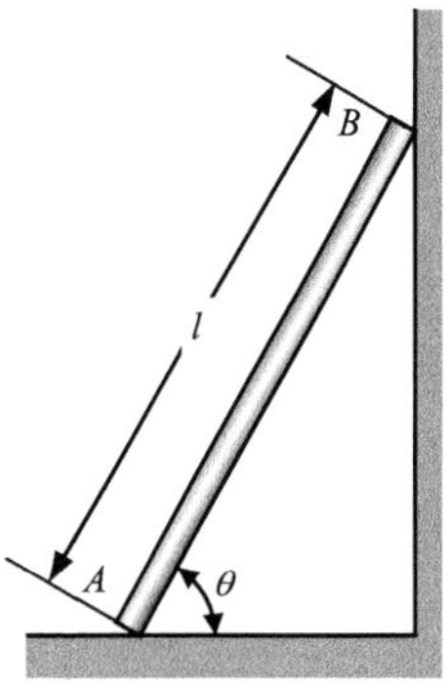

图 P4.2

4.2 长度为 l、质量为 m 的均质梯子 AB 靠在墙上。假设 A 处静摩擦系数为 μ_s，B 处静摩擦系数为零，求梯子在 $\theta=60°$ 保持平衡时 μ_s 的最小值。

4.3 重量为 W 的环 B 与弹簧 AB 连接后沿杆运动。弹簧刚度 $k=1.8\ \text{kN/m}$，当 $\theta=0$ 时弹簧未伸长。已知环与杆之间的静摩擦系数 $\mu_s=0.25$，点 A、B 之间的水平距离 $l=0.6\ \text{m}$，求当 $\theta=30°$ 环保持平衡时 W 值的范围。

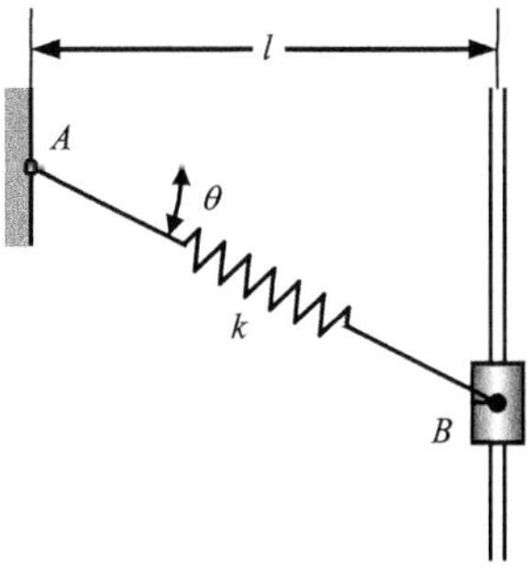

图 P4.3

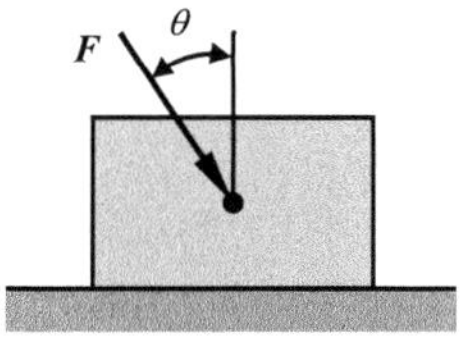

图 P4.4

4.4 $F=100\ \text{N}$ 的力作用于质量为 m 的物块上，物块放在水平面上。已知物块与平面之间的摩擦系数 $\mu_s=0.3$ 和 $\mu_k=0.2$，仅考虑 θ 小于或等于 90°，求当(a) $m=15\ \text{kg}$ 和(b) $m=30\ \text{kg}$ 物块处于向右滑动临界状态时的最小 θ 值。

4.5 $F=150\ \text{N}$ 的力作用于质量 $m=10\ \text{kg}$ 的物块上。已知物块与平面之间的摩擦系数 $\mu_s=0.3$ 和 $\mu_k=0.2$，求物块保持平衡时 θ 值的范围。

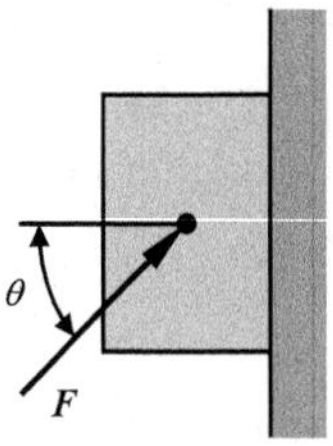

图 P4.5

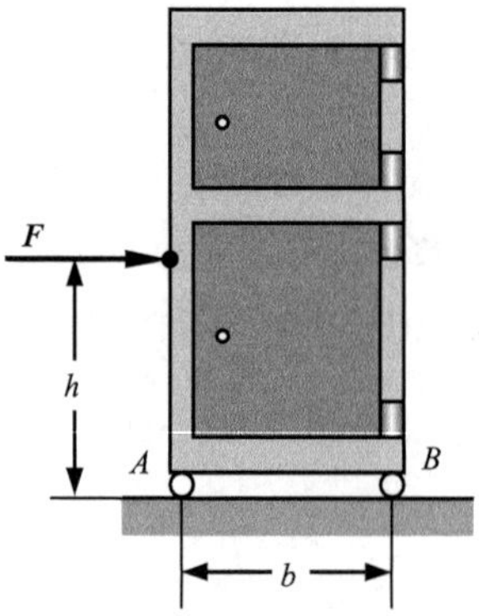

图 P4.6

4.6　质量 $m=50$ kg 的橱柜安装在小脚轮上，小脚轮能够锁定以阻止其转动。地板与小脚轮之间的静摩擦系数 $\mu_s=0.3$。已知 $b=500$ mm 和 $h=650$ mm，求在下列三种情形下橱柜具有向右临界运动时力 $\boldsymbol{F}$ 的大小：(a)所有小脚轮都锁定；(b)B 处小脚轮锁定，A 处小脚轮可以自由转动；(c)A 处小脚轮锁定，B 处小脚轮可以自由转动。

第 5 章　质点运动学

沿直线运动的质点称为在做直线运动，而沿曲线运动的质点称为在做曲线运动。

5.1　矢 量 表 示

如图 5.1 所示，考虑质点 P 沿曲线运动，质点在时间 t 的位置可由连接 O 和 P 的位置矢量 $\boldsymbol{r}$ 表示，其中 O 为在空间选择的参考点。当质点运动时，$\boldsymbol{r}$ 为 t 的函数，即

$$\boldsymbol{r}=\boldsymbol{r}(t) \tag{5.1}$$

如图 5.2 所示，假设矢量 $\boldsymbol{r}'$ 定义质点在时间 $t+\Delta t$ 的位置，那么 $\Delta\boldsymbol{r}=\boldsymbol{r}'-\boldsymbol{r}$ 表示 $\boldsymbol{r}$ 在时间间隔 Δt 内的变化。因此，质点在 t 的速度可表示为

$$\boldsymbol{v}=\lim_{\Delta t\to 0}\frac{\Delta\boldsymbol{r}}{\Delta t}=\dot{\boldsymbol{r}} \tag{5.2}$$

速度是矢量，其方向始终与运动轨迹相切，如图 5.3 所示。

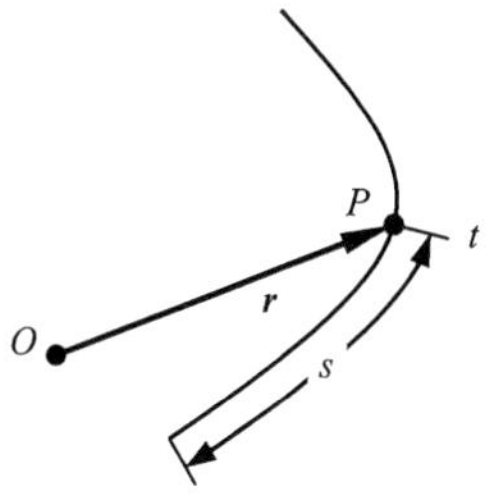

图 5.1

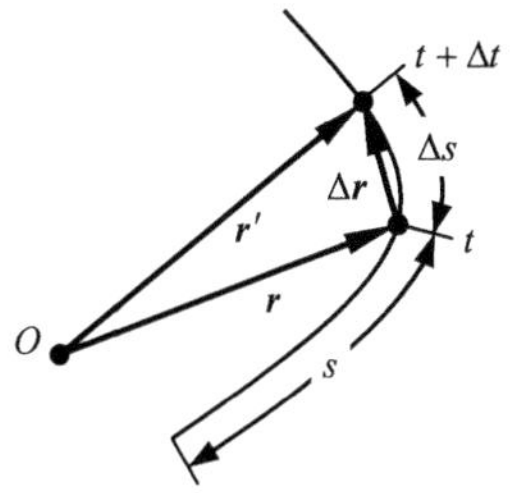

图 5.2

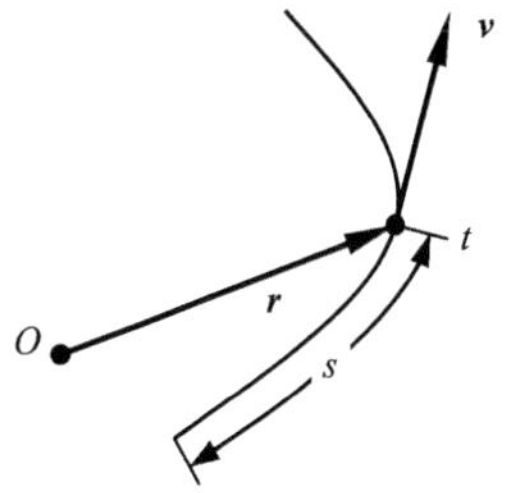

图 5.3

如图 5.4 所示，假设用矢量 $\boldsymbol{v}'$ 定义质点在时间 $t+\Delta t$ 的速度，那么矢量 $\Delta\boldsymbol{v}=\boldsymbol{v}'-\boldsymbol{v}$ 表示速度 $\boldsymbol{v}$ 在 Δt 内的变化。因此，质点在 t 的加速度可表示为

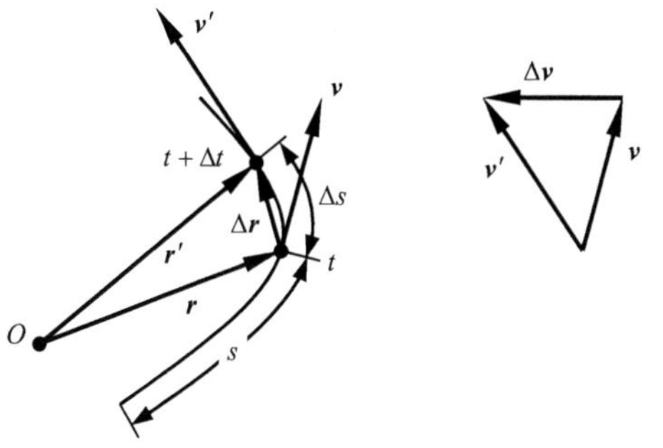

图 5.4

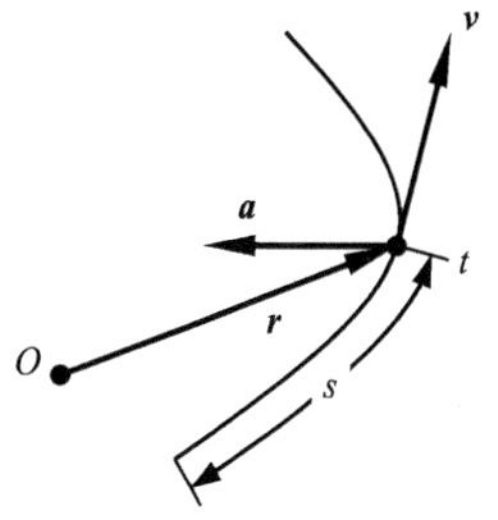

图 5.5

$$\boldsymbol{a} = \lim_{\Delta t \to 0} \frac{\Delta \boldsymbol{v}}{\Delta t} = \dot{\boldsymbol{v}} = \ddot{\boldsymbol{r}} \tag{5.3}$$

加速度是矢量，其方向通常不与运动轨迹相切，如图 5.5 所示。

5.2 直 角 分 量

建立图 5.6 所示直角坐标系 $Oxyz$，则质点 P 的位置矢量 $\boldsymbol{r}$ 可写为

$$\boldsymbol{r} = x\boldsymbol{i} + y\boldsymbol{j} + z\boldsymbol{k} \tag{5.4}$$

式中，$\boldsymbol{i}$、$\boldsymbol{j}$ 和 $\boldsymbol{k}$ 分别为沿正坐标轴的单位矢量，$x = x(t)$、$y = y(t)$ 和 $z = z(t)$ 分别为 $\boldsymbol{r}$ 在三个坐标轴上的标量分量。

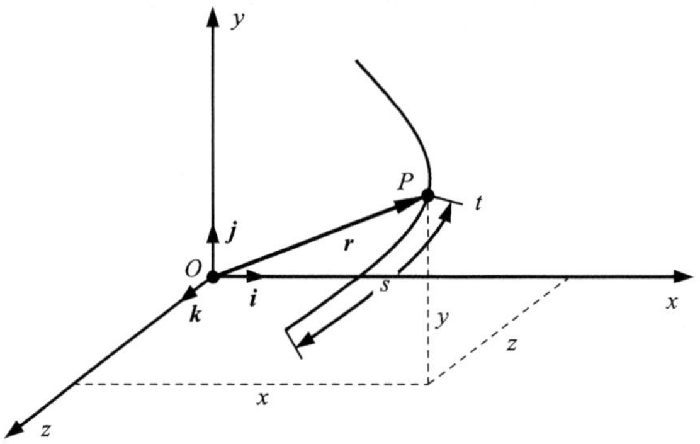

图 5.6

因为 $\boldsymbol{i}$、$\boldsymbol{j}$ 和 $\boldsymbol{k}$ 的大小和方向均保持不变，因此质点 P 在时间 t 的速度和加速度可分别表示为

$$\boldsymbol{v} = \dot{\boldsymbol{r}} = \dot{x}\boldsymbol{i} + \dot{y}\boldsymbol{j} + \dot{z}\boldsymbol{k} \tag{5.5}$$

$$\boldsymbol{a} = \ddot{\boldsymbol{r}} = \ddot{x}\boldsymbol{i} + \ddot{y}\boldsymbol{j} + \ddot{z}\boldsymbol{k} \tag{5.6}$$

式中，$\dot{x}$，$\dot{y}$，$\dot{z}$ 和 $\ddot{x}$，$\ddot{y}$，$\ddot{z}$ 分别表示 $\boldsymbol{v}$ 和 $\boldsymbol{a}$ 的标量分量，即

$$v_x = \dot{x}, v_y = \dot{y}, v_z = \dot{z} \tag{5.7}$$

$$a_x = \ddot{x}, a_y = \ddot{y}, a_z = \ddot{z} \tag{5.8}$$

例 5.1　质点的运动由位置矢量 $\boldsymbol{r} = A(\cos t + t\sin t)\boldsymbol{i} + A(\sin t - t\cos t)\boldsymbol{j}$ 定义，其中 t 的单位为 s。求当位置矢量与加速度矢量分别(a)垂直和(b)平行时的 t 值。

解　根据 $\boldsymbol{r} = A(\cos t + t\sin t)\boldsymbol{i} + A(\sin t - t\cos t)\boldsymbol{j}$，有

$$\boldsymbol{v} = \dot{\boldsymbol{r}} = A(t\cos t)\boldsymbol{i} + A(t\sin t)\boldsymbol{j}$$

$$\boldsymbol{a} = \dot{\boldsymbol{v}} = A(\cos t - t\sin t)\boldsymbol{i} + A(\sin t + t\cos t)\boldsymbol{j}$$

(a)当位置矢量与加速度矢量垂直，有

$$\boldsymbol{r} \cdot \boldsymbol{a} = 0$$

由上式得

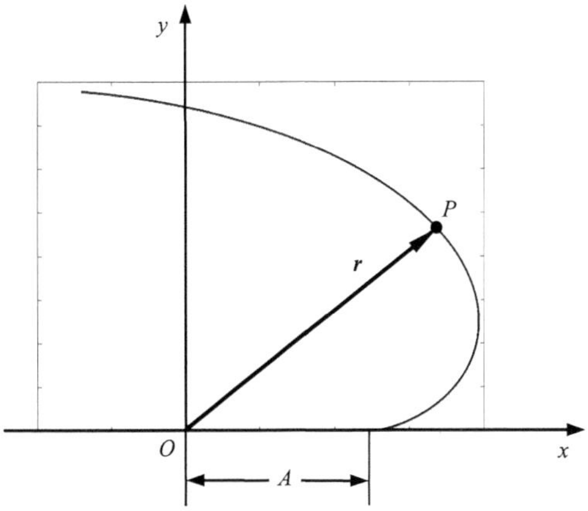

图 E5.1

$$t = 1\ \mathrm{s}$$

(b) 当位置矢量与加速度矢量平行，有

$$\boldsymbol{r} \times \boldsymbol{a} = 0$$

由上式得

$$t = 0$$

5.3　切向和法向分量

如图 5.7 所示，在沿曲线运动的质点 P 上建立自然坐标系，并定义三个单位矢量 $\boldsymbol{e}_{\mathrm{t}}$、$\boldsymbol{e}_{\mathrm{n}}$ 和 $\boldsymbol{e}_{\mathrm{b}}$，其中 $\boldsymbol{e}_{\mathrm{t}}$ 为指向运动方向的与运动轨迹相切的切向单位矢量，$\boldsymbol{e}_{\mathrm{n}}$ 为指向运动轨迹曲率中心的与运动轨迹垂直的主法向单位矢量，$\boldsymbol{e}_{\mathrm{b}}$ 为指向 $\boldsymbol{e}_{\mathrm{t}} \times \boldsymbol{e}_{\mathrm{n}}$ 方向(通过右手法则确定)的与包含 $\boldsymbol{e}_{\mathrm{t}}$ 和 $\boldsymbol{e}_{\mathrm{n}}$ 的平面垂直的副法向单位矢量，即 $\boldsymbol{e}_{\mathrm{b}} = \boldsymbol{e}_{\mathrm{t}} \times \boldsymbol{e}_{\mathrm{n}}$。包含 $\boldsymbol{e}_{\mathrm{t}}$ 和 $\boldsymbol{e}_{\mathrm{n}}$ 的平面称为密切面。

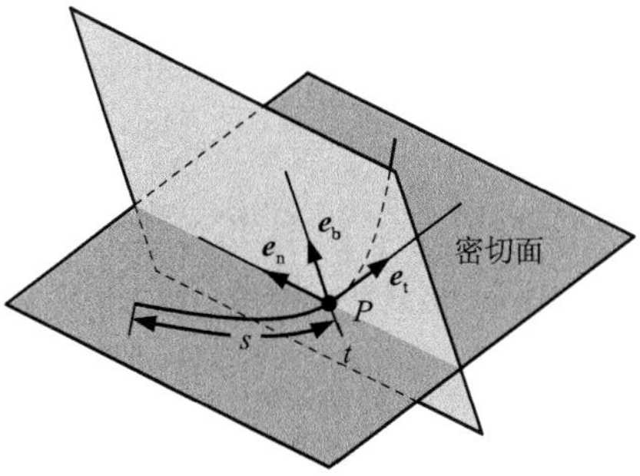

图 5.7

在自然坐标系中，质点 P 在时间 t 的速度矢量 $\boldsymbol{v}$ 可写为

$$\boldsymbol{v} = v\boldsymbol{e}_{\mathrm{t}} \tag{5.9}$$

式中，$\boldsymbol{e}_{\mathrm{t}}$ 为切向单位矢量。利用式(5.9)，质点 P 在时间 t 的加速度矢量 $\boldsymbol{a}$ 可写为

$$\boldsymbol{a} = \dot{\boldsymbol{v}} = \dot{v}\boldsymbol{e}_{\mathrm{t}} + v\dot{\boldsymbol{e}}_{\mathrm{t}} \tag{5.10}$$

利用 $\dot{\boldsymbol{e}}_{\mathrm{t}} = \dfrac{v}{\rho}\boldsymbol{e}_{\mathrm{n}}$，其中 $\boldsymbol{e}_{\mathrm{n}}$ 为主法向单位矢量，ρ 为运动轨迹的曲率半径，则质点 P 在时间 t 的加速度矢量 $\boldsymbol{a}$ 可重写为

$$\boldsymbol{a} = \dot{v}\boldsymbol{e}_{\mathrm{t}} + \frac{v^2}{\rho}\boldsymbol{e}_{\mathrm{n}} \tag{5.11}$$

式中，$\dot{v}$ 和 $\dfrac{v^2}{\rho}$ 分别表示 $\boldsymbol{a}$ 的切向和法向分量，即

$$a_{\mathrm{t}} = \dot{v},\ a_{\mathrm{n}} = \frac{v^2}{\rho} \tag{5.12}$$

式(5.12)表明，加速度的切向分量 a_{t} 等于质点速率的变化率，而法向分量 a_{n} 等于速率平方除以运动轨迹的曲率半径。如果质点速率增加，a_{t} 为正，$\boldsymbol{a}_{\mathrm{t}}$ 与运动方向相同；如果质点速率减小，a_{t} 为负，$\boldsymbol{a}_{\mathrm{t}}$ 与运动方向相反。然而，a_{n} 始终为正，$\boldsymbol{a}_{\mathrm{n}}$ 始终指向运动轨迹的曲率中心。

由此得出结论，加速度的切向分量反映质点速度的大小变化，而法向分量则反映质点速度的方向变化。

例 5.2　质点的运动由位置矢量 $\boldsymbol{r} = [(2\sin 4t)\boldsymbol{i} + (2\cos 4t)\boldsymbol{j} + (4t)\boldsymbol{k}]$ m 定义，其中 t 的单位为 s。求质点运动轨迹的曲率半径。

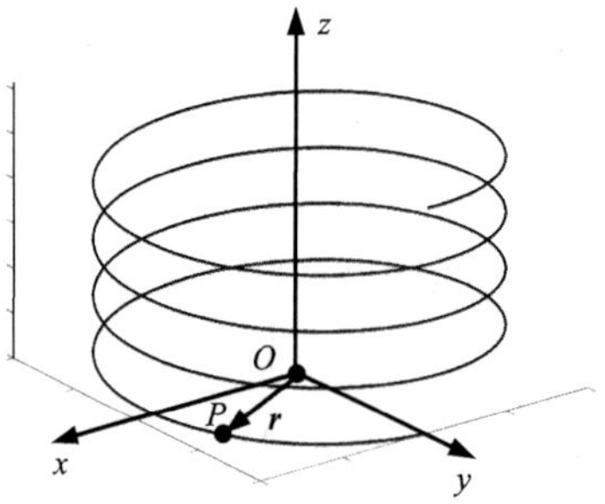

图 E5.2

解　轨迹 $\boldsymbol{r} = [(2\sin 4t)\boldsymbol{i} + (2\cos 4t)\boldsymbol{j} + (4t)\boldsymbol{k}]$ m，有

$$\boldsymbol{v} = \dot{\boldsymbol{r}} = [(8\cos 4t)\boldsymbol{i} + (-8\sin 4t)\boldsymbol{j} + (4)\boldsymbol{k}]\ \mathrm{m/s}$$

$$\boldsymbol{a} = \dot{\boldsymbol{v}} = [(-32\sin 4t)\boldsymbol{i} + (-32\cos 4t)\boldsymbol{j}]\ \mathrm{m/s^2}$$

因此，得

$$v = \sqrt{(8\cos 4t)^2 + (-8\sin 4t)^2 + (4)^2} = 4\sqrt{5}\ \mathrm{m/s}$$

$$a = \sqrt{(-32\sin 4t)^2 + (-32\cos 4t)^2} = 32\ \mathrm{m/s^2}$$

利用 $a^2 = a_t^2 + a_n^2$，其中 $a_t = \dot{v} = 0$ 和 $a_n = \dfrac{v^2}{\rho}$，得

$$\rho = \frac{v^2}{a_n} = \frac{v^2}{a} = 2.5\ \text{m}$$

习　题

5.1　质点的运动由位置矢量 $\boldsymbol{r} = (4t^2 - 3t)\boldsymbol{i} + t^3\boldsymbol{j}$ 定义，其中 r 和 t 的单位分别为 m 和 s。求(a) $t = 0.2$ s 和(b) $t = 1$ s 时质点的速度和加速度。

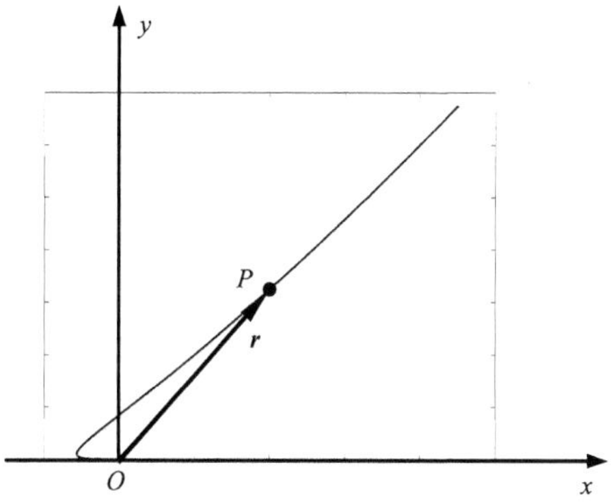

图 P5.1

5.2　质点 P 沿抛物线轨迹 $y = \dfrac{1}{20}x^2$ 运动，其中 x 和 y 的单位均为 m。当 $x = 6$ m，质点速率为 6 m/s，速率增加率为 2 m/s^2。求该瞬时速度的方向以及加速度的大小和方向。

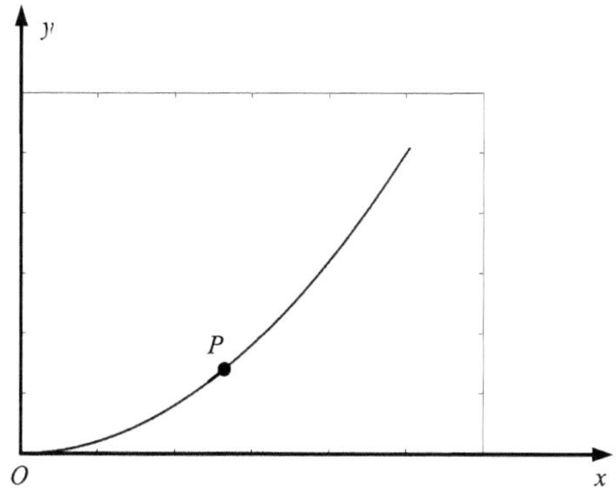

图 P5.2

5.3　质点的运动由位置矢量 $\boldsymbol{r} = \dfrac{3}{4}[1 - 1/(t+1)]\boldsymbol{i} + [\dfrac{1}{2}\exp(-\pi t/2)\cos 2\pi t]\boldsymbol{j}$ 定义，其中 r 和 t 的单位分别为 m 和 s。求(a) $t = 0.5$ s 和(b) $t = 1$ s 时质点的位置、速度和加速度。

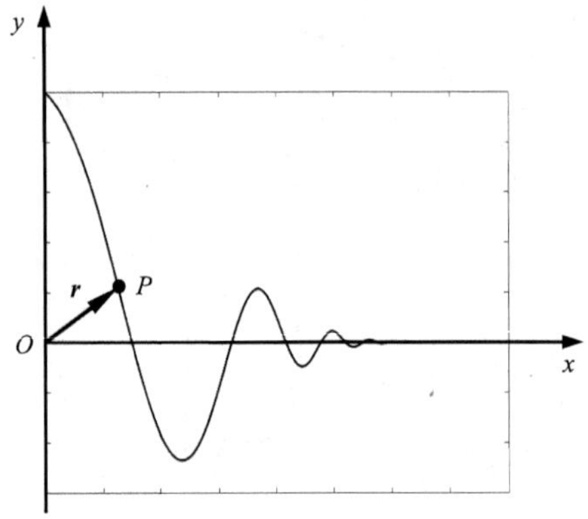

图 P5.3

5.4　质点的运动由位置矢量 $\boldsymbol{r}=[(\omega t-\sin\omega t)\boldsymbol{i}+(1-\cos\omega t)\boldsymbol{j}]$ m 定义，其中 ω 和 t 的单位分别为 rad/s 和 s。求在时间 t 质点的切向和法向加速度。

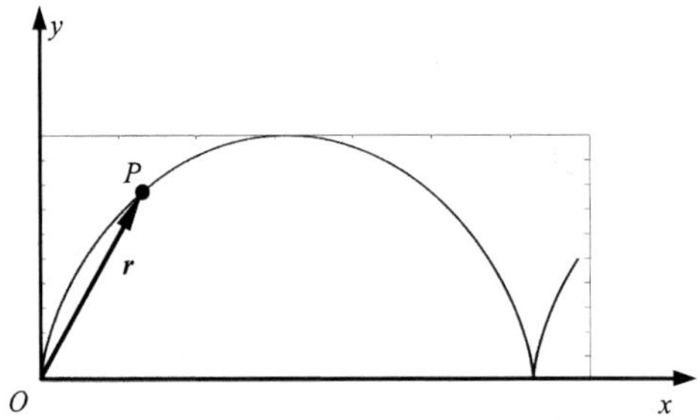

图 P5.4

第 6 章　刚体平面运动学

当刚体上所有质点都沿距某一固定平面等距离的轨迹移动，则这种运动称为平面运动。存在三种类型的平面运动，即平移、定轴转动和一般平面运动。

6.1　平　　移

如果刚体内任何直线在刚体运动期间保持方向不变，则这种运动称为平移。当刚体平移时，刚体内所有质点都沿平行运动轨迹移动。如果运动轨迹是直线，则称为直线平移，如图 6.1a 所示；如果运动轨迹是曲线，则称为曲线平移，如图 6.1b 所示。

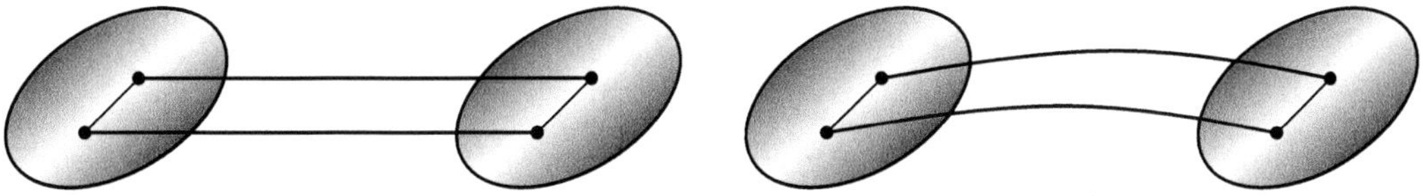

图 6.1

如图 6.2 所示，设 A 和 B 是平移刚体内任意两点，$\boldsymbol{r}_A$ 和 $\boldsymbol{r}_B$ 是质点 A 和 B 相对固定参考系 $Oxyz$ 的位置矢量，$\boldsymbol{r}_{A/B}$ 是质点 A 相对质点 B 的位置矢量。由图 6.2 得

$$\boldsymbol{r}_A = \boldsymbol{r}_B + \boldsymbol{r}_{A/B} \tag{6.1}$$

式(6.1)对时间 t 进行微分，得

$$\boldsymbol{v}_A = \boldsymbol{v}_B + \boldsymbol{v}_{A/B} \tag{6.2}$$

式中，$\boldsymbol{v}_A = \dot{\boldsymbol{r}}_A$ 和 $\boldsymbol{v}_B = \dot{\boldsymbol{r}}_B$ 分别为质点 A 和 B 的速度，$\boldsymbol{v}_{A/B} = \dot{\boldsymbol{r}}_{A/B}$ 为质点 A 相对质点 B 的速度。当刚体发生平移，$\boldsymbol{r}_{A/B}$ 为常量，即 $\dot{\boldsymbol{r}}_{A/B} = 0$，因此

$$\boldsymbol{v}_A = \boldsymbol{v}_B \tag{6.3}$$

微分，得

$$\boldsymbol{a}_A = \boldsymbol{a}_B \tag{6.4}$$

由此得到结论，刚体发生平移时刚体上所有质点在任意给定时刻具有相同的速度和加速度，如图 6.3 所示。

对于曲线平移，速度和加速度的方向不断改变，然而，对于直线平移，速度和加速度的方向保持不变。

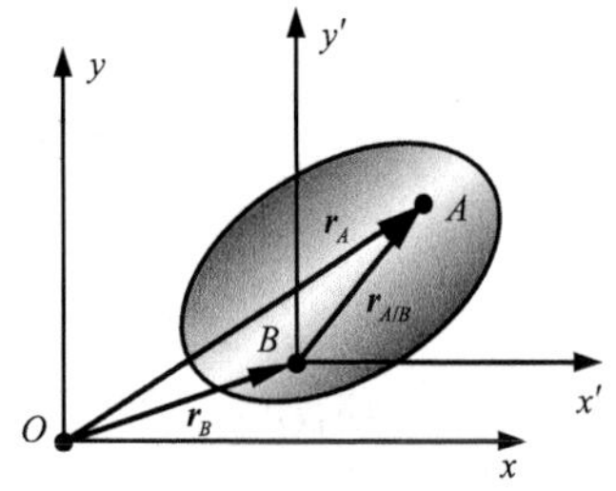

图 6.2

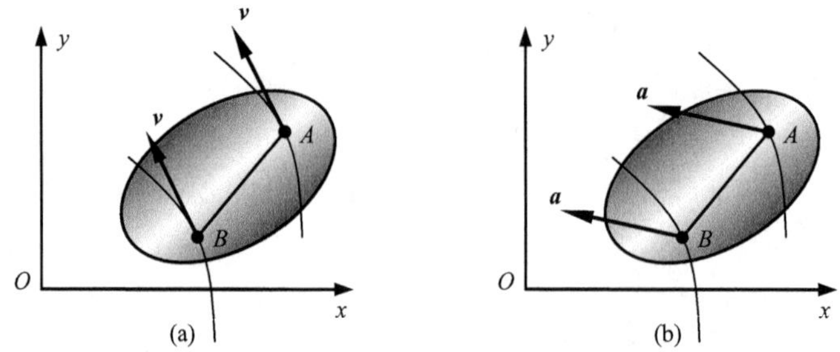

图 6.3

6.2　定 轴 转 动

当刚体绕某一固定轴旋转，则除转动轴上的质点之外，刚体上的所有质点都沿垂直于转动轴的圆形轨迹运动，如图 6.4 所示。

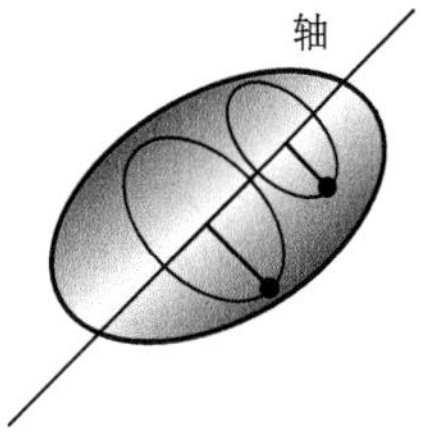

图 6.4

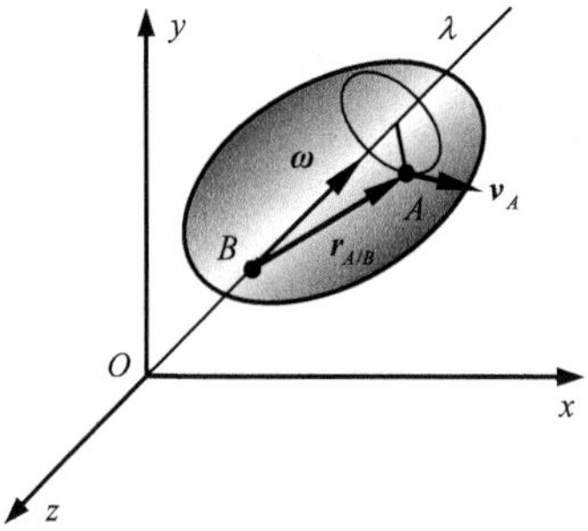

图 6.5

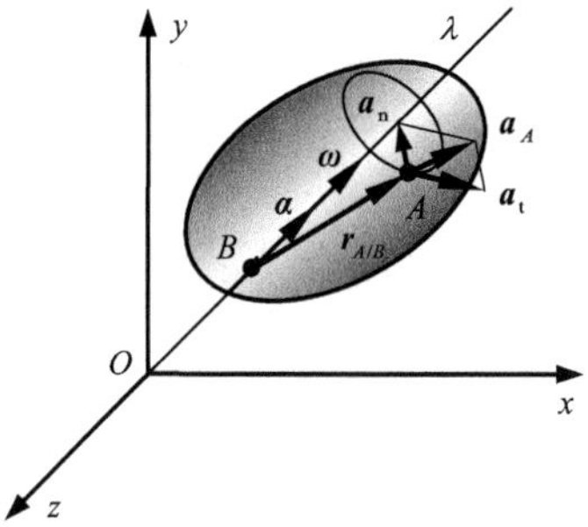

图 6.6

如图 6.5 所示，考虑刚体绕固定轴 λ 转动，A 是刚体上任意一点，$\boldsymbol{r}_{A/B}$ 是质点 A 相对转动轴上任意点 B 的位置矢量。由图 6.5 得

$$\boldsymbol{v}_A = \boldsymbol{\omega} \times \boldsymbol{r}_{A/B} \tag{6.5}$$

式中，$\boldsymbol{\omega}$ 为刚体角速度。角速度的方向沿转动轴，角速度的大小等于角坐标的变化率。

式(6.5)对时间 t 微分，得

$$\boldsymbol{a}_A = \boldsymbol{\alpha} \times \boldsymbol{r}_{A/B} + \boldsymbol{\omega} \times \boldsymbol{v}_A \tag{6.6}$$

式中，$\boldsymbol{\alpha} = \dot{\boldsymbol{\omega}}$ 为刚体角加速度。对于刚体绕定轴转动，角加速度 $\boldsymbol{\alpha}$ 的方向沿转动轴，大小等于角速度变化率。利用式(6.5)，式(6.6)可写为

$$\boldsymbol{a}_A = \boldsymbol{\alpha} \times \boldsymbol{r}_{A/B} + \boldsymbol{\omega} \times (\boldsymbol{\omega} \times \boldsymbol{r}_{A/B}) \tag{6.7}$$

也可表示为

$$\boldsymbol{a}_A = \boldsymbol{a}_{\mathrm{t}} + \boldsymbol{a}_{\mathrm{n}} \tag{6.8}$$

式中，$\boldsymbol{a}_{\mathrm{t}} = \boldsymbol{\alpha} \times \boldsymbol{r}_{A/B}$ 为加速度 $\boldsymbol{a}_A$ 的切向分量，沿轨迹切向，$\boldsymbol{a}_{\mathrm{n}} = \boldsymbol{\omega} \times \boldsymbol{v}_A = \boldsymbol{\omega} \times (\boldsymbol{\omega} \times \boldsymbol{r}_{A/B})$ 为加速度 $\boldsymbol{a}_A$ 的法向分量，指向轨迹圆心。式(6.8)表明，加速度 $\boldsymbol{a}_A$ 等于切向分量 $\boldsymbol{a}_{\mathrm{t}}$ 和法向

分量 $\boldsymbol{a}_{\mathrm{n}}$ 的矢量和，如图 6.6 所示。

刚体定轴转动可简化为板的面内转动，即使板面垂直刚体转动轴并绕板面与刚体转动轴的交点 O 发生面内转动，如图 6.7 所示。

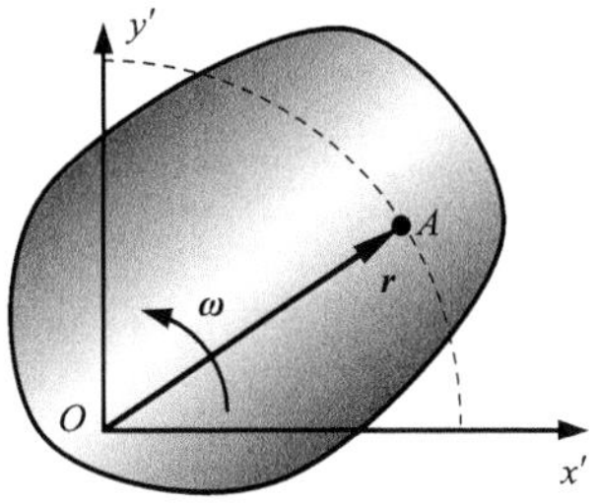

图 6.7

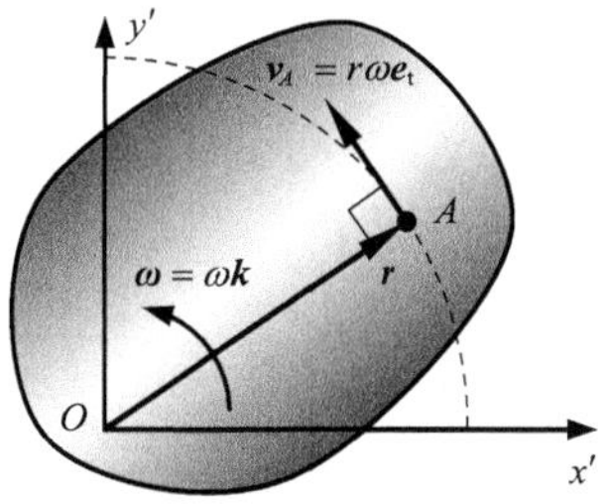

图 6.8

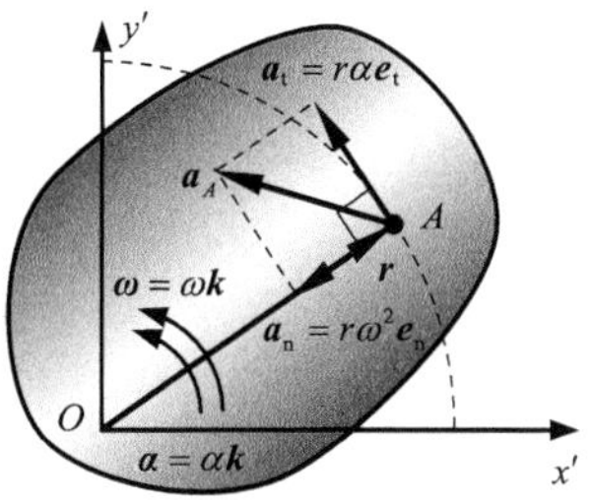

图 6.9

对于板绕定点 O 的面内转动，角速度 $\boldsymbol{\omega}$ 可表示为

$$\boldsymbol{\omega} = \omega\boldsymbol{k} \tag{6.9}$$

式中，$\boldsymbol{k}$ 为垂直板面向外的单位矢量，ω 为角速度大小，逆时针转动为正，顺时针转动为负。

利用式(6.5)，得图 6.8 所示质点 A 的速度

$$\boldsymbol{v}_A = \boldsymbol{\omega} \times \boldsymbol{r} = r\omega\boldsymbol{e}_{\mathrm{t}} \tag{6.10}$$

式中，$\boldsymbol{e}_{\mathrm{t}}$ 为指向运动方向的切向单位矢量。因此，速度大小等于

$$v_A = r\omega \tag{6.11}$$

方向由 $\boldsymbol{r}$ 沿转动方向旋转 90° 确定。

利用式(6.7)和(6.8)，得图 6.9 所示质点 A 的加速度

$$\boldsymbol{a}_A = \boldsymbol{a}_{\mathrm{t}} + \boldsymbol{a}_{\mathrm{n}} = \boldsymbol{\alpha} \times \boldsymbol{r} + \boldsymbol{\omega} \times (\boldsymbol{\omega} \times \boldsymbol{r}) = r\alpha\boldsymbol{e}_{\mathrm{t}} + r\omega^2\boldsymbol{e}_{\mathrm{n}} \tag{6.12}$$

式中，$\boldsymbol{e}_{\mathrm{n}}$ 为指向圆心的单位矢量。如果 α 为正，则 $\boldsymbol{a}_{\mathrm{t}} = r\alpha\boldsymbol{e}_{\mathrm{t}}$ 指向逆时针方向，否则指向顺时针方向。$\boldsymbol{a}_{\mathrm{n}} = r\omega^2\boldsymbol{e}_{\mathrm{n}}$ 始终指向圆心。

例 6.1　图 E6.1 所示结构由杆 AE、CE 和矩形板 $ABCD$ 焊接而成。结构绕轴 AE 以不变角速度 $\omega = 5$ rad/s 旋转。已知从点 A 观察时转动为逆时针，求点 B 的速度和加速度。

解　利用 $\boldsymbol{r}_{A/E} = (\boldsymbol{i} - 0.45\boldsymbol{j} + 0.6\boldsymbol{k})$ m，有

$$\boldsymbol{\omega} = \omega\frac{\boldsymbol{r}_{A/E}}{r_{A/E}} = (4\boldsymbol{i} - 1.8\boldsymbol{j} + 2.4\boldsymbol{k})\ \text{rad/s}$$

利用 $\boldsymbol{r}_{B/A} = (-0.6\boldsymbol{k})$ m，得

$$\boldsymbol{v}_B = \boldsymbol{\omega} \times \boldsymbol{r}_{B/A} = \begin{vmatrix} \boldsymbol{i} & \boldsymbol{j} & \boldsymbol{k} \\ 4 & -1.8 & 2.4 \\ 0 & 0 & -0.6 \end{vmatrix} = (1.08\boldsymbol{i} + 2.4\boldsymbol{j})\ \text{m/s}$$

由 $\boldsymbol{\alpha} = 0$，得

$$\boldsymbol{a}_B = \boldsymbol{\alpha} \times \boldsymbol{r}_{B/A} + \boldsymbol{\omega} \times \boldsymbol{v}_B = \begin{vmatrix} \boldsymbol{i} & \boldsymbol{j} & \boldsymbol{k} \\ 4 & -1.8 & 2.4 \\ 1.08 & 2.4 & 0 \end{vmatrix} = (-5.76\boldsymbol{i} + 2.59\boldsymbol{j} + 11.54\boldsymbol{k})\ \text{m/s}^2$$

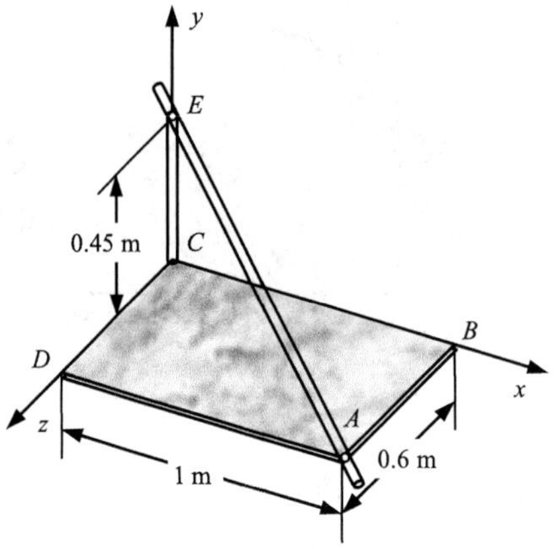

图 E6.1

6.3 一般平面运动

物体一般运动可以看成是参考面内的平移和绕参考面垂直轴的转动的合成，如图 6.10 所示。

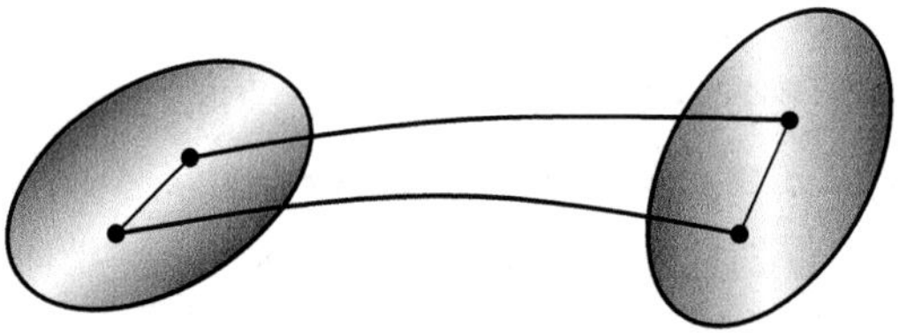

图 6.10

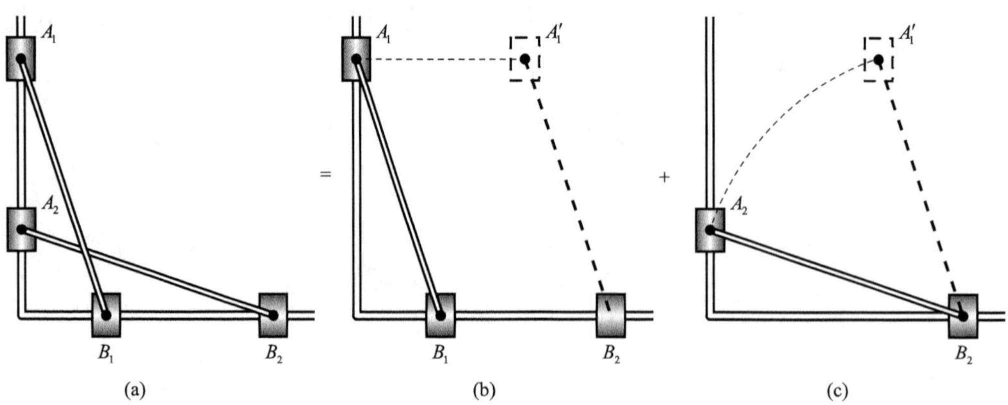

图 6.11

一般平面运动既不是平移，也不是转动，但它始终可以看成为平移和转动的合成。如图 6.11a 所示，两端分别在水平和垂直轨道中滑动的杆 AB 做一般平面运动。杆从 A_1B_1 到 A_2B_2 的运动可以分解为随基点 B 从 A_1B_1 到 $A_1'B_2$ 的向右平移(如图 6.11b 所示)和绕基点 B 从 $A_1'B_2$ 到 A_2B_2 的逆时针转动(如图 6.11c 所示)。

如图 6.12a 所示，水平面上只滚不滑的轮子的运动也为一般平面运动。轮子从 A_1B_1 到 A_2B_2 的运动也可以分解为随基点 B 从 A_1B_1 到 $A_1'B_2$ 的向右平移(如图 6.12b 所示)和绕基点 B 从 $A_1'B_2$ 到 A_2B_2 的顺时针转动(如图 6.12c 所示)。

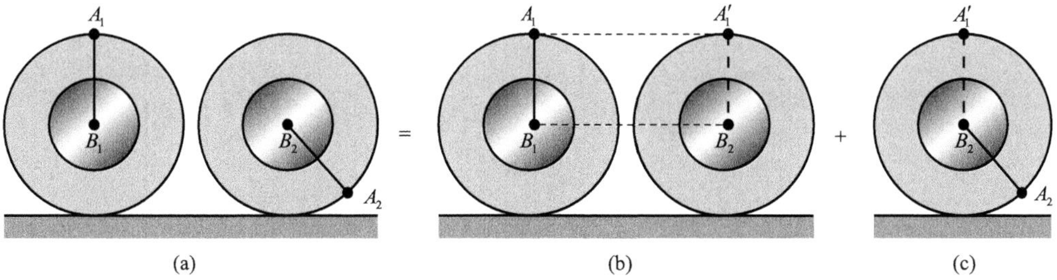

图 6.12

1. 基点法求速度

因为任何一般平面运动都可分解为随基点的平移和绕基点的转动，因此图 6.13a 所示刚体上 A 的速度 $\boldsymbol{v}_A$ 可表示为

$$\boldsymbol{v}_A = \boldsymbol{v}_B + \boldsymbol{v}_{A/B} \tag{6.13}$$

式中，$\boldsymbol{v}_B$ 为基点 B 的速度，如图 6.13b；$\boldsymbol{v}_{A/B} = \boldsymbol{\omega} \times \boldsymbol{r}_{A/B}$ 为 A 相对基点 B 的速度，如图 6.13c。应该注意，尽管基点可以任意选择，但为了方便通常选择运动已知点作为基点。

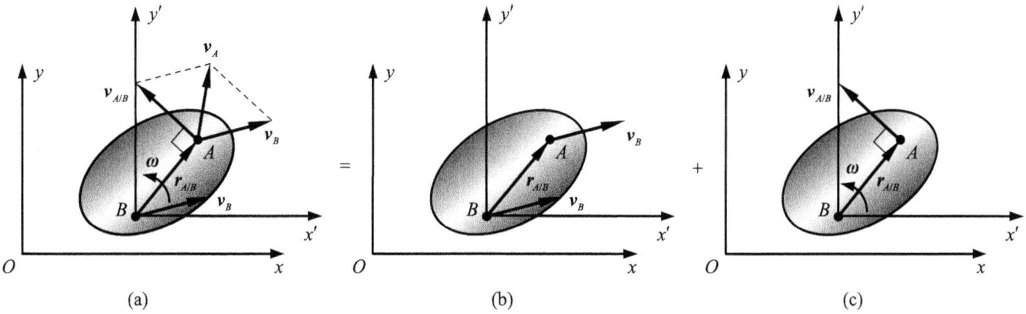

图 6.13

例 6.2　如图 E6.2a 所示，轴环 A 以不变速度 $v_A = 1\ \text{m/s}$ 向下移动。图示瞬时 $\theta = 30°$，求：(a)杆 AB 的角速度；(b)轴环 B 的速度。

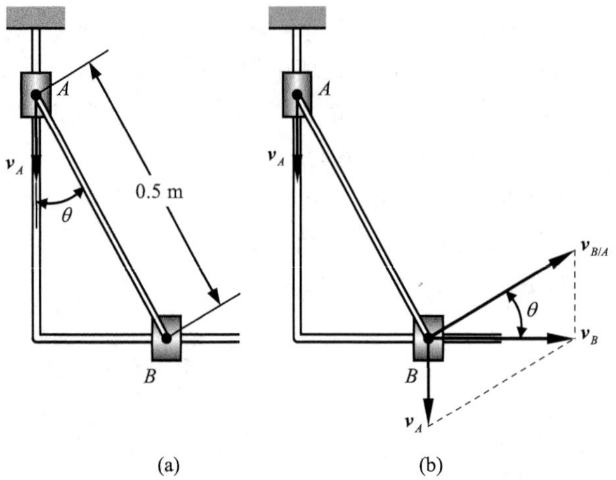

图 E6.2

解　选点 A 作为基点，则点 B 的速度为

$$\boldsymbol{v}_B = \boldsymbol{v}_A + \boldsymbol{v}_{B/A}$$

利用 $\boldsymbol{v}_B$ 水平向右和 $\boldsymbol{v}_{B/A}$ 垂直杆向上，则速度平行四边形如图 E6.2b 所示。根据平行四边形，则在该瞬时有

$$v_B = \frac{v_A}{\tan\theta} = 1.73\ \text{m/s},\ v_{B/A} = \frac{v_A}{\sin\theta} = 2\ \text{m/s}$$

利用 $v_{B/A} = AB \cdot \omega$，可得

$$\omega = \frac{v_{B/A}}{AB} = 4\ \text{rad/s (逆时针)}$$

2. 瞬心法求速度

在一般平面运动速度分析中，可选刚体上或刚体外瞬时速度为零的点作为基点。假设图 6.14 所示瞬时点 I 处速度为零，I 称为速度瞬心，则点 A 的速度 $\boldsymbol{v}_A$ 可写为

$$\boldsymbol{v}_A = \boldsymbol{\omega} \times \boldsymbol{r} \tag{6.14}$$

式中，$\boldsymbol{\omega}$ 和 $\boldsymbol{r}$ 分别为刚体的角速度和 A 相对速度瞬心 I 的位置矢量。应该注意，速度瞬心的加速度通常不为零。

在给定瞬时，速度瞬心可通过下述方法确定：

(1) $\boldsymbol{v}_A$ 和 $\boldsymbol{\omega}$ 已知，速度瞬心可按图 6.15 确定。

(2) 已知互不平行速度 $\boldsymbol{v}_A$ 和 $\boldsymbol{v}_B$，速度瞬心可按图 6.16 确定。

(3) 已知平行速度 $\boldsymbol{v}_A$ 和 $\boldsymbol{v}_B$，速度瞬心可按图 6.17 确定。图 6.17a 的特殊情况应该引起注意，即如果某一瞬时有 $\boldsymbol{v}_A = \boldsymbol{v}_B$，则速度瞬心将在无穷远处。这种特殊情况称为瞬时平移。

例 6.3　如图 E6.3a 所示，轴环 A 以不变速度 $v_A = 1$ m/s 向下移动。图示瞬时 $\theta = 30°$，求：(a) 杆 AB 的角速度；(b) 轴环 B 的速度。

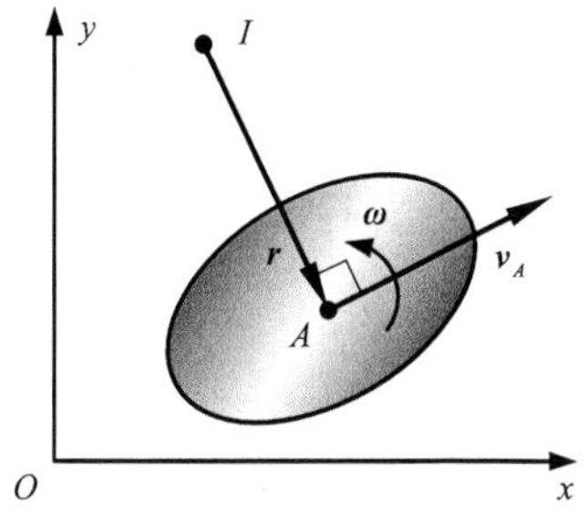

图 6.14

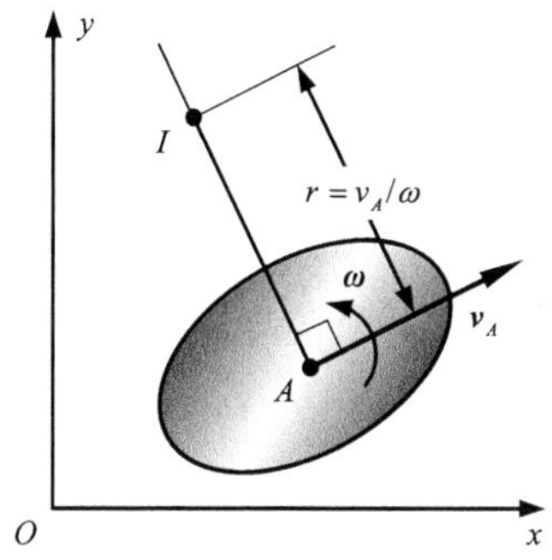

图 6.15

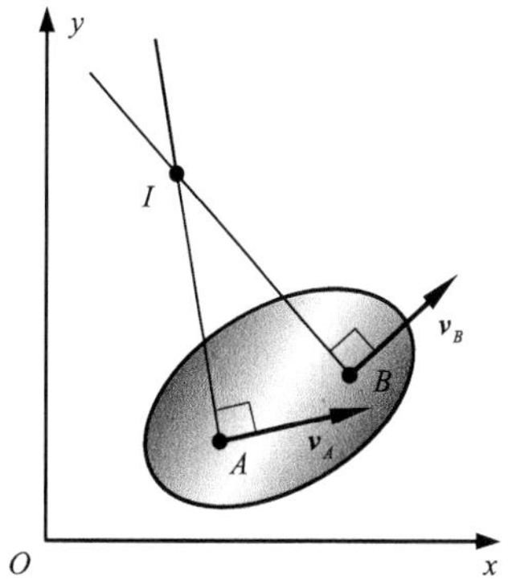

图 6.16

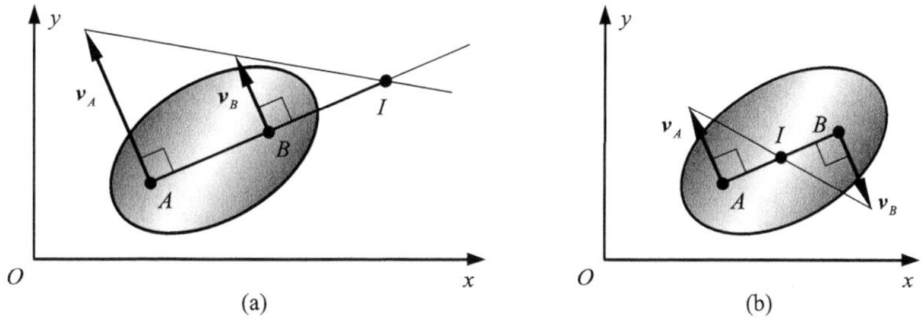

图 6.17

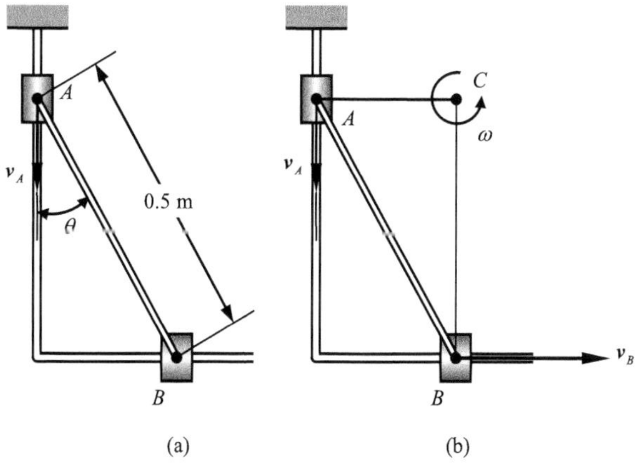

图 E6.3

解　在图示给定瞬时，速度瞬心在 C 处，则有

$$\omega = \frac{v_A}{AC} = \frac{v_A}{AB \cdot \sin\theta} = 4 \text{ rad/s}$$

$$v_B = BC \cdot \omega = (AB \cdot \cos\theta)\omega = 1.73 \text{ m/s}$$

3. 基点法求加速度

因为任何一般平面运动都可分解为随基点的平移和绕基点的转动，因此图 6.18a 所

示 A 处加速度 $\boldsymbol{a}_A$ 可表示为

$$\boldsymbol{a}_A = \boldsymbol{a}_B + \boldsymbol{a}_{A/B} = \boldsymbol{a}_B + (\boldsymbol{a}_{A/B})_\mathrm{t} + (\boldsymbol{a}_{A/B})_\mathrm{n} \tag{6.15}$$

式中，$\boldsymbol{a}_B$ 为基点加速度，$(\boldsymbol{a}_{A/B})_\mathrm{t} = \boldsymbol{\alpha} \times \boldsymbol{r}_{A/B}$ 和 $(\boldsymbol{a}_{A/B})_\mathrm{n} = \boldsymbol{\omega} \times (\boldsymbol{\omega} \times \boldsymbol{r}_{A/B})$ 分别为 A 相对基点 B 的切向和法向加速度，$\boldsymbol{a}_{A/B} = (\boldsymbol{a}_{A/B})_\mathrm{t} + (\boldsymbol{a}_{A/B})_\mathrm{n}$ 为 A 相对基点 B 的加速度。

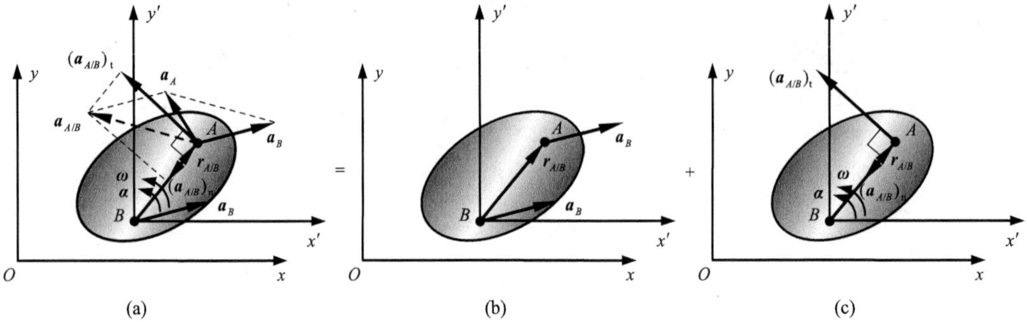

图 6.18

例 6.4　如图 E6.4a 所示，轴环 A 以不变速度 $v_A = 1$ m/s 向下移动。图示瞬时 $\theta = 30°$，求：(a) 杆 AB 的角加速度；(b) 轴环 B 的加速度。

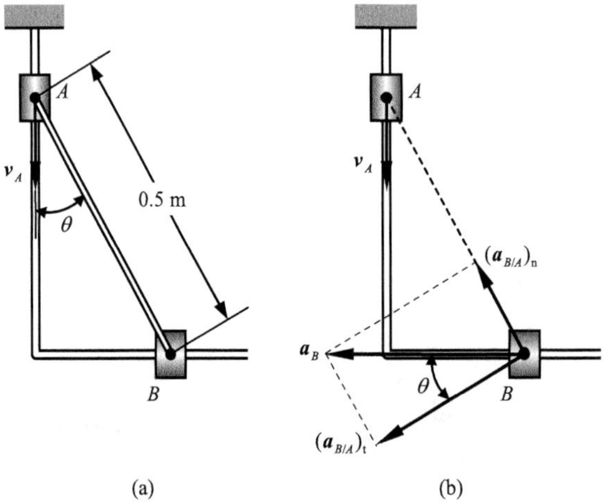

图 E6.4

解　选 A 点为基点，则 B 点加速度为

$$\boldsymbol{a}_B = \boldsymbol{a}_A + (\boldsymbol{a}_{B/A})_\mathrm{t} + (\boldsymbol{a}_{B/A})_\mathrm{n}$$

利用 $\boldsymbol{a}_B$ 水平、$\boldsymbol{a}_A$ 恒等于零，则加速度平行四边形如图 E6.4b 所示。根据平行四边形，则在该瞬时有

$$a_B = \frac{(a_{B/A})_\mathrm{n}}{\sin\theta},\ (a_{B/A})_\mathrm{t} = \frac{(a_{B/A})_\mathrm{n}}{\tan\theta}$$

利用 $\omega = 4$ rad/s（参考例 6.2 或 6.3），即 $(a_{B/A})_\mathrm{n} = AB \cdot \omega^2 = 8$ m/s^2（指向 A 点），则得

$$a_B = 16\ \text{m/s}^2,\ (a_{B/A})_t = 13.86\ \text{m/s}^2$$

根据 $(a_{B/A})_t = AB \cdot \alpha$，有

$$\alpha = \frac{(a_{B/A})_t}{AB} = 27.72\ \text{rad/s}^2\ (\text{顺时针})$$

例 6.5　如图 E6.5a 所示，杆 BD 在 B 和 D 处与连杆 AB 和 CD 相连。已知图示瞬时连杆 AB 以不变角速度 $\omega = 2$ rad/s 顺时针旋转，求(a)杆 BD 和(b)连杆 CD 的角速度和角加速度。

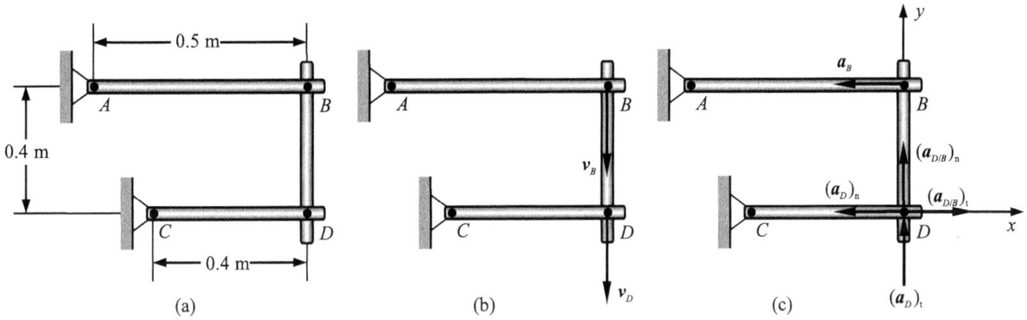

图 E6.5

解　由连杆 AB 绕 A 点顺时针旋转，则 B 点速度垂直向下，如图 E6.5b。同理，因连杆 CD 绕 C 点顺时针旋转，故 D 点速度也垂直向下。在图示瞬时，杆 BD 在 B 和 D 处具有相同速度(包括大小和方向)，因此图示瞬时杆 BD 做瞬时平移，由此得

$$v_D = v_B = AB \cdot \omega_{AB} = 1\ \text{m/s},\ \omega_{BD} = 0,\ \omega_{CD} = \frac{v_D}{CD} = 2.5\ \text{rad/s}\ (\text{顺时针})$$

选杆 BD 上的 B 点为基点，那么 D 点加速度为

$$(\boldsymbol{a}_D)_t + (\boldsymbol{a}_D)_n = \boldsymbol{a}_B + (\boldsymbol{a}_{D/B})_t + (\boldsymbol{a}_{D/B})_n$$

如图 E6.5c 建参考系 xy，设 $\boldsymbol{i}$ 和 $\boldsymbol{j}$ 分别为沿 x 和 y 方向的单位矢量，则有

$$(a_D)_t \boldsymbol{j} - (a_D)_n \boldsymbol{i} = -a_B \boldsymbol{i} + (a_{D/B})_t \boldsymbol{i} + (a_{D/B})_n \boldsymbol{j}$$

或

$$-(a_D)_n = -a_B + (a_{D/B})_t,\ (a_D)_t = (a_{D/B})_n$$

利用 $(a_D)_n = CD \cdot \omega_{CD}{}^2 = 2.5\ \text{m/s}^2$，$a_B = AB \cdot \omega_{AB}{}^2 = 2\ \text{m/s}^2$ 和 $(a_{D/B})_n = BD \cdot \omega_{BD}{}^2 = 0$，得

$$(a_{D/B})_t = a_B - (a_D)_n = -0.5\ \text{m/s}^2\ (\text{向左}),\ (a_D)_t = 0$$

在利用 $(a_{D/B})_t = BD \cdot \alpha_{BD}$ 和 $(a_D)_t = CD \cdot \alpha_{CD}$，有

$$\alpha_{BD} = -1.25\ \text{rad/s}^2\ (\text{顺时针}),\ \alpha_{CD} = 0$$

习　　题

6.1　图示结构由杆 AE, CE 和矩形板 $ABCD$ 焊接而成。结构绕轴 AE 以角速度 $\omega=5$ rad/s 和角加速度 $\alpha=10$ rad/s^2 旋转。已知从点 A 观察时角速度和角加速度为逆时针，求点 P 的速度和加速度。

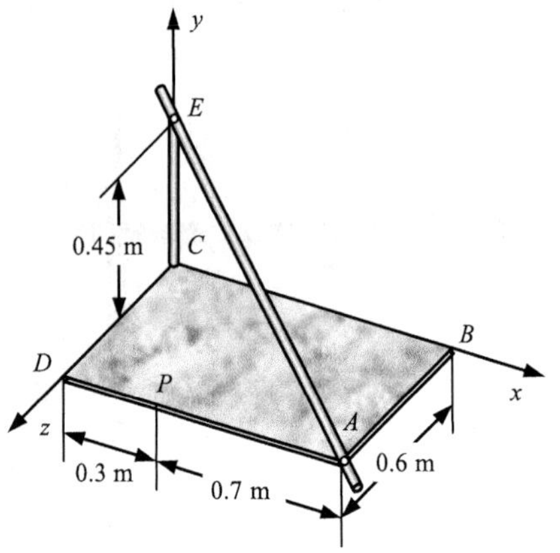

图 P6.1

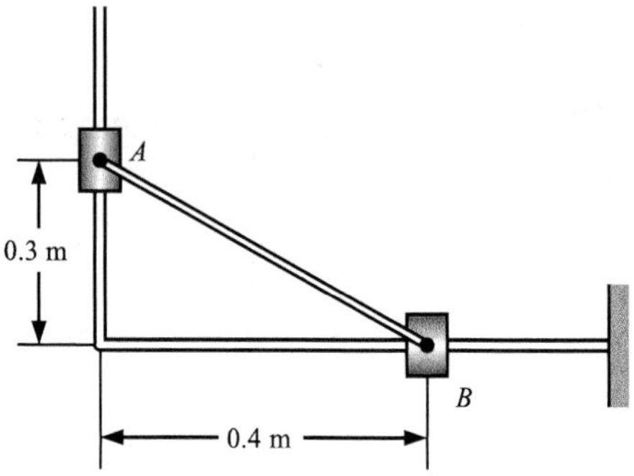

图 P6.2

6.2　轴环 A 以不变速度 $v_A=0.4$ m/s 向上移动。在图示瞬时，求：(a) 杆 AB 的角速度和轴环 B 的速度；(b) 杆 AB 的角加速度和轴环 B 的加速度。

6.3　杆 BDE 在 B 和 D 处与连杆 AB 和 CD 相连。已知图示瞬时连杆 AB 以不变角速度 $\omega=2$ rad/s 顺时针旋转，求点 E 的速度和加速度。

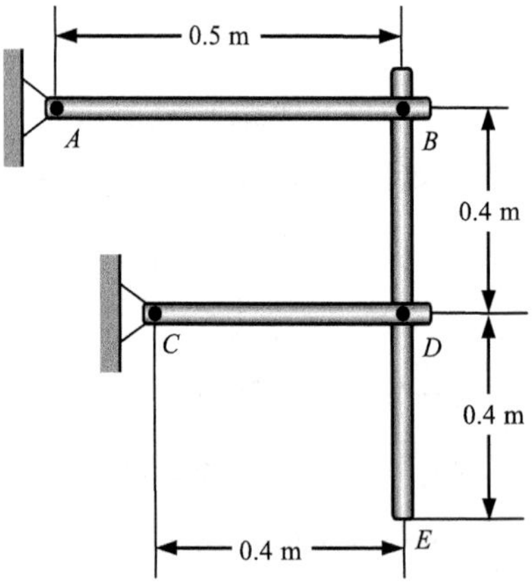

图 P6.3

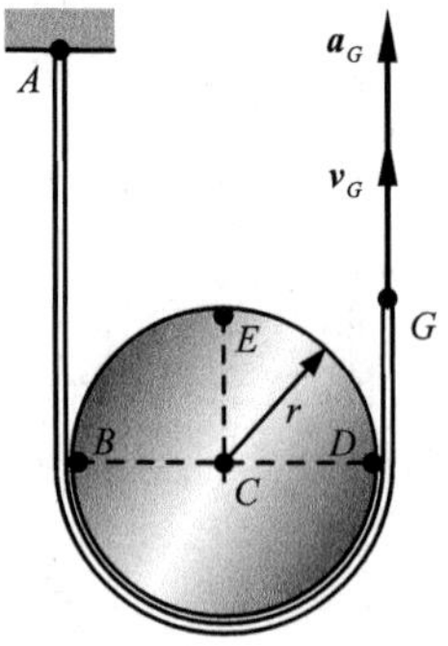

图 P6.4

6.4　半径为 r 的圆柱的运动通过缆绳 AG 控制。已知 $v_G=0.3$ m/s（向上），$a_G=0.5$ m/s^2（向上）和 $r=0.15$ m，求：(a) B 和 D 点的加速度；(b) E 点的速度和加速度。

第 7 章　质点合成运动

利用单一参考系已经分析了质点的运动。然而，在很多情况下质点相对单一参考系的运动轨迹十分复杂，因此需要通过两个或更多参考系进行质点运动分析。例如，当飞机在飞行的时候，首先观察飞机相对地面的运动，然后通过平行四边形定律叠加螺旋桨端点相对飞机的运动，可更容易描述螺旋桨端点质点的运动。

与地球相连的参考系称为固定参考系(简称定系)，而相对地球运动的参考系则称为运动参考系(简称动系)。动点相对定系的运动称为绝对运动，而动点相对动系的运动则称为相对运动。

7.1　矢量时间导数

假设 $Oxyz$ 为定系， $O'x'y'z'$ 为动系，如图 7.1 所示。

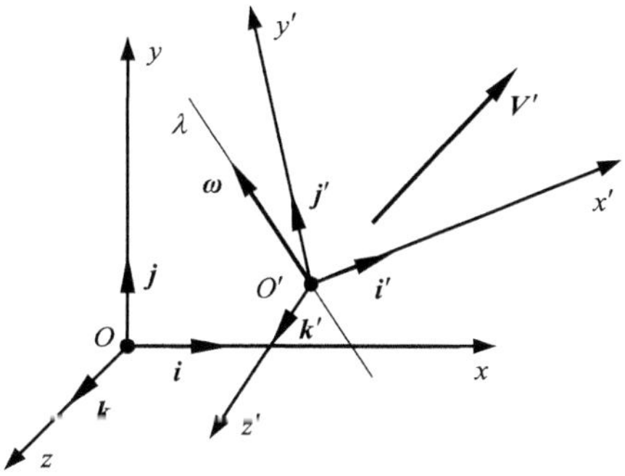

图 7.1

如果 $O'x'y'z'$ 以瞬时角速度 $\boldsymbol{\omega}$ 绕轴 λ 旋转，则矢量 $\boldsymbol{V}'$ 可表示为

$$\boldsymbol{V}' = V_x'\boldsymbol{i}' + V_y'\boldsymbol{j}' + V_z'\boldsymbol{k}' \tag{7.1}$$

式中，V_x'，V_y'，V_z' 和 $\boldsymbol{i}'$，$\boldsymbol{j}'$，$\boldsymbol{k}'$ 分别为 $\boldsymbol{V}'$ 在动系中的直角分量和用于动系的单位矢量。在动系中，$\boldsymbol{i}'$，$\boldsymbol{j}'$ 和 $\boldsymbol{k}'$ 的大小和方向均保持不变，因此有

$$\dot{\boldsymbol{V}}' = \dot{V}_x'\boldsymbol{i}' + \dot{V}_y'\boldsymbol{j}' + \dot{V}_z'\boldsymbol{k}' \tag{7.2}$$

然而，在定系中，$\boldsymbol{i}'$，$\boldsymbol{j}'$ 和 $\boldsymbol{k}'$ 的大小不变，但方向发生变化，因此得

$$\{\dot{\boldsymbol{V}}\}_O = \dot{V}_x'\boldsymbol{i}' + \dot{V}_y'\boldsymbol{j}' + \dot{V}_z'\boldsymbol{k}' + V_x'\dot{\boldsymbol{i}}' + V_y'\dot{\boldsymbol{j}}' + V_z'\dot{\boldsymbol{k}}' \tag{7.3}$$

利用 $\dot{\boldsymbol{i}}' = \boldsymbol{\omega}\times\boldsymbol{i}'$， $\dot{\boldsymbol{j}}' = \boldsymbol{\omega}\times\boldsymbol{j}'$ 和 $\dot{\boldsymbol{k}}' = \boldsymbol{\omega}\times\boldsymbol{k}'$ ，得

$$\{\dot{\boldsymbol{V}}'\}_O = \dot{V}_x'\boldsymbol{i}' + \dot{V}_y'\boldsymbol{j}' + \dot{V}_z'\boldsymbol{k}' + \boldsymbol{\omega}\times(V_x'\boldsymbol{i}' + V_y'\boldsymbol{j}' + V_z'\boldsymbol{k}') \tag{7.4}$$

式(7.1)和(7.2)代入式(7.4)，得

$$\{\dot{\boldsymbol{V}}'\}_O = \dot{\boldsymbol{V}}' + \boldsymbol{\omega} \times \boldsymbol{V}' \tag{7.5}$$

7.2 速 度 合 成

如图 7.2 所示，假设 $Oxyz$ 为定系，$O'x'y'z'$ 为动系。考虑沿空间曲线运动的动点 P，假设牵连点(即动系上与 P 的瞬时重合点)由 M 表示，则有

$$\boldsymbol{r}_P = \boldsymbol{r}_M + \boldsymbol{r}_{P/M} \tag{7.6}$$

式中，$\boldsymbol{r}_P$ 为动点 P 的位置矢量，$\boldsymbol{r}_M$ 为牵连点 M 的位置矢量，$\boldsymbol{r}_{P/M}$ 为动点相对牵连点的位置矢量。

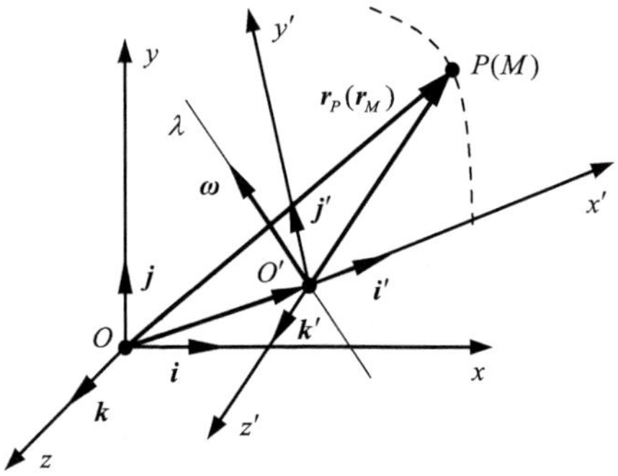

图 7.2

在定系 $Oxyz$ 中，式(7.6)对时间进行微分，得

$$\dot{\boldsymbol{r}}_P = \dot{\boldsymbol{r}}_M + \{\dot{\boldsymbol{r}}_{P/M}\}_O \tag{7.7}$$

设动系 $O'x'y'z'$ 以瞬时角速度 $\boldsymbol{\omega}$ 绕轴 λ 旋转，利用式(7.5)，得

$$\{\dot{\boldsymbol{r}}_{P/M}\}_O = \dot{\boldsymbol{r}}_{P/M} + \boldsymbol{\omega} \times \boldsymbol{r}_{P/M} \tag{7.8}$$

把式(7.8)代入式(7.7)，得

$$\dot{\boldsymbol{r}}_P = \dot{\boldsymbol{r}}_M + \dot{\boldsymbol{r}}_{P/M} + \boldsymbol{\omega} \times \boldsymbol{r}_{P/M} \tag{7.9}$$

式中，$\dot{\boldsymbol{r}}_P = \boldsymbol{v}_P$ 为动点 P 的速度，$\dot{\boldsymbol{r}}_M = \boldsymbol{v}_M$ 为牵连点 M 的速度，$\dot{\boldsymbol{r}}_{P/M} = \boldsymbol{v}_{P/M}$ 为动点相对牵连点的速度。因此，式(7.9)可重写为

$$\boldsymbol{v}_P = \boldsymbol{v}_M + \boldsymbol{v}_{P/M} + \boldsymbol{\omega} \times \boldsymbol{r}_{P/M} \tag{7.10}$$

利用 $\boldsymbol{r}_{P/M} = 0$，则式(7.10)可简化为

$$\boldsymbol{v}_P = \boldsymbol{v}_M + \boldsymbol{v}_{P/M} \tag{7.11}$$

由此得到结论，在任何瞬时绝对速度 $\boldsymbol{v}_P$ 等于牵连速度 $\boldsymbol{v}_M$ 和相对速度 $\boldsymbol{v}_{P/M}$ 的矢量和。

例 7.1 飞机 A 和 B 在相同高度飞行，假设飞机 A 以速度 $v_A = 400$ km/h 向南飞行，飞机 B 以速度 $v_B = 500$ km/h 向东偏北 30° 方向飞行，求飞机 B 相对飞机 A 的速度。

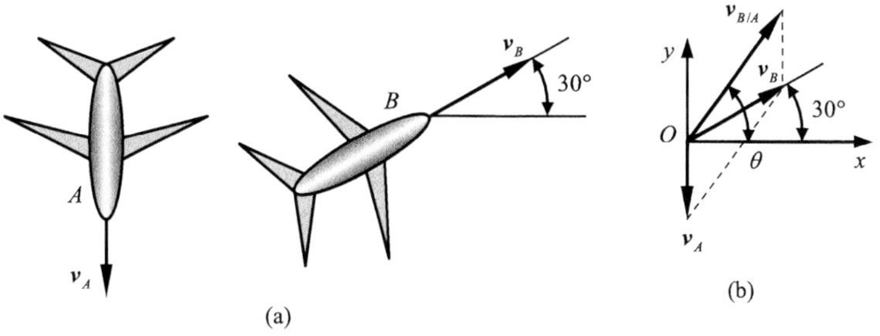

图 E7.1

解　假设地球选为定系，飞机 A 为动系，飞机 B 为动点。根据 $\boldsymbol{v}_P=\boldsymbol{v}_M+\boldsymbol{v}_{P/M}$，则有

$$\boldsymbol{v}_B=\boldsymbol{v}_A+\boldsymbol{v}_{B/A}$$

利用如图 E7.1b 所示坐标系，得

$$(v_B\cos 30°)\boldsymbol{i}+(v_B\sin 30°)\boldsymbol{j}=(-v_A)\boldsymbol{j}+(v_{B/A}\cos\theta)\boldsymbol{i}+(v_{B/A}\sin\theta)\boldsymbol{j}$$

式中，$\boldsymbol{i}$ 和 $\boldsymbol{j}$ 为分别对应 x 和 y 方向的单位矢量，θ 为 $\boldsymbol{v}_{B/A}$ 与 x 轴的夹角。代入 $v_A=400\ \text{km/h}$ 和 $v_B=500\ \text{km/h}$，得

$$v_{B/A}=781.0\ \text{km/h},\ \theta=56.33°$$

因此，飞机 B 相对飞机 A 的速度大小为 781.0 km/h，方向为东偏北 56.33°。

7.3　加速度合成

如图 7.2 所示，在定系 $Oxyz$ 中由式(7.10)对时间进行微分，得

$$\dot{\boldsymbol{v}}_P=\dot{\boldsymbol{v}}_M+\{\dot{\boldsymbol{v}}_{P/M}\}_O+\dot{\boldsymbol{\omega}}\times\boldsymbol{r}_{P/M}+\boldsymbol{\omega}\times\{\dot{\boldsymbol{r}}_{P/M}\}_O \tag{7.12}$$

利用式(7.5)，有

$$\{\dot{\boldsymbol{v}}_{P/M}\}_O=\dot{\boldsymbol{v}}_{P/M}+\boldsymbol{\omega}\times\boldsymbol{v}_{P/M} \tag{7.13}$$

和

$$\{\dot{\boldsymbol{r}}_{P/M}\}_O=\dot{\boldsymbol{r}}_{P/M}+\boldsymbol{\omega}\times\boldsymbol{r}_{P/M} \tag{7.14}$$

式(7.13)和(7.14)代入式(7.12)，得

$$\dot{\boldsymbol{v}}_P=\dot{\boldsymbol{v}}_M+\dot{\boldsymbol{v}}_{P/M}+\boldsymbol{\omega}\times\boldsymbol{v}_{P/M}+\dot{\boldsymbol{\omega}}\times\boldsymbol{r}_{P/M}+\boldsymbol{\omega}\times(\dot{\boldsymbol{r}}_{P/M}+\boldsymbol{\omega}\times\boldsymbol{r}_{P/M}) \tag{7.15}$$

再利用 $\dot{\boldsymbol{v}}_P=\boldsymbol{a}_P$，$\dot{\boldsymbol{v}}_M=\boldsymbol{a}_M$，$\dot{\boldsymbol{v}}_{P/M}=\boldsymbol{a}_{P/M}$，$\dot{\boldsymbol{\omega}}=\boldsymbol{\alpha}$ 和 $\dot{\boldsymbol{r}}_{P/M}=\boldsymbol{v}_{P/M}$，并假设 $\boldsymbol{a}_\text{C}=2\boldsymbol{\omega}\times\boldsymbol{v}_{P/M}$，得

$$\boldsymbol{a}_P=\boldsymbol{a}_M+\boldsymbol{a}_{P/M}+\boldsymbol{a}_\text{C}+\boldsymbol{\alpha}\times\boldsymbol{r}_{P/M}+\boldsymbol{\omega}\times(\boldsymbol{\omega}\times\boldsymbol{r}_{P/M}) \tag{7.16}$$

式中，$\boldsymbol{a}_P$ 为动点 P 的加速度，$\boldsymbol{a}_M$ 为牵连点 M 的加速度，$\boldsymbol{a}_{P/M}$ 为动点 P 相对牵连点 M 的加速度，$\boldsymbol{a}_\text{C}=2\boldsymbol{\omega}\times\boldsymbol{v}_{P/M}$ 为科氏加速度，$\boldsymbol{\alpha}$ 为动系角加速度。利用 $\boldsymbol{r}_{P/M}=0$，式(7.16)可简化为

$$\boldsymbol{a}_P=\boldsymbol{a}_M+\boldsymbol{a}_{P/M}+\boldsymbol{a}_\text{C} \tag{7.17}$$

由此得到结论，在任何瞬时绝对加速度 $\boldsymbol{a}_P$ 等于牵连加速度 $\boldsymbol{a}_M$、相对加速度 $\boldsymbol{a}_{P/M}$ 和科氏

加速度 $\boldsymbol{a}_C$ 的矢量和。

例 7.2　已知图 E7.2a 所示瞬时，杆 *BD* 以不变角速度 $\omega_{BD}=2$ rad/s 逆时针旋转，求：(a) 杆 *AD* 的角速度和轴环 *D* 相对杆 *BD* 的速度；(b) 杆 *AD* 的角加速度和轴环 *D* 相对杆 *BD* 的加速度。

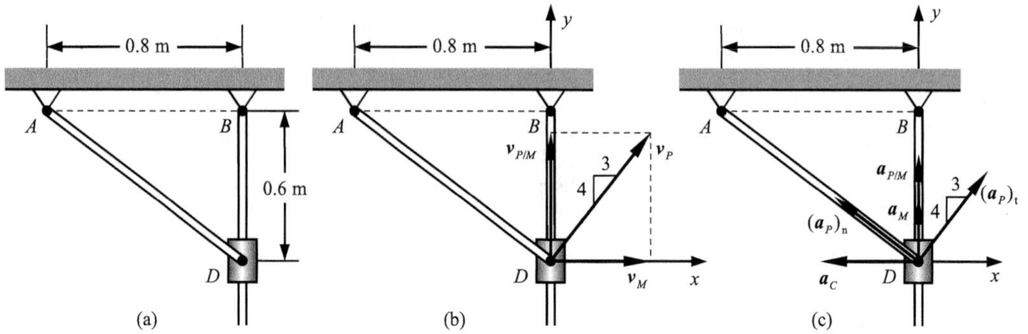

图 E7.2

解　取地球为定系、杆 *BD* 为动系和轴环 *D* 为动点。

(a) 利用 $\boldsymbol{v}_P=\boldsymbol{v}_M+\boldsymbol{v}_{P/M}$，如图 E7.2b，得

$$\frac{3}{5}v_P\boldsymbol{i}+\frac{4}{5}v_P\boldsymbol{j}=v_M\boldsymbol{i}+v_{P/M}\boldsymbol{j}$$

式中，$\boldsymbol{i}$ 和 $\boldsymbol{j}$ 为单位矢量。解上式，并利用 $v_M=BD\cdot\omega_{BD}=1.2\text{ m/s}$，得

$$v_P=\frac{5}{3}v_M=2\text{ m/s},\ v_{P/M}=\frac{4}{3}v_M=1.6\text{ m/s}$$

再利用 $v_P=AD\cdot\omega_{AD}$，得

$$\omega_{AD}=\frac{v_P}{AD}=2\text{ rad/s}\ (逆时针)$$

(b) 利用 $(\boldsymbol{a}_P)_\text{t}+(\boldsymbol{a}_P)_\text{n}=\boldsymbol{a}_M+\boldsymbol{a}_{P/M}+\boldsymbol{a}_\text{C}$，如图 E7.2c，得

$$[\frac{3}{5}(a_P)_\text{t}\boldsymbol{i}+\frac{4}{5}(a_P)_\text{t}\boldsymbol{j}]+[-\frac{4}{5}(a_P)_\text{n}\boldsymbol{i}+\frac{3}{5}(a_P)_\text{n}\boldsymbol{j}]=a_M\boldsymbol{j}+a_{P/M}\boldsymbol{j}-a_\text{C}\boldsymbol{i}$$

式中，$(a_P)_\text{n}=AD\cdot\omega_{AD}{}^2=4\text{ m/s}^2$，$a_M=BD\cdot\omega_{BD}{}^2=2.4\text{ m/s}^2$ 和 $a_\text{C}=2\omega_{BD}v_{P/M}=6.4\text{ m/s}^2$。解上式，得

$$(a_P)_\text{t}=-5.33\text{ m/s}^2,\ a_{P/M}=-4.27\text{ m/s}^2$$

利用 $(a_P)_\text{t}=AD\cdot\alpha_{AD}$，得

$$\alpha_{AD}=\frac{(a_P)_\text{t}}{AD}=-5.33\text{ rad/s}^2\ (顺时针)$$

例 7.3　在卡车以加速度 2 m/s^2 倒车的同时，吊杆外段 *D* 相对卡车以加速度 1 m/s^2 缩回，如图 E7.3a。求吊杆外段 *D* 的加速度。

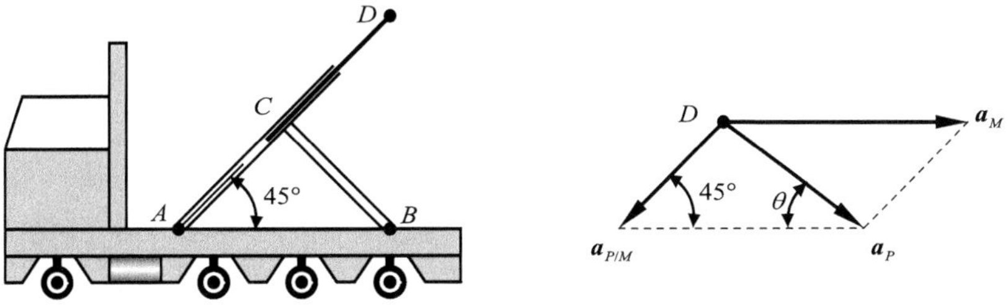

(a)　　(b)

图 E7.3

解　取地球为定系、卡车为动系、吊杆外段 D 为动点。根据 $\boldsymbol{a}_P = \boldsymbol{a}_M + \boldsymbol{a}_{P/M}$ 画加速度三角形，如图 E7.3b，得

$$a_P = \sqrt{(a_M)^2 + (a_{P/M})^2 - 2a_M a_{P/M}\cos 45°} = 1.47\ \text{m/s}^2,\ \theta = \arcsin\left(\frac{a_{P/M}}{a_P}\sin 45°\right) = 28.75°$$

因此，吊杆外段 D 的加速度大小为 $1.47\ \text{m/s}^2$，方向为东偏南 28.75°。

习　题

7.1　轴环以不变相对速率 v 沿杆 OAB 向外滑动，同时杆 OAB 以不变角速度 $\omega_B = 2$ rad/s 逆时针转动。已知 $\theta = 0$ 时轴环位于 A 处，$\theta = 90°$ 时轴环到达 B 处，求 $\theta = 30°$ 时轴环的加速度。

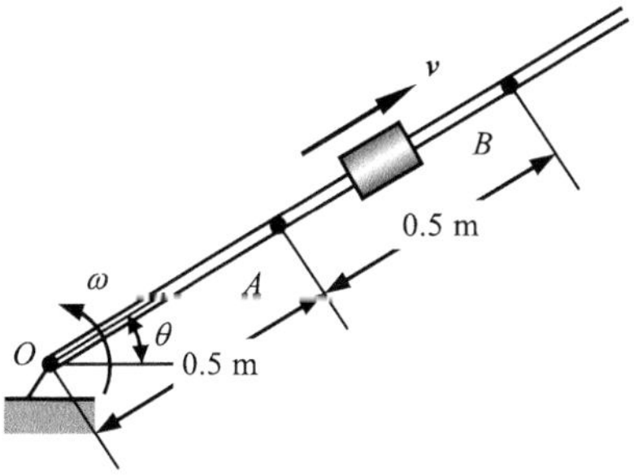

图 P7.1

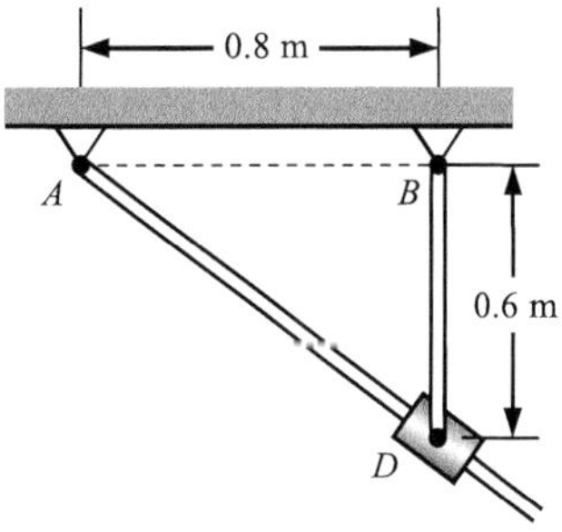

图 P7.2

7.2　已知图示瞬时杆 BD 以不变角速度 $\omega_B = 2$ rad/s 逆时针旋转，求：(a) 杆 AD 的角速度和轴环 D 相对杆 AD 的速度；(b) 杆 AD 的角加速度和轴环 D 相对杆 AD 的加速度。

7.3　在卡车以加速度 $3\ \text{m/s}^2$ 前进的同时，吊杆外段 D 相对卡车以加速度 $2\ \text{m/s}^2$ 缩回。求吊杆外段 D 的加速度。

7.4　轴环 E 沿杆 BD 滑动，并与沿垂直杆 AC 运动的轴环 F 相连。已知杆 BD 的角速度和角加速度分别为 $\omega = 6$ rad/s 和 $\alpha = 4\ \text{rad/s}^2$，两者都为顺时针，求轴环 E 的速度和加速度。

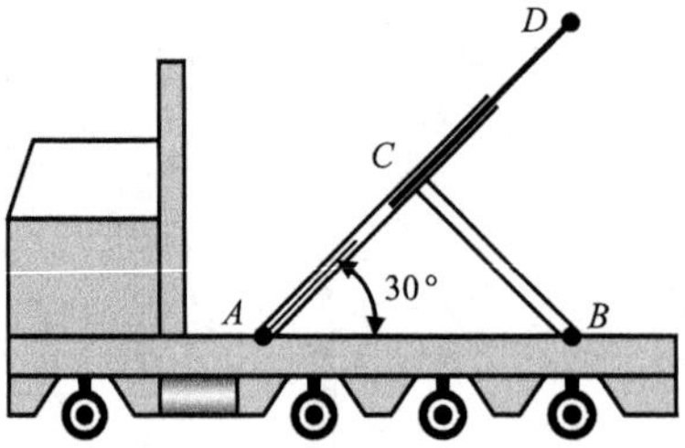

图 P7.3

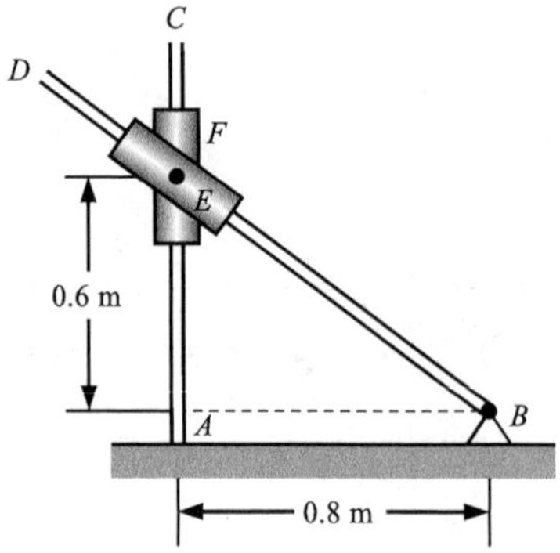

图 P7.4

7.5　沿直线水平飞行的飞机 A 在图示瞬时的速度为300 km/h，加速度为5 m/s^2 。飞机 B 在相同高度沿半径为200 m的圆形轨迹飞行。已知飞机 B 在图示瞬时的速度为400 km/h，减速度为3 m/s^2，求图示瞬时(a) B 相对 A 的速度，(b) B 相对 A 的加速度。

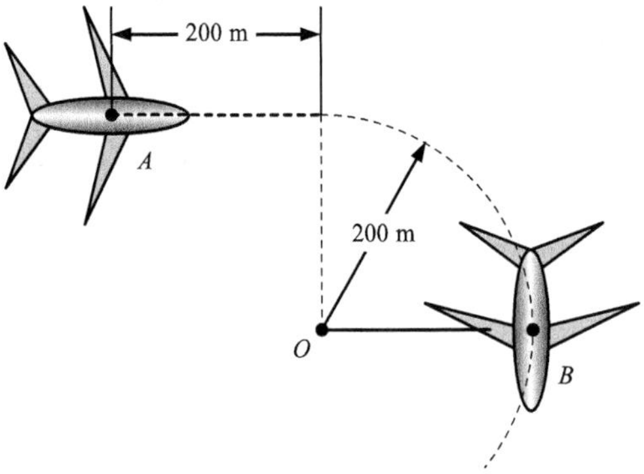

图 P7.5

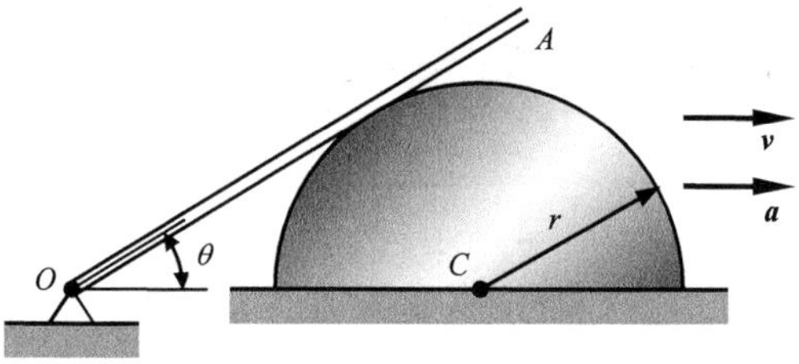

图 P7.6

7.6　杆 OA 始终与半径 $r=0.4$ m的半圆柱 C 的表面相切。已知半圆柱的速度和加速度分别为 $v=0.1$ m/s和 $a=0.2$ m/s^2，方向都向右，求 $\theta=30°$ 时杆的角速度和角加速度。

7.7　当杆 AB 的端点 A 以不变速度600 mm/s向右移动时，杆 AB 将在 C 处小轮上运动。在图示瞬时，求：(a)杆的角速度和端点 B 的速度；(b)杆的角加速度和端点 B 的加速度。

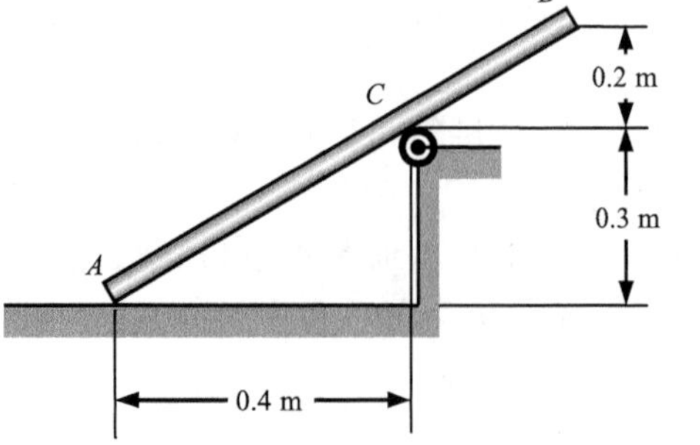

图 P7.7

第 8 章　质点动力学

8.1　运 动 方 程

根据牛顿第二定律，如果作用于质点上的合力不为零，则质点将具有与合力大小成正比并沿合力方向的加速度。数学上，该定律可表示为

$$m\boldsymbol{a} = \sum \boldsymbol{F} \tag{8.1}$$

式中，m 为质点质量，$\boldsymbol{a}$ 为质点加速度，$\sum \boldsymbol{F}$ 为合力。应该注意，牛顿第二运动定律仅在牛顿参考系或惯性参考系中成立。在质点运动问题求解中，上述矢量方程替换为由直角或自然坐标表示的等效标量方程将会给解题带来极大方便。

把加速度和力分解为直角分量，得

$$ma_x = \sum F_x,\ ma_y = \sum F_y,\ ma_z = \sum F_z \tag{8.2}$$

式中，$a_x = \dot{v}_x = \ddot{x},\ a_y = \dot{v}_y = \ddot{y},\ a_z = \dot{v}_z = \ddot{z}$ 。

把加速度和力分解为切向分量(指向运动方向)和法向分量(指向曲率中心)，得

$$ma_{\mathrm{t}} = \sum F_{\mathrm{t}},\ ma_{\mathrm{n}} = \sum F_{\mathrm{n}},\ 0 = \sum F_{\mathrm{b}} \tag{8.3}$$

式中，$a_{\mathrm{t}} = \dot{v},\ a_{\mathrm{n}} = \dfrac{v^2}{\rho}$ 。

例 8.1　如图 E8.1a，弹簧常数为 k 的弹簧 AB 与支撑 A 和质量为 m 的环 B 相连，弹簧原长为l。已知环从 $x = \sqrt{3}l$ 处静止释放，不计环与水平杆之间的摩擦，求环通过中点 O 时的速度大小。

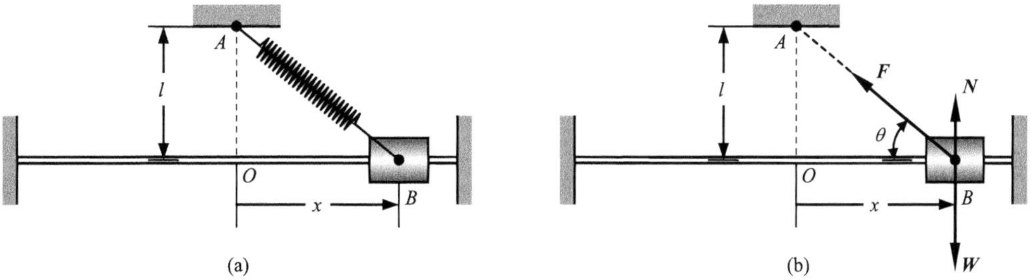

图 E8.1

解　当环在 x 处，弹簧伸长可表示为

$$\delta = \sqrt{l^2 + x^2} - l$$

相应的弹簧弹力为

$$F = k\delta = k(\sqrt{l^2 + x^2} - l)$$

由运动方程，$ma_x = \sum F_x$，得

$$ma = -F\cos\theta$$

利用 $a = \dfrac{\mathrm{d}v}{\mathrm{d}t} = \dfrac{\mathrm{d}x}{\mathrm{d}t}\dfrac{\mathrm{d}v}{\mathrm{d}x} = v\dfrac{\mathrm{d}v}{\mathrm{d}x}$ 和 $\cos\theta = \dfrac{x}{\sqrt{l^2 + x^2}}$，得

$$v\mathrm{d}v = -\frac{k}{m}(x - \frac{lx}{\sqrt{l^2 + x^2}})\mathrm{d}x$$

进行积分，得

$$\int_0^{v_O} v\mathrm{d}v = -\frac{k}{m}\int_{\sqrt{3}l}^{0}(x - \frac{lx}{\sqrt{l^2 + x^2}})\mathrm{d}x$$

即

$$v_O = l\sqrt{k/m}$$

例 8.2　如图 E8.2a，物块在 A 处的初速度 $v_A = 8\ \mathrm{m/s}$。已知物块与斜面之间的动摩擦系数 $\mu_\mathrm{k} = 0.25$，$\theta = 10°$，物块到达 B 处时静止，求物块移动距离和所用时间。

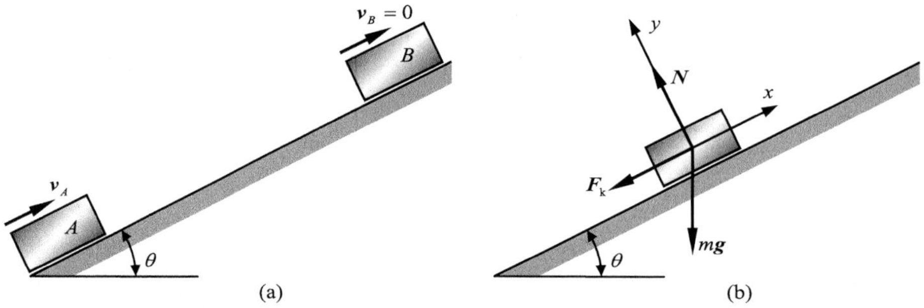

图 E8.2

解　如图 E8.2b 所示，根据运动方程，$ma_x = \sum F_x$ 和 $ma_y = \sum F_y$，得

$$ma_x = \sum F_x,\ ma = -mg\sin\theta - F_\mathrm{k}$$

$$ma_y = \sum F_y,\ 0 = N - mg\cos\theta$$

利用 $F_\mathrm{k} = \mu_\mathrm{k} N$，求解上述方程，得

$$a = -(\sin\theta + \mu_\mathrm{k}\cos\theta)g$$

（1）利用 $a = \dfrac{\mathrm{d}v}{\mathrm{d}t} = \dfrac{\mathrm{d}x}{\mathrm{d}t}\dfrac{\mathrm{d}v}{\mathrm{d}x} = v\dfrac{\mathrm{d}v}{\mathrm{d}x}$，有

$$v\mathrm{d}v = -(\sin\theta + \mu_\mathrm{k}\cos\theta)g\mathrm{d}x$$

进行积分，得

$$\int_{v_A}^{v_B} v\mathrm{d}v = \int_0^{x_{AB}} -(\sin\theta + \mu_\mathrm{k}\cos\theta)g\mathrm{d}x$$

即

$$x_{AB} = \frac{{v_A}^2 - {v_B}^2}{2(\sin\theta + \mu_k \cos\theta)g}$$

代入 $v_A = 8\ \text{m/s}$，$v_B = 0$，$\mu_k = 0.25$，$\theta = 10°$ 和 $g = 9.81\ \text{m/s}^2$，得

$$x_{AB} = 7.77\ \text{m}$$

(2) 利用 $a = \frac{dv}{dt}$，得

$$dv = -(\sin\theta + \mu_k \cos\theta)g dt$$

积分，得

$$\int_{v_A}^{v_B} dv = \int_0^{t_{AB}} -(\sin\theta + \mu_k \cos\theta)g dt$$

即

$$t_{AB} = \frac{v_A - v_B}{(\sin\theta + \mu_k \cos\theta)g}$$

代入 $v_A = 8\ \text{m/s}$，$v_B = 0$，$\mu_k = 0.25$，$\theta = 10°$ 和 $g = 9.81\ \text{m/s}^2$，得

$$t_{AB} = 1.94\ \text{s}$$

8.2 惯 性 力 法

回到式(8.1)，并假设 $\boldsymbol{F}_\text{I} = -m\boldsymbol{a}$，则牛顿第二定律可重写为

$$\sum \boldsymbol{F} + \boldsymbol{F}_\text{I} = 0 \tag{8.4}$$

式中，$\boldsymbol{F}_\text{I}$ 称为惯性力，其大小等于 ma，方向与 $\boldsymbol{a}$ 相反。这种方法由达朗贝尔提出用于分析质点的运动，通常称为惯性力法或达朗贝尔原理。应该注意，惯性力是假想的力，实际并不存在。

惯性力法可以把动力学问题在形式上转换为等效的平衡问题，因而可以采用静力学方法求解动力学问题。虽然质点并不平衡，但是如果在运动质点上施加惯性力，那么即可采用静力学平衡方程研究质点的运动。

8.3 功 能 法

功能法可用于求解涉及力、质量、速度和位移的动力学问题。

1. 功能原理

式(8.1)两边点乘 $d\boldsymbol{r}$，得

$$m\boldsymbol{a} \cdot d\boldsymbol{r} = \sum \boldsymbol{F} \cdot d\boldsymbol{r} \tag{8.5}$$

利用 $\boldsymbol{a}\cdot \mathrm{d}\boldsymbol{r}=\mathrm{d}\boldsymbol{v}\cdot \boldsymbol{v}=\frac{1}{2}\mathrm{d}(\boldsymbol{v}\cdot \boldsymbol{v})=\frac{1}{2}\mathrm{d}(v^2)$，得

$$\mathrm{d}T=\mathrm{d}W \tag{8.6}$$

式中，$\mathrm{d}T=\mathrm{d}(\frac{1}{2}mv^2)$ 和 $\mathrm{d}W=\sum \boldsymbol{F}\cdot \mathrm{d}\boldsymbol{r}$ 分别是质点动能增量和力 $\sum \boldsymbol{F}$ 在位移 $\mathrm{d}\boldsymbol{r}$ 上所做的元功。从 A_1 到 A_2 积分，得

$$T_2-T_1=W_{12} \tag{8.7}$$

式中，$T_1=\frac{1}{2}m{v_1}^2$ 和 $T_2=\frac{1}{2}m{v_2}^2$ 为质点分别在 A_1 和 A_2 处的动能，$W_{12}=\sum\int_{A_1}^{A_2}\boldsymbol{F}\cdot \mathrm{d}\boldsymbol{r}$ 为质点从 A_1 移动到 A_2 过程中作用在质点上的力所做的功。

因此，由式(8.6)或(8.7)得到结论，当质点从一点移动到另一点时，质点动能的变化等于作用于质点上的力所做的功。式(8.6)或(8.7)所表示的功和能之间的关系称为功能原理。应该注意，功能原理仅在牛顿参考系成立。

例 8.3　如图 E8.3，物块在 A 处的初速度 $v_A=8\ \mathrm{m/s}$。已知物块与斜面之间的动摩擦系数 $\mu_\mathrm{k}=0.25$，$\theta=10°$，物块到达 B 处时静止，求物块移动距离。

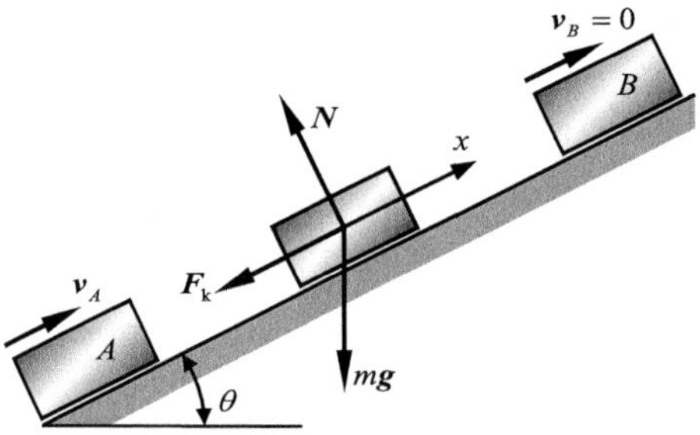

图 E8.3

解　利用功能原理，$T_B-T_A=W_{AB}$，得

$$\frac{1}{2}m{v_B}^2-\frac{1}{2}m{v_A}^2=(-F_\mathrm{k}-mg\sin\theta)x_{AB}$$

式中，$F_\mathrm{k}=\mu_\mathrm{k}N=\mu_\mathrm{k}mg\cos\theta$。

代入 $v_A=8\ \mathrm{m/s}$，$v_B=0$，$\mu_\mathrm{k}=0.25$，$\theta=10°$ 和 $g=9.81\ \mathrm{m/s^2}$，得

$$x_{AB}=\frac{{v_A}^2-{v_B}^2}{2(\sin\theta+\mu_\mathrm{k}\cos\theta)g}=7.77\ \mathrm{m}$$

2. 能量守恒

当质点从一点移动到另一点，如果作用于质点上的力所做的功与质点的移动轨迹无关，则作用于质点上的力称为保守力。重力和弹力是典型的保守力。

当质点从 A_1 移动到 A_2 时作用于质点上的保守力所做的功定义为质点在 A_1 处相对于 A_2 处的势能，即

$$V_1 - V_2 = W_{12} = \sum \int_{A_1}^{A_2} \boldsymbol{F} \cdot \mathrm{d}\boldsymbol{r} \tag{8.8}$$

式中，V_1 和 V_2 分别表示质点在 A_1 和 A_2 处的势能。

结合式(8.7)和(8.8)，得

$$T_1 + V_1 = T_2 + V_2 \tag{8.9}$$

式(8.9)表明，当质点在保守力作用下发生运动，则质点的动能和势能之和保持不变。动能和势能之和称为机械能。

例 8.4　如图 E8.4，弹簧常数为 k 的弹簧 AB 与支撑 A 和质量为 m 的环 B 相连，弹簧原长为 l。已知环从 $x=\sqrt{3}l$ 处静止释放，不计环与水平杆之间的摩擦，求环通过中点 O 时的速度大小。

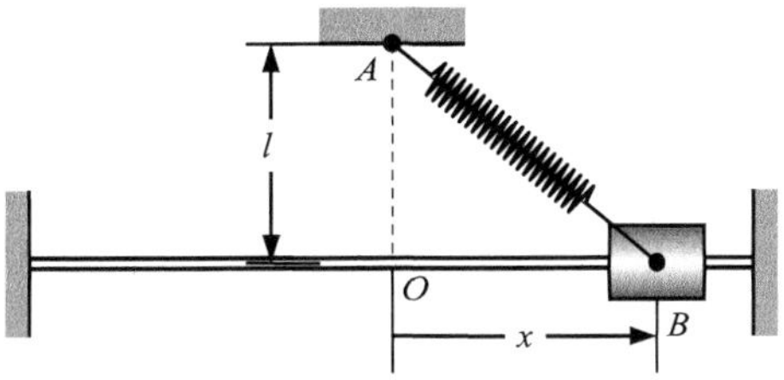

图 E8.4

解　当环在 $x=\sqrt{3}l$ 处，有

$$T_1 = 0,\ V_1 = \frac{1}{2}k\delta^2 = \frac{1}{2}k[\sqrt{l^2 + (\sqrt{3}l)^2} - l]^2 = \frac{1}{2}kl^2$$

当环在中点 O 处，有

$$T_2 = \frac{1}{2}m{v_O}^2,\ V_2 = 0$$

利用机械能守恒，$T_1 + V_1 = T_2 + V_2$，得

$$0 + \frac{1}{2}kl^2 = \frac{1}{2}m{v_O}^2 + 0$$

$$v_O = l\sqrt{k/m}$$

8.4　冲量动量法

冲量动量法可用于求解涉及力、质量、速度和时间的动力学问题。

1. 冲量动量原理

式(8.1)两边乘 $\mathrm{d}t$，得

$$m\boldsymbol{a}\mathrm{d}t = \sum \boldsymbol{F}\mathrm{d}t \tag{8.10}$$

利用 $\boldsymbol{a}\mathrm{d}t = \mathrm{d}\boldsymbol{v}$，得

$$\mathrm{d}\boldsymbol{L} = \mathrm{d}\boldsymbol{I} \tag{8.11}$$

式中，$\mathrm{d}\boldsymbol{L}=\mathrm{d}(m\boldsymbol{v})$ 和 $\mathrm{d}\boldsymbol{I}=\sum \boldsymbol{F}\mathrm{d}t$ 分别为质点动量的增量和作用于质点上的力 $\sum \boldsymbol{F}$ 在时间间隔 $\mathrm{d}t$ 内的元冲量。从 t_1 到 t_2 积分，得

$$\boldsymbol{L}_2-\boldsymbol{L}_1=\boldsymbol{I}_{12} \tag{8.12}$$

式中，$\boldsymbol{L}_1=m\boldsymbol{v}_1$ 和 $\boldsymbol{L}_2=m\boldsymbol{v}_2$ 为质点分别在 t_1 和 t_2 时的动量，$\boldsymbol{I}_{12}=\sum\int_{t_1}^{t_2}\boldsymbol{F}\mathrm{d}t$ 为时间从 t_1 到 t_2 间隔中作用在质点上的力的冲量。

因此，从式(8.11)或(8.12)得到结论，在所考虑的时间间隔内，质点动量的变化等于作用于质点上的力的冲量。式(8.11)或式(8.12)所表示的冲量和动量之间的关系称为冲量动量原理。冲量动量原理是矢量方程，因此当用于求解问题时，式(8.11)或式(8.12)需要分解为直角坐标方程。

例 8.5　如图 E8.5，物块在 A 处的初速度 $v_A=8\ \mathrm{m/s}$。已知物块与斜面之间的动摩擦系数 $\mu_\mathrm{k}=0.25$，$\theta=10°$，物块到达 B 处时静止，求所用时间。

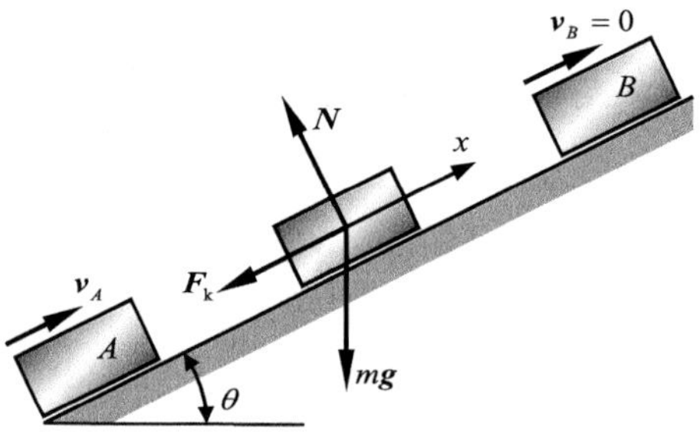

图 E8.5

解　由冲量动量原理，$\boldsymbol{L}_B-\boldsymbol{L}_A=\boldsymbol{I}_{AB}$，得

$$mv_B-mv_A=(-F_\mathrm{k}-mg\sin\theta)t_{AB}$$

式中，$F_\mathrm{k}=\mu_\mathrm{k}N=\mu_\mathrm{k}mg\cos\theta$。

代入 $v_A=8\ \mathrm{m/s}$，$v_B=0$，$\mu_\mathrm{k}=0.25$，$\theta=10°$ 和 $g=9.81\ \mathrm{m/s^2}$，得

$$t_{AB}=\frac{v_A-v_B}{(\sin\theta+\mu_\mathrm{k}\cos\theta)g}=1.94\ \mathrm{s}$$

2. 动量守恒

如果作用于质点上的力的合力为零，则式(8.12)可化简为

$$\boldsymbol{L}_1=\boldsymbol{L}_2 \tag{8.13}$$

该式表明动量守恒。

8.5　质点系运动方程

为推导由 n 个质点构成的质点系的运动方程，可把牛顿第二定律应用于系统的每一

个质点。考虑质点 P_i，质量为 m_i，外力合力为 $\boldsymbol{F}_i^{(e)}$，内力合力(即系统内所有其他质点施加在质点 P_i 上的力)为 $\boldsymbol{F}_i^{(i)}$。对质点 P_i，牛顿第二定律可表示为

$$m_i\boldsymbol{a}_i = \boldsymbol{F}_i^{(e)} + \boldsymbol{F}_i^{(i)} \quad (i=1,2,\cdots,n) \tag{8.14}$$

式中，$\boldsymbol{a}_i$ 为质点 P_i 相对牛顿参考系的加速度。式(8.14)各项对定点 O 取矩，有

$$\boldsymbol{r}_i \times m_i\boldsymbol{a}_i = \boldsymbol{r}_i \times \boldsymbol{F}_i^{(e)} + \boldsymbol{r}_i \times \boldsymbol{F}_i^{(i)} \quad (i=1,2,\cdots,n) \tag{8.15}$$

式中，$\boldsymbol{r}_i$ 为质点 P_i 相对点 O 的位置矢量。

考虑系统中的所有质点，并利用式(8.14)和(8.15)，得

$$\sum m_i\boldsymbol{a}_i = \sum \boldsymbol{F}_i^{(e)} + \sum \boldsymbol{F}_i^{(i)} \tag{8.16}$$

$$\sum(\boldsymbol{r}_i \times m_i\boldsymbol{a}_i) = \sum(\boldsymbol{r}_i \times \boldsymbol{F}_i^{(e)}) + \sum(\boldsymbol{r}_i \times \boldsymbol{F}_i^{(i)}) \tag{8.17}$$

代入 $\sum \boldsymbol{F}_i^{(i)} = 0$ 和 $\sum(\boldsymbol{r}_i \times \boldsymbol{F}_i^{(i)}) = 0$，式(8.16)和(8.17)化简为

$$\sum m_i\boldsymbol{a}_i = \sum \boldsymbol{F}_i^{(e)} \tag{8.18}$$

$$\sum(\boldsymbol{r}_i \times m_i\boldsymbol{a}_i) = \sum(\boldsymbol{r}_i \times \boldsymbol{F}_i^{(e)}) \tag{8.19}$$

质点系的线动量定义为系统中各个质点的线动量之和，即

$$\boldsymbol{L} = \sum m_i\boldsymbol{v}_i \tag{8.20}$$

式(8.20)对时间微分，有

$$\dot{\boldsymbol{L}} = \sum m_i\boldsymbol{a}_i \tag{8.21}$$

质点系对定点 O 的角动量定义为系统中各个质点对相同点 O 的角动量之和，即

$$\boldsymbol{H}_O = \sum(\boldsymbol{r}_i \times m_i\boldsymbol{v}_i) \tag{8.22}$$

式中，$\boldsymbol{r}_i \times m_i\boldsymbol{v}_i$ 为质点 P_i 对定点 O 的角动量。

式(8.22)对时间微分，得

$$\dot{\boldsymbol{H}}_O = \sum(\boldsymbol{v}_i \times m_i\boldsymbol{v}_i) + \sum(\boldsymbol{r}_i \times m_i\boldsymbol{a}_i) \tag{8.23}$$

利用 $\boldsymbol{v}_i \times \boldsymbol{v}_i = 0$，则化简为

$$\dot{\boldsymbol{H}}_O = \sum(\boldsymbol{r}_i \times m_i\boldsymbol{a}_i) \tag{8.24}$$

把式(8.21)和(8.24)代入式(8.18)和(8.19)，得

$$\dot{\boldsymbol{L}} = \sum \boldsymbol{F}_i^{(e)} \tag{8.25}$$

$$\dot{\boldsymbol{H}}_O = \sum(\boldsymbol{r}_i \times \boldsymbol{F}_i^{(e)}) = \sum \boldsymbol{M}_O(\boldsymbol{F}_i^{(e)}) \tag{8.26}$$

由此得到结论，质点系线动量的变化率和质点系对定点 O 的角动量的变化率分别等于作用于系统质点上的外力的合力和作用在系统质点上的外力对 O 的合力矩。

8.6　质点系质心运动方程

假设 $\boldsymbol{r}_C$ 为质点系质心的位置矢量，则有

$$m\boldsymbol{r}_C = \sum m_i\boldsymbol{r}_i \tag{8.27}$$

式中，$m=\sum m_i$ 为质点系总质量。

式(8.27)对时间微分两次，得

$$m\boldsymbol{a}_C=\sum m_i\boldsymbol{a}_i \tag{8.28}$$

式中，$\boldsymbol{a}_C$ 为质点系质心的加速度。

把式(8.18)代入式(8.28)，得

$$m\boldsymbol{a}_C=\sum \boldsymbol{F}_i^{(\mathrm{e})} \tag{8.29}$$

上述方程可以描述质点系的质心的运动。因此得到结论，质点系的质心的运动相当于一个质点的运动，该质点位于质点系的质心位置，并集中质点系的全部质量和作用在质点系上的所有外力。

8.7　质点系相对质心的运动方程

采用平移形心参考系研究质点系中质点的运动通常显得非常方便。设 $\boldsymbol{r}_i'$ 和 $\boldsymbol{v}_i'$ 分别为质点 P_i 相对形心 C 的位置矢量和相对形心系 $Cx'y'z'$ 的速度，则质点系相对质心的角动量定义如下：

$$\boldsymbol{H}_C'=\sum(\boldsymbol{r}_i'\times m_i\boldsymbol{v}_i') \tag{8.30}$$

同理，可定义

$$\boldsymbol{H}_C=\sum(\boldsymbol{r}_i'\times m_i\boldsymbol{v}_i) \tag{8.31}$$

式中，$\boldsymbol{v}_i$ 为在牛顿参考系 $Oxyz$ 中观察到的绝对速度。

利用 $\boldsymbol{v}_i-\boldsymbol{v}_i'=\boldsymbol{v}_C$，得

$$\boldsymbol{H}_C-\boldsymbol{H}_C'=\sum(\boldsymbol{r}_i'\times m_i\boldsymbol{v}_i)-\sum(\boldsymbol{r}_i'\times m_i\boldsymbol{v}_i')=(\sum m_i\boldsymbol{r}_i')\times\boldsymbol{v}_C \tag{8.32}$$

代入 $\sum m_i\boldsymbol{r}_i'=m\boldsymbol{r}_C'=0$，得

$$\boldsymbol{H}_C=\boldsymbol{H}_C' \tag{8.33}$$

利用 $\boldsymbol{r}_i-\boldsymbol{r}_i'=\boldsymbol{r}_C$，得

$$\boldsymbol{H}_O-\boldsymbol{H}_C=\sum(\boldsymbol{r}_i\times m_i\boldsymbol{v}_i)-\sum(\boldsymbol{r}_i'\times m_i\boldsymbol{v}_i)=\boldsymbol{r}_C\times(\sum m_i\boldsymbol{v}_i) \tag{8.34}$$

代入 $\sum m_i\boldsymbol{v}_i=m\boldsymbol{v}_C$，得

$$\boldsymbol{H}_O-\boldsymbol{H}_C=\boldsymbol{r}_C\times m\boldsymbol{v}_C \tag{8.35}$$

对时间微分，并利用 $\boldsymbol{v}_C\times\boldsymbol{v}_C=0$，得

$$\dot{\boldsymbol{H}}_O-\dot{\boldsymbol{H}}_C=\boldsymbol{r}_C\times m\boldsymbol{a}_C \tag{8.36}$$

再利用式(8.26)和(8.29)，式(8.36)可写为

$$\dot{\boldsymbol{H}}_C=\dot{\boldsymbol{H}}_O-\boldsymbol{r}_C\times m\boldsymbol{a}_C=\sum(\boldsymbol{r}_i\times\boldsymbol{F}_i^{(\mathrm{e})})-\boldsymbol{r}_C\times\sum\boldsymbol{F}_i^{(\mathrm{e})}=\sum[(\boldsymbol{r}_i-\boldsymbol{r}_C)\times\boldsymbol{F}_i^{(\mathrm{e})}] \tag{8.37}$$

利用 $\boldsymbol{r}_i-\boldsymbol{r}_C=\boldsymbol{r}_i'$，有

$$\dot{\boldsymbol{H}}_C=\sum(\boldsymbol{r}_i'\times\boldsymbol{F}_i^{(\mathrm{e})})=\sum\boldsymbol{M}_C(\boldsymbol{F}_i^{(\mathrm{e})}) \tag{8.38}$$

由此得到结论，质点系相对质心的角动量的变化率等于作用于系统质点上的外力对质心的合力矩。

习　题

8.1　承包人使用含有平台的电驱升降机把质量为 m 的物块从地面 A 运送到屋顶 B，升降机平台可以沿着固定在梯子上的导轨滑动。升降机首先以不变加速度 a_1 从 A 处由静止开始加速运动，然后再以不变减速度 a_2 做减速运动，一直运动到梯子顶部附近的 B 处停止。已知物块与平台之间的静摩擦系数 $\mu_s = 0.3$ 和梯子倾角 $\theta = 60°$，求物块不在平台上发生滑动的最大允许加速度 a_1 和最大允许减速度 a_2。

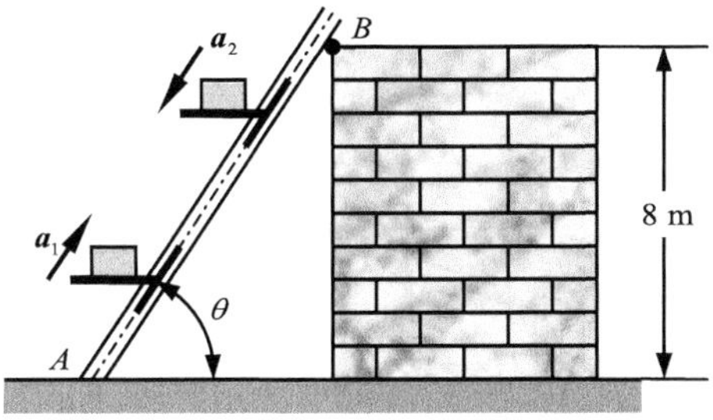

图 P8.1

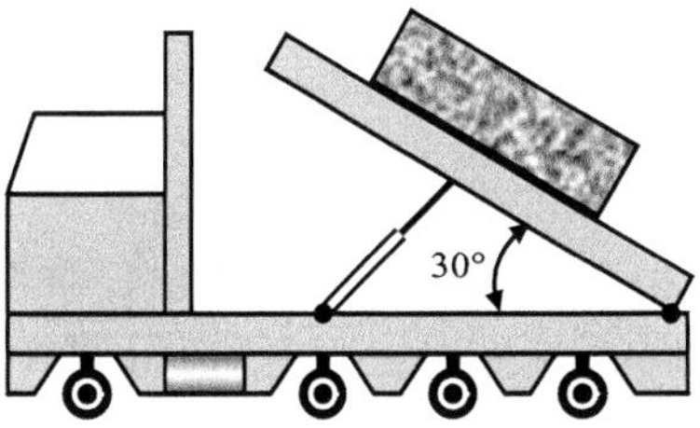

图 P8.2

8.2　为了从卡车上卸下石块，驾驶员首先倾斜卡车底板，然后从静止开始加速。已知物块与底板之间的摩擦系数 $\mu_s = 0.4$ 和 $\mu_k = 0.3$，求物块发生滑动的最小加速度。

8.3　重量 $W_B = 100$ N 的物块 B 与绳索相连后放在重量 $W_A = 300$ N 的物块 A 上面，大小 $F = 200$ N 的水平力作用于绳索。不计摩擦，求：(a) 物块 A 的加速度；(b) 物块 B 相对物块 A 的加速度。

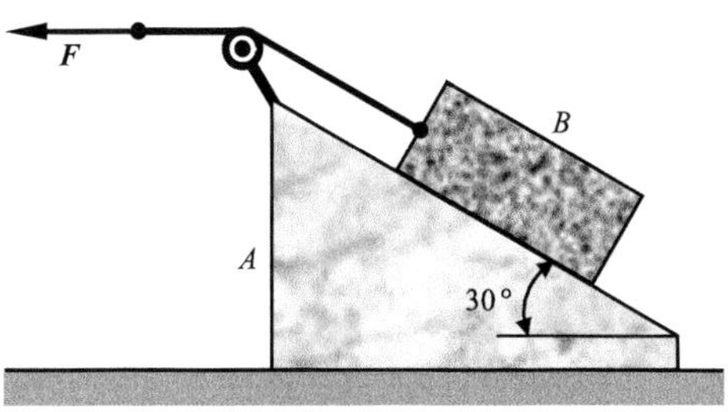

图 P8.3

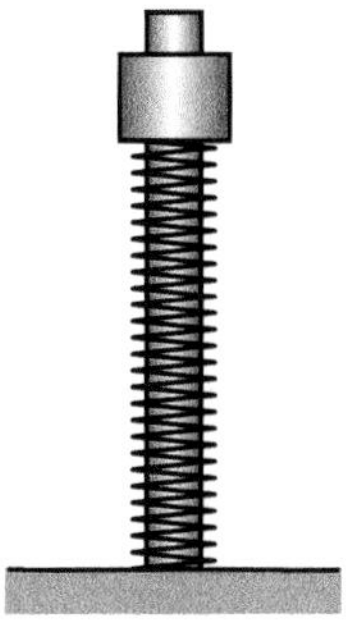

图 P8.4

8.4　重量为 30 N 的环可沿垂直杆发生无摩擦滑动，握住环使之刚好与未伸长弹簧接触。已知弹簧常数 $k = 2$ kN/m，求下列情况下弹簧的最大挠度：(a) 环慢慢释放，直到达到平衡位置；(b) 环突然释放。

8.5　质量为 2 kg 的环与弹簧相连，在垂直平面内沿圆杆发生无摩擦滑动。弹簧原长为 0.1 m，弹簧常数为 k。环在 A 处于静止状态时受到轻微推动而开始向右运动。已知环通过 B 处时将获得最大速度，求：(a) 弹簧常数 k；(b) 环的最大速度。

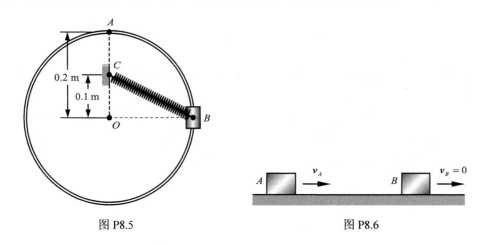

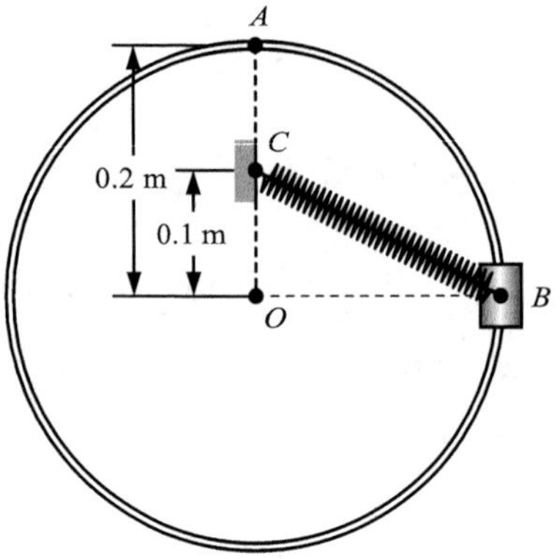

图 P8.5

图 P8.6

8.6　物块在 A 处的初速度 $v_A = 2$ m/s 。已知物块与水平面之间的动摩擦系数 $\mu_k = 0.2$ ，物块到达 B 处时静止，求物块移动距离和所用时间。

8.7　质量为 2 kg 的球通过长度为 0.5 m 的刚性绳索与固定点 O 相连。球在 A 处静止于光滑水平面上，在垂直于 OA 的方向突然给球一个速度 $\boldsymbol{v}_0$ ，球开始无摩擦运动，直到球到达 B 处而使绳索拉紧。如果作用于绳索的冲量不超过 $I = 5$ N·s，求最大允许速度。

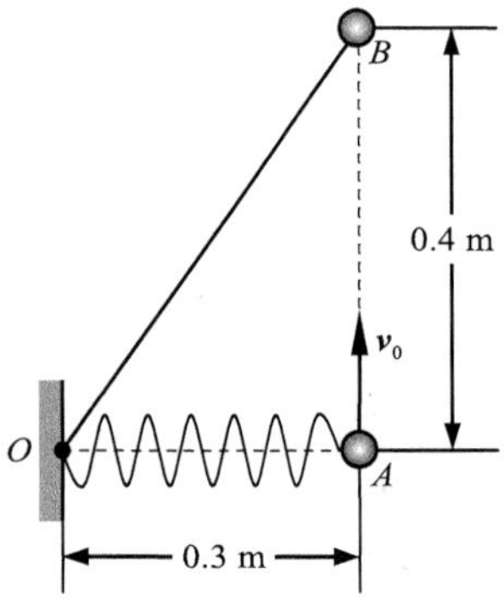

图 P8.7

图 P8.8

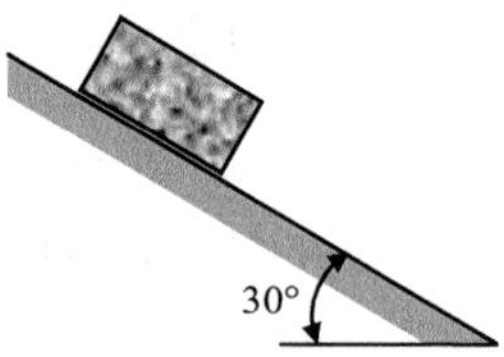

图 P8.9

8.8　三级跳远是田径运动比赛项目。假设重量为 800 N 的运动员以 8 m/s 的水平速度从左侧接近起跳线，保持与地面接触的时间为 0.2 s，并以 10 m/s 的速度沿 60°角起跳，求地面作用于脚上的平均冲力。

8.9　质量 $m = 10$ kg 的物块沿倾角 $\theta = 30°$ 的平面由静止开始向下滑动。已知物块与平面之间的动摩擦系数 $\mu_k = 0.3$ ，求物块运动 $t = 5$ s 后的速度。

第 9 章　刚体平面动力学

9.1　运 动 方 程

对平面运动刚体，刚体质心运动方程和刚体绕质心运动方程可分别写为

$$m\boldsymbol{a}_C = \sum \boldsymbol{F}, \dot{H}_C = \sum M_C(\boldsymbol{F}) \tag{9.1}$$

式中，H_C 可表示为

$$H_C = H'_C = \int r'(\mathrm{d}mv') = (\int r'^2 \mathrm{d}m)\omega = I_C\omega \tag{9.2}$$

式中，$I_C = \int r'^2 \mathrm{d}m$ 为刚体绕质心的转动惯量，ω 为刚体角速度。式(9.2)对时间微分，得

$$\dot{H}_C = I_C\dot{\omega} = I_C\alpha \tag{9.3}$$

式中，α 为刚体的角加速度。

式(9.3)代入式(9.1)，得平面运动刚体的运动方程：

$$m\boldsymbol{a}_C = \sum \boldsymbol{F}, I_C\alpha = \sum M_C(\boldsymbol{F}) \tag{9.4}$$

把上述矢量方程分解为直角坐标分量，式(9.4)可重写为

$$ma_{Cx} = \sum F_x, ma_{Cy} = \sum F_y, I_C\alpha = \sum M_C(\boldsymbol{F}) \tag{9.5}$$

例 9.1　质量为 40 kg 的均质薄板放在卡车上，A 端由光滑垂直面支撑，B 端静止放在粗糙水平面上。已知卡车减速度为 2 m/s^2，$l = 2$ m，$\theta = 60°$，求：(a) A 和 B 处反力；(b) B 处最小静摩擦系数。

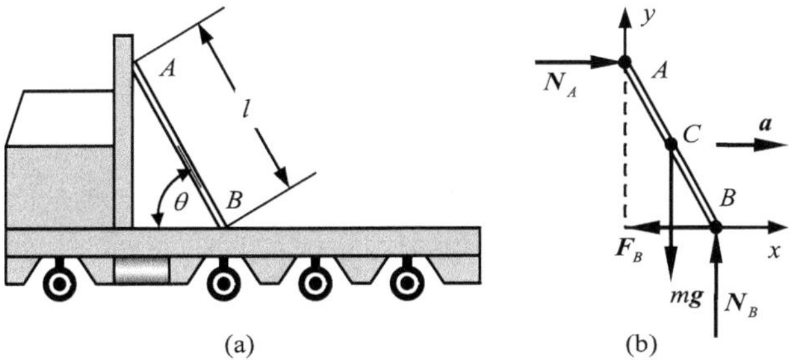

图 E9.1

解　取 AB 为自由体，画受力图，如图 E9.1b，则有

$$ma_{Cx} = \sum F_x, \quad ma = N_A - F_B$$

$$ma_{Cy} = \sum F_y, \quad 0 = N_B - mg$$

$$I_C\alpha = \sum M_C(\boldsymbol{F}),\ 0 = -N_A(\frac{1}{2}l\sin\theta) + N_B(\frac{1}{2}l\cos\theta) - F_B(\frac{1}{2}l\sin\theta)$$

解方程，并利用 $g = 9.81\ \mathrm{m/s^2}$，得

$$N_A = 153.28\ \mathrm{N},\ N_B = 392.40\ \mathrm{N},\ F_B = 73.28\ \mathrm{N}$$

利用 $F_B \leqslant \mu_s N_B$，有

$$\mu_s \geqslant 0.19$$

例 9.2 长度为 l、质量为 m 的均质杆 AB 支撑如图 E9.2a 所示。如果 B 处绳子突然断裂，求：(a)杆 AB 的角加速度；(b)支座 A 处的反力。

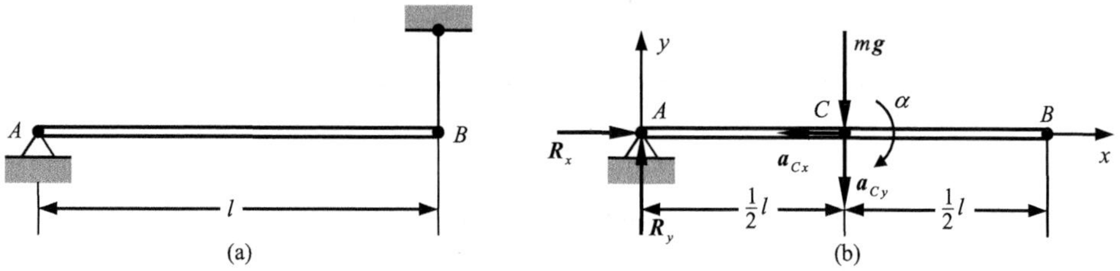

图 E9.2

解 取 AB 为自由体，画受力图，如图 E9.2b，则有

$$ma_{Cx} = \sum F_x, \quad -ma_{Cx} = R_x$$

$$ma_{Cy} = \sum F_y, \quad -ma_{Cy} = R_y - mg$$

$$I_C\alpha = \sum M_C(\boldsymbol{F}), \quad -I_C\alpha = -R_y(\frac{1}{2}l)$$

式中，$a_{Cx} = AC \cdot \omega^2 = 0$，$a_{Cy} = AC \cdot \alpha = \frac{1}{2}l\alpha$，$I_C = \frac{1}{12}ml^2$。

解上述方程，得

$$R_x = 0,\ R_y = \frac{1}{4}mg,\ \alpha = \frac{3g}{2l}$$

9.2 惯 性 力 法

惯性力法可用于分析质点运动，同样该方法也可用于刚体平面运动。对刚体平面运动，不仅需要把惯性力(大小与 $m\boldsymbol{a}_C$ 相等，方向与 $m\boldsymbol{a}_C$ 相反)加到刚体质心，而且需要把惯性力偶(大小与 $I_C\alpha$ 相等，转向与 $I_C\alpha$ 相反)加到刚体。因此，刚体平面运动方程可表示为

$$\sum \boldsymbol{F} + \boldsymbol{F}_\mathrm{I} = 0,\ \sum M_C(\boldsymbol{F}) + M_{\mathrm{I}C} = 0 \tag{9.6}$$

式中，$\boldsymbol{F}_\mathrm{I} = -m\boldsymbol{a}_C$ 和 $M_{\mathrm{I}C} = -I_C\alpha$ 分别称为惯性力和惯性力偶。

虽然所考虑的刚体并不平衡，但是平衡方程可用于分析刚体运动，只要利用基于达朗贝尔原理的惯性力法把惯性力和惯性力偶加到运动刚体。上述方程分解为直角坐标分量，得

$$\sum F_x + F_{Ix} = 0,\ \sum F_y + F_{Iy} = 0,\ \sum M_C(\boldsymbol{F}) + M_{IC} = 0 \tag{9.7}$$

式中，$F_{Ix} = -ma_{Cx}$，$F_{Iy} = -ma_{Cy}$，$M_{IC} = -I_C\alpha$。

惯性力法的优点是可以把刚体动力学问题转化为等效的平衡问题，并可对任意点取矩。因此，上述方程也可表示为

$$\sum F_x + F_{Ix} = 0,\ \sum F_y + F_{Iy} = 0,\ \sum M_A(\boldsymbol{F}) + M_A(\boldsymbol{F}_I) + M_{IC} = 0 \tag{9.8}$$

式中，A 为任意点。上述方程的两种替代形式分别为

$$\sum F_x + F_{Ix} = 0,\ \sum M_A(\boldsymbol{F}) + M_A(\boldsymbol{F}_I) + M_{IC} = 0,\ \sum M_B(\boldsymbol{F}) + M_B(\boldsymbol{F}_I) + M_{IC} = 0 \tag{9.9}$$

式中，A 和 B 连线不能垂直 x 轴，和

$$\begin{gathered}\sum M_A(\boldsymbol{F}) + M_A(\boldsymbol{F}_I) + M_{IC} = 0,\ \sum M_B(\boldsymbol{F}) + M_B(\boldsymbol{F}_I) + M_{IC} = 0,\\ \sum M_C(\boldsymbol{F}) + M_C(\boldsymbol{F}_I) + M_{IC} = 0\end{gathered} \tag{9.10}$$

式中，A、B 和 C 不能共线。

例 9.3　长度为 l、质量为 m 的均质杆 AB 支撑如图 E9.3a 所示。如果 B 处绳子突然断裂，求：(a) 杆 AB 的角加速度；(b) 支座 A 处的反力。

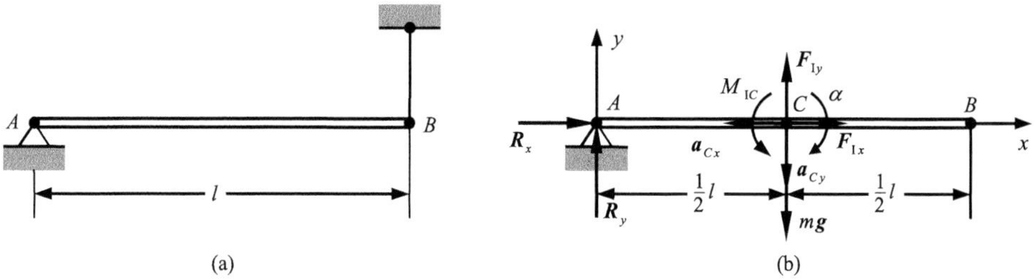

图 E9.3

解　取 AB 为自由体，画受力图，如图 E9.3b，则有

$$\sum F_x + F_{Ix} = 0,\quad R_x + ma_{Cx} = 0$$

$$\sum F_y + F_{Iy} = 0,\quad R_y - mg + ma_{Cy} = 0$$

$$\sum M_A(\boldsymbol{F}) + M_A(\boldsymbol{F}_I) + M_{IC} = 0,\ -mg(\frac{1}{2}l) + ma_{Cy}(\frac{1}{2}l) + I_C\alpha = 0$$

利用 $a_{Cx} = AC\cdot\omega^2 = 0$、$a_{Cy} = AC\cdot\alpha = \frac{1}{2}l\alpha$ 和 $I_C = \frac{1}{12}ml^2$，解方程得

$$R_x = 0,\ R_y = \frac{1}{4}mg,\ \alpha = \frac{3g}{2l}$$

9.3 功 能 法

1. 力偶矩的功

矩为 M 的力偶所做元功可表示为

$$\mathrm{d}W = M\mathrm{d}\theta \tag{9.11}$$

式中，$\mathrm{d}\theta$ 为角位移。

2. 刚体的动能

平面运动刚体的动能定义为

$$T = \int \frac{1}{2}\mathrm{d}m v^2 \tag{9.12}$$

利用 $\boldsymbol{v} = \boldsymbol{v}_C + \boldsymbol{v}'$，则式(9.12)可重写为

$$T = \int \frac{1}{2}\mathrm{d}m(\boldsymbol{v}\cdot\boldsymbol{v}) = \frac{1}{2}{v_C}^2\int \mathrm{d}m + \boldsymbol{v}_C \cdot \int \mathrm{d}m\boldsymbol{v}' + \frac{1}{2}\int \mathrm{d}m v'^2 \tag{9.13}$$

利用 $\int \mathrm{d}m\boldsymbol{v}' = m\boldsymbol{v}'_C = 0$ 和 $v' = r'\omega$，有

$$T = \frac{1}{2}m{v_C}^2 + \frac{1}{2}I_C\omega^2 \tag{9.14}$$

式中，$I_C = \int r'^2\mathrm{d}m$ 为刚体绕质心的转动惯量。该式表明，平面运动刚体的动能等于刚体随质心的平移动能和刚体相对平移形心参考系的转动动能之和。

3. 功能原理

质点的功能原理可用于刚体上每个质点。相加所有质点的动能，并考虑刚体上所有力和力偶所做功，即可得到平面运动刚体的功能原理

$$\mathrm{d}T = \mathrm{d}W \tag{9.15}$$

式中，$\mathrm{d}T$ 和 $\mathrm{d}W$ 分别为刚体动能增量和刚体上所有力和力偶所做元功。积分上式，得

$$T_2 - T_1 = W_{12} \tag{9.16}$$

式中，T_1 和 T_2 为刚体的初、末动能，W_{12} 为刚体上所有力和力偶所做功。

4. 能量守恒

如果刚体上所有做功的力都是保守力，则功能原理可替换为

$$T_1 + V_1 = T_2 + V_2 \tag{9.17}$$

该式表示机械能守恒。

例 9.4　长度为 l、质量为 m 的均质杆可绕点 O 转动，如图 E9.4a。杆从水平位置由静止状态开始释放而发生自由摆动。求：(a) 当杆通过垂直位置时杆具有最大角速度的距离 d；(b) 相应的角速度值和 O 处反力值。

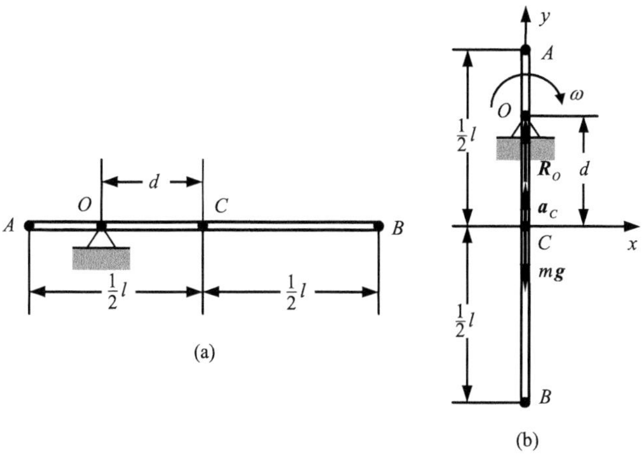

图 E9.4

解　在位置1，如图E9.4a，有

$$T_1 = V_1 = 0$$

在位置2，如图E9.4b，有

$$T_2 = \frac{1}{2}I_O\omega^2 = \frac{1}{2}(I_C + md^2)\omega^2 = \frac{1}{2}(\frac{1}{12}ml^2 + md^2)\omega^2,\ V_2 = -mgd$$

利用 $T_1 + V_1 = T_2 + V_2$，得

$$\omega = \sqrt{\frac{2g}{d + \frac{1}{12}l^2/d}}$$

显然，当 $d = \frac{1}{12}l^2/d$，即 $d = \frac{\sqrt{3}}{6}l$ 时，角速度 ω 将有最大值，即

$$\omega_{\max} = \sqrt{\frac{2g}{2\sqrt{d(\frac{1}{12}l^2/d)}}} = \sqrt[4]{12}\sqrt{\frac{g}{l}} = 1.86\sqrt{\frac{g}{l}}$$

利用在位置2处，$ma_C = \sum F_y = R_O - mg$ 和 $a_C = d\omega_{\max}^2 = g$，如图E9.4b，有

$$R_O = 2mg$$

9.4　冲量动量法

1. 冲量动量原理

考虑平面运动刚体，有

$$\mathrm{d}\boldsymbol{L} = \mathrm{d}\boldsymbol{I},\ \mathrm{d}H_C = \mathrm{d}G_C \tag{9.18}$$

式中，$\mathrm{d}\boldsymbol{L} = \mathrm{d}(m\boldsymbol{v}_C)$ 为刚体线动量增量，$\mathrm{d}\boldsymbol{I} = \sum \boldsymbol{F}\mathrm{d}t$ 为作用于刚体上力在时间间隔 $\mathrm{d}t$ 内的线冲量，$\mathrm{d}H_C = \mathrm{d}(I_C\omega)$ 为刚体绕质心 C 的角动量增量，$\mathrm{d}G_C = \sum M_C(\boldsymbol{F})\mathrm{d}t$ 为作用于刚体上力在时间间隔 $\mathrm{d}t$ 内绕质心 C 的角冲量。积分式(9.18)，得

$$\boldsymbol{L}_2-\boldsymbol{L}_1=\boldsymbol{I}_{12},\ H_{C2}-H_{C1}=G_{C12} \tag{9.19}$$

式中，$\boldsymbol{L}_1=m\boldsymbol{v}_{C1}$和$\boldsymbol{L}_2=m\boldsymbol{v}_{C2}$分别为刚体的初、末线动量，$\boldsymbol{I}_{12}=\sum\int_{t_1}^{t_2}\boldsymbol{F}\mathrm{d}t$为作用于刚体上力在$t_1$到$t_2$时间间隔内的线冲量，$H_{C1}=I_C\omega_1$和$H_{C2}=I_C\omega_2$分别为刚体绕质心$C$的初、末角动量，$G_{C12}=\sum\int_{t_1}^{t_2}M_C(\boldsymbol{F})\mathrm{d}t$为作用于刚体上力在$t_1$到$t_2$时间间隔内绕质心$C$的角冲量。

例 9.5　半径为r、重量为W的均质圆柱，具有逆时针初始角速度ω_0，放在由地板和墙壁形成的墙角，如图 E9.5a。设A和B处的动摩擦系数均为μ_k，试求圆柱停止转动时所需时间。

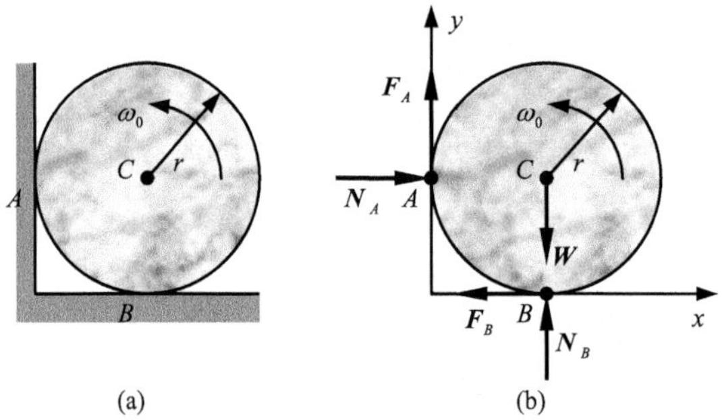

图 E9.5

解　取圆柱为自由体，画受力图，如图 E9.5b，则有

$$L_{2x}-L_{1x}=I_{12x},\quad 0-0=(N_A-F_B)t$$

$$L_{2y}-L_{1y}=I_{12y},\quad 0-0=(N_B+F_A-W)t$$

$$H_{C2}-H_{C1}=G_{C12},\quad 0-I_C\omega_0=-(F_A+F_B)rt$$

代入$F_A=\mu_k N_A$、$F_B=\mu_k N_B$和$I_C=\dfrac{1}{2}mr^2$，解方程，得

$$t=\frac{(1+\mu_k^{\ 2})I_C\omega_0}{\mu_k(1+\mu_k)Wr}=\frac{(1+\mu_k^{\ 2})r\omega_0}{2\mu_k(1+\mu_k)g}$$

式中，g为重力加速度。

2. 动量守恒

如果没有外力作用于平面运动刚体，则式(9.19)可写为

$$\boldsymbol{L}_1=\boldsymbol{L}_2,\ H_{C1}=H_{C2} \tag{9.20}$$

上式表明，刚体的线动量和绕质心C的角动量守恒。

9.5　平面运动刚体的碰撞

瞬间接触，如能引起速度突然改变，则称为碰撞。碰撞时间极短，但施加于物体的作用力却极大。因为这些力在接触点附近会产生随时间变化的变形，因此全面处理碰撞问题相当困难。然而，如下假设将使我们能够以相对简单的方式确定速度变化。这些假设主要包括：(1)碰撞过程很短，因而碰撞期间物体位移忽略不计；(2)接触点冲力很大，因而碰撞期间所有其他非冲力(如重力和弹力)忽略不计。

对平面运动刚体，碰撞期间的冲量动量原理可写为

$$m\boldsymbol{v}'_C - m\boldsymbol{v}_C = \boldsymbol{I},\ I_C\omega' - I_C\omega = M_C(\boldsymbol{I}) \tag{9.21}$$

式中，m 为刚体质量，$\boldsymbol{v}_C$ 和 $\boldsymbol{v}'_C$ 分别为刚体碰撞前后质心速度，$\boldsymbol{I}$ 为碰撞期间作用于刚体上的冲力冲量，I_C 为刚体绕质心的转动惯量，ω 和 ω' 分别为碰撞前后刚体角速度，$M_C(\boldsymbol{I})$ 为碰撞期间作用于刚体上的冲力绕质心 C 的角冲量。

例 9.6　图示质量为 m、长度为 l、与水平方向成 θ 角的均质细长杆 AB 以垂直速度 $\boldsymbol{v}$ 和零角速度撞击光滑表面。假设为完全弹性碰撞，试推导杆碰撞后的角速度、碰撞期间表面施加于杆的冲力冲量。

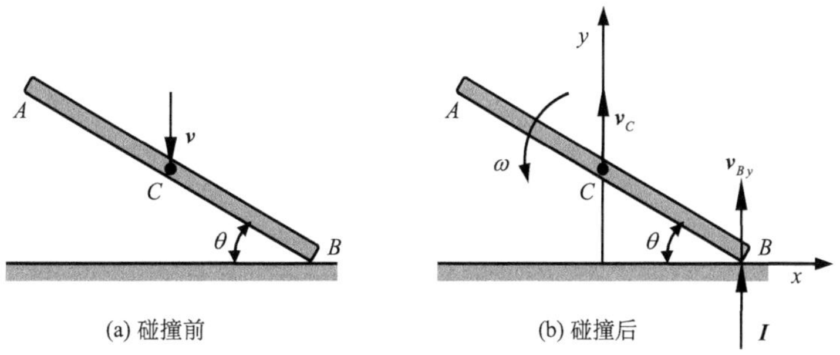

图 E9.6

解　建立图示坐标系，对杆应用冲量动量原理，并注意杆的质心速度垂直、碰撞期间作用于杆的冲力冲量也垂直，因此有

$$mv_C - m(-v) = I,\ I_C\omega - 0 = I(\frac{1}{2}l\cos\theta)$$

式中，I_C 为杆对质心的转动惯量，I 为表面作用于杆的冲力冲量值。利用 $I_C = \frac{1}{12}ml^2$，$v_{By} = v$，$v_C = v_{By} - (\frac{1}{2}l\omega)\cos\theta = v - \frac{1}{2}l\omega\cos\theta$，并求解上述方程，得

$$\omega = \frac{12v\cos\theta}{(1+3\cos^2\theta)l},\ I = \frac{2mv}{1+3\cos^2\theta}$$

例 9.7　质量为 m_1 的子弹以大小为 v 的水平速度射击进入质量为 m_2、长度为 l 的均质细长杆。已知杆最初静止，求：(a)子弹嵌入后杆的角速度；(b) A 处冲击反力的冲量；(c) A 处冲击反力为零时的距离 a。

解　对由子弹和杆件构成的系统应用冲量动量原理，并注意碰撞期间作用于杆件的冲击反力冲量位于水平方向，有

$$m_1(a\omega) + m_2(\frac{1}{2}l\omega) - m_1v = I_A,\ (m_1a^2)\omega + (\frac{1}{3}m_2l^2)\omega - a(m_1v) = 0$$

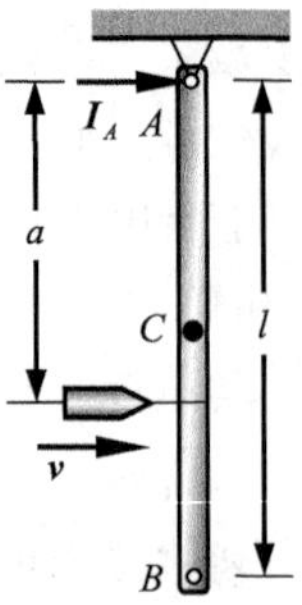

图 E147

式中，ω 为子弹刚刚嵌入后杆的角速度，I_A 为 A 处支撑施加于杆件的冲击反力的冲量。求解上述方程，得

$$\omega=\frac{3m_1av}{3m_1a^2+m_2l^2},\ I_A=\frac{(3a-2l)lm_1m_2v}{6m_1a^2+2m_2l^2}$$

由上述第 2 个表达式可以看出，如果 A 处冲击反力为零，则距离 a 等于

$$a=\frac{2}{3}l$$

杆件上距支点为 $a=2l/3$ 的点称为撞击中心。撞击中心是指：垂直碰撞发生在撞击中心时，不会在支点引起冲力。

习　　题

9.1 质量 $m=20$ kg 的均质橱柜安装在小脚轮上可以在地板上自由移动。已知 $F=100$ N、$b=500$ mm 和 $h_C=800$ mm，求：(a)橱柜加速度；(b)保证橱柜不发生翻倒的 h 值范围。

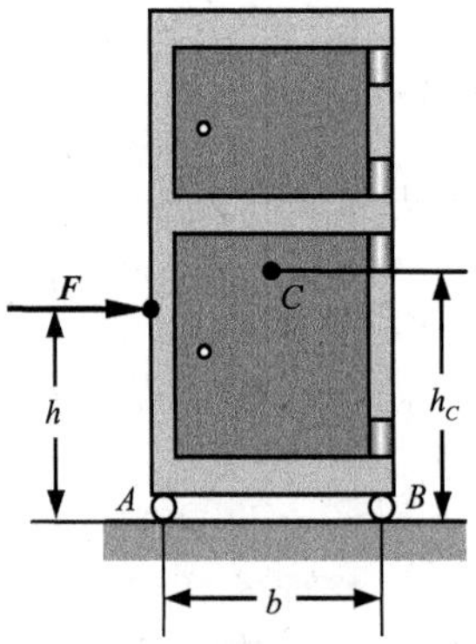

图 P9.1

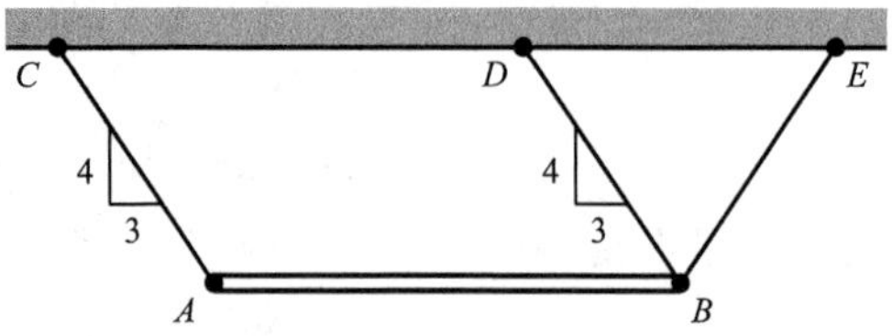

图 P9.2

9.2 质量为 2 kg 的均质细杆 AB 由长度相同的三根绳子支撑在图示位置。求绳子 BE 突然剪断后的瞬间：(a)杆 AB 的加速度；(b)每一绳子的拉力。

9.3 质量为 5 kg 的均质圆板由长度相同的三根绳子悬挂在图示位置。已知 CA 和 CB

分别水平和垂直，求绳子 BF 突然剪断后的瞬间：(a)板的加速度；(b)每一绳子的拉力。

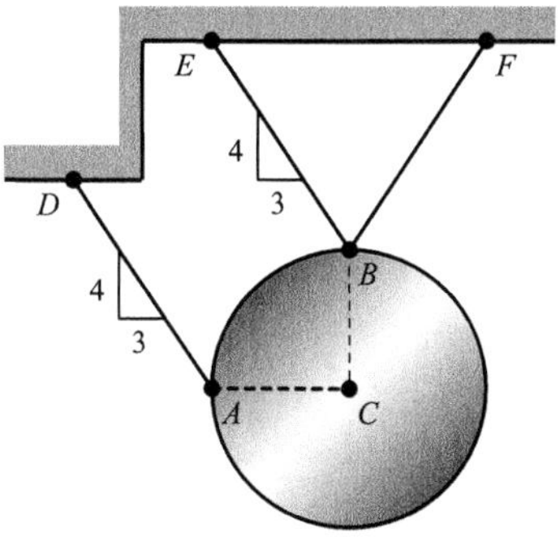

图 P9.3

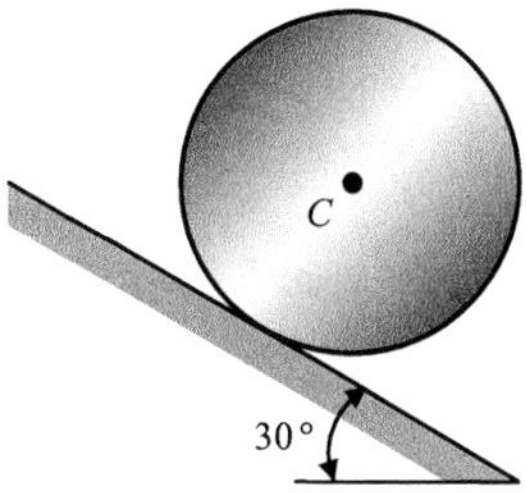

图 P9.4

9.4　质量 $m=8$ kg、半径 $r=0.15$ m 的均质圆柱沿倾角 $\theta=30°$ 的斜面向下只滚不滑。求摩擦力和质心加速度。

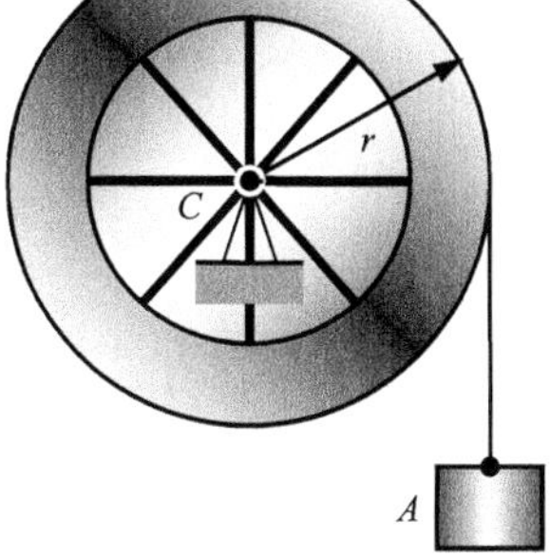

图 P9.5

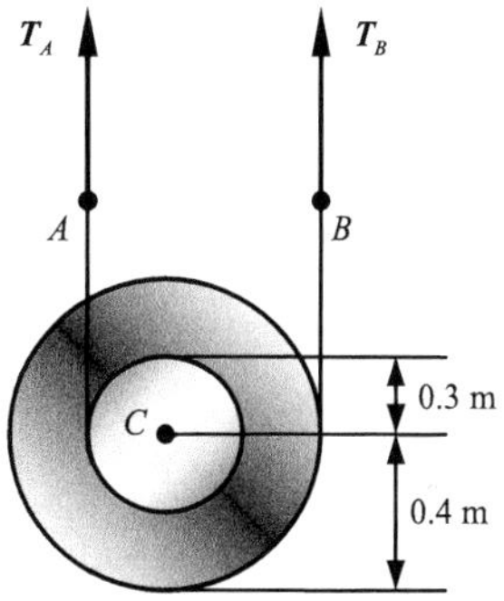

图 P9.6

9.5　图示飞轮的半径为 500 mm，质量为 150 kg，回转半径为 400 mm。质量为 20 kg 的物块 A 与绕在飞轮上的绳子相连，系统由静止释放。不计摩擦，求：(a)物块 A 的加速度；(b)物块 A 移动 2 m 后的速度。

9.6　半径为 0.4 m 的鼓轮与半径为 0.3 m 的圆盘相连。圆盘和鼓轮具有组合质量 5 kg 和组合回转半径 0.25 m，通过两根绳子悬挂在图示位置。已知 $T_A=50$ N 和 $T_B=30$ N，求绳上点 A 和 B 的加速度。

9.7　长度为 l、质量为 m 的均质杆 AB 通过图示弹簧支撑。如果 B 处弹簧突然断裂，求断裂瞬时：(a)杆 AB 的角加速度；(b)A 点的加速度。

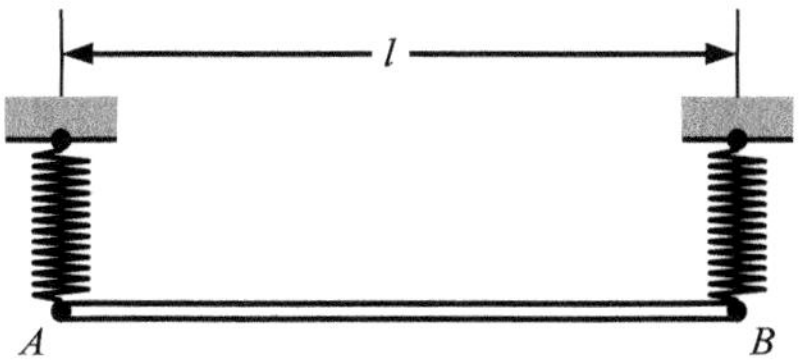

图 P9.7

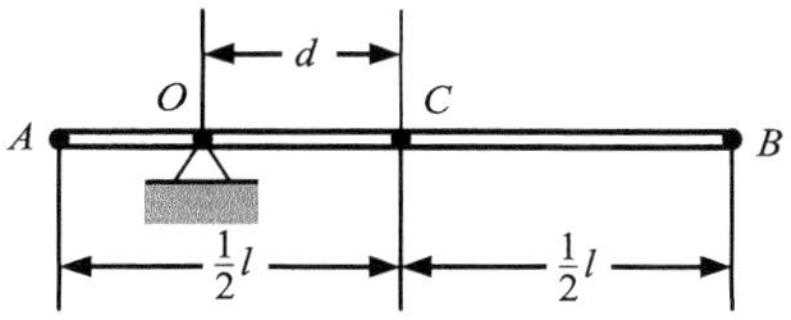

图 P9.8

9.8　长度为 l、质量为 m 的均质杆可绕点 O 转动。杆从水平位置由静止状态开始释放而发生自由摆动。已知 $d=\frac{3}{8}l$，求当杆旋转 90° 时杆的角速度和 O 处反力。

9.9　半径为 r、重量为 W 的均质圆柱，具有逆时针初始角速度 ω_0，放在由粗糙地板和光滑墙壁形成的墙角。设圆柱和地板之间的动摩擦系数均为 μ_k，试求圆柱停止转动时所需时间。

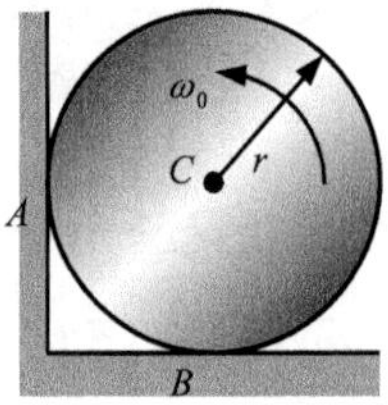

图 P9.9

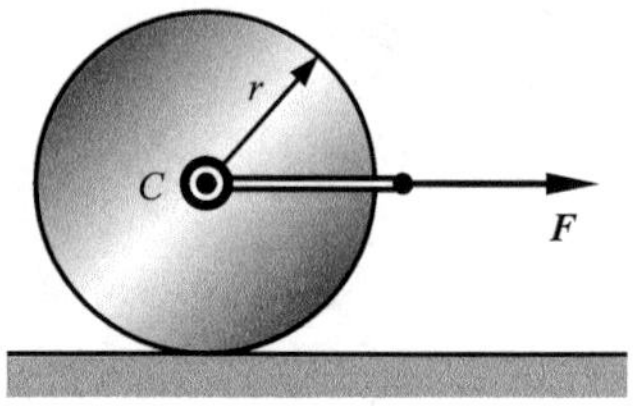

图 P9.10

9.10　质量 $m=25\text{ kg}$、半径 $r=0.2\text{ m}$ 的均质圆柱滚轮，初始静止，受 $F=120\text{ N}$ 的力作用。已知滚轮只滚不滑，求：(a) 滚轮运动 1.5 m 时的质心速度；(b) 不发生滑动所需要的摩擦力。

9.11　绳子绕在半径为 r、质量为 m 的均质圆柱上。已知圆柱由静止释放，求圆柱向下运动距离 s 后的圆柱质心的速度。

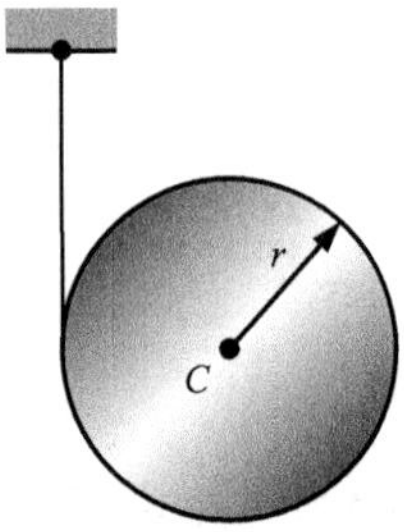

图 P9.11

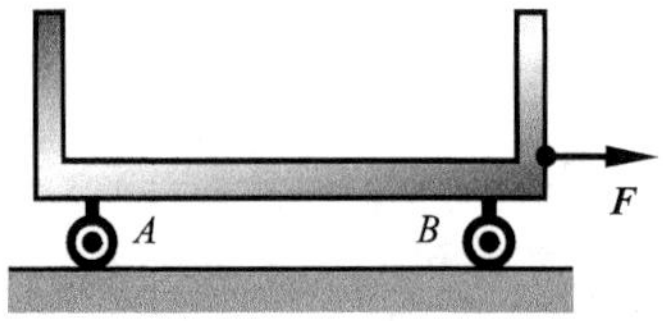

图 P9.12

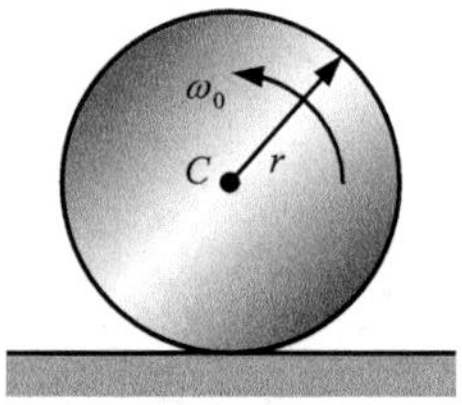

图 P9.13

9.12　质量为 10 kg 的支架，受力 $F=30\text{ N}$ 作用，由两个只滚不滑的均质圆盘支撑。每个圆盘的质量 $m=5\text{ kg}$、半径 $r=0.1\text{ m}$。已知系统初始静止，求支架运动 0.5 m 后的速度。

9.13　半径为 r、质量为 m 的均质圆柱，初始线速度为零，逆时针初始角速度为 ω_0，放在水平地板上。设圆柱和地板之间的动摩擦系数均为 μ_k，试求：(a) 圆柱开始只滚不滑的时间；(b) 圆柱开始只滚不滑时的线速度和角速度。

9.14　物块 A 和 B 通过绕在圆盘 O 表面的绳子 AB 连接。假设绳子和圆盘之间没有相对滑动，圆盘和轴承之间没有摩擦。已知图示瞬时，物块 B 以 0.4 m/s 的速度向下移动，弹簧压缩量为 0.2 m，求物块 B 下降 0.5 m 后的速度。

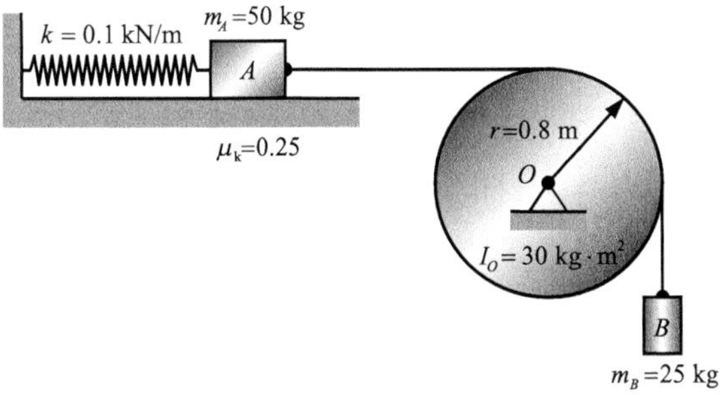

图 P9.14

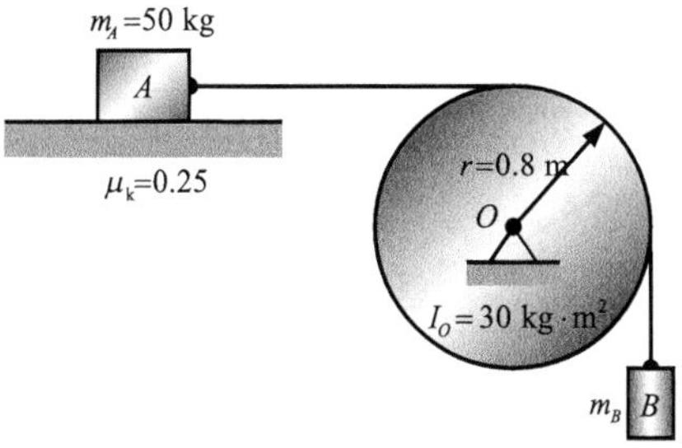

图 P9.15

9.15　物块 A 和 B 通过绕在圆盘 O 表面的绳子 AB 连接。假设绳子和圆盘之间没有相对滑动，圆盘和轴承之间没有摩擦。如果物块 A 在 6 s 内速度从 4 m/s 变为 8 m/s，求物块 B 的质量 m_B。

9.16　质量 $m_1 = 10$ g 的子弹以大小 $v = 500$ m/s 的水平速度射击进入质量 $m_2 = 5$ kg、长度 $l = 1$ m 的均质细长杆的下端。已知杆最初静止，求：(a) 子弹嵌入后杆的角速度；(b) A 处施加于杆的冲量。

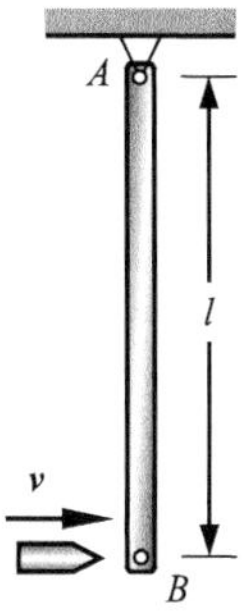

图 P9.16

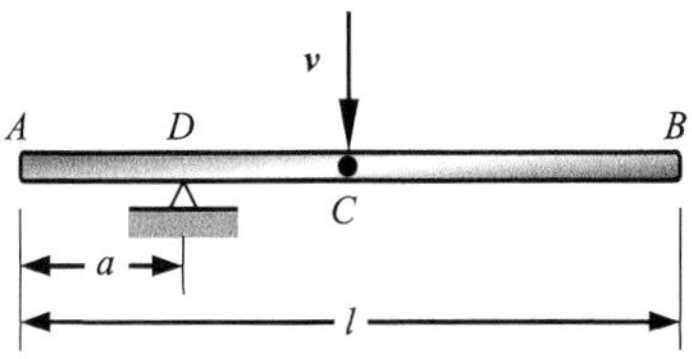

图 P9.17

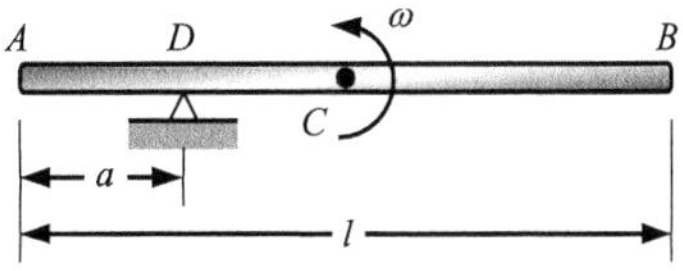

图 P9.18

9.17　质量为 m、长度为 l 的均质细长杆 AB 以大小为 v 的垂直速度和零角速度撞击光滑刚性支撑 D。已知 $a = l/5$ 和 $e = 0$，求：(a) 碰撞后杆的角速度和质心速度；(b) D 处作用于杆的冲量。

9.18　质量为 m、长度为 l 的均质细长杆 AB 以逆时针角速度 ω 和零质心速度撞击光滑刚性支撑 D。已知 $a = l/5$ 和 $e = 1$，求：(a) 碰撞后杆的角速度和质心速度；(b) D 处作用于杆的冲量。

9.19　图示质量为 m、长度为 l、与水平方向成 θ 角的均质细长杆 AB 以垂直速度 $\boldsymbol{v}$ 和零角速度撞击光滑表面。假设杆与表面之间的恢复系数为 e，试推导杆碰撞后的角速度。

9.20　图示质量为 m、长度为 l 的均质细长杆 AB 以速度 $\boldsymbol{v}$ 下落，其 B 端撞击倾角为 θ 的光滑斜面。假设碰撞为完全弹性，求碰撞后杆的角速度和质心速度。

9.21　图示质量为 m_1、长度为 l 的均质细长杆 AB 从水平位置静止释放，向下摆动到垂直位置时撞击质量为 m_2 的物块 C，物块 C 静止放置于光滑表面。假设杆与物块之间的恢复系数为e，求碰撞后物块的速度。

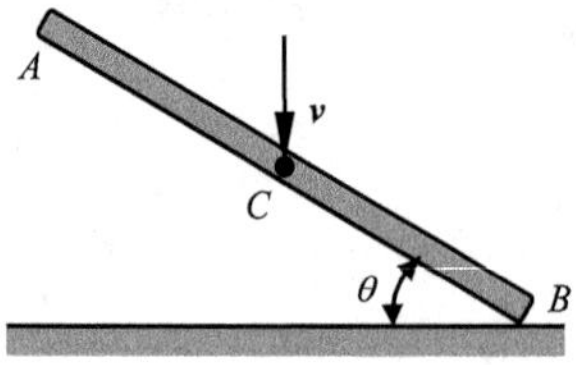

图 P9.19

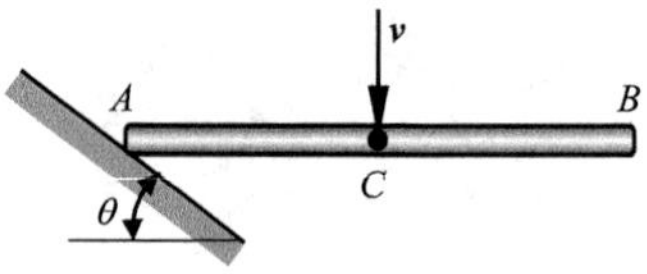

图 P9.20

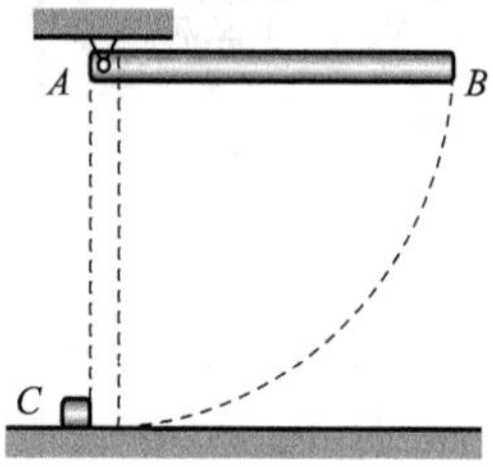

图 P9.21

第10章　分 析 力 学

10.1　约束与虚功

约束是指限制物体运动的几何或运动学条件。自由度是指完全确定给定系统位形所需的最小变量数。

物体的虚位移定义为同施加于物体上的约束相协调的假想的无限小位移。该位移可以是想象的位移，即位移没有实际发生。

虚功定义为力或力偶经历虚位移而做的功。力 $\boldsymbol{F}$ 经历虚位移 $\delta\boldsymbol{r}$，所做的虚功可表示为

$$\delta W = \boldsymbol{F}\cdot\delta\boldsymbol{r} \tag{10.1}$$

大小为 M 的力偶经历虚位移 $\delta\theta$，所做的虚功可表示为

$$\delta W = M\delta\theta \tag{10.2}$$

如果所有约束力经历任意虚位移所做的虚功都等于零，则这类约束称为理想约束。

10.2　虚 功 原 理

考虑质点系。假设 $\delta\boldsymbol{r}_i$ 是质点 i 的与约束相协调的虚位移，那么作用于质点 i 上的力所做的虚功可表示为

$$\delta W_i = (\boldsymbol{F}_i + \boldsymbol{R}_i)\cdot\delta\boldsymbol{r}_i \tag{10.3}$$

式中，$\boldsymbol{F}_i$ 是作用于质点 i 上的主动力的合力，$\boldsymbol{R}_i$ 是作用于相同质点上的约束力的合力。如果质点系处于静平衡，即 $\boldsymbol{F}_i + \boldsymbol{R}_i = 0$，那么式(10.3)可重写为

$$\delta W_i = (\boldsymbol{F}_i + \boldsymbol{R}_i)\cdot\delta\boldsymbol{r}_i = 0 \tag{10.4}$$

对 i 求和，则总虚功等于

$$\delta W = \sum\delta W_i = \sum(\boldsymbol{F}_i + \boldsymbol{R}_i)\cdot\delta\boldsymbol{r}_i = 0 \tag{10.5}$$

对理想约束系统，即 $\sum\boldsymbol{R}_i\cdot\delta\boldsymbol{r}_i = 0$，那么有

$$\delta W = \sum\boldsymbol{F}_i\cdot\delta\boldsymbol{r}_i = 0 \tag{10.6}$$

该式表示虚功原理：理想约束系统平衡的充要条件是主动力在满足约束条件任意虚位移所做的虚功等于零。

例 10.1　刚度 $k = 800$ N/m 的原长弹簧与点 I 和 J 处的销钉连接，如图所示。B 处销钉连接到构件 BCD，并可沿着固定平板上的垂直滑槽自由滑动。求当 $F = 135$ N 的水平力向右作用于点 G 时弹簧受力和点 H 的水平位移。

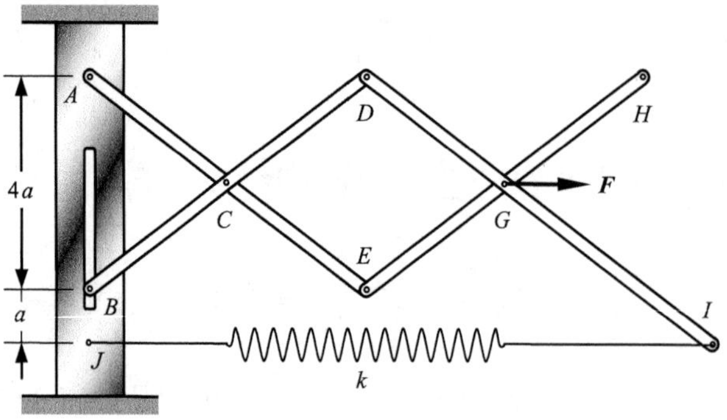

图 E10.1

解　利用 $x_G = 3x_C$, $x_H = 4x_C$, $x_I = 4.5x_C$ 有

$$\delta x_G = 3\delta x_C,\ \delta x_H = 4\delta x_C,\ \delta x_I = 4.5\delta x_C$$

根据虚功原理，$\delta W = 0$，有

$$F\delta x_G - F_{\text{spr}}\delta x_I = 0$$

即

$$F_{\text{spr}} = \frac{\delta x_G}{\delta x_I}F = \frac{3\delta x_C}{4.5\delta x_C}F = 90\ \text{N}$$

利用 $F_{\text{spr}} = k\delta x_I$，得

$$x_I = \frac{F_{\text{spr}}}{k} = 112.5\ \text{mm}$$

利用 $\delta x_H = 4\delta x_C$ 和 $\delta x_I = 4.5\delta x_C$，得

$$\delta x_H = \frac{4}{4.5}\delta x_I = 100\ \text{mm}$$

例 10.2　结构在点 G 处受力 $\boldsymbol{F}$ 作用，如图 E10.2a 所示。忽略结构重量，已知 $AB = AC = BC = CD = CE = DG = EG = a$，求点 B 处水平约束力大小。

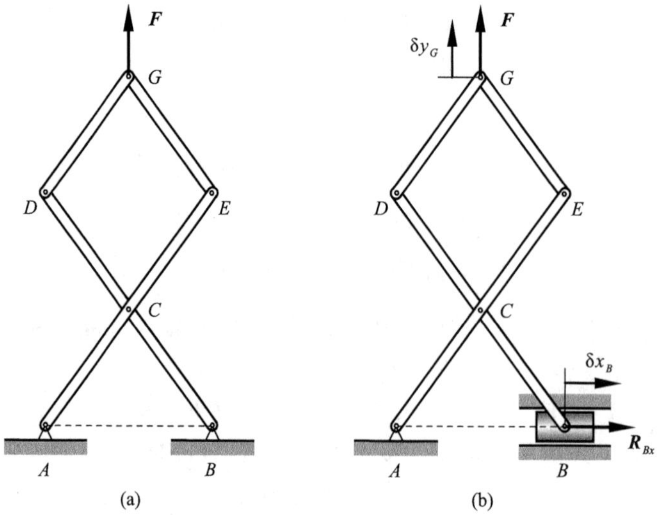

图 E10.2

解 用水平约束力$\boldsymbol{R}_{Bx}$代替点B处水平约束，并假设点B处虚位移δx_B向右为正，点G处虚位移δy_G向上为正，如图E10.2b所示。利用$x_C{}^2+y_C{}^2=a^2$，即$(\frac{x_B}{2})^2+(\frac{y_G}{3})^2=a^2$，有

$$x_B\delta x_B+\frac{4}{9}y_G\delta y_G=0$$

利用$AB=AC=BC$，即$y_C=\sqrt{3}x_C$，或$y_G=\frac{3\sqrt{3}}{2}x_B$，有

$$\delta x_B=-\frac{2\sqrt{3}}{3}\delta y_G$$

根据虚功原理，$\delta W=0$，有

$$R_{Bx}\delta x_B+F\delta y_G=0$$

即

$$R_{Bx}=-\frac{\delta y_G}{\delta x_B}F=\frac{\sqrt{3}}{2}F$$

10.3 广义坐标和广义力

1. 广义坐标

确定给定系统位形的独立变量称为广义坐标。通常广义坐标可选线坐标和角坐标。有时采用多于自由度的坐标比较合适，那么这些坐标之间一定通过某些约束而发生关联。

例如，考虑由n个质点构成的系统，假设系统的位形通过广义坐标$q_1,q_2,\cdots,q_s$进行表征，其中s是系统的自由度，那么质点i的位置矢量可表示为

$$\boldsymbol{r}_i=\boldsymbol{r}_i(q_1,q_2,\cdots,q_s,t)\quad(i=1,2,\cdots,n)\tag{10.7}$$

假设$\delta\boldsymbol{r}_i$表示满足约束的质点i的虚位移，那么$\delta\boldsymbol{r}_i$可表示为

$$\delta\boldsymbol{r}_i=\sum_{k=1}^{s}\frac{\partial\boldsymbol{r}_i}{\partial q_k}\delta q_k\quad(i=1,2,\cdots,n)\tag{10.8}$$

2. 广义力

假设质点i在主动力$\boldsymbol{F}_i$作用下发生满足约束条件的虚位移$\delta\boldsymbol{r}_i$，那么由作用于系统上的主动力所做的虚功可表示为

$$\delta W=\sum_{i=1}^{n}\boldsymbol{F}_i\cdot\delta\boldsymbol{r}_i\tag{10.9}$$

把式(10.8)代入式(10.9)，有

$$\delta W=\sum_{i=1}^{n}\boldsymbol{F}_i\cdot(\sum_{k=1}^{s}\frac{\partial \boldsymbol{r}_i}{\partial q_k}\delta q_k)=\sum_{k=1}^{s}[\sum_{i=1}^{n}(\boldsymbol{F}_i\cdot\frac{\partial \boldsymbol{r}_i}{\partial q_k})]\delta q_k=\sum_{k=1}^{s}Q_k\delta q_k \tag{10.10}$$

式中

$$Q_k=\sum_{i=1}^{n}(\boldsymbol{F}_i\cdot\frac{\partial \boldsymbol{r}_i}{\partial q_k})\quad (k=1,2,\cdots,s) \tag{10.11}$$

称为与广义坐标 q_k 对应的广义力。

对保守力系，广义力 Q_k 可表示为

$$Q_k=\sum_{i=1}^{n}(\boldsymbol{F}_i\cdot\frac{\partial \boldsymbol{r}_i}{\partial q_k})=-\frac{\partial V}{\partial q_k}\quad (k=1,2,\cdots,s) \tag{10.12}$$

式中，$V=V(q_1,q_2,\cdots,q_s)$ 是系统的势能函数。

10.4 拉格朗日方程

考虑由 n 个质点组成的质点系，受完整约束作用。假设系统位形可通过广义坐标 $q_1,q_2,\cdots,q_s$ 描述，其中 s 是系统自由度，那么质点 i 的位置矢量可表示为

$$\boldsymbol{r}_i=\boldsymbol{r}_i(q_1,q_2,\cdots,q_s,t)\quad (i=1,2,\cdots,n) \tag{10.13}$$

对时间求导，有

$$\dot{\boldsymbol{r}}_i=\sum_{k=1}^{s}\frac{\partial \boldsymbol{r}_i}{\partial q_k}\dot{q}_k+\frac{\partial \boldsymbol{r}_i}{\partial t}\quad (i=1,2,\cdots,n) \tag{10.14}$$

根据上式，得

$$\frac{\partial \dot{\boldsymbol{r}}_i}{\partial \dot{q}_k}=\frac{\partial \boldsymbol{r}_i}{\partial q_k}\quad (i=1,2,\cdots,n;k=1,2,\cdots,s) \tag{10.15}$$

点乘 $\dot{\boldsymbol{r}}_i$ 并对时间求导，有

$$\frac{\mathrm{d}}{\mathrm{d}t}(\dot{\boldsymbol{r}}_i\cdot\frac{\partial \dot{\boldsymbol{r}}_i}{\partial \dot{q}_k})=\frac{\mathrm{d}}{\mathrm{d}t}(\dot{\boldsymbol{r}}_i\cdot\frac{\partial \boldsymbol{r}_i}{\partial q_k})=\ddot{\boldsymbol{r}}_i\cdot\frac{\partial \boldsymbol{r}_i}{\partial q_k}+\dot{\boldsymbol{r}}_i\cdot\frac{\partial \dot{\boldsymbol{r}}_i}{\partial q_k}\quad (i=1,2,\cdots,n;k=1,2,\cdots,s) \tag{10.16}$$

即

$$\frac{\mathrm{d}}{\mathrm{d}t}\frac{\partial}{\partial \dot{q}_k}(\frac{1}{2}m_i\dot{\boldsymbol{r}}_i\cdot\dot{\boldsymbol{r}}_i)=(m_i\ddot{\boldsymbol{r}}_i)\cdot\frac{\partial \boldsymbol{r}_i}{\partial q_k}+\frac{\partial}{\partial q_k}(\frac{1}{2}m_i\dot{\boldsymbol{r}}_i\cdot\dot{\boldsymbol{r}}_i)\quad (i=1,2,\cdots,n;k=1,2,\cdots,s) \tag{10.17}$$

式中，m_i 是质点 i 的质量。对 i 求和，得

$$\frac{\mathrm{d}}{\mathrm{d}t}\frac{\partial}{\partial \dot{q}_k}(\sum_{i=1}^{n}\frac{1}{2}m_i\dot{\boldsymbol{r}}_i\cdot\dot{\boldsymbol{r}}_i)=\sum_{i=1}^{n}[(m_i\ddot{\boldsymbol{r}}_i)\cdot\frac{\partial \boldsymbol{r}_i}{\partial q_k}]+\frac{\partial}{\partial q_k}(\sum_{i=1}^{n}\frac{1}{2}m_i\dot{\boldsymbol{r}}_i\cdot\dot{\boldsymbol{r}}_i)\quad (k=1,2,\cdots,s) \tag{10.18}$$

即

$$\frac{\mathrm{d}}{\mathrm{d}t}\frac{\partial T}{\partial \dot{q}_k}=\sum_{i=1}^{n}(\boldsymbol{F}_i\cdot\frac{\partial \boldsymbol{r}_i}{\partial q_k})+\frac{\partial T}{\partial q_k}\quad (k=1,2,\cdots,s) \tag{10.19}$$

式中，$T=\sum_{i=1}^{n}\frac{1}{2}m_i\dot{\boldsymbol{r}}_i\cdot\dot{\boldsymbol{r}}_i$ 是系统动能，$\boldsymbol{F}_i=m_i\ddot{\boldsymbol{r}}_i$ 是作用于质点 i 上的力。利用广义力定义，

上述方程可重写为

$$\frac{\mathrm{d}}{\mathrm{d}t}\frac{\partial T}{\partial \dot{q}_k}-\frac{\partial T}{\partial q_k}=Q_k \quad (k=1,2,\cdots,s) \tag{10.20}$$

式中，$Q_k=\sum\limits_{i=1}^{n}(\boldsymbol{F}_i\cdot\frac{\partial \boldsymbol{r}_i}{\partial q_k})$ 是对应广义坐标 q_k 的广义力。上述方程称为拉格朗日方程。

对保守系统，$Q_k=-\frac{\partial V}{\partial q_k}$，则拉格朗日方程可表示为

$$\frac{\mathrm{d}}{\mathrm{d}t}\frac{\partial T}{\partial \dot{q}_k}-\frac{\partial T}{\partial q_k}=-\frac{\partial V}{\partial q_k} \quad (k=1,2,\cdots,s) \tag{10.21}$$

定义

$$L=T-V \tag{10.22}$$

式中，L 称为拉格朗日函数。利用 $\frac{\partial V}{\partial \dot{q}_k}=0$，保守系统拉格朗日方程可重写为

$$\frac{\mathrm{d}}{\mathrm{d}t}\frac{\partial L}{\partial \dot{q}_k}-\frac{\partial L}{\partial q_k}=0 \quad (k=1,2,\cdots,s) \tag{10.23}$$

如果部分广义力不是保守力，例如 Q_k'；其余部分广义力是保守力，可从势函数 V 求导得到，即

$$Q_k=Q_k'-\frac{\partial V}{\partial q_k} \quad (k=1,2,\cdots,s) \tag{10.24}$$

那么可得拉格朗日方程的最一般表达式

$$\frac{\mathrm{d}}{\mathrm{d}t}\frac{\partial L}{\partial \dot{q}_k}-\frac{\partial L}{\partial q_k}=Q_k' \quad (k=1,2,\cdots,s) \tag{10.25}$$

例 10.3　质量为 m_1 的物块沿光滑斜面滑动，质量为 m_2 的斜面沿光滑水平面滑动。求物块水平加速度和斜面加速度。

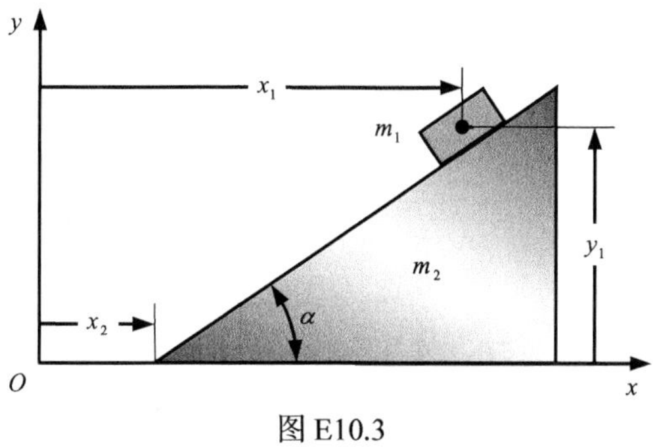

图 E10.3

解　选 x_1 和 x_2 为广义坐标，则有 $y_1=\tan\alpha(x_1-x_2)$ 和 $\dot{y}_1=\tan\alpha(\dot{x}_1-\dot{x}_2)$。因此，系统的动能和势能可分别表示为

$$T=\frac{1}{2}m_1[\dot{x}_1{}^2+\tan^2\alpha(\dot{x}_1-\dot{x}_2)^2]+\frac{1}{2}m_2\dot{x}_2{}^2,\quad V=m_1g\tan\alpha(x_1-x_2)+m_2gh$$

式中，h为常数。相应的拉格朗日函数等于

$$L=T-V=T=\frac{1}{2}m_1[\dot{x}_1{}^2+\tan^2\alpha(\dot{x}_1-\dot{x}_2)^2]+\frac{1}{2}m_2\dot{x}_2{}^2-m_1g\tan\alpha(x_1-x_2)-m_2gh$$

对质量为m_1的物块，有

$$\frac{\partial L}{\partial\dot{x}_1}=m_1[\dot{x}_1+\tan^2\alpha(\dot{x}_1-\dot{x}_2)],\quad \frac{\partial L}{\partial x_1}=-m_1g\tan\alpha$$

把上式代入拉格朗日方程，得

$$m_1[\ddot{x}_1+\tan^2\alpha(\ddot{x}_1-\ddot{x}_2)]+m_1g\tan\alpha=0$$

同理，对质量为m_2的斜面，有

$$\frac{\partial L}{\partial\dot{x}_2}=-m_1\tan^2\alpha(\dot{x}_1-\dot{x}_2)+m_2\dot{x}_2,\quad \frac{\partial L}{\partial x_2}=m_1g\tan\alpha$$

上式代入拉格朗日方程，得

$$-m_1\tan^2\alpha(\ddot{x}_1-\ddot{x}_2)+m_2\ddot{x}_2-m_1g\tan\alpha=0$$

联立求解方程组，得

$$\ddot{x}_1=-\frac{m_2g\sin\alpha\cos\alpha}{m_1\sin^2\alpha+m_2},\quad \ddot{x}_2=\frac{m_1g\sin\alpha\cos\alpha}{m_1\sin^2\alpha+m_2}$$

习　　题

10.1　图示机构受力$\boldsymbol{F}$作用。忽略机构重量，已知$AC=BC=CD=CE=DG=EG=a$和$AB=b$，推导机构平衡时所需力$\boldsymbol{R}$的大小。

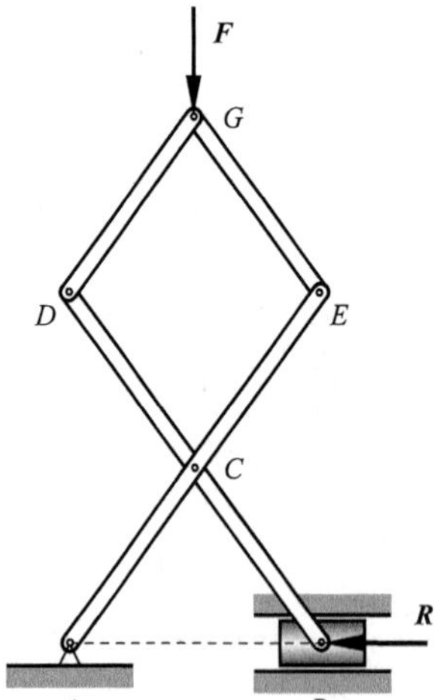

图 P10.1

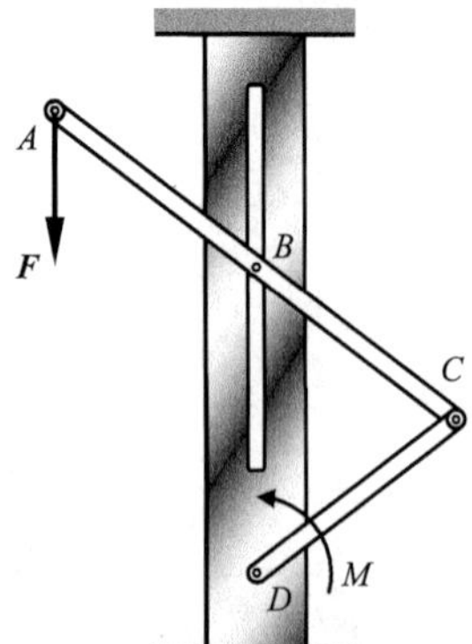

图 P10.2

10.2　销钉B与构件ABC连接，并能沿着固定平板上的滑槽滑动。忽略摩擦，已知$AB=BC=CD=a$和$BD=b$，当作用于A的力$\boldsymbol{F}$分别(a)垂直向下和(b)水平向左时，推

导维持平衡所需力偶的大小 M。

10.3　长为 l 的细长杆 AB 与环 B 连接，并静止放置于半径为 r 的半圆柱面。忽略摩擦，已知机构受力 $\boldsymbol{F}_A$ 和 $\boldsymbol{F}_B$ 作用，推导机构平衡时的 θ 表达式。

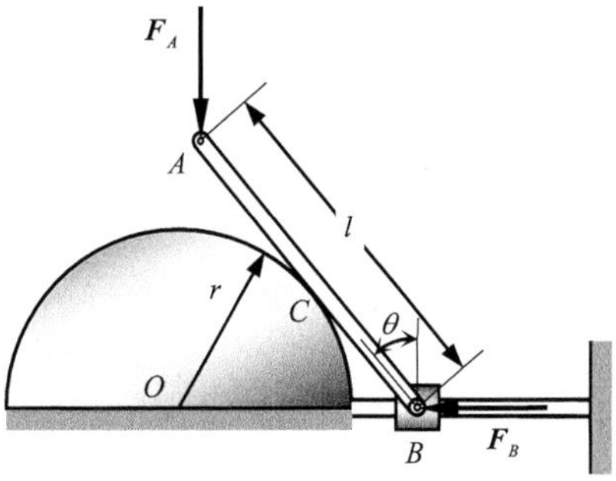

图 P10.3

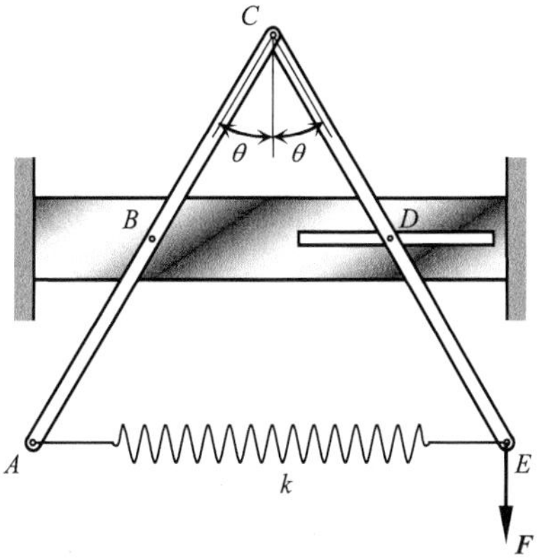

图 P10.4

10.4　两杆 ABC 和 CDE 通过销钉 C 和弹簧 AE 连接。刚度为 k，当 $\theta = 30°$时弹簧为原长。已知 $AB = BC = CD = DE = a$，推导当系统在图示载荷作用下处于平衡状态时 F、θ、l 和 k 必须满足的方程。

10.5　150 N 的水平力 $\boldsymbol{F}$ 作用于机构点 A。刚度为 k = 1.5 kN/m，当 θ = 0°时弹簧为原长。忽略机构质量，已知 a = 250 mm 和 r = 150 mm，求平衡时的 θ 值。

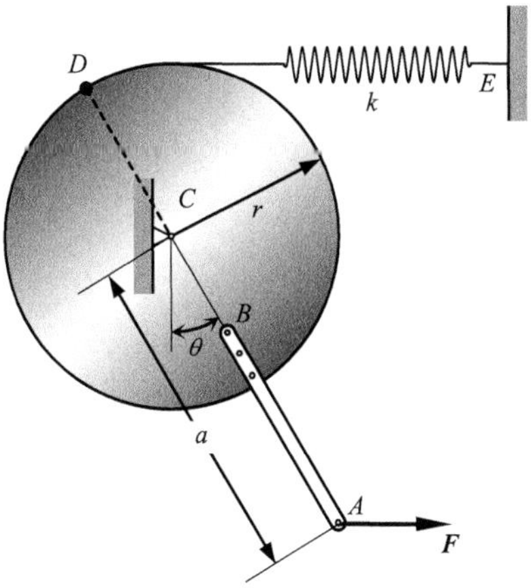

图 P10.5

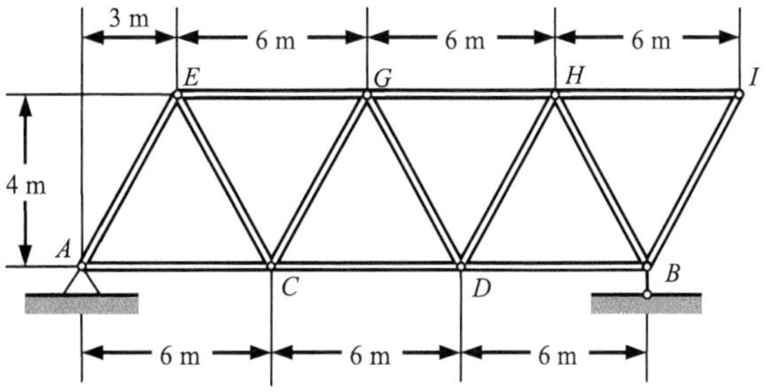

图 P10.6

10.6　如果构件 AC 的长度增加 6 mm，求节点 I 的垂直位移。

10.7　质量为 m_1 的小球与长为 l 的绳索相连后悬挂于质量为 m_2、半径为 r 的均质圆盘的边界。假设圆盘可绕 O 自由转动，求系统运动微分方程。

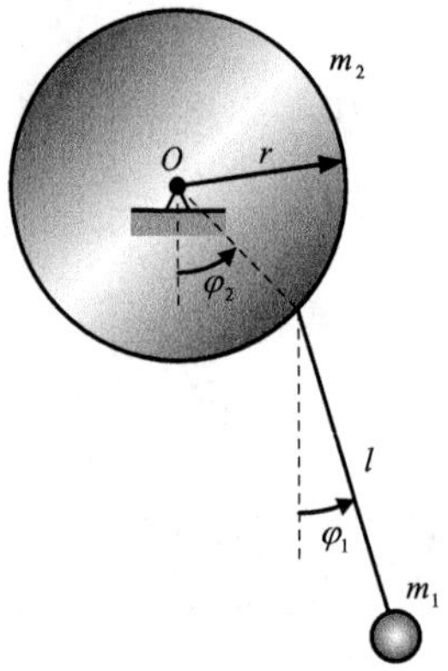

图 P10.7

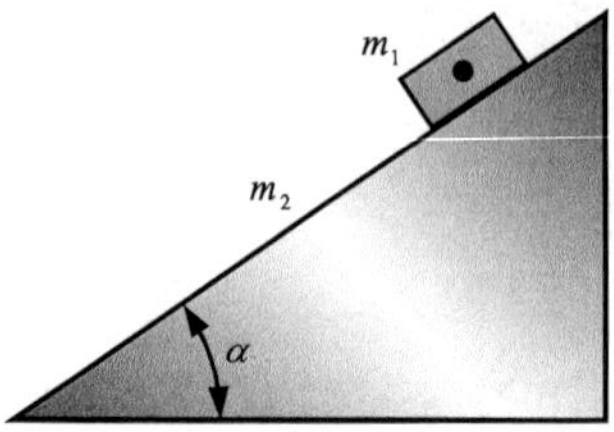

图 P10.8

10.8　质量为 m_1 的物块沿光滑斜面滑动，质量为 m_2 的斜面沿光滑水平面滑动。求物块相对斜面的相对加速度和斜面加速度。

10.9　具有 2 个自由度的无阻尼受迫振动系统由质量均为 m 的物块、刚度均为 k 的弹簧组成。系统受大小均为 $H\sin\omega_{\mathrm{f}}t$ 的周期力作用，其中 H 和 ω_{f} 分别为周期力的幅值和圆频率。求系统运动微分方程。

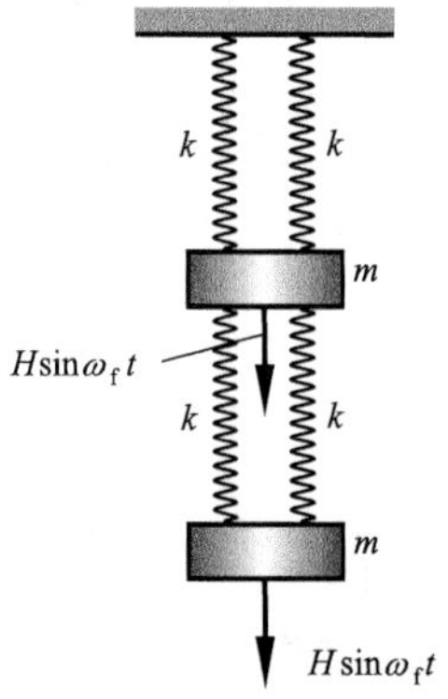

图 P10.9

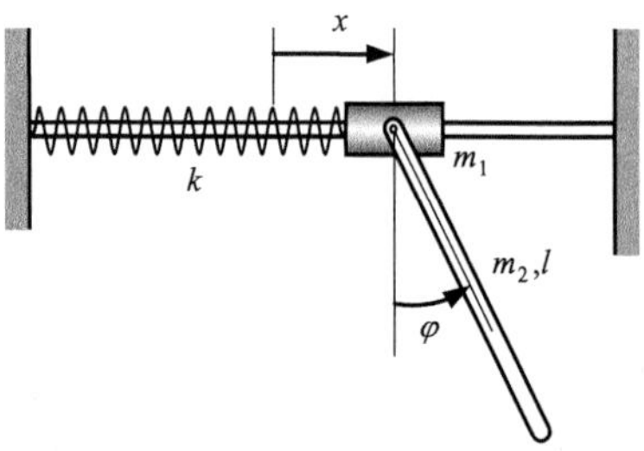

图 P10.10

10.10　质量为 m_1 的环可沿水平梁滑动，并与刚度为 k 的弹簧相连。长为 l 、质量为 m_2 的均质杆与环相连，并可在包含梁的垂直平面内摆动。不计摩擦，求系统运动的拉格朗日方程。

第 11 章　应力与应变

11.1　外　　力

作用于物体表面的外力称为表面力，简称面力。如果面力分布在物体表面的有限区域，则称为面分布载荷，如图 11.1a。如果面力作用在物体表面的狭窄区域，则称为线分布载荷，如图 11.1b。如果面力作用区域非常小，则称为集中载荷，如图 11.1c。

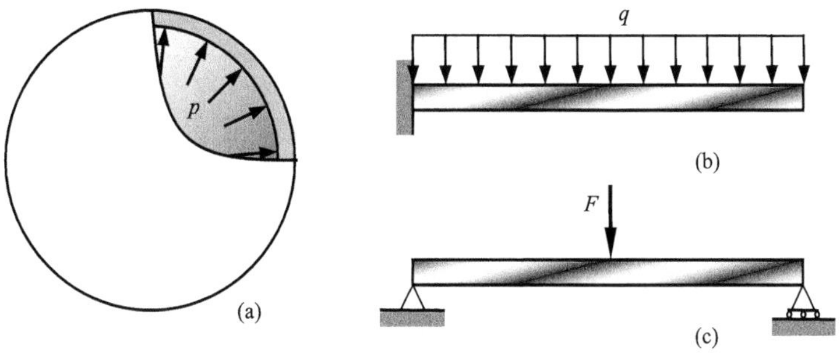

图 11.1

作用于物体内每一点的外力称为体积力，简称体力。重力是典型的体力。

11.2　内　　力

当外载荷作用于构件，那么在构件内将产生相应的分布内力。构件截面上的分布内力可通过截面法进行求解。

想象用平面，比如图 11.2a 中平面 Π，在需要求分布内力之处截开构件。移去截开平面右侧构件，用分布内力代替其对左侧构件的作用，如图 11.2b。考虑构件保留部分的平衡，则分布内力可通过平衡方程确定。虽然内力的精确分布未知，但是利用平衡方程可以确定截开截面上由分布内力引起的合力 $\boldsymbol{R}$ 和对形心 O 的合力偶 $\boldsymbol{M}_O$，如图 11.3a。

一般来说，合力 $\boldsymbol{R}$ 和合力偶 $\boldsymbol{M}_O$ 具有任意方向，即既不垂直截面，也不平行截面。然而，可以把合力和合力偶分解为四个分量，如图 11.3b：

(1) **轴力**　合力法向分量称为轴力，由 N 表示。

(2) **剪力**　合力切向分量称为剪力，由 V 表示。

(3) **扭矩**　合力偶法向分量称为扭矩，由 T 表示。

(4) **弯矩**　合力偶切向分量称为弯矩，由 M 表示。

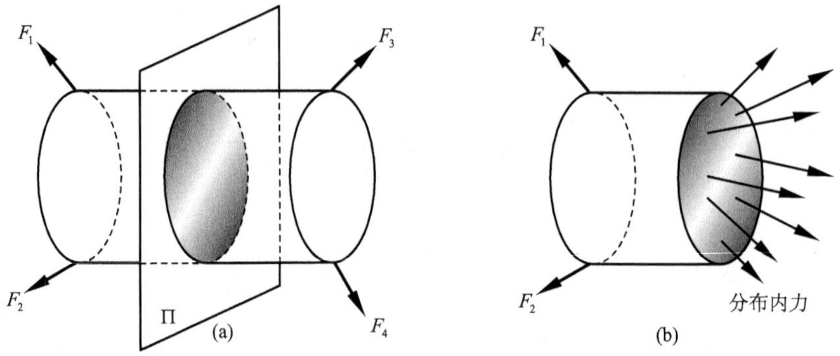

图 11.2

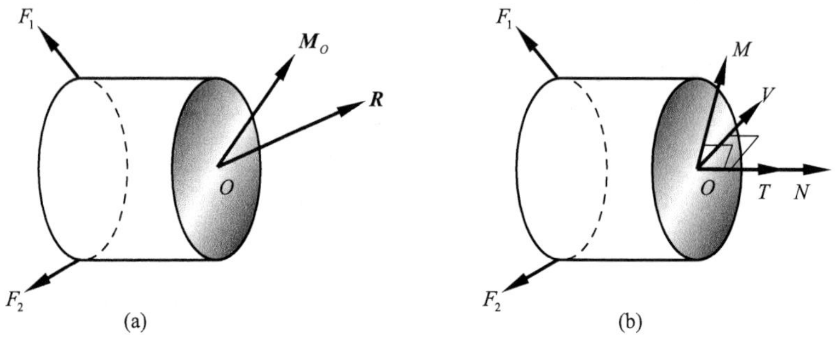

图 11.3

11.3　应　　力

当各种外载荷作用于构件，在构件内任何一点将产生分布内力。为定义截面上点 P 处应力，围绕点 P 取面元ΔA，如图 11.4a 所示，假设面元上合力为$\Delta \boldsymbol{F}$。通常$\Delta \boldsymbol{F}$ 在截面上特定点有唯一确定的方向，因而可分解为分量ΔN 和ΔV，如图 11.4b 所示。ΔN 为垂直面元ΔA 的法向分量，ΔV 为面元ΔA 内的切向分量。

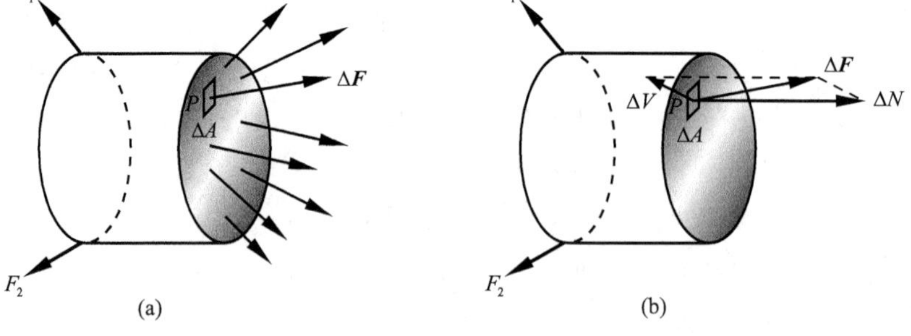

图 11.4

1. 正应力

法向力集度(即单位面积法向力)定义为正应力，用σ表示。如图 11.5 所示，点 P 处正应力表示为

$$\sigma = \lim_{\Delta A \to 0} \frac{\Delta N}{\Delta A} \tag{11.1}$$

式中，σ垂直于截面。正号用于表示拉应力，负号表示压应力。在国际单位制中，ΔN 的单位为 N，ΔA 的单位为 m^2，正应力σ的单位为 Pa。

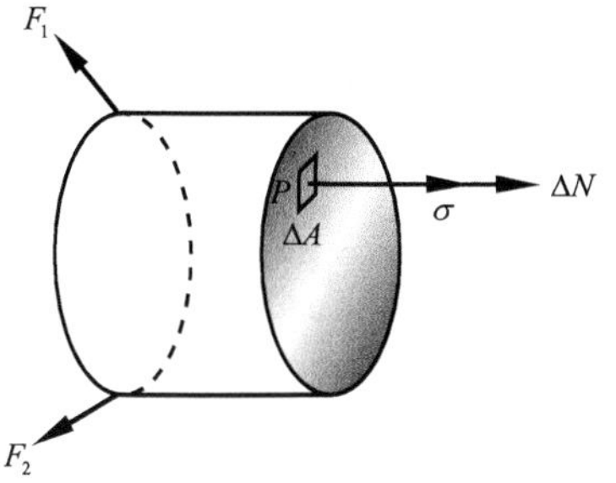

图 11.5

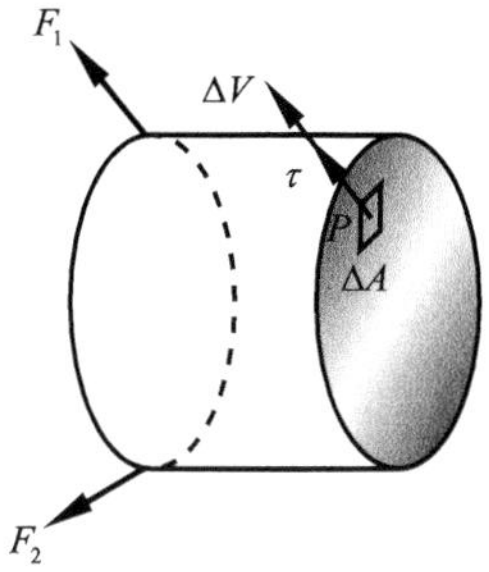

图 11.6

2. 剪应力

切向力集度(即单位面积切向力)定义为剪应力，用 τ 表示。如图 11.6 所示，点 P 处切应力写为

$$\tau = \lim_{\Delta A \to 0} \frac{\Delta V}{\Delta A} \tag{11.2}$$

式中，τ 位于截面内。在国际单位制中，剪应力τ 的单位为 Pa。

例 11.1　载荷 F 作用于图 E11.1a 所示钢杆，钢杆由开有直径为 15 mm 孔的薄板支撑。已知钢的剪应力不能超过 120 MPa，求能够作用于钢杆的最大载荷 F。

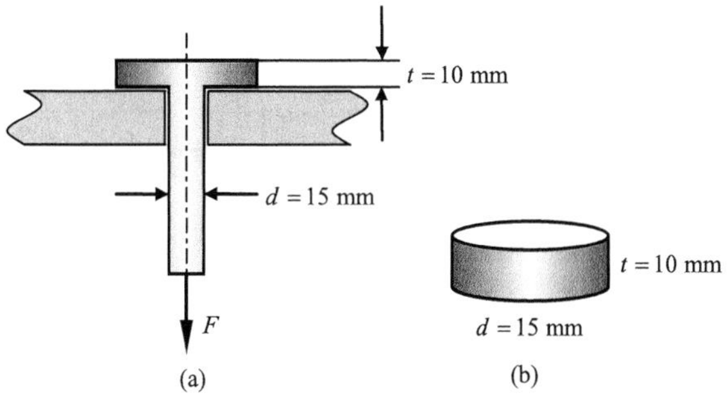

图 E11.1

解　剪切面为圆柱面，如图 E11.1b 所示，剪切面积 $A=\pi dt$。因最大剪应力 $\tau_{\max}=120\ \text{MPa}$，则利用式(11.2)，得最大载荷 $F_{\max}$ 为

$$F_{\max}=\tau_{\max}A=56.5\ \text{kN}$$

3. 挤压应力

由螺钉、销钉和铆钉连接的构件的挤压面上将产生应力。如图 11.7a 所示，考虑通过螺钉连接的两块薄板。螺钉作用于上板的力 F_{bs}(图 11.7b)与上板反作用于螺钉的力 F'_{bs}(图 11.7c)大小相等，方向相反。螺钉施加的力 F_{bs} 表示半圆柱内表面上分布力的合力。

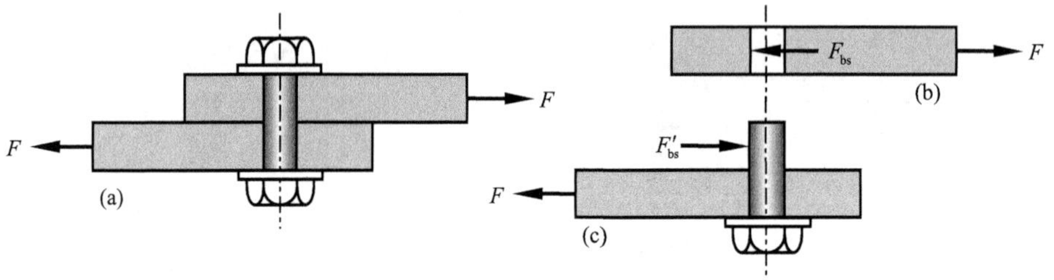

图 11.7

由于构件接触面上力的分布十分复杂，因此由 F_{bs} 除以螺钉在板上的投影面积 A_{bs} 而得到的平均应力即被认为是挤压应力 σ_{bs}。因 A_{bs} 等于 td，其中 t 为薄板厚度，d 为螺钉直径，故有

$$\sigma_{bs}=\frac{F_{bs}}{A_{bs}}=\frac{F_{bs}}{td} \tag{11.3}$$

例 11.2　载荷 F 作用于图 E11.2a 所示钢杆，钢杆由开有直径为 15 mm 孔的薄板支撑。已知钢的挤压应力不能超过 150 MPa，求能够作用于钢杆的最大载荷 F。

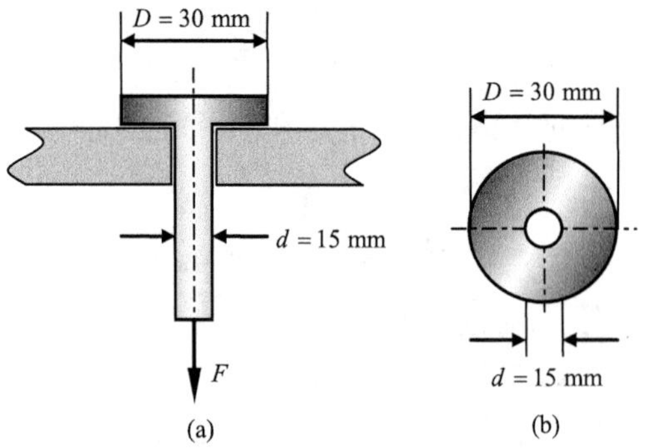

图 E11.2

解　如图 E11.2b 所示，挤压面为内径 $d=15$ mm 和外径 $D=30$ mm 的环形表面，因此挤压面积等于 $A_{bs}=\dfrac{1}{4}\pi(D^2-d^2)$。因最大挤压应力 $\sigma_{bs}=150\ \text{MPa}$，则利用式(11.3)，得最大载荷 $F_{\max}$ 为

$$F_{max} = \sigma_{bs} A_{bs} = 79.5\ \text{kN}$$

11.4　应　　变

当外载荷作用于构件时，载荷将倾向于改变构件的尺寸和形状。

1. 线应变

单位长度线段的伸长或缩短称为线应变，用ε表示。如图 11.8a 所示，考虑构件上过点 P 的线段Δl。在外载荷作用下构件发生变形，线段Δl 的长度变为$\Delta l'$，如图 11.8b 所示。线段 Δl 的变形等于$\Delta\delta = \Delta l' - \Delta l$。

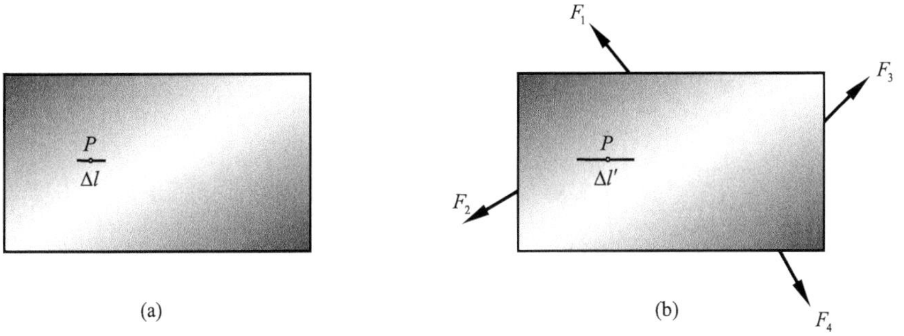

图 11.8

当Δl 趋于零，$\Delta\delta$与Δl 的商将趋于有限值。该极限值称为点 P 处沿线段方向的线应变，可表示为

$$\varepsilon = \lim_{\Delta l \to 0} \frac{\Delta\delta}{\Delta l} \tag{11.4}$$

式中，ε为无量纲量。如果线段伸长，线应变为正；如果线段缩短，线应变为负。

2. 剪应变

两垂直线段之间的角度变化称为剪应变，用γ表示。如图 11.9a 所示，考虑两垂直线段Δx 和Δy，在点 P 处相交。在外载荷作用下构件发生变形，两线段的夹角由$\frac{1}{2}\pi$ 变为θ，如图 11.9b 所示。

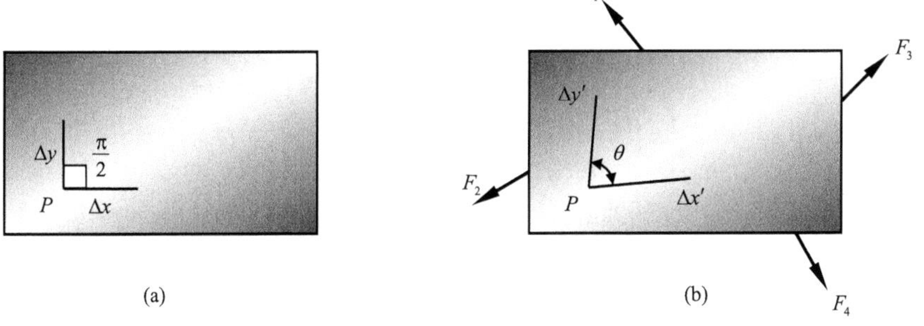

图 11.9

点 P 处在包含两线段的平面内的剪应变可写为

$$\gamma = \frac{1}{2}\pi - \lim_{\substack{\Delta x \to 0 \\ \Delta y \to 0}} \theta \tag{11.5}$$

如果 θ 小于 $\frac{1}{2}\pi$，则剪应变 γ 为正；否则，为负。剪应变 γ 的单位为 rad。

11.5　胡 克 定 律

大部分工程结构都设计为承受相对较小的变形，即仅包含应力应变曲线的直线部分。对应力应变曲线的直线部分，应力 σ 与应变 ε 成正比，则有

$$\sigma = E\varepsilon \tag{11.6}$$

上述关系称为胡克定律。E 为弹性模量或杨氏模量。由于 ε 无量纲，因此 E 与 σ 单位相同。

11.6　低碳钢拉伸性能

1. 标准试样

具有确定尺寸和形状的标准试样必须用于低碳钢拉伸试验，以便实验结果具有可比性。

图 11.10a 所示为一种常用的低碳钢标准拉伸试样。试样中间部位的横截面面积 A_0 已经精确测量，间距为 l_0 的两根标线也已经刻划在试样中间部位。距离 l_0 称为试样标距。

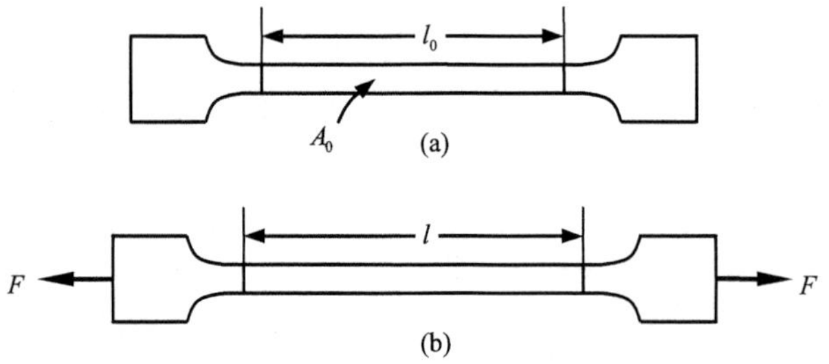

图 11.10

2. 拉伸图

把试样放入万能试验机，施加中心载荷 F。当载荷 F 增加时，两标线之间的距离 l 也随之增加，如图 11.10b 所示。对每个 F 值，记录相应的伸长量 $\delta = l - l_0$。

画载荷大小 F 与变形 δ 的曲线，得载荷变形图，如图 11.11 所示。该图称为低碳钢的拉伸图。

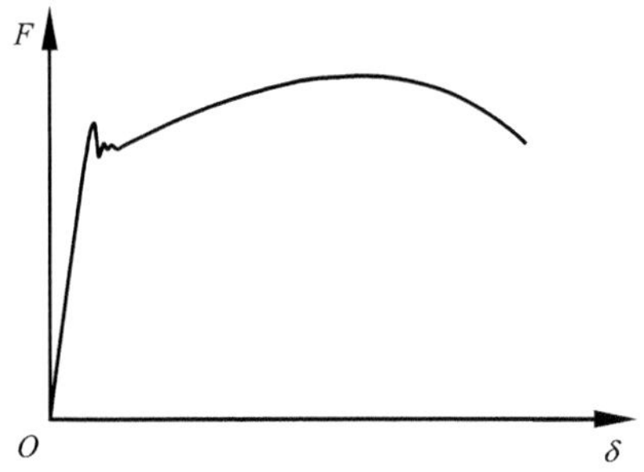

图 11.11

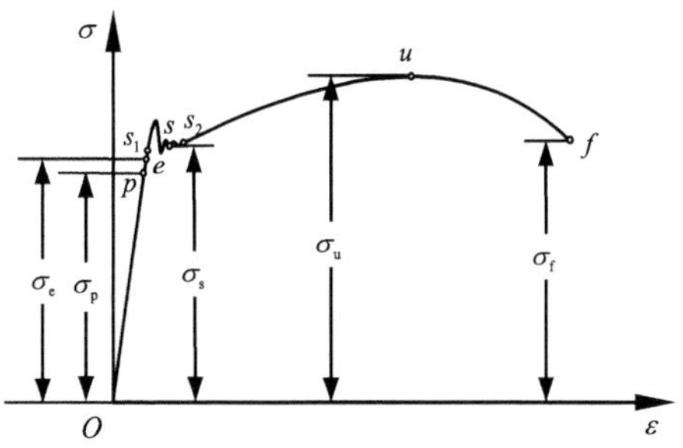

图 11.12

3. 应力应变曲线

根据每对读数 F 和δ，载荷 F 除以原始横截面面积 A_0 计算应力σ，伸长量δ 除以两标线之间的原始距离 l_0 计算应变ε。

以ε为横坐标、σ 为纵坐标绘图而得到的曲线，既能够表征低碳钢机械性能，又与所用特定试样的尺寸无关。该曲线称为低碳钢的应力应变曲线，如图 11.12 所示。根据低碳钢的应力应变曲线，可看出材料所表现出的四个不同范围。

(1) 弹性阶段 (Oe)。应力应变曲线的 Oe 区域称为弹性范围。该区域材料呈现弹性特征。对应弹性范围的最大应力称为弹性极限，用 σ_e 表示。

弹性范围 Oe 的直线区域 Op 称为线弹性范围。在该范围材料呈现线弹性特征，即应力与应变成正比。满足线性关系的最大应力称为比例极限，用 σ_p 表示。如果应力稍微超过比例极限，则材料仍然呈现弹性特征，直到应力达到弹性极限。

在弹性范围，如果卸载，则试样将恢复到其原始尺寸和形状。通常对低碳钢而言，弹性极限非常接近比例极限，因而很难确定弹性极限。

(2) 屈服阶段 (s_1s_2)。稍微高于弹性极限的应力将会引起材料失效，引起材料发生永久变形。这种现象称为屈服，屈服由应力应变曲线的 s_1s_2 区域表示。屈服开始产生所对应的应力称为屈服应力、屈服强度或屈服点，用 σ_s 表示。屈服范围产生的变形称为塑性变形。塑性变形是由材料斜截面上的剪应力引起材料沿斜截面的滑移而形成。达到屈服点后，试样在不增大载荷的情况下还会继续伸长。

对低碳钢，屈服点常常有两个值。上屈服点首先出现，紧接着承载能力突然下降到下屈服点。

(3) 强化阶段 (s_2u)。当屈服结束，更大的载荷能够作用于试样，引起曲线继续上升但变得更为平坦，直到达到最大应力。该最大应力称为极限应力或极限强度，用 σ_u 表示。以这种方式引起的曲线上升称为应变强化，应变强化由 s_2u 区域表示。

(4) 颈缩阶段 (uf)。达到极限应力后，横截面面积在试样局部区域开始急剧减小。这种现象称为颈缩，颈缩现象是由材料内部形成的滑移面所引起，实际应变是由剪应力所引起。因此，当试样进一步伸长时，横向收缩逐渐形成了颈缩区域。因为在颈缩区域横截面面积不断减小，因此越来越小面积仅能承受不断减小的载荷。所以，应力应变曲线向下弯曲直到试样达到断裂应力 σ_f 时而发生断裂。

应该注意，断裂沿着与试样横截面大约成 45° 的锥形面发生。

4. 伸长率和断面收缩率

材料塑性的标准度量方法是伸长率 δ_1，定义为

$$\delta_1 = \frac{l_1 - l_0}{l_0} \times 100\% \tag{11.7}$$

式中，l_0 和 l_1 分别为拉伸试样的原始标距和断后标距。

材料塑性的其他度量方法是断面收缩率 ψ_1，定义为

$$\psi_1 = \frac{A_0 - A_1}{A_0} \times 100\% \tag{11.8}$$

式中，A_0 和 A_1 分别为试样的原始横截面面积和断后最小横截面面积。

11.7 无明显屈服点塑性材料的应力应变曲线

许多塑性材料，屈服并不是由应力应变曲线的波动或水平部分进行表征，因为应力一直不断增大，直到达到极限强度，然后颈缩开始，最终断裂。对这类材料，屈服强度可以通过偏移方法进行定义。过横坐标 $\varepsilon = 0.2\%$ 处，画平行于应力应变曲线直线部分的倾斜直线，交应力应变曲线于 s 点，如图 11.13 所示。对应于 s 点的应力 $\sigma_{0.2}$ 定义为材料的条件屈服应力。

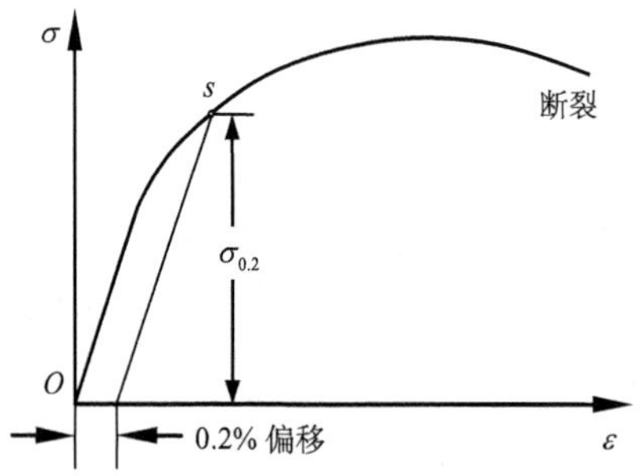

图 11.13

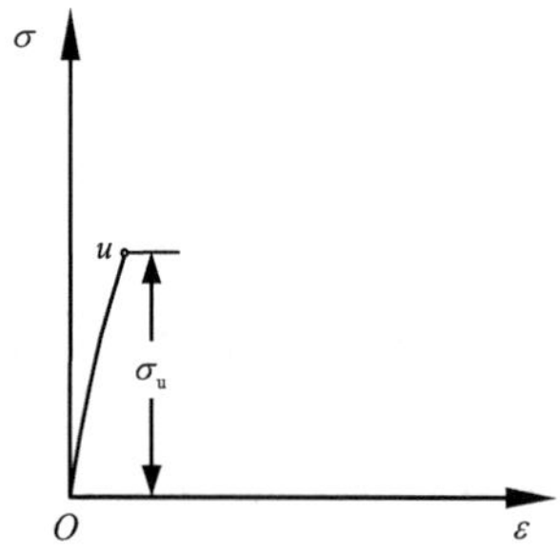

图 11.14

11.8 塑性和脆性材料

各种材料的应力应变曲线相差很大。然而，从各种材料的应力应变曲线当中可以区分一些共同特征，并根据这些特征把材料分为两大类，即塑性材料和脆性材料。

由结构钢和许多其他金属合金组成的塑性材料由其在常温时的屈服能力进行表征，如图 11.12 所示。

由铸铁、玻璃和石块组成的脆性材料由其伸长率在没有明显变化的情况下而发生的断裂进行表征，如图 11.14 所示。脆性材料试样没有颈缩阶段，断裂沿垂直于载荷的截

面发生。

11.9　材料压缩性能

当塑性材料试样压缩时，应力应变曲线的直线部分和对应屈服和强化的开头部分与塑性材料拉伸时的应力应变曲线相同。尤其值得注意的是，低碳钢的屈服强度在拉伸和压缩中相同。对较大应变，拉伸和压缩应变曲线开始分叉，应该注意压缩没有颈缩现象，如图 11.15a 所示。

对大部分脆性材料，比如铸铁，压缩极限强度远远高于拉伸极限强度，如图 11.15b 所示。

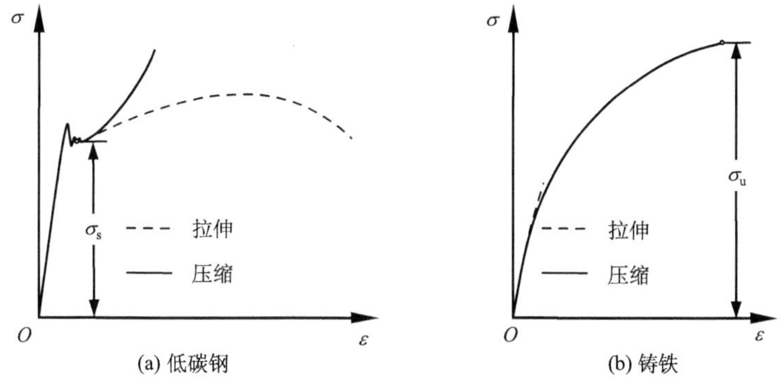

图 11.15

习　题

11.1　厚度 $t = 20$ mm、宽度 $b = 80$ mm 的两块木板通过胶合榫眼接头进行连接。已知胶的平均剪应力达到 $\tau = 1.2$ MPa 时接头失效，求切口长度 $a = 30$ mm 时的最大许可轴向载荷 F_{max}。

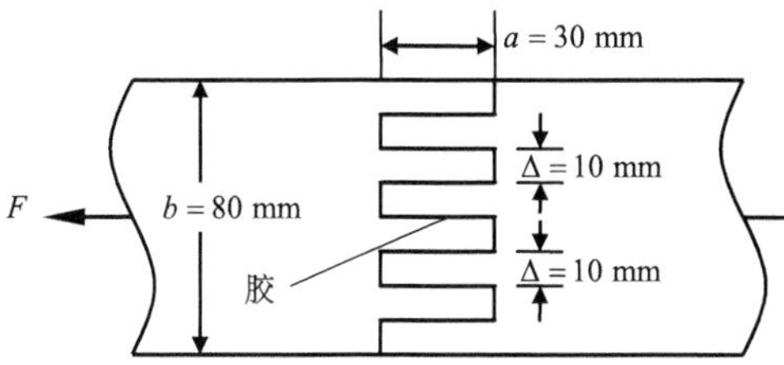

图 P11.1

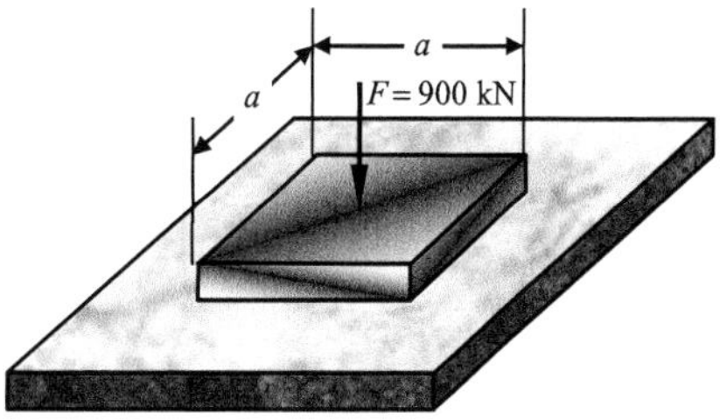

图 P11.2

11.2　载荷 F 通过方板均匀作用于混凝土地基。已知混凝土地基的挤压应力不能超过 12 MPa，求满足最经济安全设计的方板边长 a。

11.3　铝合金应力应变曲线如图 P11.3 所示，如果应力施加到 550 MPa，求卸载后的永久应变。

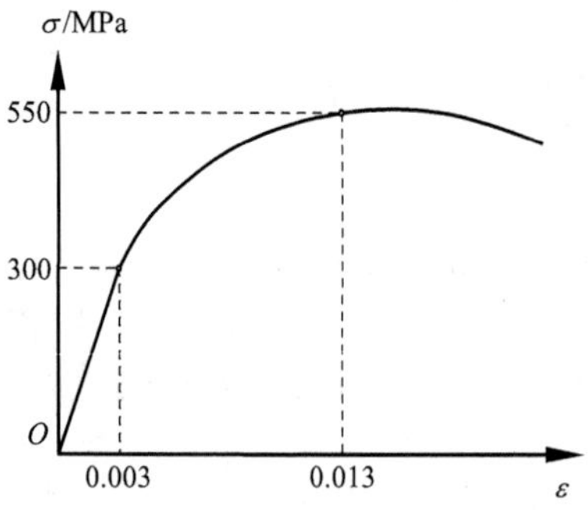

图 P11.3

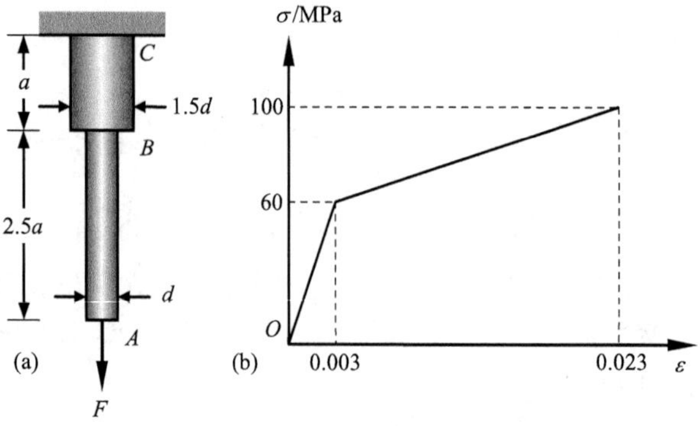

图 P11.4

11.4　图 P11.4a 所示圆截面杆受轴力 F 作用，应力应变曲线如图 P11.4b 所示，求加载后杆件伸长。已知 $a = 50\ \text{mm}$, $d = 4\ \text{mm}$, $F = 1131\ \text{N}$。

第 12 章　拉伸与压缩

考虑等截面杆，两端受等值反向并与纵向形心轴重合的一对外力作用。如果外力方向指向杆外，则杆为拉伸，如图 12.1a 所示；如果外力方向指向杆内，则杆为压缩，如图 12.1b 所示。

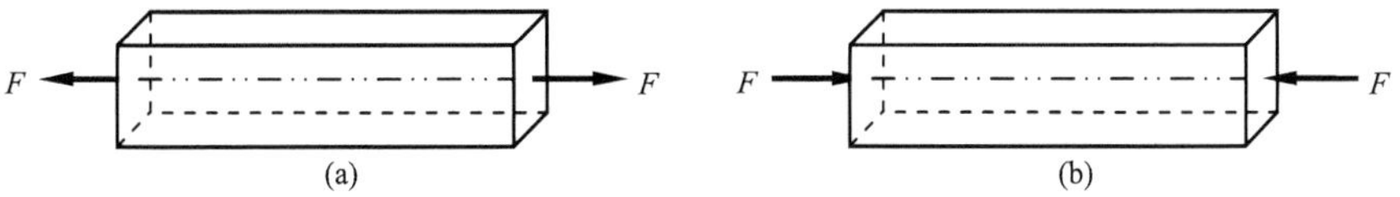

图 12.1

12.1　轴　　力

如图 12.1a 所示，在一对外力作用下，杆的横截面上将产生分布内力。这些分布内力的合力称为轴力或法向力，用 N 表示，如图 12.2 所示。

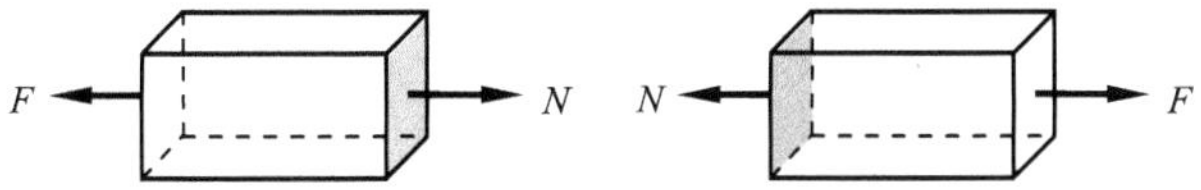

图 12.2

假想用一个平面在需要求轴力的位置沿杆的横截面把杆截开，为了得到截开面上的轴力，可移去截开面右侧(或左侧)的杆，用轴力代替对左侧(或右侧)部分的作用。通过截开横截面，相对杆的保留部分而言，原来的轴力就变成了外力。考虑杆的保留部分的平衡，轴力就可通过沿杆方向的静平衡方程求解。

12.2　横截面正应力

如图 12.3a 所示，关于拉伸或压缩杆横截面上的轴力分布，需要提出一些假设。由于外力通过截面形心，因此通常假设轴力在横截面上均匀分布，如图 12.3b 所示。因此横截面上的正应力可表示为

$$\sigma=\frac{N}{A} \tag{12.1}$$

式中，A 为横截面面积，N 为横截面上的轴力。正号表示拉应力(杆受拉)，而负号表示压应力(杆受压)。

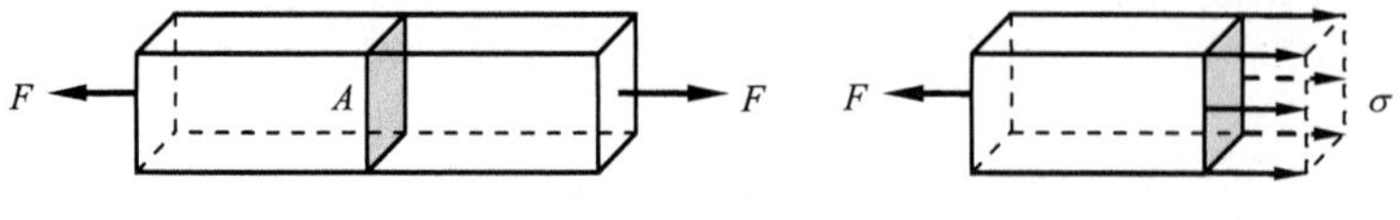

图 12.3

例 12.1 如图 E12.1a 所示，矩形截面连杆 BD 和 CE 的横截面尺寸均为 16 mm×26 mm，销钉的直径为 6 mm。求连杆(a) BD 和(b) CE 的最大平均正应力。

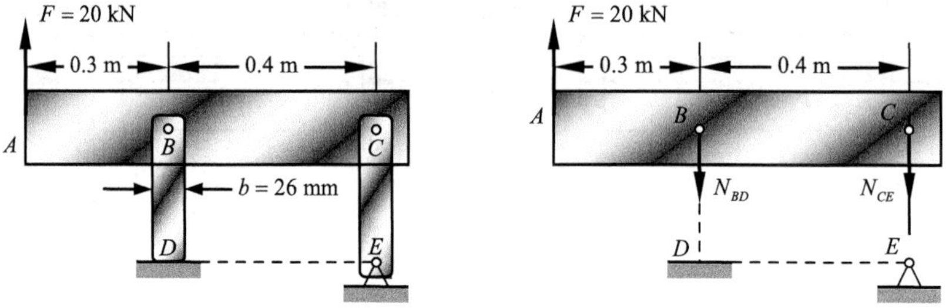

图 E12.1

解 如图 E12.1b 所示，考虑水平梁的平衡，有

$$N_{BD}=F\frac{l_{AC}}{l_{BC}}=35\text{ kN},\quad N_{CE}=F-N_{BD}=-15\text{ kN}$$

因为连杆 BD 受拉，因此其最小横截面面积等于

$$A_{BD}=t(b-d)=320\text{ mm}^2$$

根据式(12.1)，得连杆 BD 的最大平均正应力为

$$\sigma_{BD}=\frac{N_{BD}}{A_{BD}}=109.4\text{ MPa}$$

同理，连杆 CE 的最大平均正应力为

$$\sigma_{CE}=\frac{N_{CE}}{A_{CE}}=-36.1\text{ MPa}$$

12.3 斜截面正应力和剪应力

考虑图 12.4a 所示二力杆，受一对拉力 F 作用。把杆沿斜截面(与横截面成 θ 角)截开，画左侧部分的受力图。根据平衡，作用于斜截面上的力 R 必定等于 F。如图 12.4b 所示，把 R 分解为分别垂直和平行于斜截面的分量 N 和 V，则有

$$N=R\cos\theta,\ V=R\sin\theta \tag{12.2}$$

式中，N 表示斜截面上法向分布力的合力，V 表示切向分布力的合力。如图 12.4c 所示，N 和 V 分别除以斜截面面积 A_θ 可得斜截面上的正应力和剪应力

$$\sigma_\theta = N/A_\theta,\ \tau_\theta = V/A_\theta \tag{12.3}$$

利用 $A_\theta = A/\cos\theta$，其中 A 为横截面面积，得

$$\sigma_\theta = \sigma\cos^2\theta,\ \tau_\theta = \frac{1}{2}\sigma\sin 2\theta \tag{12.4}$$

式中，σ为横截面上的正应力。由式(12.4)得，当 $\theta = 0$，则 $\sigma_\theta = \sigma_{\max} = \sigma$ 和 $\tau_\theta = 0$；当 $\theta = 45°$，则 $\sigma_\theta = \tau_\theta = \tau_{\max} = \frac{1}{2}\sigma$。

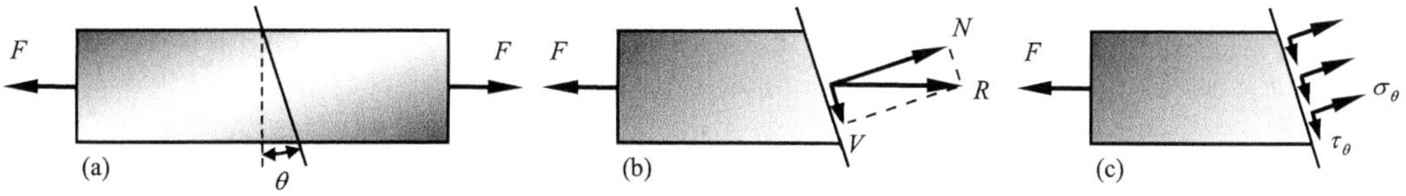

图 12.4

与抗拉能力相比，塑性材料的抗剪能力较弱。因此，塑性材料试样在受拉时，将会沿与横截面成 45°的截面发生失效，如图 12.5a 所示。然而，脆性材料的抗拉能力比抗剪能力弱。因此，脆性材料试样在受拉时，将会沿横截面发生失效，如图 12.5b 所示。

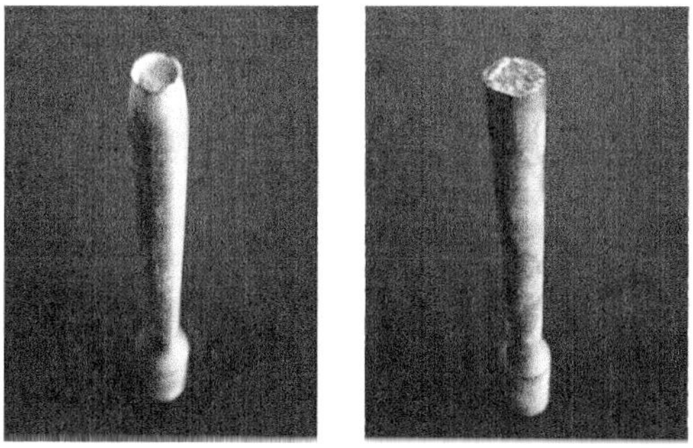

(a) (b)

图 12.5

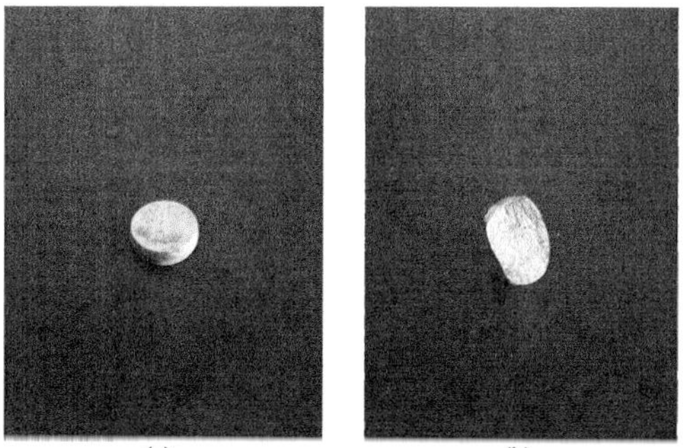

(a) (b)

图 12.6

塑性材料的抗剪能力比抗压能力弱，因此，塑性材料试样在受压时，将会沿与横截面成 45°的截面发生失效，如图 12.6a 所示。脆性材料的抗剪能力也比抗压能力弱，因此，脆性材料试样在受压时，将会沿与横截面成 45°~55°的截面发生破坏，如图 12.6b 所示。

例 12.2 具有 90 mm×120 mm 矩形横截面的两根木杆通过图示胶合接头进行连接。已知 $F = 100\ \text{kN}$，求胶合接头面上的正应力和剪应力。

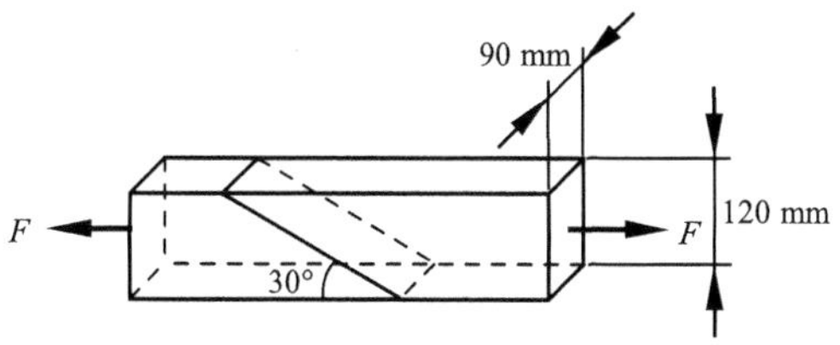

图 E12.2

解　利用式(12.4)，有

$$\sigma_{60°} = \sigma \cos^2 60° = 2.31\ \text{MPa}，\ \tau_{60°} = \frac{\sigma}{2}\sin(2\times 60°) = 4.01\ \text{MPa}$$

12.4　线 应 变

1. 纵向线应变

长度为 l、横截面面积为 A 的等截面杆，上端悬挂，如图 12.7a 所示。如果在下端施加外力 F，则杆将伸长，如图 12.7b 所示。

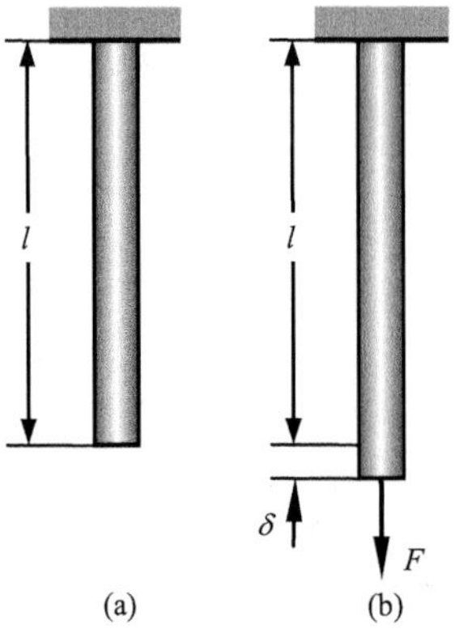

图 12.7

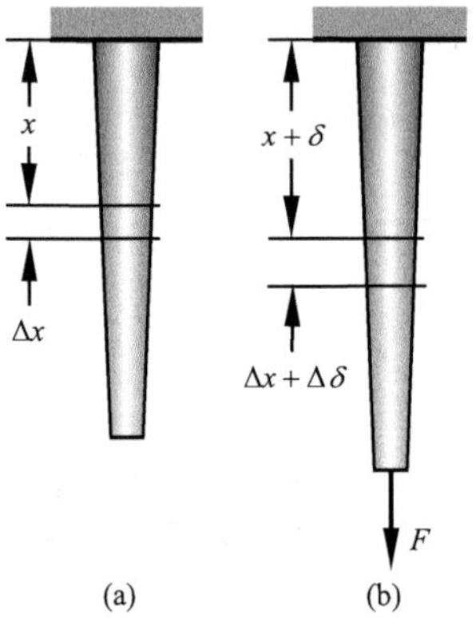

图 12.8

在轴向载荷作用下杆的纵向线应变可表示为

$$\varepsilon = \frac{\delta}{l} \tag{12.5}$$

式中，δ为杆的纵向变形，l 为杆长。

对变截面杆，横截面上的正应力沿杆将发生变化，因此有必要定义在一点处的纵向线应变。考虑未变形长度为Δx 的微段，如图 12.8a 所示。用$\Delta\delta$表示微段在载荷作用下的变形，如图 12.8b 所示。那么在 x-截面上一点处的纵向线应变可表示为

$$\varepsilon = \lim_{\Delta x \to 0} \frac{\Delta\delta}{\Delta x} = \frac{\mathrm{d}\delta}{\mathrm{d}x} \tag{12.6}$$

2. 横向线应变

当杆轴向受载，只要不超过材料的比例极限，横截面上正应力与纵向线应变将满足胡克定律。所有材料都假设为均匀和各向同性，即机械性能与位置和方向无关。因此，对图 12.9 所示载荷，有$\varepsilon_y=\varepsilon_z$，其中$\varepsilon_y$和$\varepsilon_z$分别为沿 y 和 z 轴的横向线应变。

材料的重要常数之一是泊松比，用μ表示。泊松比定义为

$$\mu = -\frac{\varepsilon_y}{\varepsilon_x} = -\frac{\varepsilon_z}{\varepsilon_x} \tag{12.7}$$

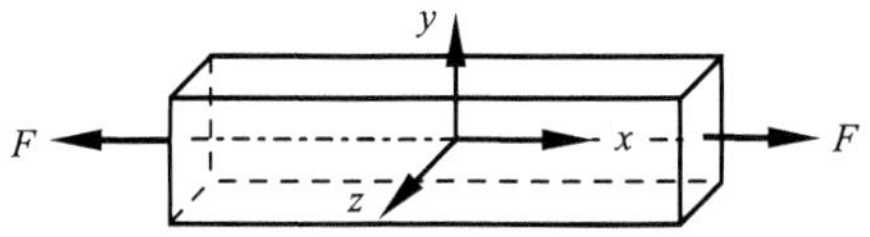

图 12.9

因此，横向线应变可表示为

$$\varepsilon_y = \varepsilon_z = -\mu\varepsilon_x \tag{12.8}$$

12.5 拉 压 变 形

考虑长度为 l、横截面面积为 A 的等截面杆，受中心轴向载荷作用，如图 12.10 所示。如果横截面上正应力不超过材料的比例极限，则胡克定律成立，即 $\sigma = E\varepsilon$。利用 $\sigma = N/A$ 和 $\varepsilon = \delta/l$，有

$$\delta = \frac{Nl}{EA} \tag{12.9}$$

式中，EA 为杆的轴向(抗拉或抗压)刚度。式(12.9)仅适用于杆满足均匀和各向同性假设、具有不变横截面和受中心轴向载荷作用。

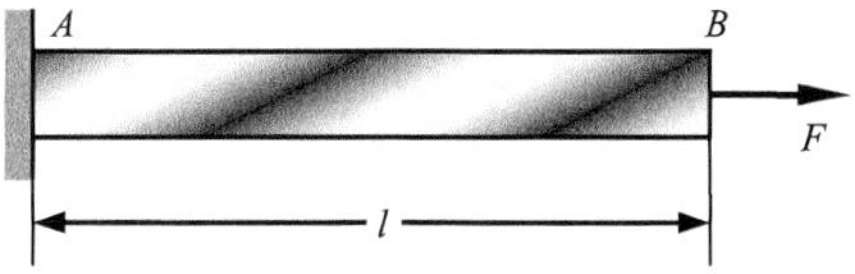

图 12.10

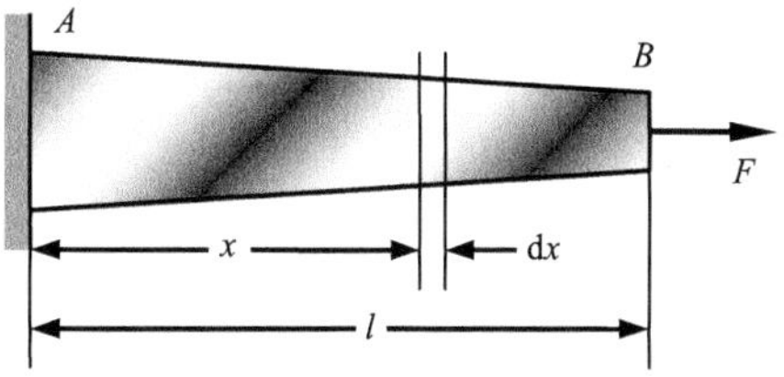

图 12.11

对变截面杆，如图 12.11 所示，则长度为 dx 的微段变形可表示为

$$\mathrm{d}\delta = \frac{N\mathrm{d}x}{EA} \tag{12.10}$$

因此，通过沿杆长积分，得杆的总变形为

$$\delta = \int \frac{N\mathrm{d}x}{EA} \tag{12.11}$$

例 12.3　直径为 4 mm 的绳子 BC 是由弹性模量 E=200 GPa 的钢制成。已知绳中最大应力不超过 190 MPa，绳的伸长不超过 6 mm，求可施加到图 E12.3a 所示结构的最大载荷 F。

解　如图 E12.3b 所示，取 AB 为自由体，考虑其平衡，有

$$\sum M_A = 0: \quad Fl_{AD} - (N_{BC}\frac{l_{AC}}{l_{BC}})l_{AB} = 0\text{，即 } F = 1.2N_{BC}$$

根据绳中应力求最大许可载荷，有

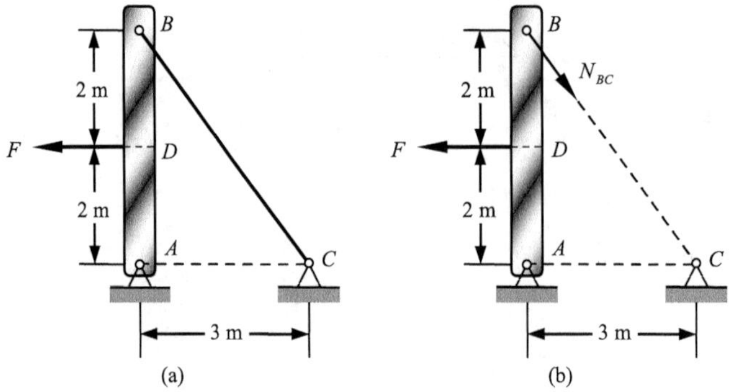

图 E12.3

$$F_{\max}^{\sigma}=1.2N_{BC}=1.2\sigma_{\max}A_{BC}=1.2\sigma_{\max}(\frac{1}{4}\pi d_{BC}^{2})=2.87\ \text{kN}$$

根据绳的变形求最大许可载荷，有

$$F_{\max}^{\delta}=1.2N_{BC}=1.2\frac{\delta EA_{BC}}{l_{BC}}=1.2\frac{\delta E(\frac{1}{4}\pi d_{BC}^{2})}{l_{BC}}=3.62\ \text{kN}$$

应该取两个最大许可载荷中的较小者，即可施加到结构的最大许可载荷 $F_{\max}$ 等于

$$F_{\max}=\min[F_{\max}^{\sigma},\ F_{\max}^{\delta}]=2.87\ \text{kN}$$

12.6　静不定拉压杆

前面各节所考虑的问题，利用静平衡方程即可求解载荷作用下杆各个部分产生的内力。然而，工程中有些问题的内力仅仅根据静平衡方程并不能进行求解。对这类问题，静平衡方程必须用满足变形协调关系的额外方程进行补充。因为静平衡方程不足以求解全部反力或内力，因此这类问题称为静不定问题。

考虑长度为 l、横截面面积为 A 和弹性模量为 E 的等截面杆 AB，受载前两端 A 和 B 固定于刚性支座，如图 12.12a 所示。

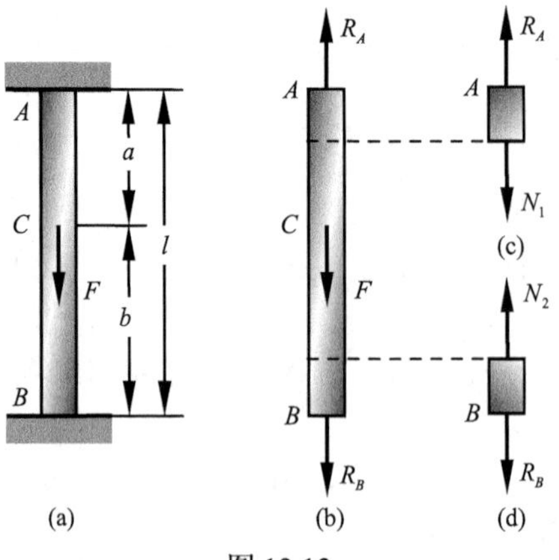

图 12.12

如图 12.12b 所示，画出杆的受力图，得平衡方程

$$\sum F_y = 0: \quad R_A - R_B - F = 0 \tag{12.12}$$

因为一个方程不足以求两个未知反力 R_A 和 R_B，因此本问题为一次静不定。

观察几何关系，得杆的总变形δ必须等于零。用δ_1和δ_2分别表示 AC 和 BC 部分的变形，有

$$\delta = \delta_1 + \delta_2 = 0 \tag{12.13}$$

用相应的内力 N_1 和 N_2 表达δ_1和δ_2，得

$$\delta_1 = \frac{N_1 a}{EA},\ \delta_2 = \frac{N_2 b}{EA} \tag{12.14}$$

根据图 12.12c 和 12.12d，有 $N_1 = R_A$ 和 $N_2 = R_B$。把这些关系代入式(12.14)，并利用式(12.13)，有

$$R_A a + R_B b = 0 \tag{12.15}$$

式(12.12)和(12.15)联立求解，得

$$R_A = \frac{Fb}{l},\ R_B = -\frac{Fa}{l} \tag{12.16}$$

12.7 拉压杆设计

在工程应用中，应力用于结构设计，以保证所设计的结构能够既安全又经济地履行规定的功能。

能够作用于杆的最大力称为极限载荷，用 F_u 表示。因为中心受载，因此极限载荷除以原始横截面面积即可得到极限应力或极限强度，可表示为

$$\sigma_u = \frac{F_u}{A} \tag{12.17}$$

在正常应用条件下，杆允许承受的工作载荷要小于极限载荷。能够施加到杆的最大工作载荷称为许可载荷，用 F_{allow} 表示。因此，施加许可载荷仅发挥杆的部分极限承载能力，剩余的承载能力予以保留以确保杆的安全。极限载荷与许可载荷的比值定义为安全因素，可表示为

$$n = \frac{F_u}{F_{allow}} \tag{12.18}$$

通过应力，安全因素还可表示为

$$n = \frac{\sigma_u}{\sigma_{allow}} \tag{12.19}$$

安全因素的选择是最重要的工程任务之一。一方面，如果安全因素太小，杆失效的可能性就会很大；另一方面，如果安全因素太大，杆的设计就会变得不经济。选择合适的安全因素需要具备基于多种考虑的工程判断能力。

习　题

12.1　实心圆柱杆 *AB* 和 *BC* 在截面 *B* 处焊接为整体，受载如图。已知 d_1=50 mm 和 d_2=30 mm，求(a)杆 *AB* 和(b)杆 *BC* 中间截面上的正应力。

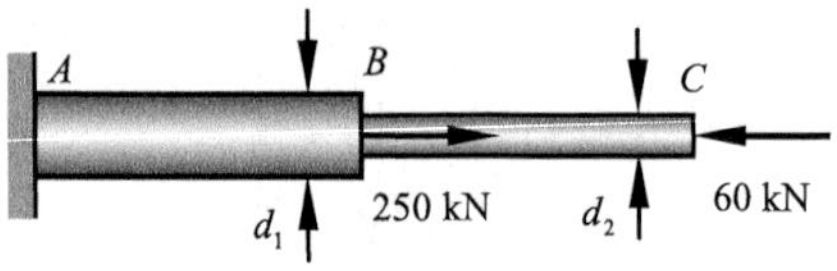

图 P12.1

12.2　直径为 *d* 的实心圆柱杆 *AB* 受两轴向载荷作用，F_1 作用于 *C* 处，F_2 作用于 *B* 处。已知 *d*=50 mm、F_1=200 kN 和 F_2=50 kN，求(a) *AC* 和(b) *BC* 段的正应力。

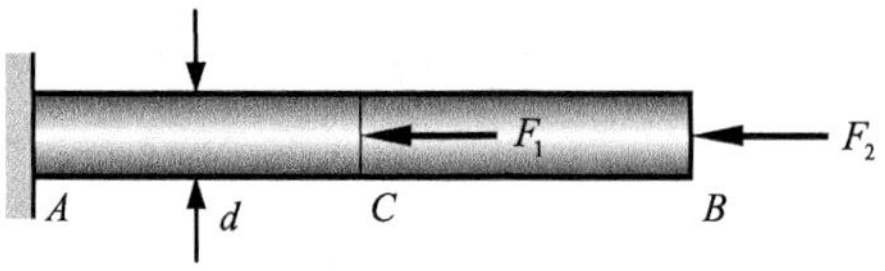

图 P12.2

12.3　宽度 *b*=50 mm、厚度 *t*=6 mm 的连杆 *AB* 用于支撑水平梁的末端。已知连杆的平均正应力为 138 MPa 和销钉的平均剪应力为 82 MPa，求(a)销钉直径 *d* 和(b)连杆的平均挤压应力。

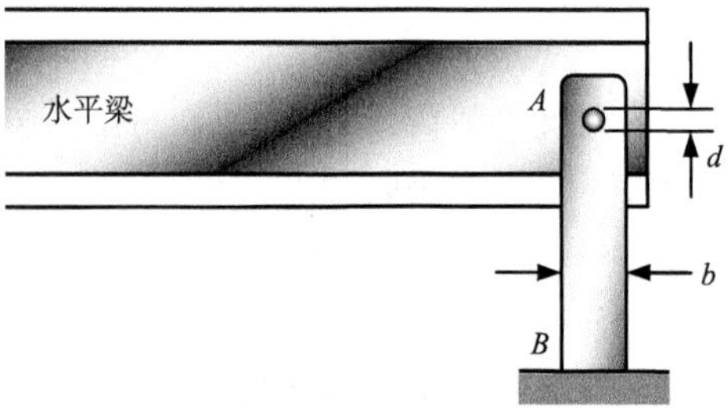

图 P12.3

12.4　矩形截面连杆 *BD* 和 *CE* 的横截面面积均为 16 mm×26 mm，销钉的直径为 6 mm。求(a) *B* 处销钉的平均剪应力、(b)连杆 *BD* 在 *B* 处的平均挤压应力和(c)梁 *ABC* 在 *B* 处的平均挤压应力，已知矩形截面梁的横截面尺寸为 10 mm×50 mm。

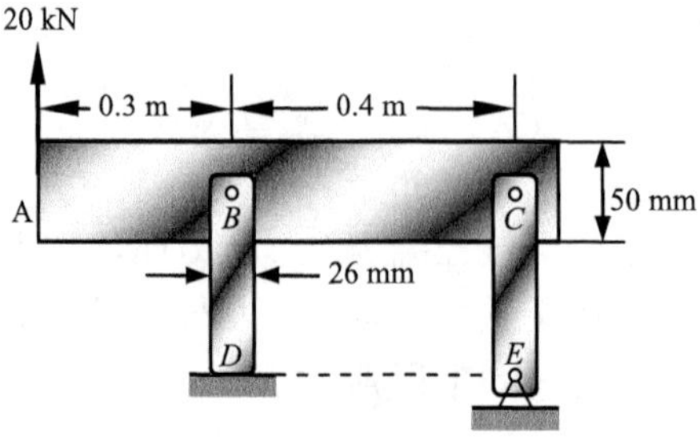

图 P12.4

12.5　轴向载荷作用于钢杆 ABC 的 C 端。已知 E=200 GPa，求当 C 点位移为 3 mm 时 BC 段的直径 d_2。

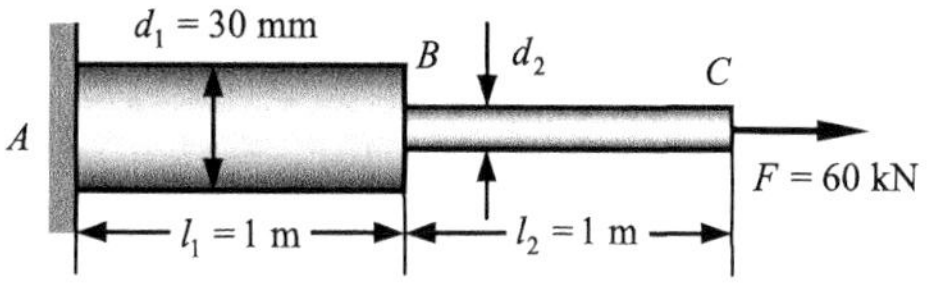

图 P12.5

12.6　考虑图示钢杆 AB(E=200 GPa)，受轴向载荷 F 作用。如果杆的应力不超过 120 MPa，杆的最大长度变化不超过杆长的 0.001 倍，求杆的最小直径。

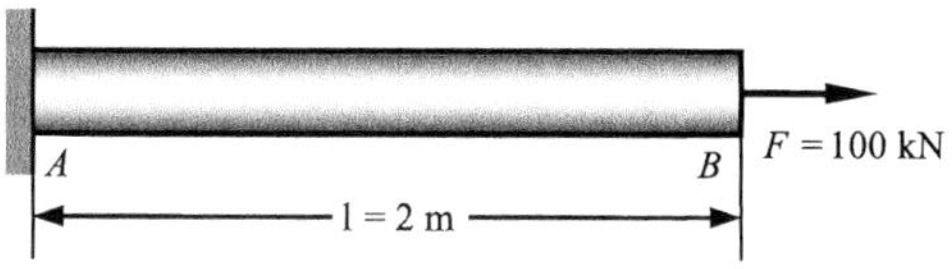

图 P12.6

12.7　杆 ABC 的两段均由黄铜(E=105 GPa)制成。已知 d_2=10 mm 和σ_{allow}=100 MPa，如果 C 点位移不超过 4 mm，求所能施加的最大载荷 F。

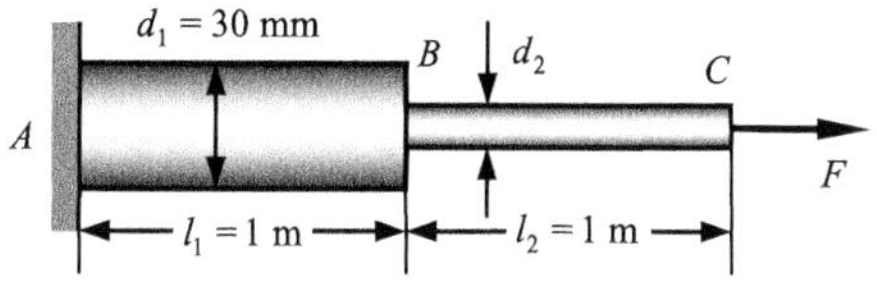

图 P12.7

12.8　连杆 AB 和 CD 由铝(E=75 GPa)制成，横截面面积为 125 mm^2。已知连杆用于支撑刚性梁 AC，求点 E 处的位移。

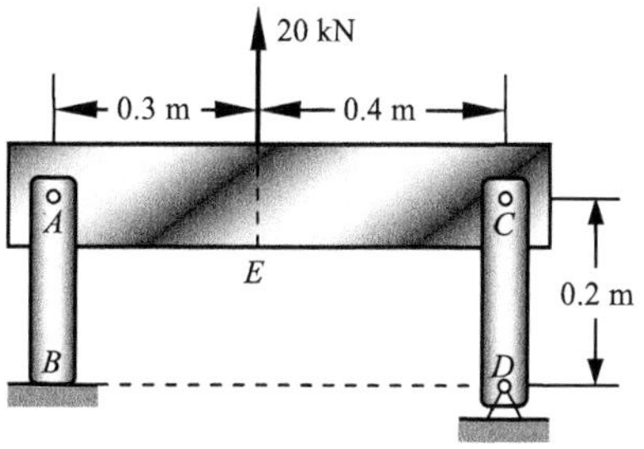

图 P12.8

12.9　刚性梁 AD 由两根直径为 1.5 mm 的钢丝(E=200 GPa)和 A 处的固定铰支座支撑。已知受载前钢丝正好拉紧，求当 1.0 kN 的载荷 F 作用于点 D 处时：(a)每根钢丝的附加拉力；(b)点 D 处的位移。

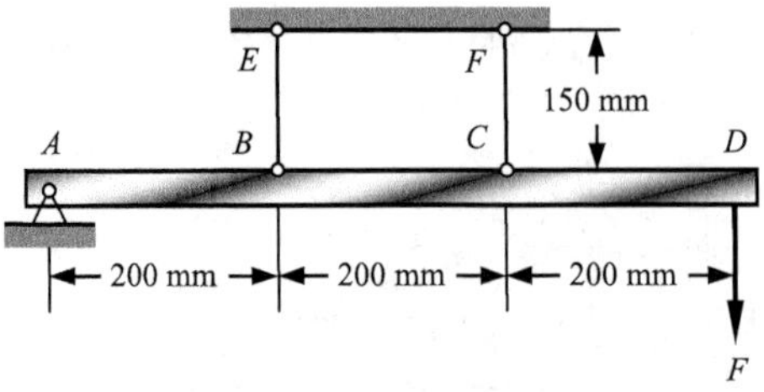

图 P12.9

12.10　刚性梁 AD 由两根直径为 1.5 mm 的钢丝（E=200 GPa）和 A 处的固定铰支座支撑。已知受载前钢丝正好拉紧，求当 1.0 kN 的载荷 F 作用于点 D 处时：(a)每根钢丝的附加拉力；(b)点 D 处的位移。

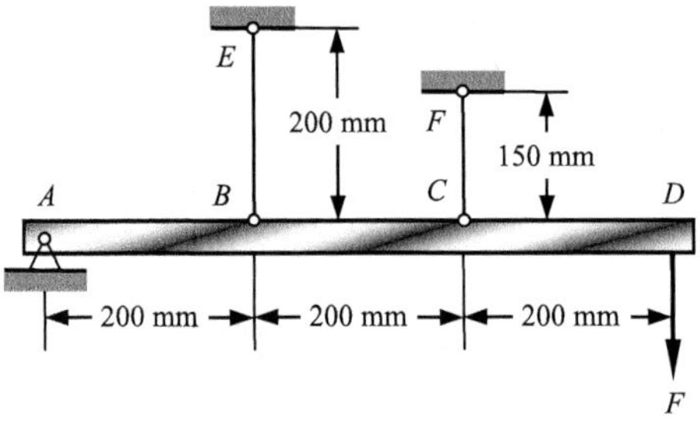

图 P12.10

第 13 章　扭　　转

如图 13.1 所示，圆截面杆两端受一对等值反向并位于纵轴垂直面内的外力偶 M_e 作用。这样的杆称为扭转轴。

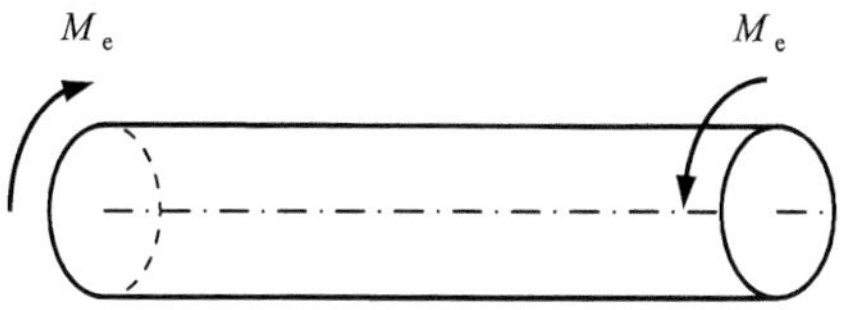

图 13.1

13.1　扭　　矩

如图 13.1 所示，在一对外力偶作用下，轴的横截面上将产生分布内力。分布内力对纵轴的合力偶称为扭矩，用 T 表示，如图 13.2 所示。

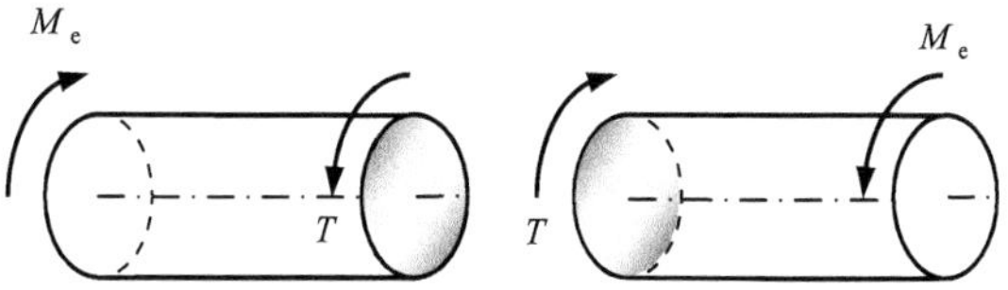

图 13.2

13.2　剪切胡克定律

考虑图 13.3a 所示纯剪切单元体，受剪应力 τ_{xy} 和 τ_{yx} 作用，τ_{xy} 和 τ_{yx} 分别垂直于 x 和 y 轴。根据单元体平衡，得

$$\tau_{xy} = \tau_{yx} \tag{13.1}$$

上述关系称为剪应力互等定理。

在纯剪应力作用下，单元体变形为菱形，如图 13.3b 所示。其中两个角从 $\frac{1}{2}\pi$ 减小为 $\frac{1}{2}\pi - \gamma_{xy}$，而另外两个角则从 $\frac{1}{2}\pi$ 增加为 $\frac{1}{2}\pi+\gamma_{xy}$，其中 γ_{xy} 为剪应变。如果剪应力不超过比例极限，对均匀各向同性材料，则有

$$\tau_{xy} = G\gamma_{xy} \tag{13.2}$$

上述关系称为剪切胡克定律，G 为剪切弹性模量，其单位与 τ_{xy} 相同。

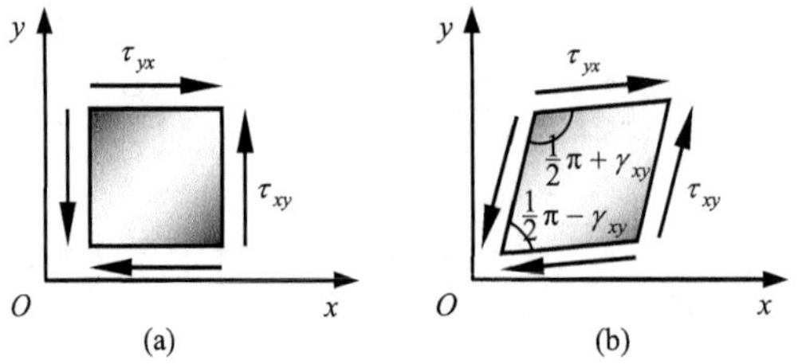

图 13.3

均匀各向同性材料的拉压和剪切弹性模量满足如下方程：

$$G = \frac{E}{2(1+\mu)} \tag{13.3}$$

式中，μ 为泊松比。

13.3　横截面剪应力

1. 平衡条件

考虑受扭圆轴 AB，在任意位置 C 处沿横截面把圆轴截开，如图 13.4a 所示。截面 C 上各点必存在垂直于半径的微剪力 dF，如图 13.4b 所示。这些微剪力是轴发生扭转时由 BC 段施加与 AC 段。

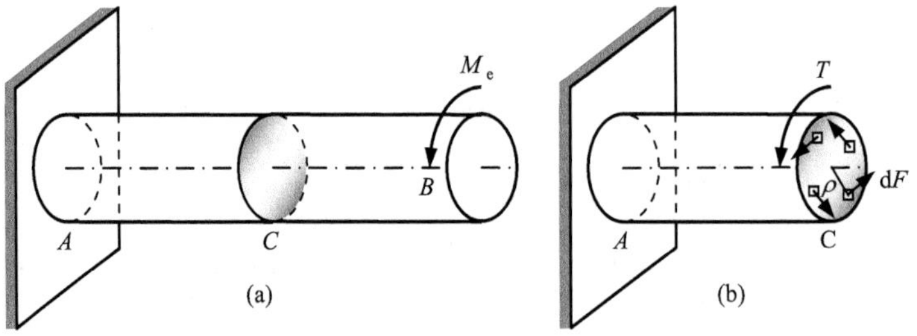

图 13.4

根据静平衡条件，微剪力对纵向形心轴的合力偶等于横截面上的扭矩 T。用ρ表示微剪力 dF 到纵向形心轴的距离，则有

$$T = \int \rho \mathrm{d}F = \int \rho\tau \mathrm{d}A \tag{13.4}$$

式中，τ 为面元 dA 上的剪应力。

2. 变形条件

如图 13.5a 所示，圆轴 AB 受外力偶 M_e 作用时将发生扭转，每个横截面保持平面。考虑长度为 dx 的微段 CD，具有相对扭转角 dφ，如图 13.5b 所示。

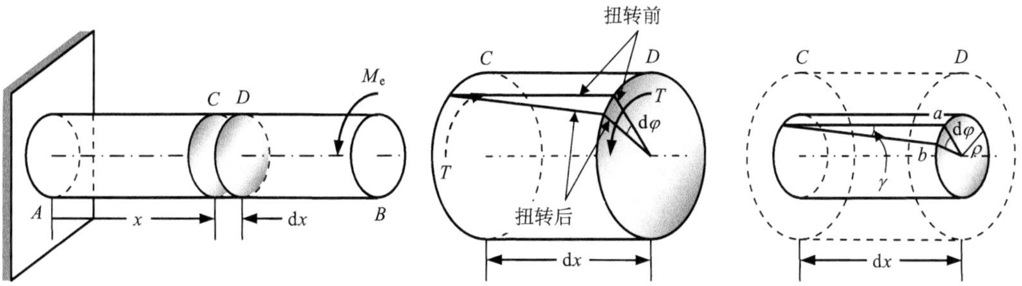

(a) (b) (c)

图 13.5

从微段 CD 中取出半径为 ρ 的圆柱，如图 13.5c 所示。对小变形，得 $\gamma \mathrm{d}x = \rho \mathrm{d}\varphi$，即

$$\gamma = \frac{\mathrm{d}\varphi}{\mathrm{d}x}\rho \tag{13.5}$$

式中，γ 和 $\mathrm{d}\varphi$ 的单位为 rad。上式表明，圆轴的剪应变与到纵向形心轴的距离 ρ 成正比。

3. 应力应变关系

假设圆轴的剪应力不超过比例极限，根据剪切胡克定律 $\tau = G\gamma$，有

$$\tau = G\frac{\mathrm{d}\varphi}{\mathrm{d}x}\rho \tag{13.6}$$

式中，G 为剪切弹性模量。式(13.6)表明，只要不超过比例极限，圆轴的剪应力与到纵向形心轴的距离 ρ 成正比。图 13.6a 和图 13.6b 分别表示了实心圆轴和空心圆轴的剪应力分布。

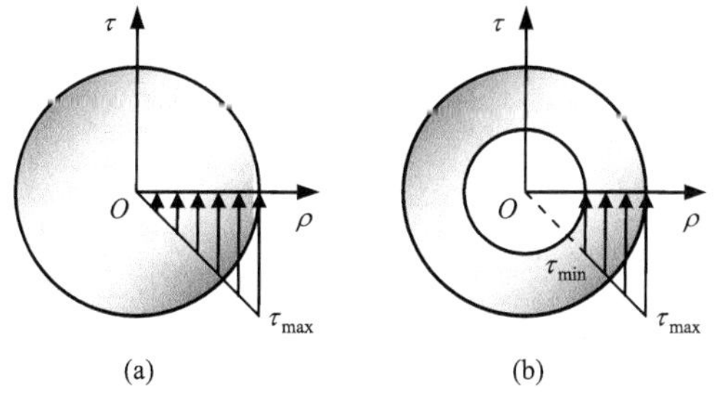

(a) (b)

图 13.6

式(13.6)代入式(13.4)，得

$$T = \int \rho\tau \mathrm{d}A = G\frac{\mathrm{d}\varphi}{\mathrm{d}x}\int \rho^2 \mathrm{d}A = GI_{\mathrm{p}}\frac{\mathrm{d}\varphi}{\mathrm{d}x} \tag{13.7}$$

式中，$I_{\mathrm{p}} = \int \rho^2 \mathrm{d}A$ 为横截面对形心 O 的极惯性矩。对直径为 d 的实心圆轴，$I_{\mathrm{p}} = \frac{1}{32}\pi d^4$；对内外径分别为 d 和 D 的空心圆轴，$I_{\mathrm{p}} = \frac{1}{32}\pi(D^4 - d^4) = \frac{1}{32}\pi D^4(1-\alpha^4)$，其中 $\alpha = d/D$。

利用式(13.6)和(13.7)，有

$$\tau = \frac{T\rho}{I_p} \tag{13.8}$$

由式(13.8)可以看出，最大剪应力可表示为

$$\tau_{max} = \frac{T_{max}\rho_{max}}{I_p} \tag{13.9}$$

注意，比值I_p/ρ_{max}仅与截面几何尺寸有关。该比值称为极截面模量或抗扭截面系数，用S_p表示。因此，式(13.9)可写为

$$\tau_{max} = \frac{T_{max}}{S_p} \tag{13.10}$$

对直径为 d 的实心圆轴，$S_p = \frac{1}{16}\pi d^3$；对内外径分别为$d$和$D$的空心圆轴，$S_p = \frac{1}{16}\pi D^3(1-\alpha^4)$，其中$\alpha = d/D$。

例 13.1　如图 E13.1 所示，外力偶M_B和M_C分别作用于变截面实心轴ABC的截面B和C处。已知D=46 mm、d=30 mm、M_B=400 N·m 和M_C=300 N·m，求(a)轴AB和(b)轴BC的最大剪应力。

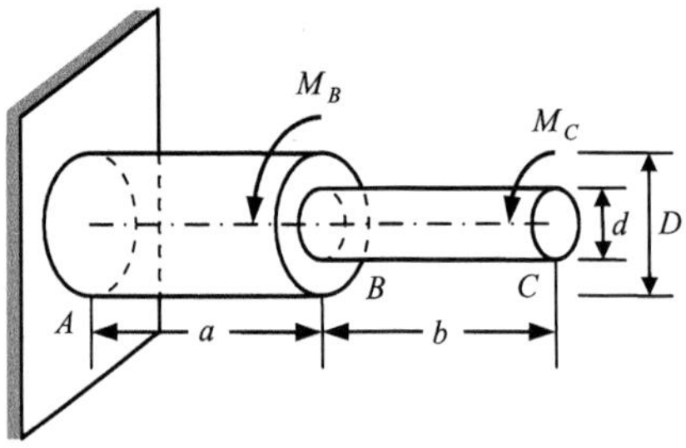

图 E13.1

解　利用截面法，得扭矩

$$T_{AB} = M_B + M_C = 700\ \text{N} \cdot \text{m},\quad T_{BC} = M_C = 300\ \text{N} \cdot \text{m}$$

对轴AB，由式(13.10)，得

$$(\tau_{max})_{AB} = \frac{T_{AB}}{(S_p)_{AB}} = \frac{T_{AB}}{\frac{1}{16}\pi D^3} = 36.6\ \text{MPa}$$

同理，对轴BC，有

$$(\tau_{max})_{BC} = \frac{T_{BC}}{(S_p)_{BC}} = \frac{T_{BC}}{\frac{1}{16}\pi d^3} = 56.6\ \text{MPa}$$

13.4　斜截面正应力和剪应力

前面考虑了横截面上的剪应力，下面考虑斜截面上的应力。如图 13.7 所示，在受扭

圆轴的前表面取三个单元体 a、b 和 c。单元体 a 的面分别平行和垂直于纵轴，因此单元体上只有剪应力，并且剪应力达到最大。单元体 b 的面与纵轴成任意角，因此单元体上同时有正应力和剪应力。单元体 c 的面与纵轴成 45°，因此单元体上只有正应力。

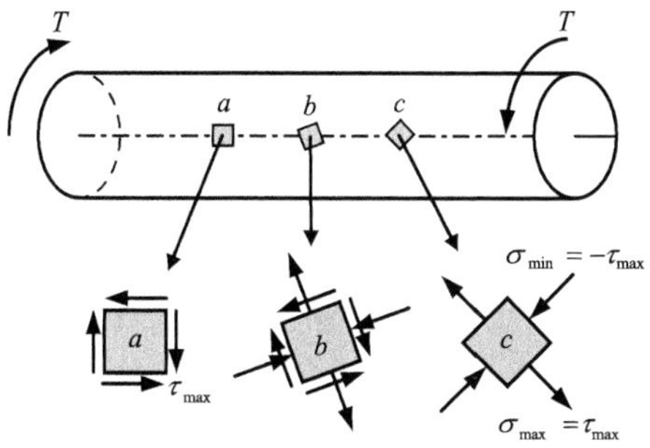

图 13.7

塑性材料的抗剪能力较差，因此塑性材料受扭时将沿横截面破坏，如图 13.8a 所示。脆性材料抗拉能力较差，因此脆性材料受扭时将沿垂直最大拉应力的截面破坏，即沿与横截面成 45°的截面破坏，如图 13.8b 所示。

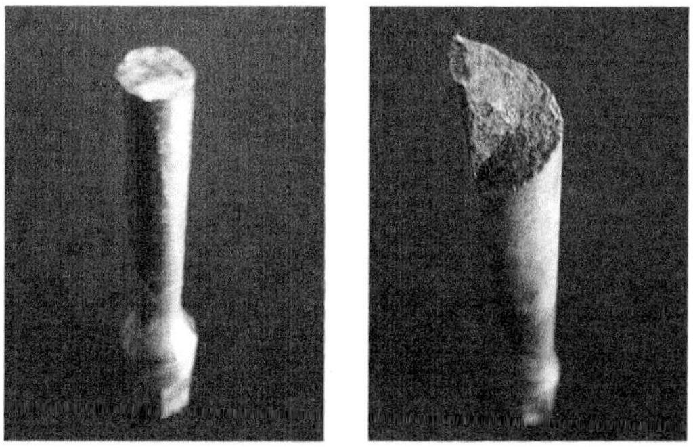

(a) (b)

图 13.8

13.5 扭 转 角

如图 13.9 所示，考虑圆轴 AB 自由端受外力偶 M_e 作用，假设扭矩用 T 表示，并且圆轴保持线弹性，利用式(13.7)，则微段 CD 的扭转角 $d\varphi$ 可写为

$$d\varphi = \frac{T dx}{GI_p} \tag{13.11}$$

式中，GI_p 为抗扭刚度。式(13.11)表明，在线弹性范围，扭转角 $d\varphi$ 与扭矩 T 成正比。积分得扭转角为

$$\varphi = \int \frac{T\mathrm{d}x}{GI_{\mathrm{p}}} \tag{13.12}$$

对等截面圆轴受不变扭矩作用，则式(13.12)简化为

$$\varphi = \frac{Tl}{GI_{\mathrm{p}}} \tag{13.13}$$

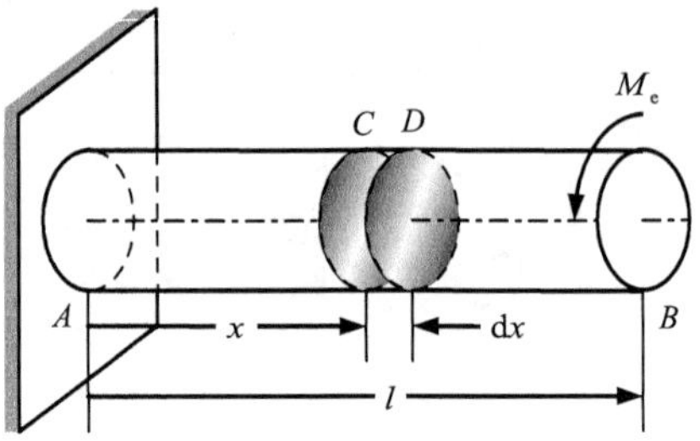

图 13.9

例 13.2　如图 E13.2 所示，外力偶 M_B 和 M_C 分别作用于变截面实心铝(G=77 GPa)轴 ABC 的截面 B 和 C 处。已知 D=46 mm、d=30 mm、a=750 mm、b=900 mm、M_B=400 N·m 和 M_C=300 N·m，求(a)截面 B、C 之间的扭转角和(b)截面 C 的扭转角。

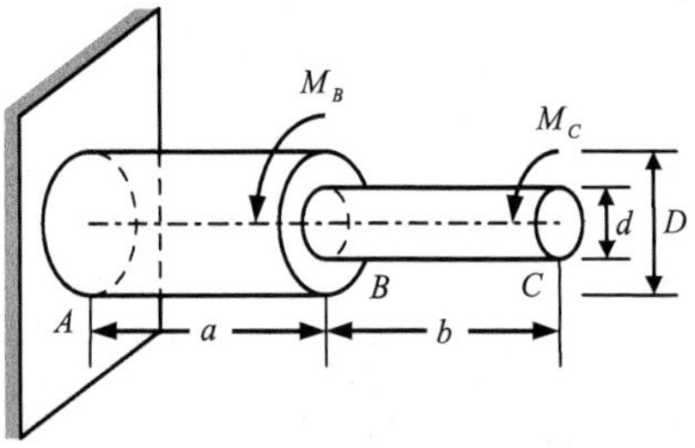

图 E13.2

解　利用式(13.13)，得截面 C 相对截面 B 的扭转角

$$\varphi_{C/B} = \frac{T_{BC}l_{BC}}{G(I_{\mathrm{p}})_{BC}} = \frac{T_{BC}l_{BC}}{G(\frac{1}{32}\pi d^4)} = 4.41\times10^{-2}\ \mathrm{rad} = 2.53°$$

和截面 C 的绝对扭转角

$$\varphi_C = \varphi_{C/A} = \varphi_{C/B} + \varphi_{B/A} = \frac{T_{BC}l_{BC}}{G(\frac{1}{32}\pi d^4)} + \frac{T_{AB}l_{AB}}{G(\frac{1}{32}\pi D^4)} = 5.96\times10^{-2}\ \mathrm{rad} = 3.41°$$

13.6　静 不 定 轴

如图 13.10a 所示，考虑圆轴 AB，两端固支，在截面 C 处受外力偶 M_{e} 作用。反力包含两个未知量，而平衡条件却只有一个，因此圆轴为一次静不定。

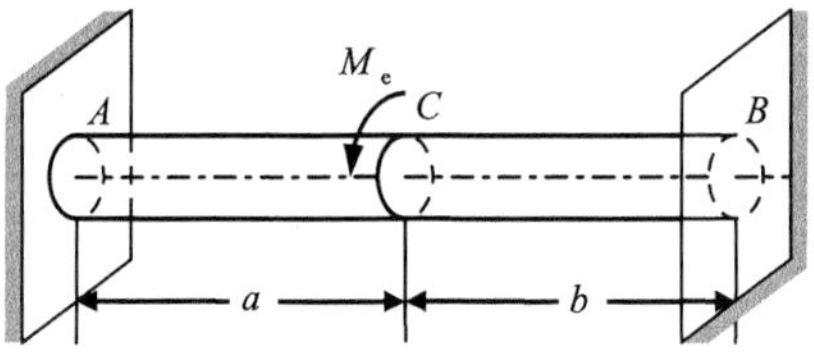

(a)

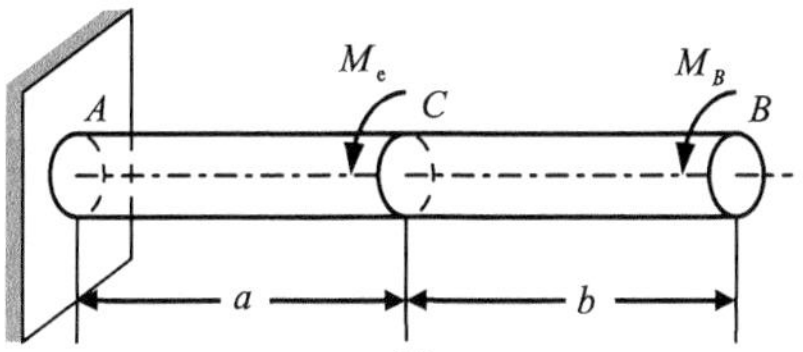

(b)

图 13.10

取支座 B 处的反力为多余反力，消除对应的支座 B，如图 13.10b 所示。多余反力可作为未知载荷 M_B，连同其余载荷 M_e，共同作用所产生的变形必须与原始支座 B 协调。利用圆轴在截面 B 处的扭转角必须等于零即可求得多余反力 M_B。

分别考虑 B 处多余反力 M_B 所产生的扭转角 $(\varphi_B)_{M_B}$ 和外力偶 M_e 在同一截面所产生的 $(\varphi_B)_{M_e}$ 即可得到解。根据式(13.13)，得 $(\varphi_B)_{M_B}=\dfrac{M_B(a+b)}{GI_p}$，$(\varphi_B)_{M_e}=(\varphi_C)_{M_e}=\dfrac{M_e a}{GI_p}$。$B$ 处扭转角必须为零，即 $\varphi_B=(\varphi_B)_{M_B}+(\varphi_B)_{M_e}=0$ 或 $\dfrac{M_B(a+b)}{GI_p}+\dfrac{M_e a}{GI_p}=0$。求解得 $M_B=-\dfrac{M_e a}{a+b}$。

13.7 扭转轴设计

扭转轴设计的主要规范是传输的功率以及轴的转速。选择合适的材料和截面尺寸以保证轴在以规定的转速传输所要求的功率时不超过许用应力。

旋转轴以转速 ω(rad/s) 或 n(r/min) 传输功率 P(kW) 时所产生的扭矩 T 可表示为

$$T(\mathrm{N\cdot m})=\frac{P(\mathrm{W})}{\omega(\mathrm{rad/s})} \text{ 或 } T(\mathrm{N\cdot m})=9549\frac{P(\mathrm{kW})}{n(\mathrm{r/min})} \tag{13.14}$$

得到扭矩 T 和选择材料后，将许用剪应力代入式(13.9)和许可扭转角代入式(13.12)，得

$$\tau_{\max}=\frac{T_{\max}\rho_{\max}}{I_p}\leqslant\tau_{\text{allow}}\text{，}\quad \varphi=\int\frac{T\mathrm{d}x}{GI_p}\leqslant\varphi_{\text{allow}} \tag{13.15}$$

式中，τ_{allow} 为许用剪应力，φ_{allow} 为许可扭转角。

例 13.3 长度为 1.8 m、外径为 45 mm 的空心圆截面钢轴的许用剪应力 τ_{allow}=75 MPa 和剪切弹性模量 G=77 GPa。已知钢轴受 M_e=900 N·m 的外力偶作用时的扭转角不能超过 5°，求最大内径 d。

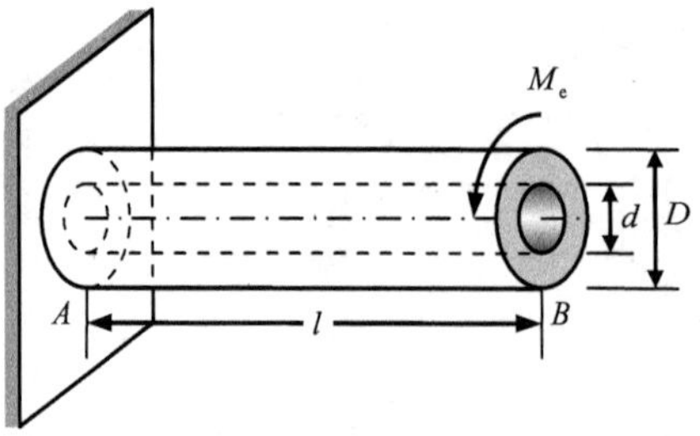

图 E13.3

解　根据$\tau_{max}=\dfrac{T\rho_{max}}{I_p}\leqslant\tau_{allow}$，有

$$(I_p)_\tau\geqslant\frac{T\rho_{max}}{\tau_{allow}}=270\times10^{-9}\ \text{m}^4$$

根据$\varphi=\dfrac{Tl}{GI_p}\leqslant\varphi_{allow}$，有

$$(I_p)_\varphi\geqslant\frac{Tl}{G\varphi_{allow}}=241\times10^{-9}\ \text{m}^4$$

应该选用较大的I_p值，因此得

$$I_p=\max[(I_p)_\tau,\ (I_p)_\varphi]\geqslant270\times10^{-9}\ \text{m}^4$$

对内径为d、外径为D的空心圆轴，利用$I_p=\dfrac{1}{32}\pi(D^4-d^4)$，有

$$d\leqslant\sqrt[4]{D^4-\frac{32I_p}{\pi}}=34\ \text{mm}$$

因此，能够用于设计的最大内径d等于 34 mm。

习　　题

13.1　已知d=30 mm 和D=40 mm，求空心轴产生 52 MPa 最大剪应力时所施加的外力偶M_e。

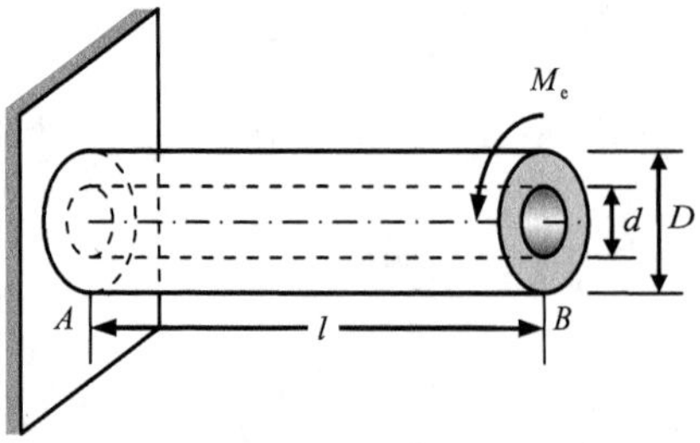

图 P13.1

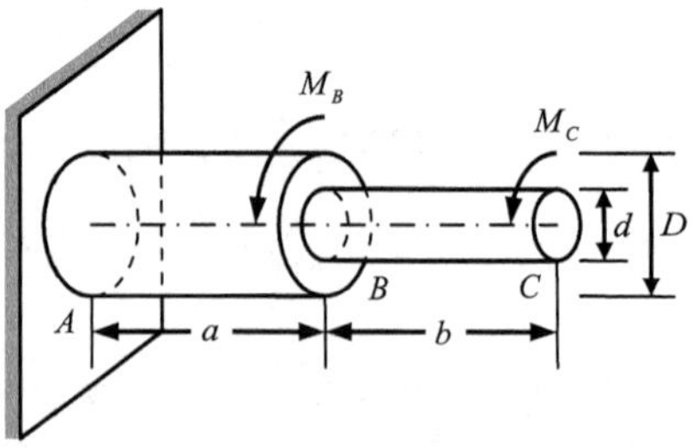

图 P13.2

13.2　外力偶 M_B 和 M_C 分别作用于变截面实心轴 ABC 的截面 B 和 C 处。已知 D = 46 mm、d=30 mm、M_B=400 N·m 和 M_C=300 N·m，为了降低轴的总质量，在轴的最大剪应力不增大的情况下，求轴 AB 的最小直径。

13.3　空心管 AB 的外径为 90 mm、壁厚为 6 mm，实心杆 BC 的直径为 60 mm。已知管和杆均由钢制成，许用剪应力为 95 MPa，求在截面 C 处所能施加的最大外力偶 M_e。

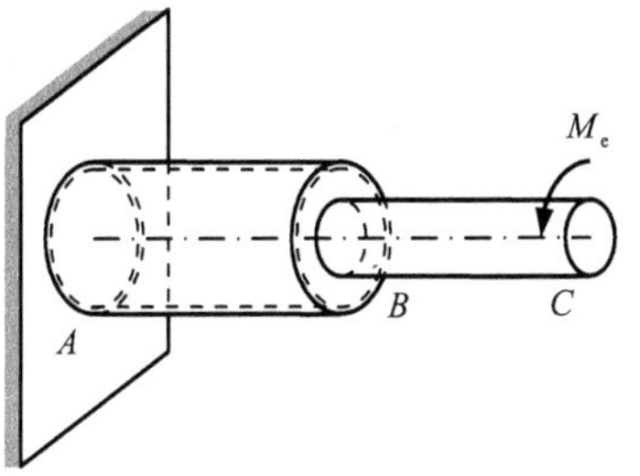

图 P13.3

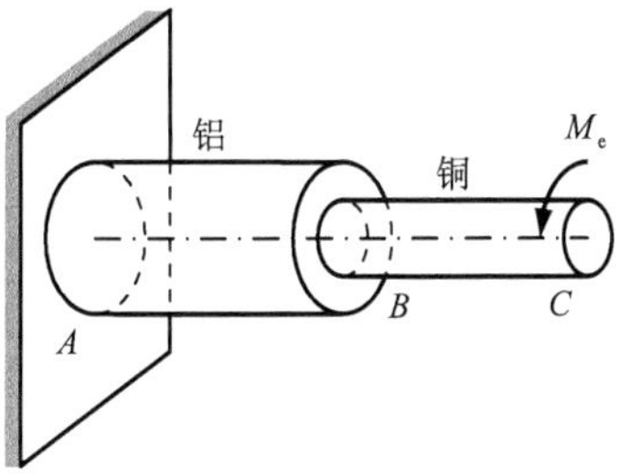

图 P13.4

13.4　铝杆 AB 的许用剪应力为 25 MPa，铜杆的许用剪应力为 50 MPa。已知 M_e= 1.5 kN·m，求(a)杆 AB 和(b)杆 BC 的所需直径。

13.5　铝杆如图所示。已知 G=27 GPa、l=1.1 m、d=10 mm 和 D=20 mm，求：(a)产生 5°扭转角的外力偶 M_e；(b)相同长度和相同横截面面积的实心圆轴在相同外力偶 M_e 作用下的扭转角。

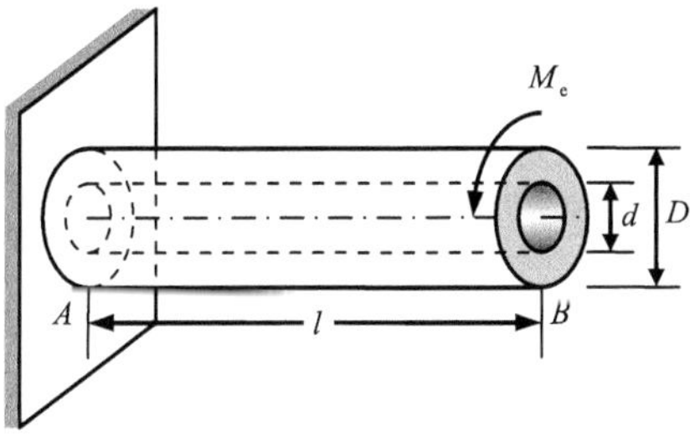

图 P13.5

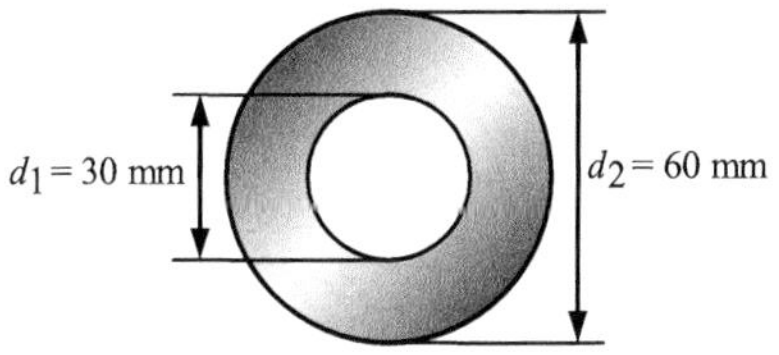

图 P13.6

13.6　当图示横截面的钢杆以 120 r/min 的角速度转动时，频闪测量显示 4 m 长度范围的扭转角为 2°。已知 G=77 GPa，求传输的功率。

13.7　长度为 1.8 m、内径为 45 mm 的空心圆截面钢轴的许用剪应力 τ_{allow}=75 MPa 和剪切弹性模量 G=77 GPa。已知钢轴受 M_e=900 N·m 的外力偶作用时的扭转角不能超过 5°，求最小外径。

13.8　机器构件的设计要求为外径 D=38 mm 的轴传输 45 kW 功率。(a)如果转速为 800 r/min，求图 P13.8a 所示实心轴的最大剪应力。(b)如果转速增加 50%达到 1200 r/min，但保持最大剪应力不变，求图 P13.8b 所示空心轴的最大内径。

13.9　长度为 1.5 m、外径 D 为 40 mm、内径 d 为 30 mm 的空心圆截面钢轴(G = 77 GPa)传输 120 kW 的功率。已知许用剪应力为 65 MPa 和扭转角不超过 3°，求钢轴的最小转速。

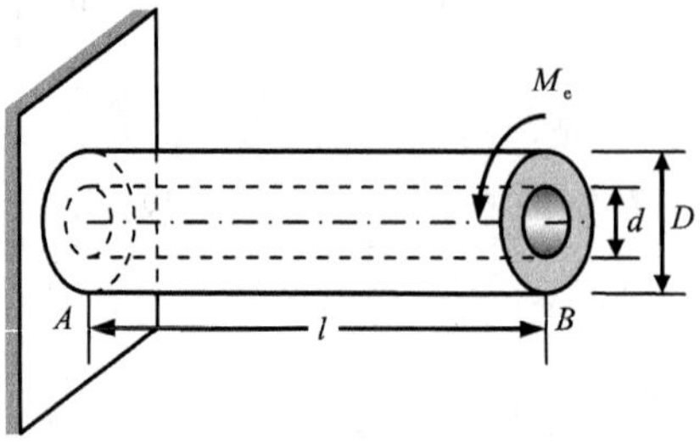

图 P13.7

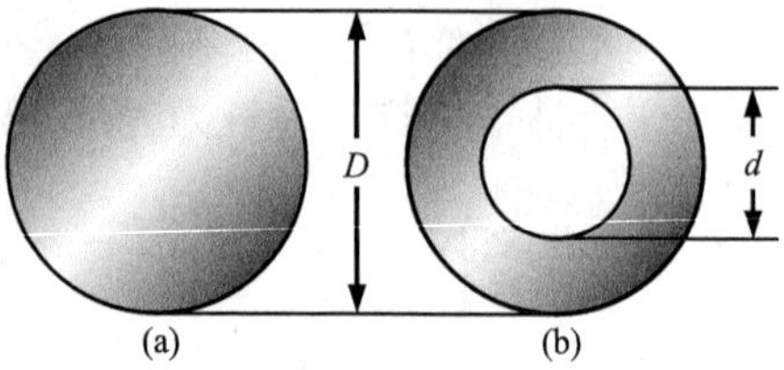

图 P13.8

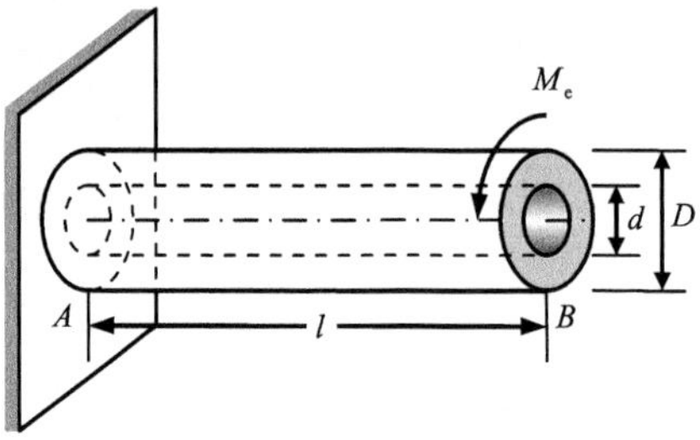

图 P13.9

第 14 章　弯 曲 内 力

受垂直杆纵轴的外力或杆纵轴所在平面内的外力偶作用的杆称为弯曲梁。

梁通过支撑方式进行分类。图 14.1 所示为几种类型的常用梁。简支梁为一端固定铰支，另一端可动铰支，如图 14.1a。外伸梁为两点支撑，梁的一端或两端越过支座，如图 14.1b。悬臂梁为一端固支，另一端自由，如图 14.1c。作用于梁的外载荷包括集中力(图 14.1a)、分布力(图 14.1b)和集中力偶(图 14.1c)。

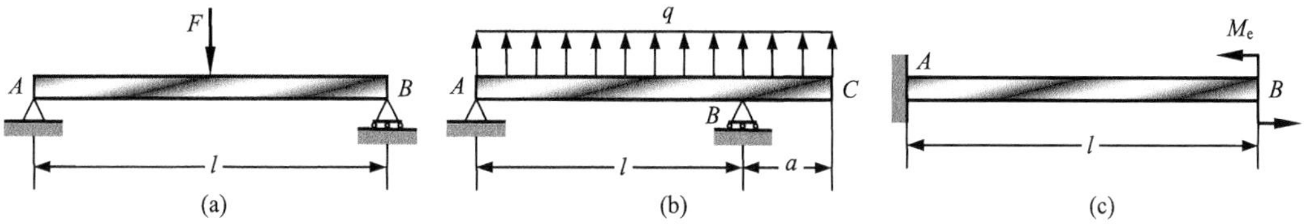

图 14.1

上述梁的支座反力总共有三个未知量，因此通过平衡方程即可确定全部未知量。这类梁称为静定梁。如果梁的支座反力有三个以上的未知量， 则不能仅仅通过平衡方程确定全部未知量，这类梁称为静不定梁。图 14.2 所示为一次静不定梁。

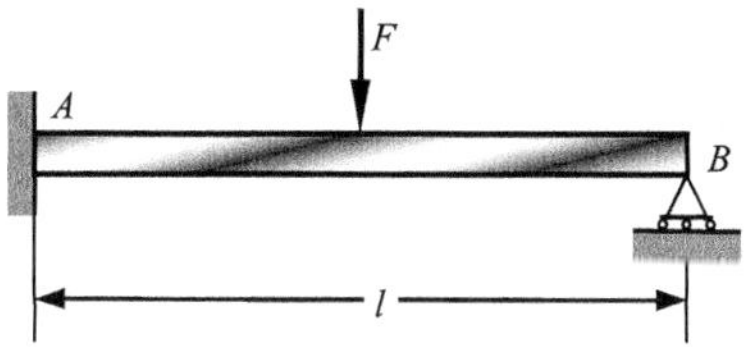

图 14.2

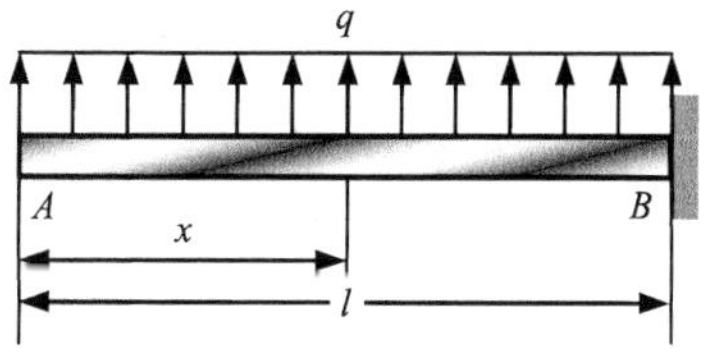

图 14.3

14.1　剪力图和弯矩图

当外载荷作用于梁，则梁将产生剪力和弯矩，如图 14.3。一般而言，剪力和弯矩是截面位置的函数，可表示为

$$V = V(x),\ M = M(x) \tag{14.1}$$

上述关系分别称为剪力和弯矩方程。根据上述方程，梁的剪力和弯矩可以用图形表示，该图形分别称为剪力图和弯矩图。

为了绘制剪力图和弯矩图，有必要首先确立剪力和弯矩的符号规则。虽然符号规则的选择具有任意性，但是为了方便，将采用常用的符号规则。引起梁段顺时针转动的剪力为正，如图 14.4a；引起梁段上侧受压的弯矩为正，如图 14.4b。

图 14.4

梁上任何截面的剪力和弯矩可通过截面法确定，即在需求剪力和弯矩的位置把梁解开(如图 14.5a)，并考虑截面左侧梁段的平衡(如图 14.5b)或截面右侧梁段的平衡(如图 14.5c)。

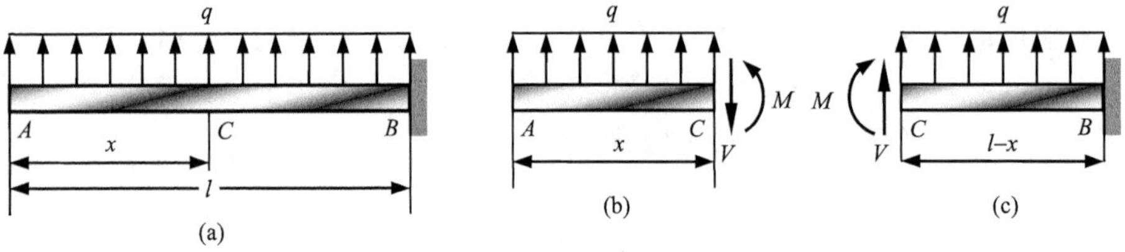

图 14.5

如图 14.5b，考虑左段梁 AC 的平衡，并利用平衡方程，得

$$\begin{aligned} &\sum F_y = 0: \quad qx - V = 0 \qquad (0 \leqslant x < l) \\ &\sum M_C = 0: \ M - \frac{1}{2}qx^2 = 0 \ (0 \leqslant x < l) \end{aligned} \tag{14.2}$$

解方程，得

$$\begin{aligned} &V = qx \qquad (0 \leqslant x < l) \\ &M = \frac{1}{2}qx^2 \ \ (0 \leqslant x < l) \end{aligned} \tag{14.3}$$

得到剪力和弯矩方程后，接着可以根据方程绘制剪力和弯矩图。图 14.5a 所示梁的剪力和弯矩图分别如图 14.6b 和 14.6c 所示。

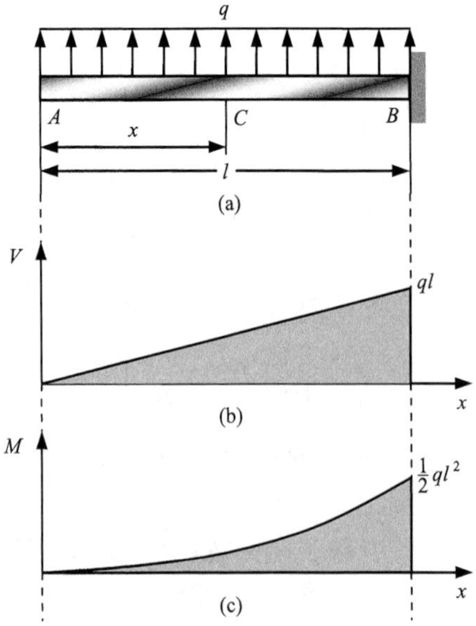

图 14.6

由图 14.6b 和 14.6c 可以看出，截面 B 是危险截面，因为该截面有最大剪力和最大弯矩。

例 14.1　绘制图 E14.1a 所示结构的弯矩图，并求最大弯矩。

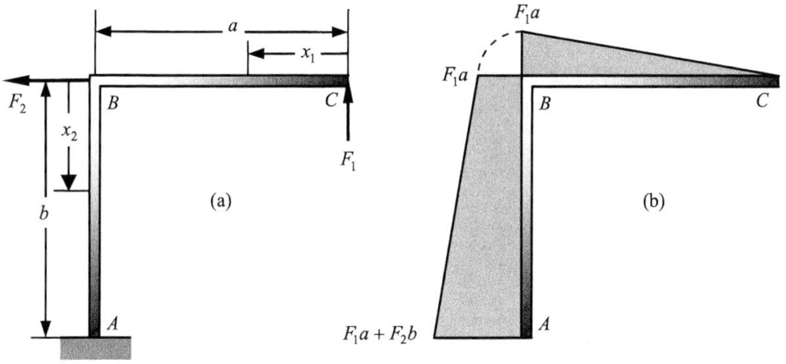

图 E14.1

解　利用截面法和平衡条件，可得弯矩方程

$$M_1 = F_1x_1 \quad (0 \leqslant x_1 \leqslant a)$$

$$M_2 = F_1a + F_2x_2 \quad (0 \leqslant x_2 < b)$$

由弯矩方程可以看出，结构每段的弯矩图均为倾斜直线。整个结构的弯矩图如图 E14.1b 所示，最大弯矩位于截面 A 处，其最大值等于

$$M_{\max} = F_1a + F_2b$$

例 14.2　绘制图 E14.2a 所示结构的弯矩图，并求最大弯矩。

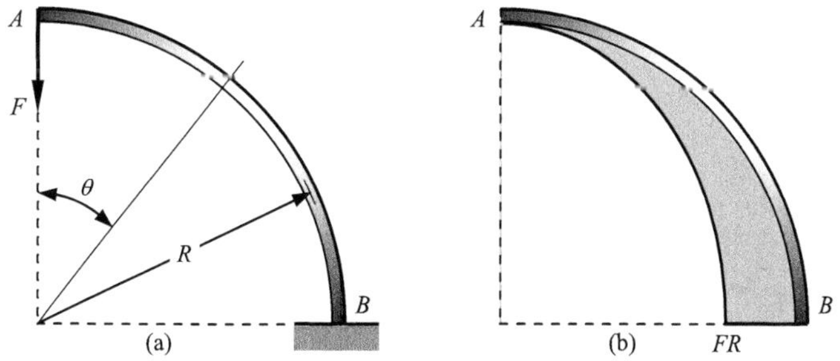

图 E14.2

解　利用截面法和平衡条件，可得弯矩方程

$$M = FR\sin\theta \quad (0 \leqslant \theta < \frac{1}{2}\pi)$$

结构的弯矩图如图 E14.2b 所示，固定端的最大弯矩为

$$M_{\max} = FR$$

14.2　分布载荷、剪力和弯矩之间的关系

当梁承受多个外载荷作用时，前节讨论的绘制剪力和弯矩图的方法就显得非常麻烦。如果考虑分布载荷、剪力和弯矩之间的关系，那么剪力和弯矩图绘制将显得非常方便。

如图 14.7a 所示，考虑简支梁 AB，受分布载荷 q 作用（q 为分布载荷集度，即沿梁轴单位长度上的作用力），设 C、D 为梁上相距 $\mathrm{d}x$ 的两截面。截取梁段 CD 并画受力图，如图 14.7b 所示。

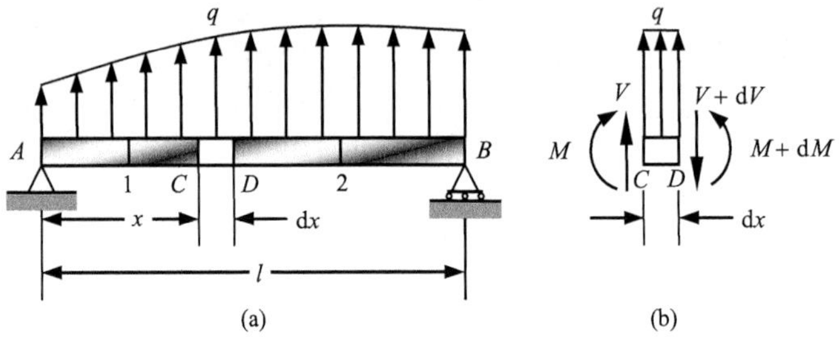

图 14.7

根据垂直方向力平衡条件，有

$$\sum F_y = 0:\quad V-(V+\mathrm{d}V)+q\mathrm{d}x=0,\ 即\ \mathrm{d}V = q\mathrm{d}x \tag{14.4}$$

除以 $\mathrm{d}x$，得

$$\frac{\mathrm{d}V}{\mathrm{d}x} = q \tag{14.5}$$

在截面 1 和 2 之间进行积分，有

$$V_2 = V_1 + \int_{x_1}^{x_2} q\mathrm{d}x \tag{14.6}$$

式中，$\int_{x_1}^{x_2} q\mathrm{d}x$ 为截面 1 和 2 之间分布载荷的合力。

根据对 D 的矩平衡条件，得

$$\sum M_D = 0:\quad (M+\mathrm{d}M)-M-V\mathrm{d}x-\frac{1}{2}q(\mathrm{d}x)^2=0,\ 即\ \mathrm{d}M = V\mathrm{d}x+\frac{1}{2}q(\mathrm{d}x)^2 \tag{14.7}$$

忽略二阶无穷小量 $(\mathrm{d}x)^2$，并除以 $\mathrm{d}x$，得

$$\frac{\mathrm{d}M}{\mathrm{d}x} = V \tag{14.8}$$

在截面 1 和 2 之间进行积分，得

$$M_2 = M_1 + \int_{x_1}^{x_2} V\mathrm{d}x \tag{14.9}$$

式中，$\int_{x_1}^{x_2} V\mathrm{d}x$ 为截面 1 和 2 之间剪力图的面积。

利用上述关系可以快速绘制在分布载荷作用下梁的剪力和弯矩图。

例 14.3　绘制图 E14.3a 所示简支梁的剪力和弯矩图，并求剪力和弯矩的最大值。

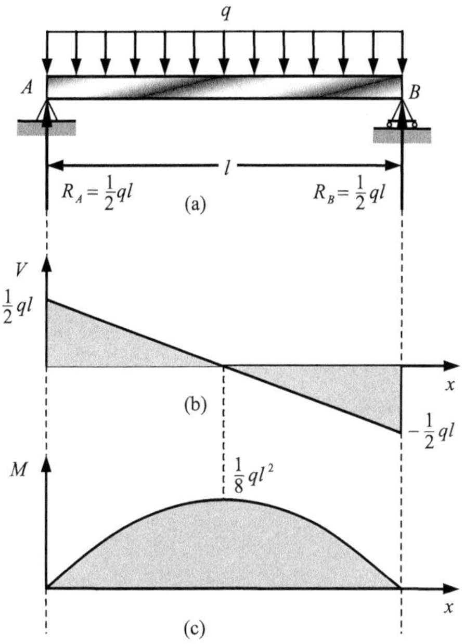

图 E14.3

解　(1) A 和 B 处反力分别为 $R_A = R_B = \dfrac{1}{2}ql$。

(2) 截面 A 右侧的剪力为 $V_A^{\mathrm{R}} = R_A = \dfrac{1}{2}ql$。利用式(14.6)，得任意截面 x 处的剪力

$$V = V_A^{\mathrm{R}} + \int_0^x (-q)\mathrm{d}x = \frac{1}{2}ql - qx \quad (0 < x < l)$$

显然，剪力图是倾斜直线，如图 E14.3b 所示。在 A 或 B 处的剪力达到最大绝对值。

(3) 利用 $M_A = 0$，根据式(14.9)，任意截面 x 处的弯矩为

$$M = M_A + \int_0^x V\mathrm{d}x = \frac{1}{2}qlx - \frac{1}{2}qx^2 \quad (0 \leqslant x \leqslant l)$$

显然，弯矩图是抛物线，如图 E14.3c 所示。在中间截面处有最大正弯矩。

14.3　集中载荷、剪力和弯矩之间的关系

如图 14.8a 所示，考虑简支梁 AB，受集中力 F 和集中力偶 M_{e} 作用，设 C、D 为梁上相距 $\mathrm{d}x$ 的两截面。假设集中力和集中力偶作用于梁段 CD 中点，截取梁段 CD 并画受力图，如图 14.8b 所示。

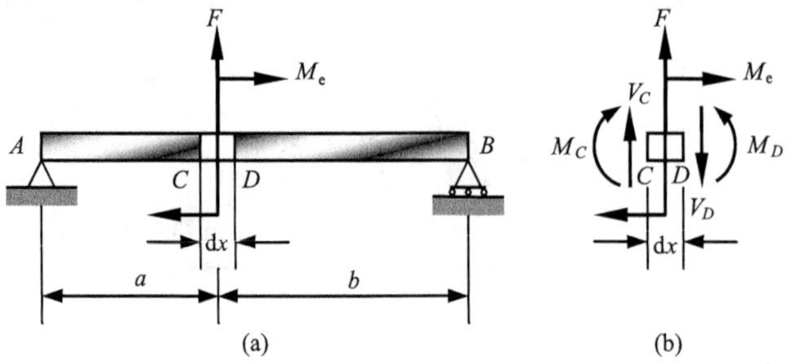

图 14.8

根据垂直方向力平衡条件，有

$$\sum F_y = 0: \quad V_C - V_D + F = 0 \tag{14.10}$$

即

$$V_D = V_C + F \tag{14.11}$$

显然，剪力图在集中力作用的截面将发生突变。

根据对 D 的矩平衡条件，得

$$\sum M_D = 0: \quad M_D - M_C - V_C \mathrm{d}x - M_{\mathrm{e}} - F(\frac{1}{2}\mathrm{d}x) = 0 \tag{14.12}$$

忽略无穷小量 $\mathrm{d}x$，得

$$M_D = M_C + M_{\mathrm{e}} \tag{14.13}$$

显然，弯矩图在集中力偶作用的截面将发生突变。

利用上述关系可以快速绘制在集中载荷作用下梁的剪力和弯矩图。

例 14.4　绘制图 E14.4a 所示简支梁的剪力和弯矩图，并求剪力和弯矩的最大值。

解　(1)根据梁的平衡条件，有

$$R_A = \frac{3}{4}F, \ R_B = \frac{1}{4}F$$

(2)截面 A 右侧的剪力等于 $V_A^{\mathrm{R}} = R_A = \frac{3}{4}F$。因没有外载荷作用于 AC 段，故 AC 段上剪力为常量，该段剪力图为水平直线。利用 $V_C^{\mathrm{L}} = V_A^{\mathrm{R}} = \frac{3}{4}F$ 和式(14.11)，得截面 C 右侧的剪力为

$$V_C^{\mathrm{R}} = V_C^{\mathrm{L}} + (-F) = -\frac{1}{4}F$$

同理，BC 段上剪力为常量，因此该段剪力图也为水平直线。

整段梁的剪力图如图 E14.4b 所示，最大剪力位于 AC 段，其最大值等于 $V_{\max} = \frac{3}{4}F$。

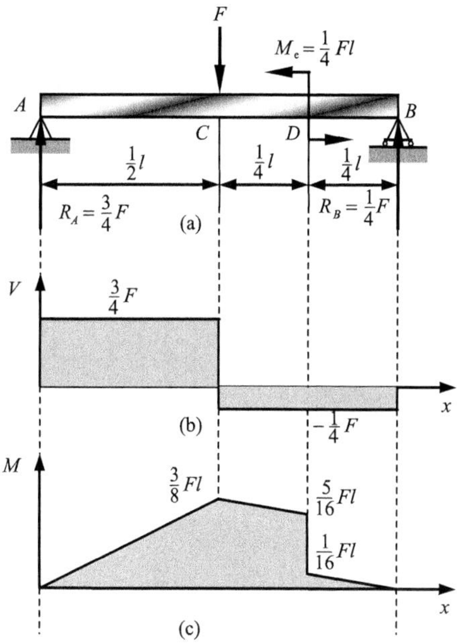

图 E14.4

(3)利用 $M_A=0$，截面 C 处弯矩为

$$M_C=M_A+\int_{x_A}^{x_C}V\mathrm{d}x=\frac{3}{4}F\cdot\frac{1}{2}l=\frac{3}{8}Fl$$

因 AC 段上剪力为常量，故相应的弯矩图为倾斜直线。同理，CD 段上的弯矩图也为倾斜直线，截面 D 左侧的弯矩为

$$M_D^{\mathrm{L}}=M_C+\int_{x_C}^{x_D}V\mathrm{d}x=\frac{5}{16}Fl$$

在截面 D 由集中力偶，因此弯矩在该截面将有突变。利用式(14.13)，有

$$M_D^{\mathrm{R}}=M_D^{\mathrm{L}}+(-M_{\mathrm{e}})=\frac{1}{16}Fl$$

同理，可绘制 DB 段的弯矩图，并且截面 B 的弯矩为

$$M_B=M_D^{\mathrm{R}}+\int_{x_D}^{x_B}V\mathrm{d}x=0$$

整段梁的弯矩图如图 E14.4c 所示，最大弯矩位于截面 C，其最大值等于 $M_{\max}=\frac{3}{8}Fl$。

习 题

14.1-14.18 绘制图示梁在图示载荷作用下的剪力和弯矩图。

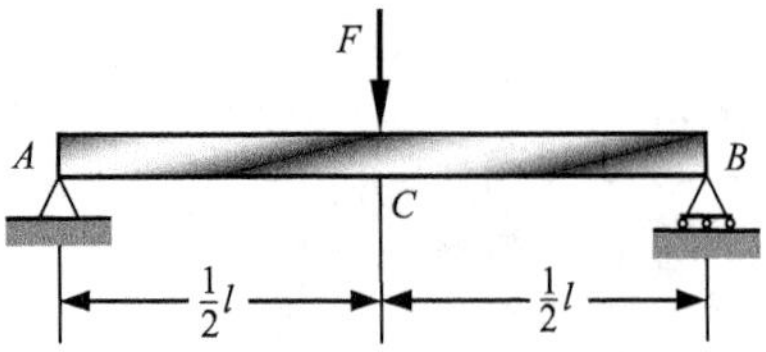

图 P14.1

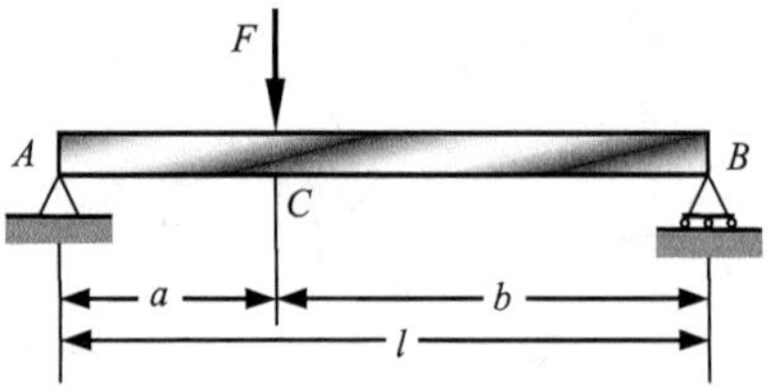

图 P14.2

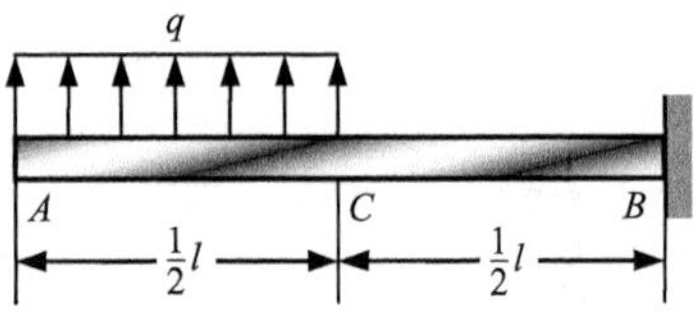

图 P14.3

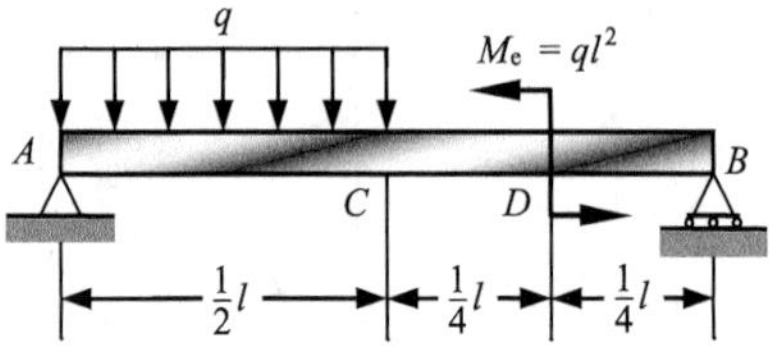

图 P14.4

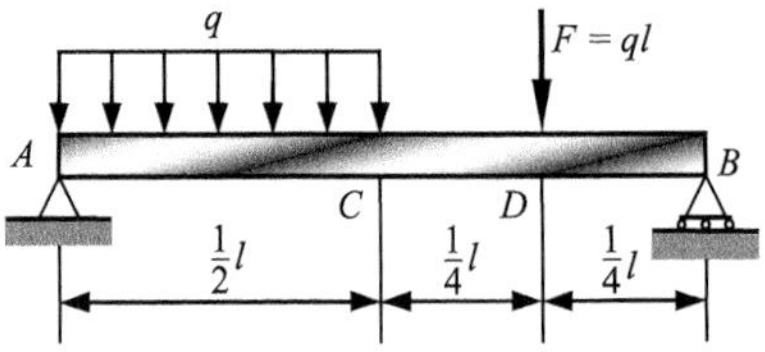

图 P14.5

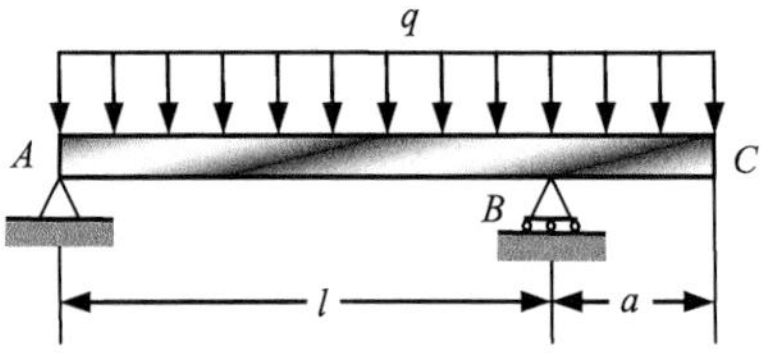

图 P14.6

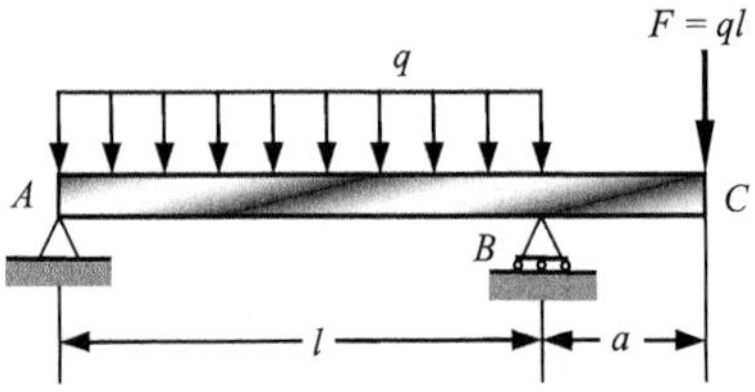

图 P14.7

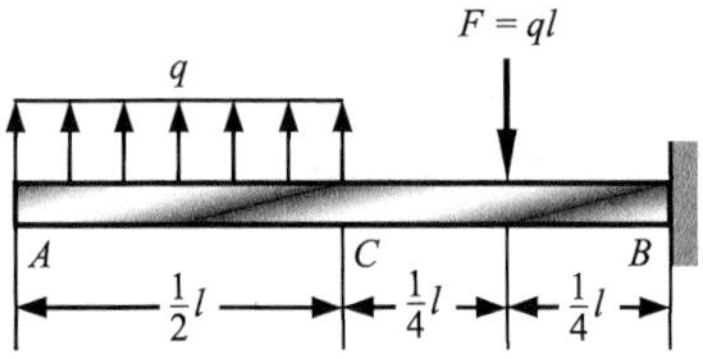

图 P14.8

第15章　弯 曲 应 力

作用于梁的外载荷包括垂直梁轴的外力和梁轴所在平面内的外力偶。这些作用于梁上的外力和外力偶产生的作用之一就是在垂直于梁轴的横截面上产生正应力和剪应力。

如果仅外力偶作用于梁，则称为纯弯曲，如图 15.1a 所示。承受纯弯曲的梁在横截面上仅存在正应力，而没有剪应力。

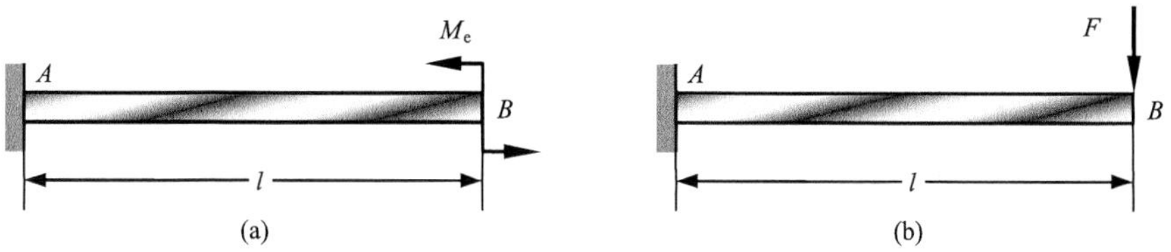

图 15.1

由不能构成力偶的外力产生的弯曲称为横力弯曲，如图 15.1b 所示。承受横力弯曲的梁在横截面上既有正应力，又有剪应力。

15.1　纯弯曲横截面正应力

考虑等截面梁，具有纵向对称面，承受纵向对称面内等值反向外力偶作用，如图 15.2 所示。

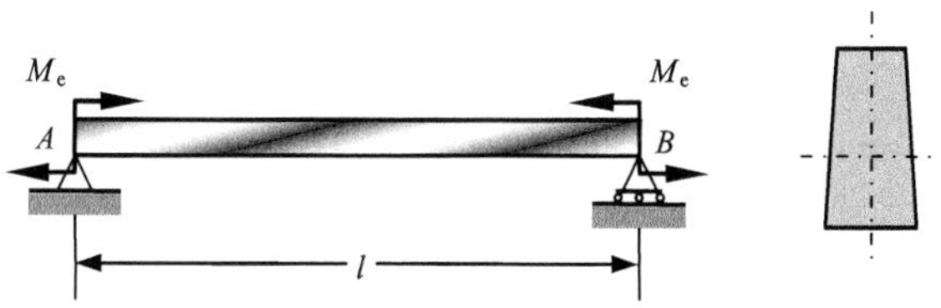

图 15.2

把梁分成许多单元体，如图 15.3a 所示，则当梁发生纯弯曲时这些单元体将产生如图 15.3b 所示的变形。

单元体上唯一不为零的应力分量为垂直于纵轴的单元体面上的正应力。因此，纯弯曲梁的每一点都处于单向应力状态。若弯矩为正，则梁的上下表面分别缩短和伸长，即梁的上部压缩，下部拉伸。

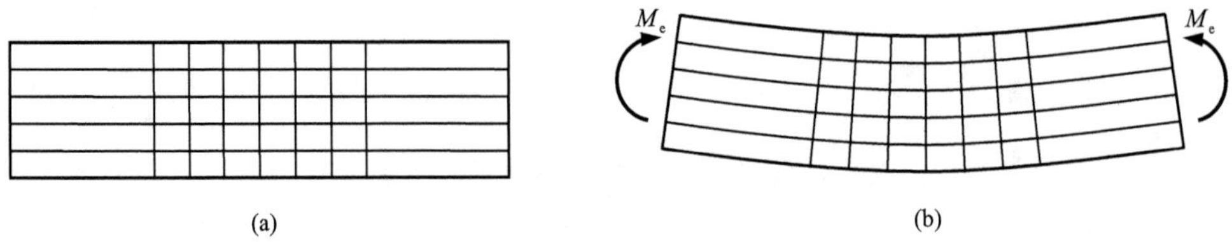

图 15.3

上述分析表明，在梁的内部必定存在与上下表面平行的线应变和正应力为零的面。该面称为中性层，如图 15.4a 所示。中性层与横截面的交线称为中性轴，如图 15.4b 所示。坐标原点选在中性层上，以便任何一点到中性层的距离可通过坐标 y 进行表示。

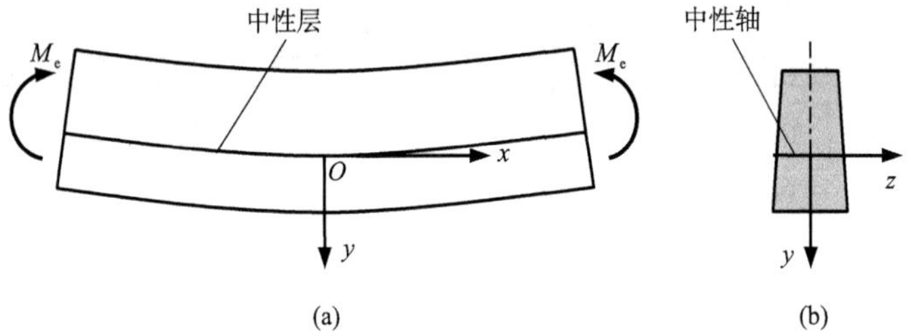

图 15.4

1. 变形条件

用 ρ 表示中性层 aa 的曲率半径，θ 表示中性层 aa 所对应的中心角，如图 15.5 所示。

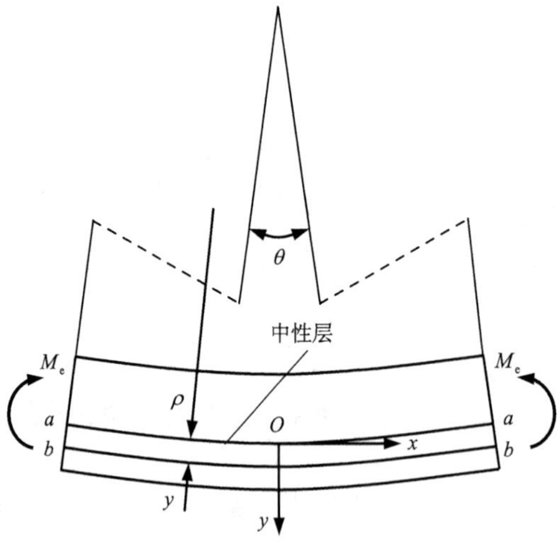

图 15.5

观察表明，变形后中性层 aa 的长度等于变形前中性层的长度 l，故有

$$l = \rho\theta \tag{15.1}$$

现考虑中性层下方 y 处的弧线 bb，变形后其长度为

$$l'_{bb}=(\rho+y)\theta \tag{15.2}$$

因弧线 bb 原长为 $l_{bb}=l$，则弧线 bb 变形可表示为

$$\delta=l'_{bb}-l_{bb}=y\theta \tag{15.3}$$

弧线 bb 的纵向线应变 ε 可通过弧线变形 δ 除以弧线原长 l_{bb} 而得到，即

$$\varepsilon=\frac{\delta}{l_{bb}}=\frac{y}{\rho} \tag{15.4}$$

式(15.4)表明，纵向线应变 ε 与到中性层的距离 y 成线性关系。

2. 应力应变关系

假设正应力不超过比例极限，则胡克定律成立，即 $\sigma=E\varepsilon$。从式(15.4)，得

$$\sigma=\frac{E}{\rho}y \tag{15.5}$$

式中，E 为弹性模量。式(15.5)表明，在线弹性范围，正应力随到中性层的距离线性变化，如图 15.6 所示。

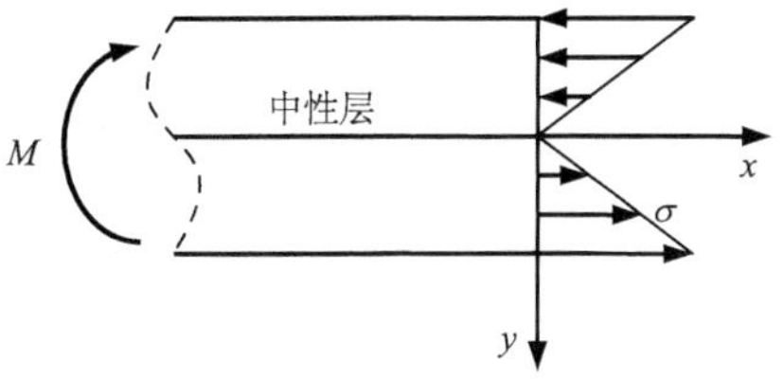

图 15.6

3. 平衡条件

根据 x 方向平衡条件，得

$$\sum F_x=0:\quad N=\int\sigma\mathrm{d}A=\frac{E}{\rho}\int y\mathrm{d}A=0 \tag{15.6}$$

即

$$\int y\mathrm{d}A=0 \tag{15.7}$$

式(15.7)表明，对纯弯曲梁，只要应力在线弹性范围，则中性轴通过截面形心。

根据绕中性轴的平衡条件，得

$$\sum M_z=0:\quad M=\int y\sigma\mathrm{d}A=\frac{E}{\rho}\int y^2\mathrm{d}A \tag{15.8}$$

即

$$\frac{1}{\rho}=\frac{M}{E\int y^2\mathrm{d}A}=\frac{M}{EI} \tag{15.9}$$

式中，$I=\int y^2\mathrm{d}A$ 为横截面对形心轴的惯性矩。把式(15.9)代入式(15.5)，得

$$\sigma=\frac{My}{I} \tag{15.10}$$

当弯矩为正时，则中性轴上方的正应力为压应力；当弯矩为负时，则中性轴上方的正应力为拉应力。

根据式(15.10)，最大正应力等于 $\sigma_{\max}=\frac{My_{\max}}{I}$。比值 $I/y_{\max}$ 仅与截面几何尺寸有关。该比值称为截面模量或抗弯截面系数，用 S 表示。根据式(15.10)，得

$$\sigma_{\max}=\frac{My_{\max}}{I}=\frac{M}{S} \tag{15.11}$$

对宽为 b，高为 h 的矩形截面梁，有 $S=\frac{1}{6}bh^2$。

例 15.1 已知外力偶作用于竖直平面内，求梁中间截面 C 上 a 和 b 点的应力。

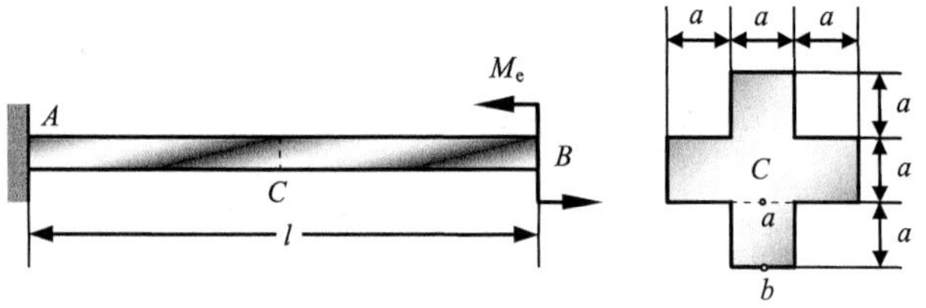

图 E15.1

解 从中间截面 C 把梁截开，考虑右侧梁的平衡，得截面 C 的弯矩 $M=M_e$。

截面对中性轴的惯性矩为

$$I=\frac{1}{12}a^4+\frac{1}{12}a(3a)^3+\frac{1}{12}a^4=\frac{29}{12}a^4$$

利用式(15.10)，得 a 和 b 点的应力

$$\sigma_a=\frac{My_a}{I}=\frac{M_e(\frac{1}{2}a)}{\frac{29}{12}a^4}=\frac{6M_e}{29a^3},\quad \sigma_b=\frac{My_b}{I}=\frac{M_e(\frac{3}{2}a)}{\frac{29}{12}a^4}=\frac{18M_e}{29a^3}$$

15.2 横力弯曲横截面正应力和剪应力

作用于梁的横向载荷将在横截面上产生正应力和剪应力。正应力由弯矩引起，而剪应力则由剪力引起。

1. 正应力

求纯弯曲梁横截面上正应力公式对横力弯曲细长梁仍然有效，如图 15.7 所示，因此横力弯曲梁的正应力为

$$\sigma = \frac{My}{I} \tag{15.12}$$

式中，弯矩 M 随截面不同而变化，即 $M = M(x)$ 是位置 x 的函数。

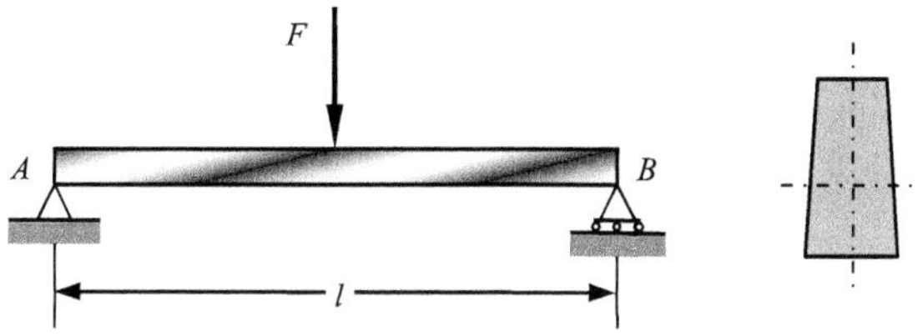

图 15.7

横力弯曲梁的最大正应力为

$$\sigma_{\max} = \frac{M_{\max} y_{\max}}{I} = \frac{M_{\max}}{S} \tag{15.13}$$

2. 剪应力

正应力在梁的设计中显得十分重要。然而，在粗短梁的设计中，剪应力也很重要。

考虑具有竖直对称面的等截面梁，受外载荷(集中力 F、集中力偶 M_e 和分布载荷 q)作用。在距 A 端为 x 处取长度为 dx 的单元体 $abcd$，设单元体的上表面到中性轴的距离为 y，如图 15.8 所示。

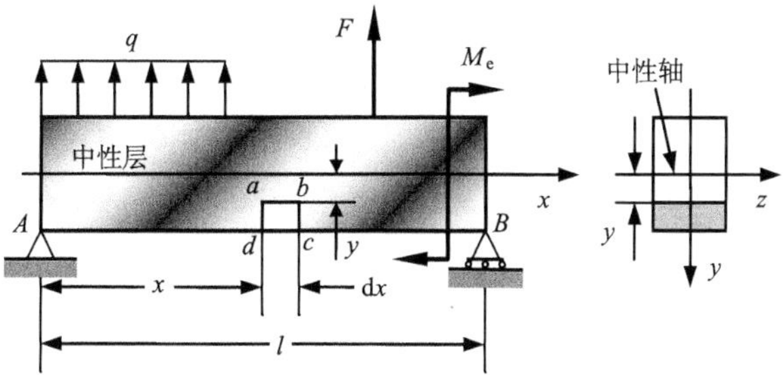

图 15.8

当 dx 趋于零，单元体左右表面上的剪应力大小相等。只要梁的横截面的宽度比高度小，那么可认为剪应力沿截面宽度均匀分布。假设单元体左表面(或右表面)上到中性轴的距离为 y 处的剪应力用 τ 表示，则根据剪应力互等定理单元体上表面的剪应力也等于 τ，如图 15.9a 所示。

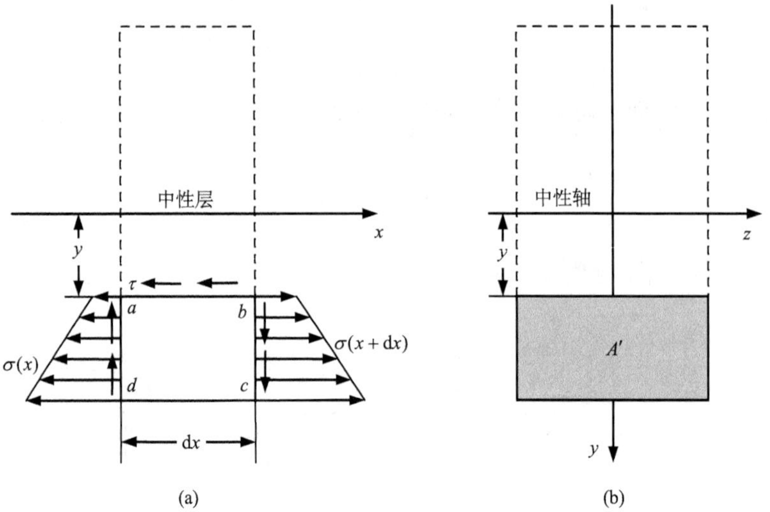

图 15.9

考虑单元体 $abcd$ 在水平方向的平衡，则有

$$\sum F_x = 0:\quad \int[\sigma(x+\mathrm{d}x)-\sigma(x)]\mathrm{d}A' - \tau(b\mathrm{d}x) = 0 \tag{15.14}$$

式中，b 为梁的横截面宽度，A' 为距中性轴为 y 的直线下方截面的面积，如图 15.9b 所示。

利用 $\sigma(x+\mathrm{d}x)-\sigma(x) = \dfrac{M(x+\mathrm{d}x)y'}{I} - \dfrac{M(x)y'}{I} = \dfrac{\mathrm{d}My'}{I} = \dfrac{(V\mathrm{d}x)y'}{I}$，解式(15.14)，有

$$\tau = \frac{V}{Ib}\int y'\mathrm{d}A' = \frac{VQ'}{Ib} \tag{15.15}$$

式中，$Q' = \int y'\mathrm{d}A'$ 为距中性轴为 y 的直线下方(或上方)截面对中性轴的静矩，I 为整个截面的形心惯性矩，V 为截面上的剪力。

如图 15.10a 所示，对宽为 b、高为 h 的矩形截面梁，$I = \frac{1}{12}bh^3$，以及

$$Q' = A'\overline{y}' = [b(\frac{1}{2}h - y)][\frac{1}{2}(\frac{1}{2}h + y)] = \frac{1}{8}b(h^2 - 4y^2) \tag{15.16}$$

则有

$$\tau = \frac{3}{2}\frac{V}{bh}[1-(\frac{2y}{h})^2] \tag{15.17}$$

式(15.17)表明，矩形截面梁横截面上的剪应力满足抛物线分布，如图 15.10b 所示。在上下表面($y = \pm\frac{1}{2}h$)，剪应力为零。设 $y = 0$，横截面上剪应力的最大值为

$$\tau_{\max} = \frac{3V}{2bh} = \frac{3V}{2A} \tag{15.18}$$

式中，A 为矩形截面梁的横截面面积。

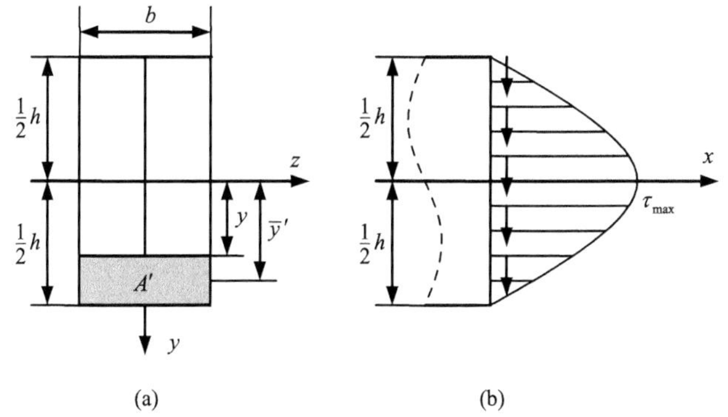

图 15.10

如图 15.11 所示，对工字截面梁(简称工字梁)，腹板上的剪应力在截面竖直方向仅发生微小变化，并且全部剪力几乎全由腹板承担。因此，剪力除以腹板横截面面积即可得到剪应力的最大值

$$\tau_{\max}=\frac{V}{A_{\text{web}}}=\frac{V}{b_0h_0} \tag{15.19}$$

式中，b_0 和 h_0 分别为工字梁腹板的宽度和高度。

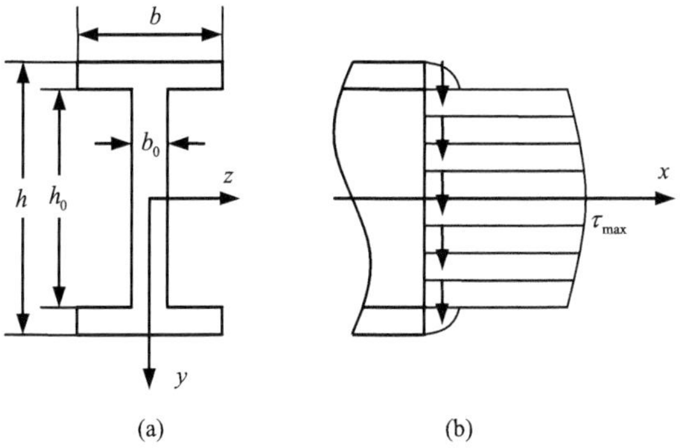

图 15.11

如图 15.12 所示，对实心圆截面梁，最大剪应力位于中性轴上，最大值可表示为

$$\tau_{\max}=\frac{4V}{3A} \tag{15.20}$$

式中，A 为实心圆截面梁的横截面面积。

如图 15.13 所示，对空心圆截面梁，中性轴上有最大剪应力，最大值可表示为

$$\tau_{\max}=\frac{2V}{A} \tag{15.21}$$

式中，A 为空心圆截面梁的横截面面积。

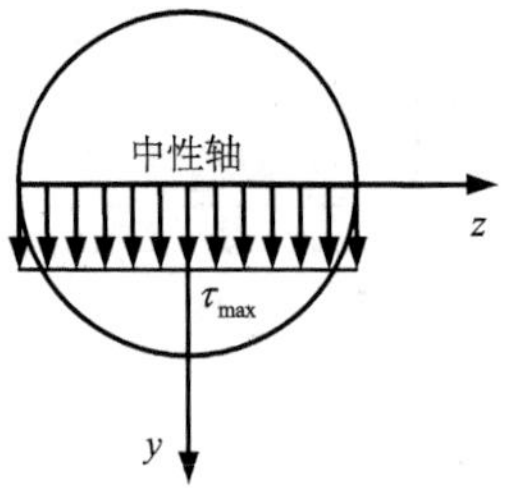

图 15.12

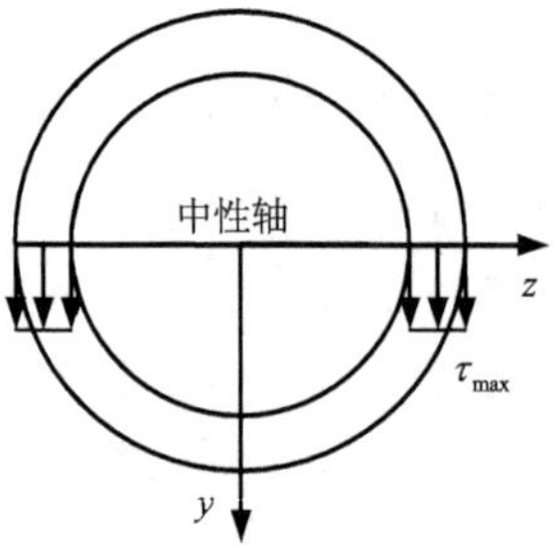

图 15.13

例 15.2　已知外力作用于竖直面内，求中间截面 C 上点 a 处的剪应力。

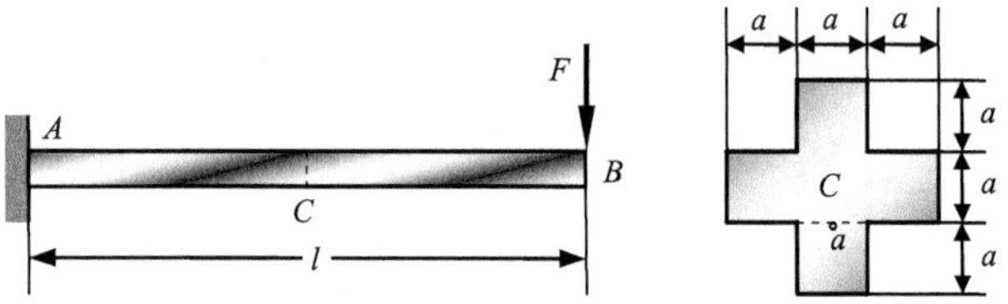

图 E15.2

解　根据截面法，剪力等于截面 B 处的外力，即 V=F。截面对中性轴的惯性矩为 $I=\dfrac{29}{12}a^4$，过点 a 水平线下方截面对中性轴的静矩为 $Q'=A'\overline{y}'=a^3$。因此点 a 处的剪应力等于

$$\tau_a=\frac{VQ'}{Ib}=\frac{Fa^3}{(\frac{29}{12}a^4)a}=\frac{12F}{29a^2}$$

15.3　弯曲梁设计

弯曲梁设计通常由梁中最大弯矩控制。最大正应力位于危险截面上下表面，可表示为

$$\sigma_{\max}=\frac{M_{\max}y_{\max}}{I}=\frac{M_{\max}}{S} \tag{15.22}$$

安全设计要求满足 $\sigma_{\max}\leqslant\sigma_{\text{allow}}$，其中 σ_{allow} 为许用应力。把 σ_{allow} 代入式(15.22)，得

$$S\geqslant\frac{M_{\max}}{\sigma_{\text{allow}}} \tag{15.23}$$

在梁的设计中，合理设计应该满足经济原则，即在类型和材料相同的梁中，当其他因素相同时，应该选择重量最轻的梁，以降低梁的费用。

例 15.3　已知钢的许用正应力为 165 MPa，选择支撑图 E15.3 所示外载荷的工字钢梁。

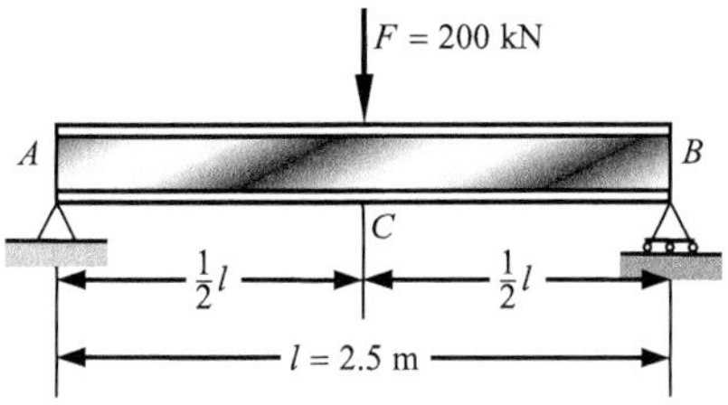

图 E15.3

解　弯矩在截面 C 处达到最大，等于 $M_{\max}=\frac{1}{4}Fl=125\ \text{kN}\cdot\text{m}$。因此，最小截面模量为

$$S_{\min}=\frac{M_{\max}}{\sigma_{\text{allow}}}=757.58\ \text{cm}^3$$

查阅附录 IV 中型钢表，选择截面模量不小于 $S_{\min}$ 的一组工字钢梁，如表 15.1 所示。

表 15.1

型号	重量/(kg/m)	截面模量/cm^3
32c	62.7	760
36a	60.0	875
36b	65.7	919

根据表 15.1，可选 32c 或 36a 工字钢。然而，最经济设计应该选择 36a 工字钢，因为与 32c 工字钢相比，36a 工字钢尽管具有较大截面模量，但其重量却只有 60.0 kg/m。

所选梁的重量为 1.47 kN。该重量与 200 kN 外载荷相比显得很小，因而可忽略不计。

习　题

15.1　图示外力偶作用于竖直面内，求中间截面 C 上(a)点 P_1 和(b)点 P_2 处的应力。

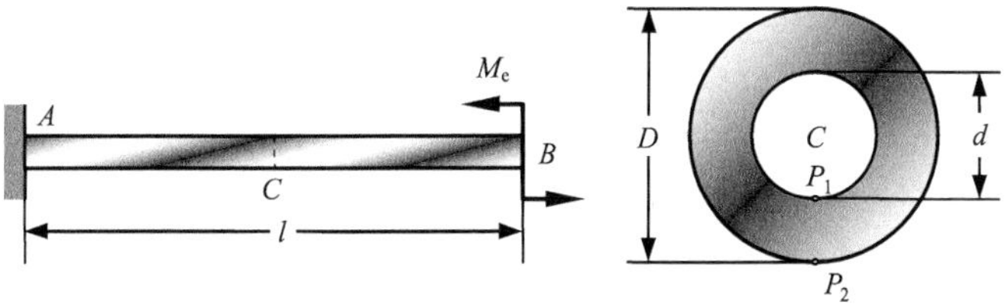

图 P15.1

15.2　已知图示工字截面钢梁的许用应力σ_{allow}=150 MPa，求所能施加的最大力偶。忽略倒角效应。

15.3　外力偶作用于图示 T 型截面梁。求梁的最大拉应力和最大压应力。忽略倒角效应。

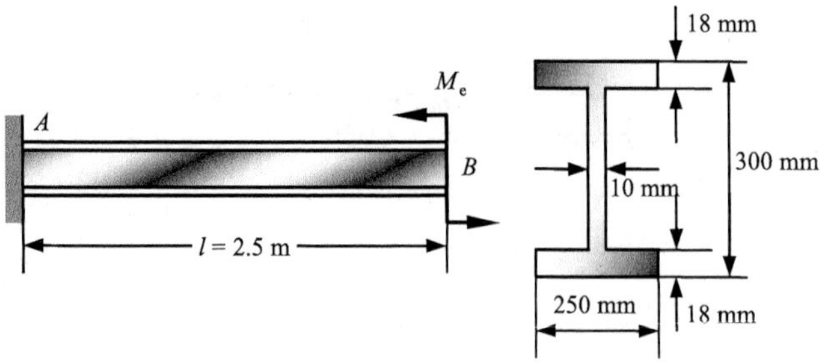

图 P15.2

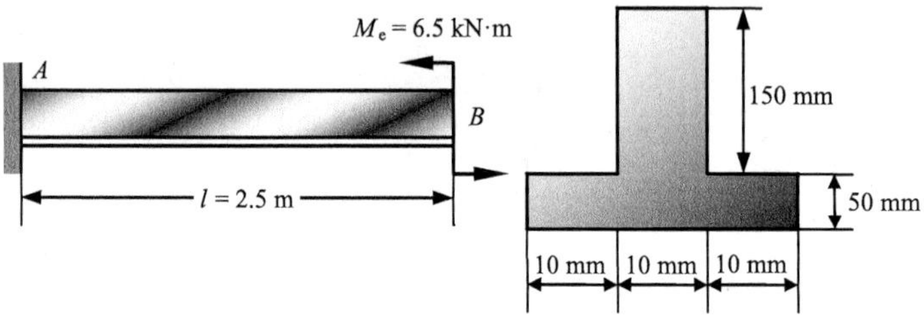

图 P15.3

15.4-15.6　已知图示外力作用于竖直面内，求中间截面 C 上点 P 处的剪应力。

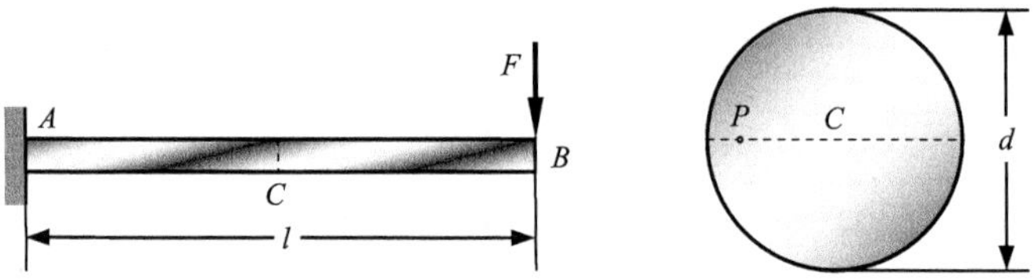

图 P15.4

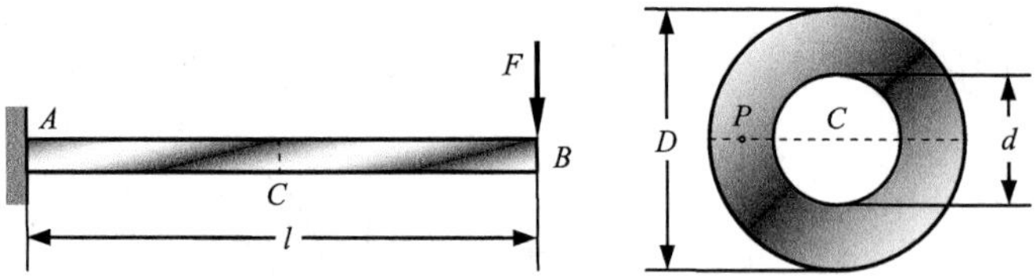

图 P15.5

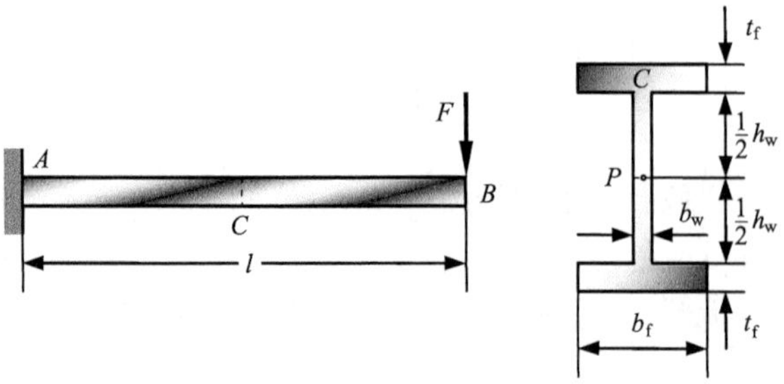

图 P15.6

15.7 已知钢的许用正应力为 165 MPa，选择工字钢梁以支撑图示外载荷。

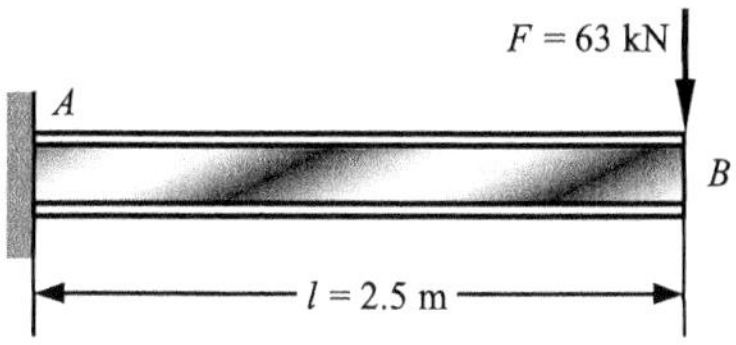

图 P15.7

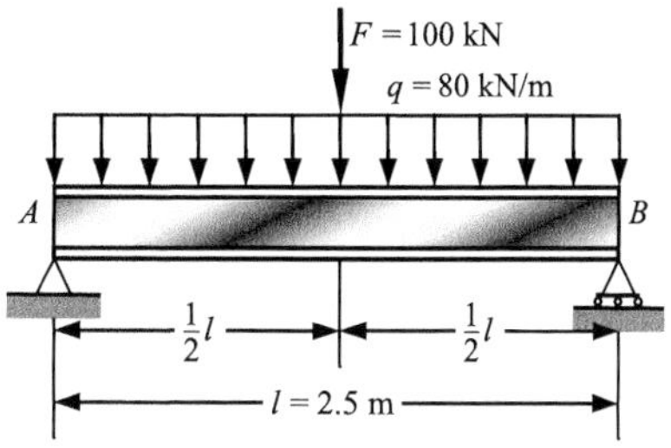

图 P15.8

15.8 已知钢的许用正应力为 165 MPa，选择最经济的工字梁以支撑图示外载荷。

第 16 章 弯曲变形

如图 16.1 所示，纯弯曲梁在变形后将弯成圆弧。在线弹性范围，中性层的曲率可表示为

$$\frac{1}{\rho}=\frac{M}{EI} \tag{16.1}$$

式中，弯矩 M 为常数，EI 为梁的抗弯刚度。

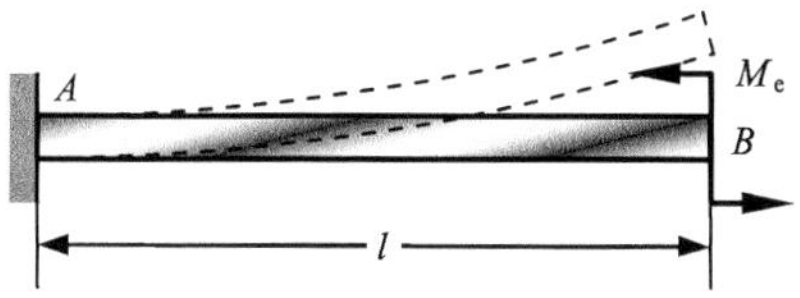

图 16.1

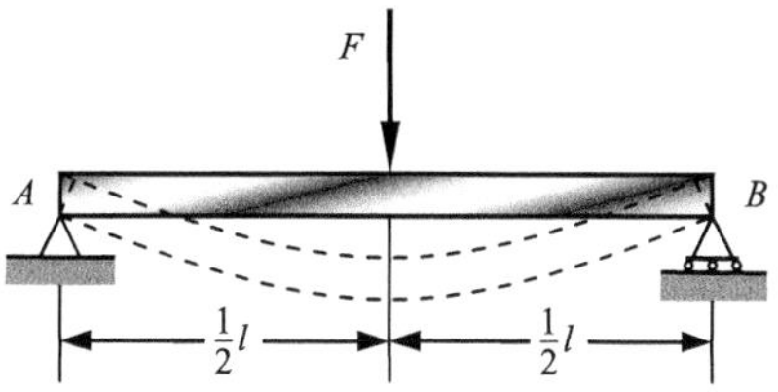

图 16.2

当图 16.2 所示细长梁承受横向载荷时，式(16.1)仍然成立。不过弯矩和中性层曲率将会随着截面的不同而发生变化。设 x 表示截面到左端的距离，则有

$$\frac{1}{\rho}=\frac{M(x)}{EI} \tag{16.2}$$

式中，弯矩 $M(x)$ 为 x 的函数。

梁在变形后中性层的曲率可表示为

$$\frac{1}{\rho}=\frac{w''}{[1+(w')^2]^{3/2}} \tag{16.3}$$

式中， $w=w(x)$ 表示挠度函数， w' 和 w'' 分别为函数 $w=w(x)$ 的一阶和二阶导数。对小变形梁，与 1 相比， w' 非常小，因而其平方可忽略，得

$$\frac{1}{\rho}=w'' \tag{16.4}$$

把式(16.4)代入式(16.2)，得

$$w''=\frac{M(x)}{EI} \tag{16.5}$$

上式称为弯曲梁的挠曲线微分方程。

16.1　积　分　法

积分式(16.5)，得

$$w' = \int \frac{M(x)}{EI} \mathrm{d}x + C_1 \tag{16.6}$$

式中，C_1为积分常数。设$\theta(x)$为变形后中性层切线与水平方向的夹角，如图16.3所示，考虑到夹角很小，则有

$$w' = \tan\theta \approx \theta \tag{16.7}$$

因此，式(16.6)可重写为

$$\theta = \int \frac{M(x)}{EI} \mathrm{d}x + C_1 \tag{16.8}$$

积分，得

$$w = \int \left(\int \frac{M(x)}{EI} \mathrm{d}x\right) \mathrm{d}x + C_1 x + C_2 \tag{16.9}$$

式中，C_2也是积分常数。

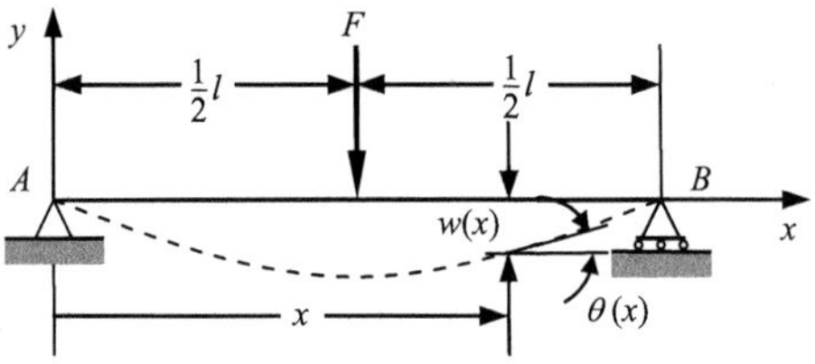

图 16.3

积分常数C_1和C_2可通过边界条件确定。三类常用支撑及其边界条件如图16.4所示。

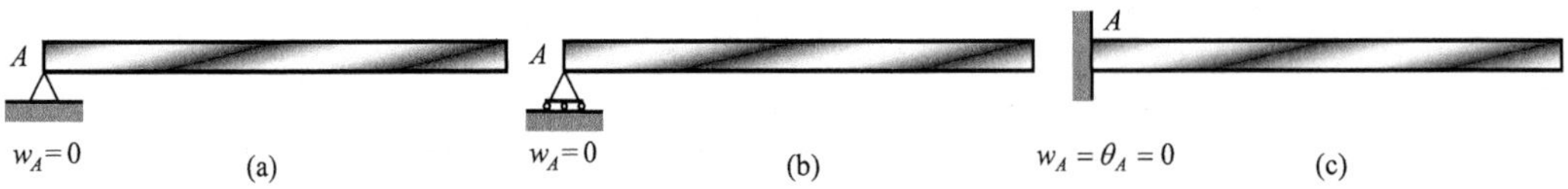

图 16.4

例 16.1　悬臂梁受图E16.1所示载荷作用，求：(a)梁的挠曲线方程；(b)自由端的挠度和转角。

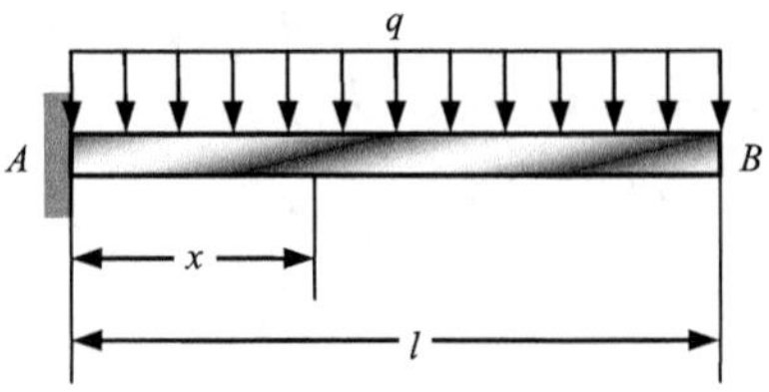

图 E16.1

解 考虑梁的 x 截面，如图 E16.1 所示，则弯矩为 $M(x)=-\frac{1}{2}q(l-x)^2$ 和挠曲线微分方程为 $w''=-\frac{q(l-x)^2}{2EI}$。利用积分法，得

$$\theta=\int[-\frac{q(l-x)^2}{2EI}]\mathrm{d}x+C_1=-\frac{q}{6EI}(3l^2x-3lx^2+x^3)+C_1$$

$$w=\int\{\int[-\frac{q(l-x)^2}{2EI}]\mathrm{d}x\}\mathrm{d}x+C_1x+C_2=-\frac{q}{24EI}(6l^2x^2-4lx^3+x^4)+C_1x+C_2$$

当 $x=0$ 时，$\theta=w=0$，则得 $C_1=0$ 和 $C_2=0$。因此挠曲线方程为

$$w=-\frac{q}{24EI}(6l^2x^2-4lx^3+x^4)$$

利用上述方程，得自由端挠度和转角分别为

$$w_B=-\frac{ql^4}{8EI},\ \theta_B=-\frac{ql^3}{6EI}$$

16.2 叠 加 法

当梁受多个载荷作用时，首先分别计算每个载荷单独作用时的挠度和转角，然后采用叠加原理，把对应各个载荷的挠度和转角相加，即可得到多个载荷同时作用时的挠度和转角。大部分工程手册都包含梁在各种载荷作用和支撑条件下的挠度和转角列表，如表 16.1 所示。

考虑图 16.5a 所示简支梁，受集中载荷和分布载荷同时作用，则梁在任意截面的挠度和转角等于图 16.5b 所示集中载荷和图 16.5c 所示分布载荷分别引起的挠度和转角之和。因此，受两个载荷同时作用时，梁的挠度和转角可表示为

$$w=w_F+w_q,\ \theta=\theta_F+\theta_q \tag{16.10}$$

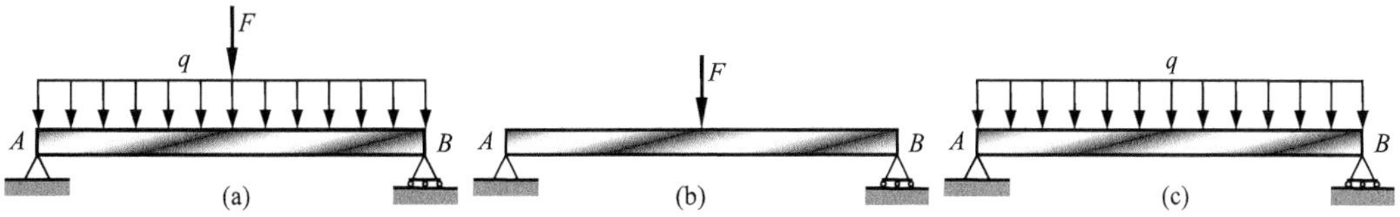

图 16.5

例 16.2 图 E16.2 所示梁和载荷，求截面 C 的挠度和截面 B 的转角。

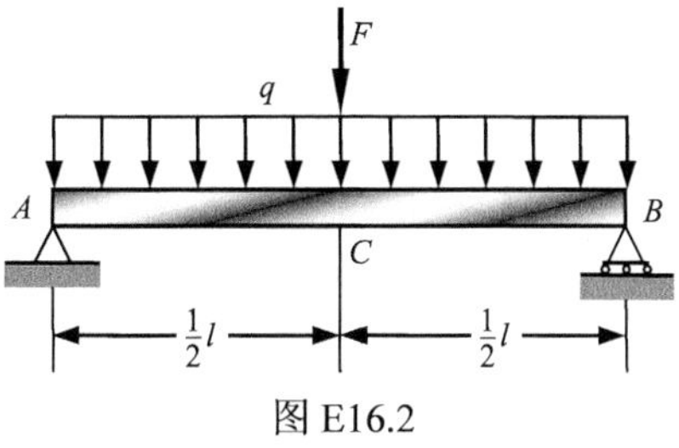

图 E16.2

表 16.1　挠度曲线

梁和载荷	挠曲线方程	临界挠度	临界转角
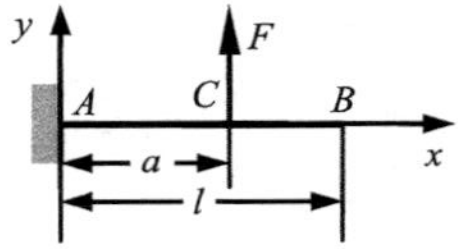	$w=\frac{Fx^2}{6EI}(3a-x)\ (0\leqslant x\leqslant a)$ $w=\frac{Fa^2}{6EI}(3x-a)\ (a\leqslant x\leqslant l)$	$w_{\max}=w_B=\frac{Fa^2}{6EI}(3l-a)$ $w_C=\frac{Fa^3}{3EI}$	$\theta_{\max}=\theta_B=\theta_C=\frac{Fa^2}{2EI}$
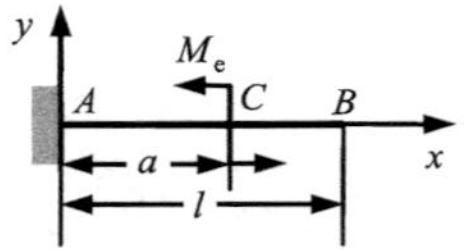	$w=\frac{M_ex^2}{2EI}\ (0\leqslant x\leqslant a)$ $w=\frac{M_ea}{2EI}(2x-a)\ (a\leqslant x\leqslant l)$	$w_{\max}=w_B=\frac{M_ea}{2EI}(2l-a)$ $w_C=\frac{M_ea^2}{2EI}$	$\theta_{\max}=\theta_B=\theta_C=\frac{M_ea}{EI}$
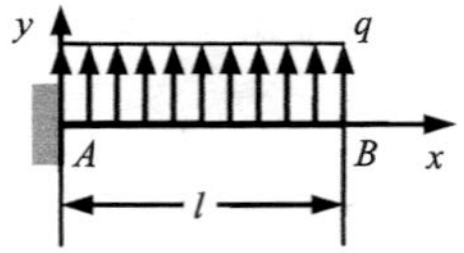	$w=\frac{qx^2}{24EI}(6l^2-4lx+x^2)$	$w_{\max}=w_B=\frac{ql^4}{8EI}$	$\theta_{\max}=\theta_B=\frac{ql^3}{6EI}$
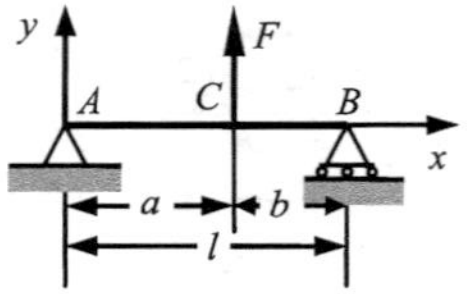	$w=\frac{Fbx}{6EIl}(l^2-b^2-x^2)\ (0\leqslant x\leqslant a)$ $w=\frac{F}{6EIl}[l(x-a)^3+b(l^2-b^2)x-bx^3]$ $(a\leqslant x\leqslant l)$	设$a\geqslant b$: $w_{\max}=w[\sqrt{\frac{1}{3}(l^2-b^2)}]=\frac{Fb\sqrt{(l^2-b^2)^3}}{9\sqrt{3}EIl}$ $w(\frac{1}{2}l)=\frac{Fb(3l^2-4b^2)}{48EI}$	设$a\geqslant b$: $\theta_A=\frac{Fab(l+b)}{6EIl}$ $\theta_{\max}=-\theta_B=\frac{Fab(l+a)}{6EIl}$
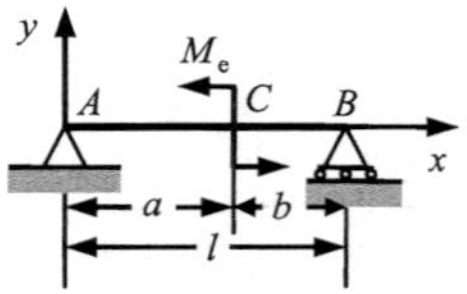	$w=\frac{M_ex}{6EIl}(x^2+3b^2-l^2)\ (0\leqslant x\leqslant a)$ $w=\frac{M_e}{6EIl}[x^3-3l(x-a)^2-(l^2-3b^2)x]$ $(a\leqslant x\leqslant l)$	设$a\geqslant b$: $w_{\max}=-w[\sqrt{\frac{1}{3}(l^2-3b^2)}]=\frac{M_e\sqrt{(l^2-3b^2)^3}}{9\sqrt{3}EIl}$ $w(\frac{1}{2}l)=-\frac{M_e(l^2-4b^2)}{16EI}$	$\theta_A=-\frac{M_e}{6EIl}(l^2-3b^2)$ $\theta_B=-\frac{M_e}{6EIl}(l^2-3a^2)$ $\theta_{\max}=\theta_C=\frac{M_e}{6EIl}[3(a^2+b^2)-l^2]$
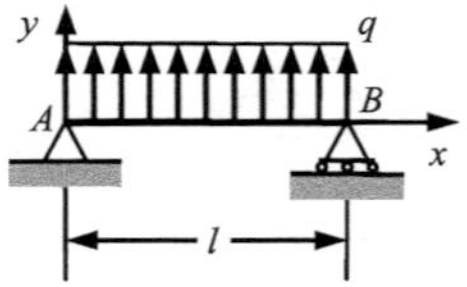	$w=\frac{qx}{24EI}(l^3-2lx^2+x^3)$	$w_{\max}=w(\frac{1}{2}l)=\frac{5ql^4}{384EI}$	$\theta_{\max}=\theta_A=-\theta_B=\frac{ql^3}{24EI}$

解　对集中载荷 F，查表 16.1，得

$$(w_C)_F = -\frac{Fl^3}{48EI}，\ (\theta_B)_F = \frac{Fl^2}{16EI}$$

同理，对分布载荷 q，查表 16.1，得

$$(w_C)_q = -\frac{5ql^4}{384EI}，\ (\theta_B)_q = \frac{ql^3}{24EI}$$

利用叠加原理，得截面 C 的挠度和截面 B 的转角为

$$w_C = (w_C)_F + (w_C)_q = -\left(\frac{Fl^3}{48EI} + \frac{5ql^4}{384EI}\right)，\ \theta_B = (\theta_B)_F + (\theta_B)_q = \frac{Fl^2}{16EI} + \frac{ql^3}{24EI}$$

16.3　静 不 定 梁

考虑图 16.6 所示等截面梁，A 端固支，B 端可动铰支。梁有四个未知反力，但只有三个平衡条件，因此梁为一次静不定。

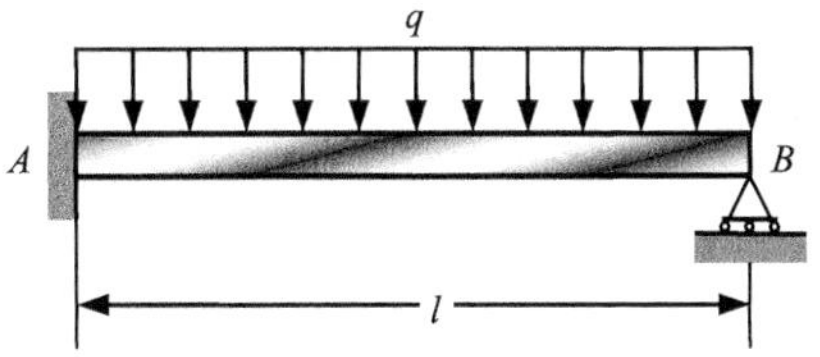

图 16.6

对图 16.6 所示一次静不定梁，把一个反力(如 B 处反力)看成多余反力，并消除多余反力处的支撑。多余反力作为未知载荷，连同其余载荷，产生与原支撑相协调的变形。

例 16.3　图 16.6 所示梁受载荷作用，求可动铰支处的反力。

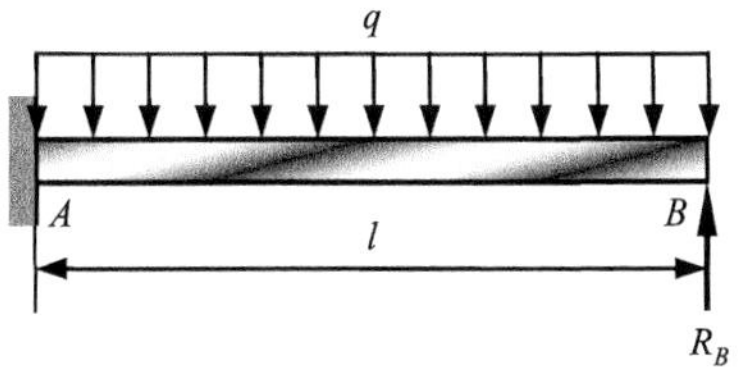

图 E16.3

解　如图 E16.3a 所示，考虑 B 处为多余反力，并释放 B 处支撑，则多余反力 R_B 可根据 B 处挠度等于零而求得。

查表 16.1，得 B 处由多余反力 R_B 引起的挠度为

$$(w_B)_{R_B} = \frac{R_B l^3}{3EI}$$

同理，B 处由均布载荷 q 引起的挠度为

$$(w_B)_q = -\frac{ql^4}{8EI}$$

利用叠加原理，得 B 处挠度

$$w_B = (w_B)_{R_B} + (w_B)_q = \frac{R_B l^3}{3EI} - \frac{ql^4}{8EI}$$

因 B 处挠度为零，则得

$$\frac{R_B l^3}{3EI} - \frac{ql^4}{8EI} = 0$$

解上式，得 B 处反力为

$$R_B = \frac{3}{8}ql$$

习　　题

16.1-16.2　图示载荷，求：(a)悬臂梁的挠曲线方程；(b)自由端的挠度和转角。

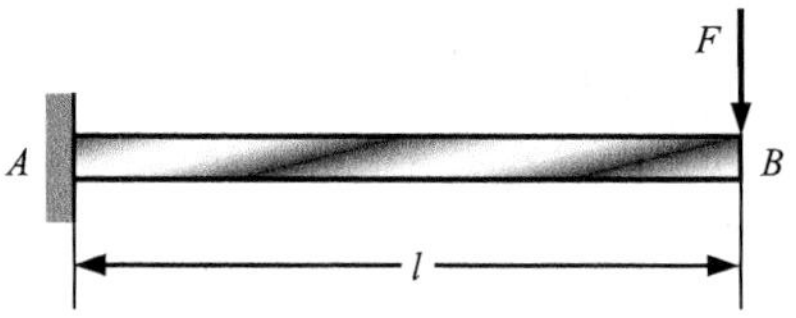

图 P16.1

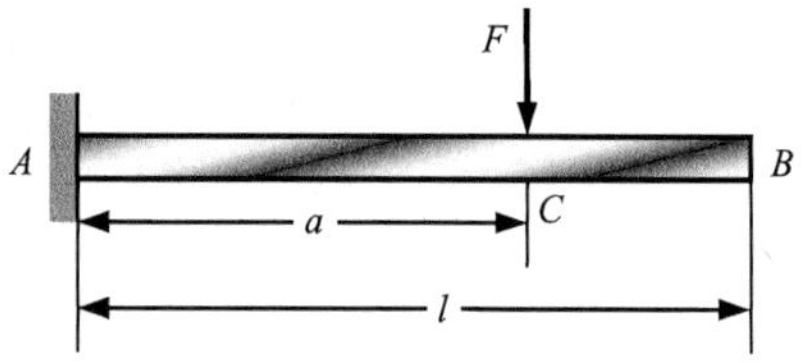

图 P16.2

16.3-16.4　图示载荷，求：(a)简支梁的挠曲线方程；(b)中间截面 C 的挠度和截面 B 的转角。

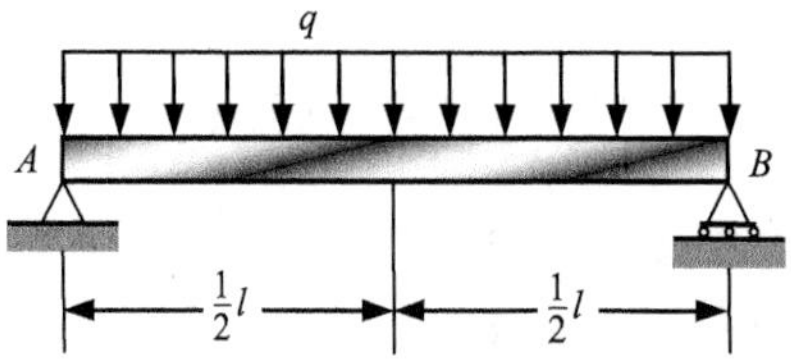

图 P16.3

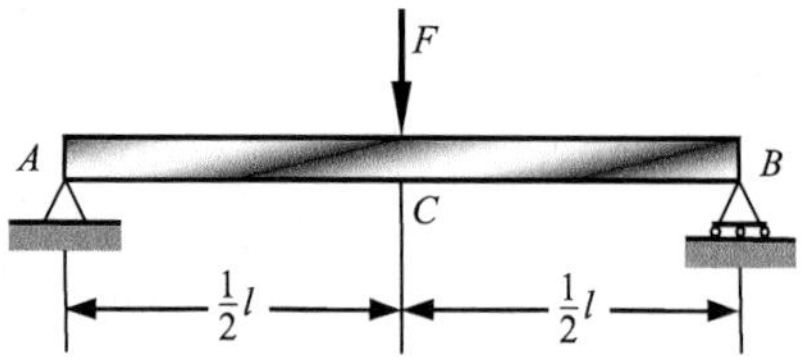

图 P16.4

16.5　图示梁和载荷，求：(a)中间截面 C 的挠度；(b)截面 B 的转角。

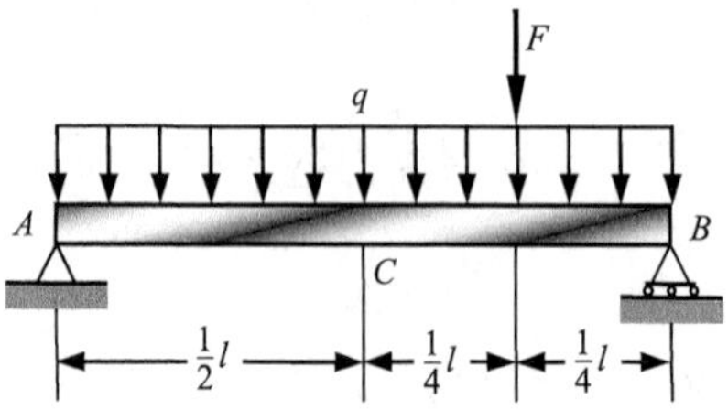

图 P16.5

16.6-16.9　图示悬臂梁和载荷，求截面 B 的挠度和转角。

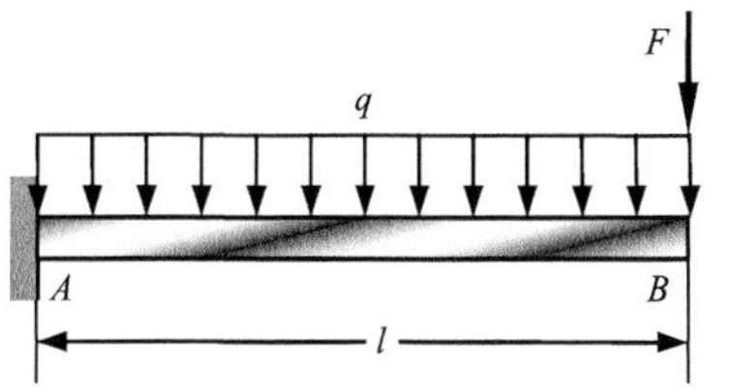

图 P16.6

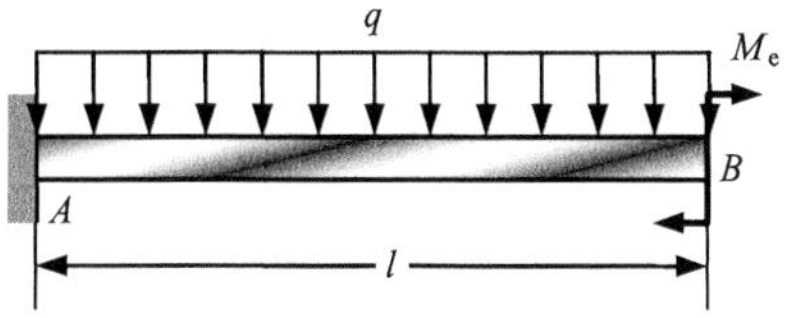

图 P16.7

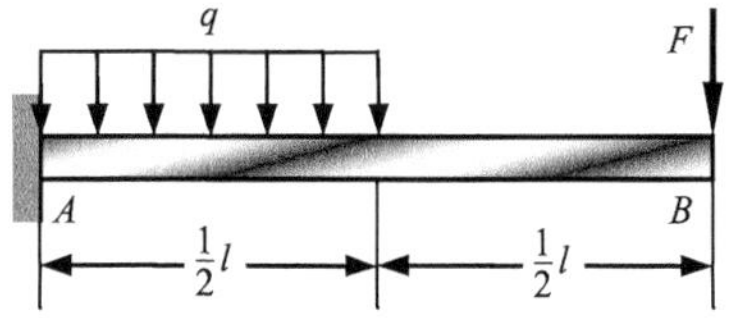

图 P16.8

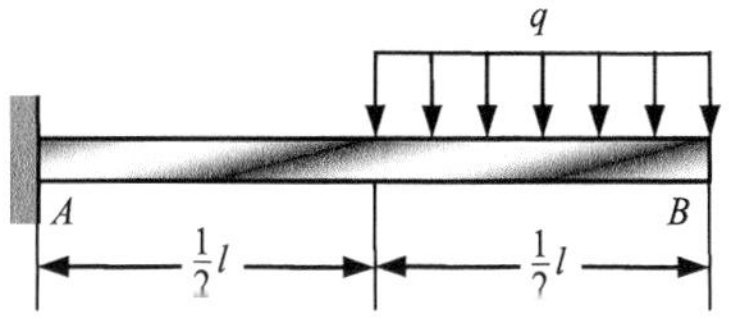

图 P16.9

16.10　图示载荷，求：(a)中间截面 C 的挠度；(b)截面 B 的转角。

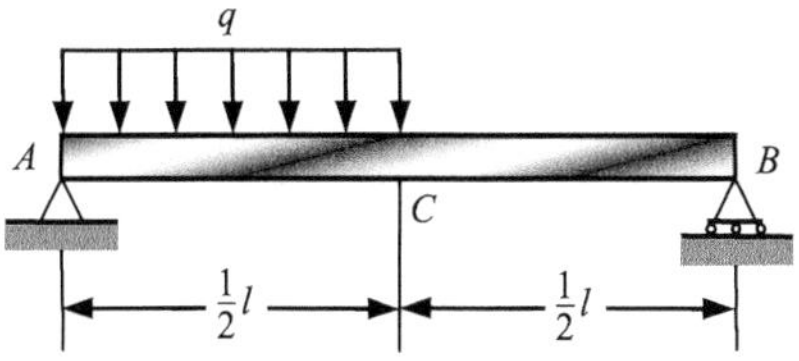

图 P16.10

16.11-16.12　图示梁和载荷，求可动铰支 B 处反力。

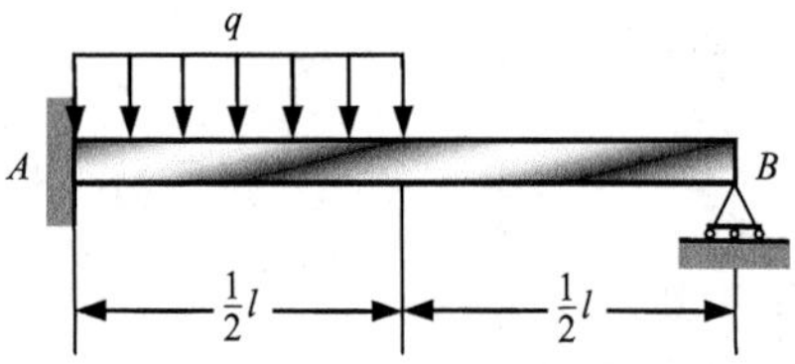

图 P16.11

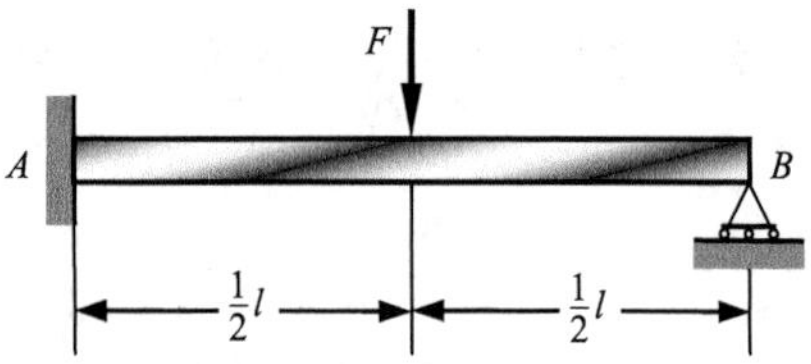

图 P16.12

第 17 章　应力分析与强度理论

给定点 P 的最一般应力状态可通过六个应力分量表示，其中三个分量 σ_x、σ_y 和 σ_z 为单元体面上的正应力，而其余三个分量 $\tau_{xy}(\tau_{yx}=\tau_{xy})$、$\tau_{yz}(\tau_{zy}=\tau_{yz})$ 和 $\tau_{zx}(\tau_{xz}=\tau_{zx})$ 为剪应力，如图 17.1 所示。如果坐标轴旋转，则相同的应力状态可通过不同的分量进行表示。应力状态的讨论主要涉及平面应力，即单元体的两个平行面上不受应力作用，如图 17.2 所示。

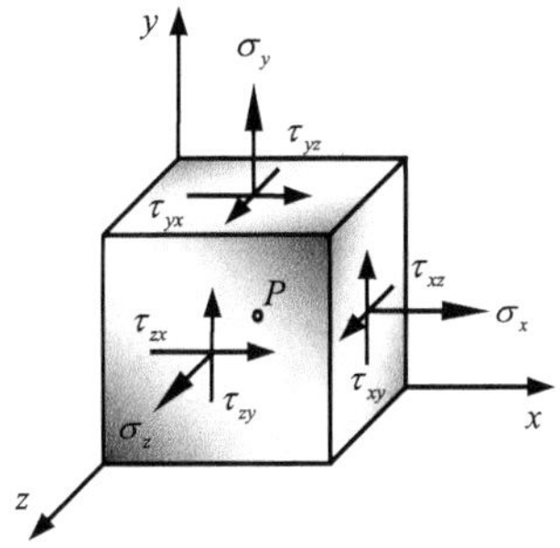

图 17.1

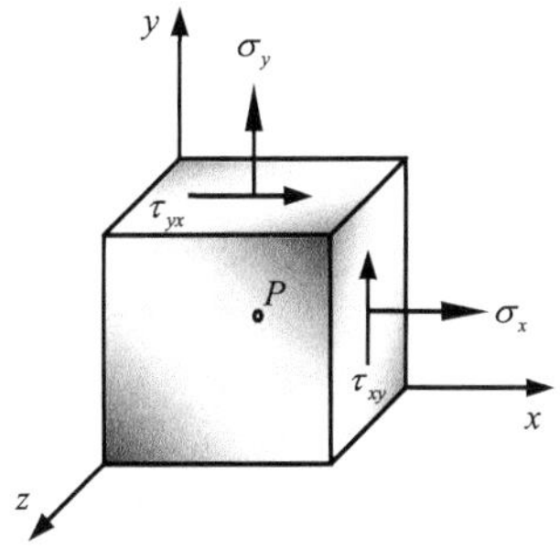

图 17.2

17.1　应 力 变 换

考虑点 P 处于平面应力状态，应力分量由 σ_x、σ_y 和 $\tau_x(\tau_y=\tau_x)$ 表示，如图 17.3a 所示。下面求单元体绕 z 轴逆时针旋转 θ 后的应力分量 $\sigma_{x'}$、$\sigma_{y'}$ 和 $\tau_{x'}(\tau_{y'}=\tau_{x'})$，如图 17.3b 所示。

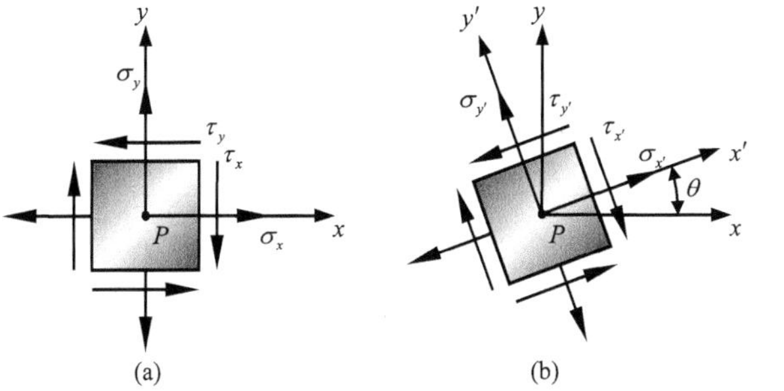

图 17.3

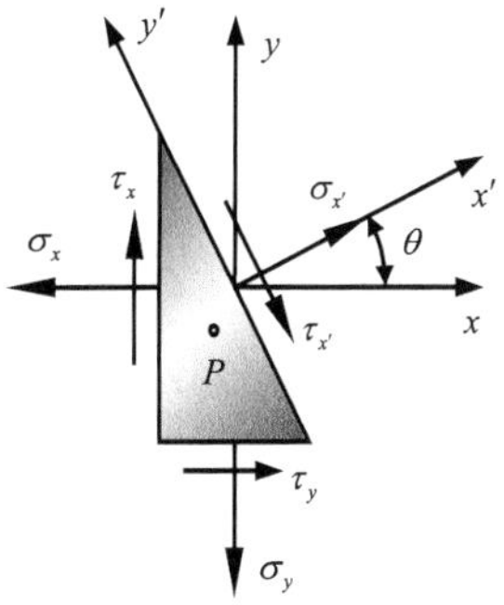

图 17.4

考虑表面分别垂直于 x、y 和 x' 的单元体，如图 17.4 所示。假设斜面面积为 ΔA，则垂直面和水平面的面积分别等于 $\Delta A\cos\theta$ 和 $\Delta A\sin\theta$。

利用 x' 和 y' 方向的静平衡条件，得

$$\begin{aligned}\sum F_{x'}=0:\quad &\sigma_{x'}\Delta A-\sigma_x(\Delta A\cos\theta)\cos\theta+\tau_x(\Delta A\cos\theta)\sin\theta-\\&\sigma_y(\Delta A\sin\theta)\sin\theta+\tau_y(\Delta A\sin\theta)\cos\theta=0\\\sum F_{y'}=0:\quad &-\tau_{x'}\Delta A+\sigma_x(\Delta A\cos\theta)\sin\theta+\tau_x(\Delta A\cos\theta)\cos\theta-\\&\sigma_y(\Delta A\sin\theta)\cos\theta-\tau_y(\Delta A\sin\theta)\sin\theta=0\end{aligned}\tag{17.1}$$

解上述方程，并利用 $\tau_y=\tau_x$、$\cos^2\theta=\frac{1}{2}(1+\cos 2\theta)$、$\sin^2\theta=\frac{1}{2}(1-\cos 2\theta)$ 和 $2\sin\theta\cos\theta=\sin 2\theta$，得

$$\begin{aligned}\sigma_{x'}&=\frac{1}{2}(\sigma_x+\sigma_y)+\frac{1}{2}(\sigma_x-\sigma_y)\cos 2\theta-\tau_x\sin 2\theta\\\tau_{x'}&=\frac{1}{2}(\sigma_x-\sigma_y)\sin 2\theta+\tau_x\cos 2\theta\end{aligned}\tag{17.2}$$

如果需要求 $\sigma_{y'}$，则用 $\theta+\frac{1}{2}\pi$ 代替 θ 代入上述方程，得

$$\sigma_{y'}=\frac{1}{2}(\sigma_x+\sigma_y)-\frac{1}{2}(\sigma_x-\sigma_y)\cos 2\theta+\tau_x\sin 2\theta\tag{17.3}$$

例 17.1　图 E17.1a 所示应力状态，求单元体顺时针旋转 20°后的正应力和剪应力。

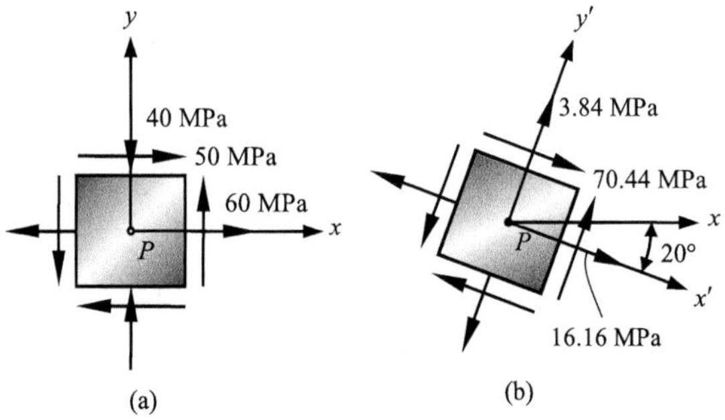

图 E17.1

解　根据图 E17.1a 所示参考系，则有

$$\sigma_x=60\text{ MPa},\ \sigma_y=-40\text{ MPa},\ \tau_x=-50\text{ MPa},\ \theta=-20°$$

利用式(17.2)和(17.3)，得

$$\sigma_{x'}=\frac{1}{2}(\sigma_x+\sigma_y)+\frac{1}{2}(\sigma_x-\sigma_y)\cos 2\theta-\tau_x\sin 2\theta=16.16\text{ MPa}$$

$$\sigma_{y'}=\frac{1}{2}(\sigma_x+\sigma_y)-\frac{1}{2}(\sigma_x-\sigma_y)\cos 2\theta+\tau_x\sin 2\theta=3.84\text{ MPa}$$

$$\tau_{x'}=\frac{1}{2}(\sigma_x-\sigma_y)\sin 2\theta+\tau_x\cos 2\theta=-70.44\text{ MPa}$$

图 E17.1b 所示为单元体顺时针旋转 20°后的正应力和剪应力。

17.2 主 应 力

设 $\sigma_{avg}=\frac{1}{2}(\sigma_x+\sigma_y)$ 和 $R=\sqrt{[\frac{1}{2}(\sigma_x-\sigma_y)]^2+\tau_x^2}$ ，则利用式(17.2)，得

$$(\sigma_{x'}-\sigma_{avg})^2+\tau_{x'}^2=R^2 \tag{17.4}$$

式(17.4)表示圆心在 $C(\sigma_{avg},0)$ 半径为 R 的圆，如图 17.5 所示。该圆称为莫尔圆或应力圆。

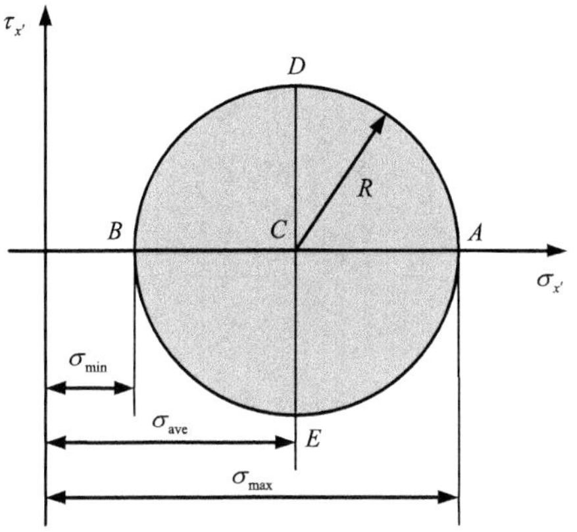

图 17.5

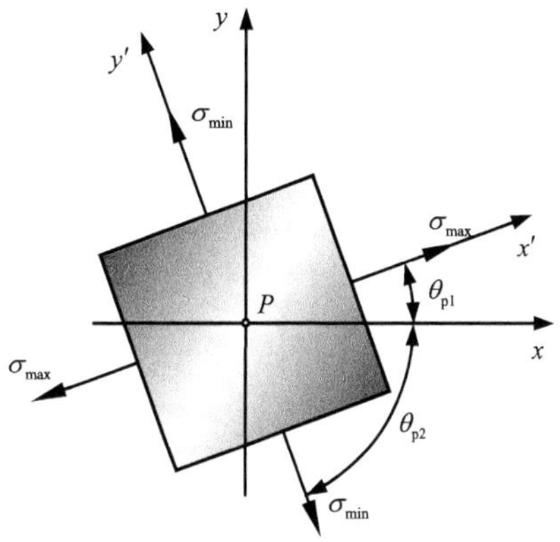

图 17.6

应力圆与水平轴的交点 A 和 B 是特别感兴趣的点：A 点对应最大正应力 σ_{max} ，而 B 点对应最小正应力 σ_{min} 。同时，这两点对应的剪应力为零。通过在式(17.2)中设 $\tau_{x'}=0$ 可得分别对应 A 和 B 的最大和最小正应力的方向，即

$$\tan 2\theta_p=\frac{-2\tau_x}{\sigma_x-\sigma_y} \tag{17.5}$$

式(17.5)定义了相差 90°的两个值 θ_{p1} 和 θ_{p2} 。通过任何一个值即可确定相应单元体的方位，如图 17.6 所示。

包含最大或最小正应力的平面称为主平面，相应的最大或最小正应力称为面内主应力。显然，没有剪应力作用于主平面。根据式(17.4)得

$$\sigma_{max,min}=\sigma_{avg}\pm R=\frac{1}{2}(\sigma_x+\sigma_y)\pm\sqrt{[\frac{1}{2}(\sigma_x-\sigma_y)]^2+\tau_x^2} \tag{17.6}$$

均匀各向同性材料内任何一点在平面应力状态下都存在三个相互垂直的主应力。如果 $\sigma_{min}\geqslant 0$ ，三个主应力可表示为 $\sigma_1=\sigma_{max}$ 、 $\sigma_2=\sigma_{min}$ 和 $\sigma_3=0$ ；如果 $\sigma_{max}\geqslant 0\geqslant\sigma_{min}$ ，则 $\sigma_1=\sigma_{max}$ 、 $\sigma_2=0$ 和 $\sigma_3=\sigma_{min}$ ；如果 $\sigma_{max}\leqslant 0$ ，则 $\sigma_1=0$ 、 $\sigma_2=\sigma_{max}$ 和 $\sigma_3=\sigma_{min}$ 。

例 17.2　图 E17.2a 所示应力状态，求主应力和相应主平面。

解　参考图 E17.2a 所示坐标系，得

$$\sigma_x=60\text{ MPa},\ \sigma_y=-40\text{ MPa},\ \tau_x=-50\text{ MPa}$$

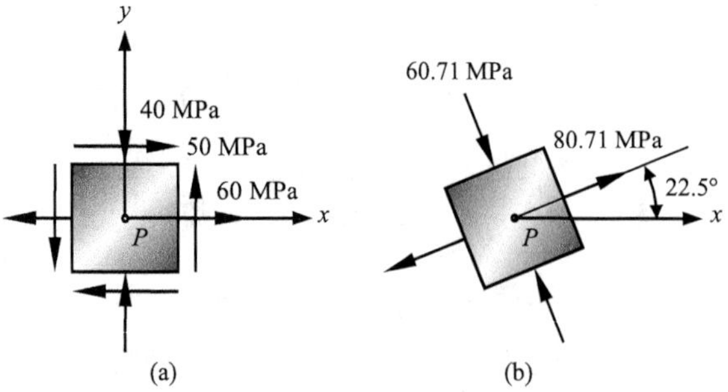

图 E17.2

利用式(17.6)，得面内主应力

$$\sigma_{\max}=\frac{1}{2}(\sigma_x+\sigma_y)+\sqrt{[\frac{1}{2}(\sigma_x-\sigma_y)]^2+\tau_x^2}=80.71\ \text{MPa}$$

$$\sigma_{\min}=\frac{1}{2}(\sigma_x+\sigma_y)-\sqrt{[\frac{1}{2}(\sigma_x-\sigma_y)]^2+\tau_x^2}=-60.71\ \text{MPa}$$

利用式(17.5)，得面内主应力的方位

$$\theta_p=\frac{1}{2}\arctan\frac{-2\tau_x}{\sigma_x-\sigma_y}=\begin{matrix}22.5° & (\text{对应}\ \sigma_{\max})\\ -67.5° & (\text{对应}\sigma_{\min})\end{matrix}$$

主应力和相应主平面如图 E17.2b 所示。因此，三个主应力分别为$\sigma_1=80.71\ \text{MPa}$、$\sigma_2=0$ 和$\sigma_3=-60.71\ \text{MPa}$。

17.3　最大剪应力

参考图 17.5 所示应力圆，位于垂直直径上的 D 和 E 点对应最大和最小剪应力。因 D 和 E 的横坐标均为$\sigma_{avg}=\frac{1}{2}(\sigma_x+\sigma_y)$，故通过在式(17.2)中设$\sigma_{x'}=\frac{1}{2}(\sigma_x+\sigma_y)$，则对应 D 和 E 的最大和最小剪应力的方向为

$$\tan 2\theta_s=\frac{\sigma_x-\sigma_y}{2\tau_x}\tag{17.7}$$

式(17.7)定义了相差 90°的两个值θ_{s1}和θ_{s2}。通过任何一个值即可确定对应于最大面内剪应力的单元体的方位，如图 17.7 所示。

由式(17.4)，得

$$(\tau_{\max})_{\text{in-plane}}=R=\sqrt{[\frac{1}{2}(\sigma_x-\sigma_y)]^2+\tau_x^2}\tag{17.8}$$

以及最大面内剪应力面上的正应力为

$$\sigma_{avg}=\frac{1}{2}(\sigma_x+\sigma_y)\tag{17.9}$$

根据式(17.5)和(17.7)，有$\tan 2\theta_s\cdot\tan 2\theta_p=-1$，因此，最大剪应力所在面与主平面成 45°角。

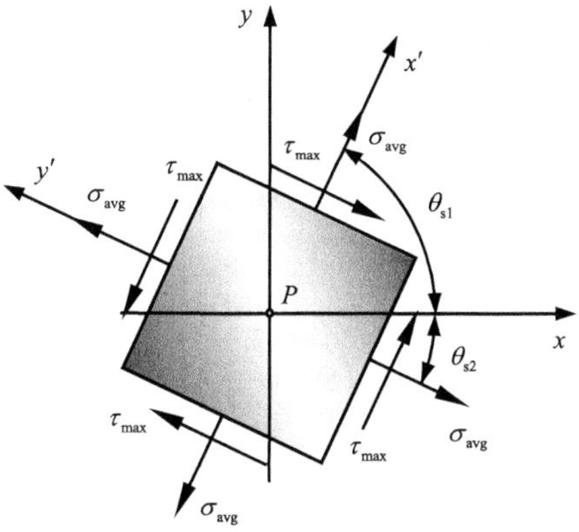

图 17.7

应该注意，最大剪应力 $\tau_{\max}=\frac{1}{2}(\sigma_1-\sigma_3)$ 可能会比由式(17.8)定义的面内最大剪应力 $(\tau_{\max})_{\text{in-plane}}$ 大。这种情况会发生在由式(17.6)定义的主应力具有相同正负号，即都是拉应力或都是压应力。

例 17.3　图 E17.3a 所示应力状态，求：(a)最大面内剪应力；(b)最大剪应力面的方位；(c)最大剪应力面上的正应力。

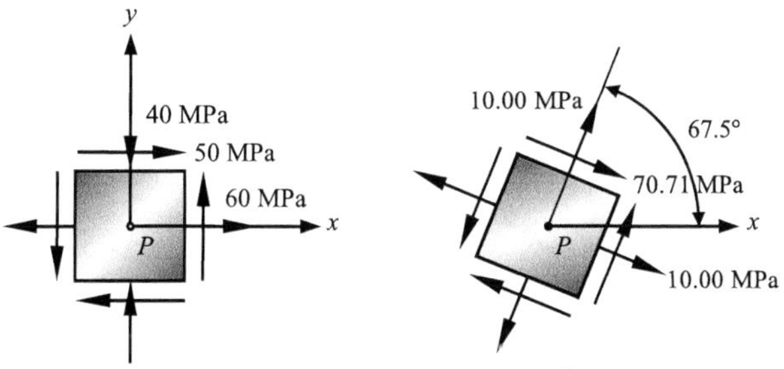

(a)　(b)

图 E17.3

解　对图 E17.3a 所示坐标系，有

$$\sigma_x=60\ \text{MPa},\ \sigma_y=-40\ \text{MPa},\ \tau_x=-50\ \text{MPa}$$

(a)利用式(17.8)，得最大面内剪应力

$$(\tau_{\max})_{\text{in-plane}}=\sqrt{[\frac{1}{2}(\sigma_x-\sigma_y)]^2+\tau_x^2}=70.71\ \text{MPa}$$

(b)利用式(17.7)，得最大剪应力面的方位

$$\theta_{\text{s}}=\frac{1}{2}\arctan\frac{\sigma_x-\sigma_y}{2\tau_x}=\begin{matrix}67.5^\circ\\-22.5^\circ\end{matrix}$$

(c)利用式(17.9)，得最大剪应力面上的正应力

$$\sigma_{\text{avg}}=\frac{1}{2}(\sigma_x+\sigma_y)=10.00\ \text{MPa}$$

最大面内剪应力和最大面内剪应力面上的正应力以及这些应力的方向如图 E17.3b 所示。

17.4　压力容器

薄壁压力容器为平面应力提供了重要应用。薄壁压力容器的应力分析仅考虑两种常见的容器类型：柱形压力容器和球形压力容器。

1. 柱形压力容器

如图 17.8 所示，柱形压力容器的内径为 d，壁厚为 t，内压为 p。下面求壁单元体上的应力，单元体的边分别平行和垂直于圆柱纵轴。

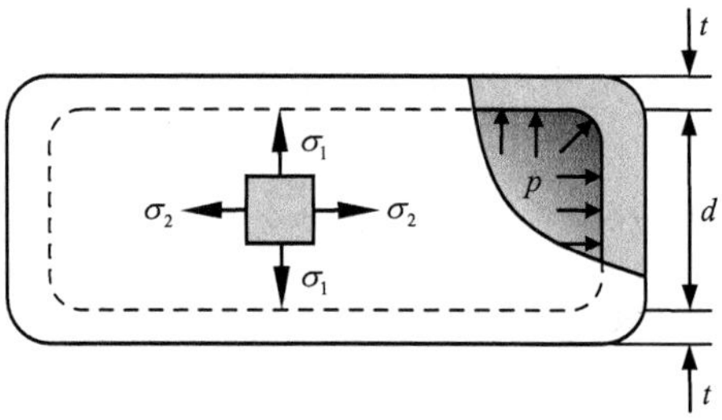

图 17.8

显然，没有剪应力作用于单元体，单元体上的正应力因此是主应力 σ_1 和 σ_2。σ_1 称为环向应力或周向应力，σ_2 称为纵向应力或轴向应力。

为了求纵向应力 σ_2，考虑图 17.9 所示自由体(或隔离体)的平衡，则有

$$\sum F_x = 0: \quad \sigma_2(\pi dt) - p(\frac{1}{4}\pi d^2) = 0 \tag{17.10}$$

解方程，得

$$\sigma_2 = \frac{pd}{4t} \tag{17.11}$$

为了求环向应力 σ_1，分离一部分容器，如图 17.10 所示。利用平衡条件，得

$$\sum F_z = 0: \quad \sigma_1(2t\Delta x) - p(d\Delta x) = 0 \tag{17.12}$$

解方程，得

$$\sigma_1 = \frac{pd}{2t} \tag{17.13}$$

利用式(17.11)和(17.13)，得最大剪应力

$$(\tau_{\max})_{\text{in-plane}} = \frac{1}{2}(\sigma_1 - \sigma_2) = \frac{pd}{8t}, \quad \tau_{\max} = \frac{1}{2}(\sigma_1 - \sigma_3) = \frac{pd}{4t} \tag{17.14}$$

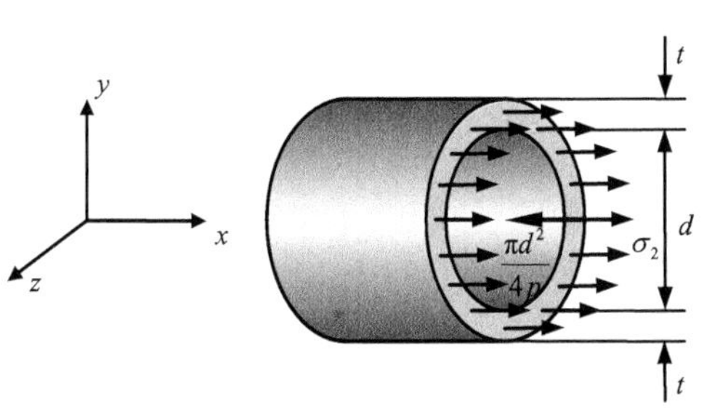

图 17.9

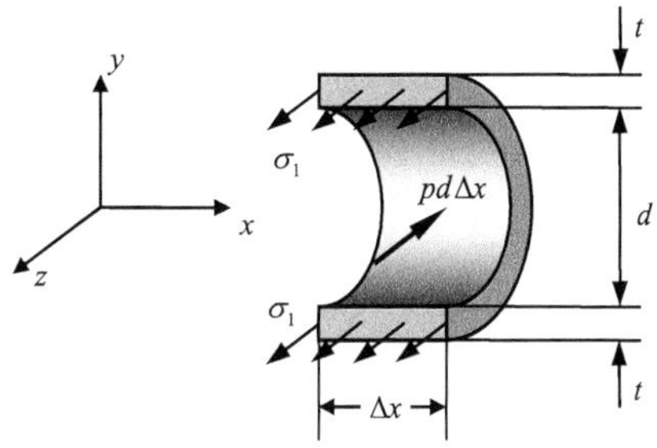

图 17.10

2. 球形压力容器

如图 17.11a 所示，球形压力容器的内径为 d，壁厚为 t，内压为 p。利用容器对称性，两个面内主应力一定相等，即 $\sigma_1=\sigma_2$。

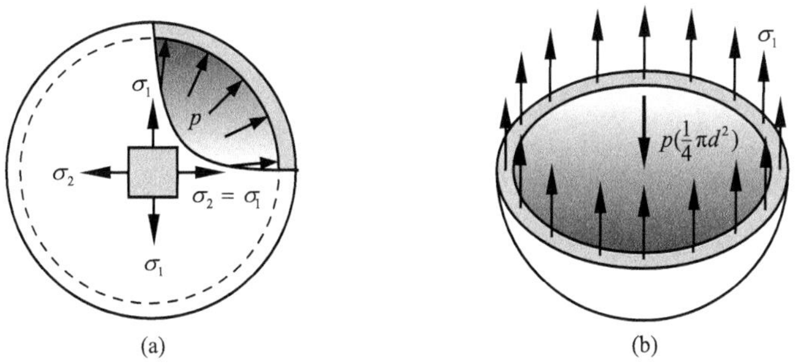

图 17.11

为了求应力 σ_1（或 σ_2），考虑图 17.11b 所示自由体。根据自由体在垂直方向的平衡，得

$$\sigma_1(\pi dt)-p(\frac{1}{4}\pi d^2)=0 \tag{17.15}$$

解方程，得

$$\sigma_1=\sigma_2=\frac{pd}{4t} \tag{17.16}$$

利用式(17.16)，得最大剪应力

$$(\tau_{\max})_{\text{in-plane}}=\frac{1}{2}(\sigma_1-\sigma_2)=0,\quad \tau_{\max}=\frac{1}{2}(\sigma_1-\sigma_3)=\frac{pd}{8t} \tag{17.17}$$

17.5　广义胡克定律

如图 17.12a 所示，考虑边长为 $\mathrm{d}x$、$\mathrm{d}y$ 和 $\mathrm{d}z$ 的单元体。当单元体受主应力 σ_1、σ_2 和 σ_3 作用，则单元体将变形为边长为 $(1+\varepsilon_1)\mathrm{d}x$、$(1+\varepsilon_2)\mathrm{d}y$ 和 $(1+\varepsilon_3)\mathrm{d}z$ 的新的单元体，如

图 17.12b 所示。其中，ε_1、ε_2 和 ε_3 为对应于主应力 σ_1、σ_2 和 σ_3 的线应变。

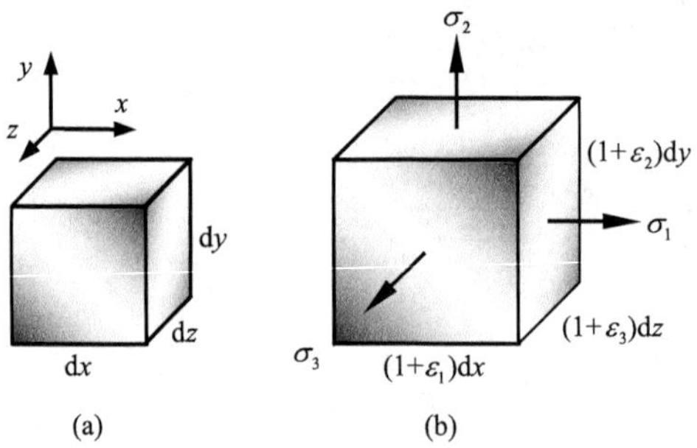

图 17.12

假设材料满足均匀、各向同性和线弹性，则通过利用叠加原理可以确定应力应变关系。

首先考虑由每个正应力引起的沿 x 方向的线应变。当 σ_1 作用时，单元体沿 x 方向伸长，由 σ_1 引起的线应变 ε_1' 为

$$\varepsilon_1' = \frac{\sigma_1}{E} \tag{17.18}$$

作用 σ_2 时，单元体在 x 方向收缩，线应变 ε_1'' 表示为

$$\varepsilon_1'' = -\mu\frac{\sigma_2}{E} \tag{17.19}$$

同理，σ_3 也引起 x 方向收缩，线应变 ε_1''' 写为

$$\varepsilon_1''' = -\mu\frac{\sigma_3}{E} \tag{17.20}$$

利用叠加原理，由 σ_1、σ_2 和 σ_3 共同引起的沿 x 方向的线应变 ε_1 等于

$$\varepsilon_1 = \varepsilon_1' + \varepsilon_1'' + \varepsilon_1''' = \frac{1}{E}[\sigma_1 - \mu(\sigma_2 + \sigma_3)] \tag{17.21}$$

同理，通过叠加原理，可以得到由 σ_1、σ_2 和 σ_3 共同引起的沿 y 和 z 方向的线应变。因此三个主应力作用下的应力应变关系可表示为

$$\varepsilon_1 = \frac{1}{E}[\sigma_1 - \mu(\sigma_2 + \sigma_3)],\ \varepsilon_2 = \frac{1}{E}[\sigma_2 - \mu(\sigma_3 + \sigma_1)],\ \varepsilon_3 = \frac{1}{E}[\sigma_3 - \mu(\sigma_1 + \sigma_2)] \tag{17.22}$$

上述关系称为均匀各向同性材料处于主应力状态下的广义胡克定律。只要应力不超过比例极限所得结果均有效。

例 17.4　在大型钢压力容器的侧面划出边长为 40 mm 的正方形。加压后正方形处于二向应力，如图 E17.4 所示。已知 E=200 GPa 和 μ=0.30，求边 AB、边 BC 和对角线 AC 的长度改变。

解　已知 $\sigma_1 = 60\ \text{MPa},\ \sigma_2 = 30\ \text{MPa},\ \sigma_3 = 0$，则利用广义胡克定律，得

$$\varepsilon_1 = \frac{1}{E}[\sigma_1 - \mu(\sigma_2 + \sigma_3)] = 255\times10^{-6},\ \varepsilon_2 = \frac{1}{E}[\sigma_2 - \mu(\sigma_3 + \sigma_1)] = 60\times10^{-6}$$

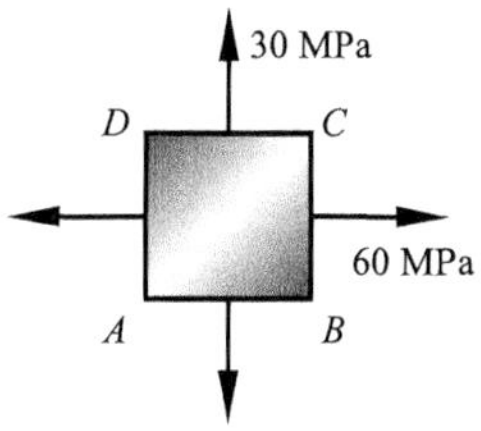

图 E17.4

利用 $\varepsilon_{AB}=\delta_{AB}/l_{AB}=\varepsilon_1$，得

$$\delta_{AB}=\varepsilon_1 l_{AB}=10.2\times10^{-3}\ \text{mm}$$

同理，得

$$\delta_{BC}=\varepsilon_2 l_{BC}=2.4\times10^{-3}\ \text{mm}$$

利用 $\delta_{AC}=\sqrt{(l_{AB}+\delta_{AB})^2+(l_{BC}+\delta_{BC})^2}-\sqrt{l_{AB}{}^2+l_{BC}{}^2}$，得

$$\delta_{AC}=8.9\times10^{-3}\ \text{mm}$$

把叠加原理用于图 17.13 所示应力状态，如果应力不超过比例极限，得一般应力状态下的广义胡克定律

$$\begin{aligned}&\varepsilon_x=\frac{1}{E}[\sigma_x-\mu(\sigma_y+\sigma_z)],\ \varepsilon_y=\frac{1}{E}[\sigma_y-\mu(\sigma_z+\sigma_x)],\ \varepsilon_z=\frac{1}{E}[\sigma_z-\mu(\sigma_x+\sigma_y)]\\&\gamma_{xy}=\frac{\tau_{xy}}{G},\ \gamma_{yz}=\frac{\tau_{yz}}{G},\ \gamma_{zx}=\frac{\tau_{zx}}{G}\end{aligned}\tag{17.23}$$

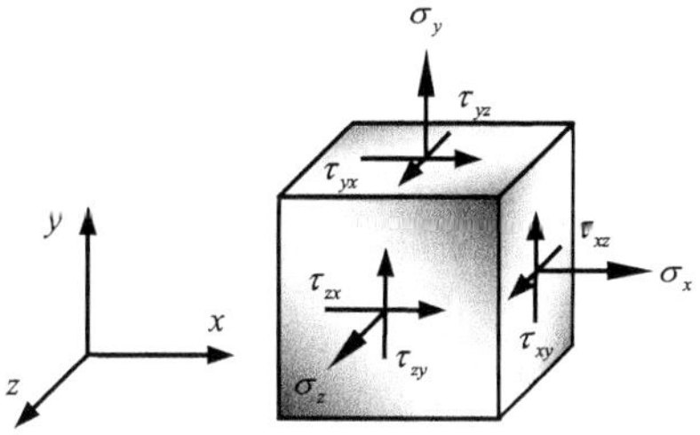

图 17.13

17.6 强 度 理 论

构件通常被设计为在预期加载条件下材料不能发生失效。对于塑性材料失效通常是指屈服，而对于脆性材料失效是指断裂。

对塑性材料只要满足 $\sigma<\sigma_s$（σ_s 为屈服强度）或对脆性材料只要满足 $\sigma<\sigma_u$（σ_u 为极限强度），则单向应力状态下的构件满足安全要求，如图 17.14 所示。

当构件处于平面应力状态或二向应力状态，不可能通过试验预测构件是否失效，如图 17.15 所示。因此对于复杂应力状态，需要建立与材料失效有关的理论。这类理论称为强度理论或失效理论。

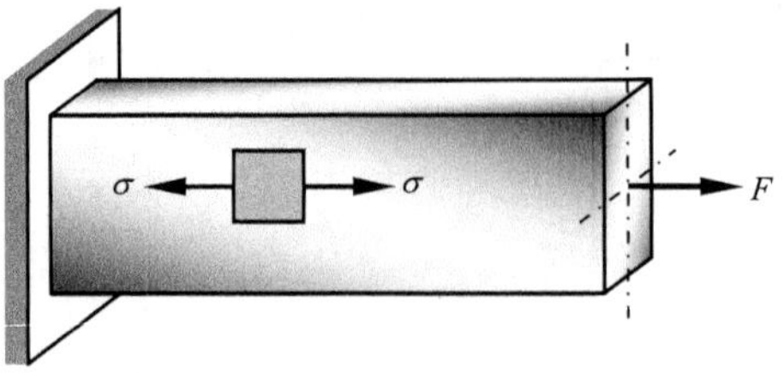

图 17.14

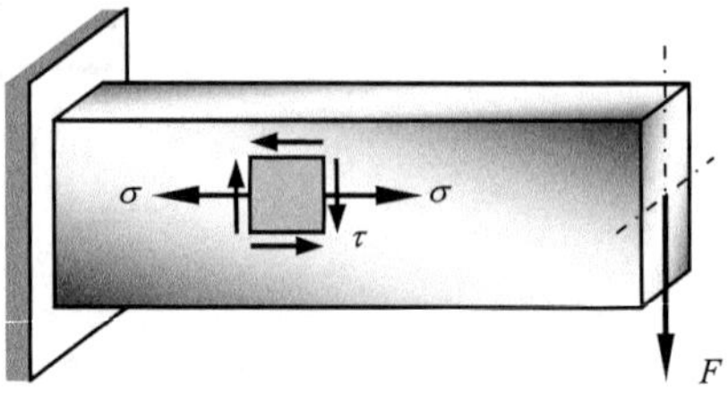

图 17.15

存在两类强度理论：一类适用于脆性断裂，另一类适用于塑性屈服。

1. 最大正应力理论

该理论认为，当材料处于二向或三向应力状态时，只要最大正应力达到相同材料单向拉伸失效正应力值，材料即发生失效。该理论是基于事实：脆性断裂是由最大正应力引起。

假设三个主应力由 σ_1、σ_2 和 σ_3 表示，并按照 $\sigma_1 \geqslant \sigma_2 \geqslant \sigma_3$ 进行排序，则失效准则可表示为

$$\sigma_{\mathrm{eq1}} = \sigma_1 = \sigma_{\mathrm{u}} \tag{17.24}$$

式中，σ_{eq1} 为对应最大正应力理论的相当应力，σ_{u} 为材料单向拉伸的极限强度。

假设安全因数为 n，则二向或三向应力状态下的强度条件可表示为

$$\sigma_{\mathrm{eq1}} = \sigma_1 \leqslant \sigma_{\mathrm{allow}} = \frac{\sigma_{\mathrm{u}}}{n} \tag{17.25}$$

式中，σ_{allow} 为许用应力。最大正应力理论与脆性断裂实验结果吻合。

例 17.5　图 E17.5 所示平面应力状态出现在由铸铁制成的构件中，已知铸铁的拉伸极限强度为 σ_{ut}=160 MPa，压缩极限强度为 σ_{uc}=320 MPa。采用最大正应力理论，判断断裂是否发生。如果断裂不发生，求安全因数。

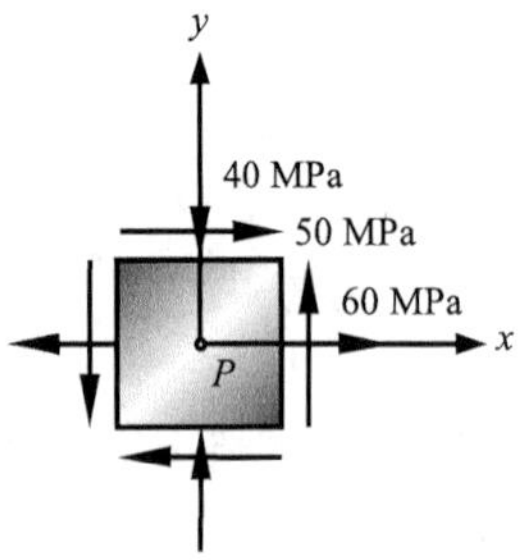

图 E17.5

解　参考例 17.2，有

$$\sigma_1 = 80.71\ \mathrm{MPa},\ \sigma_2 = 0,\ \ \sigma_3 = -60.71\ \mathrm{MPa}$$

因 $\sigma_{eq1}=\sigma_1=80.71\ \text{MPa}<\sigma_{ut}=160\ \text{MPa}$，则断裂不发生，相应的安全因数为

$$n=\frac{\sigma_{ut}}{\sigma_{eq1}}=1.98$$

2. 最大线应变理论

该理论认为，当材料处于二向或三向应力状态时，只要最大线应变达到相同材料单向拉伸失效拉应变值，材料即发生失效。该理论是基于事实：脆性断裂是由最大线应变引起。

失效准则可表示为

$$\varepsilon_1=\frac{1}{E}[\sigma_1-\mu(\sigma_2-\sigma_3)]=\frac{1}{E}\sigma_b \tag{17.26}$$

式中，ε_1 为最大线应变，μ 为泊松比。该准则也可表示为

$$\sigma_{eq2}=\sigma_1-\mu(\sigma_2-\sigma_3)=\sigma_b \tag{17.27}$$

式中，σ_{eq2} 为对应最大线应变理论的相当应力。因此，二向或三向应力状态下的强度条件可表示为

$$\sigma_{eq2}=\sigma_1-\mu(\sigma_2-\sigma_3)\leqslant\sigma_{allow}=\frac{\sigma_b}{n} \tag{17.28}$$

式中，σ_{allow} 为许用应力，n 为安全因数。目前，最大线应变理论已经很少使用。

3. 最大剪应力理论

该理论认为，当材料处于二向或三向应力状态时，只要最大剪应力达到相同材料单向拉伸失效剪应力值，材料即发生失效。该理论是基于事实：塑性屈服是由最大剪应力引起。

失效准则可表示为

$$\tau_{max}=\frac{1}{2}(\sigma_1-\sigma_3)=\frac{1}{2}\sigma_s \tag{17.29}$$

式中，τ_{max} 为最大剪应力，σ_s 为单向拉伸材料的屈服强度。该准则也可表示为

$$\sigma_{eq3}=\sigma_1-\sigma_3=\sigma_s \tag{17.30}$$

式中，σ_{eq3} 为对应最大剪应力理论的相当应力。因此，二向或三向应力状态下的强度条件可表示为

$$\sigma_{eq3}=\sigma_1-\sigma_3\leqslant\sigma_{allow}=\frac{\sigma_s}{n} \tag{17.31}$$

式中，σ_{allow} 为许用应力，n 为安全因数。最大剪应力理论广泛应用于塑性屈服。

例 17.6　图 E17.6 所示平面应力状态出现在由钢制成的构件中，已知屈服强度 $\sigma_s=310\ \text{MPa}$。采用最大剪应力理论，判断屈服是否发生。如果屈服不发生，求安全因数。

解　参考例 17.2，有

$$\sigma_1=80.71\ \text{MPa},\ \sigma_2=0,\ \sigma_3=-60.71\ \text{MPa}$$

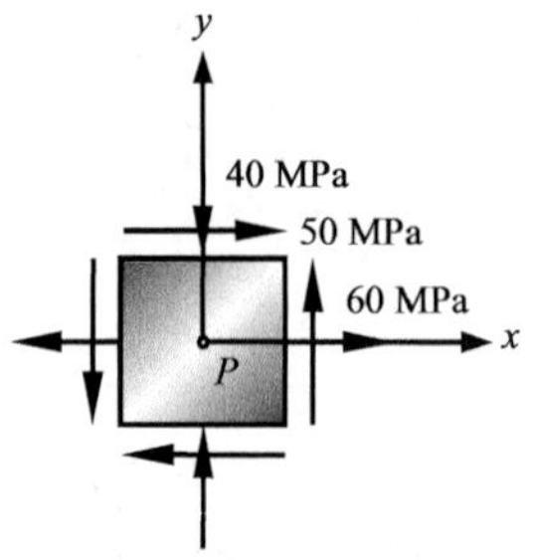

图 E17.6

因 $\sigma_{\text{eq3}} = \sigma_1 - \sigma_3 = 141.42\ \text{MPa} < \sigma_{\text{s}} = 310\ \text{MPa}$，则屈服不发生，相应的安全因数为

$$n = \frac{\sigma_{\text{s}}}{\sigma_{\text{eq3}}} = 2.19$$

4. 最大畸变能理论

该理论认为，当材料处于二向或三向应力状态时，只要畸变能密度(即单位体积畸变能)等于或超过相同材料单向拉伸屈服畸变能密度，材料即发生失效。该理论是基于事实：塑性屈服是由与材料形状改变有关的最大畸变能引起。失效准则可表示为

$$u_{\text{d}} = \frac{(1+\mu)}{6E}[(\sigma_1 - \sigma_2)^2 + (\sigma_2 - \sigma_3)^2 + (\sigma_3 - \sigma_1)^2] = \frac{(1+\mu)}{6E} 2\sigma_{\text{s}}{}^2 \tag{17.32}$$

式中，u_{d} 为对应二向或三向应力状态的畸变能密度，E 为弹性模量。上述准则也可表示为

$$\sigma_{\text{eq4}} = \sqrt{\frac{1}{2}[(\sigma_1 - \sigma_2)^2 + (\sigma_2 - \sigma_3)^2 + (\sigma_3 - \sigma_1)^2]} = \sigma_{\text{s}} \tag{17.33}$$

式中，σ_{eq4} 为对应最大畸变能理论的相当应力。因此，二向或三向应力状态下的强度条件可表示为

$$\sigma_{\text{eq4}} = \sqrt{\frac{1}{2}[(\sigma_1 - \sigma_2)^2 + (\sigma_2 - \sigma_3)^2 + (\sigma_3 - \sigma_1)^2]} \leqslant \sigma_{\text{allow}} = \frac{\sigma_{\text{s}}}{n} \tag{17.34}$$

式中，σ_{allow} 为许用应力，n 为安全因数。最大畸变能理论与塑性屈服非常吻合。

例 17.7 图示平面应力状态出现在由钢制成的构件中，已知屈服强度 σ_{s}=310 MPa。采用最大畸变能理论，判断屈服是否发生。如果屈服不发生，求安全因数。

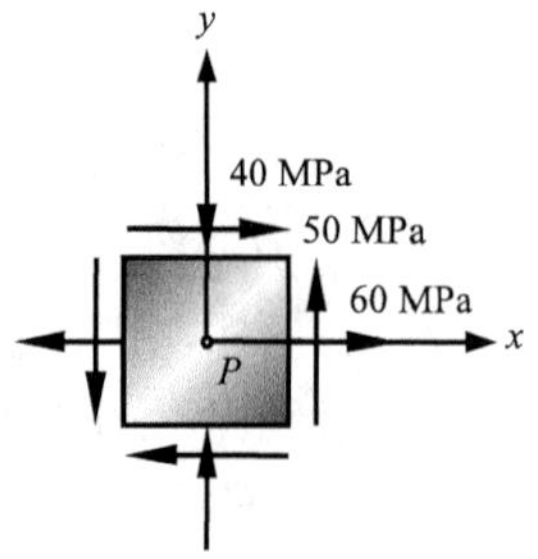

图 E17.7

解　参考例 17.2，有

$$\sigma_1 = 80.71\ \text{MPa},\ \sigma_2 = 0,\ \sigma_3 = -60.71\ \text{MPa}$$

因 $\sigma_{eq4} = \sqrt{\frac{1}{2}[(\sigma_1-\sigma_2)^2+(\sigma_2-\sigma_3)^2+(\sigma_3-\sigma_1)^2]} = 122.88\ \text{MPa} < \sigma_s = 310\ \text{MPa}$，则屈服不发生，相应的安全因数为

$$n = \frac{\sigma_s}{\sigma_{eq4}} = 2.52$$

习　　题

17.1　图示应力状态，求单元体逆时针旋转 30°后的正应力和剪应力。

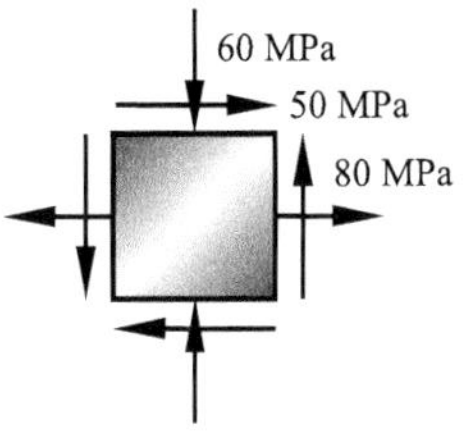

图 P17.1

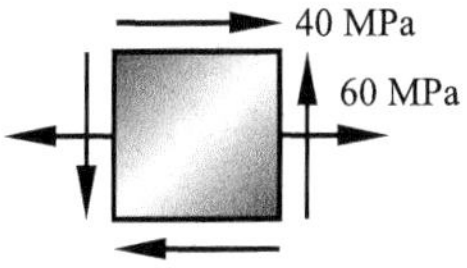

图 P17.2

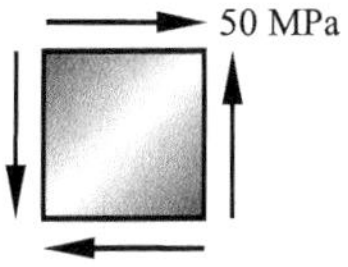

图 P17.3

17.2　图示应力状态，求单元体顺时针旋转 15°后的正应力和剪应力。

17.3　图示应力状态，求单元体顺时针旋转 55°后的正应力和剪应力。

17.4　图示应力状态，求单元体逆时针旋转 45°后的正应力和剪应力。

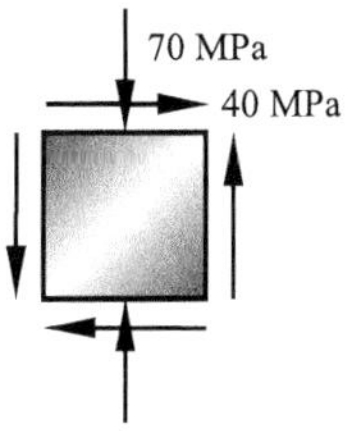

图 P17.4

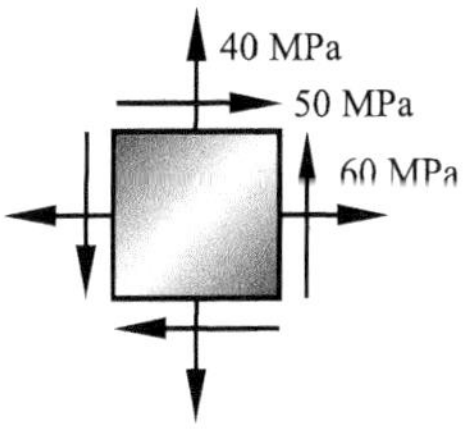

图 P17.5

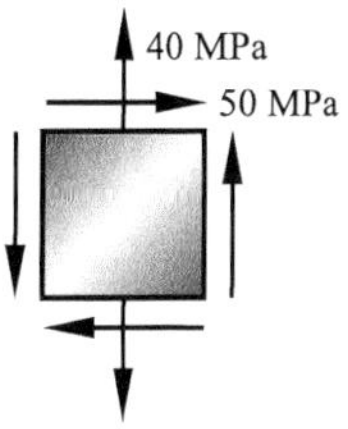

图 P17.6

17.5-17.8　图示应力状态，求主应力和相应的主平面。

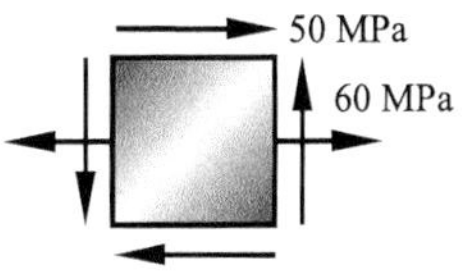

图 P17.7

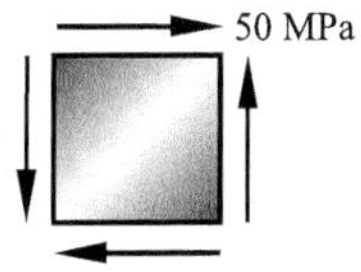

图 P17.8

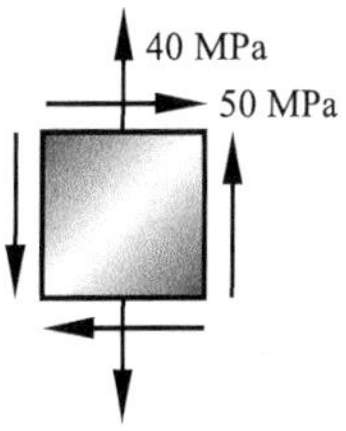

图 P17.9

17.9　图示应力状态，求：(a)最大面内剪应力；(b)最大剪应力面的方位；(c)最大剪应力面上的正应力。

17.10　边长为 40 mm 的方板受图示平面应力作用。已知 E=200 GPa 和μ=0.30，求边 AB、边 BC 和对角线 AC 的长度改变。

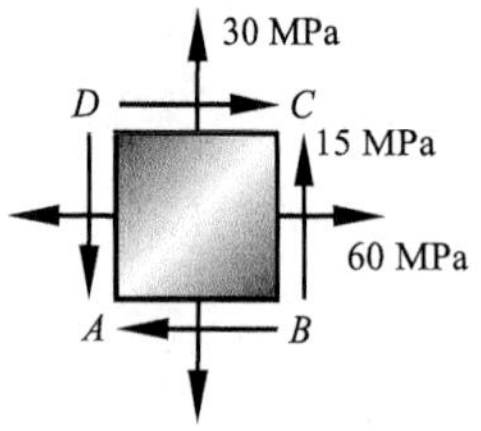

图 P17.10

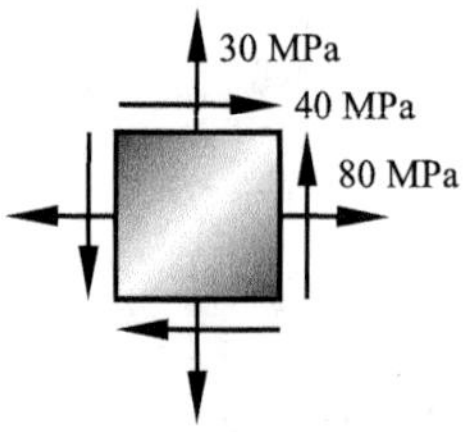

图 P17.11

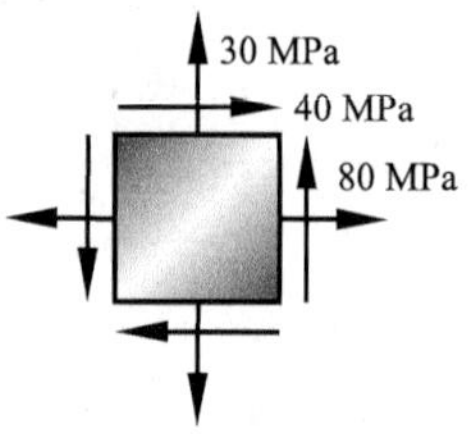

图 P17.12

17.11　图示平面应力状态出现在由钢制成的构件中，已知屈服强度σ_s=310 MPa。采用最大剪应力理论，判断屈服是否发生。如果屈服不发生，求安全因数。

17.12　图示平面应力状态出现在由钢制成的构件中，已知屈服强度σ_s=310 MPa。采用最大畸变能理论，判断屈服是否发生。如果屈服不发生，求安全因数。

第18章 组合载荷

前面各章讨论了受单一内力作用的构件，例如分析了拉压杆、扭转轴和弯曲梁，同时阐明了每一内力作用下的应力和应变求解方法。然而，在许多结构中，构件常常受两种或两种以上内力，这些同时作用于构件的内力称为组合载荷或组合加载。

组合载荷作用下构件的应力分布可以通过叠加原理进行确定。首先求解每个内力单独作用引起的应力分布，然后将这些分布应力进行叠加即可得到合成分布应力。只要在应力和载荷之间存在线性关系，那么叠加原理都能使用。另外，在使用叠加原理时，构件受载时在几何方面仅发生小变形，以保证每个载荷所产生的应力之间相互独立。常用结构都满足上述条件，因此叠加原理广泛应用于受组合载荷作用的工程结构的应力应变分析。

18.1 偏心拉压

如果外载荷作用线通过截面形心，则轴向加载杆横截面上的应力可以假设均匀分布。这种加载方式称为中心加载。

现在考虑外载荷作用线不通过截面形心时的应力分布。当外载荷作用线不通过截面形心时，称为偏心加载，如图18.1所示。

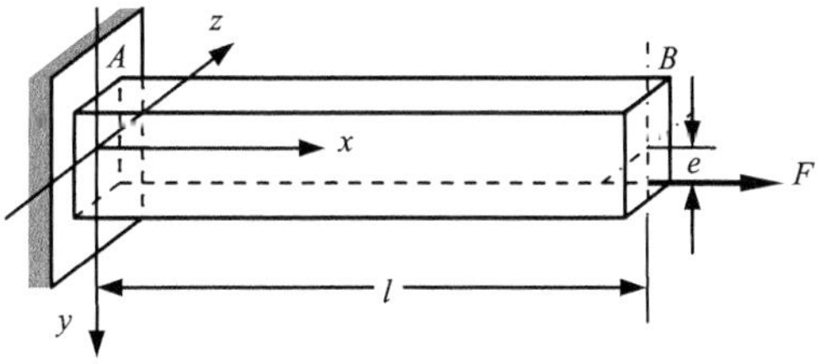

图18.1

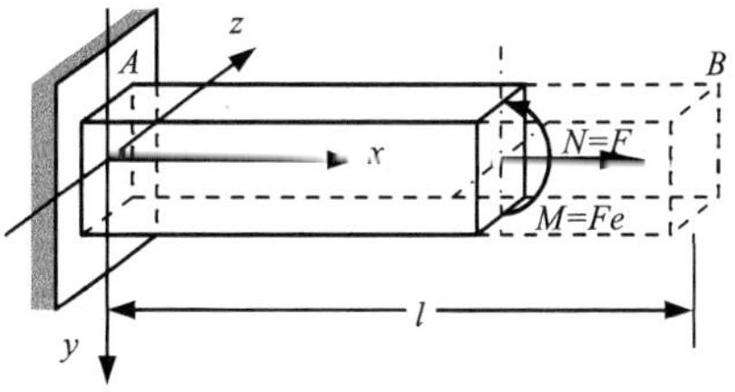

图18.2

假设外载荷作用于杆件对称面，设偏心距为 e，那么作用于任何横截面上的内力包括通过截面形心的轴力 $N = F$ 和杆件对称面内的弯矩 $M = Fe$，如图18.2所示。

如果材料满足线弹性，并发生小变形，那么由于偏心加载引起的应力分布可以通过叠加对应于轴力 N 的均匀应力分布 σ_N 和对应于弯矩 M 的线性应力分布 σ_M 获得

$$\sigma = \sigma_N + \sigma_M = \frac{N}{A} + \frac{My}{I} \tag{18.1}$$

式中，A 为横截面面积，I 为横截面的形心惯性矩，y 为横截面上点到形心轴的距离。

根据横截面几何和外载荷偏心，合成应力可以为正和负，或有相同符号。根据式(18.1)，截面上存在一根线，满足 $\sigma = 0$。这根线表示截面的中性轴。需要注意的是中

性轴与形心轴不再重合。

合成应力在上下表面达到最大，最大值可表示为

$$\sigma_{\max} = \frac{N}{A} + \frac{My_{\max}}{I} = \frac{N}{A} + \frac{M}{S} \tag{18.2}$$

式中，$y_{\max}$ 为上下表面点到形心轴的距离，$S = I/y_{\max}$ 为相对形心轴的截面模量。

上述分析表明，截面上下表面点具有最大应力。假设许用应力为 σ_{allow}，则偏心拉压杆的强度条件为

$$\sigma_{\max} \leqslant \sigma_{\text{allow}} \tag{18.3}$$

例 18.1　F=50 kN 的两个外载荷作用于 20a 工字钢的自由端。已知 e=80 mm，求梁下表面的应力。

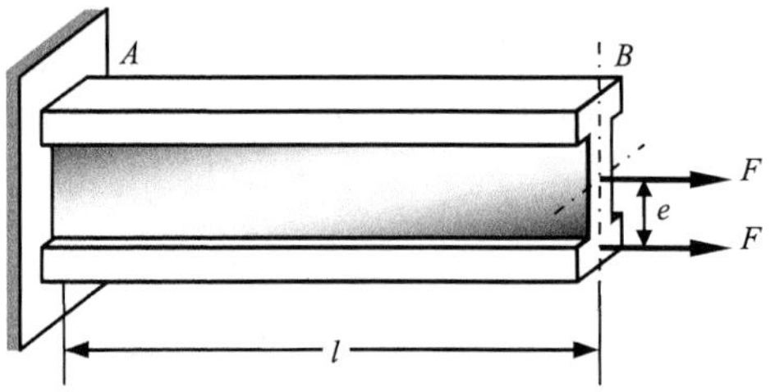

图 E18.1

解　梁截面上的轴力和弯矩分别表示为

$$N = 2F = 100\ \text{kN},\quad M = Fe = 4\ \text{kN}\cdot\text{m}$$

查阅附录 IV 型钢表，20a 工字钢的横截面面积和截面模量分别为

$$A = 35.56\ \text{cm}^2,\quad S = 237\ \text{cm}^3$$

利用式(18.2)，得梁下表面的应力

$$\sigma = \sigma_N + \sigma_M = \frac{N}{A} + \frac{M}{S} = 45.00\ \text{MPa}$$

18.2　工字梁横力弯曲

根据梁截面形状和弯矩取最大值处临界截面(危险截面)上的剪力值，最大应力可能不在截面的上下表面，而是在截面上其他点。工字钢腹板和翼缘交界处有较大正应力和较大剪应力，当这两个应力同时作用时，则在交界处产生的应力往往大于工字梁表面的应力。

考虑图 18.3 所示受横向载荷作用的工字梁，整个梁上剪力保持不变，而弯矩在固定端达到最大，因此危险截面位于固定端，并且危险截面上的内力可表示为

$$V = F,\ M_{\max} = Fl \tag{18.4}$$

在危险截面上，剪力引起的剪应力在工字梁中性轴上达到最大值，并且最大剪应力可表示为

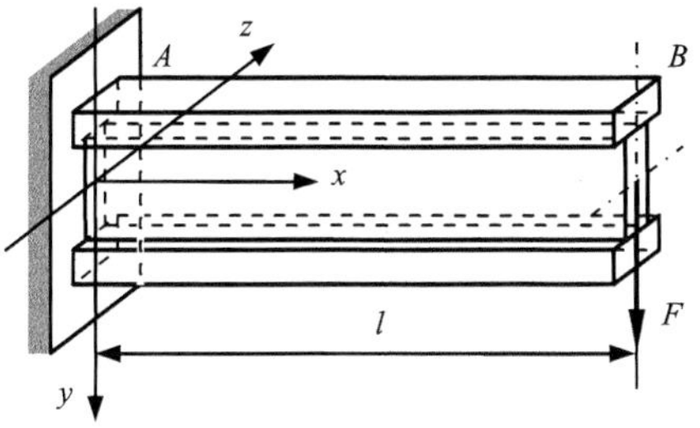

图 18.3

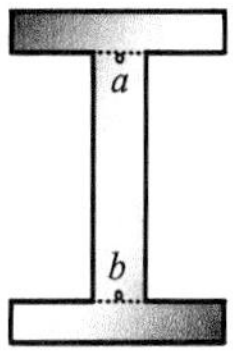

图 18.4

$$\tau_{\max} = \frac{VQ'_{\max}}{Ib} = \frac{V}{A_w} \tag{18.5}$$

式中，I 为工字型截面对中性轴的惯性矩，$Q'_{\max}$ 为中性轴上方(或下方)截面面积对中性轴的静矩，b 为腹板宽度，A_w 为腹板横截面面积。

在危险截面上，弯矩引起的正应力在上下表面达到最大，最大应力可表示为

$$\sigma_{\max} = \frac{M_{\max} y_{\max}}{I} = \frac{M_{\max}}{S} \tag{18.6}$$

式中，$y_{\max}$ 为上下表面点到中性轴的距离，$S = I/y_{\max}$ 为对中性轴的截面模量。

在危险截面的任何其他点，材料同时受正应力和剪应力作用，其值分别为

$$\sigma = \frac{M_{\max} y}{I} \tag{18.7}$$

$$\tau = \frac{VQ'}{Ib} \tag{18.8}$$

式中，y 为到中性轴的距离，Q' 为需要计算应力的点的上侧或下侧的截面对中性轴的静矩，b 为需要计算应力的点所在位置的截面宽度。

通过应力变换，可得到危险截面上任一点的主应力。对工字型截面梁，较大的剪应力和较大的主应力都出现在腹板和翼缘交界处，因此在交界处主应力会大于上下表面处的应力和中性轴上的应力。在工字梁设计中，我们特别要注意这种可能性，因此应该计算腹板和翼缘交界处 a 点和 b 点的主应力，如图 18.4 所示。

利用应力变换公式，危险截面上 a 点或 b 点的主应力可表示为

$$\sigma_{1,3} = \frac{\sigma}{2} \pm \sqrt{\left(\frac{\sigma}{2}\right)^2 + \tau^2} \tag{18.9}$$

对塑性材料，应该采用最大剪应力理论或最大畸变能理论进行工字梁设计。

如果采用最大剪应力理论，则 a 点或 b 点的相当应力可写为

$$\sigma_{eq3} = \sigma_1 - \sigma_3 = \sqrt{\sigma^2 + 4\tau^2} \tag{18.10}$$

式中，σ_{eq3} 为对应最大剪应力理论的相当应力。假设材料许用应力为 σ_{allow}，则横力弯曲工字梁上交接点处的最大剪应力强度条件可表示为

$$\sigma_{eq3} \leqslant \sigma_{allow} \tag{18.11}$$

如果采用最大畸变能理论，则横力弯曲工字梁上交接点处的强度条件可表示为

$$\sigma_{eq4}=\sqrt{\frac{1}{2}[(\sigma_1-\sigma_2)^2+(\sigma_2-\sigma_3)^2+(\sigma_3-\sigma_1)^2]}=\sqrt{\sigma^2+3\tau^2}\leqslant\sigma_{allow} \qquad (18.12)$$

式中，σ_{eq4} 为对应最大畸变能理论的相当应力。

例 18.2 F=150 kN 的外载荷施加到工字型钢梁的自由端，钢的屈服应力为σ_s = 235 MPa。利用最大剪应力理论，确定屈服是否发生。如果屈服不发生，求安全因数。忽略倒角效应。

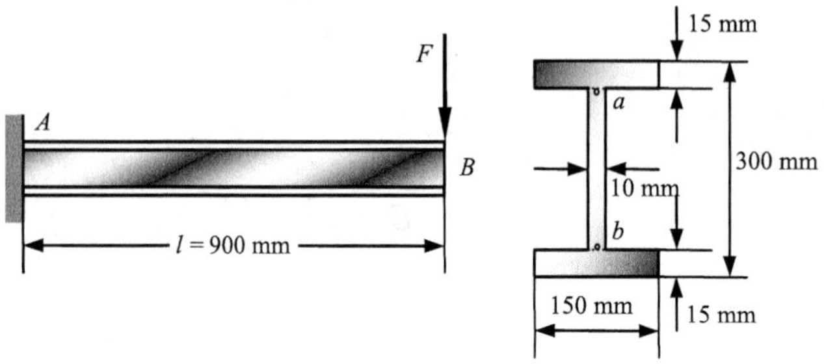

图 E18.2

解 危险截面位于固定端，弯矩和剪力分别为

$$M_{max}=Fl=135\ \text{kN}\cdot\text{m}\ ,\ \ V=F=150\ \text{kN}$$

截面惯性矩等于

$$I=\frac{1}{12}(150)(300)^3-\frac{1}{12}(140)(270)^3=1.07865\times10^8\ \text{mm}^4$$

固定端截面上最大正应力和剪应力为

$$\sigma_{max}=\frac{M_{max}y_{max}}{I}=187.73\ \text{MPa}\ ,\ \ \tau_{max}=\frac{V}{A_w}=55.56\ \text{MPa}$$

固定端截面上 a 点(或 b 点)的正应力和剪应力为

$$\sigma_a=\frac{y_a}{y_{max}}\sigma_{max}=168.96\ \text{MPa}\ ,\ \ \tau_a\approx\tau_{max}=\frac{V}{A_w}=55.56\ \text{MPa}$$

对应于最大剪应力理论的相当应力为

$$\sigma_{eq3}=\sqrt{\sigma_a{}^2+4\tau_a{}^2}=202.23\ \text{MPa}$$

由 $\sigma_{eq3}>\sigma_{max}$，得危险点位于 a 点(或 b 点)。因 $\sigma_{eq3}<\sigma_s=235\ \text{MPa}$，则不会发生屈服，相应的安全因数等于

$$n=\frac{\sigma_s}{\sigma_{eq3}}=1.16$$

18.3 拉压与弯曲

考虑如图 18.5 所示矩形截面梁，只要材料保持线弹性和发生小变形，则可通过叠加原理求组合载荷作用下梁的合成应力分布。

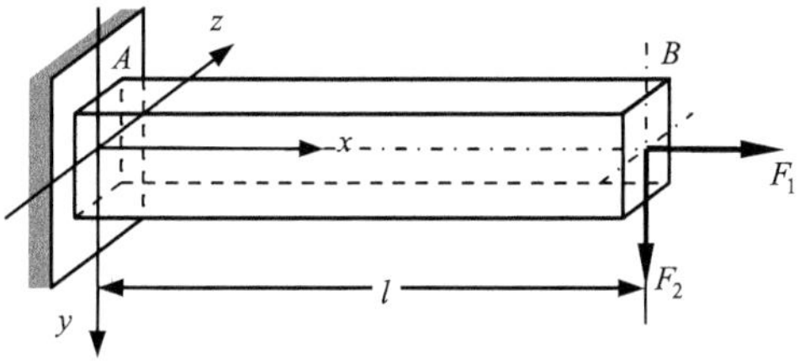

图 18.5

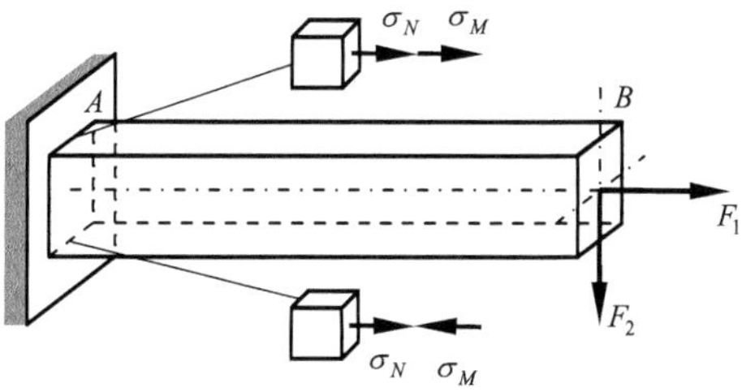

图 18.6

危险截面位于固定端，危险截面上的内力为

$$V = F_2,\ M_{\max} = F_2 l,\ N = F_1 \tag{18.13}$$

危险截面上由剪力引起的剪应力在矩形截面形心轴上将达到最大值，最大剪应力为

$$\tau_{\max} = \frac{VQ'_{\max}}{Ib} = \frac{3V}{2A} \tag{18.14}$$

式中，I 为截面对形心轴的惯性矩，$Q'_{\max}$ 为形心轴上方截面对形心轴的静矩，b 为矩形截面梁的宽度，A 为矩形截面梁的横截面面积。剪力引起的应力与轴力或弯矩引起的应力相比非常小，因而可以忽略。

危险截面上由弯矩引起的正应力在上下表面达到最大，最大值可表示为

$$\sigma_M = \frac{M_{\max} y_{\max}}{I} = \frac{M_{\max}}{S} \tag{18.15}$$

式中，$y_{\max}$ 为上下表面点到形心轴的距离，$S = I/y_{\max}$ 为对形心轴的截面模量。

危险截面上由轴力引起的正应力均匀分布，可表示为

$$\sigma_N = \frac{N}{A} \tag{18.16}$$

式中，A 为矩形截面梁的横截面面积。

综上所述，危险截面的上下表面点将有最大或最小应力，如图 18.6 所示。最大或最小值为

$$\sigma_{\max} = \sigma_N + \sigma_M = \frac{N}{A} + \frac{M_{\max}}{S} \tag{18.17}$$

如果许用应力为σ_{allow}，则强度条件可表示为

$$\sigma_{\max} \leqslant \sigma_{\text{allow}} \tag{18.18}$$

例 18.3 F_1=100 kN 和 F_2=4 kN 的两个外载荷作用于 20a 工字钢梁的自由端。已知 l=1 m，求梁中最大拉应力。

解 最大拉应力出现在梁的固定端上表面。固定端的轴力和弯矩为

$$N = F_1 = 100\ \text{kN},\ M_{\max} = F_2 l = 4\ \text{kN}\cdot\text{m}$$

对于 20a 工字钢梁，有

$$A = 35.56\ \text{cm}^2, S = 237\ \text{cm}^3$$

利用式(18.17)，得

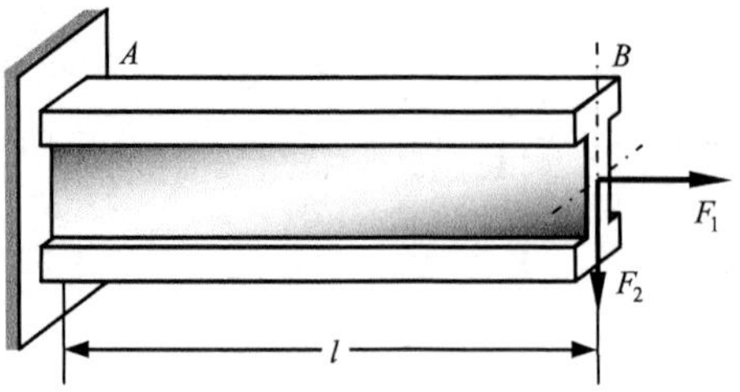

图 E18.3

$$\sigma_{\max}^{t} = \sigma_N + \sigma_M = \frac{N}{A} + \frac{M_{\max}}{S} = 45.00 \text{ MPa}$$

18.4　扭转与弯曲

考虑图 18.7 所示圆轴，设材料保持线弹性和发生小变形，那么可以通过叠加原理求组合载荷作用轴的合成应力。

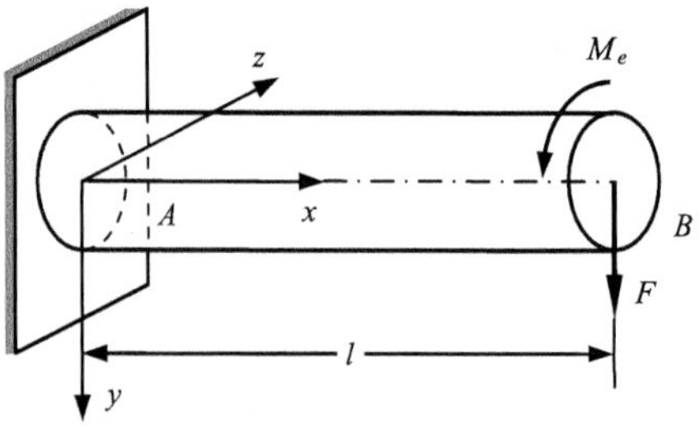

图 18.7

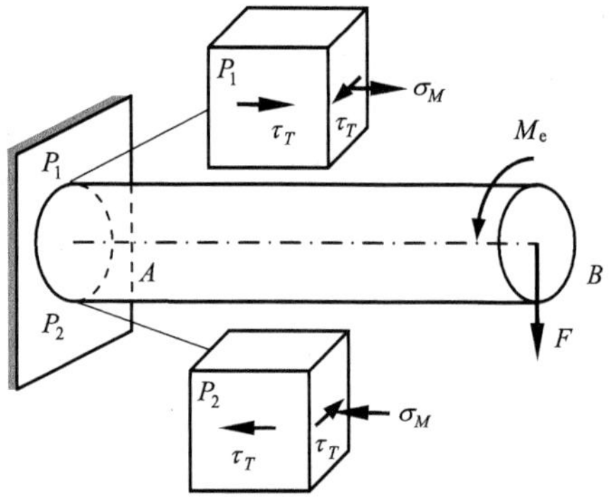

图 18.8

危险截面 A 处内力为

$$V = F, \ M_{\max} = Fl, \ T = M_e \tag{18.19}$$

危险截面上由剪力引起的剪应力在圆截面中性轴上有最大值

$$\tau_V = \frac{VQ'_{\max}}{Ib} = \frac{4V}{3A} \tag{18.20}$$

式中，I 为截面对中性轴的惯性矩，$Q'_{\max}$ 为中性轴上方截面对中性轴的静矩，b 为圆轴直径，A 为圆轴横截面面积。剪力引起的应力可以忽略不计。因此，仅需要考虑由扭矩和弯矩引起的应力。

危险截面上由弯矩引起的正应力在上下表面有最大值

$$\sigma_M = \frac{M_{\max} y_{\max}}{I} = \frac{M_{\max}}{S} \tag{18.21}$$

式中，$y_{\max}$ 为上下表面点到中性轴的距离，$S = I/y_{\max}$ 为截面模量。

危险截面上由扭矩引起的剪应力在圆轴表面达到最大值

$$\tau_T = \frac{T_{\max}\rho_{\max}}{I_p} = \frac{T_{\max}}{S_p} \tag{18.22}$$

式中，I_p 为横截面对形心的极惯性矩，$\rho_{\max}$ 为横截面半径，$S_p = I_p/\rho_{\max}$ 为极截面模量。

显然，危险点处于平面应力状态，位于危险截面上下表面，如图 18.8 所示。

危险截面上下表面点的主应力可表示为

$$\sigma_{1,3} = \frac{\sigma_M}{2} \pm \sqrt{(\frac{\sigma_M}{2})^2 + {\tau_T}^2} \tag{18.23}$$

对塑性材料，如果采用最大剪应力理论，则相当应力为

$$\sigma_{eq3} = \sigma_1 - \sigma_3 = \sqrt{{\sigma_M}^2 + 4{\tau_T}^2} \tag{18.24}$$

式中，σ_{eq3} 为对应最大剪应力理论的相当应力。

利用式(18.21)和式(18.22)，以及 $S_p = 2S$，得

$$\sigma_{eq3} = \frac{\sqrt{{M_{\max}}^2 + {T_{\max}}^2}}{S} \tag{18.25}$$

式中，对直径为 d 的实心圆轴，$S = \frac{1}{32}\pi d^3$；对外径为 D，内径为 d 的空心圆轴，$S = \frac{1}{32}\pi D^3(1-\alpha^4)$，其中 $\alpha = d/D$。

假设许用应力为 σ_{allow}，则最大剪应力强度条件可表示为

$$\sigma_{eq3} \leqslant \sigma_{allow} \tag{18.26}$$

对塑性材料，如果采用最大畸变能理论，则强度条件为

$$\sigma_{eq4} = \sqrt{{\sigma_M}^2 + 3{\tau_T}^2} = \frac{\sqrt{{M_{\max}}^2 + \frac{3}{4}{T_{\max}}^2}}{S} \leqslant \sigma_{allow} \tag{18.27}$$

式中，σ_{eq4} 为对应最大畸变能理论的相当应力。

例 18.4　图 E18.4 所示钢管，屈服强度 σ_s=325 MPa，长度 l=200 mm，外径 D=80 mm，壁厚 t=5 mm，受载荷 F=6 kN 和 M_e=1.5 kN·m 作用。采用最大剪应力理论，判断屈服是否发生。如果不发生屈服，求安全因数。

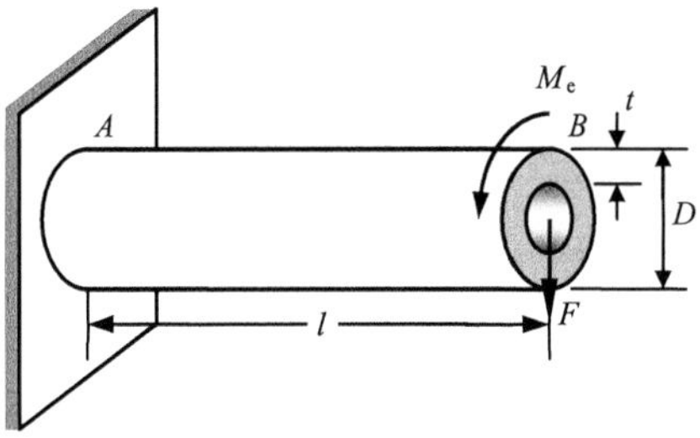

图 E18.4

解 危险点位于固定端上下表面点。固定端的弯矩和扭矩为

$$M_{\max} = Fl = 1.2\ \text{kN}\cdot\text{m}\ ,\ \ T = M_{\text{e}} = 1.5\ \text{kN}\cdot\text{m}$$

利用式(18.26)，得

$$\sigma_{\text{eq3}} = \frac{\sqrt{{M_{\max}}^2 + T^2}}{S} = \frac{\sqrt{{M_{\max}}^2 + T^2}}{\dfrac{1}{32}\pi D^3\{1-[(D-2t)/D]^4\}} = 92.35\ \text{MPa}$$

因 $\sigma_{\text{eq3}} < \sigma_{\text{s}}$，则不会发生屈服，相应的安全因数为

$$n = \frac{\sigma_{\text{s}}}{\sigma_{\text{eq3}}} = 3.52$$

习　　题

18.1　F=80 kN 的外载荷作用于 22a 工字钢梁的自由端。已知偏心距 e=80 mm，求梁上下表面的应力。

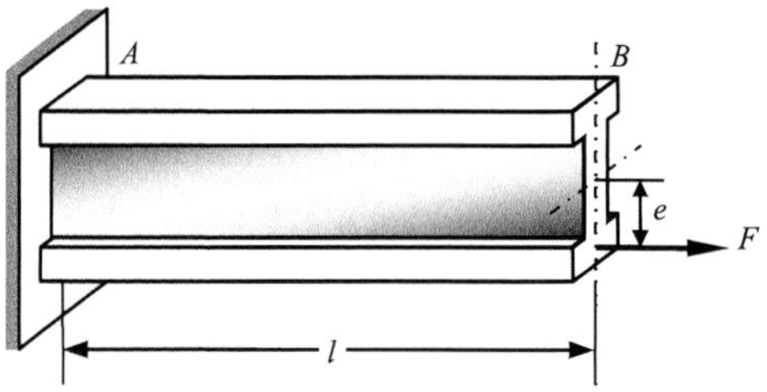

图 P18.1

18.2　F=80 kN 的外载荷作用于 22a 工字钢梁的自由端。已知偏心距 e=80 mm，求梁中最大拉、压应力。

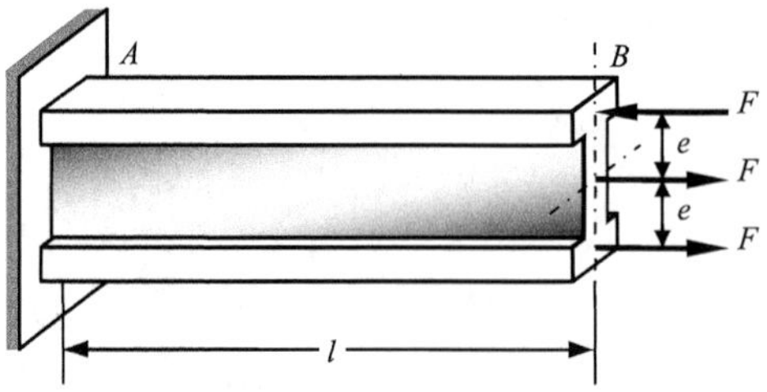

图 P18.2

18.3　F=180 kN 的外载荷作用于宽缘工字钢梁的自由端。已知屈服应力 σ_{s} = 235 MPa，采用最大畸变能理论判断屈服是否发生。如果不发生屈服，求安全因数。

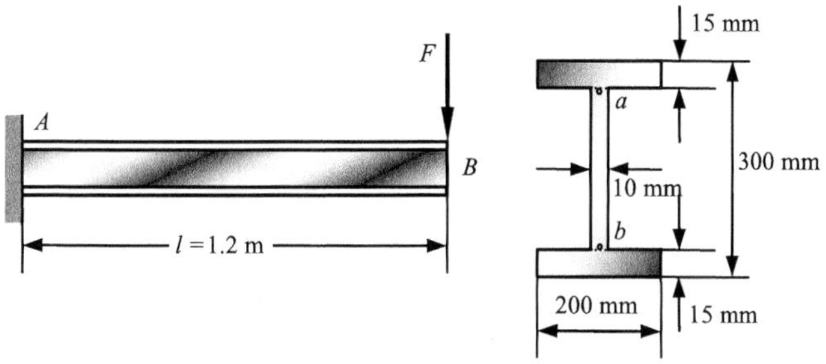

图 P18.3

18.4　外载荷 F_1=100 kN 和 F_2=20 kN 作用于 36a 工字钢梁的自由端。已知屈服应力σ_s=235 MPa，利用最大剪应力理论，确定屈服是否发生。如果屈服不发生，求安全因数。

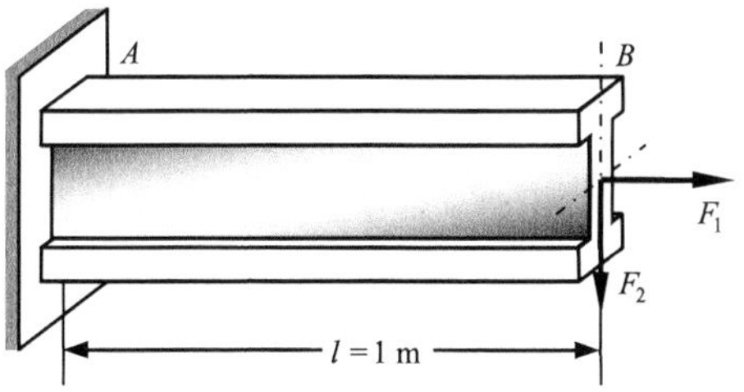

图 P18.4

18.5　图示钢管，屈服强度σ_s=325 MPa，长度 l=200 mm，外径 D=80 mm，壁厚 t=5 mm，承受载荷 F=6 kN 和 M_e=1.5 kN·m 作用。采用最大畸变能理论，判断屈服是否发生。如果屈服不发生，求安全因数。

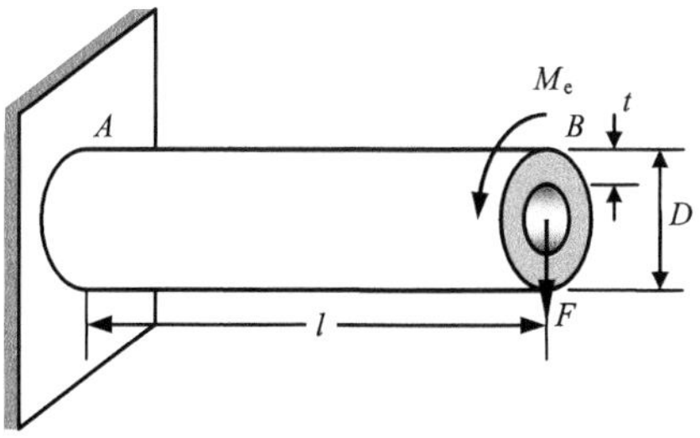

图 P18.5

18.6　悬臂工字钢梁承受图示外载荷。已知 F=200 kN 和σ_{allow}=180 MPa，求：(a)梁中最大正应力；(b)腹板和翼缘交界处最大主应力；(c)图示工字钢梁是否合适。

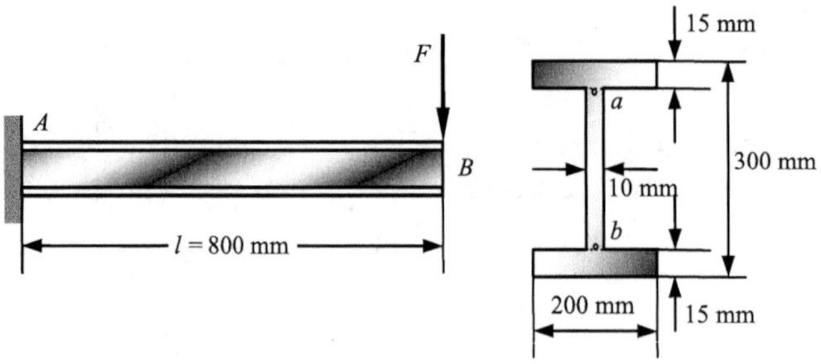

图 P18.6

第 19 章　压 杆 稳 定

承受轴向压缩的细长构件称为压杆。压杆失效是通过失稳(即突然横向挠曲)而发生。压杆失稳甚至在最大应力远小于屈服极限和强度极限时也能发生。

压杆失稳常常导致结构突然失效，因此需要特别关注压杆设计问题以便所设计的压杆能够安全支撑给定载荷而不发生失稳现象。

19.1　两端铰支细长压杆临界载荷

压杆失稳前所能承受的最大轴向压力称为临界载荷，用 F_{cr} 表示。图 19.1a 所示压杆两端铰支受轴向载荷作用。压杆在受载后会发生失稳，即发生突然横向挠曲，如图 19.1b 所示。

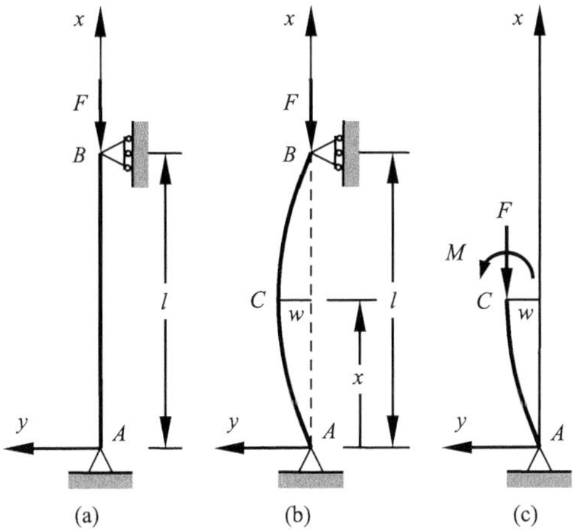

图 19.1

考虑如图 19.1c 所示 AC 段平衡，则 C 截面上的弯矩可表示为 $M=-Fw$。利用 $w''=M/EI$，得

$$w''+\frac{F}{EI}w=0 \tag{19.1}$$

式(19.1)是常系数线性二阶齐次微分方程。设

$$k^2=\frac{F}{EI} \tag{19.2}$$

则

$$w''+k^2w=0 \tag{19.3}$$

上述方程的通解可表示为

$$w = A\sin kx + B\cos kx \tag{19.4}$$

利用 $x=0,\ w=0$，得 $B=0$；再利用 $x=l,\ w=0$，得

$$A\sin kl = 0 \tag{19.5}$$

上述方程满足的条件是 $A=0$ 或 $\sin kx=0$。如果 $A=0$，得到 $w\equiv 0$，即压杆始终是直杆。如果 $\sin kx=0$，得 $kl=n\pi$，即

$$F = \frac{n^2\pi^2 EI}{l^2} \tag{19.6}$$

式(19.6)的最小值对应于 $n=1$，因此，得临界载荷为

$$F_{cr} = \frac{\pi^2 EI}{l^2} \tag{19.7}$$

式(19.7)称为欧拉公式。对于圆形或方形横截面，相对于任意形心轴的惯性矩 I 都相同，因此压杆会围绕任意形心轴而发生失稳。然而，对于其他截面形状，上式中的惯性矩是指最小惯性矩，即取 $I = I_{min}$ 计算临界载荷。

19.2　其他支撑细长压杆临界载荷

欧拉公式通过两端铰支压杆推导而来，下面确定其他支撑条件下的临界载荷。考虑如图 19.2a 所示一端固支一端自由压杆。该压杆等效于图 19.2b 所示两端铰支压杆的上半部分，因此图 19.2a 所示压杆的临界载荷与图 19.2b 所示压杆相同，其临界载荷可取二倍杆长通过欧拉公式进行计算而得到。由此可得临界载荷

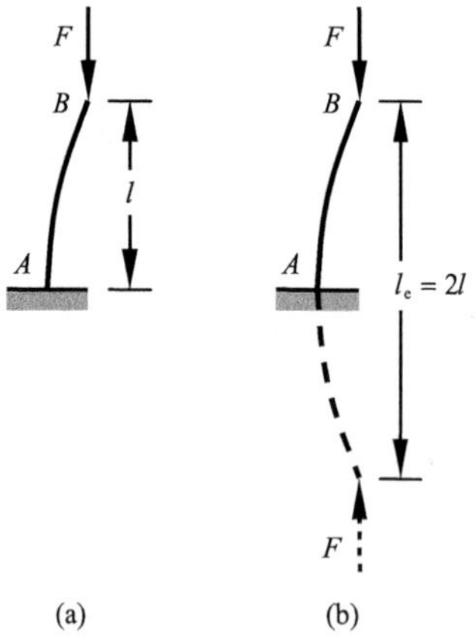

图 19.2

$$F_{cr} = \frac{\pi^2 EI}{(2l)^2} = \frac{\pi^2 EI}{(\mu l)^2} = \frac{\pi^2 EI}{l_e^{\,2}} \tag{19.8}$$

式中，μ 为有效长度因数，$l_e = \mu l$ 为有效长度。式(19.8)是欧拉公式的延伸，因此也称

为欧拉公式。

对应各种支撑的有效长度因数和有效长度如图19.3所示。

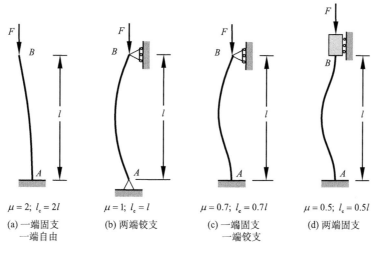

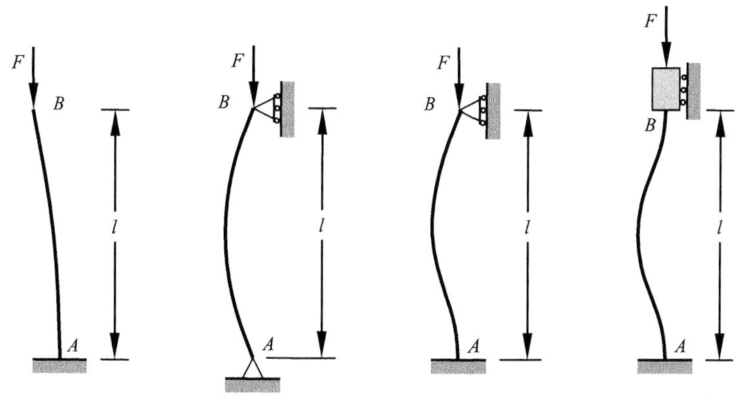

$\mu=2;\ l_e=2l$ (a) 一端固支 一端自由

$\mu=1;\ l_e=l$ (b) 两端铰支

$\mu=0.7;\ l_e=0.7l$ (c) 一端固支 一端铰支

$\mu=0.5;\ l_e=0.5l$ (d) 两端固支

图 19.3

19.3 细长压杆临界应力

对应临界载荷的应力称为临界应力，由σ_{cr}表示。利用欧拉公式并设$i^2=I/A$ (其中A为横截面面积，i为惯性半径)，得

$$\sigma_{cr}=\frac{F_{cr}}{A}=\frac{\pi^2E}{(\mu l/i)^2}=\frac{\pi^2E}{\lambda^2} \tag{19.9}$$

式中，$\lambda=\mu l/i$为压杆的长细比。式(19.9)也称为欧拉公式。欧拉公式表明，临界应力正比于弹性模量，而反比于长细比的平方。

欧拉公式是在临界应力不超过比例极限的前提之下推导而来，即

$$\sigma_{cr}=\frac{\pi^2E}{\lambda^2}\leqslant\sigma_p \tag{19.10}$$

定义$\lambda_p=\sqrt{\pi^2E/\sigma_p}$，则上述关系可表示为

$$\lambda\geqslant\lambda_p \tag{19.11}$$

式中，λ_p是临界长细比。式(19.11)表明，欧拉公式仅当$\lambda\geqslant\lambda_p$时才成立。对于低碳钢(Q235)，$E$=206 GPa和$\sigma_p$=200 MPa，则$\lambda_p$=100.8；对于铝合金，$E$=70 GPa和$\sigma_p$=175 MPa，则$\lambda_p$=62.8。

例 19.1 外径和内径分别为30 mm和20 mm的细长压杆由钢管制成，如图E19.1所示。已知l=1.5 m和E=200 GPa，求图示支撑条件下临界载荷和临界应力。

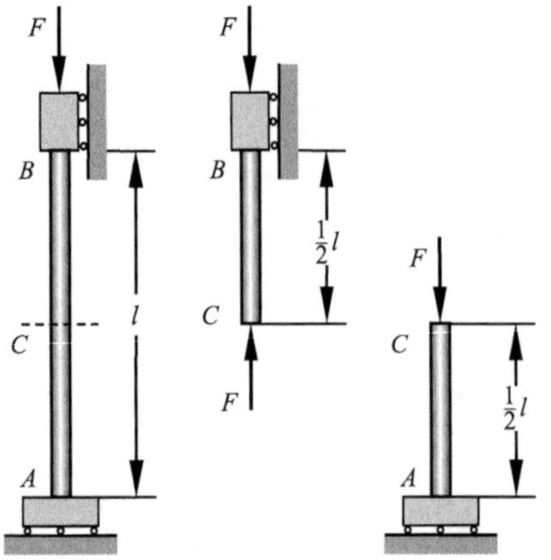

图 E19.1

解　从中间截面把压杆截开，则压杆分为 BC 和 AC 两段，每段都是一端固支一端自由的压杆。因此，图示压杆的临界载荷就等于任一段压杆的临界载荷，利用 $\mu_{BC}=\mu_{AC}=2$ 和 $l_{BC}=l_{AC}=\dfrac{1}{2}l$，得

$$F_{cr}=(F_{cr})_{BC}=(F_{cr})_{AC}=\frac{\pi^2EI}{(\mu_{AC}l_{AC})^2}=\frac{\pi^2E[\frac{1}{64}\pi(D^4-d^4)]}{l^2}=27.99\text{ kN}$$

利用式(19.9)，临界应力等于

$$\sigma_{cr}=\frac{F_{cr}}{A}=\frac{F_{cr}}{\frac{1}{4}\pi(D^2-d^2)}=71.28\text{ MPa}$$

19.4　中长压杆临界应力

对于细长压杆，长细比较大，压杆失效由欧拉公式确定，临界应力与弹性模量有关，而与屈服强度或极限强度无关。

对于粗短压杆，压杆失效是由屈服或断裂引起。对塑性材料，临界应力等于屈服强度；而对脆性材料，临界应力等于极限强度。

对于中长压杆，压杆失效与弹性模量和屈服极限(或强度极限)有关。中长压杆的失效是非常复杂的现象，因而通常采用经验公式进行中心受载中长压杆的设计。最简单的经验公式是线性公式，可表示为

$$\sigma_{cr}=a-b\lambda \tag{19.12}$$

式中，a 和 b 是与所用材料有关的常数。对低碳钢，$a=304$ MPa 和 $b=1.12$ MPa；对铸铁，$a=332.2$ MPa 和 $b=1.454$ MPa。

例 19.2　20a 工字钢压杆在自由端承受中心压缩载荷。已知 σ_p=200 MPa、E=206 GPa、a=304 MPa 和 b=1.12 MPa，求压杆长度分别为(a) l=1.2 m、(b) l=1.0 m 时的临界载荷。

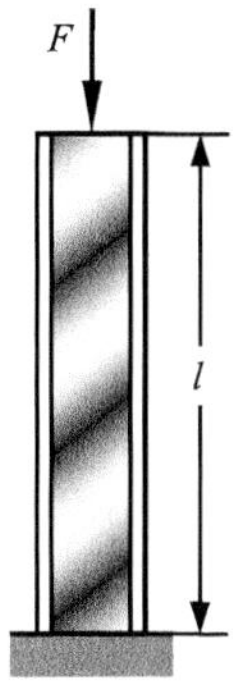

图 E19.2

解　对给定压杆，有效长度因数是 $\mu = 2.0$，临界长细比为

$$\lambda_p = \sqrt{\frac{\pi^2 E}{\sigma_p}} = 101$$

查阅附录 IV 型钢表，20a 工字钢的横截面面积和最小惯性矩分别为

$$A = 35.56\ \text{cm}^2, \quad I_{\min} = 158\ \text{cm}^4$$

(a) 长细比为

$$\lambda = \frac{\mu l}{i} = \frac{\mu l}{\sqrt{I_{\min}/A}} = 114$$

因 $\lambda > \lambda_p$，那么应该选用欧拉公式求解临界载荷，即

$$F_{cr} = A\sigma_{cr} = A\frac{\pi^2 E}{\lambda^2} = 556\ \text{kN}$$

(b) 长细比为

$$\lambda = \frac{\mu l}{\sqrt{I_{\min}/A}} = 95$$

因 $\lambda < \lambda_p$，那么应该选用经验公式求解临界载荷，即

$$F_{cr} = A\sigma_{cr} = A(a - b\lambda) = 703\ \text{kN}$$

19.5　压 杆 设 计

直线经验公式通常用于中长压杆的设计，而欧拉公式则用于细长压杆的设计。因此，用于细长、中长和粗短压杆设计的三个临界应力公式表示如下：

$$\begin{aligned} \sigma_{cr} &= \frac{\pi^2 E}{\lambda^2} \qquad (\lambda \geqslant \lambda_p) \\ \sigma_{cr} &= a - b\lambda \qquad (\lambda_s \leqslant \lambda < \lambda_p) \\ \sigma_{cr} &= \sigma_s \qquad (\lambda < \lambda_s) \end{aligned} \tag{19.13}$$

式中，$\lambda_s=(a-\sigma_s)/b$，$\lambda_p=\sqrt{\pi^2 E/\sigma_p}$。

临界应力σ_{cr}与安全因数n的比值定义为许用应力σ_{allow}。如果压杆稳定，则中心受载压杆的工作应力σ应该满足如下条件：

$$\sigma \leqslant \sigma_{allow}=\frac{\sigma_{cr}}{n} \tag{19.14}$$

例 19.3　外径和壁厚分别为D=120 mm 和t=15 mm 的空心圆形钢杆在自由端承受中心压缩载荷F作用。求压杆长度分别为(a) 1.8 m、(b) 1.2 m 和(c) 0.6 m 时的临界载荷。已知σ_p=280 MPa、σ_s=350 MPa、E=210 GPa、a=461 MPa 和b=2.568 MPa。

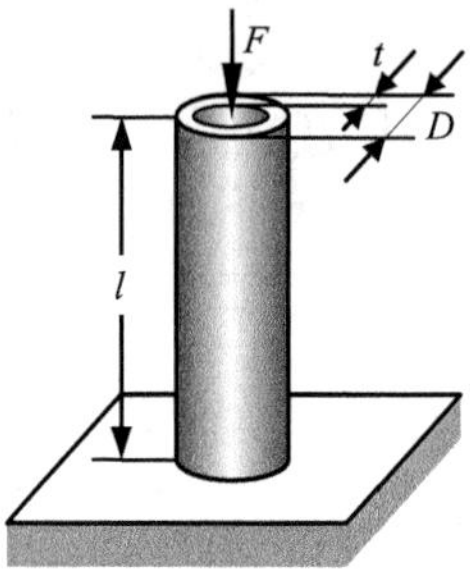

图 E19.3

解　对图 E19.3 所示压杆，有效长度因数$\mu=2.0$。对应比例和屈服极限的临界长细比分别等于

$$\lambda_p=\sqrt{\frac{\pi^2 E}{\sigma_p}}=86,\ \lambda_s=\frac{a-\sigma_s}{b}=43$$

(a) 长细比为

$$\lambda=\frac{\mu l}{i}=\frac{\mu l}{\sqrt{I/A}}=\frac{\mu l}{\sqrt{[\frac{1}{64}\pi(D^4-d^4)]/[\frac{1}{4}\pi(D^2-d^2)]}}=\frac{4\mu l}{\sqrt{D^2+d^2}}=96$$

因$\lambda>\lambda_p$，则应采用欧拉公式计算临界载荷。相应的临界载荷为

$$F_{cr}=A\sigma_{cr}=\frac{1}{4}\pi(D^2-d^2)\frac{\pi^2 E}{\lambda^2}=1113\ \text{kN}$$

(b) 长细比为

$$\lambda=\frac{4\mu l}{\sqrt{D^2+d^2}}=64$$

因$\lambda_s<\lambda<\lambda_p$，则应采用经验公式计算临界载荷。相应的临界载荷为

$$F_{cr}=A\sigma_{cr}=\frac{1}{4}\pi(D^2-d^2)(a-b\lambda)=1468\ \text{kN}$$

(c) 长细比为

$$\lambda = \frac{4\mu l}{\sqrt{D^2 + d^2}} = 32$$

因 $\lambda < \lambda_s$ 则应采用强度公式计算临界载荷。相应的临界载荷为

$$F_{cr} = A\sigma_{cr} = \frac{1}{4}\pi(D^2 - d^2)\sigma_s = 1732 \text{ kN}$$

例 19.4 平面结构 ABC 由杆件 AB、AC 和 BC 分别在节点 A、B 和 C 处铰接而成，杆件 AB、AC 和 BC 的直径均为 32 mm，长度分别为 1 m 、0.6 m 和 0.8 m，节点 C 受集中力 F 作用。已知 $\sigma_p = 280$ MPa 、$\sigma_s = 350$ MPa 、$E = 210$ GPa 、$a = 461$ MPa 、$b =$ 2.568 MPa 和 $0 < \theta < 90°$，试求临界载荷 F_{cr} 及其对应临界度 θ_{cr}。(注：仅考虑面内失稳。)

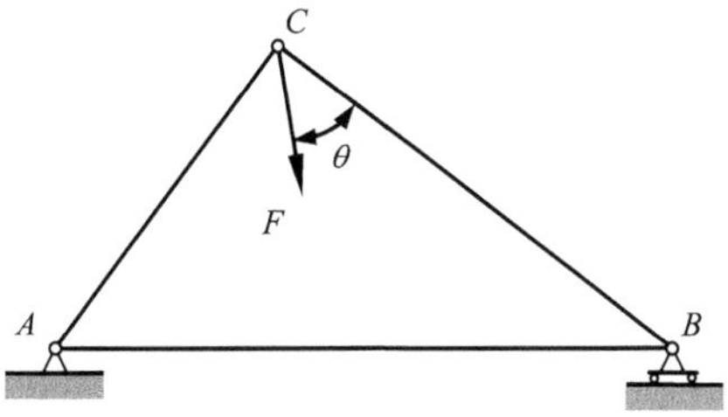

图 E19.4

解 $\lambda_p = \sqrt{\dfrac{\pi^2 E}{\sigma_p}} = 86$，$\lambda_s = \dfrac{a - \sigma_s}{b} = 43$，$\lambda_{AC} = \dfrac{4\mu l_{AC}}{d} = 75$，$\lambda_{BC} = \dfrac{4\mu l_{BC}}{d} = 100$

由于 $\lambda_s < \lambda_{AC} < \lambda_p < \lambda_{BC}$，即 AC 和 BC 分别为中长杆和细长杆，因此应该采用经验公式和欧拉公式分别计算 AC 和 BC 杆的临界载荷，即

$$(F_{AC})_{cr} = (\frac{1}{4}\pi d^2)(a - b\lambda_{AC}) = 215.86 \text{ kN},\quad (F_{BC})_{cr} = (\frac{1}{4}\pi d^2)\frac{\pi^2 E}{\lambda_{BC}{}^2} = 166.69 \text{ kN}$$

所以临界载荷和相应临界角可表示为

$$F_{cr} = \sqrt{(F_{AC})_{cr}{}^2 + (F_{BC})_{cr}{}^2} = 272.73 \text{ kN},\quad \theta_{cr} = \arctan\frac{(F_{AC})_{cr}}{(F_{BC})_{cr}} = 52.32°$$

习 题

19.1 四根细长压杆由直径 30 mm 的钢杆制成。已知 l=1.5 m 和 E=200 GPa，求每一种支撑条件下的临界载荷 F_{cr}。

19.2 四根细长压杆由外径 30 mm 和壁厚 5 mm 的铝管制成。利用 l=1.2 m、E=70 GPa 和安全因数 n=2.3，求每一种支撑条件下的许可载荷 F_{allow}。

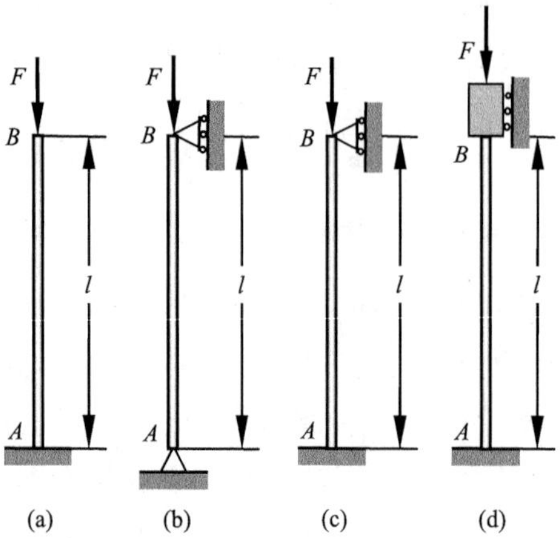

图 P19.1

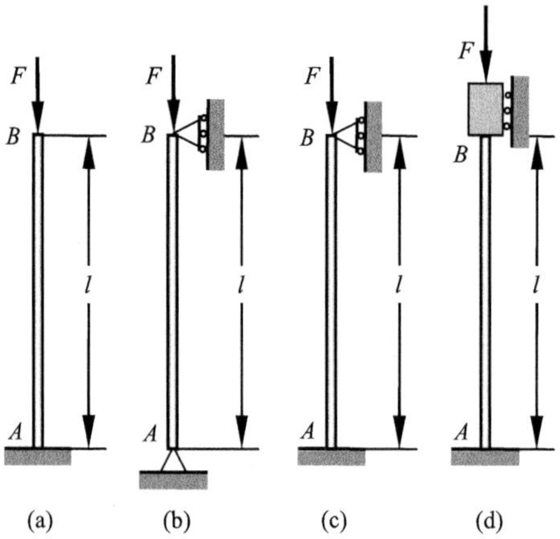

图 P19.2

19.3　直径 d=150 mm 的实心圆钢杆在自由端承受中心压缩载荷，求杆长分别为(a) 2.0 m、(b) 1.2 m 和(c) 0.5 m 时的临界载荷。已知σ_p=280 MPa、σ_s=350 MPa、E=210 GPa、a=461 MPa 和 b=2.568 MPa。

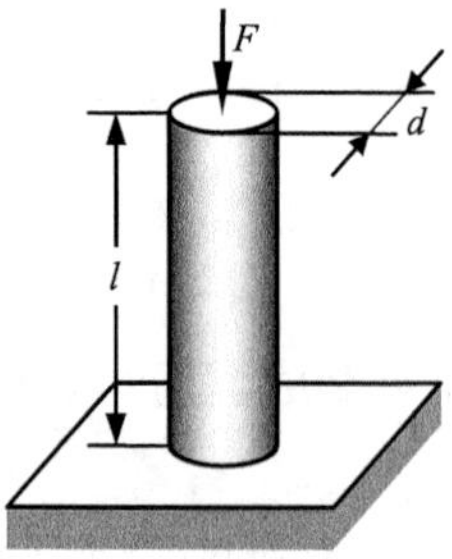

图 P19.3

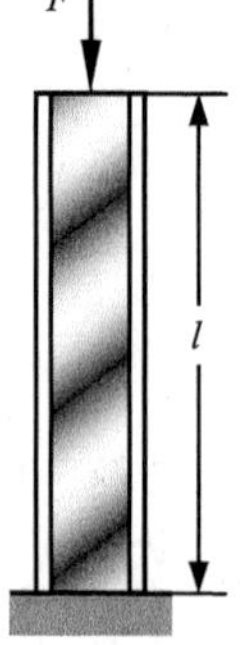

图 P19.4

19.4　25a 工字钢压杆在自由端承受中心压缩载荷。已知 σ_p=200 MPa、E=206 GPa、a=304 MPa 和 b=1.12 MPa，求杆长分别为 l=1.5 m 和 l=1.0 m 时的临界载荷。

第 20 章　能 量 方 法

20.1　外　　功

1. 外力功

如图 20.1 所示，当作用于物体的外力从零缓慢增加到终值 F，则力作用点沿力方向的相应位移也缓慢从零增加到终值δ。假设材料处于线弹性范围，那么外力对物体所做功可表示为

$$W_{e}=\frac{1}{2}F\delta \tag{20.1}$$

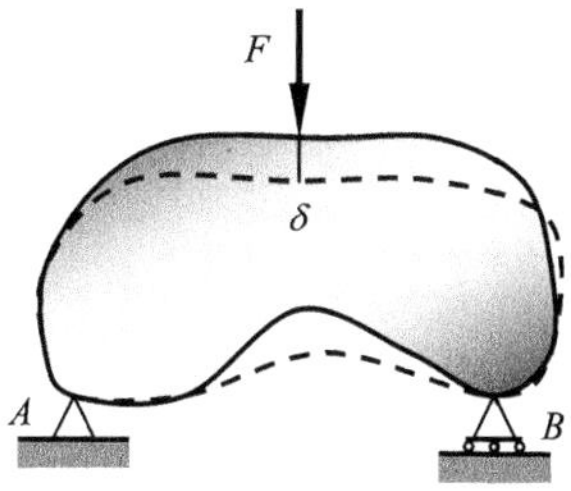

图 20.1

2. 外力偶功

当作用于物体的外力偶从零缓慢增加到终值 M_e，力偶的相应角位移也缓慢从零增加到终值θ。如果材料处于线弹性范围，则图 20.2 所示外力偶所做功可表示为

$$W_{e}=\frac{1}{2}M_{e}\theta \tag{20.2}$$

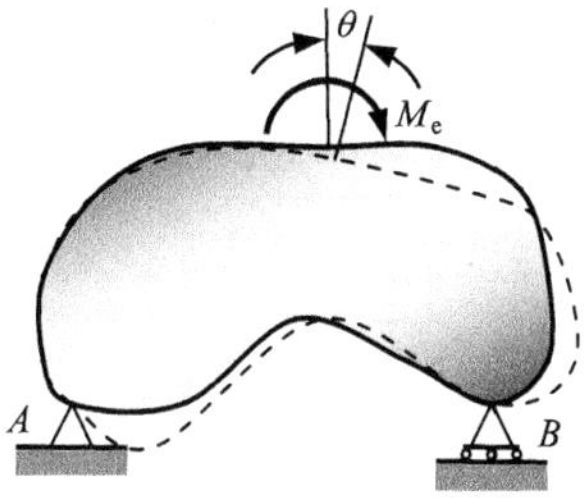

图 20.2

3. 多外载荷功

图 20.3 所示物体受外力 $F_1, F_2, \cdots, F_i, \cdots, F_n$ 作用，当外力由零开始缓慢作用于物体，直至达到终值，则力作用点沿力方向的相应位移也从零缓慢增加到终值。假设材料处于线弹性范围，则外力所做功可表示为

$$W_e = \sum \frac{1}{2} F_i \Delta_i \tag{20.3}$$

式中，Δ_i 是力 $F_1, F_2, \cdots, F_i, \cdots, F_n$ 的函数，即 $\Delta_i = \Delta_i(F_1, F_2, \cdots, F_i, \cdots, F_n)$ $(i = 1, 2, \cdots, n)$。应该注意，F_i 是广义力，Δ_i 是相应的广义位移。

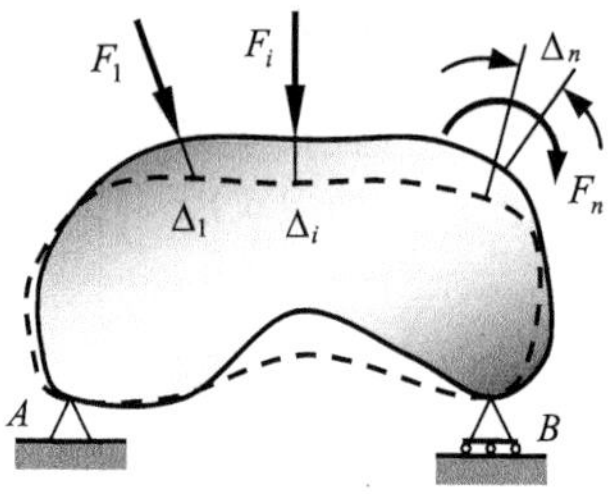

图 20.3

4. 组合载荷功

图 20.4 所示细长圆轴受组合载荷作用。如果是小变形，则组合载荷所做功可表示为

$$W_e = \sum \frac{1}{2} F_i \delta_i \tag{20.4}$$

式中，δ_i 是 F_i 的函数，即 $\delta_i = \delta_i(F_i)$ $(i = 1, 2, 3)$。

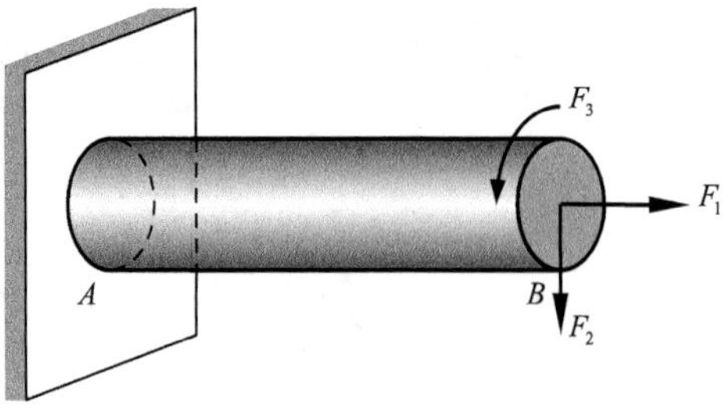

图 20.4

20.2　应变能密度

1. 单向应力应变能密度

如图 20.5 所示，当构件受单向应力时，应变能密度等于应力应变曲线下方的面积，

并可表示为

$$u = \int \sigma \mathrm{d}\varepsilon \tag{20.5}$$

式中，ε 是与 σ 对应的线应变。当 σ 位于线弹性范围，则有 $\sigma = E\varepsilon$，其中 E 是材料的弹性模量。把 σ 代入式(20.5)，得

$$u = \frac{1}{2}E\varepsilon^2 = \frac{1}{2}\sigma\varepsilon = \frac{\sigma^2}{2E} \tag{20.6}$$

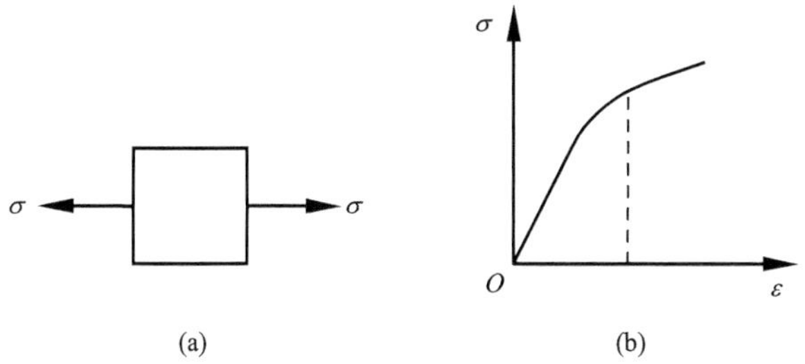

图 20.5

2. 纯剪应力应变能密度

当构件受纯剪应力，如图 20.6 所示，则应变能密度可表示为

$$u = \int \tau \mathrm{d}\gamma \tag{20.7}$$

式中，γ 是与 τ 对应的剪应变。根据式(20.7)，应变能密度 u 等于应力应变曲线下方的面积。当 τ 位于线弹性范围，即 $\tau = G\gamma$，其中 G 是材料的剪切弹性模量，则有

$$u = \frac{1}{2}G\gamma^2 = \frac{1}{2}\tau\gamma = \frac{\tau^2}{2G} \tag{20.8}$$

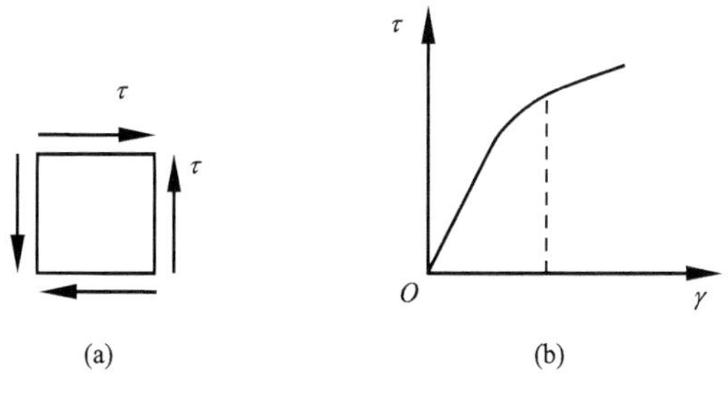

图 20.6

20.3 应 变 能

1. 轴向拉压应变能

轴向拉杆受缓慢增加的与杆轴线重合的外力 F 作用，如图 20.7 所示。

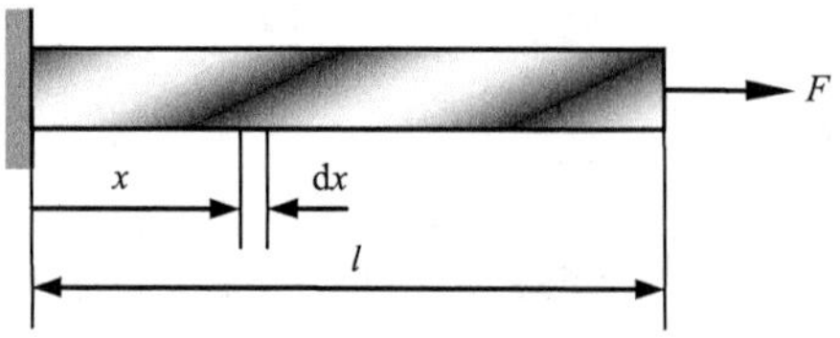

图 20.7

根据式(20.6)，杆的应变能等于

$$V_{\varepsilon}=\int\frac{\sigma^{2}}{2E}A\mathrm{d}x \tag{20.9}$$

利用$\sigma=\dfrac{N}{A}$，其中N为轴力，A为横截面面积，得

$$V_{\varepsilon}=\int\frac{N^{2}}{2EA}\mathrm{d}x \tag{20.10}$$

若面积A均匀，轴力N不变，则式(20.10)可重写为

$$V_{\varepsilon}=\frac{N^{2}l}{2EA} \tag{20.11}$$

2. 扭转应变能

圆轴受缓慢增加的外力偶M_{e}作用，外力偶作用面垂直纵轴，如图 20.8。

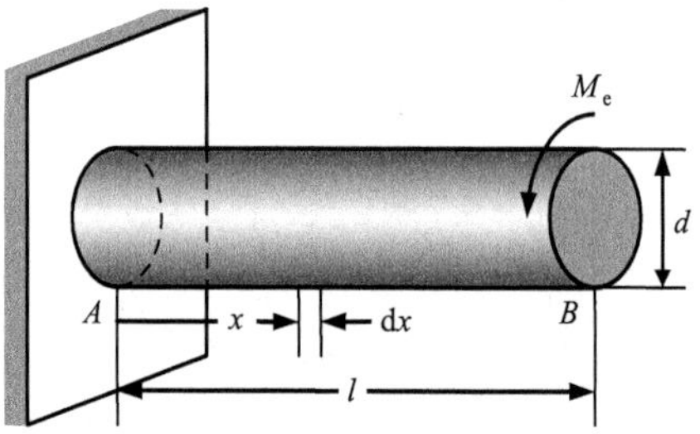

图 20.8

根据式(20.8)，轴的应变能可表示为

$$V_{\varepsilon}=\int(\int\frac{\tau^{2}}{2G}\mathrm{d}A)\mathrm{d}x \tag{20.12}$$

因$\tau=\dfrac{T\rho}{I_{\mathrm{p}}}$，其中$T$为扭矩、$I_{\mathrm{p}}$为极惯性矩、$\rho$为到截面形心的距离，则有

$$V_{\varepsilon}=\int(\frac{T^{2}}{2GI_{\mathrm{p}}{}^{2}}\int\rho^{2}\mathrm{d}A)\mathrm{d}x \tag{20.13}$$

利用$I_{\mathrm{p}}=\int\rho^{2}\mathrm{d}A$，则式(20.13)可重写为

$$V_\varepsilon = \int \frac{T^2}{2GI_p} dx \tag{20.14}$$

如果极惯性矩 I_p 和扭矩 T 均为常数，则式(20.14)可表示为

$$V_\varepsilon = \frac{T^2 l}{2GI_p} \tag{20.15}$$

3. 弯曲应变能

弯曲梁受缓慢增加的外力 F 作用，外力作用于纵向对称面，如图 20.9 所示。对细长梁，与正应力相比，剪应力引起的应变能通常很小，可忽略不计，因此梁的应变能可表示为

$$V_\varepsilon = \int \left(\int \frac{\sigma^2}{2E} dA\right) dx \tag{20.16}$$

利用 $\sigma = \frac{My}{I}$，其中 M 为弯矩、 I 为惯性矩、 y 为到中性轴的距离，则有

$$V_\varepsilon = \int \left(\frac{M^2}{2EI^2} \int y^2 dA\right) dx \tag{20.17}$$

把 $I = \int y^2 dA$ 代入上式，则式(20.17)可重写为

$$V_\varepsilon = \int_0^l \frac{M^2}{2EI} dx \tag{20.18}$$

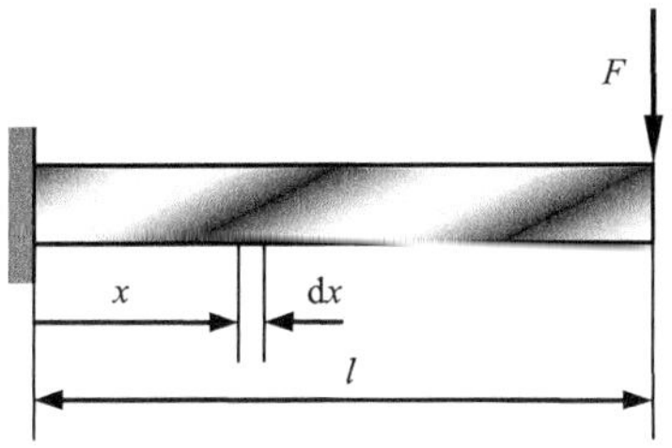

图 20.9

对纯弯曲梁，如图 20.10 所示，弯矩 M 为常数，如果惯性矩 I 也保持不变，则上式可表示为

$$V_\varepsilon = \frac{M^2 l}{2EI} \tag{20.19}$$

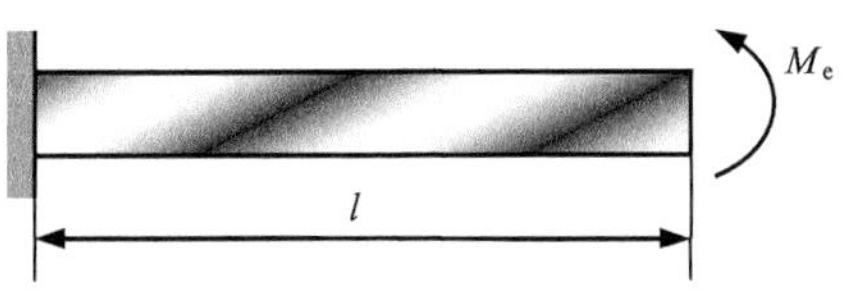

图 20.10

4. 组合载荷应变能

受组合载荷作用的圆截面杆件，如图 20.11 所示。对细长杆件，剪应力引起的应变能可忽略不计，则组合载荷作用下杆件的应变能可表示为

$$V_{\varepsilon} = \int \frac{N^2}{2EA} \mathrm{d}x + \int \frac{T^2}{2GI_{\mathrm{p}}} \mathrm{d}x + \int \frac{M^2}{2EI} \mathrm{d}x \tag{20.20}$$

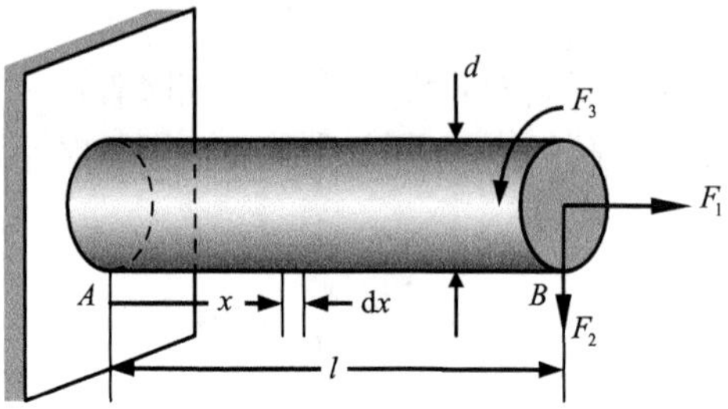

图 20.11

20.4 功 能 原 理

如果不计能量损失，则根据能量守恒，外力对物体所做功将会全部转化为物体应变能，数学上可表示为

$$V_{\varepsilon} = W_{\mathrm{e}} \tag{20.21}$$

式中，V_{ε}为应变能，W_{e}为外功。上式称为功能原理，可用于计算结构变形。

例 20.1　已知弯曲刚度 EI，利用功能原理计算悬臂梁 AB 自由端 B 处挠度。

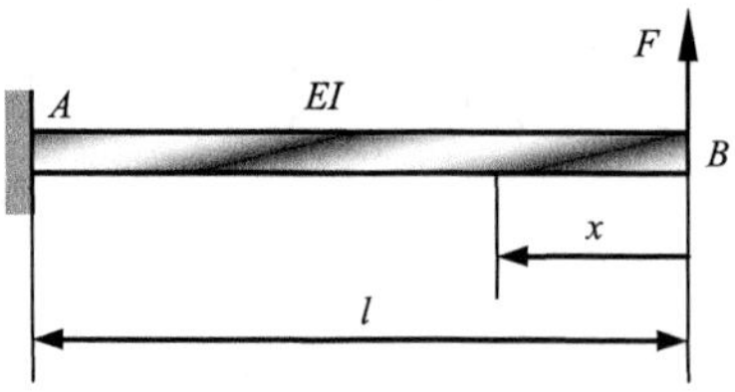

图 E20.1

解　外功和应变能分别为

$$W_{\mathrm{e}} = \frac{1}{2} F \delta_B，\ V_{\varepsilon} = \int_0^l \frac{M^2}{2EI} \mathrm{d}x = \int_0^l \frac{(Fx)^2}{2EI} \mathrm{d}x = \frac{F^2 l^3}{6EI}$$

利用功能原理，即$W_{\mathrm{e}} = V_{\varepsilon}$，得

$$\frac{1}{2} F \delta_B = \frac{F^2 l^3}{6EI}，\ \delta_B = \frac{Fl^3}{3EI}$$

例 20.2　利用功能原理，计算悬臂梁 AB 自由端 B 处转角。已知弯曲刚度 EI。

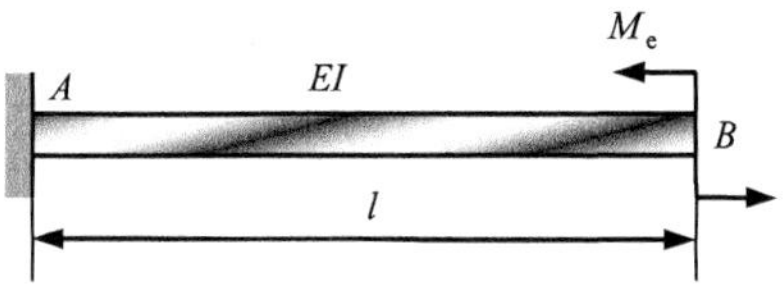

图 E20.2

解 $W_e=\frac{1}{2}M_e\theta_B$，$V_\varepsilon=\int_0^l\frac{M^2}{2EI}dx=\int_0^l\frac{M_e^2}{2EI}dx=\frac{M_e^2l}{2EI}$

根据 $W_e=V_\varepsilon$，得

$$\frac{1}{2}M_e\theta_B=\frac{M_e^2l}{2EI},\quad \theta_B=\frac{M_el}{EI}$$

20.5 互 等 定 理

考虑图 20.12 所示线弹性体受广义力作用，设先加 F_1 和 F_3，再加 F_2 和 F_4，则物体的应变能可表示为

$$V_{\varepsilon1}=\frac{1}{2}F_1\Delta_1+\frac{1}{2}F_3\Delta_3+\frac{1}{2}F_2\Delta_2+\frac{1}{2}F_4\Delta_4+F_1\Delta_1'+F_3\Delta_3' \tag{20.22}$$

式中，Δ_1（或 Δ_3）是由 F_1 和 F_3 引起的 F_1（或 F_3）作用点沿 F_1（或 F_3）方向的广义位移，Δ_2（或 Δ_4）是由 F_2 和 F_4 引起的 F_2（或 F_4）作用点沿 F_2（或 F_4）方向的广义位移，Δ_1'（或 Δ_3'）是由 F_2 和 F_4 引起的 F_1（或 F_3）作用点沿 F_1（或 F_3）方向的广义位移。

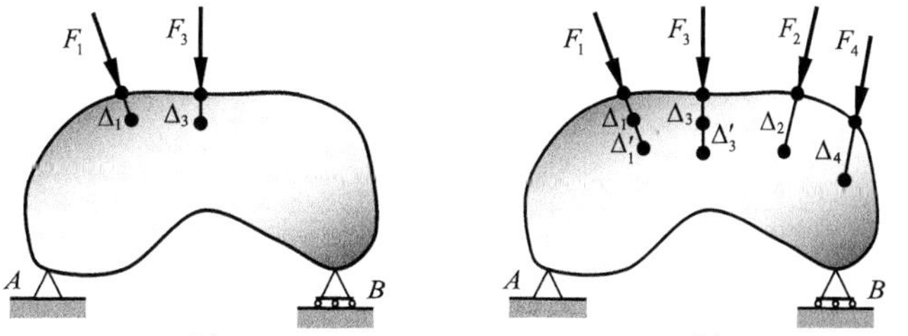

图 20.12

当广义力 F_2 和 F_4 首先作用于物体，然后广义力 F_1 和 F_3 再作用于物体，则物体的应变能为

$$V_{\varepsilon2}=\frac{1}{2}F_2\Delta_2+\frac{1}{2}F_4\Delta_4+\frac{1}{2}F_1\Delta_1+\frac{1}{2}F_3\Delta_3+F_2\Delta_2'+F_4\Delta_4' \tag{20.23}$$

式中，Δ_2'（或 Δ_4'）是由 F_1 和 F_3 引起的 F_2（或 F_4）作用点沿 F_2（或 F_4）方向的广义位移。

比较式(20.22)和(20.23)，并利用 $V_{\varepsilon1}=V_{\varepsilon2}$，得

$$F_1\Delta_1'+F_3\Delta_3'=F_2\Delta_2'+F_4\Delta_4' \tag{20.24}$$

由此得到结论，第 1 组广义力在第 2 组广义力引起的广义位移上所做功等于第 2 组广义力在第 1 组广义力引起的广义位移上所做功。上述关系称为功的互等定理。

假设 $F_3 = F_4 = 0$ 和 $F_1 = F_2$，则式(20.24)可简化为

$$\Delta_1' = \Delta_2' \tag{20.25}$$

由此得到结论，当两个广义力数值相等，则由第 2 个广义力引起的第 1 个广义力作用点沿第 1 个广义力方向的广义位移等于由第 1 个广义力引起的第 2 个广义力作用点沿第 2 个广义力方向的广义位移。该关系称为位移互等定理。

例 20.3　如图 E20.3a 所示，线弹性球受等值反向径向拉伸载荷作用。已知球的弹性模量 E、泊松比 μ、直径 d，求球的体积变化。

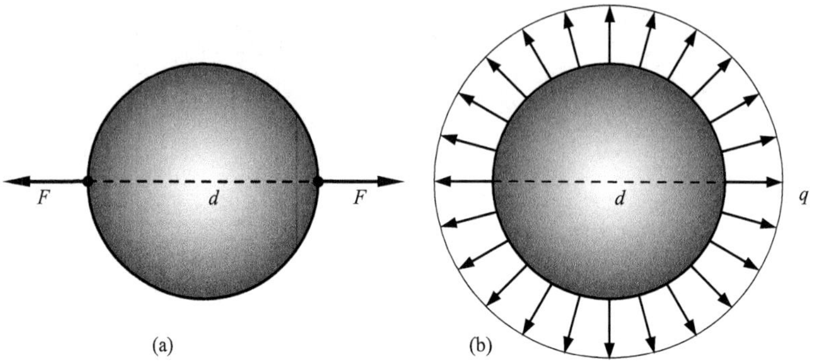

图 E20.3

解　假设球受三向均匀拉伸应力作用，如图 E20.3b 所示，则有

$$\varepsilon_1 = \frac{1}{E}[\sigma_1 - \mu(\sigma_2 + \sigma_3)] = \frac{1-2\mu}{E}q$$

利用功的互等定理，得

$$F(\varepsilon_1 d) = q\Delta V$$

即

$$\Delta V = \frac{F(\varepsilon_1 d)}{q} = \frac{1-2\mu}{E}Fd$$

20.6 卡 氏 定 理

如图 20.13 所示，变形体受外力 $F_1, F_2, \cdots, F_i, \cdots, F_n$ 作用。当作用于物体的外力由零开始缓慢增加到终值，则力作用点沿力方向的相应位移也从零缓慢增加到终值。如果材料处于线弹性范围，则作用于物体的外力所做功可表示为

$$W_e = \sum \frac{1}{2} F_i \Delta_i \tag{20.26}$$

式中，Δ_i 是外力 $F_1, F_2, \cdots, F_i, \cdots, F_n$ 的线性齐次函数。式(20.26)表明，外功 W_e 是外力 $F_1, F_2, \cdots, F_i, \cdots, F_n$ 的二次齐次函数。利用功能原理，即 $W_e = V_\varepsilon$，应变能也是外力 $F_1, F_2, \cdots, F_i, \cdots, F_n$ 的二次齐次函数，并可表示为

$$V_\varepsilon = V_\varepsilon(F_1, F_2, \cdots, F_i, \cdots, F_n) \tag{20.27}$$

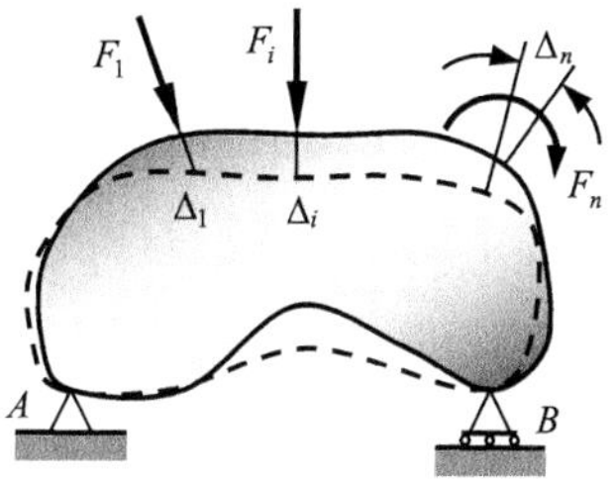

图 20.13

如果任何一个外力，比如外力 F_i，增加 $\mathrm{d}F_i$，而其余外力保持不变，那么应变能也将增加 $\mathrm{d}V_\varepsilon$，并可表示为

$$\mathrm{d}V_\varepsilon = \frac{\partial V_\varepsilon}{\partial F_i}\mathrm{d}F_i \tag{20.28}$$

因此，总应变能可写为

$$V_{\varepsilon 1} = V_\varepsilon + \mathrm{d}V_\varepsilon = \sum \frac{1}{2}F_i\Delta_i + \frac{\partial V_\varepsilon}{\partial F_i}\mathrm{d}F_i \tag{20.29}$$

总应变能 $V_{\varepsilon 1}$ 应该与外力加载次序无关，例如可以先加 $\mathrm{d}F_i$，然后再加 $F_1, F_2, \cdots, F_i, \cdots, F_n$，那么总应变能应该不变。当先加 $\mathrm{d}F_i$，外力 $\mathrm{d}F_i$ 所做功 $\mathrm{d}W'_\mathrm{e}$ 可写为

$$\mathrm{d}W'_\mathrm{e} = \frac{1}{2}\mathrm{d}F_i\mathrm{d}\Delta_i \tag{20.30}$$

式中，$\mathrm{d}\Delta_i$ 是对应 $\mathrm{d}F_i$ 的位移。当再加 $F_1, F_2, \cdots, F_i, \cdots, F_n$，则外力 $F_1, F_2, \cdots, F_i, \cdots, F_n$ 和 $\mathrm{d}F_i$ 所做功可表示为

$$W'_\mathrm{e} = \sum \frac{1}{2}F_i\Delta_i + \mathrm{d}F_i\Delta_i \tag{20.31}$$

因此，总功等于

$$W_{\mathrm{e}2} = W'_\mathrm{e} + \mathrm{d}W'_\mathrm{e} = \sum \frac{1}{2}F_i\Delta_i + \mathrm{d}F_i\Delta_i + \frac{1}{2}\mathrm{d}F_i\mathrm{d}\Delta_i \tag{20.32}$$

根据功能原理，总应变能等于

$$V_{\varepsilon 2} = W_{\mathrm{e}2} = \sum \frac{1}{2}F_i\Delta_i + \mathrm{d}F_i\Delta_i + \frac{1}{2}\mathrm{d}F_i\mathrm{d}\Delta_i \tag{20.33}$$

利用 $V_{\varepsilon 1} = V_{\varepsilon 2}$，则有

$$\sum \frac{1}{2}F_i\Delta_i + \frac{\partial V_\varepsilon}{\partial F_i}\mathrm{d}F_i = \sum \frac{1}{2}F_i\Delta_i + \mathrm{d}F_i\Delta_i + \frac{1}{2}\mathrm{d}F_i\mathrm{d}\Delta_i \tag{20.34}$$

忽略高阶小量，得

$$\Delta_i = \frac{\partial V_\varepsilon}{\partial F_i} \tag{20.35}$$

式中，F_i 是广义力，Δ_i 是相应广义位移。如果 F_i 是力偶，则 Δ_i 是与力偶对应的转角。式(20.35)称为卡氏第二定理或卡氏定理。该定理可表述为：广义力 F_i 作用点处沿 F_i 方向的广义位

移 Δ_i 等于应变能相对广义力 F_i 的一阶偏导数。

卡氏定理用于组合载荷，则可表示为

$$\Delta_i = \frac{\partial V_\varepsilon}{\partial F_i} = \int \frac{N}{EA}\frac{\partial N}{\partial F_i}\mathrm{d}x + \int \frac{T}{GI_\mathrm{p}}\frac{\partial T}{\partial F_i}\mathrm{d}x + \int \frac{M}{EI}\frac{\partial M}{\partial F_i}\mathrm{d}x \tag{20.36}$$

例 20.4　已知弯曲刚度 EI，利用卡氏定理求悬臂梁自由端挠度。

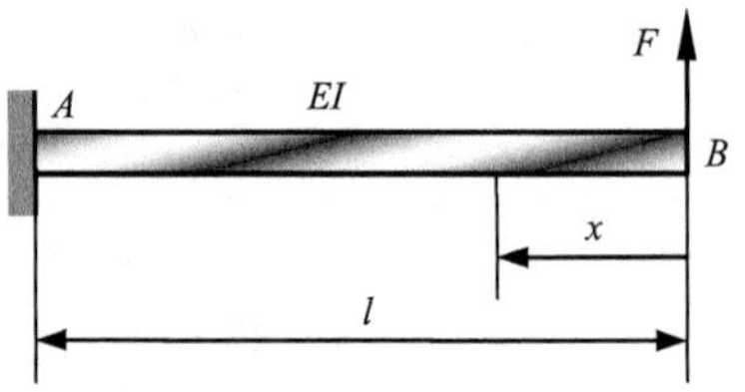

图 E20.4

解　应变能等于

$$V_\varepsilon = \int_0^l \frac{M^2}{2EI}\mathrm{d}x = \int_0^l \frac{(Fx)^2}{2EI}\mathrm{d}x = \frac{F^2l^3}{6EI}$$

利用卡氏定理，得

$$\delta_B = \frac{\partial V_\varepsilon}{\partial F} = \frac{\partial}{\partial F}\left(\frac{F^2l^3}{6EI}\right) = \frac{Fl^3}{3EI}$$

例 20.5　利用卡氏定理，求悬臂梁自由端转角。已知弯曲刚度 EI。

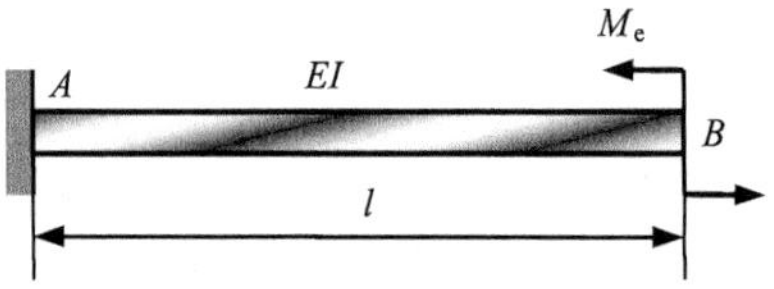

图 E20.5

解　应变能等于

$$V_\varepsilon = \int_0^l \frac{M^2}{2EI}\mathrm{d}x = \int_0^l \frac{M_\mathrm{e}^2}{2EI}\mathrm{d}x = \frac{M_\mathrm{e}^2 l}{2EI}$$

利用 $\Delta_i = \dfrac{\partial V_\varepsilon}{\partial F_i}$，得

$$\theta_B = \frac{\partial V_\varepsilon}{\partial M_\mathrm{e}} = \frac{\partial}{\partial M_\mathrm{e}}\left(\frac{{M_\mathrm{e}}^2 l}{2EI}\right) = \frac{M_\mathrm{e} l}{EI}$$

卡氏定理能用于求外力作用处的位移。如果需求无外力作用处的位移，则需要在待求位移处施加虚拟力。应变能相对虚拟力求导，进而得到虚拟力处位移，该位移是由实际力和虚拟力共同产生。通过设置虚拟力等于零，即可求得实际力产生的位移。

例 20.6　已知弯曲刚度 EI，利用卡氏定理求悬臂梁自由端转角，如图 E20.6a 所示。

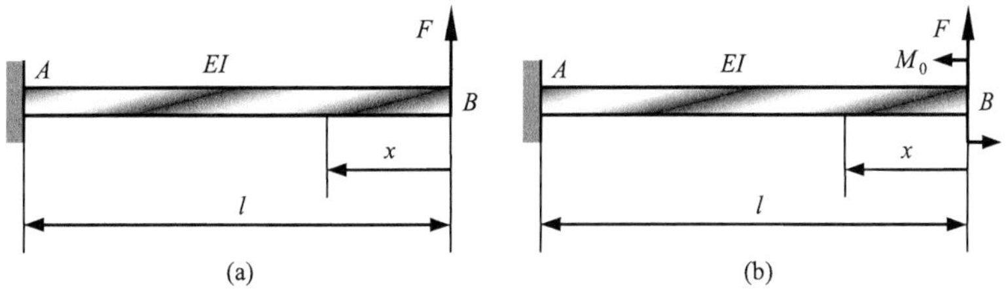

图 E20.6

解 在梁自由端施加虚拟力偶 M_0，如图 E20.6b 所示，则截面 x 处弯矩等于

$$M = Fx + M_0 \quad (0 \leqslant x < l)$$

因此，应变能可表示为

$$V_\varepsilon = \int_0^l \frac{M^2}{2EI}\mathrm{d}x = \int_0^l \frac{(Fx+M_0)^2}{2EI}\mathrm{d}x = \frac{\frac{1}{3}F^2l^3 + FM_0l^2 + M_0{}^2l}{2EI}$$

利用卡氏定理，即 $\Delta_i = \dfrac{\partial V_\varepsilon}{\partial F_i}$，得

$$\theta_B = \left(\frac{\partial V_\varepsilon}{\partial M_0}\right)_{M_0=0} = \left(\frac{Fl^2 + 2M_0l}{2EI}\right)_{M_0=0} = \frac{Fl^2}{2EI}$$

20.7 虚功原理

变形体在载荷 F 、q 和 M_e 作用下处于平衡状态，如图 20.14 所示。

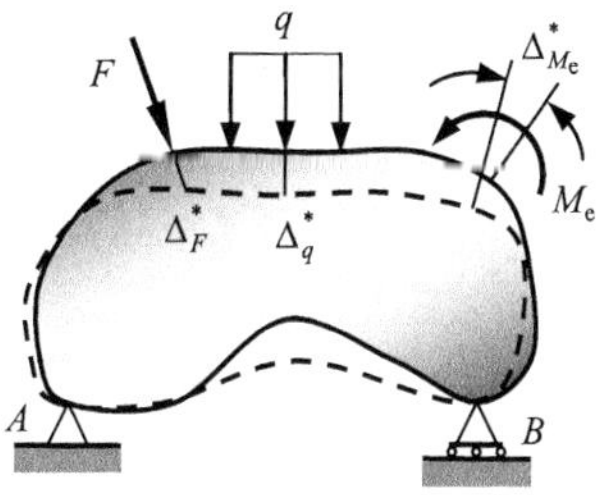

图 20.14

当因其他因素，如载荷变化或温度波动，在物体内引起虚变形时，则由载荷 F 、q 和 M_e 在虚变形上所做虚功可表示为

$$W_\mathrm{e} = F\Delta^*_F + \int q\Delta^*_q \mathrm{d}x + M_\mathrm{e}\Delta^*_{M_\mathrm{e}} \tag{20.37}$$

式中，Δ^*_F 、Δ^*_q 和 $\Delta^*_{M_\mathrm{e}}$ 分别是 F 、q 和 M_e 作用点沿 F 、q 和 M_e 方向的虚位移。

当物体发生虚变形，则作用于物体的载荷 F 、q 和 M_e 产生的内力所做虚功可表示为

$$W_i = \int N\mathrm{d}\delta^* + \int M\mathrm{d}\theta^* + \int V\mathrm{d}\lambda^* + \int T\mathrm{d}\varphi^* \tag{20.38}$$

式中，N、M、V和T分别是由载荷F、q和M_e产生的轴力、弯矩、剪力和扭矩，δ^*、θ^*、λ^*和φ^*分别是与N、M、V和T对应的虚变形。

利用$W_i = W_e$，　得

$$\int N\mathrm{d}\delta^* + \int M\mathrm{d}\theta^* + \int V\mathrm{d}\lambda^* + \int T\mathrm{d}\varphi^* = F\Delta_F^* + \int q\Delta_q^*\mathrm{d}x + M_e\Delta_{M_e}^* \tag{20.39}$$

这就是虚功原理，可表述如下：内力在虚变形上所做虚功等于外力在虚位移上所做虚功。虚功原理既可用于线性弹性体，也可用于非线性弹性体。

例 20.7　如图 E20.7a 所示，桁架 ABC 由杆 AC 和 BC 构成，每杆截面积均为常数 A，在节点C受集中力 F 作用。杆 AC 和 BC 的应力应变关系分别是$\sigma = E\varepsilon$和$\sigma^2 = k\varepsilon$，其中E和k均为常数。利用虚功原理，求C处水平和垂直位移分量。(注：结构稳定性无需考虑。)

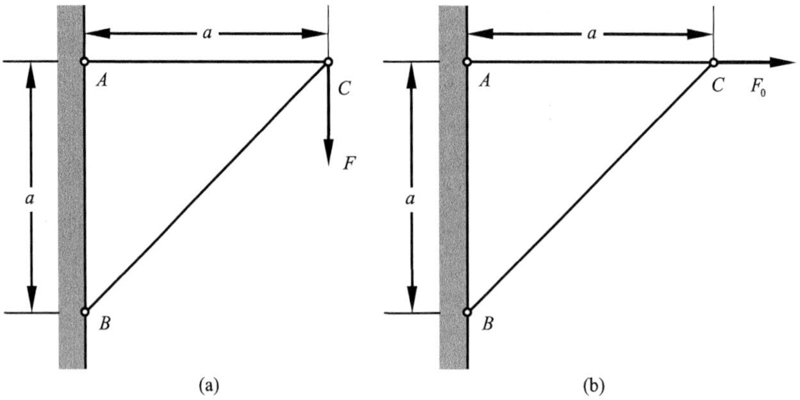

图 E20.7

解　如图 E20.7b 所示，当方向向右的水平力F_0作用于桁架节点C时，则杆 AC 和 BC 中产生的内力可表示为

$$N_{AC} = F_0,\ N_{BC} = 0$$

当原载荷F作用于结构时，在 AC 和 BC 中产生的变形可看着虚变形，并可表示为

$$\delta_{AC}^* = \frac{Fa}{EA},\ \delta_{BC}^* = [-\frac{(-\sqrt{2}F)^2}{kA^2}](\sqrt{2}a) = -\frac{2\sqrt{2}F^2a}{kA^2}$$

在C处产生的位移可看着虚位移，并等于

$$\delta_C^* = \Delta_H$$

式中，Δ_H是C处待求水平位移分量。

利用虚功原理，得

$$F_0\Delta_H = N_{AC}\frac{Fa}{EA} + N_{BC}(-\frac{2\sqrt{2}F^2a}{kA^2}),\text{即}\ \Delta_H = \frac{Fa}{EA}$$

如果方向向下的垂直力作用于桁架节点C，同理可得C处垂直位移分量

$$\Delta_V = \frac{Fa}{EA} + (-\sqrt{2})(-\frac{2\sqrt{2}F^2 a}{kA^2}) = \frac{Fa}{EA} + \frac{4F^2 a}{kA^2}$$

20.8 单位载荷法

变形体在载荷 F 、q 和 M_e 作用下处于平衡状态，如图 20.15a 所示。

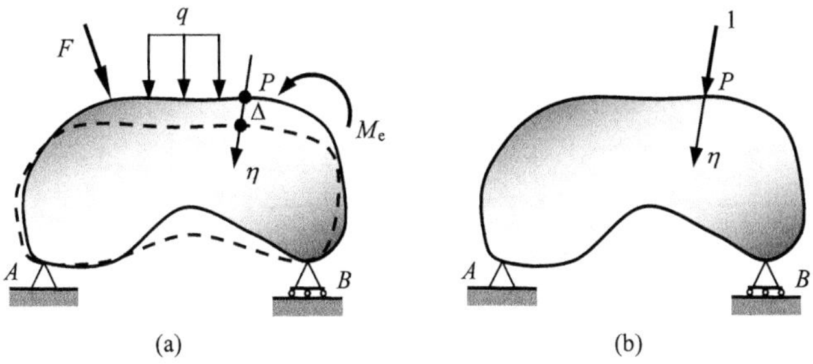

图 20.15

为求 P 处沿 η 方向的位移 Δ，假设单位载荷沿 η 方向作用于物体 P 处，如图 20.15b 所示，然后原载荷也作用于物体。如果把原载荷引起的变形看着虚变形，那么根据虚功原理，得

$$\Delta = \int \bar{N} \mathrm{d}\delta + \int \bar{M} \mathrm{d}\theta + \int \bar{V} \mathrm{d}\lambda + \int \bar{T} \mathrm{d}\varphi \tag{20.40}$$

式中，$\bar{N}$ 、$\bar{M}$ 、$\bar{V}$ 和 $\bar{T}$ 分别是由单位在和引起的轴力、弯矩、剪力和扭矩，δ 、θ 、λ 和 φ 分别是由载荷 F 、q 和 M_e 引起的与轴力、弯矩、剪力和扭矩对应的变形。上述关系称为单位载荷法，可用于确定结构上任意点沿任意方向的位移。如果转角需要确定，那么单位力偶需要施加于待求转角的位置。单位载荷法可用线性弹性体和非线性弹性体。

对受组合载荷作用的线弹性构件，式(20.40)可重写为

$$\Delta = \int \frac{N(x)\bar{N}(x)}{EA} \mathrm{d}x + \int \frac{M(x)\bar{M}(x)}{EI} \mathrm{d}x + \int \frac{T(x)\bar{T}(x)}{GI_p} \mathrm{d}x \tag{20.41}$$

式中，N 、M 和 T 分别是由组合载荷引起的轴力、弯矩和扭矩，EA 、EI 和 GI_p 分别是组合载荷作用构件的拉压刚度、弯曲刚度和扭转刚度。

例 20.8 已知弯曲刚度 EI，利用单位载荷法求悬臂梁自由端转角，如图 E20.8a 所示。

解 如图 E20.8a 所示，原载荷弯矩可表示为

$$M(x) = Fx \quad (0 \leqslant x < l)$$

单位载荷弯矩，如图 E20.8b，等于

$$\bar{M}(x) = M_0 = 1 \quad (0 < x < l)$$

根据单位载荷法，得

$$\theta_B = \int_0^l \frac{M(x)\bar{M}(x)}{EI} \mathrm{d}x = \frac{Fl^2}{2EI}$$

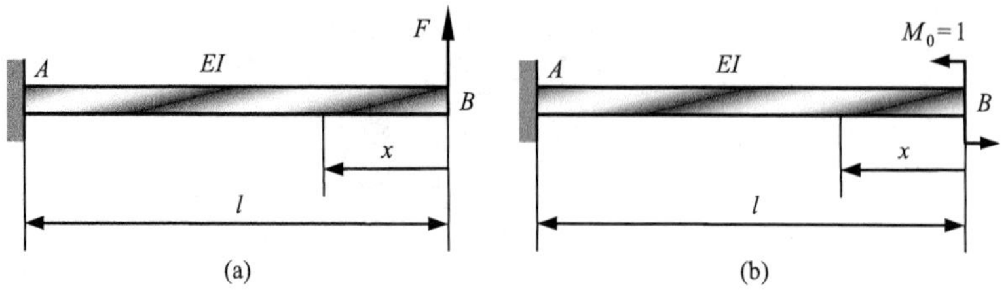

图 E20.8

例 20.9　弯曲刚度为 EI 的杆 AB 和 BC 焊接于 B 处，如图 E20.9a 所示。对图示载荷，采用单位载荷法，求点 C 处垂直挠度和截面 C 处转角。

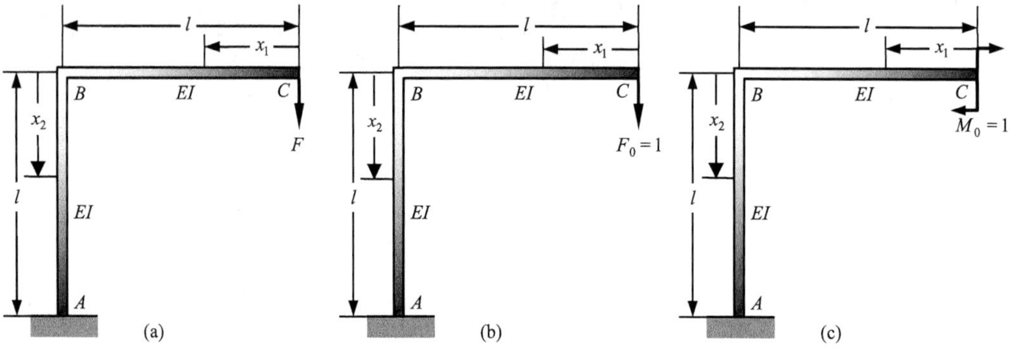

图 E20.9

解　图 E20.9a 所示原载荷引起的弯矩可表示为

$$M(x_1)=-Fx_1\ \ (0\leqslant x_1\leqslant l),\ \ M(x_2)=-Fl\ \ (0\leqslant x_2<l)$$

为求点 C 的垂直挠度，在该点垂直方向施加单位载荷 $F_0=1$，如图 E20.9b 所示，则单位载荷引起的弯矩为

$$\bar{M}(x_1)=-x_1\ \ (0\leqslant x_1\leqslant l),\ \ \bar{M}(x_2)=-l\ \ (0\leqslant x_2<l)$$

利用单位载荷法，得

$$w_{Cy}=\int_0^l\frac{M(x_1)\bar{M}(x_1)}{EI}\mathrm{d}x_1+\int_0^l\frac{M(x_2)\bar{M}(x_2)}{EI}\mathrm{d}x_2=\frac{4Fl^3}{3EI}$$

为求截面 C 的转角，在该截面施加单位力偶 $M_0=1$，如图 E20.9c 所示，显然单位力偶引起的弯矩为

$$\bar{M}(x_1)=-1\ \ (0\leqslant x_1\leqslant l),\ \ \bar{M}(x_2)=-1\ \ (0\leqslant x_2<l)$$

利用单位载荷法，得

$$\theta_C=\int_0^l\frac{M(x_1)\bar{M}(x_1)}{EI}\mathrm{d}x_1+\int_0^l\frac{M(x_2)\bar{M}(x_2)}{EI}\mathrm{d}x_2=\frac{3Fl^2}{2EI}$$

例 20.10　弯曲刚度为 EI 的杆件 AB 和 BC 焊接于 B 处，如图 E20.10a 所示，点 A 固定，点 C 铰支。对图示载荷和支撑，采用单位载荷法，求点 C 处反力。

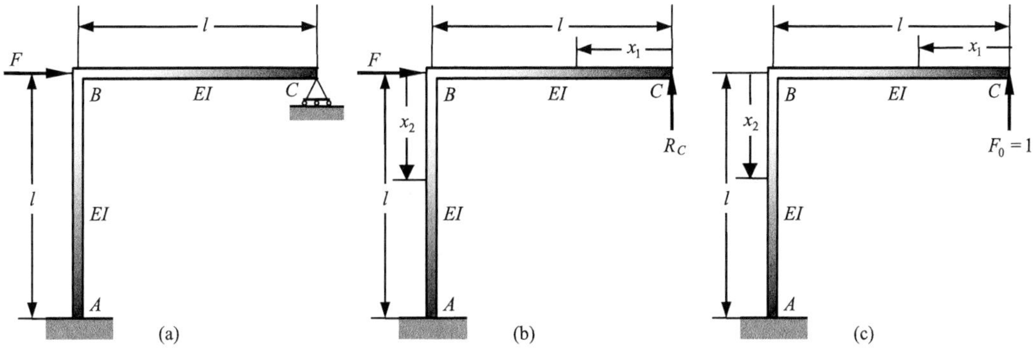

图 E20.10

解　图 E20.10a 所示为一次静不定结构。设 R_C 为点 C 处多余反力，消除相应支撑，如图 E20.10b 所示。

参考图 E20.10b，弯曲表示为

$$M(x_1) = R_C x_1 \quad (0 \leqslant x_1 \leqslant l), \qquad M(x_2) = R_C l - F x_2 \quad (0 \leqslant x_2 < l)$$

为求点 C 处垂直挠度，在该点垂直方向作用单位载荷 $F_0 = 1$，如图 E20.10c 所示，单位载荷引起的弯矩表示为

$$\bar{M}(x_1) = x_1 \quad (0 \leqslant x_1 \leqslant l), \qquad \bar{M}(x_2) = l \quad (0 \leqslant x_2 < l)$$

利用单位载荷法，得

$$w_{Cy} = \int_0^l \frac{M(x_1)\bar{M}(x_1)}{EI} \mathrm{d}x_1 + \int_0^l \frac{M(x_2)\bar{M}(x_2)}{EI} \mathrm{d}x_2 = \frac{(8R_C - 3F)l^3}{6EI}$$

因 $w_{Cy} = 0$，则得

$$R_C = \frac{3}{8} F$$

20.9　冲 击 载 荷

前面讨论的载荷都是静载荷，即载荷缓慢作用于构件直至终值，达到终值后保持不变。然而，也有载荷是动载荷，即载荷突然作用于构件。这些动载荷称为冲击载荷。当物体撞击构件，撞击期间在构件内将会产生很大应力。

1. 垂直冲击

图 20.16 所示物块弹簧系统，重量为 P、动能为 T 的物块撞击不计质量弹簧，假设物块停止运动时弹簧压缩Δ_d，如果不计冲击期间能量损失，那么能量守恒表明，物块动能和势能将会完全转化为弹簧内的应变能，或者说，动能和势能之和等于弹簧自由端移动Δ_d所做功。

根据能量守恒，有

$$T + P\Delta_\mathrm{d} = \frac{1}{2} F_\mathrm{d} \Delta_\mathrm{d} \tag{20.42}$$

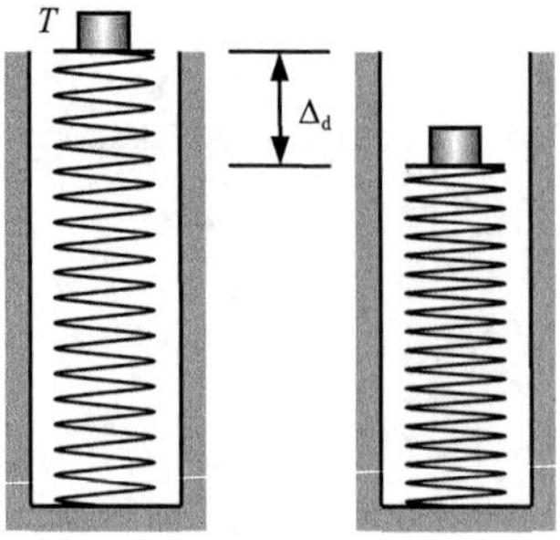

图 20.16

式中，F_d 是当弹簧压缩 Δ_d 时在物块和弹簧之间产生的作用力。假设弹簧处于线弹性范围，即 $\dfrac{F_d}{\Delta_d}=\dfrac{P}{\Delta_{st}}$，其中 Δ_{st} 为当物块静止作用于弹簧自由端时在弹簧自由端产生的静位移，则有

$$\frac{F_d}{P}=\frac{\Delta_d}{\Delta_{st}}=K_d \tag{20.43}$$

式中，K_d 动荷因数。把式(20.43)代入式(20.42)，得

$$\frac{1}{2}K_d^{\,2}-K_d-\frac{T}{P\Delta_{st}}=0 \tag{20.44}$$

求解 K_d，得

$$K_d=1+\sqrt{1+\frac{2T}{P\Delta_{st}}} \tag{20.45}$$

假设物块由静止释放，下落距离 h 后撞击弹簧，那么动荷因数可表示为

$$K_d=1+\sqrt{1+\frac{2h}{\Delta_{st}}} \tag{20.46}$$

K_d 确定后，弹簧的动应力和动变形可分别表示为

$$\sigma_d=K_d\sigma_{st}，\ \delta_d=K_d\delta_{st} \tag{20.47}$$

式中，σ_{st} 和 δ_{st} 分别是当载荷静止作用于弹簧时而产生的静应力和静变形。

如果物块从刚接触弹簧时由静止状态突然释放，即 h=0，那么根据式(20.45)或式(20.46)，动荷因数 K_d=2 和动应力 σ_d=2σ_{st}，即当物块从弹簧顶端由静止突然下落(突加载)，动应力是物块静止放在弹簧上(静加载)引起的静应力的两倍。

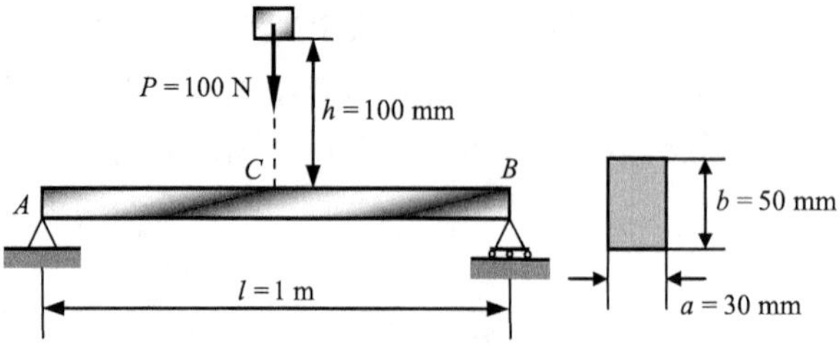

图 E20.11

例 20.11 重量为 P 的物块，初始静止，从高度 h 处落下，撞击简支梁中点。已知弹性模量 E=105 GPa，求(a)梁的最大挠度，(b)梁中最大正应力。

解 (a)查表 16.1，得

$$\Delta_{\rm st} = \frac{Pl^3}{48EI} = \frac{Pl^3}{48E(\frac{1}{12}ab^3)} = 6.35\times10^{-2}\ \text{mm}$$

利用式(20.46)，得

$$K_{\rm d} = 1 + \sqrt{1 + \frac{2h}{\Delta_{\rm st}}} = 57.13$$

因此，最大动挠度为

$$\Delta_{\rm d} = K_{\rm d}\Delta_{\rm st} = 3.63\ \text{mm}$$

(b)根据式(15.11)，得最大静应力

$$\sigma_{\rm st} = \frac{M}{S} = \frac{\frac{1}{4}Pl^2}{\frac{1}{6}ab^2} = 2.00\ \text{MPa}$$

利用式(20.47)，得最大动应力

$$\sigma_{\rm d} = K_{\rm d}\sigma_{\rm st} = 114.26\ \text{MPa}$$

例 20.12 质量为 m 的环圈从图示位置由静止释放，撞击固定在垂直杆 AB(直径为 d)自由端 B 处的平板。已知 E=70 GPa，求(a)杆的最大伸长，(b)杆内最大正应力。(g = 9.81 m/s^2)

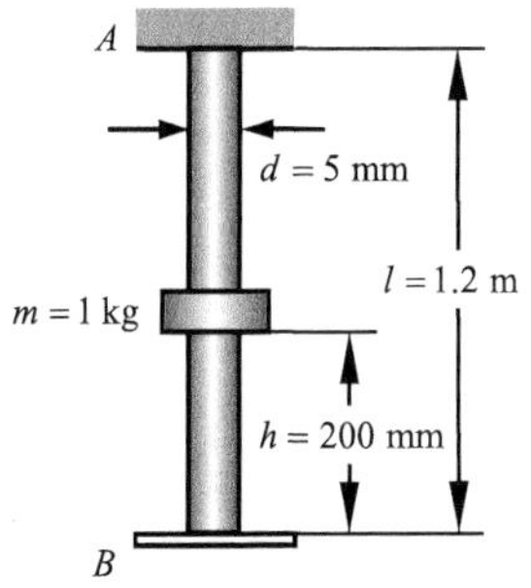

图 E20.12

解 (a)杆的最大伸长

$$\Delta_{\rm st} = \frac{mgl}{EA} = \frac{mgl}{E(\frac{1}{4}\pi d^2)} = 8.56\times10^{-3}\ \text{mm}, K_{\rm d} = 1 + \sqrt{1 + \frac{2h}{\Delta_{\rm st}}} = 217.17,\ \Delta_{\rm d} = K_{\rm d}\Delta_{\rm st} = 1.86\ \text{mm}$$

(b)杆内最大正应力

$$\sigma_{\rm st} = \frac{mg}{A} = \frac{mg}{\frac{1}{4}\pi d^2} = 0.50\ \text{MPa},\ \sigma_{\rm d} = K_{\rm d}\sigma_{\rm st} = 108.58\ \text{MPa}$$

例 20.13 平面机构 $ABCD$ 由物块 A 和 D、光滑定滑轮 B 和 C、弹性绳索 $ABCD$ 构

成，绳索长度为0.6 m，横截面面积为50 mm^2，物块 A 和 D 的重量均为120 N，分别向下和向上匀速运动，速率均为56 mm/s。已知 $E=10$ GPa，假设物块 D 突然卡住而停止运动，试求绳索最大应力。

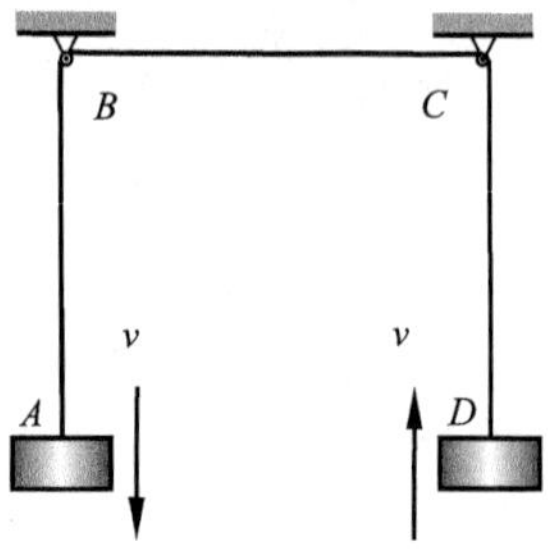

图 E20.13

解

$$\Delta_{st}=\frac{P_A l}{EA}=144\ \mu m,\ K_d=1+\sqrt{\frac{2T_A}{P_A\Delta_{st}}}=2.49,\ \sigma_{st}=\frac{P_A}{A}=2.4\ MPa,\ \sigma_d=K_d\sigma_{st}=5.98\ MPa$$

2. 水平冲击

如图 20.17 所示，物块弹簧系统置于光滑水平面。重量为 P、动能为 T 的物块沿水平面滑动而撞击弹簧，假设物块达到最大位移时弹簧压缩量等于Δ_d。如果不计冲击能量损失和弹簧质量，并设弹簧处于线弹性范围，则物块运动静止时其动能将完全转化为弹簧的应变能。

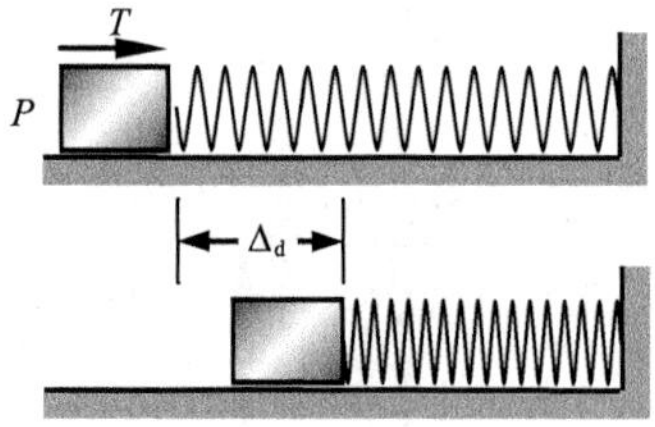

图 20.17

根据能量守恒，得

$$T=\frac{1}{2}F_d\Delta_d=\frac{1}{2}P\Delta_{st}K_d^{\ 2} \tag{20.48}$$

式(20.48)的正根可表示为

$$K_d=\sqrt{\frac{2T}{P\Delta_{st}}} \tag{20.49}$$

式中，Δ_{st}是当与物块重量 P 相等的等效力作用于弹簧自由端时而在弹簧自由端产生的静

位移。利用$T=\frac{1}{2}\frac{P}{g}v^2$，得

$$K_{\mathrm{d}}=\sqrt{\frac{v^2}{g\Delta_{\mathrm{st}}}} \tag{20.50}$$

根据式(20.49)或(20.50)，弹簧中动应力和弹簧的动变形分别等于

$$\sigma_{\mathrm{d}}=K_{\mathrm{d}}\sigma_{\mathrm{st}},\quad \delta_{\mathrm{d}}=K_{\mathrm{d}}\delta_{\mathrm{st}} \tag{20.51}$$

式中，σ_{st}和δ_{st}分别是当与物块重量 P 相等的等效力静止作用于弹簧时而产生的弹簧静应力和静变形。

例 20.14　质量为 m 的小球向右运动以速度 v 正碰立柱 AB(直径为 d)自由端 A 处。已知 E=200 GPa，求(a)立柱最大挠度，(b)立柱最大正应力。

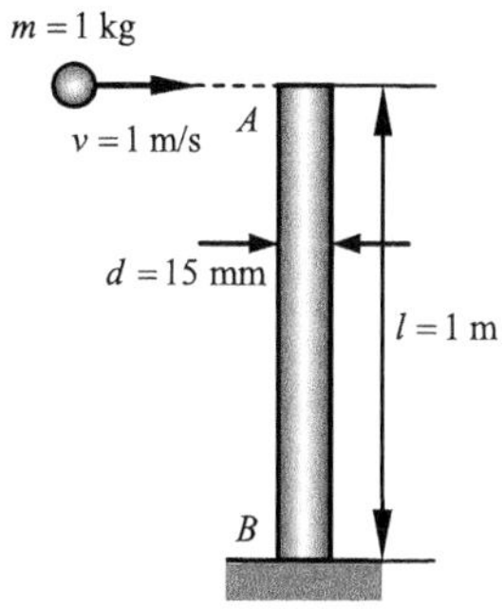

图 E20.14

解　(a)立柱最大挠度

$$\Delta_{\mathrm{st}}=\frac{mgl^3}{3EI}=\frac{mgl^3}{3E(\frac{1}{64}\pi d^4)}=6.58\ \mathrm{mm},\quad K_{\mathrm{d}}=\sqrt{\frac{v^2}{g\Delta_{\mathrm{st}}}}=3.94,\quad \Delta_{\mathrm{d}}=K_{\mathrm{d}}\Delta_{\mathrm{st}}=25.9\ \mathrm{mm}$$

(b)立柱最大正应力

$$\sigma_{\mathrm{st}}=\frac{M}{W}=\frac{mgl}{\frac{1}{32}\pi d^3}=29.6\ \mathrm{MPa},\quad \sigma_{\mathrm{d}}=K_{\mathrm{d}}\sigma_{\mathrm{st}}=117\ \mathrm{MPa}$$

例 20.15　质量为 m 的小球以速度 v 正碰杆 AB 自由端 A。已知 E=100 GPa，求(a)点 A 处位移，(b)杆内最大正应力。

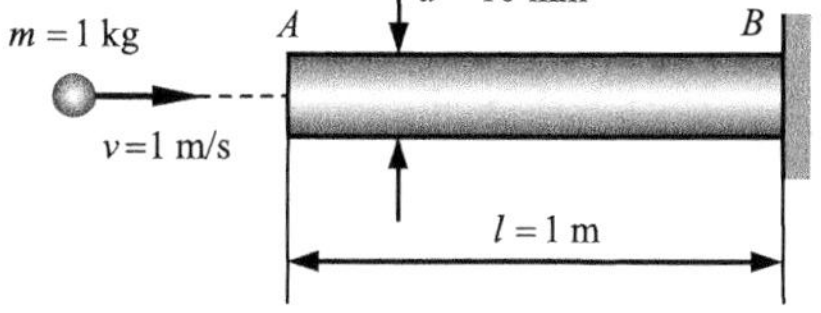

图 E20.15

解　(a)点 A 处位移

$$\Delta_{st}=\frac{mgl}{EA}=\frac{mgl}{E(\frac{1}{4}\pi d^2)}=1.25\times10^{-6}\ \text{m}，\ K_d=\sqrt{\frac{v^2}{g\Delta_{st}}}=285.57，\ \Delta_d=K_d\Delta_{st}=0.357\ \text{mm}$$

(b)杆内最大正应力

$$\sigma_{st}=\frac{mg}{A}=\frac{mg}{\frac{1}{4}\pi d^2}=0.125\ \text{MPa}，\ \sigma_d=K_d\sigma_{st}=35.7\ \text{MPa}$$

习　　题

20.1　利用功能原理，求悬臂梁自由端挠度。已知 AB 段弯曲刚度为 $2EI$，BC 段弯曲刚度为 EI。

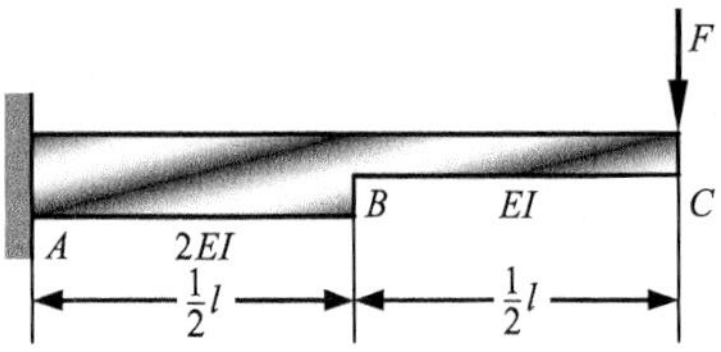

图 P20.1

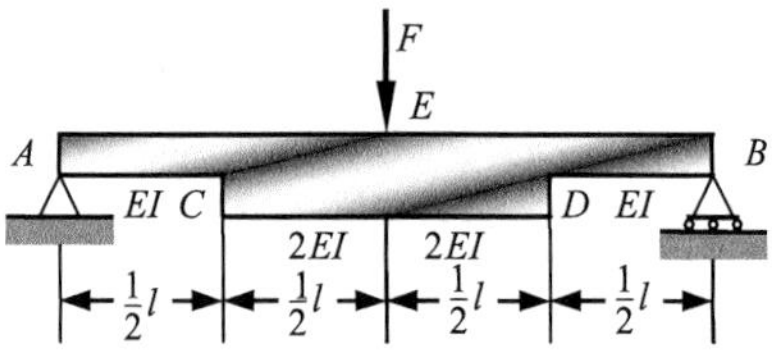

图 P20.2

20.2　利用功能原理，求简支梁中间截面挠度。已知 CD 段弯曲刚度为 $2EI$，AC 和 BD 段弯曲刚度为 EI。

20.3　利用卡氏定理，求悬臂梁自由端挠度和转角。已知弯曲刚度为 EI。

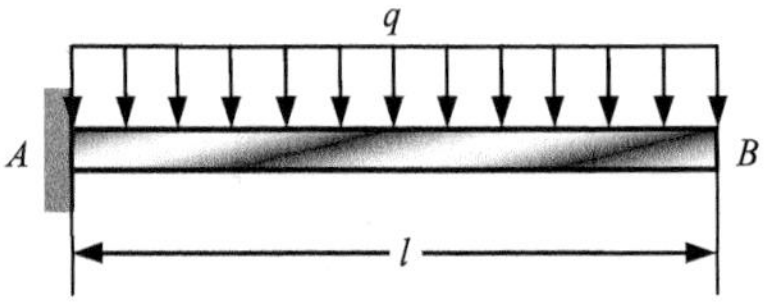

图 P20.3

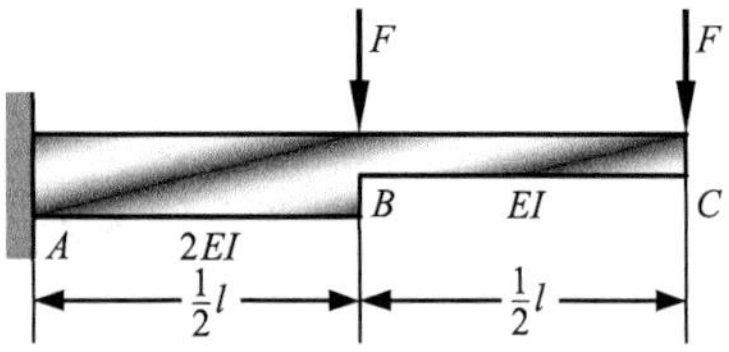

图 P20.4

20.4　利用卡氏定理，求悬臂梁自由端挠度。已知 AB 段弯曲刚度 $2EI$，BC 段弯曲刚度 EI。

20.5　利用卡氏定理，求简支梁中间截面挠度。已知弯曲刚度 EI。

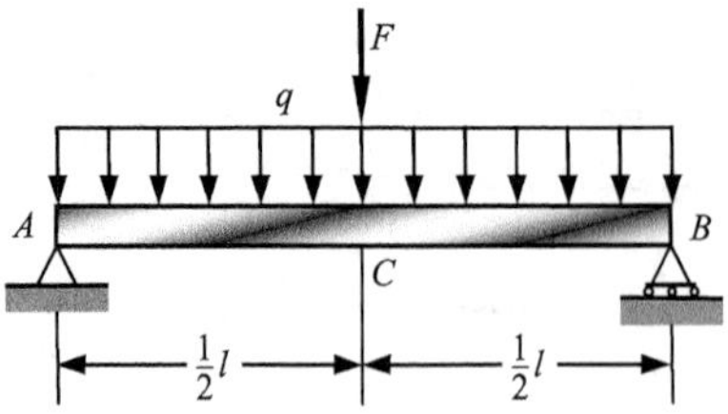

图 P20.5

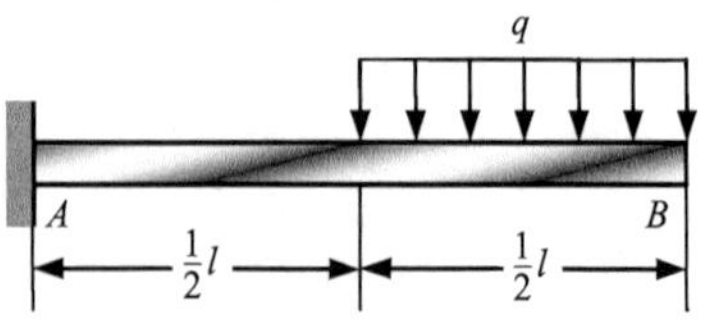

图 P20.6

20.6 利用卡氏定理，求点 B 处挠度和转角。已知弯曲刚度 EI。

20.7 利用虚功原理，求悬臂梁自由端挠度和转角。已知应力应变关系 $\sigma^2 = k\varepsilon$，其中 k 是常数。

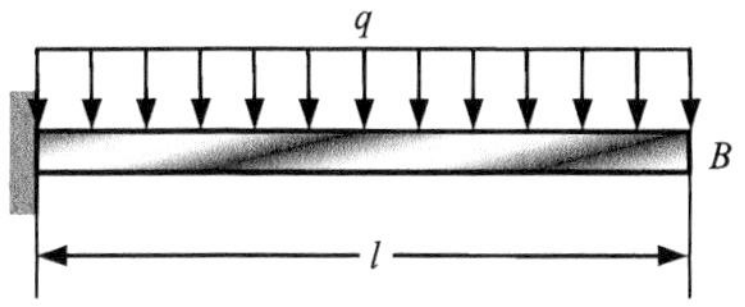

图 P20.7

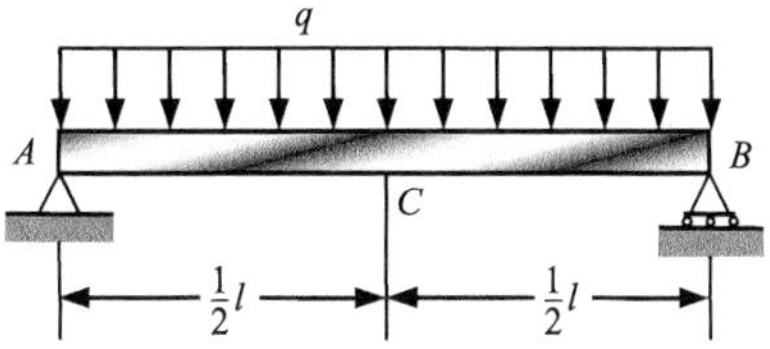

图 P20.8

20.8 利用虚功原理，求简支梁中间截面挠度。已知应力应变关系 $\sigma^2 = k\varepsilon$，其中 k 是常数。

20.9 利用单位载荷法，求悬臂梁自由端挠度和转角。已知弯曲刚度为 EI。

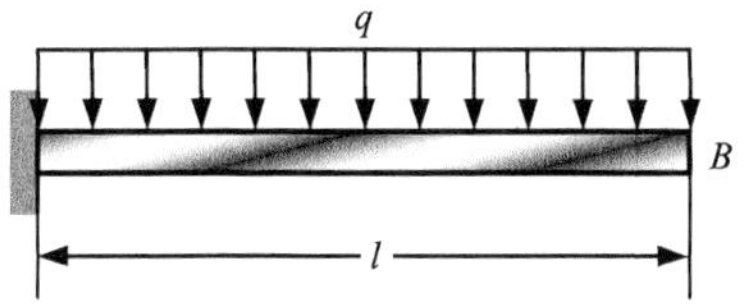

图 P20.9

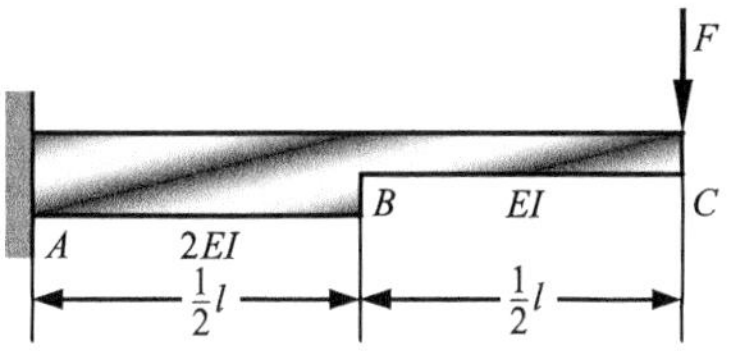

图 P20.10

20.10 利用单位载荷法，求悬臂梁自由端挠度和转角。已知 AB 段弯曲刚度为 $2EI$，BC 段弯曲刚度为 EI。

20.11-20.12 利用单位载荷法，求简支梁中间截面 C 处挠度和截面 B 处转角。已知弯曲刚度 EI。

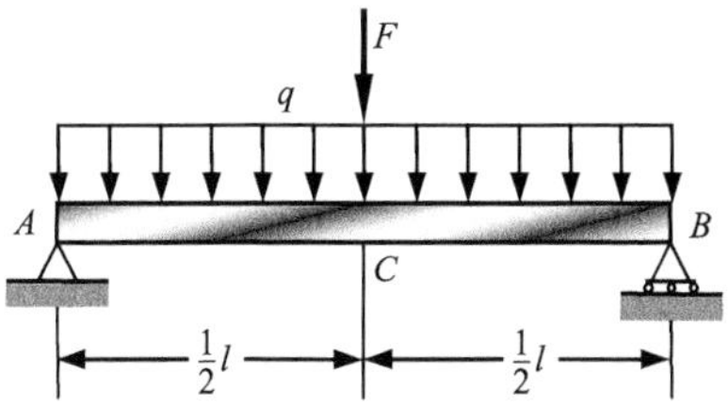

图 P20.11

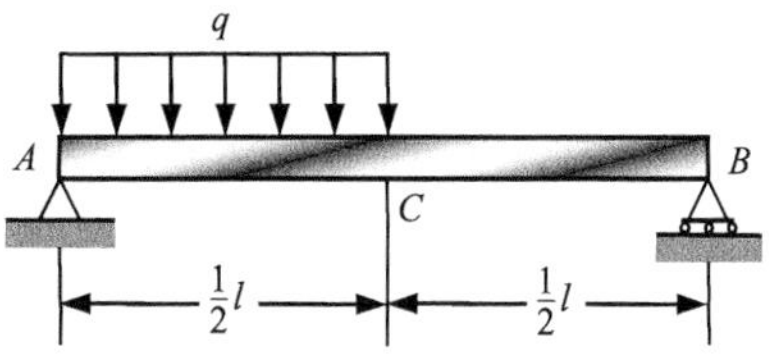

图 P20.12

20.13 利用单位载荷法，求点 A 处水平和垂直挠度。已知弯曲刚度 EI。

20.14-20.15 弯曲刚度为 EI 的杆件 AB 和 BC 焊接于 B 处。对图示载荷，采用单位载荷法，求点 C 处水平和垂直挠度以及转角。

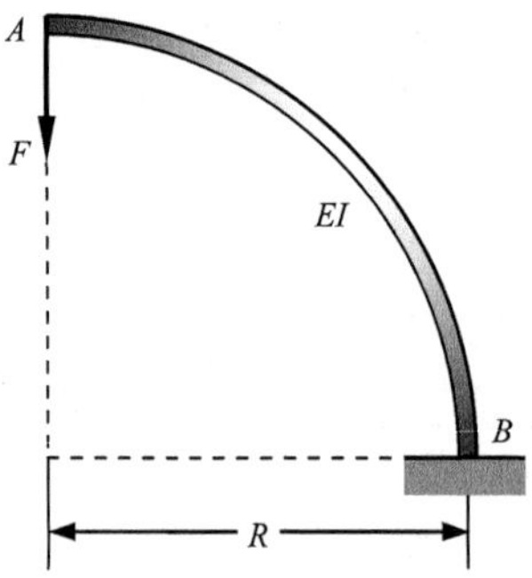

图 P20.13

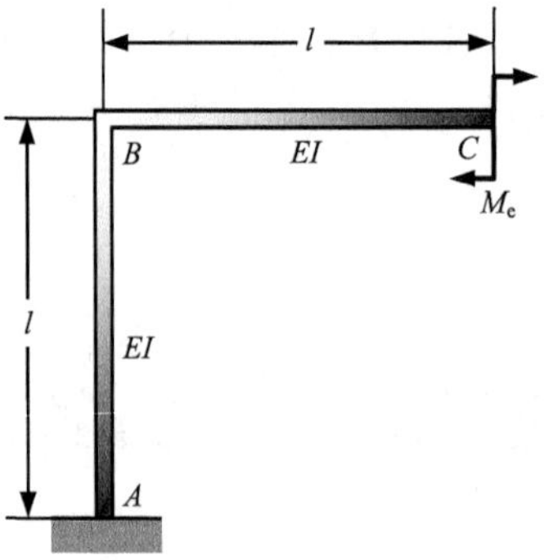

图 P20.14

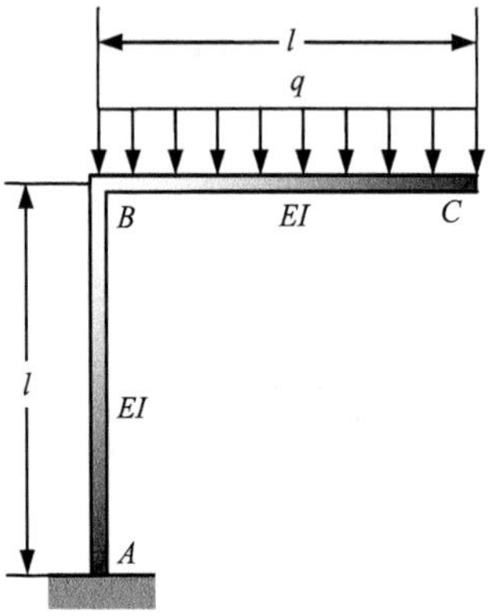

图 P20.15

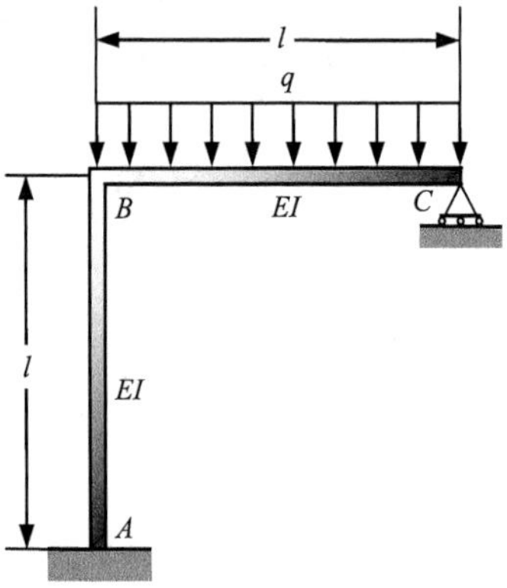

图 P20.16

20.16　弯曲刚度为 EI 的杆件 AB 和 BC 焊接于 B 处，点 A 固定，点 C 铰支。对图示载荷，采用单位载荷法，求点 C 处反力。

20.17　质量为 m 的物块，初始静止，从高度 h 处落下，撞击悬臂梁自由端。已知弹性模量 E=206 GPa，求(a)梁的最大挠度，(b)梁中最大正应力。

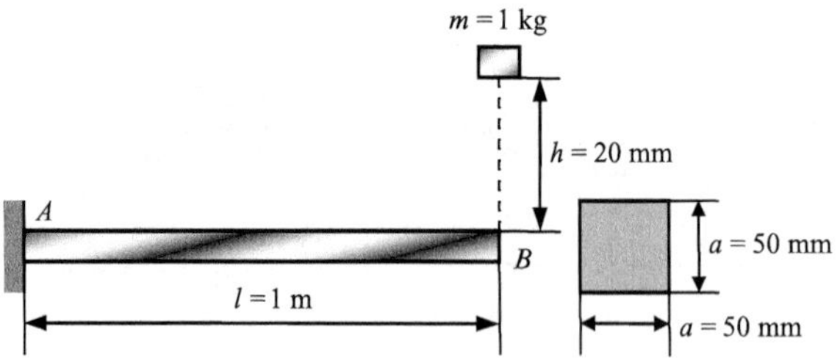

图 P20.17

20.18　质量为 m 的物块，初始静止，从高度 h 处落下，撞击悬臂梁中点。已知弹性模量 E=206 GPa，求(a)梁的最大挠度，(b)梁中最大正应力。

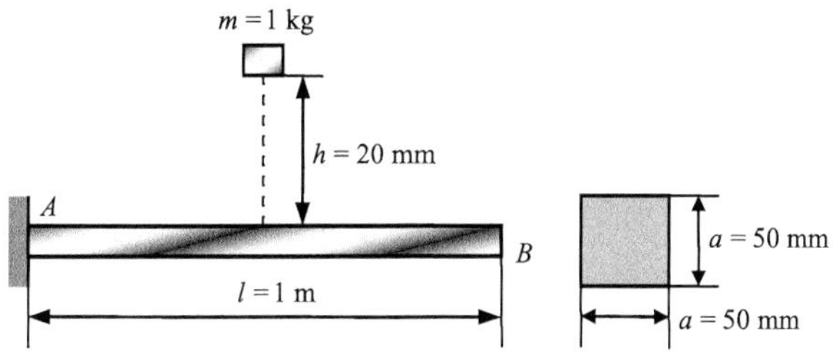

图 P20.18

20.19 质量为 m 的物块，初始静止，从高度 h 处落下，撞击直径为 d 的外伸梁自由端。已知弹性模量 E=73 GPa，求(a)点 C 处挠度，(b)梁中最大正应力。

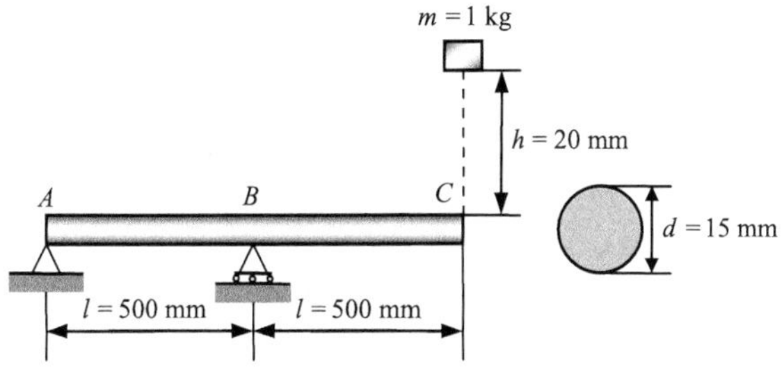

图 P20.19

20.20 质量为 m 的圆柱从图示位置由静止释放，撞击垂直杆 AB(直径为 d)的自由端 A。已知 E=200 GPa，求杆中最大正应力。

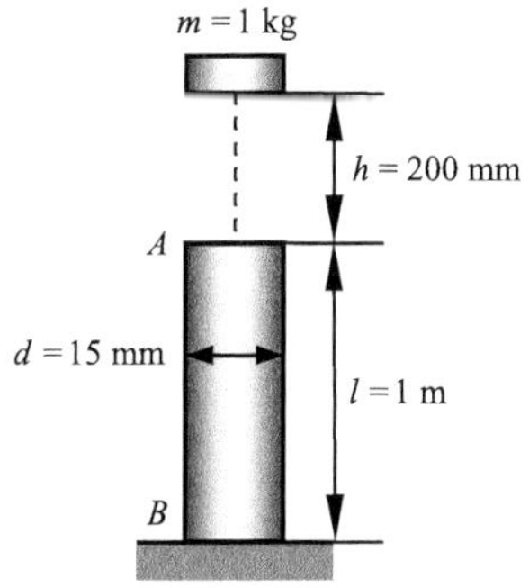

图 P20.20

20.21 质量为 m 的小球以速度 v 正碰立柱 AB(直径为 d)的中点 C。已知 E=200 GPa，求(a)点 A 的挠度，(b)立柱中最大正应力。

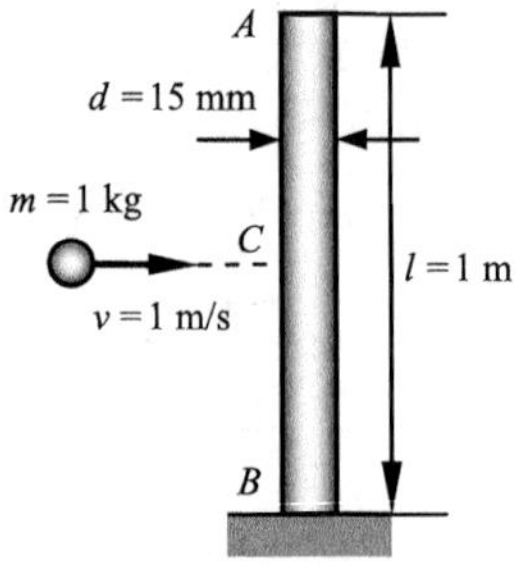

图 P20.21

20.22　质量为 m 的物块以速度 v 正碰非均匀杆 ABC 的自由端 A。已知 E=100 GPa，d_1=10 mm，d_2=20 mm，求(a)点 A 处位移，(b)杆中最大正应力。

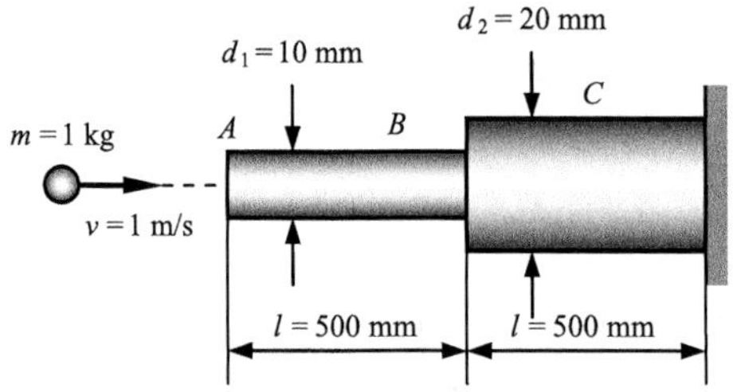

图 P20.22

参 考 文 献

哈尔滨工业大学理论力学教研室. 理论力学(I)[M]. 8版. 北京: 高等教育出版社, 2016.
哈尔滨工业大学理论力学教研室. 理论力学(II)[M]. 8版. 北京: 高等教育出版社, 2016.
刘鸿文. 材料力学(I)[M]. 6版. 北京: 高等教育出版社, 2017.
刘鸿文. 材料力学(II)[M]. 6版. 北京: 高等教育出版社, 2017.

附录I 重　　心

I.1 二 维 物 体

考虑图 I.1 所示二维物体，其重心定义为

$$\bar{x}=\frac{\int x\mathrm{d}W}{W},\ \bar{y}=\frac{\int y\mathrm{d}W}{W} \tag{I.1}$$

式中，$W=\int \mathrm{d}W$ 为二维物体的重量。

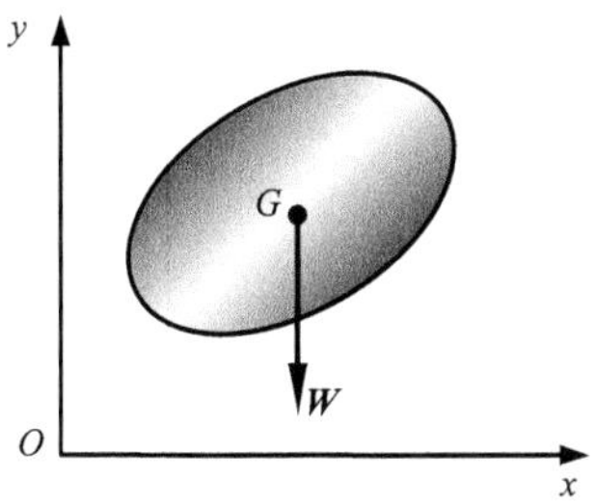

图 I.1

如果二维物体具有均匀密度ρ和均匀厚度t，则二维物体重心G将与其形心C重合，如图 I.2。利用$W=\rho gAt$，则二维物体形心可表示为

$$\bar{x}=\frac{\int x\mathrm{d}A}{A},\ \bar{y}=\frac{\int y\mathrm{d}A}{A} \tag{I.2}$$

式中，$A=\int \mathrm{d}A$ 为二维物体的面积。

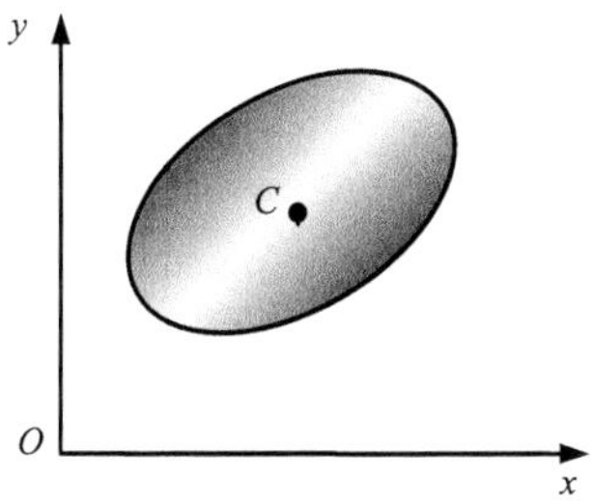

图 I.2

I.2　二维组合物体

考虑图 I.3 所示二维组合物体，则二维组合物体的重心可表示为

$$\overline{X}=\frac{\sum\overline{x}_iW_i}{W},\ \overline{Y}=\frac{\sum\overline{y}_iW_i}{W} \tag{I.3}$$

式中，$W=\sum W_i$ 为二维组合物体的重量。

如果二维组合物体具有均匀密度 ρ 和均匀厚度 t，则二维组合物体的重心 G 将与其形心 C 重合，如图 I.4 所示。利用 $W_i=\rho gA_it$，则二维组合物体的形心可表示为

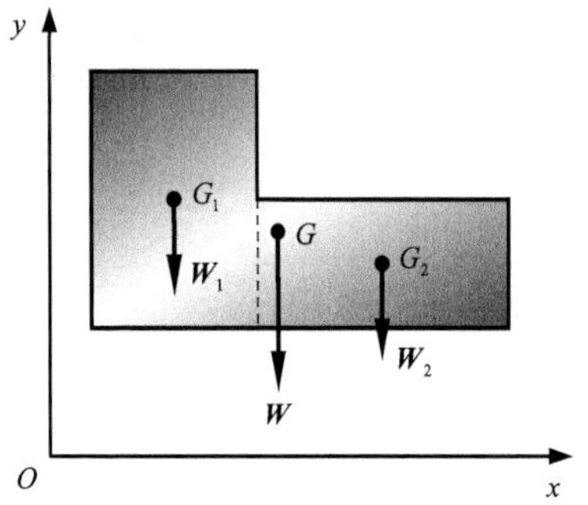

图 I.3

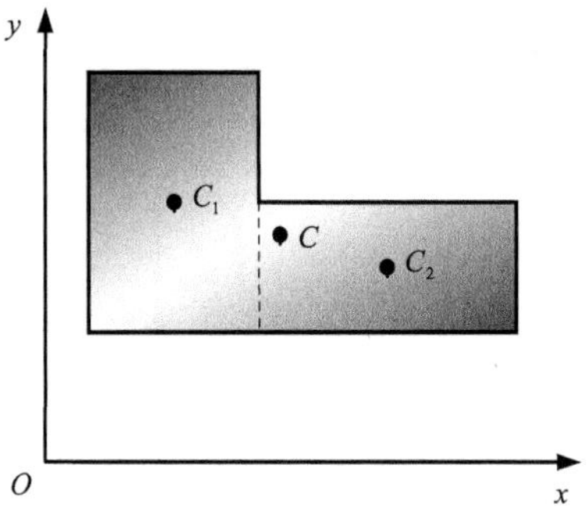

图 I.4

$$\overline{X}=\frac{\sum\overline{x}_iA_i}{A},\ \overline{Y}=\frac{\sum\overline{y}_iA_i}{A} \tag{I.4}$$

式中，$A=\sum A_i$ 为二维组合物体的面积。

I.3　三 维 物 体

考虑图 I.5 所示三维物体，其重心定义为

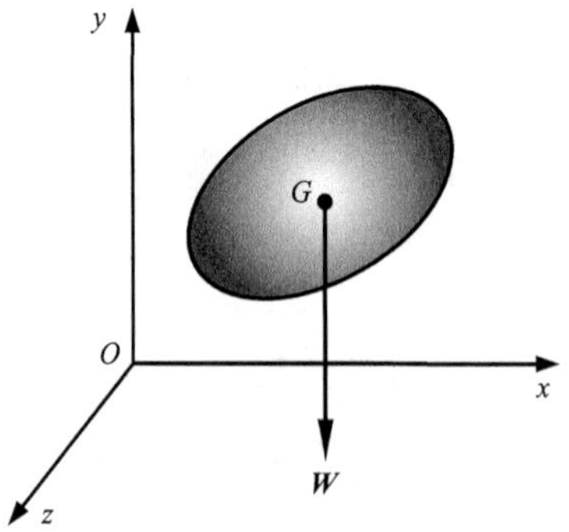

图 I.5

$$\bar{x}=\frac{\int x\mathrm{d}W}{W},\ \bar{y}=\frac{\int y\mathrm{d}W}{W},\ \bar{z}=\frac{\int z\mathrm{d}W}{W} \tag{I.5}$$

式中，$W=\int \mathrm{d}W$ 为三维物体的重量。

如果三维物体具有均匀密度 ρ，则三维物体的重心 G 将与其形心 C 重合，如图 I.6。利用 $W=\rho gV$，则三维物体的形心可表示为

$$\bar{x}=\frac{\int x\mathrm{d}V}{V},\ \bar{y}=\frac{\int y\mathrm{d}V}{V},\ \bar{z}=\frac{\int z\mathrm{d}V}{V} \tag{I.6}$$

式中，$V=\int \mathrm{d}V$ 为三维物体的体积。

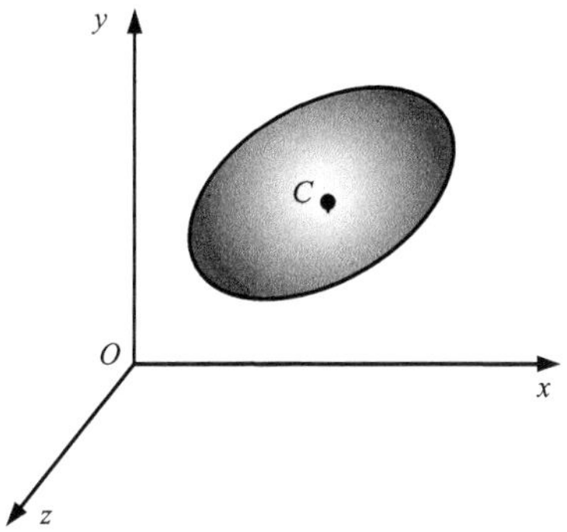

图 I.6

I.4 三维组合物体

考虑图 I.7 所示三维组合物体，则三维组合物体的重心定义为

$$\bar{X}=\frac{\sum \bar{x}_i W_i}{W},\ \bar{Y}=\frac{\sum \bar{y}_i W_i}{W},\ \bar{Z}=\frac{\sum \bar{z}_i W_i}{W} \tag{I.7}$$

式中，$W=\sum W_i$ 为三维组合物体的重量。

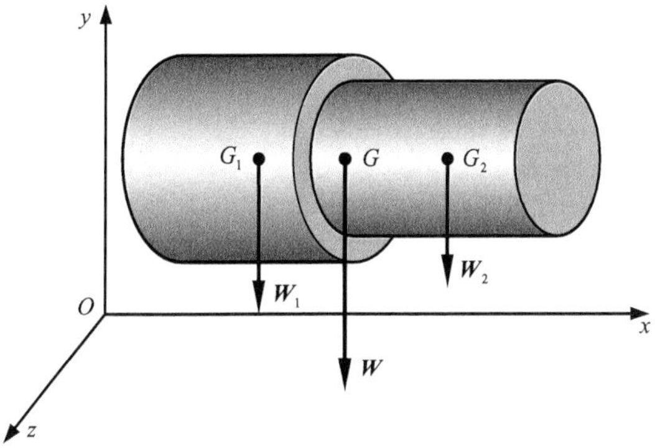

图 I.7

如果三维组合物体具有均匀密度 ρ，则三维组合物体的重心 G 将与其形心 C 重合，如图 I.8 所示。利用 $W_i = \rho g V_i$，则三维组合物体的形心可表示为

$$\overline{X} = \frac{\sum \overline{x}_i V_i}{V},\ \overline{Y} = \frac{\sum \overline{y}_i V_i}{V},\ \overline{Z} = \frac{\sum \overline{z}_i V_i}{V} \tag{I.8}$$

式中，$V = \sum V_i$ 为三维组合物体的体积。

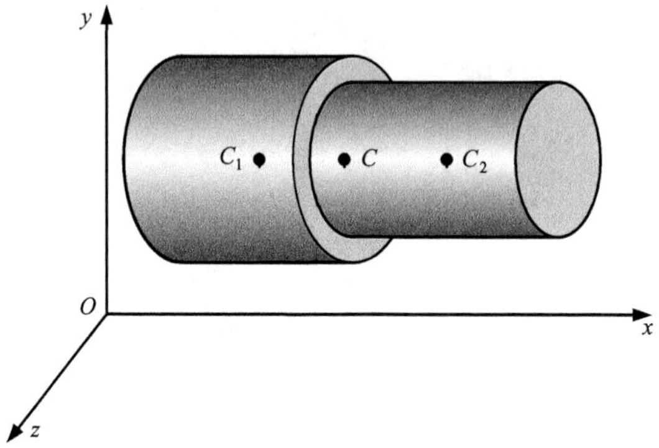

图 I.8

附录II 转 动 惯 量

II.1 转动惯量与回转半径

考虑图 II.1 所示质量为 m 的三维物体，则物体对 x、y 和 z 轴的转动惯量分别定义为

$$I_x = \int (y^2 + z^2)\mathrm{d}m,\ I_y = \int (z^2 + x^2)\mathrm{d}m,\ I_z = \int (x^2 + y^2)\mathrm{d}m \tag{II.1}$$

物体对 x、y 和 z 轴的回转半径分别定义为

$$i_x = \sqrt{\frac{I_x}{m}},\ i_y = \sqrt{\frac{I_y}{m}},\ i_z = \sqrt{\frac{I_z}{m}} \tag{II.2}$$

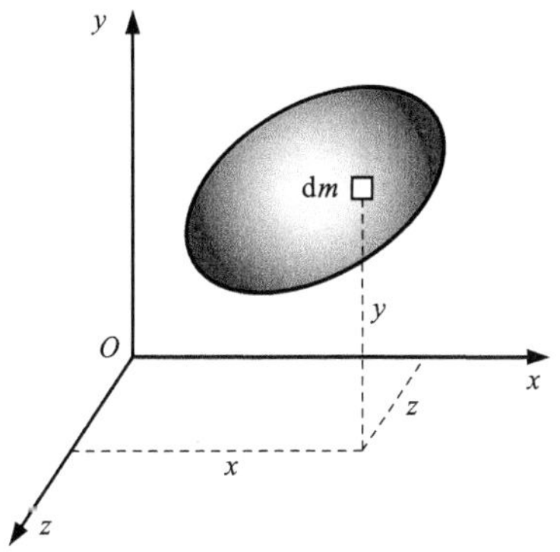

图 II.1

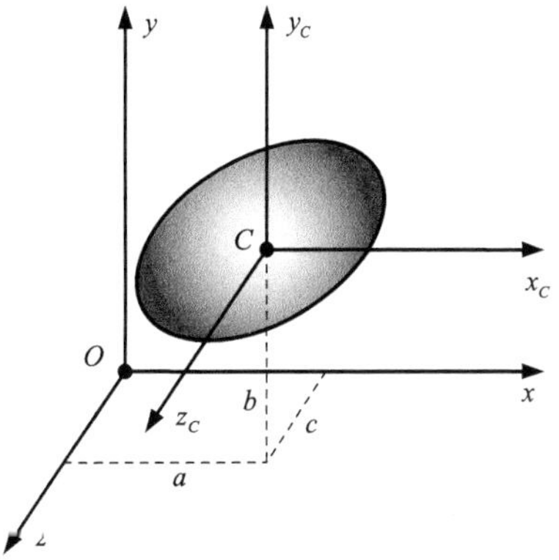

图 II.2

II.2 平行移轴定理

考虑图 II.2 所示质量为 m 的三维物体，物体对 x 和 x_C 轴的转动惯量分别表示为

$$I_x = \int (y^2 + z^2)\mathrm{d}m,\ I_{x_C} = \int ({y_C}^2 + {z_C}^2)\mathrm{d}m \tag{II.3}$$

代入 $y = y_C + b$ 和 $z = z_C + c$，得

$$I_x = \int (y^2 + z^2)\mathrm{d}m = \int ({y_C}^2 + {z_C}^2)\mathrm{d}m + 2\int (by_C + cz_C)\mathrm{d}m + (b^2 + c^2)\int \mathrm{d}m \tag{II.4}$$

利用 $\int (by_C + cz_C)\mathrm{d}m = 0$，上述方程可简化为

$$I_x = I_{x_C} + m(b^2 + c^2) \tag{II.5}$$

同理，得

$$I_y = I_{y_C} + m(c^2 + a^2),\ I_z = I_{z_C} + m(a^2 + b^2) \tag{II.6}$$

方程(II.5)和(II.6)所表示的关系称为平行移轴定理。当已知物体对形心轴的转动惯量时，上述关系常用于计算物体对任意轴的转动惯量。

附录III 截 面 性 质

III.1 静 矩

考虑图 III.1 所示阴影截面 A，截面对 z 和 y 轴的静矩分别定义为

$$Q_z = \int y \mathrm{d}A, \quad Q_y = \int z \mathrm{d}A \tag{III.1}$$

截面的形心定义为

$$\bar{y} = \frac{\int y \mathrm{d}A}{A}, \quad \bar{z} = \frac{\int z \mathrm{d}A}{A} \tag{III.2}$$

比较方程(III.1)和(III.2)，得

$$Q_z = A\bar{y}, \quad Q_y = A\bar{z} \tag{III.3}$$

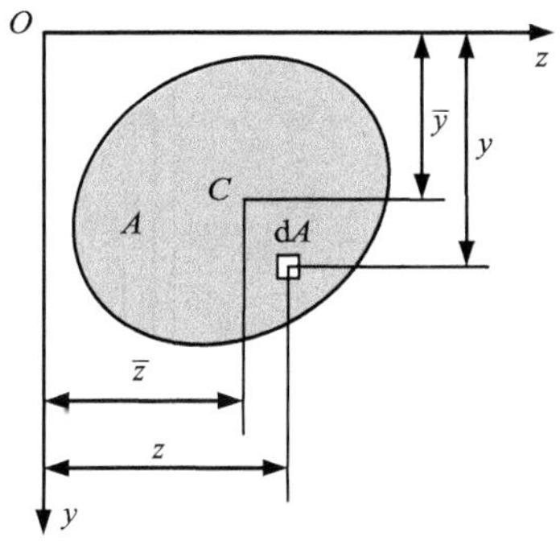

图 III.1

对于图 III.2 所示组合截面，其由截面 A_1 和 A_2 组成，则组合截面对 z 和 y 轴的静矩分别表示为

$$Q_z = \sum (Q_z)_i = \sum A_i \bar{y}_i, \quad Q_y = \sum (Q_y)_i = \sum A_i \bar{z}_i \tag{III.4}$$

组合截面的形心可写为

$$\bar{y} = \frac{\sum A_i \bar{y}_i}{\sum A_i}, \quad \bar{z} = \frac{\sum A_i \bar{z}_i}{\sum A_i} \tag{III.5}$$

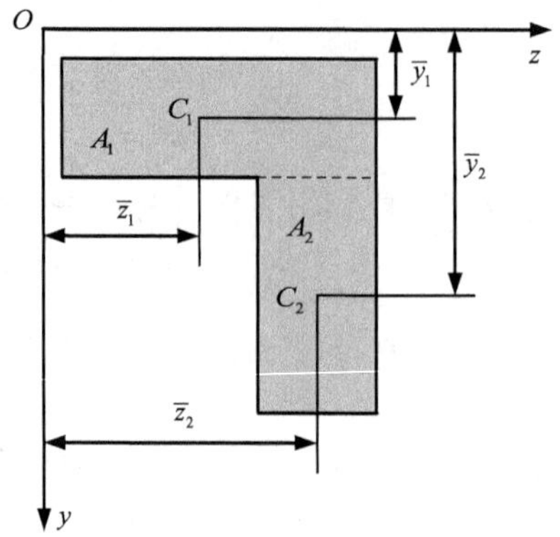

图 III.2

III.2　惯性矩与极惯性矩

1. 惯性矩

考虑图 III.3 所示阴影截面 A，截面对 z 和 y 轴的惯性矩分别定义为

$$I_z = \int y^2 \mathrm{d}A，\ I_y = \int z^2 \mathrm{d}A \tag{III.6}$$

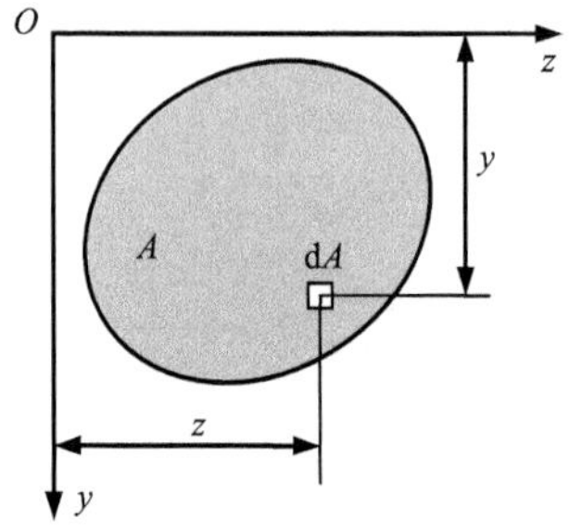

图 III.3

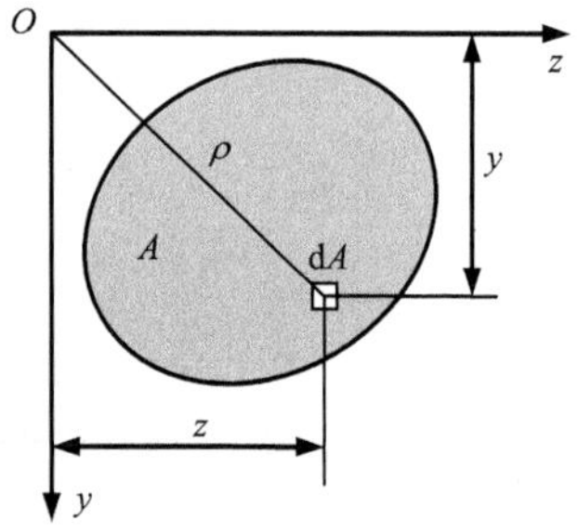

图 III.4

2. 极惯性矩

考虑图 III.4 所示阴影截面 A，截面对原点 O 的极惯性矩定义为

$$I_{\mathrm{p}} = \int \rho^2 \mathrm{d}A \tag{III.7}$$

利用 $\rho^2 = y^2 + z^2$，得

$$I_{\mathrm{p}} = \int \rho^2 \mathrm{d}A = \int y^2 \mathrm{d}A + \int z^2 \mathrm{d}A = I_z + I_y \tag{III.8}$$

III.3　惯性半径与极惯性半径

1. 惯性半径

考虑图 III.3 所示阴影截面 A，截面对 z 和 y 轴的惯性半径分别定义为

$$i_z = \sqrt{\frac{I_z}{A}}, \quad i_y = \sqrt{\frac{I_y}{A}} \tag{III.9}$$

2. 极惯性半径

考虑图 III.4 所示阴影截面 A，截面对原点 O 的极惯性半径定义为

$$i_{\mathrm{p}} = \sqrt{\frac{I_{\mathrm{p}}}{A}} \tag{III.10}$$

III.4　惯　性　积

考虑图 III.3 所示阴影截面 A，截面的惯性积定义为

$$I_{zy} = \int zy\mathrm{d}A \tag{III.11}$$

III.5　平行移轴定理

考虑图 III.5 所示阴影截面 A，截面对 z 和 z_C 轴的惯性矩分别表示为

$$I_z = \int y^2\mathrm{d}A, \quad I_{z_C} = \int {y_C}^2\mathrm{d}A \tag{III.12}$$

代入 $y = y_C + b$，有

$$I_z = \int y^2\mathrm{d}A = \int (y_C + b)^2\,\mathrm{d}A = \int {y_C}^2\mathrm{d}A + 2b\int y_C\mathrm{d}A + b^2\int \mathrm{d}A \tag{III.13}$$

利用 $\int y_C\mathrm{d}A = 0$，则上述方程可简化为

$$I_z = I_{z_C} + Ab^2 \tag{III.14}$$

同理，得

$$I_y = I_{y_C} + Aa^2, \quad I_{zy} = I_{z_C y_C} + Aab, \quad I_{\mathrm{p}} = (I_{\mathrm{p}})_C + A(a^2 + b^2) \tag{III.15}$$

方程(III.14)和(III.15)所表示的关系称为平行移轴定理。当已知截面对形心轴的惯性矩和极惯性矩时，上述关系常用于计算截面对任意轴的惯性矩和极惯性矩。

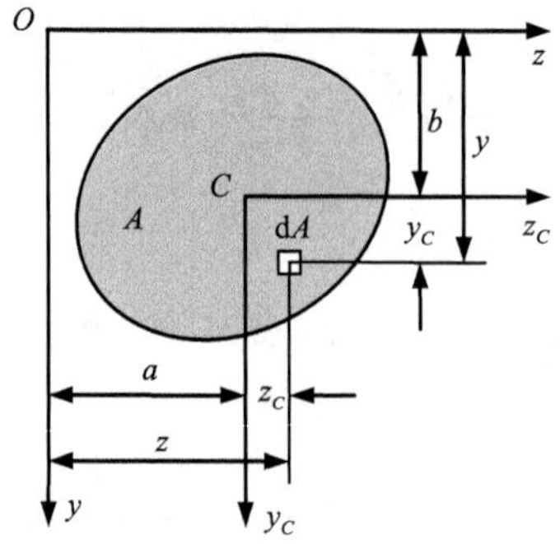

图 III.5

附录IV 型 钢

IV.1 工 字 钢

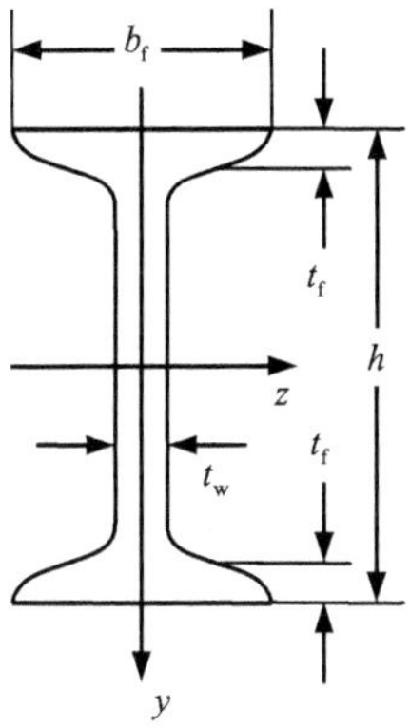

型号	重量/(kg/m)	高度 h/mm	翼缘 宽度 b_f/mm	翼缘 厚度 t_f/mm	腹板 厚度 t_w/mm	面积 A/cm^2	惯性矩 I_z/cm^4	惯性矩 I_y/cm^4	截面模量 S_z/cm^3	截面模量 S_y/cm^3
10	11.3	100	68	7.6	4.5	14.33	245	33.0	49.0	9.72
12	14.0	120	74	8.4	5.0	17.80	436	46.9	72.7	12.7
12.6	14.2	126	74	8.4	5.0	18.10	488	46.9	77.5	12.7
14	16.9	140	80	9.1	5.5	21.50	712	64.4	102	16.1
16	20.5	160	88	9.9	6.0	26.11	1130	93.1	141	21.2
18	24.1	180	94	10.7	6.5	30.74	1660	122	185	26.0
20a	27.9	200	100	11.4	7.0	35.56	2370	158	237	31.5
20b	31.1		102		9.0	39.55	2500	169	250	33.1
22a	33.1	220	110	12.3	7.5	42.10	3400	225	309	40.9
22b	36.5		112		9.5	46.50	3570	239	325	42.7
24a	37.5	240	116	13.0	8.0	47.71	4570	280	381	48.4
24b	41.2		118		10.0	52.51	4800	297	400	50.4
25a	38.1	250	116		8.0	48.51	5020	280	402	48.3
25b	42.0		118		10.0	53.51	52800	309	423	52.4
27a	42.8	270	122	13.7	8.5	54.52	6550	345	485	56.6
27b	47.0		124		10.5	59.92	6870	366	509	58.9
28a	43.5	280	122		8.5	55.39	7110	345	508	56.6
28b	47.9		124		10.5	60.97	7480	379	534	61.2
30a	48.1	300	126	14.4	9.0	61.22	8950	400	597	63.5
30b	52.8		128		11.0	67.22	9400	422	627	65.9
30c	57.5		130		13.0	73.22	9850	445	657	68.5

续表

型号	重量/(kg/m)	高度 h/mm	翼缘 宽度 b_f/mm	翼缘 厚度 t_f/mm	腹板 厚度 t_w/mm	面积 A/cm^2	惯性矩 I_z/cm^4	惯性矩 I_y/cm^4	截面模量 S_z/cm^3	截面模量 S_y/cm^3
32a	52.7	320	130	15.0	9.5	67.12	11100	460	692	70.8
32b	57.7		132		11.5	73.52	11600	502	723	76.0
32c	62.7		134		13.5	79.92	12200	544	760	81.2
36a	60.0	360	136	15.8	10.0	76.42	15800	552	875	81.2
36b	65.7		138		12.0	83.64	16500	582	919	84.3
36c	71.3		140		14.0	90.84	17300	612	962	87.4
40a	67.6	400	142	16.5	10.5	86.07	21700	660	1090	93.2
40b	73.8		144		12.5	94.07	22800	692	1140	96.2
40c	80.1		146		14.5	102.1	23900	727	1190	99.6
45a	80.4	450	150	18.0	11.5	102.4	32200	855	1430	114
45b	87.4		152		13.5	111.4	33800	894	1500	118
45c	94.5		154		15.5	120.4	35300	938	1570	122
50a	93.6	500	158	20.0	12.0	119.2	46500	1120	1860	142
50b	101		160		14.0	129.2	48600	1170	1940	146
50c	109		162		16.0	139.2	50600	1220	2080	151
55a	105	550	166	21.0	12.5	134.1	62900	1370	2290	164
55b	114		168		14.5	145.1	65000	1420	2390	170
55c	123		170		16.5	156.1	68400	1480	2490	175
56a	106	560	166		12.5	135.4	65600	1370	2340	165
56b	115		168		14.5	146.6	68500	1490	2450	174
56c	124		170		16.5	157.8	71400	1560	2550	183
63a	121	630	176	22.0	13.0	154.6	93900	1700	2980	193
63b	131		178		15.0	167.2	98100	1810	3160	204
63c	141		180		17.0	179.8	102000	1920	3300	214

IV.2 槽　钢

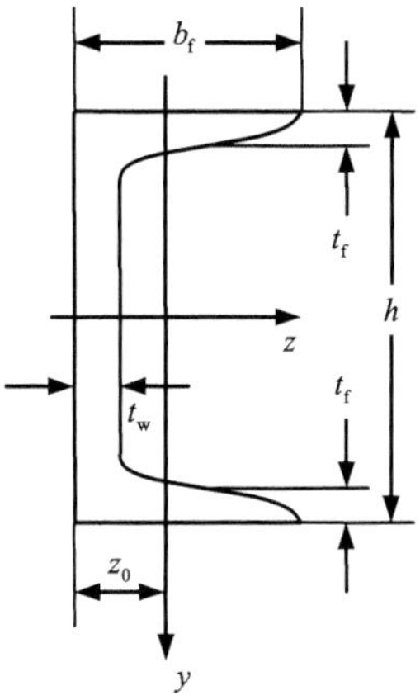

型号	重量 /(kg/m)	高度 h/mm	翼缘		腹板	面积 A/cm²	惯性矩		截面模量		z_0/cm
			宽度 b_f/mm	厚度 t_f/mm	厚度 t_w/mm		I_z/cm⁴	I_y/cm⁴	S_z/cm³	S_y/cm³	
5	5.44	50	37	7.0	4.5	6.925	26.0	8.3	10.4	3.55	1.35
6.3	6.63	63	40	7.5	4.8	8.446	50.8	11.9	16.1	4.50	1.36
6.5	6.51	65	40	7.5	4.3	8.292	55.2	12.0	17.0	4.59	1.38
8	8.04	80	43	8.0	5.0	10.24	101	16.6	25.3	5.79	1.43
10	10.0	100	48	8.5	5.3	42.74	198	25.6	39.7	7.8	1.52
12	12.1	120	53	9.0	5.5	15.36	346	37.4	57.7	10.2	1.62
12.6	12.3	126	53	9.0	5.5	15.69	391	38.0	62.1	10.2	1.59
14a	14.5	140	58	9.5	6.0	18.51	564	53.2	80.5	13.0	1.71
14b	16.7		60		8.0	21.31	609	61.1	87.1	14.1	1.67
16a	17.2	160	63	10.0	6.5	21.95	866	73.3	108	16.3	1.80
16b	19.8		65		8.5	25.15	935	83.4	117	17.6	1.75
18a	20.2	180	68	10.5	7.0	25.69	1270	98.6	141	20.0	1.88
18b	23.0		70		9.0	29.29	1370	111	152	21.5	1.84
20a	22.6	200	73	11.0	7.0	28.83	1780	128	178	24.2	2.01
20b	25.8		75		9.0	32.83	1910	144	191	25.9	1.95
22a	25.0	220	77	11.5	7.0	31.83	2390	158	218	28.2	2.10
22b	28.5		79		9.0	36.23	2570	176	234	30.1	2.03
24a	26.9	240	78	12.0	7.0	34.21	3050	174	254	30.5	2.1
24b	30.6		80		9.0	39.01	3280	194	274	32.5	2.03
24c	34.4		82		11.0	43.81	3510	213	293	34.4	2.00
25a	27.4	250	78	12.0	7.0	34.91	3370	176	270	30.6	2.07
25b	31.3		80		9.0	39.91	3530	196	282	32.7	1.98
25c	35.3		82		11.0	44.91	3690	218	295	35.9	1.92

续表

型号	重量 /(kg/m)	高度 h/mm	翼缘		腹板	面积 A/cm^2	惯性矩		截面模量		z_0/cm
			宽度 b_f/mm	厚度 t_f/mm	厚度 t_w/mm		I_z/cm^4	I_y/cm^4	S_z/cm^3	S_y/cm^3	
27a	30.8	270	82	12.5	7.5	39.27	4360	216	323	35.5	2.13
27b	35.1		84		9.5	44.67	4690	239	347	37.7	2.06
27c	39.3		86		11.5	50.07	5020	261	372	39.8	2.03
28a	31.4	280	82		7.5	40.02	4760	218	340	35.7	2.10
28b	35.8		84		9.5	45.62	5130	242	366	37.9	2.02
28c	40.2		86		11.5	51.22	5500	268	393	40.3	1.95
30a	34.5	300	85	13.5	7.5	43.89	6050	260	4.3	41.1	2.17
30b	39.2		87		9.5	49.89	6500	289	433	44.0	2.13
30c	43.9		89		11.5	55.89	6950	316	463	46.4	2.09
32a	38.1	320	88	14.0	8.0	48.50	7600	305	475	46.5	2.24
32b	43.1		90		10.0	54.90	8140	336	509	49.2	2.16
32c	48.1		92		12.0	61.30	8690	374	543	52.6	2.09
36a	47.8	360	96	16.0	9.0	60.89	11900	455	660	63.5	2.44
36b	53.5		98		11.0	68.09	12700	497	703	66.9	2.37
36c	59.1		100		13.0	75.29	13400	536	746	70.0	2.34
40a	58.9	400	100	18.0	10.5	75.04	17600	592	879	78.8	2.49
40b	65.2		102		12.5	83.04	18600	640	932	82.5	2.44
40c	71.5		104		14.5	91.04	19700	688	986	86.2	2.42

IV.3 等 边 角 钢

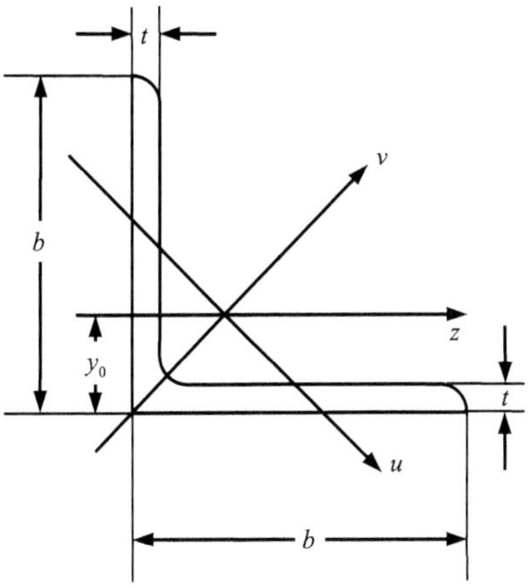

型号	重量/(kg/m)	宽度 b/mm	厚度 t/mm	面积 A/cm²	惯性矩			y_0/cm
					I_z/cm⁴	I_v/cm⁴	I_u/cm⁴	
2	0.89	20	3	1.132	0.40	0.63	0.17	0.60
	1.15		4	1.459	0.50	0.78	0.22	0.64
2.5	1.12	25	3	1.432	0.82	1.29	0.34	0.73
	1.46		4	1.859	1.03	1.62	0.43	0.76
3.0	1.37	30	3	1.749	1.46	2.31	0.61	0.85
	1.79		4	2.276	1.84	2.92	0.77	0.89
3.6	1.66	36	3	2.109	2.58	4.09	1.07	1.00
	2.16		4	2.756	3.29	5.22	1.37	1.04
	2.65		5	3.382	3.95	6.24	1.65	1.07
4.0	1.85	40	3	2.359	3.59	5.69	1.49	1.09
	2.42		4	3.086	4.60	7.29	1.91	1.13
	2.98		5	3.792	5.53	8.76	2.30	1.17
4.5	2.09	45	3	2.659	5.17	8.20	2.14	1.22
	2.74		4	3.486	6.65	10.6	2.75	1.26
	3.37		5	4.292	8.04	12.7	3.33	1.30
	3.99		6	5.077	9.33	14.8	3.89	1.33
5	2.33	50	3	2.971	7.18	11.4	2.98	1.34
	3.06		4	3.897	9.26	14.7	3.82	1.38
	3.77		5	4.803	11.2	17.8	4.64	1.42
	4.46		6	5.688	13.1	20.7	5.42	1.46
5.6	2.62	56	3	3.343	10.2	16.1	4.24	1.48
	3.45		4	4.390	13.2	20.9	5.46	1.53
	4.25		5	5.415	16.0	25.4	6.61	1.57
	5.04		6	6.420	18.7	29.7	7.73	1.61
	5.81		7	7.404	21.2	33.6	8.82	1.64
	6.57		8	8.367	23.6	37.4	9.89	1.68

续表

型号	重量 /(kg/m)	宽度 b/mm	厚度 t/mm	面积 A/cm^2	惯性矩			y_0/cm
					I_z/cm^4	I_v/cm^4	I_u/cm^4	
6	4.58	60	5	5.829	19.9	31.6	8.21	1.67
	5.43		6	6.914	23.4	36.9	9.60	1.70
	6.26		7	7.977	26.4	41.9	11.0	1.74
	7.08		8	90.20	29.5	46.7	12.3	1.78
6.3	3.91	63	4	4.978	19.0	30.2	7.89	1.70
	4.82		5	6.143	23.2	36.8	9.57	1.74
	5.72		6	7.288	27.1	43.0	11.2	1.78
	6.60		7	8.412	30.9	49.0	12.8	1.82
	7.47		8	9.515	34.5	54.6	14.3	1.85
	9.15		10	11.66	41.1	64.9	17.3	1.93
7	4.37	70	4	5.570	26.4	41.8	11.0	1.86
	5.40		5	6.876	32.2	51.1	13.3	1.91
	6.41		6	8.160	37.8	59.9	15.6	1.95
	7.40		7	9.424	43.1	68.4	17.8	1.99
	8.37		8	10.67	48.2	76.4	20.0	2.03
7.5	5.82	75	5	7.412	40.0	63.3	16.6	2.04
	6.91		6	8.797	47.0	74.4	19.5	2.07
	7.98		7	10.16	53.6	85.0	22.2	2.11
	9.03		8	11.50	60.0	95.1	24.8	2.15
	10.1		9	12.83	66.1	105	27.5	2.18
	11.1		10	14.13	72.0	114	30.1	2.22
8	6.21	80	5	7.912	48.8	77.3	20.3	2.15
	7.38		6	9.397	57.4	91.0	23.7	2.19
	8.53		7	10.86	65.6	104	27.1	2.23
	9.66		8	12.30	73.5	117	30.4	2.27
	10.8		9	13.73	81.1	129	33.6	2.31
	11.9		10	15.13	88.4	140	36.8	2.35
9	8.35	90	6	10.64	82.8	131	34.3	2.44
	9.66		7	12.30	94.8	150	39.2	2.48
	10.9		8	13.94	106	169	44.0	2.52
	12.2		9	15.57	118	187	48.7	2.56
	13.5		10	17.17	129	204	53.3	2.59
	15.9		12	20.31	149	236	62.2	2.67
10	9.37	100	6	11.93	115	182	47.9	2.67
	10.8		7	13.80	132	209	54.7	2.71
	12.3		8	15.64	148	235	61.4	2.76
	13.7		9	17.46	164	260	68.0	2.80
	15.1		10	19.26	180	285	74.4	2.84
	17.9		12	22.80	209	331	86.8	2.91
	20.6		14	26.26	237	374	99.0	2.99
	23.3		16	29.63	263	414	111	3.06
11	11.9	110	7	15.20	177	281	73.4	2.96
	13.5		8	17.24	199	316	82.4	3.01
	16.7		10	21.26	242	384	100	3.09
	19.8		12	25.20	283	448	117	3.16
	22.8		14	29.06	321	508	133	3.24

续表

型号	重量 /(kg/m)	宽度 b/mm	厚度 t/mm	面积 A/cm²	惯性矩			y_0/cm
					I_z/cm⁴	I_v/cm⁴	I_u/cm⁴	
12.5	15.5	125	8	19.75	297	471	123	3.37
	19.1		10	24.37	362	574	149	3.45
	22.7		12	28.91	423	671	175	3.53
	26.2		14	33.37	482	764	200	3.61
			16	37.74	537	851	224	3.68
14	21.5	140	10	27.37	515	817	212	3.82
	25.5		12	32.51	604	959	249	3.90
	29.5		14	37.57	689	1090	284	3.98
	33.4		16	42.54	770	1220	319	4.06
15	18.6	150	8	23.75	521	827	215	3.99
	23.1		10	29.37	638	1010	262	4.08
	27.4		12	34.91	749	1190	308	4.15
	31.7		14	40.37	856	1360	352	4.23
	33.8		15	43.06	907	1440	374	4.27
	35.9		16	45.74	958	1520	395	4.31
16	24.7	160	10	31.50	780	1240	322	4.31
	29.4		12	37.44	917	1460	377	4.39
	34.0		14	43.30	1050	1670	432	4.47
	38.5		16	49.07	1180	1870	485	4.55
18	33.2	180	12	42.24	1320	2100	543	4.89
	38.4		14	48.90	1510	2410	622	4.97
	43.5		16	55.47	1700	2700	699	5.05
	48.6		18	61.96	1880	2990	762	5.13
20	42.9	200	14	54.64	2100	3340	864	5.46
	48.7		16	62.01	2370	3760	971	5.54
	54.4		18	69.30	2620	4160	1080	5.62
	60.1		20	76.51	2870	4550	1180	5.69
	71.2		24	90.66	3340	5290	1380	5.87
22	53.9	220	16	68.67	3190	5060	1310	6.03
	60.3		18	76.75	3540	5620	1450	6.11
	66.5		20	84.76	3870	6150	1590	6.18
	72.8		22	92.68	4200	6670	1730	6.26
	78.9		24	100.5	4520	7170	1870	6.33
	85.0		26	108.3	4830	7690	2000	6.41
25	69.0	250	18	87.84	5270	8370	2170	6.84
	76.2		20	97.05	5780	9180	2380	6.92
	83.3		22	106.2	6280	9970	2580	7.00
	90.4		24	115.2	6770	10700	2790	7.07
	97.5		26	124.2	7240	11500	2980	7.15
	104		28	133.0	7700	12200	3180	7.22
	111		30	141.8	8160	12900	3380	7.30
	118		32	150.5	8600	13600	3570	7.37
	128		35	163.4	9240	14600	3850	7.48

IV.4　不等边角钢

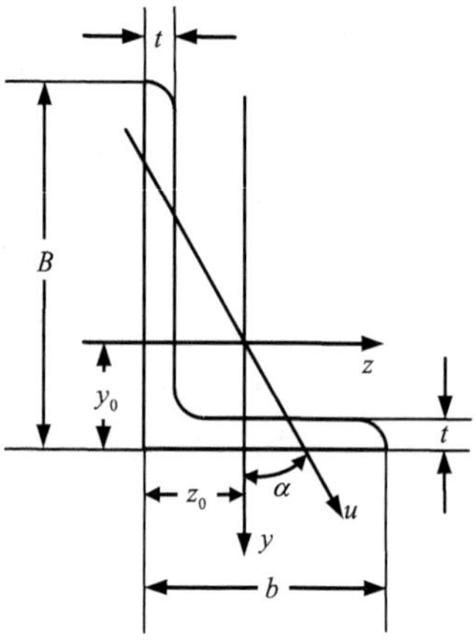

型号	重量 /(kg/m)	宽度		厚度	面积	惯性矩			z_0/cm	y_0/cm	tanα
		B/mm	b/mm	t/mm	A/cm^2	I_z/cm^4	I_y/cm^4	I_u/cm^4			
2.5/1.6	0.91	25	16	3	1.162	0.70	0.22	0.14	0.42	0.86	0.392
	1.18			4	1.499	0.88	0.27	0.17	0.46	0.90	0.381
3.2/2	1.17	32	20	3	1.492	1.53	0.46	0.28	0.49	1.08	0.382
	1.52			4	1.939	1.93	0.57	0.35	0.53	1.12	0.374
4/2.5	1.48	40	25	3	1.890	3.08	0.93	0.56	0.59	1.32	0.385
	1.94			4	2.467	3.93	1.18	0.71	0.63	1.37	0.381
4.5/2.8	1.69	45	28	3	2.149	4.45	1.34	0.80	0.64	1.47	0.383
	2.20			4	2.806	5.69	1.70	1.02	0.68	1.51	0.380
5/3.2	1.91	50	32	3	2.431	6.24	2.02	1.20	0.73	1.60	0.404
	2.49			4	3.177	8.02	2.58	1.53	0.77	1.65	0.402
5.6/3.6	2.15	56	36	3	2.743	8.88	2.92	1.73	0.80	1.78	0.408
	2.82			4	3.590	11.5	3.76	2.23	0.85	1.82	0.408
	3.47			5	4.415	13.9	4.49	2.67	0.88	1.87	0.404
6.3/4	3.19	63	40	4	4.058	16.5	5.23	3.12	0.92	2.04	0.398
	3.92			5	4.993	20.0	6.31	3.76	0.95	2.08	0.396
	4.64			6	5.908	23.4	7.29	4.34	0.99	2.12	0.393
	5.34			7	6.802	26.5	8.24	4.97	1.03	2.15	0.389
7/4.5	3.57	70	45	4	4.553	23.2	7.55	4.40	1.02	2.24	0.410
	4.40			5	5.609	28.0	9.13	5.40	1.06	2.28	0.407
	5.22			8	6.644	32.5	10.6	6.35	1.09	2.32	0.404
	6.01			7	7.658	37.2	12.0	7.16	1.13	2.36	0.402
7.5/5	4.81	75	50	5	6.126	34.9	12.6	7.41	1.17	2.40	0.435
	5.70			6	7.260	41.1	14.7	8.54	1.21	2.44	0.435
	7.43			8	9.467	52.4	18.5	10.9	1.29	2.52	0.429
	9.10			10	11.59	62.7	22.0	13.1	1.36	2.60	0.423
8/5	5.00	80	50	5	6.376	42.0	12.8	7.66	1.14	2.60	0.388
	5.93			6	7.560	49.5	15.0	8.85	1.18	2.65	0.387
	6.85			7	8.724	56.2	17.0	10.2	1.21	2.69	0.384
	7.75			8	9.867	62.8	18.9	11.4	1.25	2.73	0.381

续表

型号	重量 /(kg/m)	宽度		厚度 t/mm	面积 A/cm^2	惯性矩			z_0/cm	y_0/cm	$\tan\alpha$
		B/mm	b/mm			I_z/cm^4	I_y/cm^4	I_u/cm^4			
9/5.6	5.66	90	56	5	7.212	60.5	18.3	11.0	1.25	2.91	0.385
	6.72			6	8.557	71.0	21.4	12.9	1.29	2.95	0.384
	7.76			7	9.881	81.0	24.4	14.7	1.33	3.00	0.382
	8.78			8	11.18	91.0	27.2	16.3	1.36	3.04	0.380
10/6.3	7.55	100	63	6	9.618	99.1	30.9	18.4	1.43	3.24	0.394
	8.72			7	11.11	113	35.3	21.0	1.47	3.28	0.394
	9.88			8	12.58	127	39.4	23.5	1.50	3.32	0.391
	12.1			10	15.47	154	47.1	28.3	1.58	3.40	0.387
10/8	8.35	100	80	6	10.64	107	61.2	31.7	1.97	2.95	0.627
	9.66			7	12.30	123	70.1	36.2	2.01	3.00	0.626
	10.9			8	13.94	138	78.6	40.6	2.05	3.04	0.625
	13.5			10	17.17	167	94.7	49.1	2.13	3.12	0.622
11/7	8.35	110	70	6	10.64	133	42.9	25.4	1.57	3.53	0.403
	9.66			7	12.30	153	49.0	29.0	1.61	3.57	0.402
	10.9			8	13.94	172	54.9	32.5	1.65	3.62	0.401
	13.5			10	17.17	208	65.9	39.2	1.72	3.07	0.397
12.5/8	11.1	125	80	7	14.10	228	74.4	43.8	1.80	4.01	0.408
	12.6			8	15.99	257	83.5	49.2	1.84	4.06	0.407
	15.5			10	19.71	312	101	59.5	1.92	4.14	0.404
	18.3			12	23.35	364	117	69.4	2.00	4.22	0.400
14/9	14.2	140	90	8	18.04	366	121	70.8	2.04	4.50	0.411
	17.5			10	22.26	446	140	85.8	2.12	4.58	0.409
	20.7			12	26.40	522	170	100	2.19	4.66	0.406
	23.9			14	30.46	594	192	114	2.27	4.74	0.403
15/9	14.8	150	90	8	18.84	442	123	74.1	1.97	4.92	0.364
	18.3			10	23.26	539	14	89.9	2.05	5.01	0.362
	21.7			12	27.60	532	173	105	2.12	5.09	0.359
	25.0			14	31.86	721	196	120	2.20	5.17	0.356
	26.7			15	33.95	764	207	127	2.24	5.21	0.354
	28.3			16	36.03	806	217	134	2.27	5.25	0.352
16/10	19.9	160	100	10	25.32	669	205	122	2.28	5.24	0.390
	23.6			12	30.05	785	239	142	2.36	5.32	0.388
	27.2			14	34.71	896	271	162	2.43	5.40	0.385
	30.8			16	39.28	1000	302	183	2.51	5.48	0.382
18/11	22.3	180	110	10	28.37	956	278	167	2.44	5.89	0.376
	26.5			12	33.71	1120	325	195	2.52	5.98	0.374
	30.5			14	38.97	1290	370	222	2.59	6.06	0.372
	34.6			16	44.14	1440	412	249	2.67	6.14	0.369
20/12.5	29.8	200	125	12	37.91	1570	483	286	2.83	6.54	0.392
	34.4			14	43.87	1800	551	327	2.91	6.62	0.390
	39.0			16	49.74	2020	615	366	2.99	6.70	0.388
	43.6			18	55.53	2240	677	405	3.06	6.78	0.385